MW00564507

Vitiligo

Mauro Picardo
Alain Taïeb (Eds.)

Vitiligo

Dr. Mauro Picardo
Istituto Dermatologico San Gallicano
via Elio Chianesi 53
00144 Roma
Italy
picardo@ifo.it

Prof Alain Taïeb
Service de Dermatologie
De Dermatologie Pédiatrique
Centre de référence des maladies rares de la Peau
Hôpital Saint-André
1 rue Jean Burguet
33000 Bordeaux
France
alain.taieb@chu-bordeaux.fr

ISBN: 978-3-540-69360-4 e-ISBN: 978-3-540-69361-1

DOI: 10.1007/978-3-540-69361-1

Springer Heidelberg Dordrecht London New York

Library of Congress Control Number: 2009929162

Cover design: eStudio Calamar, Figueres/Berlin

Printed on acid-free paper

Springer is part of Springer Science+Business Media (www.springer.com)

Preface

Vitiligo has been, until recently, a rather neglected area in dermatology and medicine. Patients complain about this situation, which has offered avenues to quacks, and has led to the near orphan status of the disease. The apparently, simple and poorly symptomatic presentation of the disease has been a strong disadvantage to its study, as compared to other common chronic skin disorders such as psoriasis and atopic dermatitis. Vitiligo is still considered by doctors as a non disease, a simple aesthetic problem. A good skin-based angle of attack is also lacking because generalized vitiligo is clearly epitomizing the view of skin diseases as simple targets of a systemic unknown dysregulation (diathesis), reflecting the Hippocratic doctrine. This view has mostly restricted vitiligo to the manifestation of an auto-immune diathesis in the past 30 years. Thus, skin events, which are easily detected using skin biospies in most other situations, have not been precisely recorded, with the argument that a clinical diagnosis was sufficient for the management (or most commonly absence of management) of the patient.

This book is an international effort to summarize the information gathered about this disorder at the clinical, pathophysiological and therapeutic levels. Its primary aim is to bridge current knowledge at the clinical and investigative level, to point to the many unsolved issues, and to delineate future priorities for research. Its impetus was also to provide the best guidelines for integrated patient care, which is currently possible at a very limited number of places around the world, especially for surgical procedures.

A striking feature in the vitiligo field was, until recently, the absence of consensus on definitions, nomenclature, and outcome measures. With a group of European dermatologists, who had a strong interest in vitiligo and pigment cell research, we had launched some years ago, the Vitiligo European Task Force (VETF). The VETF has addressed those issues as a priority. This group, joined by other colleagues from the rest of the world also involved in the vitiligo research community, has communicated its experience in this book. We have tried to pilot the editing of the book according to consistent principles based on discussions held at VETF meetings and international IPCC (international pigment cell conference) workshops. However, some areas remain controversial and we have highlighted the existing conflicting issues and uncertainties.

After reviewing the field, much needs to be done. In particular, besides basic research based on the many hypotheses raised, new unbiased epidemiological, clinical, histopathological, natural history, and therapeutic data are clearly needed. They should be confronted by genetics and other investigative variables to better define the disease and its subsets. We hope that the combined efforts of all participating authors will prove useful to bring more attention to this field, and we are confident that both the research community (the mystery of melanocyte loss in vitiligo is a true scientific challenge) and the drug industry (the potential market is large) will be stimulated to bring in new treatment strategies to this large number of patients with unmet needs.

Alain Taïeb — Bordeaux, France
Mauro Picardo — Rome, Italy

Contents

Contributors

A. Alomar, M.D Department of Dermatology; Respiratory Medicine, Hospital de la Santa Creu I Sant Pau,Barcelona,Spain
aalomar@santpau.es

Flávia Pretti Aslanian M.D, Ph.D Federal University of Rio de Janeiro, Rio de Janeiro, Brasil
fpretti@cyberwal.com.br

Marco Ardigò, M.D San Gallicano Dermatological Institute, IRCCS, Rome, Italy
ardigo@ifo.it

Ratnam Attili Visakha Institute of Skin and Allergy, Visakhapatnam, India
vrattili@hotmail.com

Laila Benzekri, M.D, Ph.D Department of Dermatology, CHU Ibn Sina, Rabat, Maroc
lbenzekri2005@yahoo.fr

Markus Böhm, M.D Department of Dermatology, University of Münster, Münster, Germany
bohmm@uni-muenster.de

Raymond E. Boissy, Ph.D University of Cincinnati College of Medicine, Cincinnati, OH, USA,
boissy.raymond@uc.edu

Barbara Boone, Ph.D Department of Dermatology, University Hospital of Ghent, Ghent, Belgium
barbara.boone@ugent.be

Valeria Brazzelli, M.D Institute of Dermatology, University of Pavia, Fondazione IRCCS Policlinico S.Matteo, Pavia, Italy
vbrazzelli@libero.it

Muriel Cario-André, Ph.D Inserì U876, Centre de référence des maladies rares de la peau, Université V Segalen Bordeaux 2, Bordeaux, France
muriel.cario-andre@dermatol.u-bordeaux2.fr

Tullia Cuzzi M.D. Ph.D Federal University of Rio de Janeiro, Rio de Janeiro, Brasil
tullia@terra.com.br

Maria Lucia Dell'Anna, M.D Laboratory of Cutaneous Physiopathology, San Gallicano Dermatological Institute IFO, Rome, Italy
citolab@ifo.it

Alida DePase, Associazione Ricerca Informazione per la Vitiligine (ARIV), Cernusco Lombardone (LC), Italy
depase@arivonlus.it

Dipankar De, M.D Department of Dermatology, Postgraduate Institute of Medical Education and Research, India
chhinu75@yahoo.co.in

Gisela F. Erf, M.D Center of Excellence for Poultry Science, University of Arkansas, Fayetteville, Arkansas, USA
gferf@uark.edu

Khaled Ezzedine, M.D, Ph.D. Service de dermatologie, Hôspital St André, CHU de Bordeaux, France
khaled.ezzedine@chu-bordeaux.fr

Rafael Falabella, M.D Universidad del Valle and Centro Medico Imbanaco, Cali, Colombia
rfalabella@uniweb.net.co

Absalom Lima Filgueira M.D, Ph.D Department of Dermatology, Federal University of Rio de Janeiro, Leblon, Rio de Janeiro, Brasil
absalom@uol.com.br

Yvon Gauthier, M.D Service de Dermatologie, Hôpital Saint-André, CHU de Bordeaux, France
yvongauthier@free.fr

DJ Gawkrodger, M.D Department of Dermatology, Royal Hallamshire Hospital, Sheffield, UK
david.gawkrodger@sth.nhs.uk

Giampiero Girolomoni, M.D Clinica Dermatologica, Università di Verona, Verona, Italy.
Giampiero.girolomoni@univr.it

Somesh Gupta, M.D Department of Dermatology and Venereology, All India Institute of Medical Science, New Delhi, India
someshgupta@hotmail.com

Seung-Kyung Hann, M.D Korea Institute of Vitiligo Research, Drs. Woo & Hann's skin clinic, Yongsan-Gu, Seoul, South Korea
skhann@paran.com

Genji Imokawa Ph.D Tokyo University of Technology, School of Bioscience and Biotechnology, Tokyo, Japan
imokawag@dream.ocn.ne.jp

Thomas Jouary, M.D Service de Dermatologie, Hôpital Saint André, CHU de Bordeaux, France
thomas.jouary@chu-bordeaux.fr

E. Helen Kemp, Ph.D School of Medicine and Biomedical Sciences, University of Sheffield, Sheffield, UK
e.h.kemp@sheffield.ac.uk

Panagiota Kostopoulou, M.D Service de Dermatologie, Hôspital St André, CHU de Bordeaux, France

Prasad Kumarasinghe, M.D Consultant Dermatologist, Department of Dermatology, Royal Perth Hospitalm, Western Australia, Australia
prasadkumarasinghe@yahoo.com

Cheng-Che Eric Lan Department of Dermatology, Kaohsiung Medical University, 100 Shih-Chuan 1st Rd, Kaohsiung, Taiwan
laneric@kmu.edu.tw

Jo Lambert, M.D Department of Dermatology, Ghent University Hospital, Ghent, Belgium
jo.lambert@ugent.be

Giovanni Leone, M.D Phototherapy Unit, San Gallicano Dermatological Institute, IFO, Rome, Italy
gleone@ifo.it

I. Caroline Le Poole Ph.D Loyola University Chicago, Maywood, Illinois
ilepool@lumc.edu

Sébastien Lepreux Department of Pathology, Hôpital Pellegrin, CHU de Bordeaux, France
sebastien.lepreux@chu-bordeaux.fr

Torello M. Lotti, M.D Department of Dermatology, University of Florence, Florence, Italy
torello.lotti@unifi.it

Juliette Mazereeuw-Hautier, M.D Service de Dermatologie, CHU Purpan, Toulouse, France
mazereeuw-hautier.j@chu-toulouse.fr

Ilse Mollet, M.D Department of Dermatology, Ghent University Hospital, Ghent, Belgium
ilse.mollet@ugent.be

Silvia Moretti, M.D Department of Dermatological Sciences, University of Florence, Florence, Italy
silvia.moretti@unifi.it

Fanny Morice-Picard, M.D Departments of Medical Genetics and Dermatology, Hôpital Pellegrin-Enfants, Bordeaux, France
fanny.morice@free.fr

Francesca Muzio, M.D Department of Human and Hereditary Pathology, Institute of Dermatology, University of Pavia, Fondazione IRCCS Policlinico S.Matteo, Pavia, Italy

L.Nieuweboer-Krobotova, M.D Netherlands Institute for Pigment Disorders, and Department of Dermatology, Academic Medical Centre, University of Amsterdam, Amsterdam, The Netherlands
l.krobotova@amc.uva.nl

David A. Norris, M.D Department of Dermatology, University of Colorado Denver School of Medicine, Anschutz Medical Campus, Aurora, Colorado, USA
david.norris@uchsc.edu

Mats J. Olsson, M.D Department of Medical Sciences, Dermatology and Venereology, Uppsala University, Sweden
mats.olsson@medsci.uu.se

Jean-Paul Ortonne, M.D Department of Dermatology, Archet-2 hospital, Nice, France
Jean-Paul.ORTONNE@unice.fr

Alessia Pacifico, M.D Phototherapy Unit, San Gallicano Dematological Institute, IFO, Rome, Italy
alessia.pacifico@tiscali.it

Davinder Parsad, M.D Department of Dermatology, Postgraduate Institute of Medical Education & Research, Chandigarh, India
parsad@mac.com

Thierry Passeron, M.D, Ph.D Department of Dermatology, University Hospital of Nice, Nice, France
passeron@unice.fr

Anna Peroni, M.D Dermatological Clinic, University of Verona,Verona, Italy
anna.peroni@univr.it

Mauro Picardo, M.D Laboratory of Cutaneous Physiopathology, San Gallicano Dermatological Institute, IFO, Via San Gallicano, 25, 00153 Rome, Italy
picardo@ifo.it

Julien Seneschal, M.D Service de dermatologie, Hôpital St André, CHU de Bordeaux, France
julien.seneschal@free.fr

Richard Spritz, M.D Human Medical Genetics Program, University of Colorado Denver, Aurora, Colorado, USA
richard.spritz@uchsc.edu

Alain Taïeb, M.D Service de Dermatologie, Hôpital St André, CHU de Bordeaux, France
alain.taieb@chu-bordeaux.fr

Adrian Tanew, M.D Phototherapy Unit, Division of Special and Environmental Dermatology, Medical University of Vienna, Vienna, Austria
adrian.tanew@meduniwien.ac.at

J.P.W. van der Veen, M.D Netherlands Institute for Pigment Disorders, and Department of Dermatology, Academic Medical Centre, University of Amsterdam, Amsterdam, The Netherlands
j.p.vanderveen@amc.uva.nl

Nanny Van Geel, M.D Department of Dermatology, Ghent University Hospital, Ghent, Belgium
nanny.vangeel@ugent.be

Béatrice Vergier, M.D Department of Pathology, Hôpital du Haut Lévêque, Bordeaux, France
beatrice.vergier@chu-bordeaux.fr

Bas S. Wind Netherlands Institute for Pigment Disorders, and Department of Dermatology, Academic Medical Centre, University of Amsterdam, Amsterdam, The Netherlands

Maxine Eloise Whitton Wanstead, UK
mewhitton@aol.com

Ching-Shuang Wu Faculty of Biomedical Laboratory Science, Kaohsiung Medical University, 100 Shih-Chuan 1st Rd, Kaohsiung, Taiwan
m785034@kmu.edu.tw

Hsin-Su Yu Department of Dermatology, Kaohsiung Medical University, 100 Shih-Chuan 1st Rd, Kaohsiung, Taiwan
dermyu@kmu.edu.tw

Section 1

Defining the Disease

Historical Aspects

1.1

Yvon Gauthier and Laila Benzekri

Contents

Y. Gauthier (✉)
Service de Dermatologie, Hôpital Saint-André,
CHU de Bordeaux, France
e-mail: yvongauthier@free.fr

Core Messages

- The term vitiligo was introduced in the first century of our era.
- The mistaking of leprosy for vitiligo in the Old Testament under Zoorat or Zaraath is an important cause for the social stigma attached to white spots on the skin.
- Modern photochemotherapy (PUVA) was an improvement of photochemotherapy practised in the ancient world with herbals containing furocoumarins.
- The stigma historically associated with the disease has not disappeared, and information and educational programmes should be implemented to fight this problem, especially the fear of contagion.

The physician must know what his predecessors have known, if he does not deceive both himself and others

Hippocrates

1.1.1 Before Vitiligo: Understanding Old Terms Meaning White Skin Spots

Even though the term vitiligo appeared in the first century of our era, descriptions of the disease now known as vitiligo can be found in the ancient medical classics of the second millennium BC [17, 27, 28, 34]. The earliest reference to the disease was found in 2200 BC in the period of Aushooryan, according to the ancient literature of Iran "Tarkh-e-Tibble" [26]. In the Ebers

M. Picardo and A. Taïeb (eds.), *Vitiligo*,
DOI 10.1007/978-3-540-69361-1_1.1, © Springer-Verlag Berlin Heidelberg 2010

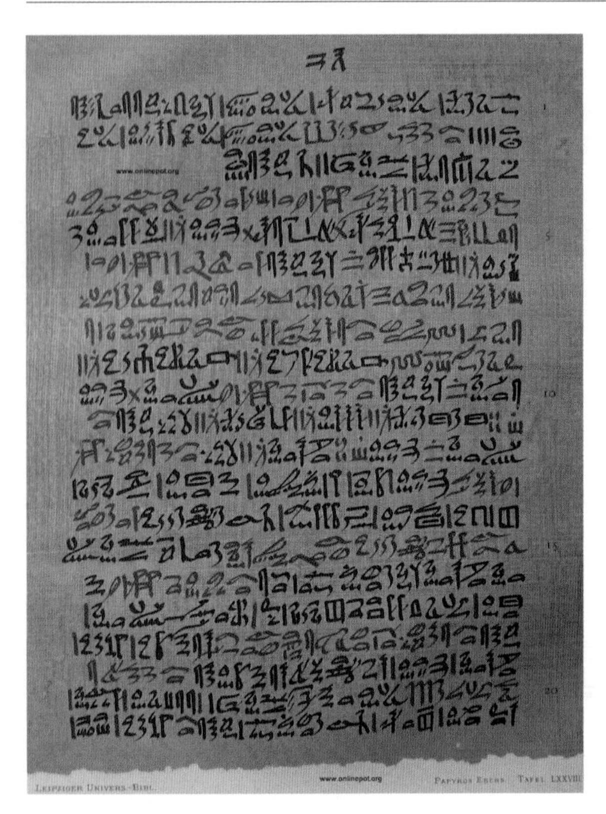

Fig. 1.1.1 Ebers papyrus (Leipzig Museum)

Papyrus [8] (ca. 1500 BC), there is a mention of two types of diseases affecting the colour of the skin. One of them is associated with swellings and recommended to be left alone "Thou shall anything to-do it" is probably leprosy. The other one, manifested with only changes of colour is likely to be vitiligo, as this is said to have been treated (Fig. 1.1.1).

Hundreds of years before Christ, vitiligo was present in ancient Indian sacred books, such as the Atharva Veda (1400 BC) and the Buddhist sacred book Vinay Pitak (224–544 BC). Atharva Veda gives a description of vitiligo and many achromic or hypochromic diseases under different names such as "Kilasa", "Sveta Khista", "Charak". The sanskrit word "Kilas" is derived from kil meaning white. So "kilas" means "which throw away colour"; the term "Sveta Khista" was meaning white leprosy and vitiligo was probably confused with macular leprosy. The term "Charak" used by villagers means, "which spreads or is secret". In the Buddhist "Vinay Pitak", the term "kilas" was also used to describe the white spots on the skin [7, 27, 28, 31, 41].

Descriptions of vitiligo and other leukodermas can also be found in other ancient Indian medical writings such as the Charaka Samhita (800 BC) (medicine), Manus mriti (200 BC) (law) and Amorkasha (600 AC) [2]. There are also references to a disease with a whitening of the skin in the early classics of the Far East. In "Makataminoharai", a collection of Shinto prayers (dating back to about 1200 BC), there is mention of a disease "shira-bito" which literally means white man. This also could have been vitiligo [27, 28].

Though vitiligo, the disease with white spots, was recognized in the ancient times, it was frequently confused with leprosy. Even Hippocrates (460–355 BC) did not differentiate vitiligo and leprosy, and included lichen, leprosy, psoriasis and vitiligo under the same category. In the Old Testament of the Holy Bible, the Hebrew word "Zora'at" is referring to a group of skin diseases that were classified into five categories [17]: (1) "White spots" per se, interpreted as vitiligo or post-inflammatory leukoderma; (2) White spots associated with inflammation; (3) White spots associated with scaling; (4) White spots associated with atrophy; (4) White spots associated with the regrowth of white hairs.

Ptolomy II (250 BC) asked for a translation of the Hebrew Bible into Greek so that more people could understand the Bible. The word "Zora'at'" was unfortunately translated in Greek versions of the Bible as "white leprosy" (Leviticus Chap. XIII). According to modern dermatologists and theologians, biblical leprosy represented not a specific illness, but psoriasis or leukoderma and other disorders perceived to be associated with spiritual uncleanliness. Consequently, persons with white spots, independent of the cause, were isolated from the healthy ones [17, 26, 27, 28, 34, 35].

"Bohak", "baras", "alabras" are the Arabic names used to describe vitiligo. "Baras" means "white skin". In the Koran (Surah The family of Imran Chap. 3, verse 48, and Surah The table spread Chap. 5 verse109) we can read (Fig. 1.1.2): *"In accord with God' will Jesus was able to cure patients with vitiligo"*. "Alabras" was translated as leprosy in many languages. However, Tahar Ben Achour [3] in 1973 explained that "alabras" meant vitiligo. For this theologian, the aetiology of vitiligo in the Koran was unknown but inherited and not contagious.

The name vitiligo was first used by the famous Roman physician Celsus at the second century BC in his medical classic "De Medicina". The word vitiligo has often been said to have derived from "vitium" (defect or blemish) rather than "vitellus" meaning calf [34].

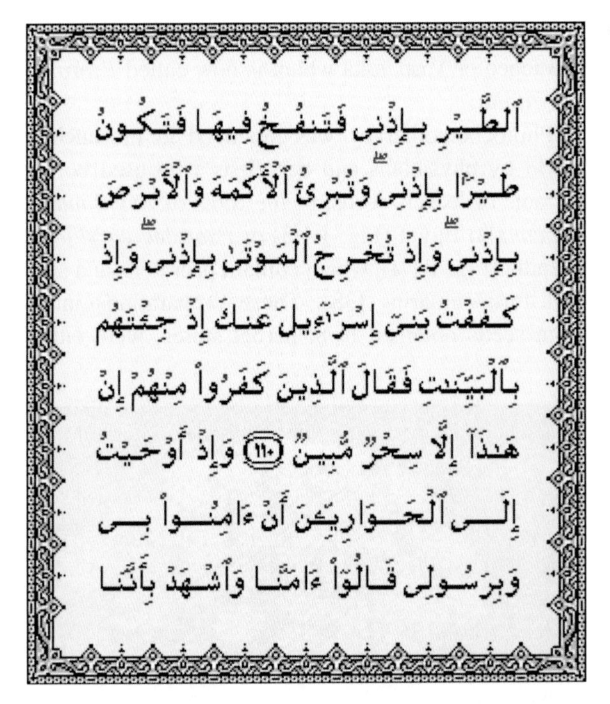

ٱلطَّيْرِ بِإِذْنِي فَتَنفُخُ فِيهَا فَتَكُونُ
طَيْرًا بِإِذْنِي وَتُبْرِئُ ٱلْأَكْمَهَ وَٱلْأَبْرَصَ
بِإِذْنِي وَإِذْ تُخْرِجُ ٱلْمَوْتَىٰ بِإِذْنِي وَإِذْ
كَفَفْتُ بَنِىٓ إِسْرَٰٓءِيلَ عَنكَ إِذْ جِئْتَهُم
بِٱلْبَيِّنَٰتِ فَقَالَ ٱلَّذِينَ كَفَرُوا۟ مِنْهُمْ إِنْ
هَٰذَآ إِلَّا سِحْرٌ مُّبِينٌ ﴿١١٠﴾ وَإِذْ أَوْحَيْتُ
إِلَى ٱلْحَوَارِيِّـۧنَ أَنْ ءَامِنُوا۟ بِى
وَبِرَسُولِى قَالُوٓا۟ ءَامَنَّا وَٱشْهَدْ بِأَنَّنَا

Fig. 1.1.2 In the Koran: Surah the Table Chap. 5 verse 109

1.1.2 From Celsus to the Modern Period

Several studies in the late nineteenth century reported that there was a propensity for depigmentation in traumatized clinically normal skin of vitiligo subjects. There appeared to be a minimal threshold of injury, required for a depigmented patch to occur [19], a phenomenon later named after Koebner, who reported it in psoriasis (Sect. 2.2.2.1). Over the last century and up to now, the alternative use of the terms of "vitiligo" and "leukoderma" has perpetuated confusion in the medical literature. Pearson, just after the end of the nineteenth Century described under the name leukoderma, a disease which seems to be vitiligo [cited in 33]. In the late twentieth century, the elusive nature of vitiligo has led to prudent definitions excluding disorders of established aetiology. "*Vitiligo can be described as an acquired primary, usually progressive, melanocyte loss of still mysterious aetiology, clinically characterized by circumscribed achromic macules often associated with leucotrichia, and progressive disappearance of melanocytes in the involved skin*" [38]. However, controversies still remain about achromic macules occurring during the evolution of malignant melanoma and lymphomas. Are they "vitiligoid lesions" (Chap. 1.2.1) or true vitiligo?

Simultaneously, there have been many attempts to classify vitiligo, often with confusing results. Classifications have been based either on the distribution or localization of hypopigmented macules (focal, segmental, generalized), [22] or on functional sweat stimulation studies: type A, or non dermatomal, considered as an autoimmune disease; and type B, or dermatomal, with sympathetic dysfunction [20]. Presence of leukotrichia has been accorded special significance. The absence of hair depigmentation was considered as a favourable prognostic sign [32, 33].

Until now,specific pathomechanisms of vitiligo have not been identified, and the only highly characteristic feature, studied by light and electron microscopic study, was considered to be the partial or total absence of melanocytes in the epidermis. Although the initial cause of vitiligo remains unknown, there have been significant advances in our understanding of this disorder. Three main theories have been proposed successively. The neurogenic hypothesis suggested that a chemical could be released from nerve endings in the skin, which would induce melanocyte destruction [6]. The self-destruction theory contended that melanocytes could be destroyed by a failure of the normal protective mechanism that usually removes toxic chemicals generated by melanogenesis and oxidoreduction process [23]. The more recent autoimmune theory was based on a frequent association with autoimmune disorders and the common occurrence of autoantibodies in these patients [4]. It is uncertain whether vitiligo is a one disease entity with a specific pathogenesis or a final common pathway of several different processes, such as that recently reported as a primary chronic loss of melanocytes due to poor anchoring of melanocytes to the basal lamina – the melanocytorrhagic hypothesis [16].

1.1.3 Social Status of Vitiligo Patients Across the Ages

The history of vitiligo reveals much information regarding the social stigma of patients suffering from this disease. In the Buddhist literature (624–544 BC) we can read "*men and women suffering from the disease named Kilasa were not eligible to get ordainment*". The social implication of this disease is also well documented in "Rigveda" (Indian book) "*persons suffering from switra and their progeny are disqualified from marrying others*" [7, 31].

Herodotus (484–425 BC) in his book "Clio," has written "*if a Persian has leprosy or white sickness he is not allowed to enter into a city or to have dealings with other Persians, he must have sinned against the sun. Foreigners suffering from this disorder are forced to leave the country; even white pigeons are often driven away, as guilty of the same offence*".

In Leviticus (XIII, 34), white spot diseases are considered as a punishment sent by God: "*Anyone with these skin affections must wear torn clothes and have his hair dishevelled, he must conceal his upper lip, and call out unclean, unclean". So long the disease persists, he is to be considered virtually unclean and live alone outside the camp*". During many centuries, the stigma of leprosy was strengthened by old edicts and cruel laws [17].

In 1943, according to the initiative of Pope Pius XII and the American Catholic office, the church added the following note with reference to Leviticus XIII: "*Various kinds of skin blemishes are treated here, which were not contagious but simply disqualified their subjects from associations with others, until they were declared ritually clean. The Hebrew term used does not refer to Hansen's disease, currently called leprosy*" [17, 34].

Islamic theologians consider that baras is a defect in the couple. Thus, the husband or the wife has the choice to divorce. Arabic kings talked to vitiligo patients only behind a screen as reported by the poet Harrith bin Hilliza [3].

The long lasting misleading association of contagion with vitiligo emphasizes the need for patients (or their parents if children) to be reassured about the non contagious nature of the disease. Unfortunately, patients with vitiligo are still regarded as social outcasts in some countries.

1.1.4 Precursors to Phototherapy of Vitiligo

The Ebers Papyrus (ca. 1550 BC), the Atharva Veda (ca. 1400 BC), other Indian and Buddhist medical literature (ca. 200 AC), and Chinese manuscript from the Sung period (ca. 700 AC), make reference to the treatment of vitiligo with black seeds from the plant Bavachee or Vasuchika which is now called *Psoralea corylifolia*.

Photochemotherapy was practiced in the ancient world by physicians and herbalists who used boiled extracts of leaves, seeds, or the roots of *Ammi majus linneaus* in Egypt (Fig. 1.1.3) or *Psoralea corylifolia* in India (Fig. 1.1.4), which contain psoralens and several furocoumarins [38]. These preparations made from seeds obtained from herbal stores, were either

Fig. 1.1.3 *Ammi majus linneaus* (AML), which grows throughout the Nile Valley as a weed. The seeds of AML were used as a therapy for leucoderma

Fig. 1.1.4 *Psoralea corylifolia*, which grows in India. Preparations for the treatment of vitiligo were made from the seeds of this plant

applied to the skin or ingested before solar exposure. In the ancient "Ayurvedic system of medicine", the use of *Psoralea corylifolia* (Leguminosae family) for inducing the repigmentation of vitiligo is carefully recorded. In "Charaka Samhita" both topical and systemic use of figs is also recommended before sun exposure. Figs are known to contain both psoralens and glucosides of psoralens. In Atharva Veda (2000–1400 BC) and many other writings of this period, the treatment of vitiligo with "bavacheree" (*Psoralia corylifolia*) is described, associated with exposure to solar radiation and worshipping. This is illustrated in Atharva Veda by the following poem to one of the plant used [5, 28, 42].

> Born by night art thou, O plant
> Dark, black, sable, do thou
> That art rich in colour, stain
> This leprosy and white grey spots.
> Even colour is the name of thy mother,
> Even colour is the name of thy father.
> Thou, O plant produced even colour,
> Render this spot to even colour

According to Indian medical books, the two most commonly applied and effective herbs were "Malapu" (*Ficus hispida*) and "Bawachi" (*Psoralea corylifolia*) [7, 25, 40, 41]. These herbs were administered orally or topically followed by exposure to sunlight until sweating was observed. Blisters were then produced and after their rupture, repigmentation occurred. In Ayurveda, the association of internal treatment consisting of dried ginger, black-pepper, pippali and leadwort root fermented in cow's urine and local ointment with a paste made of several medicinal herbs including *Psoralea corylifolia*, was proposed to induce repigmentation of vitiligo [7].

Another important plant *Ammi majus linnaeus* [7, 39] which grows throughout the Nile Valley as a weed, has been used many centuries BC as a treatment for leukoderma. Ibn el Bitar in his thirteenth century book "Mofradat Al Adwiya," described the treatment of vitiligo "baras" with the seeds of Aatrillal (*Ammi majus*) and sunlight [18]. Aatrillal seemed to be a common therapy for leukoderma in "Ben Shoerb", a Berberian tribe living in the North-western African desert. Egyptian herbalists used these herbs in the form of a powder [10]. Through ages, in Egypt, topical or oral use of seeds from the plant *Ammi majus* followed by sunlight exposure, as mentioned earlier in India, was the mainstay of treatment. Natural psoralens have been found in more than thirty plants, including lime, lemon, bergamot, celery, fig and clove [26].

1.1.5 Modern History of Some Treatments of Vitiligo

- Phototherapies

In 1938, Kuske investigated phytophotodermatitis occurring on areas of the skin which have been in contact with certain plants and also exposed to sunlight [21, 36, 37]. Extensive research began in 1941 in Cairo, Egypt, in the laboratory of Fahmi and his group. Three crystalline compounds were obtained from *Ammi majus l* and named ammoidin, ammidin, and majudin [10]. El Mofty [9] was the first, in the early 1940s, to use crystallin methoxsalen followed by exposure to sunlight for the treatment of vitiligo. The results were found to be good and shortly afterwards two of the compounds, 8-methoxypsoralen and 8-isoamyleneoxypsoralen were produced by an Egyptian firm in Cairo. In 1974, a new high intensity UVA light source was initially used in combination with either oral 8 MOP or 4–5–8 trimethyl psoralen for the treatment of vitiligo. It was the beginning of PUVA therapy which permitted to induce a good repigmentation on hairy skin using the melanocytes from the follicular reservoir. The period from 1974 to 1988 was the period of photomedicine that established the therapeutic effectiveness of psoralens in combination with UVA in the treatment of vitiligo, psoriasis and various other skin diseases [14, 27, 29]. The use of PUVA has declined over the last 20 years due to the evidence of photodamage and photocarcinogenesis due to this technique, and to the parallel development of narrow-band UVB irradiation devices (Part 3.3).

- Surgical treatments

Many investigators have tried to induce repigmentation in patients with vitiligo not responding to medical therapies, with grafting and transplantation techniques. Dermoepidermal grafts were introduced by Behl [1]. Epidermal grafting with suction blisters was successfully used for the first time in 1971 by Falabella [11] on segmental vitiligo and post burn depigmentation. Minigrafting was proposed in 1983 by Falabella in three patients with segmental vitiligo [12]. In 1989, epidermis with melanocytes was grown for the first time for repigmentation purposes by Falabella [12] and Falabella et al. [13]. The use of epidermal suspensions for repigmentation of stable vitiligo macules was initially described in 1992 by Gauthier and Surlève Bazeille [15]. This method was improved in 1998 by Olsson and Juhlin [30]. In 1992, pure melanocyte suspensions were cultured and implanted in patients with vitiligo by Olsson and Juhlin [31]. A new approach to simplify the delivery of cultured melanocytes to patients via a cell delivery carrier, which could also be used for transport of cells over considerable distances, has been proposed by Mac Neil and Eves [24].

1.1.6 Conclusions

Vitiligo, under different names, was recognized and feared over several centuries. Even though modern phototherapies represent an improvement of the herbal and sun, ancient world techniques, our current therapies, including surgical techniques, fail in a large number of cases. The stigma historically associated with the disease has not disappeared, and information and educational programmes should be implemented to fight this problem, especially the fear of contagion.

References

1. Behl PN (1964) Treatment of vitiligo with homologous thin Thiersch skin grafts. Curr Med Pract 8:218–221
2. Behl PN, Arora RB, Srivastava G (1992) Traditional Indian dermatology. Concepts of past and present. Skin Institute and School of Dermatology, New Delhi, pp 62–72
3. Ben Achour T (1973) Tafsir Attahir wa Tancuir. Edn Dar Sahoune Liannachr, Tunis, Tunisie
4. Betterle C, Caretto A, De Zio A et al (1985) Incidence and significance of organ-specific autoimmune disorders. Dermatologica 171:419–430
5. Bloomfield M (1897) Sacred books of the east hymns of Atharva veda, Vol XLII. Clarendon, Oxford
6. Chanco-Turner ML, Lerner AB (1965) Physiologic changes in vitiligo. Arch Dermatol 91:190–196
7. Donata SR, Kesavan M, Austin SR (1990) Clinical trial of certain ayurveda medicines indicated in vitiligo. Ancient Sci of life 4:202–206
8. Ebbel B (1937) The papyrus Ebers. Levin and Munskgaard, Copenhagen
9. El Mofty AM (1948) A preliminary clinical report on the treatment of leukodermias with *Ammi Majus* Linn. J Egypt Med Assoc 31:651–656
10. Fahmy IR, Abn-Shady H (1947) Pharmacological study and isolation of cristalline constituents: Amoïdin. Quart J Pharm Pharmacol 20:281–286
11. Falabella R (1971) Epidermal grafting: an original technique and its application in achromic and granulating areas. Arch Dermatol 104:592–600
12. Falabella R (1984) Repigmentation of leucoderma by autologous epidermal grafting. J Dermatol Surg Oncol 10: 136–144
13. Falabella R, Escobar C, Borrero I (1992) Treatment of refractory and stable vitiligo by transplantation of in vitro cultured epidermal autografts bearing melanocytes. J Am Acad Dermatol 26(2 Pt 1):230–236
14. Fitzpatrick TB, Pathak MA (1959) Historical aspects of methoxalen and other furocoumarins. J Invest Dermatol 32:229–231
15. Gauthier Y, Surlève-Bazeille JE (1992) Autologous grafing with non cultured melanocytes: a simplified method for treatment of depigmented lesions. J Am Acad Dermatol 26:191–194
16. Gauthier Y, Cario André M, Taieb A (2003) A critical appraisal of vitiligo etiologic theories. Is melanocyte loss a melanocytorragy? Pigment Cell Res 16:322–333
17. Goldman L, Richard S, Moraites R (1966) White spots in biblical times. Arch Derm 93:744–753
18, Fahmy IR (1947). "Ibn El Bitar Mofradat el Adwija" Quart J Pharmacol 20:744–753
19. Kaposi M (1891) Pathologie et traitement des maladies de la peau [translated by Besnier and Barthelemy], Masson, Paris, France, pp 105–110
20. Koga M (1977) Vitiligo: a new classification and therapy. Br J Dermatol 97:255–261
21. Kuske H (1938) Experimentelle Untersuchung zur Photosensibilisierung der Haut durch planzliche Wirstoffs. Arch Dermatol Syphil 178:112–113
22. Lerner AB (1959) Vitiligo. J Invest Dermatol 32:285–310
23. Lerner AB (1971) On the etiology of vitiligo and gray hair. Am J Med 51:141–147
24. Mac Neil S, Eves P (2007) Simplifying the delivery of cultured melanocytes. In: Gupta S, Olsson MJ, Kanwar A (eds) Surgical management of vitiligo. Blackwell, Oxford, pp 191–201
25. Nair BK (1978) Vitiligo: a retrospect. Int J Dermatol 17: 755–757
26. Najamabadi M (1934) Tarikh -e- Tibbe-Iran, Vol I, Shamsi, Teheran, Iran
27. Njoo MD,Westerhof W (1997) Vitiligo: a review. Hautartz 48:677–693
28. Njoo MD (2000) Treatment of vitiligo. Thela Thesis, Amsterdam, pp 17–20
29. Njoo M, Spuls PI, Bos J, et al (1998) Non surgical repigmentation therapies in vitiligo. Arch Dermatol 134: 1532–1540
30. Olsson M, Juhlin M (1998) Leucoderma treated by transplantaton of a basal cell layer enriched suspension. Br J Dermatol 138:644–648
31. Olsson M, Juhlin M (1992) Melanocyte transplantation in vitiligo. Lancet 34(8825):981
32. Ortonne JP, Mac Donald DM, Micoud A, Thivolet J (1978) Repigmentation of vitiligo induced by oral photochemotherapy. Histoenzymological and ultrastructural study. Ann Dermatol Venereol 105:939–940

33. Ortonne JP (1983) Vitiligo and other hypomelanoses of hair and skin. In: Ortonne JP, Mosher DP, Fitzpatrick TB (eds) Topics in dermatology. Plenum Medical School, New York, pp 129–132
34. Panda AK (2005) The medicohistorical perspective of vitiligo. Bull Ind Hist Med 25:41–46
35. Parrish JA, Fitzpatrick TB, Pathak A (1974) Photochemotherapy of psoriasis with oral methoxalen and long wave ultraviolet light. N Engl J Med 291:1207–1211
36. Pathak M, Daniels F, Fitzpatrick TB (1962) The presently known distribution of furocoumarins (psoralens) in plants. J Invest Dermatol 39:225–239
37. Pathak M, Fitzpatrick TB (1992) The evolution of photochemotherapy with psoralens and UVA (PUVA) 2000 BC to 1992 AD. J Photochem Photobiol 14:3–22
38. Prasad PV, Bhatnagar VK (2003) Medico-historical study of "Kilasa" (vitiligo/leucoderma) a common skin disorder. Bull Ind Inst Hist Med 33:113–127
39. Sidi E, Bourgeois-Gavardin XX (1952) The treatment of vitiligo with *Ammi Majus* Linn. J Invest Dermatol 18:391–396
40. Srivastava G (1994) Introduction vitiligo update. Asian Clin Dermatol 1:1–4
41. Sushruta S, Bhishagratna KL (1963) Second edition volI. Chapter 46 published by Chaukhamba Sanskritt series office Gopal Mandir Lane, printed by Vidya Vikas, Varanasi India
42. Whitney WD (1905) Atharva veda samhita (translation and notes). Harvard Oriental Series, Vol 7. Harvard University Press, Cambridge, MA

Clinical Overview 1.2

Epidemiology, Definitions and Classification

1.2.1

Alain Taïeb and Mauro Picardo

Contents

Core Messages

- Vitiligo occurs worldwide with an estimated overall prevalence of less than 0.5% in population-based studies.
- Vitiligo vulgaris/NSV (non-segmental vitiligo) is an acquired chronic pigmentation disorder characterized by white patches, often symmetrical, which usually increase in size with time, corresponding to a substantial loss of functioning epidermal, and sometimes hair follicle melanocytes.
- Segmental vitiligo (SV) is defined descriptively as for NSV, except for a unilateral distribution ("asymmetric vitiligo") that may totally or partially match a cutaneous segment such as a dermatome, but not necessarily.
- NSV and SV may coexist, and in this case SV lesions are usually more refractory to treatment.

1.2.1.1 Introduction

Although vitiligo is one of the most common cutaneous disorders, difficulties arise when searching for the vitiligo literature due to a lack of consensus in definitions. In order to address this problem, the Vitiligo European Task Force (VETF) has proposed consensus definitions, which form the basis of those included in this chapter [34]. The simplest classification of vitiligo, distinguishing segmental (SV) and non-segmental (NSV) forms of disease [18], served as a pragmatic framework. For the sake of consistency, the definitions used throughout the book are those proposed and reviewed thereafter.

A. Taïeb (✉)
Service de Dermatologie, Hôpital St André,
CHU de Bordeaux, France
e-mail: alain.taieb@chu-bordeaux.fr

M. Picardo and A. Taïeb (eds.), *Vitiligo*,
DOI 10.1007/978-3-540-69361-1_1.2.1, © Springer-Verlag Berlin Heidelberg 2010

1.2.1.2 Epidemiology

Vitiligo occurs worldwide with an estimated overall prevalence of less than 0.5% in population-based studies [4,8,16,30]. Some peaks of prevalence have been noted, especially in India, which may correspond to the still poor identification of environmental (but with a clear predominance of occupational factors) or genetic factors [29]. Almost half the patients present before the age of 20 years, and nearly 70–80% before the age of 30 years. Adults and children of both sexes are equally affected, although larger number of females consult the doctor probably due to the greater psycho–social perceived impact of the disease [29]. Compared to NSV, SV has a younger peak of onset and is rare (around 1 SV against every 10 NSV cases) (Chap. 1.3.2). Vitiligo prevalence and incidence seem stable over time, at variance with allergic and autoimmune diseases [2]. Compared to other common chronic skin disorders for monozygous twin concordance, a marker of the inherited component in complex disorders, the inherited component of vitiligo is weaker: only 23% concordance for vitiligo [1], vs. 35–56% in psoriasis [6,9] and up to 72% in atopic dermatitis [28]. Familial aggregation of NSV cases takes a non-Mendelian pattern that is suggestive of polygenic, multifactorial inheritance. Segregation analyses suggest that multiple major loci contribute to vitiligo susceptibility. However, according to NSV genome-wide linkage analyses performed in populations of various ethnic backgrounds, the major inherited loci are not the same (Chap. 2.2.1). Data in European-descent populations suggest that some major genes are markers of an autoimmune diathesis and others segregate with vitiligo in isolation [33]. The risk of skin cancer in vitiligo patients is discussed in Chap. 1.3.9. There might be two co-existing modes of inheritance for vitiligo depending on age of onset (Chap. 1.3.13) and specific HLA haplotypes [10], which are strongly associated with a family history of vitiligo, severity of disease, age of onset and population geography (Chap. 2.2.1).

1.2.1.3 Vitiligo Vulgaris or Non-Segmental Vitiligo

There is still a lack of consensus among experts about several aspects of this disease though they can identify it without much difficulty in most instances. The primary defect(s) underlying common NSV remains unclear (Part 2), but there is an agreement on NSV being the clinical expression of a progressive loss of melanocytes. As a starting point of the VETF mechanisms were not included consensus definition. The rest of this definition is descriptive. There is no reference to either a single disease or to a syndrome of various causes in this definition. Its aim is to allow investigators in the field to substantiate further evidence of subgroups by providing scientific data (e.g. immune or inflammatory vs. non immune; halo-naevus associated vs. non-halo naevus associated….).

Vitiligo vulgaris/NSV is an acquired chronic pigmentation disorder characterized by white patches, often symmetrical, which usually increase in size with time, corresponding to a substantial loss of functioning epidermal and sometimes hair follicle melanocytes.

As such, the definition is not specific enough; thus, it needs to be completed by a list of disorders which may clinically overlap with NSV (the acquired generalised hypomelanoses), but which are clearly attributable to known etiologic factors.

1.2.1.4 Conditions to Exclude from the Definition of NSV

Inherited or genetically induced hypomelanoses. Usually, contrary to vitiligo, hypopigmented patches are present at birth, but in patients of low phototype, hypopigmented patches are usually discovered after the first sun exposure, sometimes in the second or third year of life. The group of genetic disorders to exclude is detailed in Table 1.2.1.1 and Figs. 1.2.1.1–1.2.1.4. As usual for monogenetic disorders, information about specific ethnic background/ consanguinity and a detailed family tree are mandatory. Piebaldism may be mistaken for vitiligo when the patient without informative family history comes with symmetrical limb patches without midline anomalies. The hyperpigmented rim at the interface of depigmented and normally pigmented skin is characteristic of vitiligo after sun exposure, but may also be seen in piebaldism, which shows, however, a ventral distribution of lesions and early onset.

On the other hand, vitiligo universalis is sometimes misdiagnosed as albinism when the history cannot be obtained properly (Chap. 1.3.3). Classic oculocutaneous

Table 1.2.1.1 Monogenic hypomelanoses

Disorder/OMIM number	Clinical presentation	Transmission/diagnosis
Piebaldism/172800	White forelock, anterior body midline depigmentation, bilateral shin depigmentation	Autosomal dominant. Skin biopsy: usually, absence of c-kit protein immunostaining in melanocytes Molecular analysis of *c-KIT* gene or *SNA1* gene
Tuberous sclerosis/191100	Small or larger (ash-leaf) white spots, seizures; other usually later cutaneous symptoms (chagreen patches, angiofibromas…)	Autosomal dominant; Brain imaging; cardiac and renal imaging Molecular analysis of TSC1 and 2
Ito's hypomelanosis/300337	Blaschkolinear distribution, uni or bilateral of hypopigmented streaks	Sporadic chromosomal or genetic mosaicism (blood or skin cells)
Waardenburg's syndrome/193500 (type I)	White forelock, hypertelorism, deafness (variable according to genotype) Possible association to congenital megacolon (Hirschprung's disease)	Autosomal dominant. Genetic testing according to phenotype (several clinical variants and six possible causative genes mutated)
Hermanski-Pudlak syndrome/203300 (type 1)	Diffuse depigmentation pattern, eye pigment dilution, haemorrhagic diathesis Specific ethnic background	Autosomal dominant, genetic heterogeneity (eight possible causative genes) Molecular testing possible according to ethnic background and phenotype
Menkès syndrome/300011	Hair and body diffuse pigment dilution; Pili torti; Neurodegenerative changes	X-linked recessive ATP7A gene mutations; affects cupper dependent enzymes. Molecular testing possible
Ziprkowski-Margolis syndrome/300700	Iris heterochromia, depigmentation (diffuse) + hyperpigmented macules can remain, neurosensorial deafness	Described in Israel in one family. X-linked recessive Mapped to Xq26.3–q27.1
Griscelli's syndrome 214450 (type 1)	Silvery hairs usually diffuse liver enlargement and symptoms related to immunodeficiency	Autosomal recessive, three types corresponding to three different causative genes Molecular testing possible

albinism with hair depigmentation and nystagmus at birth is not a consideration, but milder syndromic albinisms should be mentioned (Table 1.2.1.1). Some rare vitiligo-like or true vitiligo conditions ("syndromic vitiligo") are seen in the context of monogenic disorders. They are reviewed separately in Chap. 1.3.11.

Post-inflammatory hypomelanoses (Fig. 1.2.1.5): There are two main processes for melanin pigment to be eliminated following inflammation. First, in inflammatory disorders accompanied by an increased epidermal turn over (e.g. psoriasis, atopic dermatitis), pigment is lost upwards within eliminated cells, resulting in focal pigment dilution until cessation of inflammation and recovery of melanin production/distribution within the epidermis. Second, when an acute lichenoid/cytotoxic infiltrate attacks the epidermal basal layer (e.g. lichen planus, toxic drug reactions), there is a leakage of epidermal pigment into the superficial dermis ("pigment incontinence"), which is removed when the inflammation has stopped by macrophages ("melanophages"), a process which takes months or years. In NSV (Chap. 1.3.9 and Sect. 2.2.3.1), there is limited evidence of pigment incontinence, which may occur during the acceleration phase of depigmentation, in case of mild subclinical lichenoid inflammation. In the majority of cases, a progressive pigment dilution accompanies the disappearance of pigment cells. Distinguishing NSV from post-inflammatory hypomelanoses is usually made on clinical grounds rather than on histopathology, when the primary skin inflammatory disease can be diagnosed clinically (e.g. scalp or plaque psoriasis, flexural dermatitis for atopic dermatitis, scleroderma plaques, lichen sclerosus "white spot disease"…). However, true association may occur [3] and sometimes in similar locations, suggesting pathophysiological links such as koebnerization in the case of pruritic dermatoses followed by vitiliginous patches (see Chap. 1.3.8 and Sect. 2.2.2.1). In difficult

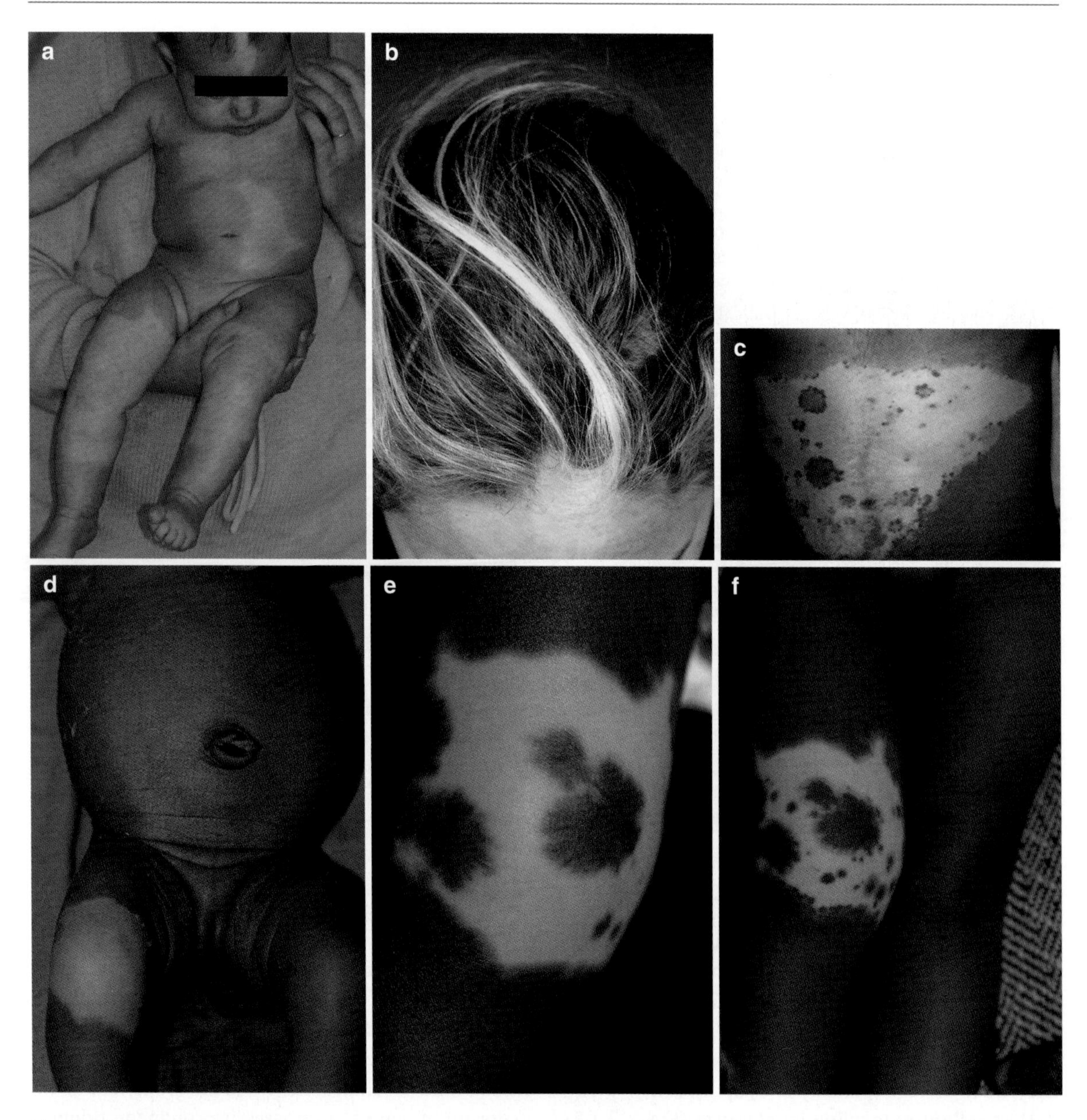

Fig. 1.2.1.1 Piebaldism: diffuse form without spontaneous repigmentation (**a**); limited form (*white forelock*) in the mother (**b**); typical spontaneous repigmentation pattern in an adult (courtesy Dr Y Gauthier) (**c**); repigmentation in the two first years of life (**d–f**), becoming clearly trichrome in (**f**)

cases, a biopsy can be useful to make an accurate diagnosis, e.g. showing spongiosis (eczema) or psorisiasiform changes with neutrophils in the stratum corneum (psoriasis) [17].

Para-malignant hypomelanoses/mycosis fungoides (Fig. 1.2.1.6). In dark-skinned patients, skin whitening may correspond to an early stage of epidermotropic T cell lymphoma. When signs of inflammation and skin infiltration are lacking, this presentation can be misleading [31]. A biopsy is able to show diagnostic changes (large size and atypia of epidermotropic lymphocytes especially in the basal layer, Chap. 1.2.2). The mechanism of pigment dilution in mycosis fungoides has been studied and a reduced expression of c-Kit has been shown. The loss of c-kit, and subsequent downstream effects on melanocyte survival, might be initiated by

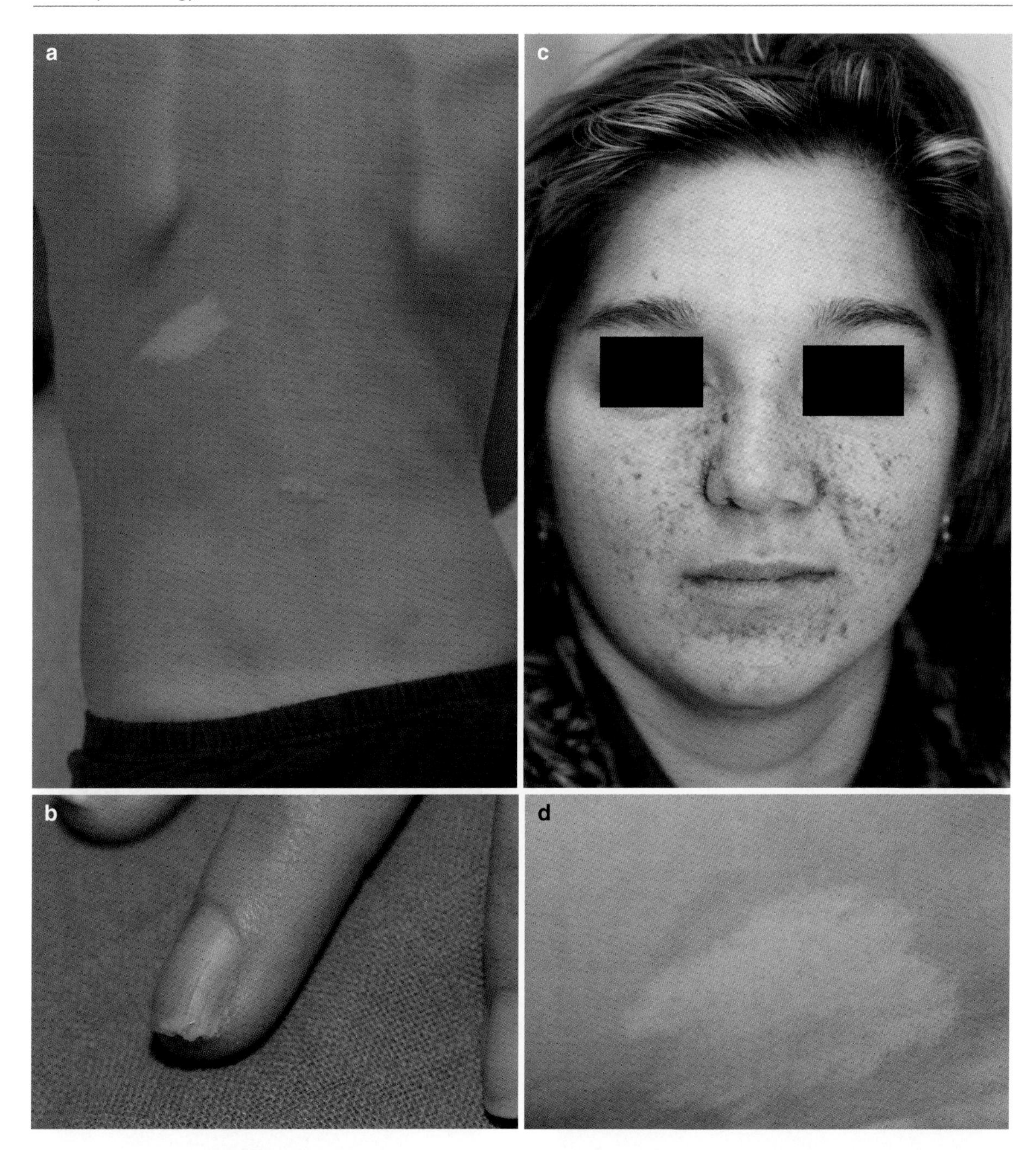

Fig. 1.2.1.2 Tuberous sclerosis (**a**) ash-leaf macules and chagreen patch on the dorsum of a young patient; Koenen's periungueal tumour in the same patient (**b**); facial angiofibromas (**c**) and large vitiligoid patch (**d**) of the abdominal belt in a 25-year-old female patient

cytotoxic effects of melanosomal-antigen-specific CD8 positive neoplastic T lymphocytes [31] (Sect. 2.2.7.4).

Para-malignant hypomelanoses/melanoma-associated depigmentation (Fig. 1.2.1.7). The vitiligoid changes associated with melanoma may result from a halo of depigmentation around a cutaneous melanoma (malignant Sutton's phenomenon) to more widespread vitiligoid changes. The margins of such vitiligoid lesions under Wood's lamp are usually less distinct than those of common vitiligo, and depigmentation is usually incomplete

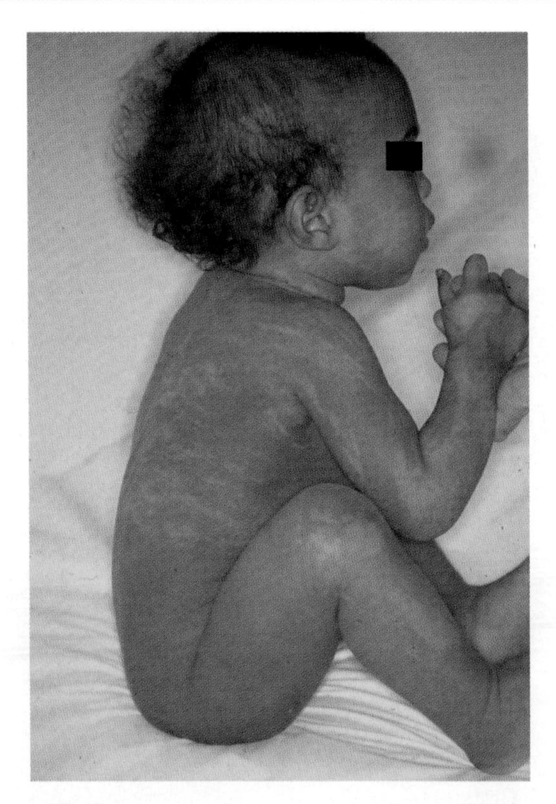

Fig. 1.2.1.3 Hypomelanosis of Ito (courtesy Dr O Enjolras, France). Note the striking blaschko-linear pattern (thin bands of depigmentation following the dorsoventral pattern of skin development)

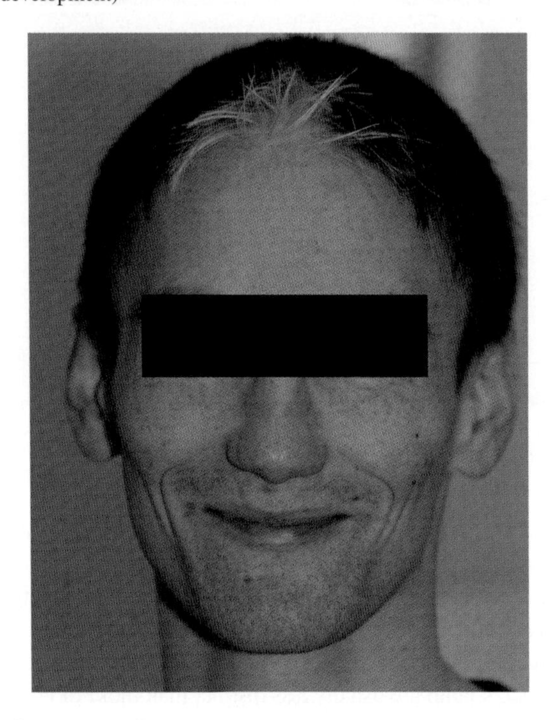

Fig. 1.2.1.4 Waardenburg's syndrome

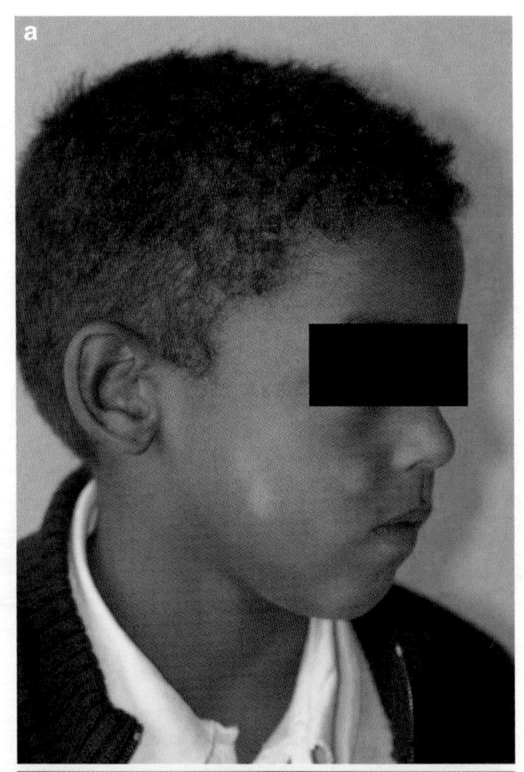

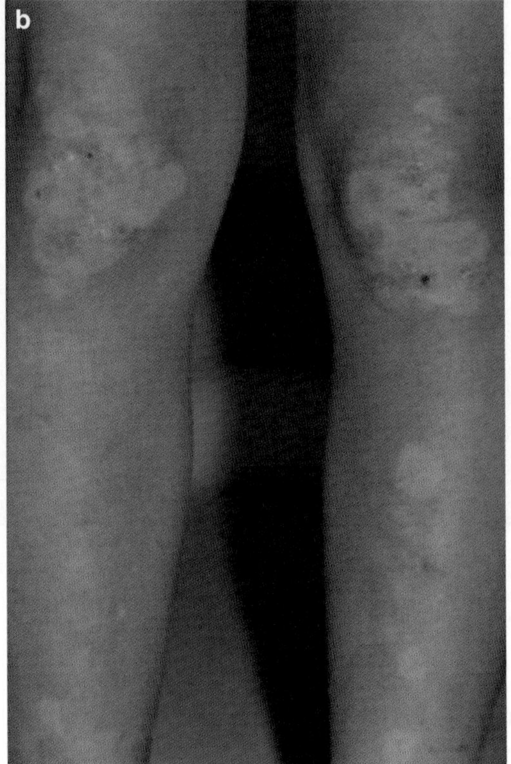

Fig. 1.2.1.5 Dermatitis with postinflammatory hypopigmentation: atopic dermatitis (**a**) (pityriasis alba); psoriasis (**b**)

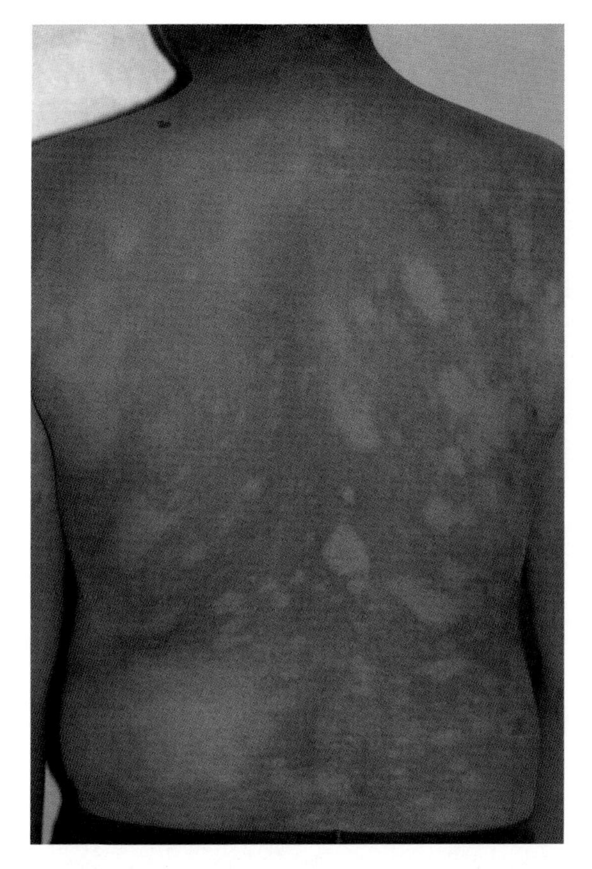

Fig. 1.2.1.6 Mycosis fungoides can present as hypopigmented patches in dark-skinned individuals. Histopathology is diagnostic (Chap. 1.2.2)

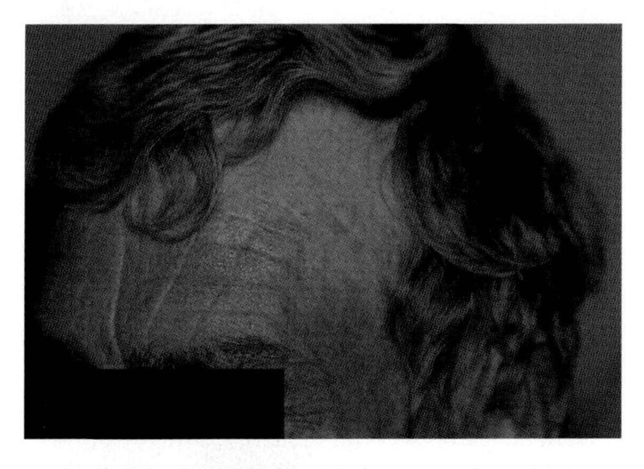

Fig. 1.2.1.7 Melanoma associated depigmentation. Melanoma-associated leucoderma is associated with spontaneous or vaccine-induced regression of a primary melanoma

[15]. The Koebner's phenomenon is usually absent. The prognostic value of depigmentation in the context of melanoma treated with interferon has been established [12].

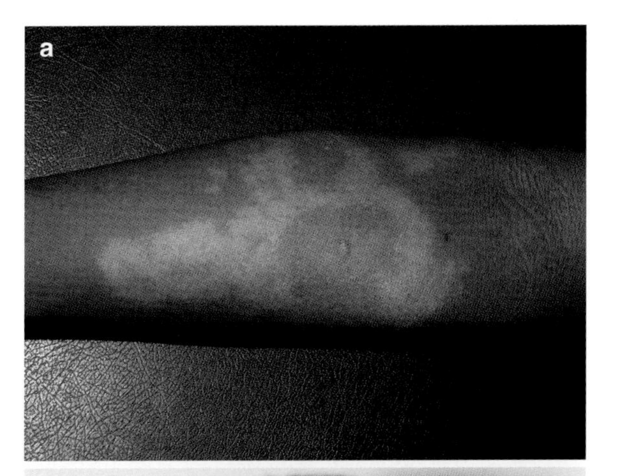

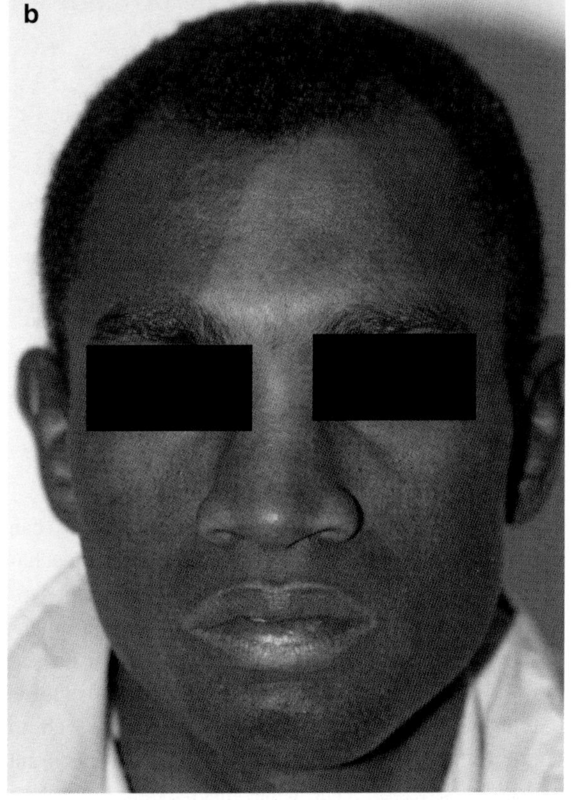

Fig. 1.2.1.8 Indeterminate leprosy (**a**, **b**) should be suspected in individuals living in endemic areas. Loss of sensitivity is usually associated. (courtesy (**a**) T. Passeron Collection)

Para-infectious hypopigmentation: Tinea versicolor can cause vitiligoid changes, generally after treatment in the absence of re-exposure to UV light. However, distribution, shape of the lesion, and some scaling and green fluorescence of untreated lesions, allow a definite diagnosis. Indeterminate leprosy (Fig. 1.2.1.8) is manifested by hypochromic patches which are hypoesthetic under

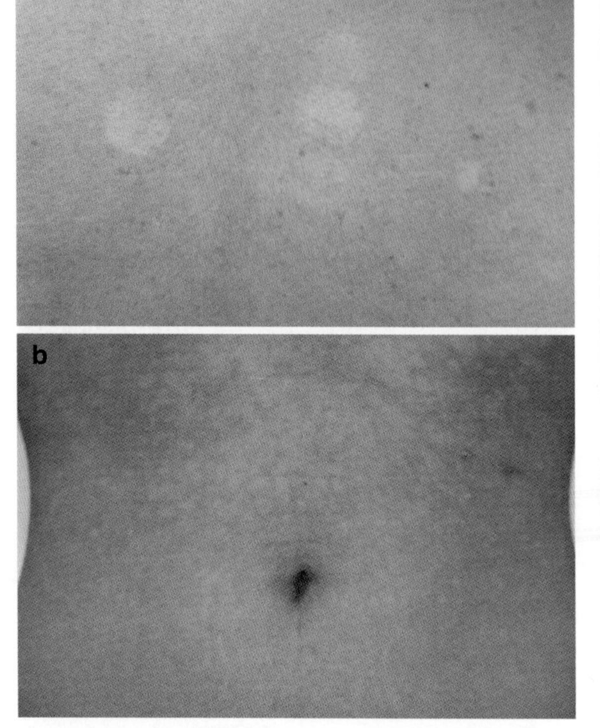

Fig. 1.2.1.9 Hypopigmented vitiligoid sequel of pityriasis versicolor (**a**) (upper back) and acquired macular hypomelanosis (**b**) (abdomen) (Guillet Westerhof syndrome)

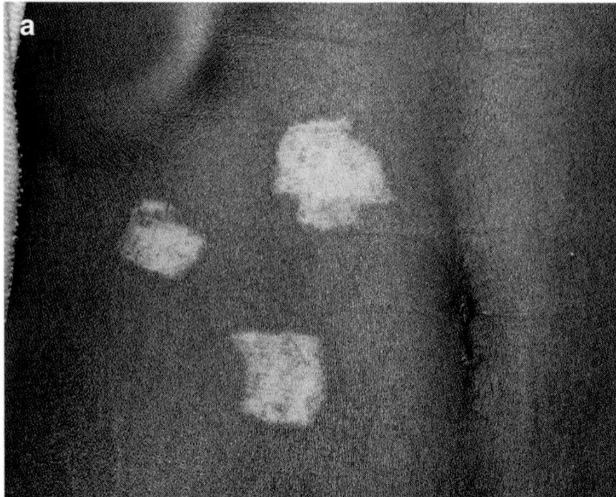

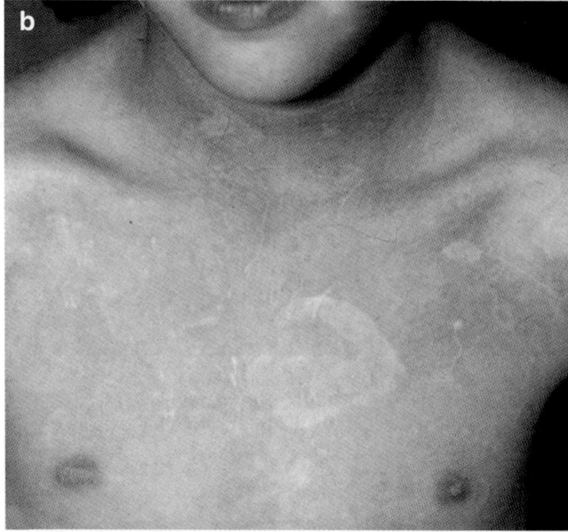

Fig. 1.2.1.10 Post traumatic depigmentation (courtesy T. Passeron Collection) (**a**). Note the unsusal square limits (**b**) definitive depigmentation following toxic epidermal necrolysis

light touch. In both cases, the infectious process can inhibit melanogenesis through largely unknown mechanisms. For tinea versicolor, toxic effects on pigment synthesis by fungal metabolites, especially tryptophan-derived metabolites of *M. furfur*, have been discussed [35] (Fig. 1.2.1.9a). Acquired macular hypomelanosis (Guillet-Westerhof disease) (Fig. 1.2.1.9b) is seen in young adults and frequently referred to as a "recalcitrant pityriasis versicolor". The white macules are present on the trunk with a reinforcement on the lower back and axillae. The role of *Propionibacterium acnes* has been based on red spots centring the macules under Wood's lamp, but this finding is not constant [13, 19, 25, 36]. This disease is largely unknown by dermatologists and usually not diagnosed.

Post traumatic leucoderma: When the melanocyte reservoir is depleted, as after deep burns or scars, which remove the hair follicles entirely or when the bulge area which contains melanocyte precursors is destroyed, the resulting wound healing process will not recapitulate pigmentation from the centre, and marginal repigmentation fails to compensate the loss. It may sometimes be difficult to distinguish some aspects from true vitiligo, when scarring is not obvious. This is the case in depigmenting sequels of toxic epidermal necrolysis (Fig. 1.2.1.10) [24].

Melasma: This common hypermelanotic disorder can be a diagnostic pitfall in the vitiligo clinic when the hyperpigmented facial lesions surround normal but hypochromic-looking skin (Fig. 1.2.1.11). Usually the pattern is different from vitiligo and the examination of other body sites allows a definitive diagnosis.

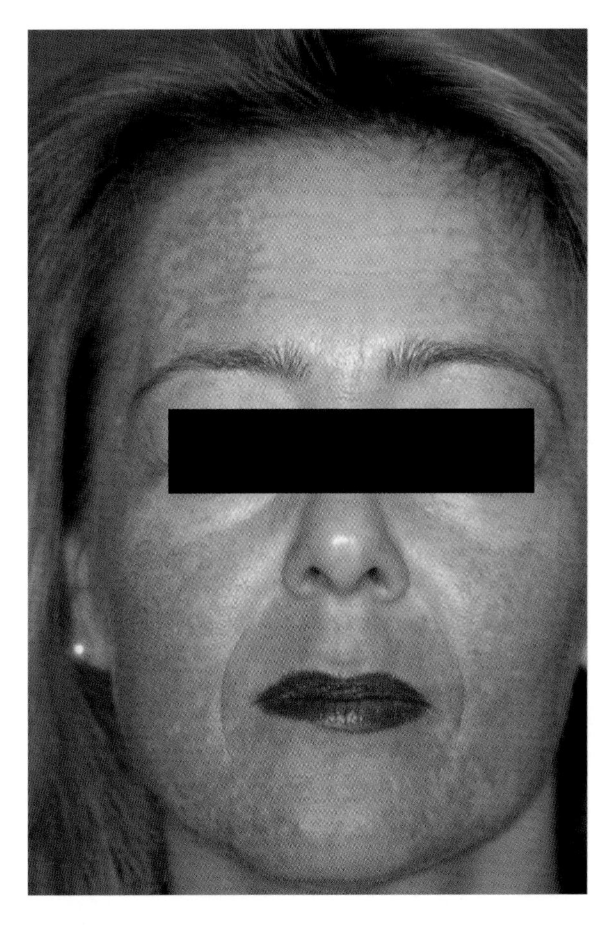

Fig. 1.2.1.11 Melasma: a complete skin examination helped by Wood's lamp examination is mandatory in this setting

Occupational and drug-induced depigmentation (Sect. 2.2.2.2): The so-called occupational vitiligo is a subset of vitiligo triggered by an occupational exposure, which evolves from contact depigmentation caused generally by a phenolic-catecholic derivative to a generalised phenomenon. When depigmentation is widespread, the limits between drug-induced depigmentation and true vitiligo are not easy to delineate. An occupational medicine advice can be helpful. For systemic drugs, except the depigmentation occurring after toxic epidermal necrolysis, which is related to a kind of superficial burn-like injury and also a rare type of universal depigmentation [32], other drugs may rarely cause vitiligoid depigmentation (chloroquine, fluphenazine, physostigmine) [5]. More recently, vitiligoid depigmentations following imatinib treatment for chronic myeloid leukaemia have been reported and are thought to be mediated via c-Kit signalling downregulation [7], as well as topically-induced vitiligoid lesions following imiquimod therapy

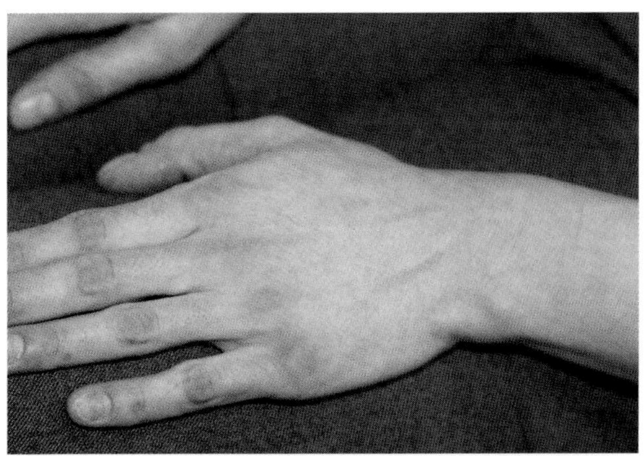

Fig. 1.2.1.12 Vitiligoid depigmentation of dorsum of hands in a patient with hand eczema treated with potent topical corticosteroids. Note persistence of pigment on knuckle pads

[22]. Long term use of potent topical corticosteroids may cause vitiligoid depigmentation (Fig. 1.2.1.12).

1.2.1.5 Segmental Vitiligo

SV is defined descriptively as for NSV except for a unilateral distribution ("asymmetric vitiligo") that may totally or partially match a cutaneous segment such as a dermatome, but not necessarily. The term *focal* is preferred for a limited lesion i.e. where the affected patch is small (10–15 cm^2) without an obvious distribution pattern. Other distribution patterns of SV can be encountered, that cross several dermatomes, or correspond to large areas delineated by Blaschko's lines. Some specific features exist such as rapid onset and hair follicle pigmentary system involvement. One unique segment is involved in most patients, but two or more segments with ipsi- or contralateral distribution is involved in rare patients (Chap. 1.3.2 and Part 2.3).

1.2.1.6 Conditions to Exclude from the Definition of SV

Although several disorders may pose problems, congenital hypomelanoses of segmental distribution named collectively naevus depigmentosus or achromic

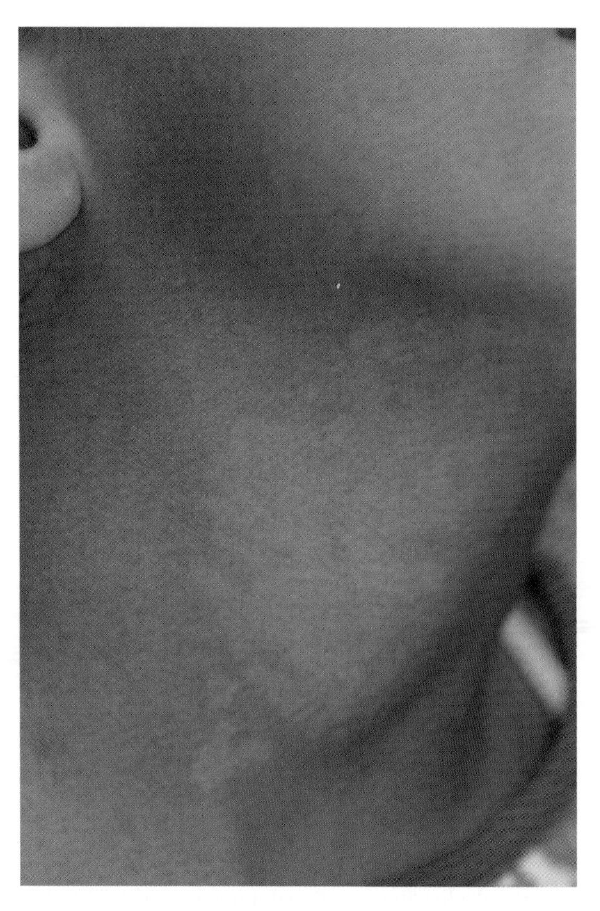

Fig. 1.2.1.13 Nevus depigmentosus, a common pitfall for the diagnosis of segmental vitiligo (see text)

nevus [14, 20], which might correspond to somatic mosaic defects of cutaneous pigmentary genes, are the most common differential diagnoses. Naevus depigmentosus is usually congenital or detectable in the first year of life and stable in size in proportion to the child's growth (Fig. 1.2.1.13). The lesion usually contains a normal or subnormal number of melanocytes compared with control perilesional skin, but the production of melanin pigment is reduced. Sun exposure may attenuate the difference in pigmentation from normal skin. In difficult cases, a biopsy is needed to differentiate naevus depigmentosus from SV. Segmental or hemicorporeal hypomelanosis of Ito (Fig. 1.2.1.3), which corresponds to another type of cutaneous mosaicism made of narrow depigmented streaks following Blaschko's lines, is not in practice, a real differential diagnosis problem for multisegmental vitiligo (Table 1.2.1.1).

1.2.1.7 Mixed Vitiligo

NSV and SV may coexist, and in this case SV lesions are usually more refractory to treatment [11], but mild segmental hypomelanoses may coexist with NSV (Part 2.3). Such refractory SV lesions can be unmasked by phototherapy. This entity has probably been underreported until recently [23, 26, 27].

1.2.1.8 Classification Issues

Acrofacial, common (*vulgaris*), generalised (non acrofacial), universal (Chap. 1.3.3), and mucosal (Chap. 1.3.4) forms of NSV have been frequently distinguished, on the basis of initial involvement, extension and major recognizable pattern. However, there is no clear scientific evidence of these being distinct diseases [21]. Vitiligo may evolve over time, e.g. the disease may be classified as acrofacial at its early debut, but it will later progress to be classified as generalized. Some cases of NSV may spare relatively the extremities (generalised non-acrofacial vitiligo). Some have a flexural tropism and others a predilection for extensor aspects. The Koebner phenomenon may or may not be prominent (Sect. 2.2.2.1). The association with halo nevi could also be an issue of clinical classification [10]. We thus recommend to classify all generalized cases, just as NSV, with the introduction of subsets if needed for specific studies or descriptive aims. In rare cases, when it is not possible to classify initially a case between the two main types SV and NSV, the category "*unclassified*" should be used. Subtypes are useful for reporting across different ethnic backgrounds. Mucosal cases are rare or undetected in white individuals. *Focal vitiligo* can belong either to SV or NSV when this classification scheme is used. It remains a kind of black hole in vitiligo, because of the limited investigations performed so far in this category of patients. Some cases may, in fact, be distinct from true vitiligo and considered merely as delayed-onset developmental pigment cell anomalies. However, most cases are, until further evidence, situated within SV as cases with limited/non systematized pattern of distribution. For SV cases, topographical classifications for the common head and neck involvement are not yet settled (Sect. 1.3.2.2). A tentative classification summary appears on Table 1.2.1.2.

Table 1.2.1.2 Classification of vitiligo

Type of vitiligo	Subtypes	Remarks
Non segmental (NSV)	(Focal),[a] mucosal, acrofacial, generalised, universal	Subtyping may not reflect a distinct nature, but useful information for epidemiologic studies
Segmental (SV)	Focal,[b] Mucosal, Unisegmental, bi- or plurisegmental	Further classification according to distribution pattern possible, but not yet standardized
Mixed (NSV + SV)	According to severity of SV	Usually the SV part in mixed vitiligo is more severe
Unclassified	Focal at onset, multifocal asymmetrical non segmental, mucosal (one site)	This category is meant to allow, after a sufficient observation time (and if necessary investigations), to make a definitive classification

[a]Possible onset of NSV
[b]See text for discussion

References

1. Alkhateeb A, Fain PR, Thody A et al (2003) Epidemiology of vitiligo and associated autoimmune diseases in Caucasian probands and their families. Pigment Cell Res 16:208–214
2. Bach JF (2002) The effect of infections on susceptibility to autoimmune and allergic diseases. N Engl J Med 347: 911–920
3. Berger TG, Kiesewetter F, Maczek C et al (2006) Psoriasis confined strictly to vitiligo areas – a Koebner-like phenomenon? J Eur Acad Dermatol Venereol 20:178–183
4. Boisseau-Garsaud AM, Garsaud P, Calès-Quist D et al (2000) Epidemiology of vitiligo in the French West Indies (Isle of Martinique). Int J Dermatol 39:18–20
5. Boissy RE, Manga P (2004) On the etiology of contact/occupational vitiligo. Pigment Cell Res 17:208–214
6. Brandrup F, Holm N, Grunnet N et al (1982) Psoriasis in monozygotic twins: variations in expression in individuals with identical genetic constitution. Acta Derm Venereol 62: 229–236
7. Cario-André M, Ardilouze L, Pain C et al (2006) Imatinib mesilate inhibits melanogenesis in vitro. Br J Dermatol 155: 493–494
8. Das SK, Majumder PP, Chakraborty R et al (1985) Studies on vitiligo. I. Epidemiological profile in Calcutta, India. Genet Epidemiol 2:71–78
9. Duffy DL, Spelman LS, Martin NG (1993) Psoriasis in Australian twins. J Am Acad Dermatol 29:428–434
10. De Vijlder HC, Westerhof W, Schreuder GM et al (2004) Difference in pathogenesis between vitiligo vulgaris and halo nevi associated with vitiligo is supported by an HLA association study. Pigment Cell Res 17:270–274
11. Gauthier Y, Cario-Andre M, Taieb A (2003) A critical appraisal of vitiligo etiologic theories. Is melanocyte loss a melanocytorrhagy? Pigment Cell Res 16:322–332
12. Gogas H, Ioannovich J, Dafni U et al (2006) Prognostic significance of autoimmunity during treatment of melanoma with interferon. N Engl J Med 354:709–718
13. Guillet G, Helenon R, Gauthier Y et al (1988) Progressive macular hypomelanosis of the trunk: primary acquired hypopigmentation. J Cutan Pathol 15:286–289
14. Hann SK, Lee HJ (1996) Segmental vitiligo: clinical findings in 208 patients. J Am Acad Dermatol 35:671–674
15. Hartmann A, Bedenk C, Keikavoussi P et al (2008) Vitiligo and melanoma-associated hypopigmentation (MAH):shared and discriminative features. J Dtsch Dermatol Ges 6:1053–1059
16. Howitz J, Brodthagen H, Schwartz M et al (1977) Prevalence of vitiligo: epidemiological survey on the Isle of Bornholm, Denmark. Arch Dermatol 113:47–52
17. Kim YC, Kim YJ, Kang HY et al (2008) Histopathologic features in vitiligo. Am J Dermatopathol 30:112–116
18. Koga M (1977) Vitiligo: a new classification and therapy. Br J Dermatol 97:255–261
19. Kumarasinghe SP, Tan SH, Thng S et al (2006) Progressive macular hypomelanosis in Singapore: a clinico-pathological study. Int J Dermatol 45:737–742
20. Lee HS, Chun YS, Hann SK (1999) Nevus depigmentosus: clinical features and histopathologic characteristics in 67 patients. J Am Acad Dermatol 40:21–26
21. Liu JB, Li M, Yang S et al (2005) Clinical profiles of vitiligo in China: an analysis of 3742 patients. Clin Exp Dermatol 30:327–331
22. Mashiah J, Brenner S (2008) Possible mechanisms in the induction of vitiligo-like hypopigmentation by topical imiquimod. Clin Exp Dermatol 33:74–76
23. Mulekar SV, Al Issa A, Asaad M et al (2006) Mixed vitiligo. J Cutan Med Surg 10:104–107
24. Oplatek A, Brown K, Sen S, et al (2006) Long-term follow-up of patients treated for toxic epidermal necrolysis. J Burn Care Res 27:26–33
25. Relyveld GN, Dingemans KP, Menke HE et al (2008) Ultrastructural findings in progressive macular hypomelanosis indicate decreased melanin production. J Eur Acad Dermatol Venereol 22:568–574
26. Schallreuter KU, Krüger C, Rokos H et al (2007) Basic research confirms coexistence of acquired Blaschkolinear Vitiligo and acrofacial Vitiligo. Arch Dermatol Res 299: 225–230
27. Schallreuter KU, Krüger C, Würfel BA et al (2008) From basic research to the bedside: efficacy of topical treatment with pseudocatalase PC-KUS in 71 children with vitiligo. Int J Dermatol 47:743–753
28. Schultz Larsen F (1993) Atopic dermatitis: a genetic-epidemiologic study in a population-based twin sample. J Am Acad Dermatol 28:719–723
29. Sehgal VN, Srivastava G (2007) Vitiligo: compendium of clinico-epidemiological features. Indian J Dermatol Venereol Leprol 73:149–156
30. Singh M, Singh G, Kanwar AJ, Belhaj MS (1985) Clinical pattern of vitiligo in Libya. Int J Dermatol 24:233–235
31. Singh ZN, Tretiakova MS, Shea CR, Petronic-Rosic VM (2006) Decreased CD117 expression in hypopigmented

mycosis fungoides correlates with hypomelanosis: lessons learned from vitiligo. Mod Pathol 19:1255–1260
32. Smith DA, Burgdorf WH (1984) Universal cutaneous depigmentation following phenytoin-induced toxic epidermal necrolysis. J Am Acad Dermatol 10:106–109
33. Spritz RA (2007) The genetics of generalized vitiligo and associated autoimmune diseases. Pigment Cell Res 20: 271–278
34. Taïeb A, Picardo M; VETF Members (2007) The definition and assessment of vitiligo: a consensus report of the Vitiligo European Task Force. Pigment Cell Res 20:27–35
35. Thoma W, Krämer HJ, Mayser P (2005) Pityriasis versicolor alba. J Eur Acad Dermatol Venereol 19:147–152
36. Westerhof W, Relyveld GN, Kingswijk MM et al (2004) Propionibacterium acnes and the pathogenesis of progressive macular hypomelanosis. Arch Dermatol 140: 210–214

Histopathology

1.2.2

Flávia Pretti Aslanian, Absalom Filgueira, Tullia Cuzzi, and Béatrice Vergier

Contents

F. P. Aslanian (✉)
Federal University of Rio de Janeiro, Rio de Janeiro, Brasil
e-mail: fpretti@cyberwal.com.br

Core Messages

- Adjacent, apparently normally pigmented skin should always be included to compare pigmented and non-pigmented skin.
- It is important to indicate to the pathologist whether the lesion is stable or progressive.
- The main histological finding in vitiligo is a marked reduction or absence of pigmentation along the basal layer of the epidermis.
- Absence or loss of melanocytes in vitiligo leads to an absence of melanin resulting in depigmentation. Histochemistry (Fontana-Masson stain) and immunohistochemistry help to demonstrate pigment and cell loss.
- Associated inflammatory changes and/or pigmentary incontinence may or may not be present, in variable intensity, generally mild, and are noted at an early stage at the border of progressing lesions.
- Histopathology is needed to rule out other causes of acquired hypopigmentation, especially mycosis fungoides and leprosy.

1.2.2.1 Introduction

Different histopathological findings have been used to support various theories concerning the development of vitiligo [17]. From the practical standpoint, vitiligo lesions can be easily overlooked by histopathologists. Vitiliginous skin shows a complete loss of melanin pigment from the epidermis and an absence of melanocytes

M. Picardo and A. Taïeb (eds.), *Vitiligo*,
DOI 10.1007/978-3-540-69361-1_1.2.2, © Springer-Verlag Berlin Heidelberg 2010

in its basal layer [32]. On the contrary, the diagnosis of vitiligo can be made readily on clinical grounds. In general, only in certain difficult cases, a skin biopsy may be required for differential diagnosis purposes [35]. The role of skin biopsies is currently debated for staging purposes (microinflammation) (Chap. 1.3.10).

Usually, the thickness of the tissue sections cut from paraffin-embedded blocks is 4 μm. It might be an additional reason for the scarcity of changes visible on routine microscopy in vitiligo [22]. Thinner (1 μm) sections may result in a more accurate histological evaluation when necessary [27]. Indeed, when carefully evaluated by light and electron microscopy, vitiligo lesions reveal striking abnormalities, such as an absence or diminished number of melanocytes and their vacuolization, keratinocyte vacuolization, absent or migrating melanocytes in the outer root sheath, changes in nerves, dermal infiltrate with T-lymphocytes, histiocytes, melanophages and erythrocytes. This chapter is focused on generalized (non-segmental) vitiligo (NSV), which represents the great majority of vitiligo cases,and can be associated with microinflammatory changes. Histopathological data on segmental vitiligo are scarce or not detailed enough to conclude [5]. However, there is evidence that lichenoid inflammation can also occur in typical segmental cases (Dr. Attili, India, personal communication to the Editors and Chap. 1.3.10).

1.2.2.2 Recommendations for a Proper Interpretation of Biopsies

As a rule, when performing a biopsy of a hypopigmented lesion suspected of being vitiligo, it is worthwhile to include some adjacent apparently normal skin in the specimen. This enables histopathologists to compare and contrast the pigmented and non-pigmented samples, with regard to features such as the number of melanocytes and amount of melanin within the epidermis [1]. The identification of the transition area between those segments, called the "lesional border", is usually possible. However, in some cases, there is a marked reduction of pigmentation in the basal layer of the adjacent clinically non-affected skin in the specimen, which is best demonstrated using the Fontana-Masson (FM) silver stain (Table 1.2.2.1). Before analysing a biopsy, it is important to know whether the disease was progressing or not when the biopsy was performed. The epidermal and dermal changes in

Table 1.2.2.1 Some histochemical stains and antibodies commonly used in dermatopathology, focusing on melanin or normal melanocytes

Stain	Purpose of stain	Comments	References
Silver nitrate	Melanin(argyrophilic)	Stains black	[25]
Fontana-Masson (FM)	Melanin (argentaffin)	Black silver precipitate	[25]
DOPA reaction	Melanocytes	Tyrosinase activity converts colourless DOPA into DOPA-melanin (dark brown to black)	[24, 25]
Antibodies	*Directed against*		
S-100(P)	Melanocytes	Most sensitive (initial screening) Lack of specificity (Langerhans cells react)	[7, 34]
Melan-A(A-103)/Mart-1 (GP100 group)	Melanocytes	High degree of specificity Melanosome-associated marker Intermediate sensitivity (60–80%)	[7, 34]
NKI-beteb	Melanocytes (100-kDa melanosome associated antigen)	High specificity	[20]
T311	Melanocytes (anti-tyrosinase)	Intermediate sensitivity (80%)	[7]
HMB-45	Melanocytes (Melanosome-associated cytoplasmic antigen)	Highly specific Lack of sensitivity	[7, 20]
MEL-5(clone TA99)	Pigment associated antigen 70–80-kDa glycoprotein	Good sensitivity	[8, 20]

vitiligo are likely to be more prominent in actively spreading vitiligo than in stable disease. By comparing specimens obtained in patients with active disease and patients with stable vitiligo, it has been shown that degenerative changes in melanocytes and keratinocytes, epidermal and dermal infiltration of lymphocytes and melanophages occur in the skin of actively spreading vitiligo, as well as in the adjacent normal-appearing skin, in a more noticeable fashion than in stable vitiligo [12]. The punctate peripheral areas of depigmented patches are frequently microinflammatory [5].

1.2.2.3 Histo- and Immunohistochemistry Techniques for Studying Vitiligo

Certain staining methods allow the visualization of melanocytes and their products, under light microscopy. Silver stains (Fontana-Masson stain) evidence the presence of melanin, due to its argentaffin property. It occurs when the ammoniated silver nitrate is reduced by melanin phenolic groups, forming a black silver precipitate [25]. The DOPA reaction (Bloch's reaction) is instructive with regard to the melanization process. Histochemical assessment of vitiliginous areas reveals a complete absence of DOPA-positive melanocytes, indicating a total absence of tyrosinase activity [16]. However, it is not used in routine diagnostic dermatopathology (Chap. 2.2.5).

Immunohistochemical detection of melanocytes can be accomplished by several methods, which have a variable sensibility and specificity for melanocytes [7, 8, 24]. This area has benefited from antibodies developed for melanoma research. Table 1.2.2.1 summarizes some frequently used methods for melanin and melanocyte detection, which are useful in vitiligo studies.

1.2.2.4 Pathological Findings

Pathologically, vitiligo is perceived as an inflammatory process, beginning with sparse superficial mononuclear cell infiltration at the dermal-epidermal interface (Fig. 1.2.2.1). Early on, melanocytes are still present at the dermo-epidermal junction but they disappear progressively from affected sites. When they disappear, no melanin is produced as demonstrated by the FM silver stain [2]. For diagnostic purposes, specimen are most frequently evaluated when the inflammatory phenomena have disappeared. The absence of marked inflammatory reaction allows to differentiate vitiligo from other lesions that produce hypochromia, such as lupus erythematosus, eczema or psoriasis (Chap. 1.2.1).

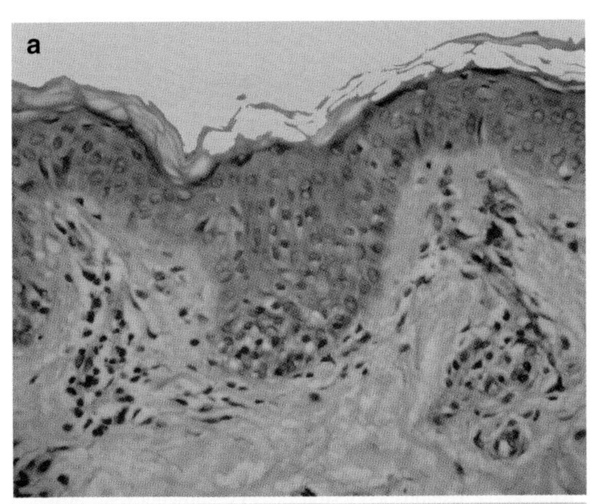

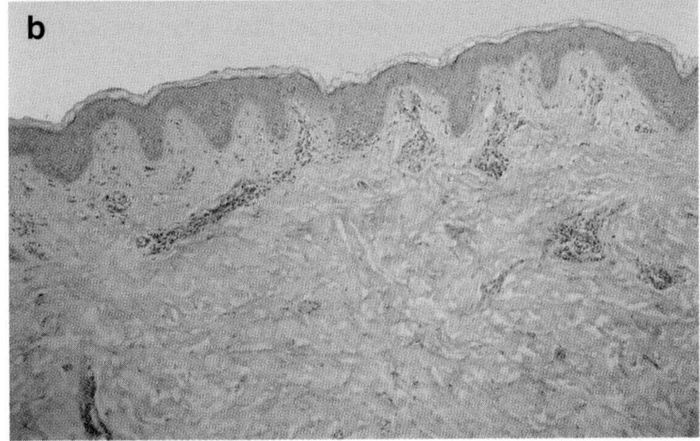

Fig. 1.2.2.1 Early stages of NSV: (**a**) superficial perivascular infiltrates of lymphocytes and presence of lymphocytes in the epidermis, with mild spongiosis, and pigment incontinence (HE 400X) (**b**) superficial perivascular cellular infiltrates and epidermal infiltrate (HE 100×)

Pathologic Stages

Histopathological findings in vitiligo differ according to the three phases of the disease: (1) early stages/progressive lesions, (2) established lesions and (3) long standing lesions [1, 11, 24]. *Established vitiligo macules* show

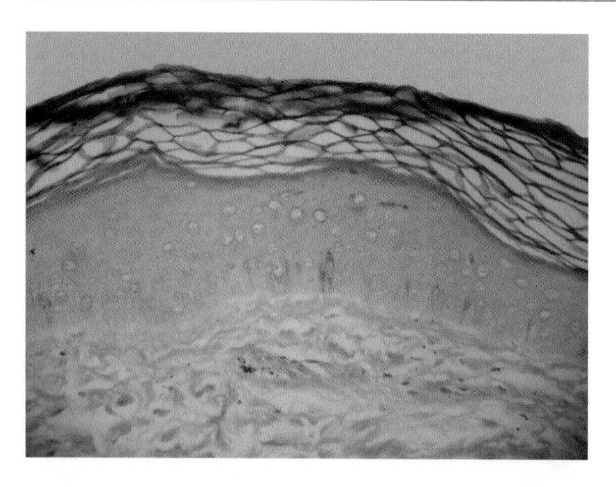

Fig. 1.2.2.2 Established lesion of NSV showing total absence of pigment in the basal layer and pigmentary incontinence (melanophages in the upper dermis) (Fontana-Masson 400×)

apparently normal skin except for the absence of pigmentation of the basal epidermal layer (Fig. 1.2.2.2). The FM silver stain confirms the absence of melanin [28]. There may be a sparse superficial perivascular infiltrate. Such changes are not diagnostic or specific for vitiligo [35]. Histochemical studies show a loss of DOPA-positive melanocytes in the basal layer. On the edge of vitiliginous macules, the melanocytes look larger than usual, and present long dendritic processes filled with melanosomes [28]. Immunohistochemical studies using antibodies panel have shown that melanocytes are commonly absent in vitiligo lesions. They are demonstrated only occasionally in the affected skin in reduced number, and sometimes presenting degenerative changes [30].

Long-standing lesions characteristically present only a marked absence of melanin in the epidermis as the main finding. Inflammatory infiltrates cannot be demonstrated [1]. In contrast, *early stage lesions* show a superficial perivascular infiltrate of lymphocytes, and sometimes a variable number of lymphocytes in the lower half of the epidermis. This aspect can be difficult to relate to vitiligo without proper clinical information, and can be mistaken for a series of other cutaneous inflammatory lichenoid/spongiotic disorders. If serial sections are examined, a lymphocyte can sometimes be found in close apposition to a melanocyte at the advancing edge of the lesion [32]. The content of melanin in the epidermal basal layer can be either reduced or not.

Pigment Incontinence

Melanophages can be better seen in the upper dermis of actively spreading lesions by the FM stain [26] either when there is a mild or marked reduction of melanin in the basal layer (Figs. 1.2.2.1a and 1.2.2.2).

Findings in Perilesional and Distant Areas in NSV

Some authors have described that the normal appearing skin, adjacent to vitiliginous skin, shows characteristic histopathological features. Degenerative changes have been documented in both keratinocytes and melanocytes, not only from the border of lesions but also from adjacent skin. Focal areas of vacuolar degeneration in the basal layer, in association with a mild mononuclear cell infiltrate, have also been observed in clinically normal pigmented skin adjacent to vitiliginous areas [6, 21]. Mosher et al. [23] have described that normally pigmented skin up to 15 cm from the amelanotic macules has basal layer vacuolization, vacuolated keratinocytes, and extracellular granular material deposits. In contrast, Gokhale and Mehta [10] mentioned that biopsy specimens from the normal skin of vitiligo patients did not show any degenerative or inflammatory changes or presence of suprabasal clear cells. Additionally, they observed mainly inflammatory changes in early lesions and prominent degenerative changes in long-standing patches, besides degenerative changes in nerves and sweat glands. Another study, in which immunohistochemistry using the melanocyte-specific antibody NKI-beteb was performed, showed a similar melanocyte distribution and pattern in the non-lesional skin of vitiligo patients when compared to that of normal controls, whereas melanocytes were absent or fragmented in the lesional segment of the perilesional epidermis, and totally depleted in lesional vitiligo skin [30].

More recently, a microscopic disappearance of melanocytes, also called "microdepigmentation", has been described in association with T-cell infiltrates in the dermoepidermal junction of the clinically normal-pigmented skin in patients with active generalised vitiligo [31]. This finding has been corroborated by the observation of a decreased epidermal basal

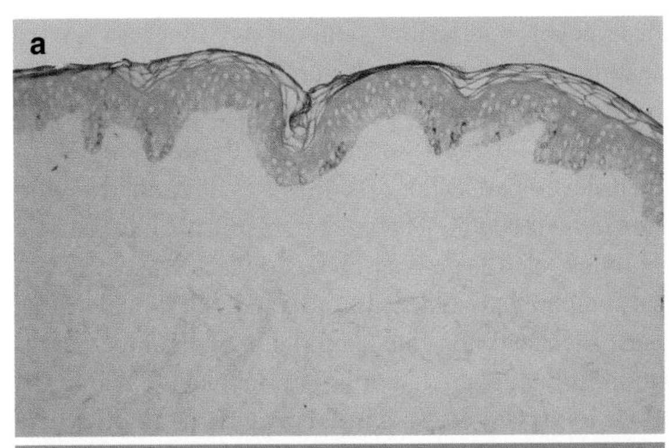

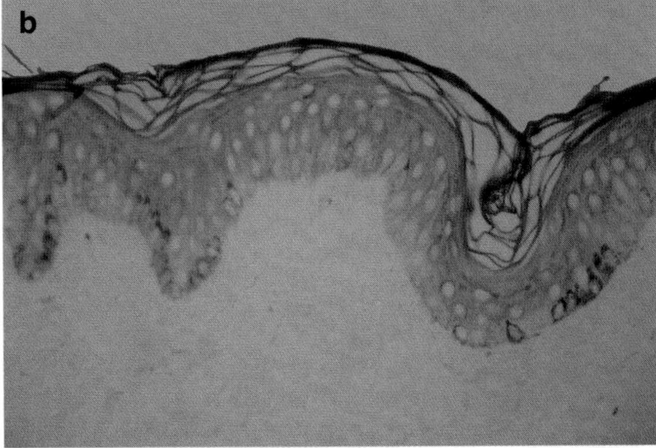

Fig. 1.2.2.3 Clinically not affected in NSV: areas of epidermal basal cell layer depigmentation (Fontana-Masson 100X (**a**) and 400X(**b**))

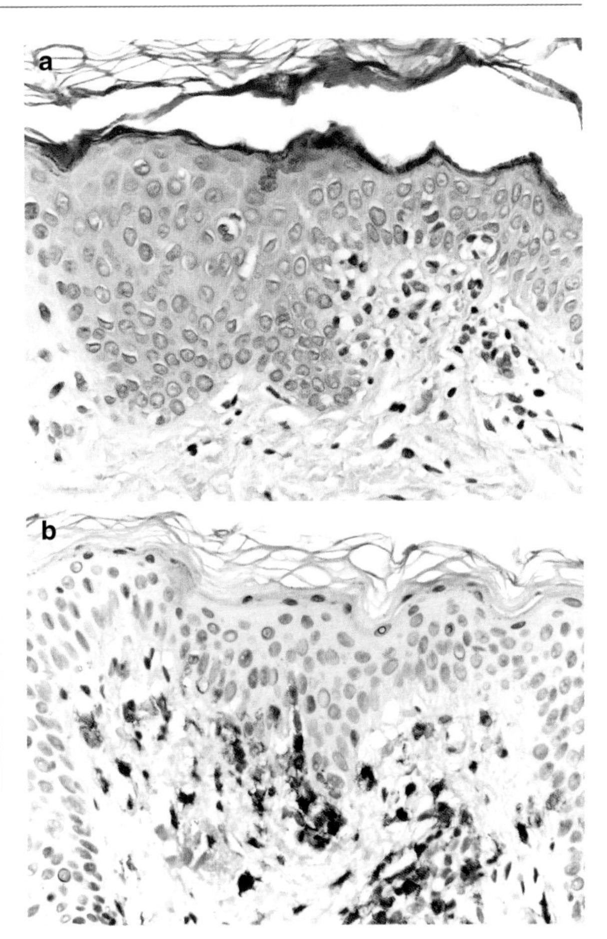

Fig. 1.2.2.4 Border of a progressing lesion in a patient with NSV showing a lichenoid and intraepidermal infiltrate (**a**), as indicated by HE stain 400×, made mostly of CD8$^+$ cells (**b**)

pigmentation in the normal-appearing skin of vitiligo patients (Fig. 1.2.2.3), which was statistically significant when compared to normal controls in skin samples stained with haematoxylin-eosin (HE) and Fontana-Masson. Surprisingly, such abnormal histological findings included the clinically pigmented skin at sites even further from the afflicted areas in those patients [26].

Nature of the Inflammatory Infiltrate

The composition of the cell infiltrate [13] in the skin of patients with vitiligo has been studied using monoclonal antibodies since the 1990s [3, 4]. Al Badri et al. [4] compared the margins of vitiligo lesions with non-lesional skin of the same patient, and controls without skin disease, and found a much greater number of CD3$^+$T lymphocytes and CLA$^+$ cells in the epidermis and dermis of the affected border than in other specimens (Fig. 1.2.2.4). In addition, the CD3$^+$CD8$^+$T lymphocytes and CLA$^+$ cells were more frequent than the number of T helper lymphocytes (CD4$^+$) in active lesions of vitiligo. The CLA$^+$T lymphocytes were more numerous in the epidermis than in the dermis, and many lymphocytes expressed MHC class II and interferon gamma, denoting their activation. Ahn et al. [3] examined the marginal skin of actively spreading and stable vitiligo using ICAM-1, HLA-DR, CD4 and CD8 monoclonal antibodies. Their main finding was that ICAM-1 was expressed on basal keratinocytes in active lesions, but not in stable lesions.

Le Poole et al. [19, 20] investigated the inflammatory infiltrate in both lesional and non-lesional skin in three cases of inflammatory vitiligo (Chap. 1.3.10), which showed exuberant histological changes. The main immunohistochemical findings were the increased number of CD68$^+$ macrophages in the skin, increased CD8/CD4 ratio, and increased expression of CLA in the cells that infiltrated lesional skin. CD8$^+$T lymphocytes were shown juxtaposed to the remaining melanocytes, suggesting the involvement of cellular immunity leading to melanocyte destruction. Although these infiltrates are much more easily characterized in rare cases of inflammatory vitiligo, these findings may be extended to the majority of cases of rapidly progressive NSV [5].

Some authors have confirmed the involvement of $CD8^{+}CLA^{+}T$ lymphocytes and macrophages in the loss of melanocytes in NSV, through single and double staining [30]. Degenerated melanocytes were found in the perilesional skin, and a total absence of NKI-beteb staining (specific antibodies to melanocytes) was found in lesional skin. $CD8^{+}T$ cells were in greater numbers in the affected skin (CD4/CD8 ratio = 0.48) and juxtaposed to the remaining melanocytes, which could be proven by double staining NKI-beteb/CD8. While the overall proportion of $CLA^{+}T$ lymphocytes was not increased compared to normal controls, these lymphocytes were found clustered at the dermo-epidermal junction, and expressed the markers of cytotoxicity granzyme-B and perforin, exactly in the place of interaction with melanocytes. There was also a reduction in the number of Langerhans cells (LC) in lesional skin in 7 out of 10 patients with progressive NSV, some of which received PUVA therapy, which can independently lead to a depletion of these cells [14, 33].

Sharquie et al. [27] demonstrated that epidermal mononuclear cell infiltration occurred in 80% of the marginal areas of lesions whose age ranged from 3 to 12 weeks, suggesting that the epidermal lymphocytic infiltration is the primary immune event in vitiligo. However, a heavy lymphocytic infiltrate in the upper dermis is a rare finding [15].

With regards to the number, changes and role of LC in vitiligo skin, data are conflicting. We agree with those who found a depletion of epidermal LC during periods of activity of the disease (personal unpublished data), and during the repigmentation period. The reappearance of these cells possibly would occur in stable vitiligo [14, 30].

1.2.2.5 Differential Diagnosis

In some cases, clinical features associated with cutaneous depigmentation may provide a clue (Chap. 1.2.1). However, the differential diagnosis with some conditions may be challenging, especially in the early stages of hypochromic lesions, without any other associated features, such as initial lesions of indeterminate leprosy with no alteration of peripheral nerve sensitivity. Indeterminate leprosy seen prior to the development of well-developed lesions, manifests as ill-defined hypopigmented macules. There are non-specific microscopic findings, such as perivascular, periadnexal and perineural lymphohistiocytic infiltrates and occasional organisms seen with the Fite stain, mainly when the nerves are infiltrated, which favour or establish the diagnosis. It is important to obtain subcutaneous tissue where infiltrates can be seen in this setting.

In hypopigmented mycosis fungoides [9, 29] epidermotropism is a clue for the diagnosis (Fig. 1.2.2.5). Post-inflammatory hypopigmentation rarely shows a

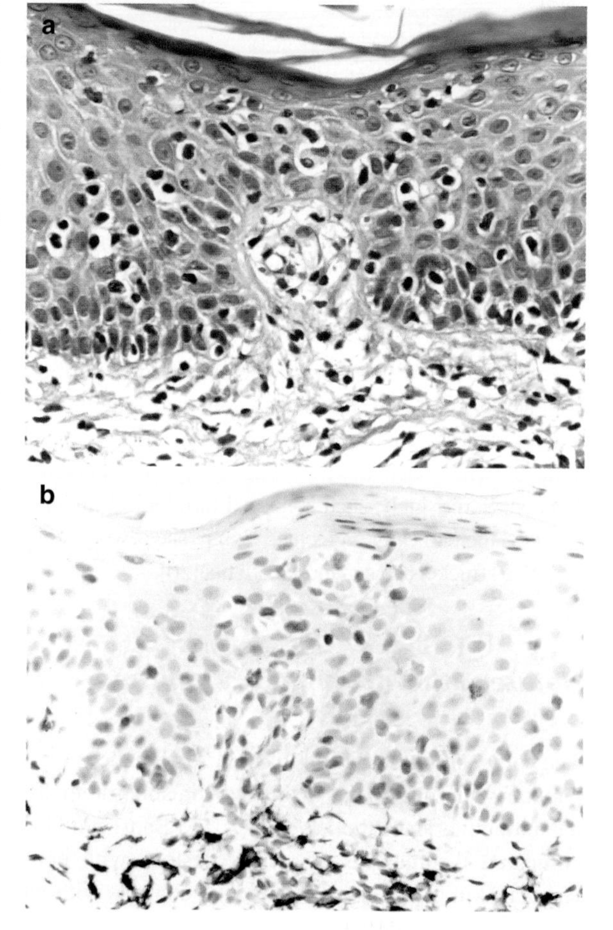

Fig. 1.2.2.5 Mycosis fungoides showing epidermotropism (**a**). Note lymphocytes with large and atypical nuclei between keratinocytes of the basal and spinous layers. They are mostly $CD4^{+}$ since CD8 stain (**b**) is negative in the epidermis – note some cells stained in the dermis

complete loss of melanocytes, and melanophages are often seen in the dermis [1], but we have demonstrated that melanophages are likely to be observed in active vitiligo lesions [26]. For hypochromic lesions without characteristic histopathology, histopathological evaluation is not helpful. Sometimes, immunohistochemical staining for melanocytes is needed for a differential diagnosis of vitiligo (Table 1.2.2.1). Nevus depigmentosus is a congenital, hypopigmented, stable macule or patch, usually on the trunk or proximal extremities, which may be clinically similar to segmental vitiligo. The microscopic feature is a normal or decreased number of melanocytes [18] in contrast with SV which shows none or rare melanocytes on microscopic examination. Using specific antibodies, Kim et al. [15] showed that their number was significantly decreased in vitiligo skin as compared with normal or nevus depigmentosus skin, although melanocytes can exist in a small number of vitiligo lesions. So, the NKI/beteb and MART-1 immunostains would be helpful to differentiate those lesions. With the FM stain it is possible to reveal the remaining melanin pigment, and the ratio of pigmented area to epidermal area. This ratio is far lower in vitiligo skin than in nevus pigmentosus.

In challenging cases, confronting clinical characteristics and histopathological findings should be considered to distinguish vitiligo from other hypopigmented lesions, since an accurate diagnosis may have important therapeutic implications [34].

References

1. Ackerman AB, Chongchinant N, Sanchez J et al (1997) Histologic diagnosis of inflammatory skin diseases. An algorithmic method based on pattern analysis, 2nd edn. Williams & Wilkins, Baltimore, MD
2. Ackerman AB, Kerl H, Sánchez J (2000) A clinical atlas of 101 common skin diseases (with histopathologic correlation), 1st edn. Ardor Scribendi, New York, pp 645–650
3. Ahn SK, Choi EH, Lee SH et al (1994) Immunohistochemical studies from vitiligo – comparison between active and inactive lesions. Yonsei Med J 35:404–410
4. Al Badri AMT, Todd PM, Garioch JJ et al (1993). An immunological study of cutaneous lymphocytes in vitiligo. J Pathol 170:149–155
5. Attili VR, Attili SK (2008) Lichenoid inflammation in vitiligo – a clinical and histopathologic review of 210 cases. Int J Dermatol 47:663–669
6. Bhawan J, Bhutani LK (1983) Keratinocyte damage in vitiligo. J Cutan Pathol 10:207–212
7. Clarkson KS, Sturdgess IC, Molyneux AJ (2001) The usefulness of tyrosinase in the immunohistochemical assessment of melanocytic lesions: a comparison of the novel T311 antibody (anti-tyrosinase) with S-100, HMB45, and A103(anti-melan-A). J Clin Pathol 54:196–200
8. Dean NR, Brennan J, Haynes J et al (2002) Immunohistochemical labelling of normal melanocytes. Appl Immunohistochem Mol Morphol 10:199–204
9. El-Darouti MA, Marzouk SA, Azzam O et al (2006) Vitiligo vs. hypopigmented mycosis fungoides (histopathological and immunohistochemical study, univariate analysis). Eur J Dermatol 16:17–22
10. Gokhale BB, Mehta LN (1983) Histopathology of vitiliginous skin. Int J Dermatol 22:477–480
11. Hann SK, Kim YS, Yoo JH (2000) Clinical and histopathologic characteristics of trichrome vitiligo. J Am Acad Dermatol 242:589–596
12. Hann SK, Park YK, Lee KG et al (1992) Epidermal changes in active vitiligo. J Dermatol 19:217–222
13. Horn TD, Abanmi A (1997) Analysis of the lymphocytic infiltrate in a case of vitiligo. Am J Dermatopathol 19:400–402
14. Kao CH, Yu HS (1990) Depletion and repopulation of Langerhans cells in nonsegmental type vitiligo. J Dermatol 17:287–96
15. Kim YC, Kim YJ, Kang HY et al (2008) Histopathologic features in vitiligo. Am J Dermatopathol 30:112–116
16. Koranne RV, Sachdeva KG (1988) Vitiligo. Int J Dermatol 27:676–681
17. Kovacs SO (1998) Vitiligo. J Am Acad Dermatol 38: 647–666
18. Lee HS, Chun YS, Hann SK (1999) Nevus depigmentosus: clinical features and histopathologic characteristics in 67 patients. J Am Acad Dermatol 40:21–26
19. Le Poole IC, Van den Wijngaard RM, Westerhof W et al (1996) Presence of T-cells and macrophages in inflammatory vitiligo skin parallels melanocyte disappearance. Am J Pathol 148:1219–1228
20. Le Poole IC, Van den Wijngaard RM, Westerhof W et al (1993) Presence or absence of melanocytes in vitiligo lesions: an immunohistochemical investigation. J Invest Dermatol 100:816–822
21. Moellmann G, Klein-Angerer S, Scollay DA et al (1982) Extracellular granular material and degeneration of keratinocytes in the normally pigmented epidermis of patients with vitiligo. J Invest Dermatol 79:321–330
22. Montes LF, Abulafia J, Wilborn WH et al (2003) Value of histopathology in vitiligo. Int J Dermatol 42:57–61
23. Mosher DB, Fitzpatrick TB, Ortonne JB et al (1999) Hypomelanoses and hypermelanoses. In: Freedberg IM, Eisen AZ, Wolff K, Austen KF, Goldsmith LA, Katz SI, Fitzpatrick TB et al (eds) Dermatology in general medicine, 5th edn. McGraw-Hill, New York, pp 945–1017
24. Mosher DB, Fitzpatrick TB, Ortonne JB et al (1999) Normal skin color and general considerations of pigmentary disorders. In: Freedberg IM, Eisen AZ, Wolff K, Austen KF, Goldsmith LA, Katz SI, Fitzpatrick TB et al (eds) Dermatology in general medicine, 5th edn. McGraw-Hill, New York, pp 936–944
25. Murphy FG (2005) Histology of the skin. In: Elder D, Elenitsas R, Jaworsky C, Johnson B (eds) Lever's histopathology of the skin, 9th edn. Lipincott Williams& Wilkins, Philadelphia, pp 16–17

26. Pretti Aslanian FM, Noe RA, Cuzzi T et al (2007) Abnormal histological findings in active vitiligo include the normal-appearing skin. Pigment Cell Res 20:144–145
27. Sharquie KE, Mehenna SH, Naji AA et al (2004) Inflammatory changes in vitiligo. Stage I and II depigmentation. Am J Dermatopathol 26:108–112
28. Spielvogel RL, Kantor GR (1997) Pigmentary disorders of the skin. In: Elder D, Elenitsas R, Jaworsky C et al (eds) Lever's histopathology of the skin, 8th edn. Lipincott-Raven, Philadelphia, pp 617–623
29. Tretiakova MS, Shea CR, Petronic-Rosic VM (2006) Decreased CD117 expression in hypopigmented mycosis fungoides correlates with hypomelanosis: lessons learned from vitiligo. Mod Pathol 19:1255–1260
30. Van den Wijngaard R, Wañkowicz-Kaliñska A, Le Poole C et al (2000) Local immune response in skin of generalized vitiligo patients. Destruction of melanocytes is associated with the proeminent presence of CLA+T-cells at the perilesional site. Lab Invest 80:1299–309
31. Wankowicz-Kalinska A, Van den Wijngaard RM, Tigges BJ et al (2003) Immunopolarization of CD4+ and CD8+ T-cells to type-1-like is associated with melanocyte loss in human vitiligo. Lab Invest 83:683–695
32. Weedon D, Strutton G (2002) Disorders of pigmentation. Skin pathology, 2nd edn. Churchill Livingston, London, pp 321–341
33. Westerhof W, Groot I, Krieg SR et al (1986) Langerhans'cell population studies with OKT6 and HLA-DR monoclonal antibodies in vitiligo patients treated with oral phenylalanine loading and UVA irradiation. Acta Derm Venereol 66: 259–262
34. Wick MR (2006) Immunohistology of melanocytic neoplasms. In: Dabbs DJ (ed). Diagnostic immunohistochemistry, 2nd edn. Churchill Livingstone, Pittsburgh, PA, pp 166–168
35. Wolff RK, Johnson RA, Surmond D (2005) Pigmentary disorders. In: Fitzpatrick's color atlas & synopsis of clinical dermatology, 5th edn. McGraw-Hill, New York, pp 336–343

1.3 Clinical aspects

Generalized Vitiligo

1.3.1

Thierry Passeron and Jean-Paul Ortonne

Contents

Core Messages

- The existence of various phenotypes among nonsegmental vitiligo (NSV) (e.g., Acrofacial or Vulgaris) is suggested by clinical observation and possibly by genetic associations.
- Half of initially "focal" vitiligo patients evolve into NSV.
- The fingers, hands and face are frequently reported to be the initial sites by the patients.
- In darker-skinned individuals, palms and soles have to be examined with Wood's lamp.
- Most of the "spontaneous" repigmentation reported by the patients is correlated to sun exposure.
- Multichrome vitiligo refers to various degrees of depigmentation within a vitiligo macule, a phenomenon noted in dark skin.

Although, there is no uniform classification of vitiligo, most authors agree to separate segmental (SV) and nonsegmental (NSV) forms, and mixed forms have been recently recognized (Chap. 1.2.1). Furthermore, some data suggest that different phenotypes of vitiligo may have different genetic background [15]. NSV usually includes acrofacial vitiligo, vitiligo vulgaris, and universal vitiligo (Chaps. 1.2.1 and 1.3.3).

1.3.1.1 Common Clinical Features

Generalized vitiligo is characterized by asymptomatic well-circumscribed milky-white macules, involving both sides of the body with usually a symmetrical pattern

T. Passeron (✉)
Department of Dermatology, University Hospital of Nice, Nice, France
e-mail: passeron@unice.fr

M. Picardo and A. Taïeb (eds.), *Vitiligo*,
DOI 10.1007/978-3-540-69361-1_1.3.1, © Springer-Verlag Berlin Heidelberg 2010

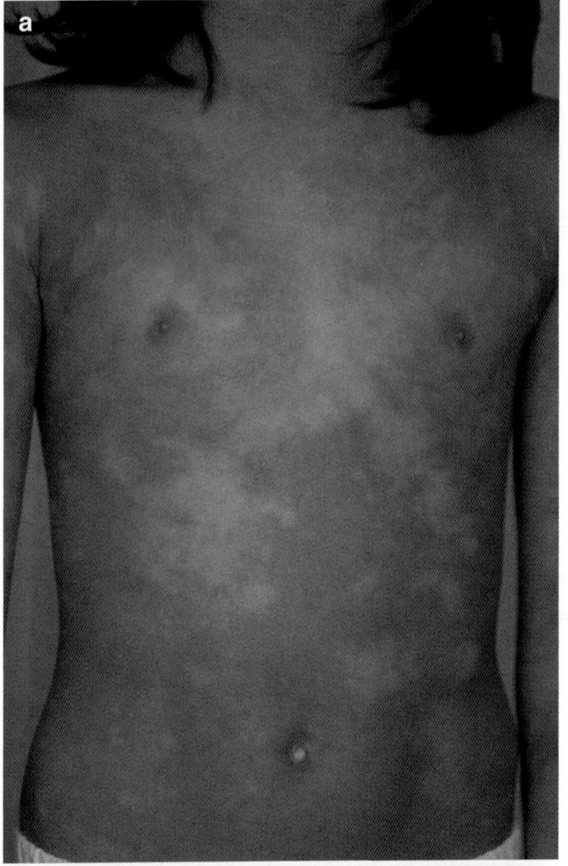

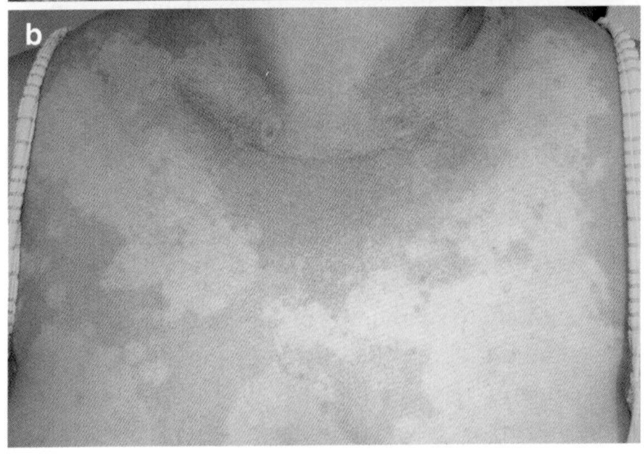

Fig. 1.3.1.1 Extensive generalized vitiligo with mostly symmetrical lesions in a child (**a**) and erythema in sun-exposed areas in an adult (**b**)

[11]. The shape of individual macules is round to oval with slightly brushed to fairly distinct margins (Fig. 1.3.1.1a). Initial spot size varies from a few to several centimeters in diameter. The depigmentation is macular, and the epidermis shows no sign of atrophy, telangiectasias, or any other signs. The presence of epidermal atrophy should suggest some other disorders, especially lichen sclerosus (Chaps. 1.2.1 and 1.3.12). A transient erythema, which can be clinically misleading, is frequently observed after ultraviolet (UV) exposure, but the history is contributive (Fig. 1.3.1.1b). Hyperpigmented lesional borders are not uncommon, especially in dark-skinned patients and after UV exposure. Hair is usually spared and remain pigmented, but in some cases hair depigmentation may also occur simultaneously. In the scalp, vitiligo usually leads to localized patches of grey or white hair, but total depigmentation of the scalp hair may occur. Depigmented body hair within vitiligo macules are considered as markers of poor repigmentation prognosis [12].

Vitiligo patches are easy to recognize on darker phototypes, but depigmentation is sometimes difficult to detect in patients with very fair skin. Wood's lamp examination is very helpful to delineate the areas involved in light-colored individuals and also to assess the remaining reservoir of melanocytes (Fig. 1.3.1.2a). Even in darker-skinned individuals, palms and soles are light colored and have to be examined with Wood's lamp. Wood's lamp examination can also show the earliest signs of repigmentation at the border of the lesion or in perifollicular areas (Fig. 1.3.1.2b).

1.3.1.2 Distribution

Generalized vitiligo can start at any site of the body, but the fingers, hands, and face are frequently reported to be the initial sites by the patients. It has been reported that when the hands are the initial site, vitiligo most commonly progresses to the face, explaining the frequency of acrofacial vitiligo in those patients [3]. The same study suggests that when the posterior trunk, hands, or feet are the initial sites, vitiligo tends to have a more widespread progression. The extensor surfaces are commonly affected, including interphalangeal joints, metacarpal/metatarsal interphalangeal joints, elbows, and knees. Other surfaces involved include volar wrists, malleoli, umbilicus, lumbosacral area, anterior tibia, and axillae. The role of the Koebner phenomenon in the distribution pattern is discussed in Chapter 2.2.2.1. Most of the time, the distribution of the lesions is clearly symmetrical. Vitiligo macules may also be periorificial and involve the skin around the eyes, nose, ears,

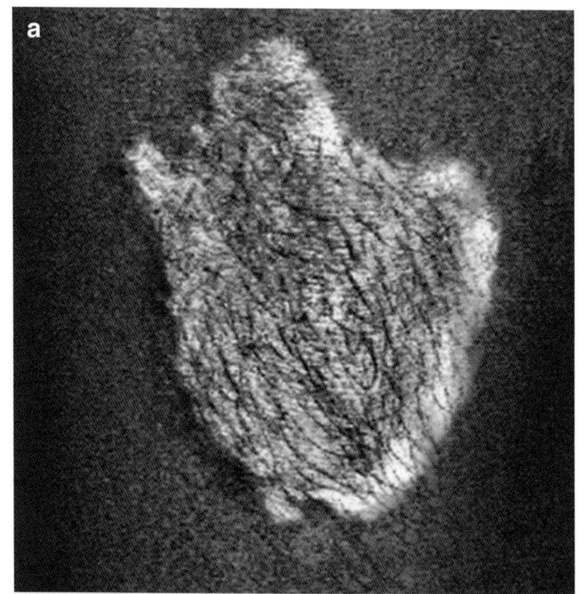

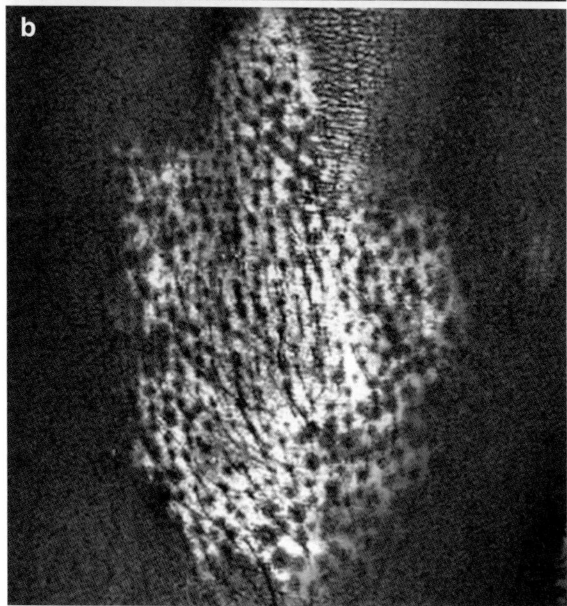

Fig. 1.3.1.2 Aspect of vitiligo lesion of the knee in Wood's lamp examination (**a**). Note the persistence of a *grey* pattern in the centre of the lesion, suggesting of a remaining melanocyte reservoir; (**b**) same lesion during after phototherapy. Note the perifollicular repigmentation

mouth, and anus. Periungual involvement may occur alone or with certain mucosal surfaces (lips, distal penis, nipples). Vitiligo can affect genital areas and in some cases almost exclusively. This location should be discussed with the patients because it causes a great concern.

1.3.1.3 Natural Course

The evolution of generalized vitiligo is unpredictable. Focal vitiligo, although stable for a time, may be a precursor of generalized vitiligo. In a large series of vitiligo patients, about three-quarters of the patients occurred as focal vitiligo, but more than half of those patients evolved to a generalized form [8]. The natural course of common vitiligo is often one of abrupt onset, followed by progression for a time; then a period of stability follows and may last for some time, even decades. This may be followed later by another period of more rapid evolution. Periods of rapid progression last often less than a year, after which there is little extension or regression. The evolution to vitiligo universalis is rare (Chap. 1.3.3). The most common course is one of gradual extension of existing macules and periodic development of new ones. Vitiligo can repigment spontaneously, but this phenomenon is rare and mostly localized to some lesions. Most of the "spontaneous" repigmentation reported by the patients is correlated to sun exposure (see also Chap. 1–6).

1.3.1.4 Clinical Subtypes

This issue has been reviewed for classification purposes in Chap. 1.2.1. *Vitiligo vulgaris* is the most common form of generalized vitiligo corresponding to several achromic patches symmetrically distributed on the body [7]. In the largest series of patients (from China), vitiligo vulgaris accounted for 41% of cases [8]. *Acrofacial vitiligo* involves distal digits and periorificial facial areas. *Mixed vitiligo* is a term used when SV and NSV lesions are associated in a same patient [9] (reviewed in Chap. 1.2.1).

1.3.1.5 Clinical Variants

Inflammatory Vitiligo

The lesions could sometimes have a raised red border, but inflammation of numerous lesions at the same time is very uncommon [1]. A mild pruritus could be associated. When the inflammation disappears, the skin becomes depigmented [2, 4, 10, 14] (Chap. 1.3.10). Such inflammation is sometimes reported with occupational vitiligo, and those patients should be questioned about chemical exposures (Chap. 2.2.2.2).

Multichrome Vitiligo

This form of vitiligo is mostly seen in darker phototypes (IV–VI). Within a vitiligo lesion, areas of depigmentation coexist with hypopigmented areas and with normal color as in surrounding skin (Fig. 1.3.1.3). In the hypopigmented area, a partial loss of melanocyte is observed. Trichrome vitiligo is commonly used to describe this pattern, but various degrees of hypopigmentation can be observed leading to trichrome, quadrichrome, or pentachrome vitiligo [5, 13]. Thus, the term "multichrome vitiligo" should be preferred.

Vitiligo Minor

Although rarely reported, this clinical pattern is not unusual in dark-skinned individuals. It is characterized by a homogeneous hypopigmented pattern (Fig. 1.3.1.4a, b).

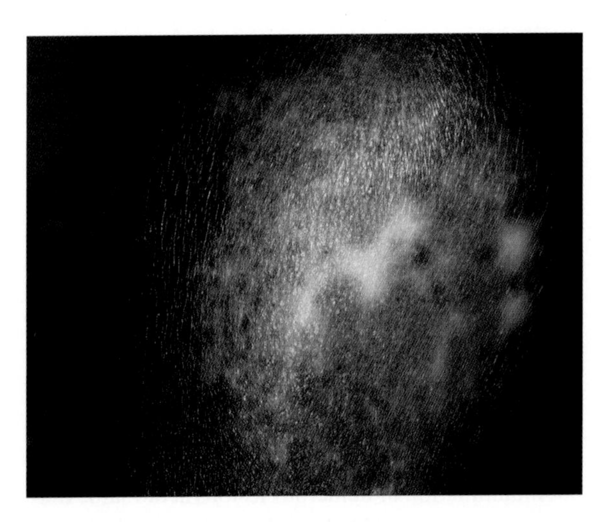

Fig. 1.3.1.3 Multichrome vitiligo of the leg

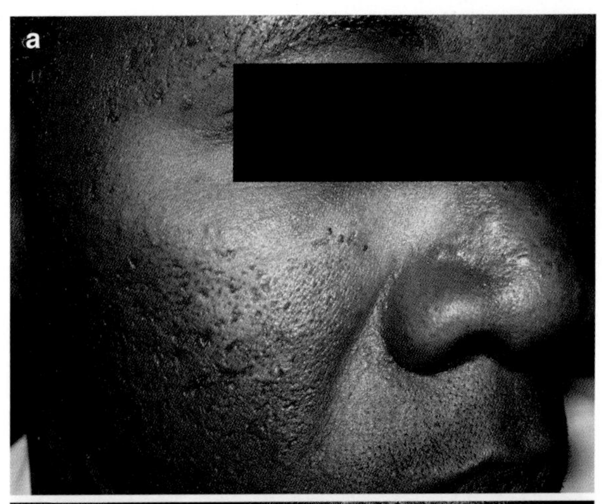

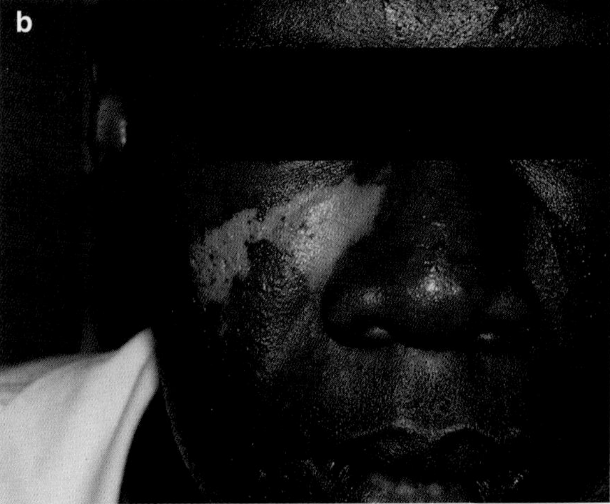

Fig. 1.3.1.4 Vitiligo minor of the face (**a**), and (**b**) evolution to complete depigmentation sixmonths later

Blue Vitiligo

Postinflammatory hyperpigmentation leading to a blue discoloration of vitiligo patches has been reported after inflammatory vitiligo [6].

References

1. Arata J, Abe-Matsuura Y (1994). Generalized vitiligo preceded by a generalized figurate erythematosquamous eruption. J Dermatol 21:438–441.
2. Buckley WR, Lobitz WC Jr (1953). Vitiligo with a raised inflammatory border. AMA Arch Derm Syphilol 67: 316–320.

3. Chun WH, Hann SK (1997). The progression of nonsegmental vitiligo: clinical analysis of 318 patients. Int J Dermatol 36:908–910.
4. Ishii M, Hamada T (1981). Ultrastructural studies of vitiligo with inflammatory raised borders. J Dermatol 8:313–322.
5. Fargnoli MC, Bolognia JL (1995). Pentachrome vitiligo. J Am Acad Dermatol 33:853–856.
6. Ivker R, Goldaber M, Buchness MR (1994). Blue vitiligo. J Am Acad Dermatol 30:829–831.
7. Le Poole C (2006). Vitiligo vulgaris. In: Nordlund J (ed) The pigmentary system. Blackwell, Oxford, pp 551–598.
8. Liu JB, Li M, Yang S, et al (2005). Clinical profiles of vitiligo in China: an analysis of 3742 patients. Clin Exp Dermatol 30:327–331.
9. Mulekar SV, Al Issa A, Asaad M, et al (2006). Mixed vitiligo. J Cutan Med Surg 10:104–107.
10. Ortonne JP, Baran R, Civatte J (1979). [Vitiligo with an inflammatory border. Apropos of 2 cases with review of the literature (18 cases)]. Ann Dermatol Venereol 106: 613–615.
11. Ortonne JP, Mosher DB, Fitzpatrick TB (1983). Vitiligo and other hypomelanoses of hair and skin. In: Parrish JA, Fitzpatrick TB (eds) Monographs in topics in dermatology. Plenum, New York, p 683.
12. Ortonne JP (2007). Vitiligo and other disorders of hypopigmentation. In: Bolognia J, Jorizzo JL, Rapini RP (eds) Dermatology. Mosby, New York, pp 913–938.
13. Sehgal VN, Srivastava G (2007). Vitiligo: compendium of clinico-epidemiological features. Indian J Dermatol Venereol Leprol 73:149–156.
14. Watzig V (1974) [Vitiligo with inflammatory marginal dam]. Dermatol Monatsschr 160:409–413.
15. Zhang XJ, Liu JB, Gui JP, et al (2004). Characteristics of genetic epidemiology and genetic models for vitiligo. J Am Acad Dermatol 51:383–390.

Segmental Vitiligo

1.3.2

Seung-Kyung Hann, Yvon Gauthier, and Laila Benzekri

Contents

Core Messages

- Segmental vitiligo (SV) shares some characteristics and can be associated with nonsegmental vitiligo (NSV).
- However, SV is currently considered as a distinct entity.
- SV usually has an early onset and it spreads rapidly in the affected dermatomal area.
- The hair follicle melanocytic compartment is frequently involved.
- The purpose of the classification of SV, especially facial, is to establish a prognosis.
- Multiple SV should be differentiated from NSV.
- Early medical and UV treatment of SV is recommended based on recent data.

There is in general no problem in defining SV (Chap. 1.2.1). However, interpretations have varied concerning its relation to nonsegmental forms. Furthermore, there is still a debate about SV clinical subclassifications, mostly according to the interpretation of the relative importance of a presupposed dermatomal (sensitive nerve distribution) and/or associated developmental anomaly (Part 2.3). We have requested for contributions from experts for this chapter, and we have left the debate open for controversial opinions such as the classification of facial SV *(The Editors)*.

S.-K. Hann (✉)
Korea Institute of Vitiligo Research,
Drs. Woo & Hann's skin clinic, Yongsan-Gu, Seoul,
South Korea
e-mail: skhann@paran.com

M. Picardo and A. Taïeb (eds.), *Vitiligo*,
DOI 10.1007/978-3-540-69361-1_1.3.2, © Springer-Verlag Berlin Heidelberg 2010

1.3.2.1 Epidemiology and Clinical Features

Yvon Gauthier, Laila Benzekri,
and Seung-Kyung Hann

It has been proposed by Koga in 1977 that two types of vitiligo exist from the pathophysiological and clinical points of view, each having a separate pathogenesis [10]. Nonsegmental vitiligo (NSV, Type A) associated with autoimmune disease and segmental vitiligo (SV, Type B) resulting from the dysfunction of sympathetic nerves in the affected area. In the two types, acquired melanocyte loss is a common feature. Based on pure clinical considerations, localized forms of vitiligo can be further subclassified into *focal*, meaning one or more macules in one area, but not clearly in a segmental distribution (Chap. 1.2.1); *segmental*, one or more macules, more or less distributed according to a zosteriform distribution; and *mucosal*, with isolated mucous membrane involvement (Chap. 1.3.4).

Epidemiology and General Features

The global incidence of vitiligo is generally acknowledged to be 0.5–1%. In published series of vitiligo, the percentage of SV varies from 5 to 7.9% [1, 6, 13] and up to 15% in our personal recruitment (Yvon Gauthier). Korean studies have shown a range between 5.5 and 16.1% [14, 15].

Koga and Tango reported that SV affects, more commonly, the young and indicated a higher (27.9%) incidence in children [9]. In Hann and Lee's report [6], SV developed before 30 years of age in 87.0% of cases, 41.3% were younger than 10 years, and the mean age of onset was 15.6 years. The earliest reported onset was immediately after birth, whereas the latest was 54 years. Most cases were less than 3 years in duration at referral, ranging from 2 months to 15 years. Delayed referral may reflect the prevailing belief among doctors that no treatment is available for vitiligo.

Clinical Features

The typical lesion is not very different from the macule observed in NSC. The most common form is a totally amelanotic macule surrounded by normal skin. The color of the macule is usually pure white or chalk white. However, as in NSV, a multichrome variation of hypopigmentation can be observed, and overall a less uniform depigmentation pattern was seen in SV compared to NSV when a decision was needed for grading a patch [16]. In some cases, such as fair skin, the lesions are not easy to see under normal light, but can be distinguishable with Wood's light examination. Similar to herpes zoster, the depigmented patches are confined to a definite dermatome [3] (partial or complete involvement), but they also commonly overlap several dermatomes and sometimes may cross the midline (Fig. 1.3.2.1).

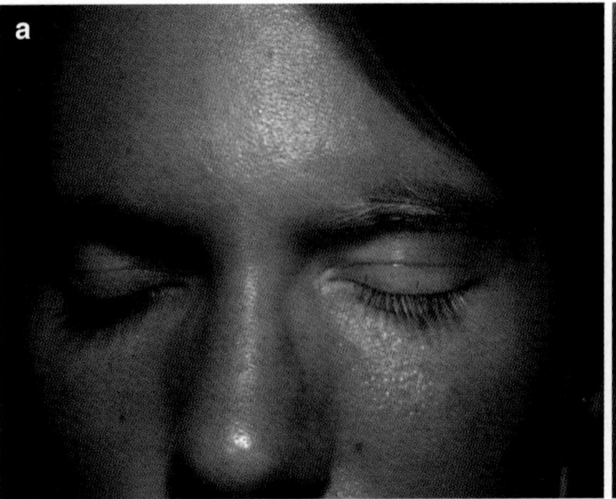

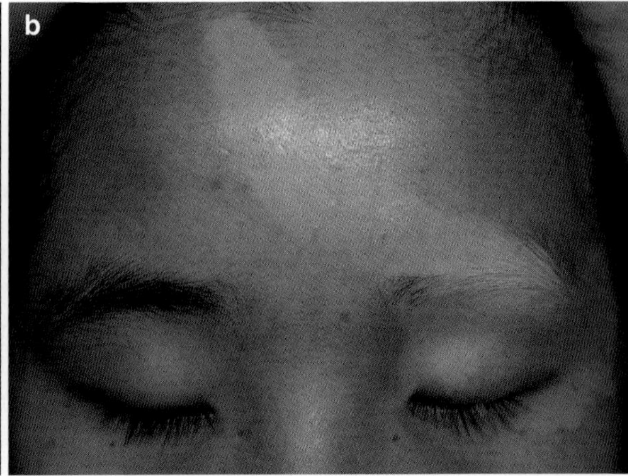

Fig. 1.3.2.1. Typical aspect of SV of the face with poliosis crossing either slightly the mid line (**a**), or more markedly (**b**). (It corresponds to Hann's type I and Gauthier's type IVC, see Figs. 1.3.2.3 and 1.3.2.4, respectively)

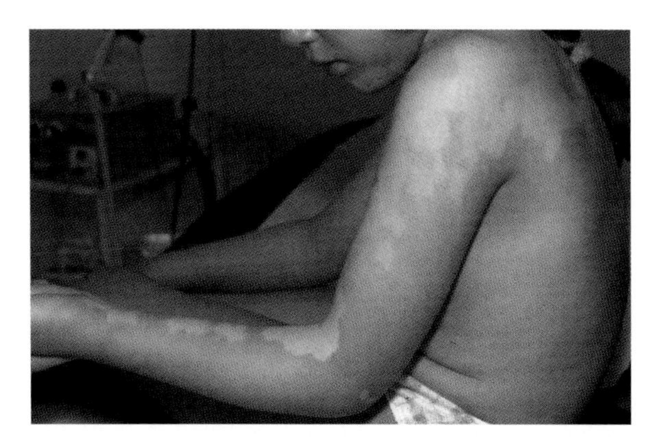

Fig. 1.3.2.2 SV of the left shoulder and upper limb

In majority of cases, depigmentation spreads within the segment in a short period of time and then stops. It leaves the skin segment partially or totally depigmented. It is rare for a patient with SV to progress to the generalized form. Such a phenomenon is probably related to the rare association of SV and NSV called "mixed vitiligo" [5]. In this case, the increased severity of SV versus NSV in response to therapy suggests that the dosing of the predisposing skin anomaly is augmented in the SV area (Part 2.3).

The head is involved in more than 50% of cases. In decreasing order of frequency, the trunk, the limbs (Fig. 1.3.2.2), the extremities, and the neck are common sites of involvement [1, 2, 4, 6, 7, 11]. In females, the neck is more frequently involved than the extremities. Lerner [12] reported that SV occurs as a single lesion in 75% of patients, a finding confirmed by the study of Hann and Lee [6] who found that 87% of patients had a single lesion.

There is no preferential distribution between right and left sides of the body. Some patients have lesions in two different unilateral dermatomes. The most commonly involved is that of the trigeminal nerve [6–8].

Depigmentation of hair (poliosis, leukotrichia) occurs in vitiliginous macules. Poliosis has been shown to occur in 48.6% of cases of SV [6, 7]. Involvement is variable; a few to many hair of a single macule may be depigmented. The eyebrow and eyelashes are commonly involved in SV located on the ophthalmic branch of the trigeminal nerve. Other involved hairy sites include scalp, pubis, and axillae.

A family history of vitiligo is present in approximately 12% of SV cases [2, 6, 7]. El Mofty [4] and Koga [9] suggested that SV is not significantly associated with other autoimmune disease. This is still debated because Park et al. [14] showed that about 9.5% of SV cases were associated with other diseases. Similarly, Hann and Lee [6] reported 6.7% of patients with associated disease, either allergic or autoimmune. Common diseases associated with SV include atopic dermatitis, common in general in this age group, and halo nevus, which is associated to SV nearly as frequently as to NSV (discussed in Chap. 1.3.5 and Part 1.7). Considering the prevalence of allergic and autoimmune disorders among the general population, whether these findings are pathogenically associated or coincidental is not established.

Precipitating Factors

The appearance of linear or macular depigmentation after scratches or traumas was not found by Koga [9] in SV outside the involved segment. However, other authors reported that sunburns, trauma, or local repeated pressures were locally aggravating factors in 4.8% of patients with SV [2].

References

1. Bang JS, Lee JW, Kim TH et al (2000) Comparative clinical study of segmental and non segmental vitiligo. Kor J Dermatol 38:1037–1044
2. Barona SK, Arruneteguy A, Falabella R (1995) An epidemiologic case-control study in a population with vitiligo. J Am Acad Dermatol 33:621–625
3. Bolognia JL, Orlow SJ, Glick SA (1994) Lines of Blasehko. J Am Acad Dermatol 31:157–190
4. El Mofty AM, El Mofty M (1980) Vitiligo: a symptom complex. Int J Dermatol 19:237–244
5. Gauthier Y, Cario Andre M, Taïeb A (2003) A critical appraisal of vitiligo etiologic theories. Is melanocyte loss a melanocytorrhagy? Pigment Cell Res 16:322–332
6. Hann SK, Lee HJ (1996) Segmental vitiligo: clinical findings in 208 patients. J Am Acad Dermatol 35:671–674
7. Hann SK, Park SK, Chan WH (1997) Clinical features of vitiligo. Clinic Dermatol 15:891–897
8. Hann SK, Chang HJ, Lee HS (2000) The classification of segmental vitiligo on the face. Yonsei Med J 41:209–212
9. Koga M, Tango T (1988) Clinical features and course of type A and type B vitiligo. Br J Dermatol 118:223–228
10. Koga M (1977) Vitiligo: a new classification and therapy. Br J Dermatol 97:255–261
11. Lee SJ, Cho SB, Hann SK (2007) Classification of vitiligo. In: Gupta S, Olsson M, Kanwar AJ, Ortonne JP (eds) Surgical management of vitiligo. Blackwell, Oxford, pp 20–30
12. Lerner AB (1959) Vitiligo. J Invest Dermatol 32:285–310

13. Ortonne JP (1983) Vitiligo and other hypomelanosis of hair and skin. Plenum, New York, pp 147–148
14. Park K, Youn JL, Lee YS (1988) A clinical study of 326 cases of vitiligo. Korean J Dermatol 26:200–205
15. Song MS, Hann SK, Ahn PS et al (1994) Clinical study of vitiligo: comparative study of type A and type B vitiligo. Ann Dermatol (Seoul) 6:22–30
16. Taïeb A, Picardo M; VETF Members (2007) The definition and assessment of vitiligo: a consensus report of the Vitiligo European Task Force. Pigment Cell Res 20:27–35

1.3.2.2 Classification, Course and Prognosis

Seung-Kyung Hann and Yvon Gauthier

Following Koga's initial report [6] suggesting that SV and NSV were distinct vitiligo subtypes, further work confirmed that a key aspect of these cases was that the lesions did not cross the midline and were distributed along a unilateral dermatome, thus enabling a prediction of the prognosis [2, 4, 7, 9]. SV does not always show classical dermatomal distribution but affects usually only one segment of the integument. The segment might be composed of several or parts of several adjacent dermatomes or have no relationship to dermatomes at all, nor to any other lines such as Blaschko's lines or acupuncture lines (see discussion in Part 2.3). The progression of SV, which is usually limited to months or a few years [2, 7], differs remarkably from the chronic progressive course of NSV [1, 4, 9]. Since the face is a commonly involved site of vitiligo and the area that causes the most psychological impact, most patients are willing to undergo intensive treatment.

Therefore, knowledge about the exact spreading pattern and prognosis is important for both patients and doctors. The two following classification schemes can help making predictions.

Classification of Segmental Vitiligo on the Face (Hann)

The distribution of SV on the face is classified into five patterns (Fig. 1.3.2.3) [5]. Type Ia represents the lesion which initiates from the right side of the forehead, crosses the midline of the face and spreads down to the eyeball, nose and cheek of the left side of the face. Type Ib appears as a mirror image of Ia. The lesion starts from the left side of the face and spreads down the right side of the face, crossing the midline. In type II, the lesion starts from the area between the nose and lip, then arches to the preauricular area. In type III, the lesion initiates from the lower lip and spreads down to the chin and neck. In type IV, the lesion originates from the right side of the forehead and spreads down to

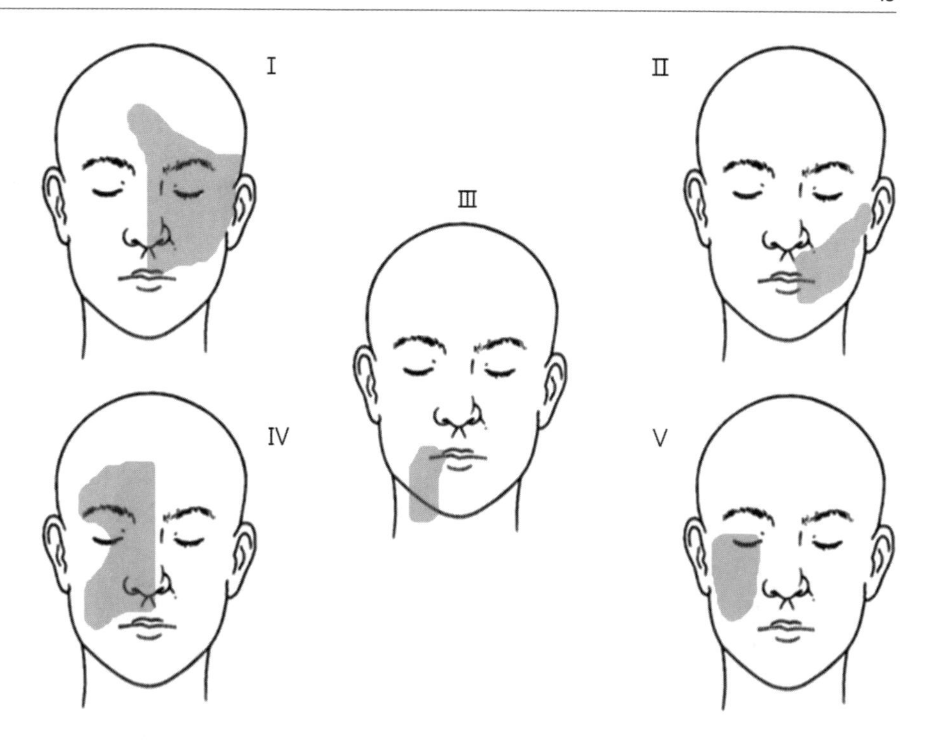

Fig. 1.3.2.3 Hann's classification of segmental vitiligo of the face

the eyeball, nose and cheek areas without crossing the midline. In type V, the lesion is confined to the left cheek area.

Some SV on the face cannot be classified by this system. Type I is the most common and type V is the least common. There are no significant differences in age, sex, duration of initial lesions, progression pattern, and clinical type among these classifications.

Classification of Segmental Vitiligo of the Face and Neck (Gauthier)

Recently, we have proposed a new simplified classification of SV of the face [10], which varies from the previous classification by Hann et al. [5]. With this new system all SV of the face can be classified. The sites involved by herpes zoster and SV were compared before establishing this new classification. In 26% of cases, SV was distributed exactly to a trigeminal dermatome: ophthalmic (V1) maxillary (V2) mandibular (V3). In 64% of cases SV did not follow exactly dermatomes and was overlapping one, two or three dermatomes, as in many cases of facial herpes zoster. A classification of facial SV according to five topographic patterns was thus proposed (Fig. 1.3.2.4): type I, corresponding to the V1 ophthalmic branch (partial or total involvement); type II, corresponding to the V2 maxillar branch (partial or total involvement); type III, corresponding to V3 mandibular branch (partial or total involvement); type IV, corresponding to mixed distribution on several dermatomes (4a = V1 + V2, 4b = V2 + V3, 4c = V1 + V2 + V3); and type V, corresponding to the cervicofacial one.

Diagnosis, Course, and Special Locations

Most often SV patches remain unchanged for the rest of the patient's life after rapid initial spreading in the affected segment [2]. However, rarely it can progress again after being quiescent for several years (Fig. 1.3.2.5). When SV progresses, it usually spreads over the predicted segment. However, in very rare cases, lesions may become generalized, a situation referred to as mixed vitiligo (Chap. 1.2.1). Early SV most often appears as a solitary oval shaped white macule or as a patch that is difficult to differentiate from "focal" vitiligo until proven by a subsequent typical distribution pattern. A white macule on the nipple

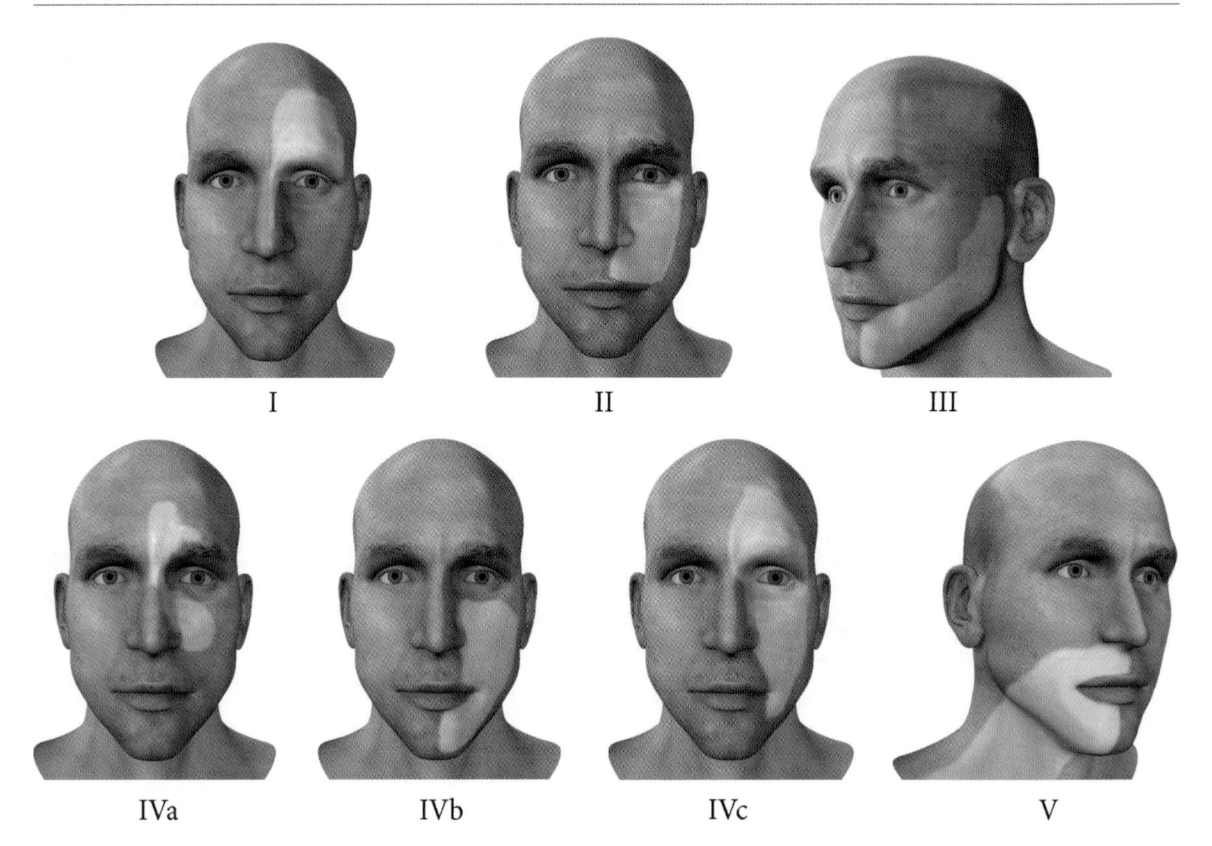

Fig. 1.3.2.4 Gauthier's classification of segmental vitiligo of the face: *I* corresponding to V1; *II* corresponding to V2; *III* corresponding to V3; *IVa* corresponding to mixed distributionV1 + V2; *IVb* corresponding to mixed distribution V2 + V3; *IVc* corresponding to mixed distribution V1 + V2 + V3; *V* corresponding to cervicofacial

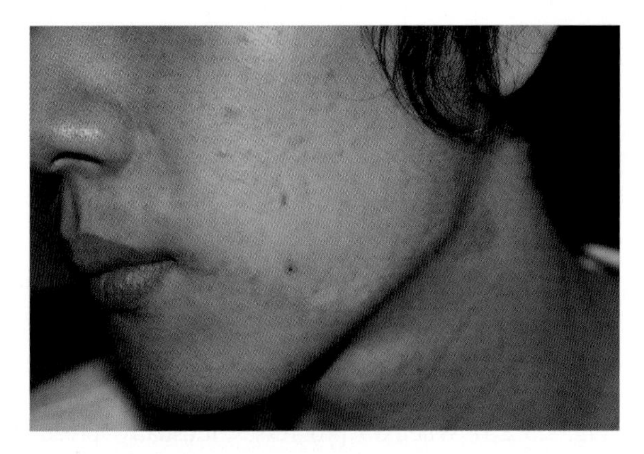

Fig. 1.3.2.5 Recurrence of vitiliginous lesions at the periphery of an autologously grafted SV site. The graft was done 10 years before the recurrence (corresponds to Hann's type II and Gauthier's type III, see Figs. 1.3.2.3 and 1.3.2.4, respectively)

or areola appearing as the initial lesion can be assumed to be an early manifestation of SV (Fig. 1.3.2.6) [4]. Nipple or areolar involvement as the initial lesion in NSV is very rare and becomes bilateral later (Fig. 1.3.2.7). The repigmentation of vitiligo on nipple or areolar skin is difficult with UV or medical therapies. Instead, surgical treatment such as epidermal grafting can lead to complete repigmentation for both SV and NSV. If SV occurs bilaterally, following the same contralateral (or different) dermatomes, it may cause difficulties in defining vitiligo type (Fig. 1.3.2.8). It may at some point be confused with NSV or, for some locations, such as white patches of both legs, to piebaldism (Chaps. 1.2.1 and 1.3.11). Lee and Hann [8] reported that 5 out of 240 patients, who had SV, exhibited two different depigmented segments on the same or

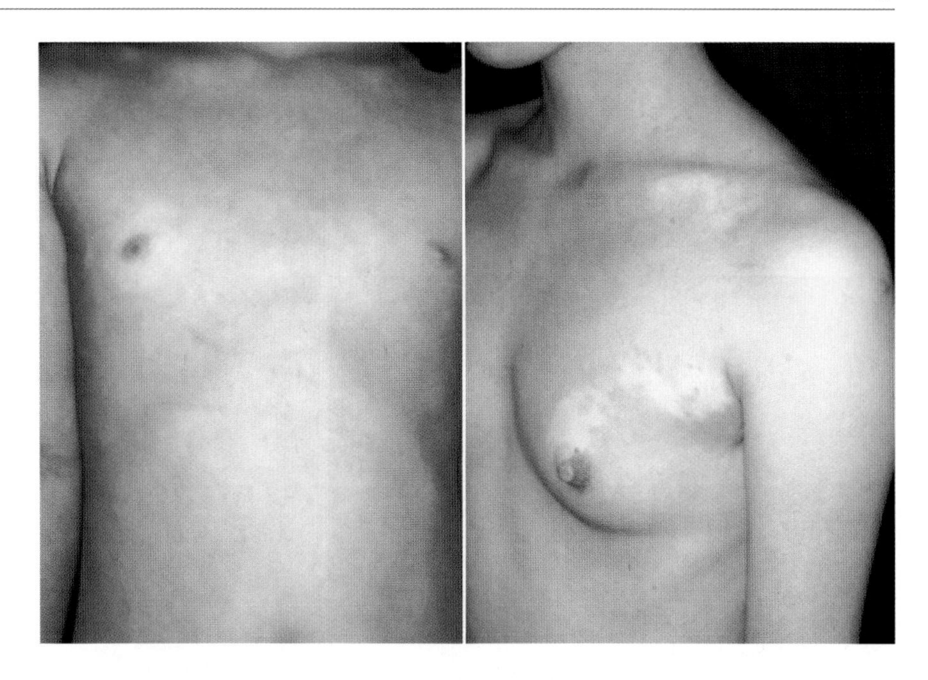

Fig. 1.3.2.6 A white macule on the areola or nipple often progresses to segmental vitiligo (*left panel* early stage, difficult to predict; *right panel*, clear segmental involvement)

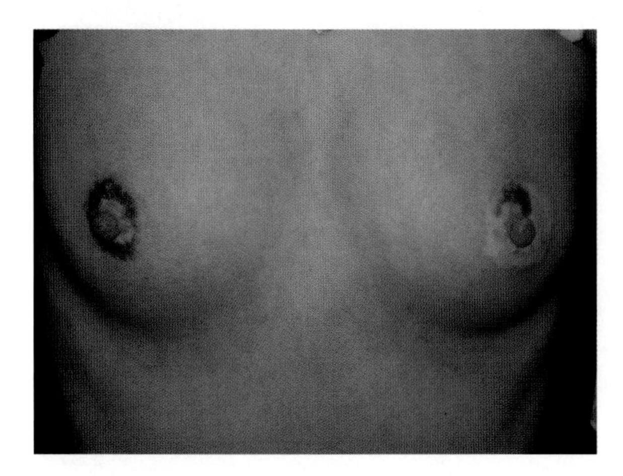

Fig. 1.3.2.7 Bilateral involvement of the nipple in NSV

opposite sites of the body. The clinical course of bilateral SV seems to be the same as unilateral SV although only five cases have been followed for up to maximum of 3 years. In our experience, PUVA therapy and topical steroid treatment can induce repigmentation or stop progression in bilateral SV.

Treatment Overview

SV was previously known to be resistant to treatment. However, recent studies reported surprisingly good results. SV at an early stage has an excellent prognosis. Since the most frequently involved site of SV is the face, it can be easily detected and treated. Effective treatment modalities at an early stage are not standardized but include topical steroids, topical calcineurin inhibitors or UV therapy, isolated or in combination (Fig. 1.3.2.9). Because SV often causes leukotrichia, it may resist standard medical therapies. Stable SV with leukotrichia can be cured successfully with epidermal grafting and subsequent UV treatment [3]. There may be also a possibility of activation and migration of epidermal melanocytes to the hair follicle. Overall, stable SV is a good indication for epidermal grafting and can be cured almost completely without recurrence (for a discussion of surgical therapies, see Part 3.7).

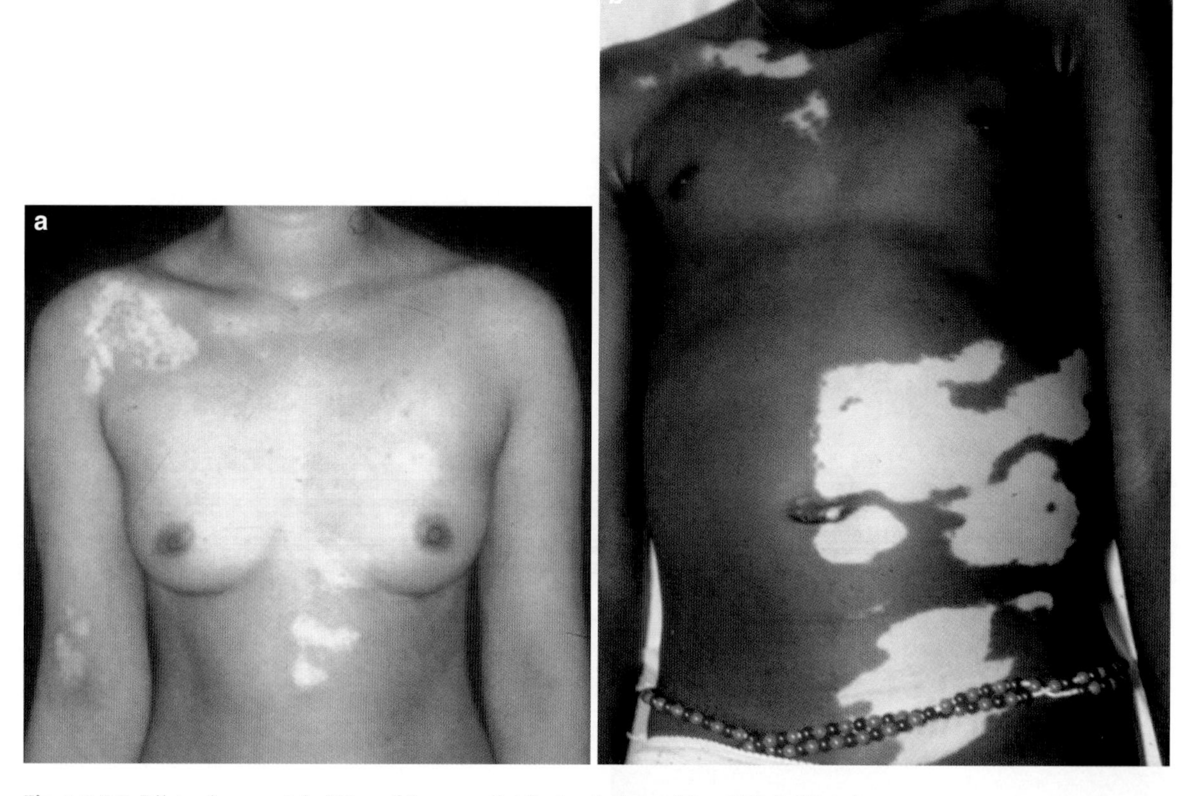

Fig. 1.3.2.8 Bilateral segmental vitiligo of the same distribution in Asian (**a**) and black (**b**) patients

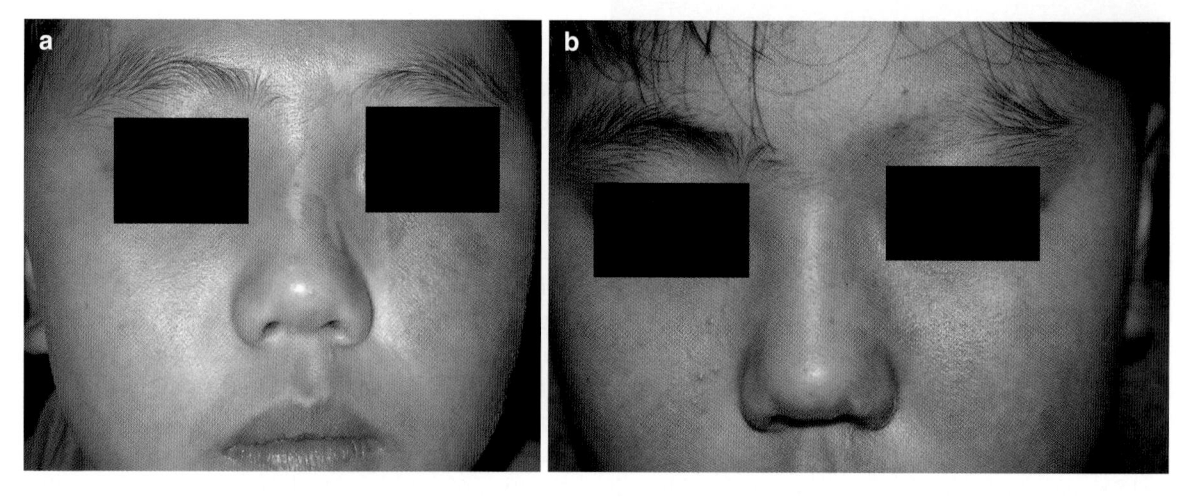

Fig. 1.3.2.9 Almost complete repigmentation of SV after 4 months of 308 nm excimer laser treatment (**a**, before; **b**, after) (corresponds to Hann's type I and Gauthier's type IVa, see Figs. 1.3.2.3 and 1.3.2.4, respectively)

References

1. Bang JS, Lee JW, Kim TH et al (2000) Comparative clinical study of segmental vitiligo and non-segmental vitiligo. Korean J Dermatol 38:1037–1044
2. Hann SK, Lee HJ (1996) Segmental vitiligo: clinical findings in 208 patients. J Am Acad Dermatol 35:671–674
3. Hann SK, Im S, Park YK, Hur W (1992) Repigmentation of leukotrichia by epidermal grafting and systemic psoralen plus UV-A. Arch Dermatol 128:998–999
4. Hann SK, Park YK, Chun WH (1997) Clinical features of vitiligo. Clinic Dermatol 15:891–897
5. Hann SK, Chang JH, Lee HS, Kim SM (2000) The classification of segmental vitiligo on the face. Yonsei Med J 41:209–212

6. Koga M (1977) Vitiligo: a new classification and therapy. Br J Dermatol 97:255–261
7. Koga M, Tango T (1988) Clinical features and courses of type A and type B vitiligo. Br J Dermatol 118:223–228
8. Lee HS, Hann SK (1998) Bilateral segmental vitiligo. Ann Dermatol (Seoul) 10:129–131
9. Song MS, Hann SK, Ahn PS et al (1994) Clinical study of vitiligo: comparative study of type A and type B vitiligo. Ann Dermatol (Seoul) 6:22–30
10. Gauthier Y, Taieb A (2006) Proposal for a new classification of segmental vitiligo of the face. Pigment Cell Res 19:515 (abstract)

Vitiligo Universalis

1.3.3

Prasad Kumarasinghe

Contents

P. Kumarasinghe
Consultant Dermatologist, Department of Dermatology,
Royal Perth Hospital,
Western Australia, Australia
e-mail: prasadkumarasinghe@yahoo.com

Core Messages

- Vitiligo universalis is the most uncommon form of vitiligo and is probably the extreme severity end of the spectrum of nonsegmental vitiligo.
- Associated autoimmune phenomena seem more common; patients have to be followed up, as associated diseases may manifest months or years after being diagnosed vitiligo universalis.
- Sun protection is important to avoid sunburns and other sequelae.
- Depigmenting creams such as monobenzyl ether of hydroquinone or laser depigmentation may be used to depigment small areas of breakthrough pigmentation.

1.3.3.1 Definition and Epidemiology

Acquired complete depigmentation or nearly complete depigmentation of the skin (and sometimes hair) is termed as vitiligo universalis (VU). It is the most extensive form of vitiligo. This form can start as common nonsegmental vitiligo (NSV) and advance to complete depigmentation of the skin and hair. True segmental vitiligo (SV) does not progress to VU. Within the group of NSV, it is not clear what factors trigger the cascade of complete or near complete depigmentation of the integument in VU.

There appears to be some minor differences in the prevalence of vitiligo among different populations, but no specific data is available on VU [19]. VU is

M. Picardo and A. Taïeb (eds.), *Vitiligo*,
DOI 10.1007/978-3-540-69361-1_1.3.3, © Springer-Verlag Berlin Heidelberg 2010

certainly the most uncommon clinical manifestation of vitiligo and appears to occur worldwide. VU is recognized more easily in dark-skinned individuals. It is not clear whether VU has the same prevalence among the fair-skinned Caucasians as compared to Asians or Africans. Although vitiligo may occur at any age, VU is very uncommon in childhood. The population prevalence of VU is not known. In a study by Song et al. [20], it has been reported that there was only one case of VU among 1315 vitiligo patients. In a retrospective study we carried out at the National Skin Centre, Singapore [22], 1.4% (4 patients out of 282 vitiligo patients) had VU. Male to female ratio of VU appears to be the same. More data are needed to confirm this observation.

1.3.3.2 Clinical Features

In VU, nearly all the skin becomes completely depigmented. However, in sun-exposed areas there can be minor perifollicular, discrete or coalescent pigmentation (Figures 1.3.3.1–1.3.3.4). Areolae and genitalia may depigment either early in the disease process or later. Some VU patients may show areas of partial depigmentation, appearing clinically as trichrome vitiligo. However, these apparently resisting areas may also lose color and become uniformly depigmented.

Some patients retain dark hair on the scalp at least in the early course of disease. Pubic and axillary hair may or may not be depigmented. Body hair often gets depigmented in affected areas. Generally, in established VU, the hair in the depigmented skin also gets depigmented.

Iris pigmentation is usually unaffected, but exceptions do occur. In some VU patients, associated uveitis may even lead to blindness. Mucosae, particularly the lips, gums, genitalia too may be partially or completely depigmented [6].

According to a Turkish study [2] minor hearing defects are not uncommon in NSV patients. These authors have reported that 14% of NSV patients have at least mild hypoacousis. Eye and ear symptoms in extensive vitiligo patients are likely to be due to the loss of melanocytes in these organs. This is a prominent feature in Vogt-Koyanagi Harada syndrome (Chap. 1.3.8).

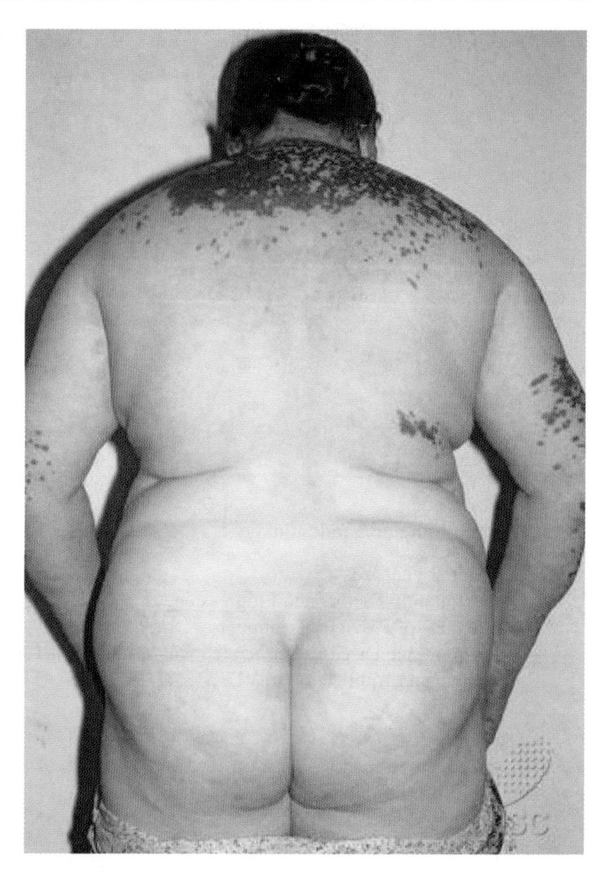

Fig. 1.3.3.1 A Singaporean woman of Indian origin with vitiligo universalis and breakthrough repigmentation on exposed areas (Photograph, courtesy National Skin Centre, Singapore)

Precipitating Factors and Progression

It is uncertain what factors (genetic or environmental) precipitate common NSV to progress to VU. Often the progression is symmetrical. Approximately 20% of vitiligo patients have a first-degree relative with some form of vitiligo [11], there is no evidence that VU incidence increases the risk of another VU case among family members. No sufficient genetic data, which would help in determining the risk of VU in a new case of NSV in a vitiligo family, is available.

Extensive NSV can evolve into VU. Sometimes this evolution can be rather rapid after a period of apparent stability (personal observation). Depigmentation may spread from the edges of existing lesion or by occurrence of new macules. Occasionally before complete depigmentation, surrounding borders of pigmented skin may show signs of inflammation, such as erythema.

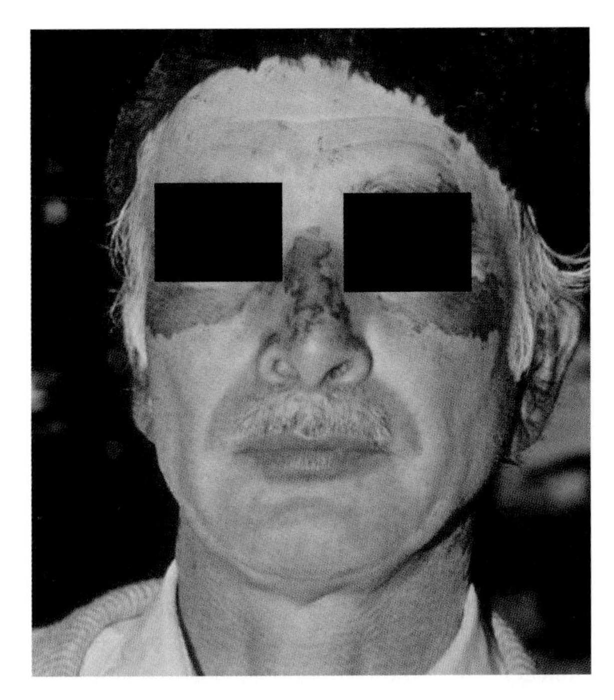

Fig. 1.3.3.2 A 65-year-old Indian male with vitiligo universalis and breakthrough repigmentation in a "lupus pattern" (Photograph courtesy of Prof Binod Khaitan, All India Institute of Medical Science, New Delhi, India)

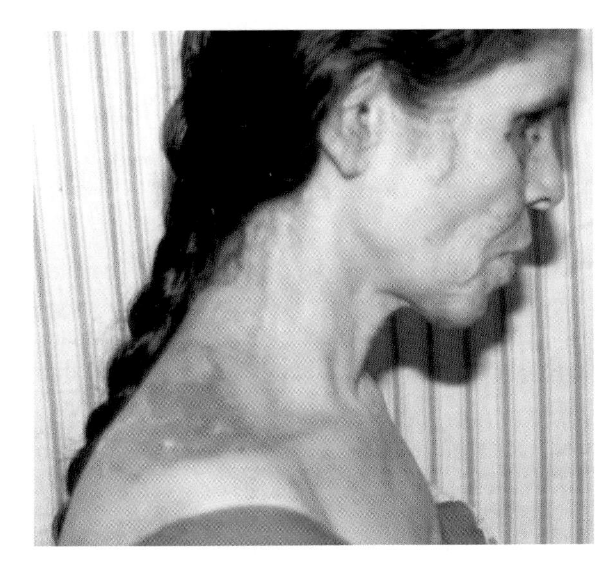

Fig. 1.3.3.3 A 58-year-old Sri Lankan woman with vitiligo universalis with an acute sunburn on the shoulders after exposure to strong sun while working in a paddy field

After the inflammation recedes, these areas also become depigmented. In one case of vitiligo (personal observation), a patient with limited, stable generalized vitiligo

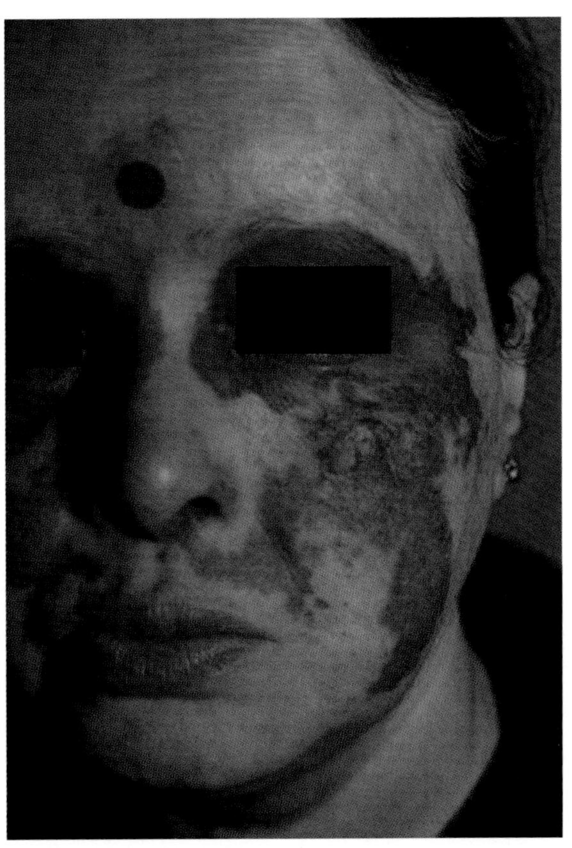

Fig 1.3.3.4 An Indian woman with vitiligo universalis and photo-induced partial repigementation, with a squamous cell carcinoma on the cheek. Carcinomas on vitiliginous skin are rare although sunburns are common (Photograph courtesy of Prof Binod Khaitan, All India Institute of Medical Science, New Delhi, India)

rapidly progressed to VU after an episode of diarrhea and vomiting, following food poisoning. Similarly, pregnancy, physical injury, malignant melanoma, and emotional stress have been incriminated as "triggers" for progression of vitiligo. But it is not clear whether these associations are true associations or chance occurrences [6]. Occasionally, chemical leukoderma, which starts focally, can spread all over the body. Following topical use of diphencyprone for extensive alopecia areata, extensive vitiligo has been reported to appear in areas beyond the contact areas [15]. Perifollicular breakthrough pigmentation on sun-exposed areas, especially the upper cheeks and exposed areas of forearms, is not uncommon (Figures 1.3.3.2 and 1.3.3.4). Pigmentation around nipples may be lost or retained. The author has observed several cases of localized repigmentation in VU cases, after a secondary event, such as lichen planus or insect-bite reactions. Dogra et al. have reported partial repigmentation of VU in a patient who was given dexamethasone and cyclophosphamide pulse therapy for pemphigus vulgaris [4]. This coincidental finding addresses an important aspect. Indeed, even in established VU, inactive, but viable melanocytes are probably present and can be reactivated by adequate signals of the cellular environment. This confirms earlier studies, where some viable melanocytes were found within established vitiliginous macules [7, 24]. Most of these melanocytes or precursors of melanocytes appear to be associated with hair follicles. This may be a reason why the perifollicular pigmentation is the most common type of breakthrough repigmentation in VU.

Influence of Previous Treatments

When a patient with VU presents to a dermatologist, he or she may have already tried several medications. There may be skin atrophy and striae due to prolonged use of potent topical steroids. Monobenzyl ether of hydroquinone (MBEHQ) may have been used to depigment the residual pigmented spots (Part 3.11). This may sometimes cause a contact dermatitis. Breakthrough pigmentation may be a problem if proper sun protection has not been adhered to. The skin may appear yellowish if high doses of oral beta-carotene have been taken. Some may have applied artificial tanning substances (dyes) on the skin (Part 3.9). A few of these tanning substances may cause allergic contact dermatitis, but it is uncommon. Scalp hair, as well as beard or moustache may have been dyed. Contact dermatitis due to PPD (*p*-phenylene diamine) containing hair dyes is not uncommon.

Autoimmune Diseases Associated with Extensive Vitiligo and VU

Many autoimmune/autoinflammatory disorders and several other conditions [1, 3, 5, 9, 14, 17, 21] have been reported in association with vitiligo (Chap. 1.3.8). These conditions may be found in patients with VU as well. These patients may have a family history of other autoimmune disorders, as well as a family history of NSV.

Some patients with Vogt-Koyanagi-Harada syndrome may also evolve into VU, and the limits between the two conditions are somewhat blurred. In this condition, visual defects, alopecia, auditory, and CNS symptoms may precede skin depigmentation. Often, they are presented to the neurologist or ophthalmologist first, due to eye or meningial symptoms.

In VU, other autoimmune disorders may evolve after the onset of vitiligo, therefore these patients have to be followed up with this in mind. In the author's experience, autoimmune thyroiditis and alopecia areata are the most common associations.

1.3.3.3 Diagnosis

Usually there is no diagnostic difficulty in VU. Diagnosis can be made based on the history and the clinical features. Skin biopsy is not required unless for a research purpose. In some situations, some forms of oculo-cutaneous albinism may be confused with vitiligo, especially if the onset of disease or history is not known. Nystagmus and photophobia are not associations of vitiligo, but of oculo-cutaneous albinism. A skin biopsy is helpful in this setting to show the presence of nonfunctioning melanocytes in albinism.

1.3.3.4 Management

Recommended investigations. In simple NSV, detailed investigations are not essential, but in VU several investigations are recommended (Table 1.3.3.1). Many patients with extensive vitiligo have serum antibodies

Table 1.3.3.1 Recommended investigations in vitiligo universalis

Basic haematology	Full cell blood count
	Fasting blood sugar
Endocrinology	Free thyroxine
	TSH
	PTH
Immunology	Anti-thyroid antibodies
	Antinuclear antibodies
Others	Ophtalmological tests
	Audiological tests

Table 1.3.3.2 Management of vitiligo universalis

Chemical depigmentation	Physical depigmentation	Camouflage	Other
MBEHQ	Q switched Ruby laser	Oral beta-carotene	Counselling patient and/or family
	Alexandrite laser	Topical dyes for skin	Psychiatric help
	NdYag laser	Topical dyes for scalp hair and eyebrows	Sun protection
			Surveillance for autoimmune and neoplastic diseases

against melanocytes [12], as well as antithyroid, antiparietal cell, and antinuclear antibodies. Antimelanocyte antibodies are not found in all cases of vitiligo, and these antibodies may be found in some other conditions as well (e.g., mucocutaneous candidiasis). Therefore, this test is not indicated as a routine test.

Therapy. Once VU has fully developed, it is unrealistic to expect complete repigmentation. Management of VU patients is mostly directed at treating the residual patches of pigmented skin, preventing breakthrough pigmentation, and protecting from sun damage while addressing camouflage issues and psychosocial aspects (Table 1.3.3.2) (Parts 3.9–3.11). Counseling is important before implementing complete depigmentation procedures [8, 20]. Occasionally, some patients are keen to attempt to retain pigmentation in some areas, notwithstanding extensive depigmentation in other areas. Q-switched ruby and Q-switched alexandrite lasers have proven to be useful in destroying residual pigmented macules in patients where topical bleaching agents have failed to depigment [13, 16, 23].

Sun Protection

Sun protection is important to avoid sunburns, as well as to prevent unwanted breakthrough and spotty repigmentation. Some studies have revealed that wild type *p53* gene is upregulated in vitiligo patients [18]. This may have a protective effect with regards to sun-induced skin malignancies. Tumors, such as squamous cell carcinomas, on vitiliginous skin are very uncommon in dark-skinned individuals. However, it can occur in rare cases (Fig 1.3.3.4). Furthermore, as melanomas are a known association of vitiligoid depigmentation, if a suspicious new pigmented lesion appears, particularly in association with a pre-existing naevus, it should be excised and sent for histological evaluation.

Psychosocial Aspects

VU causes a huge psychological impact in dark-skinned individuals, and some patients may show signs of depression at the time of presentation [10]. In certain communities and cultures, VU would be even more problematic—patients being ostracized with regards to marriage and social contacts.

Counseling (Part 3.13), and sometimes even psychiatric help may be needed, depending on the emotional impact, particularly in dark-skinned individuals. Counseling the spouse and the family members may also be necessary in some situations. Occasionally, even in the fair-skinned individuals, vitiligo may cause major psychological problems, because sun-induced pigmentation occurs only in the unaffected areas. After understanding the disease process and after realization that no treatment is 100% effective in VU, some patients accept it and continue with their lives well.

1.3.3.5 Conclusions

VU is probably the extreme severity end of the spectrum of NSV, and differentiation from VKH syndrome is not always easy. Genetic studies of families with VU in different ethnic groups would be helpful. Future research should focus on identifying and modifying the triggering factors which start the cascade of depigmentation process all over the skin.

Acknowledgments I am grateful to Prof Roy Chan, Medical Director and Dr. Goh Boon Kee, Consultant Dermatologist, of the National Skin Centre Singapore (where I worked earlier) for providing some of the photographs.

References

1. Alkhateeb A, Fain PR, Thody A et al (2003) Epidemiology of vitiligo and associated autoimmune diseases in Caucasian probands and their families. Pigment Cell Res 16:208–214
2. Aydogan K, Turan OF, Onart S et al (2006) Audiological abnormalities in patients with vitiligo. Clin Exp Dermatol 31:110–113
3. Boisseau-Garsaud AM, Vezon G et al (2000) High prevalence of vitiligo in lepromatous leprosy. Int J Dermatol 39:837–839
4. Dogra S, Kumar B (2005) Repigmentation in vitiligo universalis: role of melanocyte density, disease duration and melanocyte reservoir. Dermatol Online J 11:30
5. Gopal KV, Rama Rao GR, Kumar YH et al (2007) Vitiligo: a part of systemic autoimmune process. Indian J Dermatol Venereol Leprol 73:162–165
6. Hann SK, Nordlund JJ (eds) (2000) Clinical features of generalized vitiligo. In: Vitiligo. Blackwell, Oxford, pp 81–88
7. Husain I, Vijayan E, Ramaiah A et al (1982) Demonstration of tyrosinase in the vitiligo skin of human beings by a sensitive fluorometric method as well as by 14C(U) L tyrosine incorporation into melanin. J Invest Dermatol 78:243–252
8. Kumarasinghe SPW (1995) An optimistic approach to management of vitiligo. Ceylon Med J 40:94–96
9. McGowan JW, Long JB, Johnston CA, Lynn A (2006) Disseminated vitiligo associated with AIDS. Cutis 77:169–173
10. Mattoo SK, Handa S, Kaur I et al (2002) Psychiatric morbidity in vitiligo: prevalence and correlates in India. J Eur Acad Dermatol Venereol 16:573–578
11. Nath SK, Majumder PP, Nordlund JJ (1994) Genetic epidemiology of vitiligo: multilocus recessivity cross-validated. Am J Hum Gene 55:981–990
12. Naughton GK, Eisenger M, Bystryn JC (1983) Antibodies to normal human melanocytes in vitiligo. J Exp Med 158:246–251
13. Njoo MD, Vadegel RM, Westerhof W (2000 Depigmentation therapy in vitiligo universalis with topical 4-methoxyphenol and the Q switched ruby laser. J Am Acad Dermatol 42:760–769
14. Nordlund JJ, Ortonne JP, Le Poole C (2006) Genetic hypomelanosis: acquired depigmentation. In: Nordlund JJ, Boissy RE, Hearing VJ, King RA, Oetting WS, Ortonne JP (eds) The Pigmentary system, 2nd edn. Blackwell, Malden, pp 551–598
15. Pan JY, Goh BK, Theng C, Kumarasinghe SPW (2007) Vitiligo as a reaction to topical treatment with diphencyprone. In: Second Meeting of the Asian Society for Pigment Cell Research, Singapore, July 2007 (abstract in Pigment Cell Res 20:247)
16. Rao J, Fitzpatrick RE (2004) Use of Q switched 755nm Alexandrite laser to treat recalcitrant pigment after depigmentation therapy for vitiligo. Dermatol Surg 30: 1043–1045
17. Saban J, Rodriguez Garcia JL, Gil J et al (1991) Porphyria cutanea tarda associated with autoimmune hypothyroidism, vitiligo and alopecia universalis. Neth J Med 39:350–352
18. Schallreuter KU, Behrens-Williams S, Khaliq TP et al (2003) Increased epidermal functioning wildtype p53 expression in vitiligo. Exp Dermatol 12:268–277
19. Shah AS, Supapannachart N, Nordlund JJ (1993) Acquired hypomelanotic disorders. In: Levine N (ed) Pigmentation and pigmentary disorders. CRS, Boca Raton, pp 337–351
20. Song MS, Hann SK, Ahn PS et al (1994) Clinical study of vitiligo: comparative study of Type A and Type B vitiligo. Annals Dermatol 6:22–30
21. Spritz RA (2007) The genetics of generalized vitiligo and associated autoimmune diseases. Pigment Cell Res. 20: 271–278
22. Tan WP, Goh BK, Tee SI, Kumarasinghe SPW (2007) Clinical profile of vitiligo in Singapore. In: Second Meeting of the Asian Society for Pigment Cell Research, Singapore, July 2007 (abstract in Pigment Cell Res 20:253)
23. Thiessen M. Westerhof W (1997) Laser treatment for further depigmentation in vitiligo. Int J Dermatol 36:386–388
24. Tobin DJ, Swanson NN, Pittlekow MR et al (2000) Melanocytes are not absent in lesional skin of long duration vitiligo. J Pathol 191:407–416

Mucosal Vitiligo

1.3.4

Davinder Parsad

Contents

D. Parsad
Department of Dermatology, Postgraduate Institute of Medical Education & Research, Chandigarh, India
e-mail: parsad@mac.com

Core Messages

- Mucosal vitiligo (MV) can present as a part of involvement in vitiligo vulgaris, as an extension of perioral involvement of acrofacial vitiligo, and as pure mucosal.
- Pure MV is probably a distinct subset of this disease.

1.3.4.1 Definition and Epidemiology

Vitiligo involving the oral and/or genital mucosae has been referred to as mucosal vitiligo (MV). It can occur as a part of generalized vitiligo or as an isolated condition. The status of pure MV is not very clear in the literature, and many authors do not consider it as a different subset of vitiligo. According to the Vitiligo European Task Force (VETF) [22], subclassification of nonsegmental vitiligo into acrofacial, universal, mucosal and so on is not mandatory, as there is no pressing evidence that these are distinct disorders. Another important question which remains unanswered is whether or not all types of vitiligo including MV share a common genetic background.

1.3.4.2 Oral Mucosa

In low phototype individuals, the buccal mucosa is very lightly pigmented despite the presence of melanocytes in the oral epithelium. The number of melanocytes in the mucosa corresponds numerically to that of

M. Picardo and A. Taïeb (eds.), *Vitiligo*,
DOI 10.1007/978-3-540-69361-1_1.3.4, © Springer-Verlag Berlin Heidelberg 2010

skin; however, in the mucosa, their activity is reduced. This may explain that oral vitiligo has long been considered as quite uncommon. According to Coulan and Esquier [8], MV is very rare. Casals [3] mentioned that MV was more common in African blacks and Costa [7] reported many cases of MV from Brazil. Diascopy and Wood's lamp examination may help detecting clinically subtle macules of vitiligo.

Lip involvement is a common feature in dark-skinned vitiligo patients and its incidence has been reported to vary between 20 and 50% [6, 18, 20] (Fig. 1.3.4.1). According to a recent study, mucosa is the site of onset of vitiligo in 7.84% of patients [1]. In a study of 45 patients from Tanzania, 75% had patchy loss of pigment from gum or mucosa of inner lip [15]. In late-onset vitiligo cases, we found that 2.2% had pure mucosal involvement [9]. Besides the mucosal form of vitiligo, associated involvement of mucosae with other clinical variants (vulgaris or acrofacial) was seen in 17% patients with late onset vitiligo. The exact incidence of vitiligo of oral mucosa is not known, but patchy loss of pigment from the buccal mucosa, gingival, and gum line is invariably observed in vitiligo [10].

More commonly, vitiligo involves the vermilion and spares the wet labial mucosa [6]. An inverse distribution, which is, sparing of vermilion and band-like involvement of the labial mucosa, can occur uncommonly [5]. Another uncommon presentation is the involvement of only the most lateral part of the lips. In the acrofacial subtype of vitiligo, there is a characteristic involvement of lips along with involvement of acral parts. In some patients, herpes labialis induced

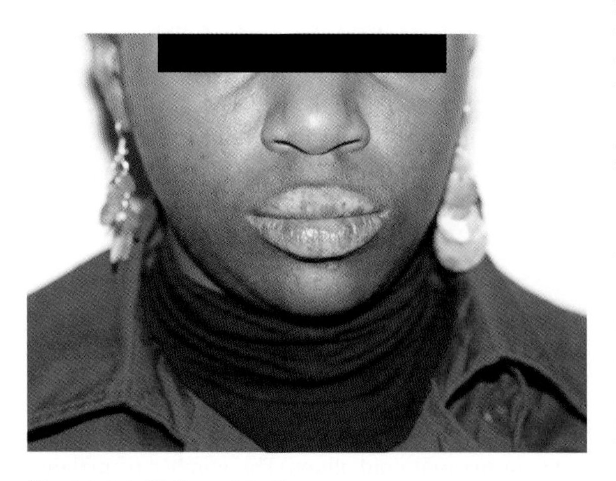

Fig. 1.3.4.1 Vitiligo of the lips

isomorphic response or koebnerization can lead to lip vitiligo [2]. This condition needs to be differentiated from recurrent herpes induced depigmentation occurring after attacks of herpes labialis on and around the lips with resulting depigmentation. Depigmentation corresponds to the area of appearance of vesicles. In the absence of vitiligo lesions elsewhere, it is debatable whether it is an independent entity or an isomorphic response. There are not many controlled studies in the literature which correlate involvement of MV with progression of the disease. However, one study mentioned significant progression of vitiligo in patients with mucosal involvement indicating that it could be a poor prognostic factor [11].

1.3.4.3 Genital Mucosa

As with lip involvement, there can be similar depigmentation involving other mucosae like glans, vaginal, or anal mucosa. In most of the subjects, usually vitiligo lesions are present over other parts of the skin. There can also be isolated involvement of these mucosae in the absence of skin lesions. In a study of 5000 consecutive new male patients examined, 22 patients had vitiligo on the genitalia [12, 17]. Involvement of vulva and vaginal mucosa is less reported. It may be possible that this is due to failure to perform genital examination on a regular basis in female vitiligo patients. Early genital lichen sclerosus et atrophicus (LSA) may be difficult to distinguish from vitiligo (Chap. 1.3.12). Extragenital lesions of LSA located on the neck, shoulders, trunk, and extremities may be present. Histopathology is useful to differentiate between MV and LSA. Simultaneous occurrence of genital LSA and vitiligo has been reported [19].

As there is paucity of studies on MV, we conducted a retrospective analysis of data of 3300 patients of vitiligo registered to our clinic over the last 10 years (Table 1.3.4.1). Family history of vitiligo was positive in 6.4% of patients. The sites of onset in decreasing order of frequency were both the lips simultaneously in 30.5%, lower lip in 25.4%, glans penis in 13.5%, prepuce and angles of mouth in 11.8% each, and labia minora in 1.6% patients. Three patients (5%) had onset simultaneously at different sites.

Table 1.3.4.1 Mucosal vitiligo at the Department of Dermatology, Postgraduate Institute of Medical Education and Research, Chandigarh

Number of patients recruited	59
Percentage of pure mucosal vitiligo	2.3% (59/3,300)
Mean age at onset	29.42 (range 1–63 years)
Male/female	40/19 (2:1)
Mucosal area involvement	
Lower lip	38.9%
Upper lip	30.5%
Both lips	30.5%
Glans penis	16.9%
Prepuce	13.5%
Angles of mouth	11.8%
Labia minora	3.3%
Hard palate and gum	3.3%

1.3.4.4 Management Principles

Depigmentation of the lips and labial mucosa is cosmetically embarrassing and socially stigmatizing in pigmented individuals. Vitiligo lesions involving the lips, oral, and genital mucosa are more resistant to medical therapies, as no melanocyte reservoir exists in these areas because of an absence of hair follicles. Therefore, treatment of MV is an arduous challenge as the medical management of lip vitiligo often results in a sluggish or poor response. In early vitiligo, we found topical tacrolimus to be effective to some extent in lip and penile vitiligo. Recently, topical pimecrolimus was found to be effective in mucosal depigmentation [21].

Micropigmentation (tattooing) gives immediate results and excellent color matching has been reported in various studies, especially in dark individuals [4, 14, 15]. The main drawback of this procedure is that with time, implanted pigment may present an unsightly and inappropriate look, which is difficult to remove even with lasers.

The success rate of various surgical procedures for lip vitiligo varies widely. The cosmetic outcome with individual procedures also varies significantly. Punch grafting has been found to be effective, but it is associated with cobblestoning [16]. Similarly, thin split thickness grafts may be associated with thickened edges and milia formation. Recently, autologous melanocyte transfer via epidermal graft has been found to be an effective and safe therapeutic option for stable vitiligo of the lips. It is cosmetically more acceptable, as there is no abnormal keratinization, which is a problem associated with dermo-epidermal grafts [13] (Part 3.7).

References

1. Al-Mutairi N, Sharma AK (2006) Profile of vitiligo in Farwaniya region in Kuwait. Kuwait Med J 38:128–131
2. Bose SK (2007) Herpes simplex viral infection in association with lip leucoderma. J Dermatol 34:280–281
3. Casals A (1943) Vitiligo. Rev Med Trop 9:15
4. Centre JM, Mancini S, Baker GI et al (1998) Management of gingival vitiligo with use of a tattoo technique. Br J Dermatol 138:359–360
5. Chitole VR (1991) Overgrafting for leukoderma of the lower lip: a new application of an already established method. Ann Plast Surg 26:289–290
6. Coondoo A, Sen N, Panja RK (1976) Leucoderma of the lips. Indian J Dermatol 21:29–33
7. Costa OG (1947) Vitiligo in Brazil. Br J Dermatol 104–108
8. Coulan J and Esquier DA (1926) Vitiligo. Bull Soc Franc Syph 33:681
9. Dogra S, Parsad D, Handa S, Kanwar AJ (2005) Late onset vitiligo: a study of 182 patients. Int J Dermatol 44:193–196
10. Dummet CO (1959) The oral tissues in vitiligo. Oral Surg Oral Med Oral Pathol 12:1073–1079
11. Dutta AK, Mandal SB (1969) Clinical study of 650 vitiligo cases and their classification. Indian J Dermatol 14:103–115
12. Gaffoor PM (1984) Depigmentation of male genitalia. Cutis 34:492–494
13. Gupta S, Sandh K, Kanwar A, Kumar B (2004) Melanocyte transfer via epidermal grafts for vitiligo of labial mucosa. Dermatol Surg 30:45–48
14. Halder RM, Pham HN, Breadon JY, Johnson BA (1989) Micropigmentation for treatment of vitiligo. J Dermatol Surg Oncol 15:1092–1098
15. Hann SK, Nordlund JJ (2000) Clinical features of generalized vitiligo. In: Hann Sk, Nordlund JJ (eds) Vitiligo. Blackwell, Oxford, pp 35–48
16. Malakar S, Lahiri K (2004) Punch grafting in lip leucoderma. Dermatology 208:125–128
17. Moss TR, Stevenson CJ (1981) Incidence of male genital vitiligo. Report of screening programme. Br J Venereal Dis 57:145–146
18. Ortonne JP (2002) Depigmentation of hair and mucous membrane. In: Hann S-K, Nordlund JJ (eds) Vitiligo, 1st edn. Blackwell, Oxford, pp 76–80
19. Osborne GE, Francis ND, Bunker CB (2000) Synchronous onset of penile lichen sclerosus and vitiligo. Br J Dermatol 143:218–219
20. Sehgal VN (1974) A clinical evaluation of 202 cases of vitiligo. Cutis 14:439–445
21. Souza Leite RM, Craveiro Leite AA (2007) Two therapeutic challenges: periocular and genital vitiligo in children successfully treated with pimecrolimus cream. Int J Dermatol 46:986–989
22. Taïeb A, Picardo M; VETF Members (2007) The definition and assessment of vitiligo: a consensus report of the Vitiligo European Task Force. Pigment Cell Res 20:27–35

Halo Nevi and Vitiligo

1.3.5

Thomas Jouary and Alain Taïeb

Contents

T. Jouary (✉)
Service de Dermatologie, Hôpital Saint André,
CHU de Bordeaux, France
e-mail: thomas.jouary@chu-bordeaux.fr

Core Messages

- The halo nevus (HN) is a common lesion, with a suspected prevalence of 1% in the general population.
- The HN incidence is apparently equally increased in both segmental (SV) and nonsegmental (NSV) vitiligo.
- There is a trend toward less extensive NSV in the case of association with HN.
- The usual explanation of the depigmented ring is that the immunologic response, first directed against the nevus melanocytic antigen(s), outgrows its first target to cause the loss of normal melanocytes of the surrounding skin.
- HN and vitiligo involve two independent pathways leading to depigmentation, but a possible overlap exists in a subset of patients, and there is an overall increased proneness to HN in both SV and NSV patients.

1.3.5.1 Definition

A halo nevus (HN) is a melanocytic nevus surrounded by a white depigmented circle. An accurate and complete depiction of the halo naevus has been portrayed by Matthias Grünewald in his painting "The Temptation of St. Anthony", which is part of the Isenheim altar piece (1512–1516) exhibited in Colmar, Alsace [1]. Sutton gave the first incomplete clinical description in 1916, under "leucoderma acquisitum centrifugum", leaving the nature of the central lesion in the dark, but since then, the term of Sutton's nevus as been used

M. Picardo and A. Taïeb (eds.), *Vitiligo*,
DOI 10.1007/978-3-540-69361-1_1.3.5, © Springer-Verlag Berlin Heidelberg 2010

among dermatologists [2]. More than 50 years later, the term of HN was used by Frank et al. [3]. Both terms are now currently used for this clinical entity.

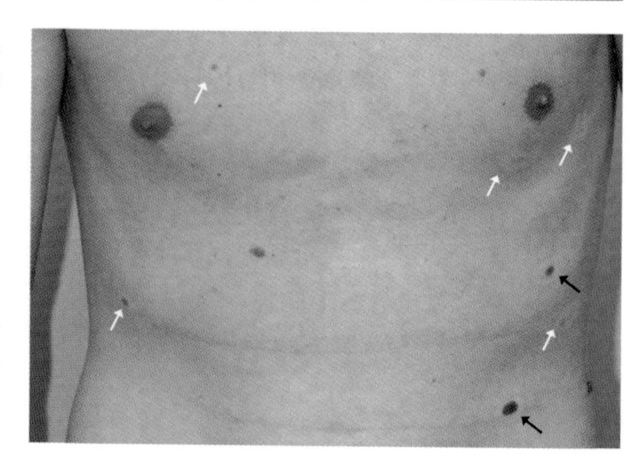

Fig. 1.3.5.2 A 32-year-old man with the sporadic atypical mole syndrome without personal or familial history of melanoma. Many halo nevi can be observed on his trunk at different stages of evolution with some well-pigmented nevi with or without clinical atypia (*black arrows*) and others with partial to complete loss of pigmentation (*white arrows*). This patient had no family history or clinical symptoms of vitiligo. Similarly, no history of autoimmune disorders was found

1.3.5.2 Clinical Features

The clinical aspect of HN is easily recognized by patients and physicians. The melanocytic lesion is usually an acquired melanocytic nevus, which can be a junctional, dermal, or compound naevus (Fig. 1.3.5.1). The diameter of the depigmented circle varies from less than one millimeter to several centimeters [4]. The size of the nevus itself is highly variable and range from a few millimeters to several centimeters. At the centre of the white circle, the presentation of the nevus itself varies according to the stage of the depigmenting process from normally pigmented to complete resolution of the pigmented lesion (Fig. 1.3.5.2): only a white circle remains without any sign of a previous melanocytic nevus. In case of multiple lesions, different stages can be observed in the same individual for both the halo and the nevi.

Epidemiology and Associated Disorders

The HN is a common lesion, with a suspected frequency of 1% in the general population [5,6]. The age of onset is usually in childhood or early adulthood.

The link between vitiligo and HN has been studied because of the striking clinical association of these disorders (Fig. 1.3.5.3). Based on epidemiological studies, the frequency of HN in vitiligo cohorts from different geographic areas varies from 0.5 to 14%. How to explain this large interval? In large series, including adults and children, the history of a previous HN was rarely recorded [7]. Thus, the frequency of HN is probably underestimated in the vitiligo population by the majority of authors. Indeed, the true frequency of this association is probably better estimated by studies of younger vitiligo patients or children, that is, at the age when HN can be observed. Recent studies on such populations have shown frequencies from 4.4 [8] to 7.2% [9]. Another report from Colombia showed that HN was associated with both segmental (SV) and nonsegmental

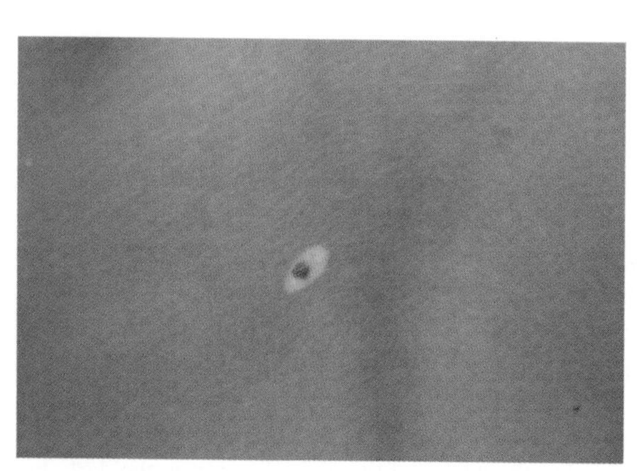

Fig. 1.3.5.1 Halo nevus, at typical halo stage

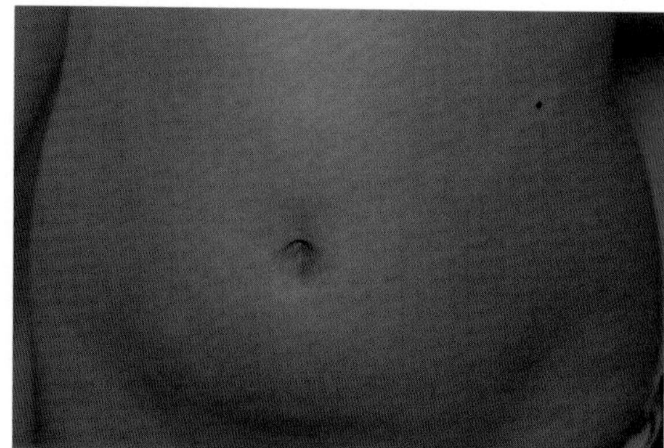

Fig. 1.3.5.3 Association of NSV and halo naevus

vitiligo (NSV) types without significant difference (8.6% in NSV vs. 6.4% in SV) [10]. Similarly, a recent study noted HN in 18.3% of vitligo patients, without significant difference between vulgaris or acrofacial nonsegmental types of vitiligo [11].

HN has been associated with Turner's 45X syndrome. In a case-control study, HN were found in 18.05% of Turner's syndrome patients as compared with only 1% in the control group. The authors concluded that HN was a typical dermatological sign of the Turner's syndrome, but that vitiligo was not clearly associated [5].

Fig. 1.3.5.4 Histopathology of the edge of a halo nevus showing an intense lymphocytic infiltrate stained by the common leukocyte antigen CD45 and peripheral normal skin corresponding to the depigmented halo (CD45 antibody stain, ×400, courtesy Prof Béatrice Vergier, Bordeaux)

Pathogenesis, Histological, Immunological, and Genetic Data

Because of the association between HN and vitiligo, some authors have proposed that HN could be a risk factor for the development of vitiligo [10], and furthermore that HN could be a clinical sign of vitiligo. However, some cases of extensive vitiligo clearly spare or avoid melanocytic nevi. The nature of the HN-depigmented ring has been investigated by several recent studies to address this question: true vitiligo or not?

The loss of pigmentation in HN is thought to be an inflammatory/immunological process. This hypothesis is based on the analysis of the T-cell infiltrate made of both CD4+ and CD8+ lymphocytes (Fig. 1.3.5.4). These lymphocytes express the homing CLA antigen, indicating a preferential circulation in the skin. It has been shown that the inflammatory infiltrate surrounding the nevus varies according to the age of the lesion. In the early stage, there is a presence of T, B lymphocytes, and macrophages; at later stages, the T-cell population is preeminent and finally disappears as the nevus vanishes. This lymphocytic infiltrate is mainly made of CD8+ cells with a minority of CD4+ cells. If the immune cellular responses seem to be more important than the humoral response, some authors have identified serum antibodies against tyrosinase and directed against antigens common to melanoma and melanocytes in patients with vitiligo, HN, and congenital giant nevi. One hypothesis was that the T-cell mediated response could be the primary event causing the release of nevocytic antigen(s) leading B cells to secrete specific IgM and IgG antibodies [12]. The hypothesis that cytotoxic T cells are responsible for the primary event, that is, nevus damage and regression, is still not formally proven.

A local cutaneous expansion of T lymphocytes in HN has been identified previously in vitro, based on the analysis of the TCR (T-cell receptor). According to Musette et al. the T-cell infiltrate was composed of both a polyclonal T-cell population and a minority of clonally expanded T cells. Those T-cell clones were identified *in vivo* by RT-PCR. In two HN patients, the T-cell clones were the same in different HN from the same patient. It would therefore be hypothesized that a common antigen is triggering lymphocytes in different skin sites in the same patient. The nature of this antigen is still unknown, but different from melanoma antigens (MAGE, BAGE, GAGE, and RAGE) [13]. The usual explanation of the depigmented ring is that the immunologic response, first directed against the nevus melanocytic antigen(s), outgrows its first target to cause the loss of normal melanocytes of the surrounding skin [14]. The origin of this autoimmune mechanism against melanocytic naevus is poorly understood so far.

The halo around the nevus is always remarkably round or oval. This geometrical figure has led some authors to suggest a phenomenon beginning at the centre where the nevus is situated, and then spreading centrifugally around [15]. Oxidative stress is thought to play a role in the pathogenesis of HN, and whether it is a cause or a consequence is still debatable. Following studies showing an accumulation of H_2O_2 and low catalase levels in vitiligo [16, 17], it has been shown *in vivo* (cutaneous and epidermal suction blisters samples) that H_2O_2, pterin-4a-carbinolamine dehydratase (PCD) and catalase levels were differently affected in HN associated with vitiligo versus vitiligo arising in isolation. Indeed, PCD was found upregulated and associated with low levels of H_2O_2, while epidermal catalase levels was only slightly decreased in HN associated with vitiligo (Chap. 2.2.6). Then, two different biochemical pathways could be involved in the depigmentation of HN associated with vitiligo and in NSV in isolation [11].

A case-control study in a Dutch population has demonstrated a positive association of vitiligo with HLA-DR4 and a negative association with HLA-DR3, whereas HN-associated vitiligo patients were not associated with these previous alleles, but with DR1, DR10, and DQ5 alleles. There was a trend toward less-extensive vitiligo in case of association with HN. The presence of other autoimmune diseases, such as thyroididis and diabetes mellitus differed between the subtypes, with less personal and familial autoimmunity in the presence of HN. The association with HLA Class II indicates a relevant involvement activation of CD4+ T cells in accordance with previously described immunological studies in both HN and NSV. However, variations across the HLA Class II repertoire may produce different genetic activation pathways in vitiligo with and without associated HN, according to the modulation of responses to autoantigens originating from normal melanocytes in vitiligo, and melanocytic nevi in halo-nevi [15].

1.3.5.3 Conclusions

HN and vitiligo are probably two distinct entities involving independent pathways leading to depigmentation, but a possible overlap exists in a subset of patients. An immune response to nevocytic antigens spread to melanocytic antigens (or the reverse) in the context of vitiligo melanocyte "fragility" can be speculated because of the frequency of HN in vitiligo patients as compared to the general population. Whether the presence of HN represents a risk factor for vitiligo onset or a sign of progression of active vitiligo is not yet settled. Similar to other clinical symptoms such as the Koebner's phenomenon (Sect. 2.2.2.1) and familial history of vitiligo or canitia (Chap. 1.3.6), HN needs to be further evaluated as a prognostic factor in prospective studies.

References

1. Happle R (1999) Grünewald nevus. Hautarzt 45:882–883
2. Sutton RL (1916) An unusual variety of vitiligo (leukoderma acquisitum centrifugum). Cutan Dis 34:797–800
3. Frank SB, Cohen HJ (1964) The halo nevus. Arch Dermatol 89:367–373
4. Wayte DM, Helwig EB (1968) Halo nevi. Cancer 22:69–90
5. Brazzelli V, Larizza D, Martinetti M et al (2004) Halo nevus, rather than vitiligo, is a typical dermatologic finding in Turner's syndrome: clinical, genetic, and immunogenetic study in 72 patients. J Am Acad Dermatol 51:354–358
6. Larsson PA, Liden S (1980) Prevalence of skin diseases among adolescents 12–16 years of age. Acta Derm Venereol 60:415–423
7. Handa S, Kaur I (1999) Vitiligo: clinical findings in 1436 patients. J Am Acad Dermatol 26:653–657
8. Handa S, Dogra S (2003) Epidemiology of childhood vitiligo: a study of 625 patients from North India. Pediatr Dermatol 20:207–210
9. Hu Z, Liu JB, Ma SS et al (2006) Profile of childhood vitiligo in China: an analysis of 541 patients. Pediatr Dermatol 23:114–116
10. Barona MI, Arrunategui A, Falabella R, Alzate A (1995) An epidemiologic case-control study in population with vitiligo. J Am Acad Dermatol 33:621–625
11. Schallreuter KU, Kothari S, Elwary S et al (2003) Molecular evidence that halo in Sutton's naevus is not vitiligo. Arch Dermatol Res 295:223–228
12. Zeff RA, Freitag A, Grin CM, Grant-Kels J (1997) The immune response in halo nevi. J Am Acad Dermatol 37: 620–624
13. Musette P, Bachelez H, Flageul B et al (1999) Immune-mediated destruction of melanocytes in halo nevi is associated with the local expansion of a limited number of T cell clones. J Immunol 162:1789–1794
14. Bergman W, Willemze R, de Graaff-Reitsma C, Ruiter DJ (1985) Analysis of major histocompatibility antigens and the mononuclear cell infiltrate in halo nevi. J Invest Dermatol 85:25–29
15. De Vijlder HC, Werterhof W, Schreuder GM et al (2004) Difference in pathogenesis between vitiligo vulgaris and halo nevi associated with vitiligo is supported by an HLA association study. Pigment Cell Res 17:270–274
16. Schallreuter KU, Wood JM, Berger J (1991) Low catalase levels in the epidermis of patients with vitiligo. J Invest Dermatol 97:1081–1085
17. Schallreuter KU, Moore J, Wood JM et al (2001) Epidermal H_2O_2 accumulation alters tetrahydrobiopterin (6BH4) recycling in vitiligo: identification of a general mechanism in regulation of all 6BH4-dependent processes? J Invest Dermatol 116:167–174

Hair Involvement in Vitiligo

1.3.6

Rafael Falabella

Contents

R. Falabella
Universidad del Valle and Centro Medico Imbanaco, Cali, Colombia
e-mail: rfalabella@uniweb.net.co

Core Messages

- Hairs become affected by vitiligo with variable frequency according to subtype, and this does not usually relate to disease progression.
- Loss of hair melanocyte reservoir means that repigmentation may be very difficult with medical therapy.
- Canities (diffuse premature hair greying) indicate a genetically and age-related failure of pigmentation mechanisms, and may also express early vitiligo in some affected patients.
- Poliosis (cluster of depigmented hair) is not specific to vitiligo.
- Repigmentation of leukotrichia may be possible with melanocyte transplantation, but methods are still at a development stage.
- Hair follicles melanocytes are frequently affected by vitiligo and determine leukotrichia, manifested clinically as permanent depigmentation of hair. Such defects may be considered by patients as unacceptable. Sometimes, a treatment can be envisaged.

1.3.6.1 Hair, Melanocytes, and Pigmentation

Hair Follicles

Human skin contains around 5 million hair follicles, 100,000 of which belong to the scalp [27]. The three types of hair follicles observed in humans, terminal

M. Picardo and A. Taïeb (eds.), *Vitiligo*,
DOI 10.1007/978-3-540-69361-1_1.3.6, © Springer-Verlag Berlin Heidelberg 2010

hair, vellous hair, and lanugo, are located in different anatomical sites and have similar basic features, although they have some structural and pigmentary differences.

The hair follicle has a dermal compartment composed of the connective tissue sheath and the dermal papilla with a fine vascular network, and an epithelial compartment with very active replicating matrix cells that originate three concentrical layers of the hair follicle's cylindrical structure, the outer, and inner root sheaths and the hair shaft [2].

In the scalp, where typical terminal hair is present, the hair follicle has a continuous proliferation during the growing cycle, the anagen, catagen, and telogen phases [8]. After telogen, hair regenerates with pluripotent stem cells in a rapid manner and during hair growth and differentiation, more than 20 different cell types are involved having different pathways and interaction with each other. Hairs in other cutaneous areas have shorter, but similar cycles to those of scalp hair.

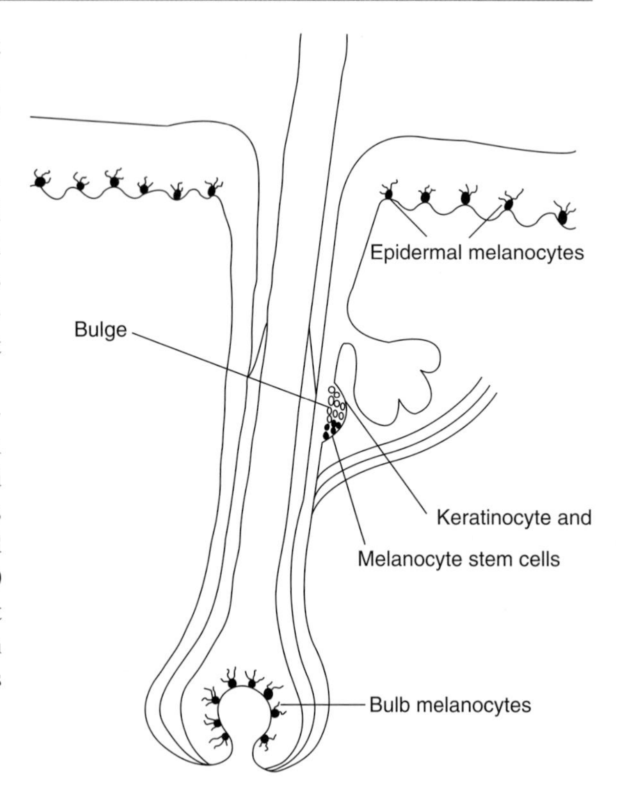

Fig. 1.3.6.1 Schematic representation of hair follicle and melanocytes. Notice keratinocyte and melanocyte stem cells present in the bulge and sub-bulge area

Hair Follicle Melanocytes, Melanocyte Reservoir, and Pigment Stem Cells

Hair follicle melanocytes in primary cultures display two morphologically and antigenically distinct types of cells: (1) pigmented, polydendritic melanocytes, present in the infundibulum and bulb (DOPA-positive); (2) small, bipolar, amelanotic melanocytes found in the outer root sheath of middle and lower hair (DOPA-negative) [31, 32]; the latter are present in the outer root sheath and cannot be identified by antibodies to tyrosinase, TRP-1, or TRP-2, or HMB45 antibody, since they are inactive and do not contain any of the enzymatic proteins necessary for melanin production [12].

The main precursor pigment stem cells of the hair follicle are known to be present within a restricted area, the bulge and sub-bulge regions, located just deep to the sebaceous gland and adjacent to the insertion site of the erector pili muscle [20]; these cells have a slow replicating cycle, but a high proliferative potential being pluripotential and self renewing, becoming active during the anagen phase (Fig. 1.3.6.1) [14].

1.3.6.2 Pigmentation in the Hair Follicle

Melanogenesis of the hair follicle involves follicular melanocytes, the transfer of melanin granules into cortical and medulla keratinocytes, and the formation of pigmented hair shafts, all steps being regulated by several enzymes, structural and regulatory proteins, transporters, receptors and their ligands, acting during development, and also directly on cells or hair follicles.

Follicular melanogenesis is tightly coupled to the anagen stage of the hair cycle, being switched-off in catagen and remaining absent through telogen; in addition, the melanocyte compartments in the upper hair follicle provide a reservoir for the repigmentation of epidermis and for the cyclic formation of new anagen hair bulbs.

Melanin synthesis and pigment transfer to bulb keratinocytes are dependent on melanin precursors; it is regulated by signal transduction pathways, with intervention of autocrine, paracrine, or intracrine mechanisms and can be modified by hormonal signals. Regulation is mediated by melanocortin-1 receptor, adrenocorticotropic hormone, melanocyte stimulating hormone, agouti protein ligands (in rodents), c-Kit, and the endothelin receptors with their ligands [28].

1.3.6.3 Hair Involvement in Vitiligo

Depigmentation in vitiligo primarily affects the epidermis which is visualised clinically as white patches of skin of involved areas. Hair is initially spared from depigmentation, a very common finding in vitiligo (Fig. 1.3.6.2), but after a prolonged course, lack of hair pigmentation within vitligo affected areas may occur

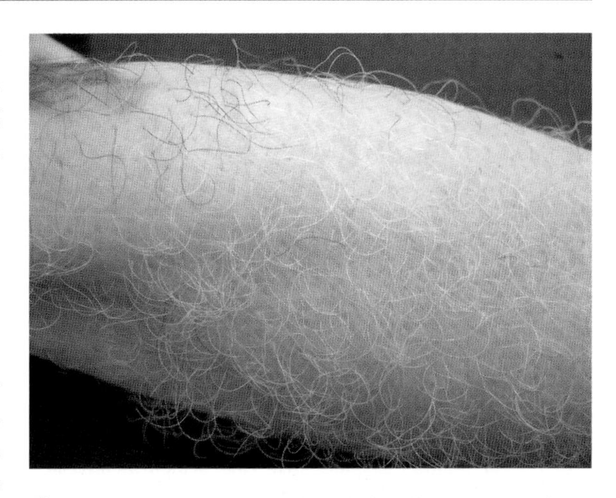

Fig. 1.3.6.3 Depigmented hairs within NSV macule. After a prolonged course leukotrichia may develop within affected skin

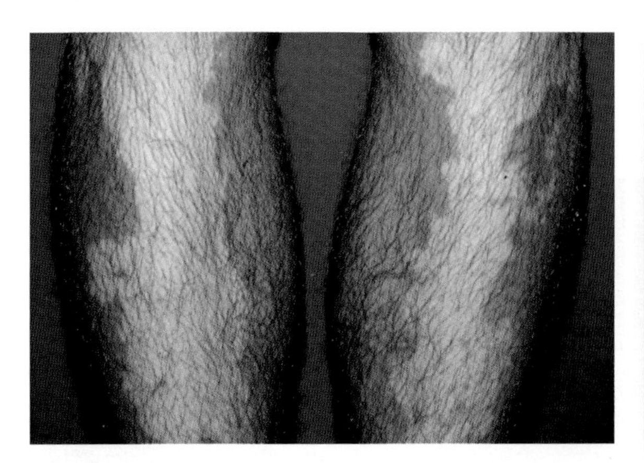

Fig. 1.3.6.2 Pigmented hairs within NSV vitiligo lesions. Hair may remain pigmented for months or years, and also indefinitely

as poliosis or leukotrichia (Fig. 1.3.6.3). This manifestation usually does not correlate with disease activity since vitiligo does not progress more rapidly in patients with leukotrichia than in patients with normally pigmented hair [11].

Leukotrichia frequently occurs in rapidly developing segmental vitiligo (SV) (Chap. 1.3.2), whereas in non-segmental vitiligo (NSV) usually occurs at a slower rate. In addition, hair depigmentation may develop randomly in affected vitiligo patches or as part of several clinical conditions associated with vitiligo as in Vogt Koyanagi and Alezzandrini syndromes (Chap. 1.3.8).

Scalp poliosis is the most frequent manifestation of hair depigmentation; in a study, 355 (27%) out of 1,350 patients with vitiligo had some type of leukotrichia, with scalp being the most frequently affected area in

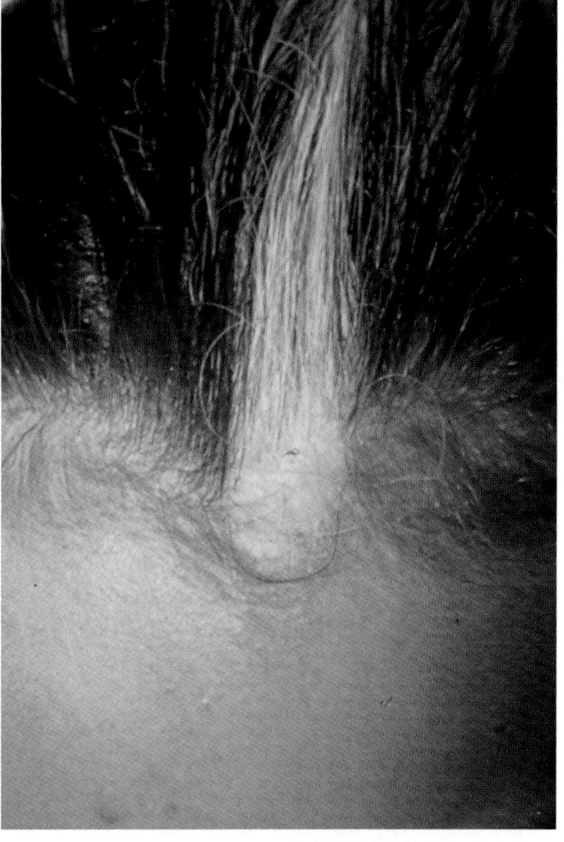

Fig. 1.3.6.4 Poliosis in scalp vitiligo. White forelock simulating piebaldism appearing in a 16-year-old girl

63.7% of patients (Fig. 1.3.6.4), followed by eyebrows in 13.8%, pubic hair in 8.4%, and axilla in 2% of affected individuals [29].

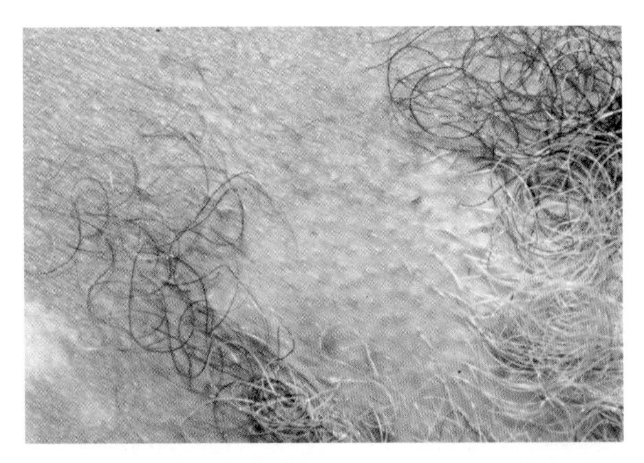

Fig. 1.3.6.5 Poliosis of pubic area. Hair depigmentation is a sign of poor prognosis for repigmentation of vitiligo because of permanent exhaustion of the melanocyte reservoir

The incidence of body leukotrichia may be as high as 60% [26] or as low as 10% [4] in the affected population. In SV, where the head and neck are frequently involved, poliosis was found in 48.6% in a group of 101 patients [10]; in this site it is frequent to observe poliosis of eyebrows and eyelashes in addition to scalp areas. Extensive hair depigmentation may occur in patients with scalp vitiligo, although in individuals so affected or with marked skin depigmentation, that is, universal vitiligo (Chap. 1.3.3), it is not infrequent to observe that most of the scalp hairs display normal pigmentation. With regard to histological and ultrastructural findings, depigmented hair in vitiligo do not show any melanocytes, whereas during greying or hair whitening due to the ageing process, abnormal melanocytes may be observed in the hair follicle [19].

In vitiligo, normally pigmented hair is the main source of repigmentation since they constitute the most important reservoir of melanocytes in this condition. When leukotrichia is present, the possibilities of repigmentation are minimal or null because hair melanocyte loss is usually permanent and repigmentation does not occur with medical therapy (Figs. 1.3.6.5 and 1.3.6.6) (Chap. 2.2.10).

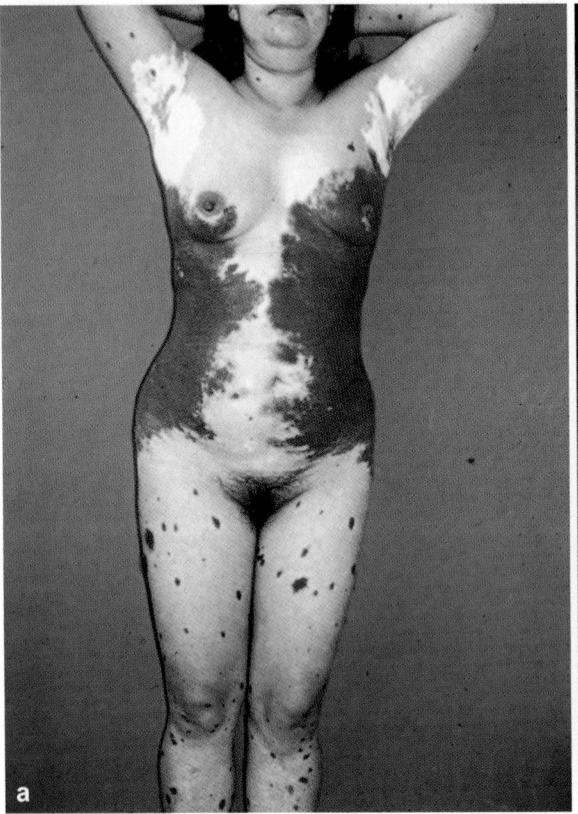

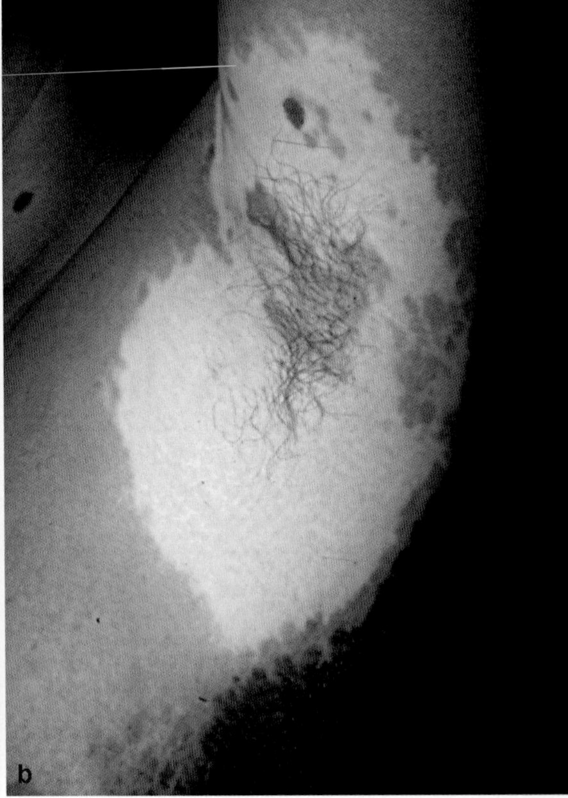

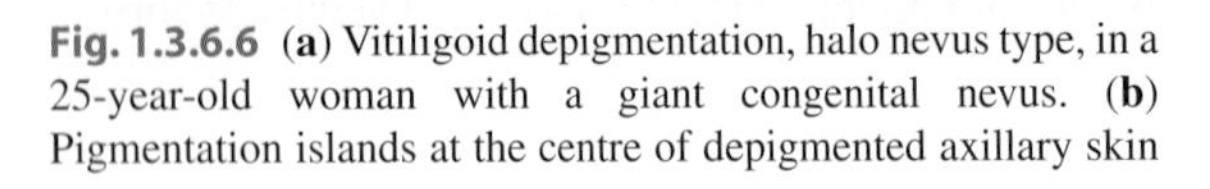

Fig. 1.3.6.6 (**a**) Vitiligoid depigmentation, halo nevus type, in a 25-year-old woman with a giant congenital nevus. (**b**) Pigmentation islands at the centre of depigmented axillary skin is provided by pigmented hairs, indicating the important role of an intact melanocyte reservoir for repigmentation in depigmented skin

1.3.6.4 Canities

Hair greying is also known as canities, a word derived from the Latin word *canus* = white. Although hair greying and whitening are the most obvious signs of ageing in humans, the aetiology of this hair manifestation is largely unknown.

As mentioned earlier, one important distinguishing feature of follicular melanogenesis, compared to the continuous melanogenesis in the epidermis, is the tight coupling of hair follicle melanogenesis to the hair-growth cycle, with periods of melanocyte proliferation (early anagen), maturation (mid to late anagen), and melanocyte death via apoptosis (early catagen). Thus, each hair cycle is associated with the reconstruction of an intact hair follicle pigmentary unit occurring during approximately the first 10 cycles, which is approximately around 40 years of age, when canities appear for the first time in most individuals. Thereafter, grey and white hair develop, suggesting an age-related, genetically regulated exhaustion of the pigmentary potential of each individual hair follicle [34]. A reduction in tyrosinase activity within hair bulbar melanocytes occurs together with inappropriate melanocyte-cortical keratinocyte interactions, and defective migration of melanocytes from the upper outer root sheath to the follicular papilla of the hair bulb, all of which will disrupt the normal function of the pigmentary unit [34]. In addition, in melanocyte-tagged transgenic mice and aging human hair follicles, hair greying has been found to be caused by defective self-maintenance of melanocyte stem cells by their selective apoptosis, not occurring in differentiated melanocytes, within the niche at the beginning of the dormant state [17].

In young patients, early greying or isolated patches of individual white hair before the age of 30 has been suggested to be a form of vitiligo [19]. It is interesting to note that premature hair greying has also been observed in up to 37% of patients with vitiligo [16], suggesting that poliosis in some patients may be an early sign of vitiligo onset.

As a differential diagnosis, canities have also been suggested as a marker for osteopenia; in a case-control study of 293 healthy postmenopausal women, premature hair greying was found to be associated with this condition, indicating that this finding might be a clinically useful risk factor for osteoporosis [18]. In another study, an association between premature greying and low bone mass was found, and a relationship with genes that control peak bone mass or factors that regulate bone turnover was suggested [23]. Nevertheless, in a larger and recent study, bone mineral density (BMD) measured at the spine, hip, and total body in 508 women and 380 men, hair greying was not significantly associated with BMD in either group [15].

Another differential diagnosis of hair hypopigmentation with iron deficiency has been reported to induce a band like segmented heterochromic hair which recovered completely after iron supplementation, coinciding with increased eumelanogenesis in repigmented hair [25].

1.3.6.5 Poliosis

Poliosis is a word derived from the Greek word *polios* = grey (med. grey or white scalp hair) and *osis* = state, condition. It is a synonym of a commonly used word that has a similar meaning, leukotrichia, from the Greek word *leukos* = white, and *thrix* = hair.

Poliosis is the name given to a localised patch of white hair, usually describing a white forelock, but it can involve a patch of white hair anywhere on the body. Poliosis can occur in otherwise healthy people, and it is also observed in association with a wide variety of conditions [20] (Table 1.3.6.1). Among autoimmune/autoinflammatory conditions, a T-cell-mediated cytotoxicity and apoptosis in the development of skin lesions has been suggested in Vogt-Koyanagi-Harada syndrome [7, 30, 33] in Alezzandrini syndrome, and in a frustrated form of a migratory variety of alopecia areata [5]. *KIT* gene mutations [22] have been described

Table 1.3.6.1 Etiology of poliosis, adapted from [20]

Inflammatory or autoimmune: vitiligo, halo nevus, alopecia areata, post-inflammatory dermatoses, post-trauma, Vogt-Koyanagi-Harada syndrome, Alezzandrini syndrome, alopecia areata
Inherited: tuberous sclerosis, piebaldism, Waardenburg syndrome, isolated forelock, isolated occipital (X linked recessive), white forelock with osteopathia striata (autosomal or X linked dominant), white forelock with multiple malformations (autosomal or X-linked recessive)
Nevoid: with nevus comedonicus, secondary to mosaicism
Drug induced: topical prostaglandin F2alpha and its analogs (latanoprost and isopropyl unoprostone)
Idiopathic

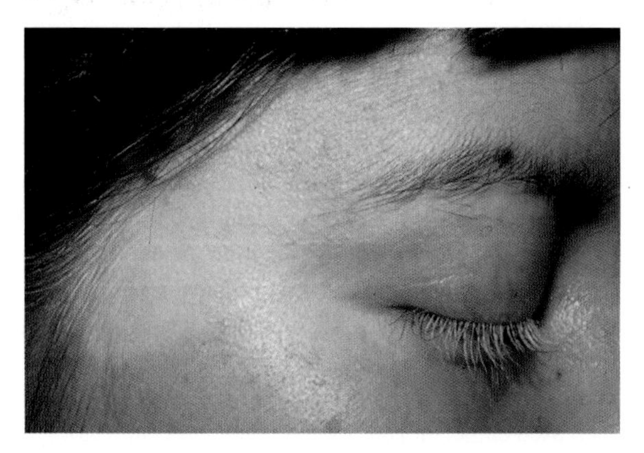

Fig. 1.3.6.7 Eyelash poliosis. This 10-year-old girl developed poliosis of eyelashes following a rapidly developing segmental vitiligo of temporal area and adjacent skin

as the major causative mechanisms involved in genetic hair depigmentation in piebaldism (see Chap. 1.3.11). Topical prostaglandin F2alpha and its analogues (latanoprost and isopropyl unoprostone), which are useful for primary open angle glaucoma in some patients have been reported to develop eyelash poliosis (Fig. 1.3.6.7) [3, 24].

1.3.6.6 Surgical Correction of Leukotrichia

Poliosis or leukotrichia contribute to the unsightly appearance of affected vitiligo skin occurring anywhere in SV or in well-established vitiligo vulgaris, where hair follicle melanocytes disappear and leukotrichia arises to remain permanently.

Although depigmented skin is more noticeable than leukotrichia, and vellous hairs are frequently unnoticeable within depigmented vitiligo skin, in certain areas with terminal, dark hair (scalp, beard skin, eyebrows, eyelashes), leukotrichia may be notorious and patients seek advice or possibilities of repigmentation. This therapy could become possible according to several reports of patients recovering from leukotrichia with surgical procedures.

In fact, during melanocyte transplantation of *in vitro* cultured epidermis, [6] epidermal grafts [9], and split thickness grafts [1] leukotrichia improved after treatment, suggesting that melanocytes could have migrated in a reversal mechanism from that seen in perifollicular repigmentation during vitiligo therapy; such melanocyte migration could occur from the grafted epidermis towards the outer root sheath of the hair follicle and from this site to the hair bulb [6]. The results achieved have shown that repigmentation of leukotrichia is possible; some areas such as the eyebrows, scalp, or beard areas responded well to surgical interventions in seven patients who were treated [1]; in addition, hairs have been observed to remain repigmented after 10 years (personal observation) in the axillary area of a patient with SV after *in vitro* transplantation of melanocytes, reported elsewhere [6]. It is important to note that for repigmentation of leukotrichia to be successful, complete disease stability is necessary.

In another study of five patients with piebaldism and who received epidermal grafts on depigmented skin, the pre-existing white hair in recipient sites became pigmented within a year after epidermal grafting. Studies using the melanocyte-specific antibody NKI/beteb revealed the presence of melanocytes in the newly pigmented hair follicles [13].

The most reasonable explanation for the surgical outcome in all mentioned reports is that melanocytes transplanted in the close vicinity of depigmented hair would have reached the follicular ostium and migrated downward through the external root sheath as described earlier. However, additional refinements of melanocyte transplantation for repigmentation of leukotrichia should be worked out in future trials to standardise these procedures, since at present surgical methods are at the stage of case reports.

A warning word of caution about epilation should be mentioned, because leukotrichia developed as a side effect in 29 of 821 patients treated with a non-coherent IPL system, with a 650 nm flashlamp filter for epilation; although spontaneous restoration of hair colour occurred in 9 patients, the remaining 20 patients had no improvement within the next 2–6 months [21]. The light absorbed and the heat produced by melanin may be sufficient enough to destroy or impair the function of melanocytes, although insufficient to damage the hair follicle cells. This side effect should be carefully considered in patients with a diagnosis of vitiligo, even with minor extension of lesions before epilation is performed for cosmetic reasons, since dissemination of vitiligo within an area devoid of hair—this is without a melanocyte reservoir—could possibly originate irreversible and refractory vitiligo.

References

1. Agrawal K, Agrawal A (1995) Vitiligo: surgical repigmentation of leukotrichia. Dermatol Surg 21:711–715
2. Bernard BA (2006 The life of human hair follicle revealed. Med Sci 22:138–143
3. Chen CS, Wells J, Craig JE (2004) Topical prostaglandin F(2alpha) analog induced poliosis. Am J Ophthalmol 137:965–966
4. Dutta AK, Mandal SB (1969) A clinical study of 650 vitiligo cases and their classification. Indian J Dermatol 14: 103–111
5. Elston DM, Clayton AS, Meffert JJ et al (2000) Migratory poliosis: a forme fruste of alopecia areata? J Am Acad Dermatol 42:1076–1077
6. Falabella R, Escobar C, Borrero I (1992) Treatment of refractory and stable vitiligo by transplantation of in vitro cultured epidermal autografts bearing melanocytes. J Am Acad Dermatol 26:230–236
7. García Hernández FJ, Ocaña Medina C, Castillo Palma MJ (2006) Vogt-Koyanagi-Harada disease. Characteristics of a series of Andalusian patients. Rev Clin Esp 206:388–391
8. Giacometti L (1965) The anatomy of the human scalp, Chapter VI. In: Montagna W (ed) Advances in biology of skin, vol VI. Pergamon, Oxford, pp 97–120
9. Hann SK, Im S, Park YK et al (1992) Repigmentation of leukotrichia by epidermal grafting and systemic psoralen plus UV-A. Arch Dermatol 128:998–999
10. Hann SK, Lee HJ (1996) Segmental vitiligo: clinical findings in 208 patients. JAm Acad Dermatol 35:671–674
11. Hann SK, Chun WH, Park YK (1997) Clinical characteristics of progressive vitiligo. Int J Dermatol 36:353–355
12. Horikawa T, Norris DA, Johnson TW et al (1996) DOPA-negative melanocytes in the outer root sheath of human hair follicles express premelanosomal antigens but not a melanosomal antigen or the melanosome-associated glycoproteins tyrosinase, TRP-1, and TRP-2. J Invest Dermatol 106:28–35
13. Horikawa T, Mishima Y, Nishino K et al (1999) Horizontal and vertical pigment spread into surrounding piebald epidermis and hair follicles after suction blister epidermal grafting. Pigment Cell Res 12:175–180
14. Loomis CA, Koss J, Chu D (2008) Embryology. In: Bolognia JL, Jorizzo JL, Rapini RP (eds) Dermatology, 2nd edn. Elsevier, Amsterdam, pp 37–47
15. Morton DJ, Kritz-Silverstein D, Riley DJ et al (2007) Premature graying, balding, and low bone mineral density in older women and men: the Rancho Bernardo study. J Aging Health 19:275–285
16. Moscher DB (1993) Vitiligo: etiology, pathogenesis, diagnosis and treatment. In: Fitzptrick TB, Eisen AZ, Wolff K, Freedberg AM, Austen KF (eds) Dermatology in general medicine, 4th edn, vol I. McGraw Hill, New York, pp 923–933
17. Nishimura EK, Granter SR, Fisher DE (2005) Mechanisms of hair graying: incomplete melanocyte stem cell maintenance in the niche. Science 307:720–724
18. Orr-Walker BJ, Evans MC, Ames RW et al (1997) Premature hair graying and bone mineral density. J Clin Endocrinol Metab 82:3580–3583
19. Ortonne JP (2000) Depigmentation of hair and mucous membranes. In: Hann SK, Nordlund JJ (eds) Vitiligo. Blackwell Science, Oxford, pp 77–80
20. Ortonne JP (2008) Vitiligo and other disorders of hypopigmentation. In: Bolognia JL, Jorizzo JL, Rapini RP (eds) Dermatology, 2nd edn, Chapter 65, Mosby Elsevier Limited
21. Radmanesh M, Mostaghimi M, Yousefi I et al (2002) Leukotrichia developed following application of intense pulsed light for hair removal. Dermatol Surg 28: 572–574
22. Richards KA, Fukai K, Oiso N et al (2001) A novel KIT mutation results in piebaldism with progressive depigmentation. J Am Acad Dermatol 44:288–292
23. Rosen CJ, Holick MF, Millard PS (1994) Premature graying of hair is a risk marker for osteopenia. J Clin Endocrinol Metab 79:854–857
24. Sasaki S, Hozumi Y, Kondo S (2005) Influence of prostaglandin F2alpha and its analogues on hair regrowth and follicular melanogenesis in a murine model. Exp Dermatol 14:323–328
25. Sato S, Jitsukawa K, Sato H et al (1989) Segmented heterochromia in black scalp hair associated with iron-deficiency anemia. Canities segmentata sideropaenica. Arch Dermatol 125:531–535
26. Seghal VN (1974) A Clinical evaluation of 202 cases of vitiligo. Cutis 14:439–445
27. Sinclair RD, Banfield CC, Dawber RP (1999) Handbook of diseases of the hair and scalp. Blackwell Science, Oxford
28. Slominski A, Wortsman J, Plonka PM et al (2005) Hair follicle pigmentation. J Invest Dermatol 124:13
29. Song MS, Hann SK, Ahn PS et al (1994) Clinical study of vitiligo: comparative study of type A and B vitiligo. Ann Dermatol 6:22–30
30. Sukavatcharin S, Tsai JH, Rao NA (2007) Vogt-Koyanagi-Harada disease in Hispanic patients. Int Ophthalmol 27:143–148
31. Takada K, Sugiyama K, Yamamoto I et al (1992) Presence of amelanotic melanocytes within the outer root sheath in senile white hair. J Invest Dermatol 99:629–633
32. Tobin DJ, Bystryn JC (1996) Different populations of melanocytes are present in hair follicles and epidermis. Pigment Cell Res 9:304–310
33. Tsuruta D, Hamada T, Teramae H et al (2001) Inflammatory vitiligo in Vogt-Koyanagi-Harada disease. J Am Acad Dermatol 44:129–131
34. Van Neste D, Tobin DJ (2004) Hair cycle and hair pigmentation: dynamic interactions and changes associated with aging. Micron 35:193–200

Non-Skin Melanocytes in Vitiligo

1.3.7

Raymond E. Boissy

Contents

Core Messages

- Although melanocyte loss in vitiligo predominantly involves the skin, alterations in the melanocytes populations of extracutaneous sites have been reported.
- Loss of ocular pigmentation can occur in the eyes of a percentage of patients with vitiligo with focal hypopigmented lesions on the iris, anterior chamber, and retinal pigment epithelium/choroid.
- Uveitis associated with central nervous system abnormalities is a feature of Vogt–Koyanagi–Harada (VKH) syndrome.
- Loss of otic melanocytes may occur in some patients with vitiligo as well as subtle abnormalities of hearing and brainstem auditory response.
- The possible involvement of leptomeningeal melanocytes has been speculated mostly in patients with VKH syndrome, even if the role of these melanocytes is not known.

1.3.7.1 Introduction

Melanocytes are most prevalent in the skin of the interfollicular epidermis and the hair follicle. However, significant populations of melanocytes occur in the eye, ear, and leptomeninges of the brain [10]. The function of melanocytes in both the eye and ear is twofold. Melanocytes, and/or the melanin they synthesise, are crucial during embryogenesis for the neuronal

R. E. Boissy
University of Cincinnati College of Medicine,
Cincinnati, OH, USA
e-mail: boissy.raymond@uc.edu

M. Picardo and A. Taïeb (eds.), *Vitiligo*,
DOI 10.1007/978-3-540-69361-1_1.3.7, © Springer-Verlag Berlin Heidelberg 2010

development between these sensory organs and their respective target centres in the brain even if these melanocyte populations facilitate the functions of these sensory organs postnatally. Although melanocyte loss in vitiligo is predominantly confined to the skin of a patient, alterations in the extracutaneous sites have been reported and/or implied for the eye and ear with associated function compromise in these sensory organs [27].

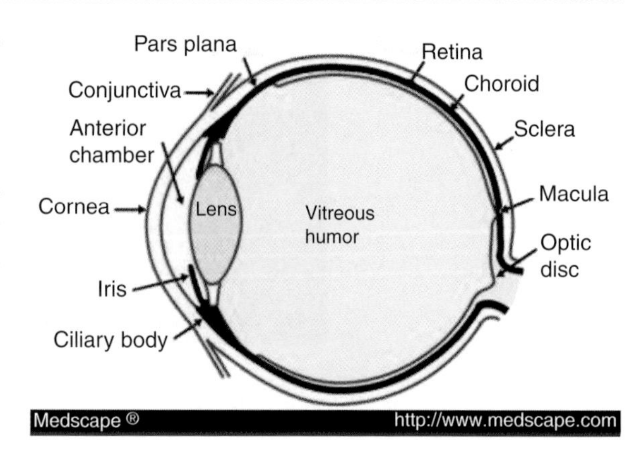

Fig. 1.3.7.2 Sagittal section of the eye showing in black the uveal tract where are found neural crest origin melanocytes

1.3.7.2 Ocular Pigmentation

The melanocyte population in the eye (Fig. 1.3.7.1) resides predominantly in the uveal tract (Fig. 1.3.7.2) that consist of the choroid, the ciliary bodies, and the iris, and is the site of most eye melanomas [10]. These melanocytes are of neural crest origin and immigrate into the uveal tract during its embryonic development. There is an additional population of melanocytes in the eye termed as the retinal pigment epithelium (RPE) (Fig. 1.3.7.3). These cells are also derived from the neural tube; however, this unicellular layer positioned between the choroid and the retina proper develops directly from the neuroectoderm neural tube of the developing forebrain without migrating as neural crest cells. The presence of the RPE and the melanin

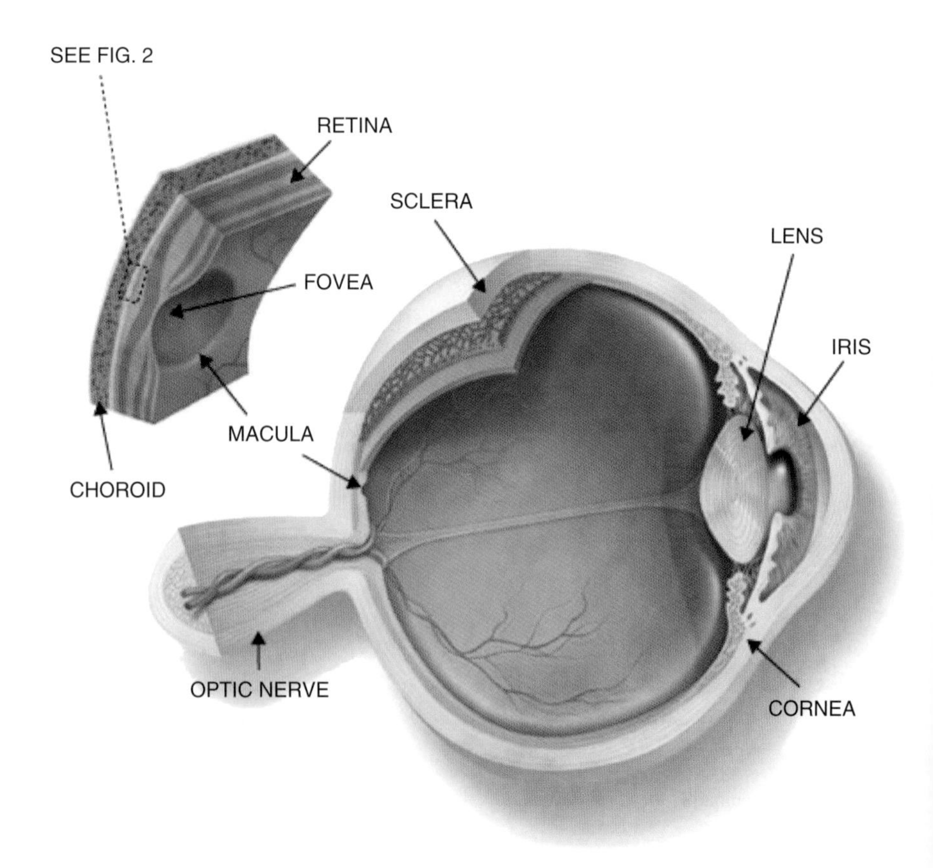

Fig. 1.3.7.1 Eye anatomy

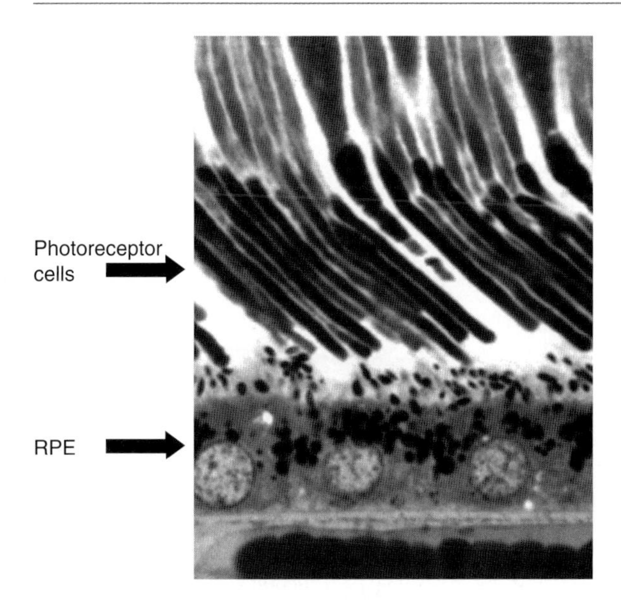

Fig. 1.3.7.3 The retinal pigment epithelium. The pigment cells originate directly from the neuroectoderm neural tube of the developing forebrain

synthesised by both the RPE and melanocytes of the uveal tract are important for the guidance of the neural projections from the retina proper to the visual cortex of the brain [17]. Specifically, neurons from the temporal half of each eye are directed to the ipsilateral visual cortex in the brain, whereas the neurons from the nasal half of each eye cross at the optic chiasma to the contralateral visual cortex of the brain [17, 21]. In the absence of melanin synthesis, as characterised by albinism, this neuronal pattern is offset so that there is an increase in neurons crossing at the optic chiasma to the contralateral geniculate, resulting in the scrambling of the neuronal input to the visual cortex of the brain [18, 24]. Once melanocytes and melanin are correctly established in the eye, melanin synthesis is substantially halted during the life of an individual, and the melanocytes of the uveal tract and RPE are relatively melanogenically dormant [10]. However, these dormant cells still provide a function to the eye. Ocular melanin absorbs incident light that enters through the retina and prevents photons from bouncing about and re-stimulating or over-stimulating the rods and cones.

Loss of ocular pigmentation can occur in the eyes of patients with vitiligo. Several studies using ophthalmoscopic evaluation have demonstrated in 18–50% of patients with vitiligo focal hypopigmented lesions on the iris, anterior chamber, and RPE/choroid, with the fundus in the latter being particularly affected [1, 2, 8, 12, 16]. Pigment loss in the skin of the eyelids and poliosis of the eyebrows and eyelashes was frequently correlated with this ocular depigmentation [2, 16]. However, a recent study demonstrated that pigmentary changes were not observed in a small group ($n = 17$) of individuals with vitiligo [6].

Inflammation of the uveal tract, that is, uveitis, can also occur in patients with vitiligo [27]. Minimal ocular inflammation has been reported in 5% of vitiligo patients and conversely idiopathic uveitis has been observed in 5% of patients with vitiligo or poliosis [41]. The Vogt–Koyanagi–Harada Syndrome (VKH) is a multi-system disorder characterised by severe uveitis, central nervous system abnormalities without a history of ocular trauma [3]. Vitiligo-like depigmentation is seen in as many as 63% of VKH patients, usually during the convalescent stage of the syndrome [28, 36]. Skin biopsies from involved areas in patients with VKH-associated vitiligo demonstrate inflammatory characteristics similar to vitiligo, with however mononuclear infiltration consisting primarily of CD4+ lymphocytes, whereas CD8+ cells are noted in common NSV [31] (Chaps. 1.2.2, 1.3.8 and 2.2.7).

Of note is the Smyth chicken, the avian animal model for vitiligo [35] (Chap. 2.2.4). The chicken had been originally bred for early and rapid development of complete vitiligo. Concurrent to this advanced stage was a high incidence of blindness [35]. The blindness was a result of an immunological removal of melanocytes from the uveal tract and a destruction of the RPE [9]. As a consequence, the photoreceptor cells and retina proper became degraded and vision was severely impaired. However, if the Smyth line chicken was bred for the development of minimally expressed vitiligo, the blindness, and ocular pathology was not readily observed.

1.3.7.3 Otic Pigmentation

The melanocyte population in the ear resides in the stria vascularis and the modiolus of the cochlea, the semi-circular ducts, the vestibular organ composed of the utricle, the ampullae, the saccule, and the endolymphatic sac [26] (Figs. 1.3.7.4 and 1.3.7.5). Otic melanocytes are of neural crest origin [22] and their embryonic development is under the regulation of Kit

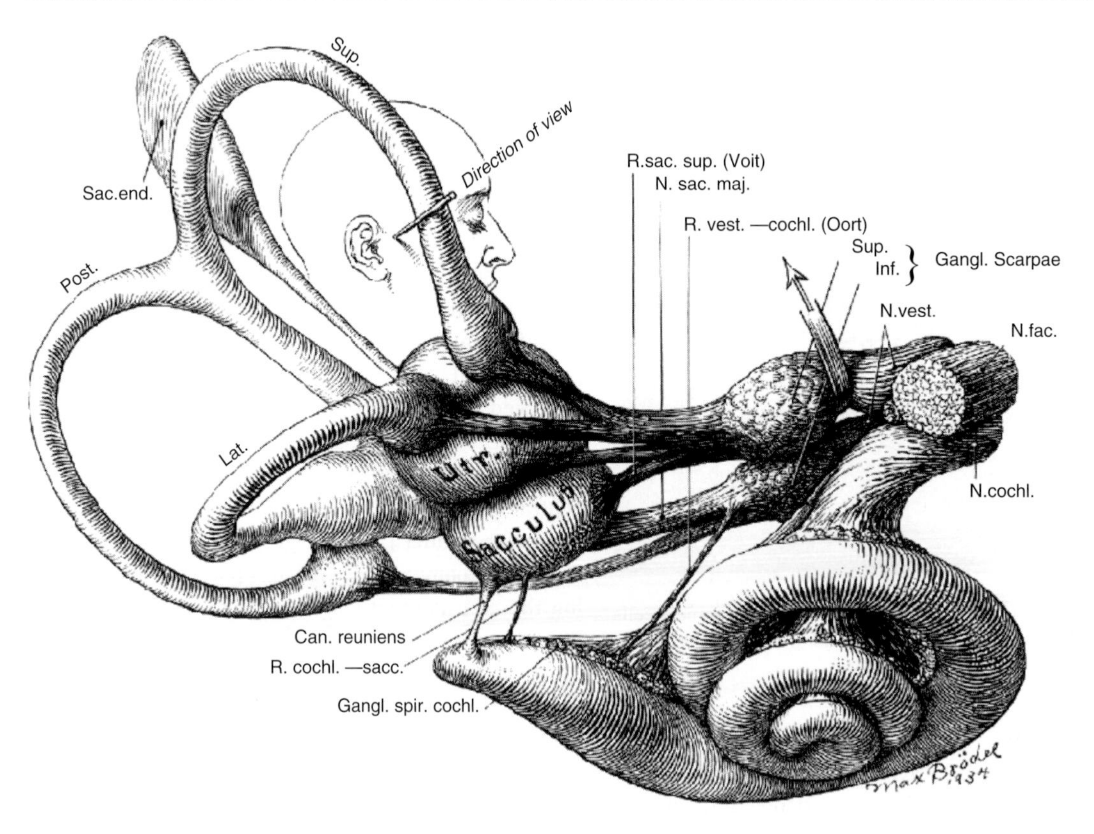

Fig. 1.3.7.4 General view of inner ear

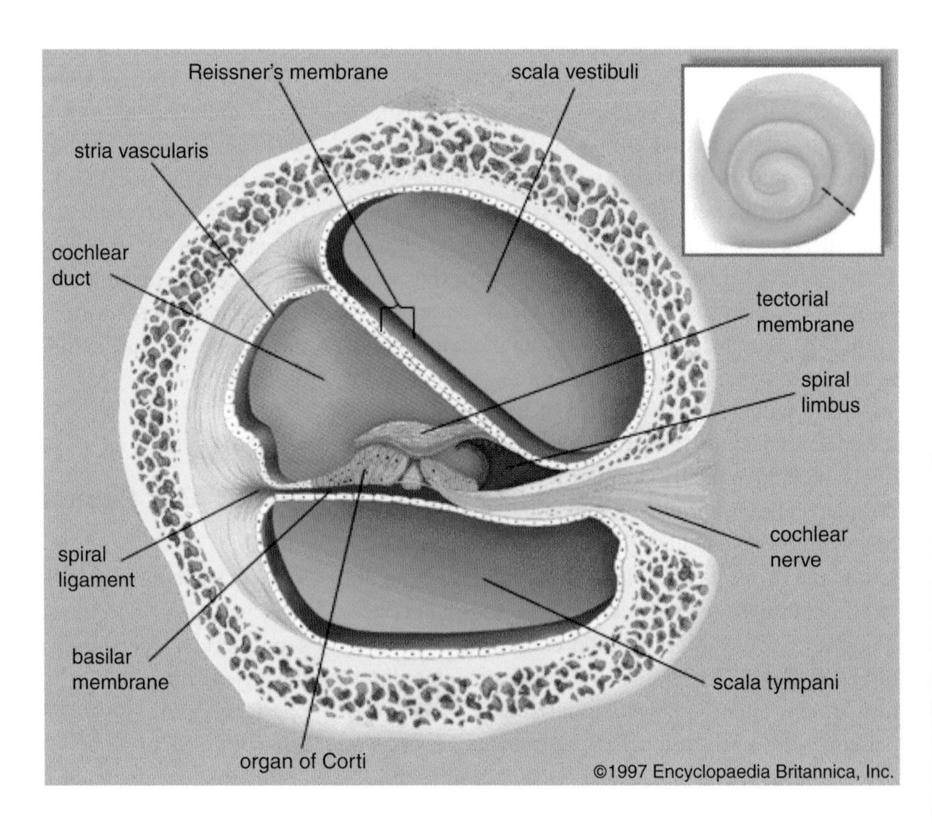

Fig. 1.3.7.5 Section of cochlea. Melanocytes are situated in stria vascularis and modiolus of the cochlea

[25] and Sox 10 [11, 33]. Once established in the ear, otic melanocytes synthesise pigmented melanosomes until congested, and then halt melanin synthesis [13]. There are several functions attributed to otic melanocytes that include [i] the embryonic development of the ear and hearing, [ii] the maintenance of endocochlear potential, and [iii] the prevention of noise- and/or toxin-induced hearing loss. In the absence of these melanocytes, neural sensory deafness can occur as exemplified by various severe piebald syndromes (i.e. Waardenburg syndromes 1–4) [34] and animal models of albinism [15]. The endocochlear potential that regulates transduction of sound waves to the hair-cell receptors can be reduced in albinos [15]. Susceptibility to noise-induced hearing loss has been demonstrated in some animals and humans with albinism [7, 14, 23].

Loss of otic melanocytes/pigment in patients with vitiligo may occur, however this loss has not been histologically confirmed. Congenital and acquired deafness has been associated with vitiligo in certain kindred, reported in the 1970s and 1980s [19, 38–40], but we know now that connexin 26 mutations causing deafness are common and these associations should be probably questioned. It has also been reported that abnormalities of hearing and brainstem auditory responses do exist in 13–16% of patients with vitiligo [19, 29, 32, 40]. Specifically, pure tone thresholds to high-frequency sounds can be significantly lower in patients with vitiligo than the normal population, particularly for men [4]. Recently, a mild degree of sensorineural hypoacusis was found in 14% of patients with active vitiligo [5]. In clinical practice, hypoacusis is rarely a complaint reported by patients.

1.3.7.4 Leptomeningeal Pigmentation

Melanocytes in the brain are confined to the leptomeninges, the covering layer of the ventrolateral surfaces of the medulla oblongata [20, 37]. The function of these melanocytes is unknown. Also, it is unknown whether this population of melanocytes is lost in vitiligo. Of interest, it has been speculated that aseptic meningitis associated with depigmentation, especially in VKH syndrome, may be a result of the loss or destruction of this melanocyte population [20, 30]. There have been a very few anecdotal cases of patients with active vitiligo developing concurrent severe headaches. It is conceivable that the destructions of the leptomeningeal melanocytes, probably by an accelerated immunologic response, may induce such a presentation. However, confirmation of this possibility is warranted.

1.3.7.5 Conclusions

The evidence of an involvement of non-cutaneous melanocytes in common forms of vitiligo is limited. There is currently no pressing need to assess the patients for hearing loss or specific visual deficiency. In case of universal vitiligo and VKH syndrome, the targets of the disease are situated beyond the skin, and there is a need for more careful evaluation.

References

1. Albert DM, Nordlund JJ, Lerner AB (1979) Ocular abnormalities occurring with vitiligo. Ophthalmology 86: 1145–1158
2. Albert DM, Wagoner MD, Pruett RC et al (1983) Vitiligo and disorders of the retinal pigment epithelium. Br J Ophthalmol 67:153–156
3. Andreoli CM, Foster CS (2006) Vogt-Koyanagi-Harada disease. Int Ophthalmol Clin 46:111–22
4. Ardic FN, Aktan S, Kara CO et al (1998) High-frequency hearing and reflex latency in patients with pigment disorder. Am J Otolaryngol 19:365–369
5. Aydogan K, Turan OF, Onart S et al (2006) Audiological abnormalities in patients with vitiligo. Clin Exp Dermatol 31:110–113
6. Ayotunde A, Olakunle G (2005) Ophthalmic assessment in black patients with vitiligo. J Natl Med Assoc 97:286–287
7. Barrenas ML, Lindgren F (1990) The influence of inner ear melanin on susceptibility to TTS in humans. Scand Audiol 19:97–102
8. Biswas G, Barbhuiya JN, Biswas MC et al (2003) Clinical pattern of ocular manifestations in vitiligo. J Indian Med Assoc 101:478–480
9. Boissy RE, Smyth JR Jr, Fite KV (1983) Progressive cytologic changes during the development of delayed feather amelanosis and associated choroidal defects in the DAM chicken line. A vitiligo model. Am J Pathol 111:197–212
10. Boissy RE, Hornyak TJ (2006) Extracutaneous melanocytes. In: Nordlund JJ, Boissy RE, Hearing VJ (eds) The pigmentary system: physiology and pathophysiology. Blackwell Scientific, Oxford
11. Bondurand N, Pingault V, Goerich DE et al (2000) Interaction among SOX10, PAX3 and MITF, three genes altered in Waardenburg syndrome. Hum Mol Genet 9:1907–1917

12. Bulbul Baskan E, Baykara M, Ercan I et al (2006) Vitiligo and ocular findings: a study on possible associations. J Eur Acad Dermatol Venereol 20:829–833
13. Cable J, Steel KP (1991) Identification of two types of melanocyte within the stria vascularis of the mouse inner ear. Pigment Cell Res 4:87–101
14. Conlee JW, Abdul-Baqi KJ, McCandless GA et al (1988) Effects of aging on normal hearing loss and noise-induced threshold shift in albino and pigmented guinea pigs. Acta Otolaryngol (Stockh) 106:64–70
15. Conlee JW, Bennett ML (1993) Turn-specific differences in the endocochlear potential between albino and pigmented guinea pigs. Hear Res 65:141–150
16. Cowan CL, Halder RM, Grimes PE et al (1986) Ocular disturbances in vitiligo. J Am Acad Dermatol 15:17–24
17. Creel D, O'Donnell FE, Witkop CJ (1978) Visual system anomalies in human ocular albinos. Science 201:931–933
18. Creel DJ, Summers CG, King RA (1990) Visual anomalies associated with albinism. Ophthalmic Paediatr Genet 11: 193–200
19. Dereymaeker AM, Fryns JP, Ars J et al (1989) Retinitis pigmentosa, hearing loss and vitiligo: report of two patients. Clin Genet 35:387–389
20. Goldgeier MH, Klein LE, Klein-Angerer S et al (1984) The distribution of melanocytes in the leptomeninges of the human brain. J Invest Dermatol 82:235–238
21. Guillery RW (1974) Visual pathways in albinos. Sci Am 230:44–55
22. Hilding DA, Ginzberg RD (1977) Pigmentation of the stria vascularis. The contribution of neural crest melanocytes. Acta Otolaryngol (Stockh) 84:24–37
23. Hood JD, Poole JP, Freedman L (1976) The influence of eye colour upon temporary threshold shift. Audiology 15: 449–464
24. King RA, Hearing VJ, Creel DJ et al (1994) Albinism. In: Scriver CR, Beaud AL, Sly WS (eds) The metabolic and molecular bases of inherited disease. McGraw Hill, New York
25. Mackenzie MA, Jordan SA, Budd PS et al (1997) Activation of the receptor tyrosine kinase kit is required for the proliferation of melanoblasts in the mouse embryo. Dev Biol 192:99–107
26. Meyer zum Gottesberge AM (1988) Physiology and pathophysiology of inner ear melanin. Pigment Cell Res 1:238–249
27. Mills MD, Albert DM (2000) Ocular and otic findings in vitiligo. In: Hann SK, Nordlund JJ (eds) Vitiligo: a monograph on the basic and clinical sciences. Blackwell Science, Oxford
28. Moorthy RS, Inomata H, Rao NA (1995) Vogt-Koyanagi-Harada syndrome (review). Surv Ophthalmol 39: 265–292
29. Nikiforidis GC, Tsambaos DG, Karamitsos DS et al (1993) Abnormalities of the auditory brainstem response in vitiligo. Scand Audiol 22:97–100
30. Nordlund JJ, Albert DM, Forget BM et al (1980) Halo nevi and the Vogt-Koyanagi-Harada syndrome: manifestations of vitiligo. Arch Dermatol 116:690–692
31. Okada T, Sakamoto T, Ishibashi T et al (1996) Vitiligo in Vogt-Koyanagi-Harada disease: immunohistological analysis of inflammatory site. Graefe's Arch Clin Exp Ophthalmol 234:359–363
32. Orecchia G, Marelli MA, Fresa D et al (1989) Audiologic disturbances in vitiligo (letter to the editor). J Am Acad Dermatol 21:1317–1318
33. Potterf SB, Gurumura M, Dunn KJ et al (2000) Transcription factor hierarchy in Waardenburg syndrome: regulation of MITF expression by SOX10 and PAX3. Hum Genet 107: 1–6
34. Read AP, Newton VE (1997) Waardenburg syndrome. J Med Genet 34:656–65
35. Smyth JR Jr (1989) The Smyth chicken: a model for autoimmune amelanosis. CRC Crit Rev Poult Biol 2:1–19
36. Sugiura S (1978) Vogt-Koyanagi-Harada disease. Jpn J Ophthalmol 22:9–35
37. Symmers W (1905) Pigmentation of the pia mater with special reference to the brain of modern Egyptians. J Anat 40: 25–27
38. Thurmon TF, Jackson J, Fowler CG (1976) Deafness and vitiligo. Birth Defects Orig Art Ser 12:315–320
39. Tosti A, Bardazzi F, De Padova MP et al (1986) Deafness and vitiligo in an Italian family. Dermatologica 172:178–179
40. Tosti A, Bardazzi F, Tosti G et al (1987) Audiologic abnormalities in cases of vitiligo. J Am Acad Dermatol 17: 230–233
41. Wagoner MD, Albert DM, Lerner AB et al (1983) New observations on vitiligo and ocular disease. Am J Ophthalmol 96:16–26

Autoimmune/Inflammatory and Other Diseases Associated with Vitiligo 1.3.8

Ilse Mollet, Nanny van Geel, and Jo Lambert

Contents

I. Mollet (✉)
Department of Dermatology, Ghent University Hospital, Ghent, Belgium
e-mail: ilse.mollet@ugent.be

Core Messages

- Although the exact aetiology of vitiligo remains unclear, vitiligo is widely considered to have an autoimmune component in its pathogenesis based on disease associations.
- According to ethnic background, large variations exist in association with autoinflammatory/autoimmune disorders.
- Some reported associations may reflect a simple coexistence of two disorders, and the repertoire of associations may help to detect shared heritable predispositions for inflammation and autoimmunity.
- Thyroid disorders are the most commonly associated to vitiligo in Caucasian populations, alopecia areata in Chinese populations, rheumatoid arthritis (RA) being more rarely associated in both.

1.3.8.1 A General Overview of Association Studies

Vitiligo remains poorly understood, but an immune component is strongly considered because of associated autoimmune/autoinflammatory disorders [61, 66, 72]. This chapter focuses on the clinical aspects of autoimmune, inflammatory, and other diseases associated with vitiligo listed in Table 1.3.8.1, which are critically reviewed for an alleged association.

Large surveys from Caucasian and Chinese-based vitiligo cohorts have provided data on the magnitude

M. Picardo and A. Taïeb (eds.), *Vitiligo*,
DOI 10.1007/978-3-540-69361-1_1.3.8,

Table 1.3.8.1 Autoimmune/inflammatory and other disorders associated or reported with vitiligo

Disorders	References	Strength of association with vitiligo
Addison's disease	[2, 31, 47, 48, 52, 100]	Demonstrated/possible
Alezzandrini's syndrome	[46]	Particular form of depigmentation, probably not vitiligo
Alopecia areata	[1, 2, 35, 52, 56, 80, 81, 87, 88]	Demonstrated/possible
Autoimmune polyendocrine syndrome	[3, 9, 10, 47, 59, 84]	Demonstrated see Chap. 1.3.11
Chronic active hepatitis	[26, 57, 74]	No clear association
Diabetes mellitus (type 1)	[2, 11, 19, 22, 32, 38, 43, 52, 56, 58]	Demonstrated/possible
Human immunodeficiency virus disease	[5, 30, 33]	Reported association see Chap. 1.3.9
Ichthyosis	[27, 56]	Reported association, ichthyosis not classified
Inflammatory bowel disease	[36, 52, 67, 73, 85]	No clear association
Lichen planus	[4, 6, 21, 76, 94]	Reported association
Malignant melanoma	[8, 14, 15, 37, 41, 50, 53, 62, 78]	Reported association
Multiple autoimmune disease	[49]	Particular/rare
Multiple sclerosis	[2, 44, 63, 96]	No clear association
Myasthenia gravis	[2, 18, 29, 51, 79, 88, 91]	No clear association
Pernicious anaemia	[2, 19, 23, 39, 52, 90]	Demonstrated/possible
Psoriasis	[2, 16, 25, 52, 56, 60, 69, 75, 82, 97]	Demonstrated/possible
Rheumatoid arthritis	[2, 11, 45, 52, 55, 56]	Demonstrated/possible
Sarcoidosis	[7, 13, 24, 57, 89]	Particular/rare
Scleroderma	[2, 12, 34, 52, 95, 99]	No clear association
Sjögren's syndrome	[2, 52, 64]	No clear association
Systemic lupus erythematosus	[2, 17, 41, 42, 56, 70]	Demonstrated/possible
Thyroid disease	[2, 19, 47, 48, 52, 54, 56, 65]	Demonstrated/possible
Urticaria	[41, 56, 83]	Reported association
Vogt-Koyanagi-Harada syndrome	[20, 40, 68, 71, 86, 92, 93, 98]	Particular form of depigmentation, probably not vitiligo

and type of associations, but their results are difficult to reconcile (Table 1.3.8.2). A large survey of more than 2,600 unselected, mainly Caucasian patients with generalized vitiligo and their close relatives from North America and the United Kingdom found significantly elevated frequencies of vitiligo itself, autoimmune thyroid disease (particularly Hashimoto's thyroiditis leading to hypothyroidism), pernicious anaemia, Addison's disease, systemic lupus erythematosus (SLE), and probably inflammatory bowel disease (IBD) in vitiligo probands and their first-degree relatives, when compared with the general population frequency of these six autoimmune diseases, suggesting that vitiligo patients and their first-degree relatives have a genetically determined susceptibility to this specific group of autoimmune diseases. About 30% of patients with generalized vitiligo were affected, with at least one additional autoimmune disease. No significant increases in the frequencies of alopecia areata, Type 1 diabetes mellitus, multiple sclerosis (MS), myasthenia gravis, psoriasis, RA, scleroderma, or Sjögren's syndrome among either vitiligo probands or their first-degree relatives were observed [2]. A survey by the same group of 133 Caucasian families, in which multiple individuals had generalized vitiligo found significantly elevated frequencies of autoimmune thyroid disease, pernicious anaemia, Addison's disease, RA, adult-onset insulin-dependent diabetes mellitus, and psoriasis. A broader repertoire of associated autoimmune diseases and earlier disease onset was found in familial generalized vitiligo compared to sporadic vitiligo, which most likely reflects a greater inherited genetic component of autoimmune susceptibility in these families. Consistent with the previous non-familial survey, the frequencies of alopecia areata, ankylosing spondylitis, scleroderma, and Sjögren's syndrome were not elevated among the family probands [52]. The results of a retrospective survey of 3,742 patients with vitiligo in China [56] could not confirm some of the earlier findings made by the Spritz's group [2] about the associated diseases of vitiligo. The frequency of thyroid

Table 1.3.8.2 Associated or reported autoimmune disorders in vitiligo probands and their first-degree relatives in populations of various ethnic background

Associated/reported autoimmune disorders	Vitiligo probands		First-degree relatives	
	Significant increase in frequency[a]	No significant increase in frequency[a]	Significant increase in frequency[a]	No significant increase in frequency[a]
Addison's disease	Caucasian[b] [2] Caucasian[b] [52]	–	Caucasian [2] Caucasian [52]	–
Alopecia areata	Chinese [56]	Caucasian [2] Caucasian [52]	–	Caucasian [2] Caucasian [52]
Diabetes mellitus (type 1)	Caucasian [52] Romanian [11]	Caucasian [2] Chinese [56]	–	Caucasian [52] Romanian [11]
Inflammatory bowel disease	Caucasian [2]	Caucasian [52]	–	Caucasian [2] Caucasian [52]
Multiple sclerosis	–	Caucasian [2]	–	Caucasian [2]
Myasthenia gravis	–	Caucasian [2]	–	Caucasian [2]
Pernicious anaemia	Caucasian [2] Caucasian [52]	–	Caucasian [2] Caucasian [52]	–
Psoriasis	Caucasian [52]	Caucasian [2] Chinese [56]	Caucasian [52]	Caucasian [2]
Rheumatoid arthritis	Caucasian [52] Chinese [56] Romanian [11]	Caucasian [2]	Caucasian [52] Romanian [11]	Caucasian [2]
Scleroderma	–	Caucasian [2] Caucasian [52]	–	Caucasian [2] Caucasian [52]
Sjögren's syndrome	–	Caucasian [2] Caucasian [52]	–	Caucasian [2] Caucasian [52]
Systemic lupus erythematosus	Caucasian [2]	Chinese [56]	Caucasian [52]	–
Thyroid disease	Caucasian [2] Caucasian [52] Romanian [11]	Chinese [56]	Caucasian [2] Caucasian [52] Romanian [11]	Chinese [56]

Numbers given in square brackets are references

[a]Comparison of frequency with population frequency of each associated autoimmune disorder

[b]For references [2] and [52] the Caucasian population studied originated from the UK and North America

diseases was lower than that of the general population. However, most of the patients studied were under the peak- onset age of commonly associated diseases. Compared with the general population, no significant difference was observed with respect to psoriasis, asthma, or SLE. Among this large cohort of vitiligo in China the significantly elevated coexisting disorders were RA, chronic urticaria, and alopecia areata and more surprisingly ichthyosis (not classified).

In addition to these studies, a high frequency of generalized vitiligo and several other associated autoimmune diseases has been observed by the Spitz's group in an isolated inbred community in Romania, confirming autoimmune thyroid disease, adult-onset Type 1 diabetes mellitus, and RA as significantly elevated among the 51 vitiligo probands and their first-degree relatives when compared with the frequencies of these autoimmune diseases in the general population [11]. A retrospective study of 182 Indian patients with late onset vitiligo revealed that 21.4% had associated autoimmune and endocrine disorders, such as diabetes mellitus, thyroid diseases, and RA. Hypertension and seizure disorders were present among other systemic diseases [28]. The Hamburg study including 321 patients found only association with thyroid disease [77].

Together, these studies indicate that generalized vitiligo is epidemiologically associated with other mostly chronic inflammatory diseases, such as autoimmune thyroiditis and RA, suggesting that pathologic variants in specific genes predispose to these disorders (for a discussion of genetic epidemiology, see Chap. 2.2.1). However, good genetic epidemiology studies are still lacking to understand the major discrepancies noted across populations of various ethnic backgrounds, including adequate control populations to check the validity of reported associations, which may reflect the simple coexistence of the two disorders, vitiligo itself not being rare. The quality of the surveys is indeed difficult to assess, especially for family history. Part 1.4 provides tools to help clinical investigators on this difficult issue, where patients are usually ignorant of the names of specific diseases.

1.3.8.2 Autoinflammatory/Autoimmune Diseases Associated with Vitiligo: Clinical Analysis and Relevance of the Association

Disorders with a Demonstrated or Possible Association with Vitiligo

- Thyroid disorders

The most common causes of autoimmune hypothyroidism is *Hashimoto's thyroiditis* or goitrous type thyroiditis and at later stages of the disease, atrophic type thyroiditis with minimal residual thyroid tissue. The late clinical manifestations of resulting hypothyroidism are currently rarely encountered due to better screening. They include dry coarse thickened skin, pallor, often with a yellow tinge to the skin, cool peripheral extremities, cold intolerance, retarded nail growth, brittle thinning hair, hair loss, weight gain, decreased appetite, constipation, fatigue, weakness, bradycardia, diastolic hypertension, impaired memory and concentration, puffy face with oedematous eyelids and non-pitting pretibial oedema (myxedema), delayed tendon reflex relaxation, carpal tunnel syndrome, decreased libido, and menstrual disturbances [47, 48]. In Caucasian NSV patients, a screening for anti- TPO antibodies is considered useful (see Part 1.4)

Graves' disease is a very common form of autoimmune hyperthyroidism. The thyroid gland is usually diffusely enlarged (goitre) to two to three times its normal size. Clinical features of Graves' disease include nervousness, hyperactivity, irritability, anxiety, fatigue, weakness, tremor, palpitation, heat intolerance, increased appetite, weight loss, tachycardy, systolic hypertension, sweating, warm, moist, smooth skin, hair loss, onycholysis, stare, eyelid retraction, diarrhoea, muscle weakness, proximal myopathy, loss of libido, and menstrual disturbances. Graves' opthalmopathy, marked by lid retraction, periorbital oedema, and proptosis may occur. The typical lesion of pretibial myxedema or thyroid dermopathy is a non-inflamed, indurated plaque with a deep pink or purple colour and an 'orange-skin' appearance, most frequent over the anterior and lateral aspects of the lower leg. Acropathy or clubbing of the nail is strongly associated with ophthalmopathy and pretibial myxedema [47, 48]. An increased incidence of thyrotoxicosis has been found in Caucasian patients with vitiligo [2, 52, 54, 65], and in their first-degree relatives [2, 52]. Vitiligo usually preceded the onset of the symptoms of Graves' disease [65]. Vitiligo has been also found more frequently in patients from United Kingdom with thyrotoxicosis [19].

- Alopecia areata

Alopecia areata (Fig. 1.3.8.1) is a common reversible patchy hair loss disease which typically begins on the scalp and sometimes progresses to complete baldness (alopecia totalis) or even total loss of all body hair (alopecia universalis). Alopecia areata may be observed at any age without sex predominance and with a worldwide occurrence. It tends to present more often in children and adolescents [35]. The nails may show typical signs of alopecia areata such as small pits, red spotted lunulae, or longitudinal ridging. Nail involvement is usually related to widespread alopecia areata [35, 88].

Some associations of alopecia areata with other autoimmune diseases such as vitiligo [80, 81], hypothyroidism, Type I diabetes mellitus, SLE, psoriasis, RA, or hypo-parathyroidism have been reported [1, 87]. When compared with the general population, significant elevated frequencies of alopecia areata were observed among vitiligo patients in China [56], whereas no significant increase in frequencies of alopecia areata was found in Caucasian patients with vitiligo and their first-degree relatives [2, 52].

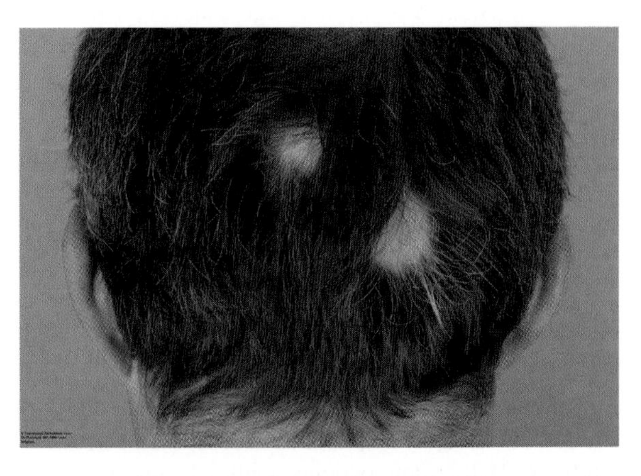

Fig. 1.3.8.1 Patches of alopecia areata with regrowing white hairs, which is common in this setting and not related to vitiligo. However, true vitiligo may coexist with alopecia areata

- Rheumatoid arthritis

The RA is a systemic inflammatory autoimmune disease characterized by persistent inflammatory synovitis, usually involving peripheral joints, especially those of the hands, wrists, knees, and feet in a symmetric pattern. The RA usually begins insidiously with fatigue, anorexia, generalized weakness, and vague musculoskeletal symptoms until the characteristic synovitis presents. Manifestations of RA include pain in affected joints aggravated by movement, morning stiffness lasting more than an hour, joint swelling, and synovial inflammation which causes swelling, tenderness, and limitation of motion. The RA most often causes symmetric arthritis with characteristic involvement of the proximal interphalangeal (PIP) and metacarpophalangeal (MCP) joints. The distal interphalangeal (DIP) joints are rarely involved. Radial deviation at the wrist with ulnar deviation of the digits is a characteristic deformity of RA. Extraarticular manifestations include rheumatoid nodules, vasculitis, and internal organ involvement. Rheumatoid nodules are subcutaneous nodules usually found on periarticular structures, extensor surfaces, or other pressure points, such as the olecranon bursa, proximal ulna, and Achilles tendon [45, 55]. When compared to the general population frequency, significantly elevated frequencies of RA were found in vitiligo patients in China [56], in vitiligo probands and their first-degree relatives in a Romanian population isolate (Table 1.3.8.2) [11], and in Caucasian affected members of the multiplex vitiligo families and their first-degree relatives [52]. In contrast, no significant increase in the frequency of RA was observed in Caucasian unselected vitiligo probands and their first-degree relatives [2].

- Pernicious anaemia

Pernicious anaemia is an autoimmune disorder characterized by the absence of intrinsic factor (IF), a cobalamin-binding protein in the gastric juice. It is the most common cause of vitamin B12 deficiency, resulting in peripheral neuropathy and several abnormalities of the digestive tract. Paresthesias and numbness are the most frequent manifestations of peripheral neuropathy, whereas diarrhoea, malabsorption, and the presence of a smooth and beefy red tongue represent gastrointestinal features of vitamin B_{12} deficiency. Two mechanisms are responsible for the malabsorption of vitamin B_{12} in patients with pernicious anaemia. First, failure of IF production can result from progressive destruction and loss of parietal cells from the gastric mucosa. Second, autoantibodies in the gastric juice can bind to the vitamin B_{12}-binding site of IF, thereby preventing the formation of the vitamin B_{12}-IF complex [90]. Pernicious anaemia has been found significantly associated with vitiligo [19, 23, 39]. The frequency of pernicious anaemia was found significantly elevated in Caucasian vitiligo probands and their relatives, when compared with the population frequency of pernicious anaemia [2, 52].

- Addison's disease

Addison's disease or chronic primary adrenal insufficiency is a condition in which inadequate secretion of corticosteroids is due to bilateral destruction of the adrenal cortex. The most common cause of Addison's disease is autoimmune adrenalitis, resulting in adrenocortical atrophy. Clinical features of Addison's disease include weakness, reduced appetite, weight loss, anorexia, abdominal pain, nausea, vomiting, postural hypotension, and skin hyperpigmentation, mostly on face, neck, and back of the hands [47, 48]. It is well-known that patients with autoimmune Addison's disease have a higher prevalence of other autoimmune disorders [100]. When compared with the population frequency, the frequency of Addison's disease is greatly increased in Caucasian vitiligo probands and their first-degree relatives [2, 52]. Autoimmune Addison's disease may occur in isolation or in association with other autoimmune disorders such as vitiligo in autoimmune polyendocrine syndrome Type 1 (APS-1) and Type 2 (APS-2) [31] (Chap. 1.3.11).

- Systemic lupus erythematosus

SLE is a chronic, usually life-long autoimmune disease, which is characterized by unpredictable disease flares followed by periods of disease inactivity. During a flare, there may be a clinical involvement of the skin, joints, kidney, brain, serosa, lung, heart, and gastrointestinal tract. Manifestations of SLE include the presence of a malar 'butterfly-shaped' rash, photosensitivity, arthritis, swelling, and tenderness of the joints, persistent proteinuria, seizures, psychosis, depression, pleuritis, pericarditis, haemolytic anaemia, leucopenia, thrombocytopenia, lymphopenia, and painless oral or nasopharyngeal ulceration. Anti-double-stranded DNA antibodies are highly specific for SLE. Other pathogenic autoantibodies in SLE are anti-Ro (SS-A), anti-La (SS-B), anti-Sm, and anti-nucleosome antibodies

[17, 42, 70]. There is a highly significant increase in the frequency of SLE in Caucasian vitiligo probands and their first-degree relatives, when compared with the population frequency of this disease [2]. No significant increase in the frequency of SLE was observed in patients with vitiligo in China [56]. The coexistence of vitiligo, malignant melanoma, lupus erythematosus, and urticaria has been reported in Turkey [41].

- Psoriasis

Psoriasis is a common and chronic inflammatory skin disorder characterised by various clinical presentations, plaque psoriasis (vulgaris) being the most common. Psoriasis fulfils many criteria of an autoinflammatory disorder, but only the identification of the putative autoantigens will finally prove its autoimmune nature [69]. Cutaneous and extracutaneous manifestations may occur in psoriasis patients [16, 60]. Extracutaneous joint involvement may be present in about one-third of the psoriasis patients. Psoriatic arthritis belongs to the spondyloarthritides group, and typically occurs only after many years of skin-disease activity [60, 97]. Significantly, elevated frequencies of psoriasis have been found in affected members of the multiplex vitiligo families and their first-degree relatives [52]. Other authors observed no significant increases in the frequencies of psoriasis among vitiligo probands [2, 56], and their first-degree relatives [2]. Colocalization of psoriasis to areas affected by vitiligo [25, 75], and of vitiligo colocalizing to areas previously affected by psoriasis have been reported in the literature [82].

- Adult onset insulin-dependent diabetes mellitus

Type I (insulin-dependent) diabetes mellitus is the most common form of diabetes mellitus among children and adolescents, although onset in adults is not uncommon. It is caused by pancreatic β-cell destruction, often autoimmune mediated, that leads to virtually loss of insulin secretion and absolute insulin deficiency. The clinical features are related to the presence of hyperglycaemia and the resulting effects on fluid and electrolyte balance [32, 43, 58]. An increased incidence of insulin-dependent diabetes mellitus in patients with vitiligo has been reported [19, 38]. An association between vitiligo and mature-onset diabetes mellitus has also been described [22]. When compared with the general population frequency, significantly elevated frequencies of adult-onset autoimmune diabetes mellitus have been found in Caucasian generalized multiplex vitiligo patients, but not in their siblings [52], and in vitiligo probands in an isolated Romanian inbred community, but not in their first-degree relatives [11]. No significant increase in the frequency of Type I diabetes mellitus has been found in Caucasian unselected vitiligo probands [2], and in Chinese vitiligo patients [56].

Disorders with No Clear Association with Vitiligo

- Inflammatory bowel disease and spondyloarthritides

The IBD is an immune-mediated chronic inflammatory condition of the gastrointestinal tract. The two major types of IBD are ulcerative colitis and Crohn's disease. The IBD is associated with multiple intestinal and extra-intestinal manifestations [36].

Ulcerative colitis is a mucosal disease which usually involves the rectum and extends proximally to involve all or part of the colon. Symptoms of ulcerative colitis include diarrhoea, bloody stools, tenesmus, passage of mucus, abdominal pain, cramping, anorexia, nausea, vomiting, fever, malaise, fatigue, and weight loss. The severity of the symptoms correlates with the extent of the disease. Compared with the general population, patients with extensive ulcerative colitis have an increased risk for colon cancer [36, 85]. *Crohn's disease* has the potential to involve any part of the gastrointestinal tract from the mouth to the anus. There are three major parts of disease distribution: disease confined to the small intestine (ileitis), disease present in the ileum and colon (ileocolitis), and disease confined to the colon (colitis). The site of the disease influences the clinical manifestations. The predominant symptoms in Crohn's disease are abdominal pain, weight loss, and diarrhoea. Fistulization is a complication of Crohn's disease [36, 85]. The most common extra-intestinal manifestation of IBD is arthritis, consisting of peripheral arthritis and axial arthritis or ankylosing spondylitis. Peripheral arthritis is asymmetric, polyarticular, migratory, and most often affects large joints of the knees, hips, ankles, wrists, and elbows. Ankylosing spondylitis is the 'prototype disease' for the spondyloarthritides (SpA), which also include Reiter's syndrome, reactive arthritis, psoriatic arthritis, enteropathic arthritis, and a variety of less clearly defined conditions known as undifferentiated spondylitis. Ankylosing spondylitis most often affects the spine and pelvis, and presents with symptoms of inflammatory diffuse low back pain, stooped posture, and morning stiffness [67, 73]. Other extra-intestinal manifestations of IBD include osteoporosis, uveitis, episcleritis, urinary tract complications, nephrolithiasis, pyoderma gangrenosum, erythema nodosum, hepatic steatosis, pericholangitis, chronic active hepatitis, cirrhosis, primary sclerosing cholangitis, and cholelithiasis [36, 85]. A statistically significant increase in the frequency of IBD (ulcerative colitis and Crohn's disease), compared with its population frequency, has been observed among unselected Caucasian vitiligo probands, but not among the probands' first-degree relatives. The frequencies of IBD among the familial vitiligo probands and their siblings were not significantly increased above their population frequencies [52]. In a survey of 234 patients with SpA and 468 control patients without SpA, 3.4% of the patients with SpA were found to have vitiligo, whereas only 1.06% of the control patients had vitiligo. The difference in the frequency of vitiligo between the two groups was statistically significant. The authors suggest that vitiligo and SpA occur together more frequently than by chance and that vitiligo should be included in the list of diseases associated with SpA [67]. We saw a 37-year old male patient affected with Bechterew's disease (ankylosing spondilytis) for several years. He developed typical vitiligo patches on dorses of both hands, and volar sides of the wrists. At that time, he was on a treatment with etanercept 2 × 25 mg SC per week. We found no associations of vitiligo and biological anti TNF treatments reports, and continued this treatment, associated with a local treatment with pimecrolimus.

- Multiple sclerosis

The MS is an inflammatory disease of the central nervous system (CNS), characterized by demyelination, axonal loss, and progressive neurological function [63]. It affects women more frequently than men. The age of onset is typically between 20 and 40 years, but the disease can present across the entire lifespan [44]. An autoimmune aetiology for MS is hypothesized and supported by studies of the immune system in MS patients and by the laboratory model of experimental allergic encephalomyelitis (EAE) [96]. Symptoms of MS are extremely varied and depend on the location and severity of lesions within the CNS. They can include sensory loss, optic neuritis, muscle weakness, paresthesias, diplopia, ataxia, vertigo, fatigue, bladder dysfunction, heat sensitivity, and cognitive dysfunction [44]. No significant increase in the frequency of MS among Caucasian vitiligo probands and their first-degree relatives has been observed [2].

- Myasthenia gravis

Myasthenia gravis is a neuromuscular disorder characterized by weakness and fatigability of skeletal muscles. Women are affected more frequently than men. Peaks of incidence occur in women in their twenties or thirties. The cranial muscles, in particular the lids and extraocular muscles, are often involved early in the course of the disease. Diplopia, ptosis, difficulty in swallowing, and weakness in chewing may occur. The muscle weakness increases during repeated use and may improve following rest or sleep. In most of the patients, the weakness becomes generalized, affecting the limb muscles as well [29]. Myasthenia gravis is associated with other autoimmune disorders such as Hashimoto's thyroiditis, Graves' disease, RA, SLE [29], and vitiligo [18, 51, 79, 88, 91]. No significant increases in the frequencies of myasthenia gravis among either Caucasian vitiligo probands or their first-degree relatives were observed [2].

- Scleroderma

Scleroderma is a chronic disorder of unknown aetiology which affects connective tissue and the microvasculature in the skin, lungs, gastrointestinal tract, kidneys, and heart. The most common form of localized scleroderma is morphea, which is characterized by circumscribed, indurated sclerotic plaques with ivory-coloured centres. A clinical variant of systemic scleroderma is CREST syndrome, so-called for its features of calcinosis cutis, Raynaud's phenomenon, oesophageal dysfunction, sclerodactyly, and telangiectasia. The skin in scleroderma is atrophic, appears tense and smooth, and has become firmly bound to the underlying structures [95, 99]. The coexistence of vitiligo and scleroderma has been described [12, 34]. No significant increase in frequencies of scleroderma has been observed among either Caucasian vitiligo probands or their first-degree relatives [2, 52].

- Sjögren's syndrome

Sjögren's syndrome is a chronic, inflammatory, autoimmune disorder characterized by dryness of the mucous membranes of the eyes, mouth, nose, and vagina. Primary Sjögren's syndrome is not associated with another underlying autoimmune disorder, whereas secondary Sjögren's syndrome occurs with other connective tissue diseases including RA, SLE, or scleroderma. The disease most commonly occurs in women. Typical Sjögren's syndrome anti-nuclear antibody patterns are anti-Ro (SSA) and anti-La (SSB) antibodies. Clinical manifestations may include xerostomia, keratoconjunctivitis sicca, xerosis, xeroderma, pruritus, fatigue, muscle and joint pain, and vaginal dryness. The most important cutaneous feature associated with Sjögren's syndrome is vasculitis, affecting small blood vessels. Morphologic lesions of this vasculitis consist of palpable and non-palpable purpura and urticaria-like lesions. Sjögren's syndrome may also affect the kidneys, lungs, liver, pancreas, and brain [64]. When compared with the population frequency of the disease, no increase in the frequencies of Sjögren's syndrome among either Caucasian vitiligo probands or their first-degree relatives has been observed [2, 52].

- Chronic active hepatitis

Autoimmune-type chronic active hepatitis is characterised by continuing hepatocellular necrosis and inflammation, usually with fibrosis, which can progress to cirrhosis and liver failure. When untreated, chronic active hepatitis may have a 6-month mortality,which can be as high as 40% [26]. The association of autoimmune-type chronic active hepatitis, vitiligo, nail dystrophy, alopecia areata, and a variant of liver kidney microsomal autoantibodies has been described [74]. Autoimmune chronic hepatitis has also been reported in association with vitiligo, autoimmune thyroiditis, and scar sarcoidosis [57].

1.3.8.3 Particular and Rare Associations

Autoimmune Polyendocrine Syndrome and Multiple Autoimmune Disease

The APS, also known as polyglandular autoimmune syndromes (PGA), are associated with vitiligo (Chap. 1.3.11). The APS have been defined as a heterogeneous group of disorders involving autoimmune diseases associated with multiple endocrine gland insufficiency. APS-1 and APS-2 have been distinguished as the two major APS [59]. Additional APS categories have been proposed [10], but are not widely accepted [84]. APS-1 and APS-2 develop Addison's disease, while APS-3 does not [10].

APS-1 (otherwise known as autoimmune polyendocrinopathy-candidiasis-ectodermaldystrophy(APECED) or Whitaker syndrome) is diagnosed when a patient presents with at least two of its three clinical features: chronic mucocutaneous candidiasis, chronic hypoparathyroidism, and Addison's disease. Autoimmune regulator (AIRE) on the long arm of Chromosome 21 is the gene mutated in this rare childhood disease [10]. Clinical features of chronic hypoparathyroidism include paresthesias, neuromuscular hyperexcitability, hypotension, malabsorption, steatorrhea, dry scaly and puffy skin, brittle nails, and coarse and sparse hair [10, 47]. Vitiligo presents in 0–25% of APS-1 cases [9].

APS-2 or Schmidt syndrome is characterized by the presence of Addison's disease (always present), autoimmune thyroid disease and/or Type 1 diabetes mellitus. It is a rare syndrome, affecting mainly adult females [10].

APS-3 is characterized by autoimmune thyroid diseases associated with other autoimmune diseases (excluding Addison's disease and/or hypoparathyroidism). APS-3A includes all endocrine diseases, while APS-3B includes the gastrointestinal autoimmune diseases. APS-3C contains skin or neuromuscular or nervous system autoimmune diseases. APS-3D comprises all collagen diseases and vasculitis [3, 10].

APS-4 includes all clinical combinations of autoimmune diseases, which cannot be included in the previous subtypes [10].

Vitiligo can be present in all subtypes of APS, but the most frequent association appears to be in APS-3 [3].

Multiple autoimmune diseases are defined by the occurrence in the same patient of three or more autoimmune diseases. It is an unusual condition in which dermatological autoimmune and especially vitiligo have an important place. Multiple autoimmune disease associating vitiligo have been described [49].

Vogt–Koyanagi–Harada and Alezzandrini's Syndrome

Vogt–Koyanagi–Harada (VKH) syndrome, also known as uveomeningitic syndrome, is a very uncommon chronic hypomelanotic multisystem autoimmune disorder principally affecting pigmented tissues in the ocular, central nervous, auditory, and integumentary systems [20, 71]. The syndrome is characterized by a bilateral granulomatous uveitis associated with meningitis, headache, vertigo, tinnitus, dysacusis, vitiligo, poliosis, and alopecia areata [46, 92, 93, 98]. Poliosis is a localized patch of white hair, which, in VKH, can involve the scalp, the eyelashes, and the eyebrows with variable extension. The alopecia in VKH very rarely totalizes, and may be subtle, diffuse, or in patches [68]. These manifestations are variable and race dependent. The VKH may occur at all ages and primarily affects certain pigmented races, such as Asians, Native Americans, Asian Indians, and Hispanics [86]. Ocular complications of VKH syndrome that lead to visual loss may include cataract, glaucoma, choroidal neovascular membrane formation, and subretinal fibrosis [71].

The VKH is currently considered to be a T-cell-mediated autoimmune disease directed against self-antigens present in melanocytes. Clinical symptoms of VKH correlate with the destruction of melanocytes in the affected body areas such as skin, eyes, ears, and CNS. Sometimes not all the characteristics of VKH syndrome are present, and many clinical signs are very similar to universal vitiligo and Alezzandrini's syndrome [68]. Therefore, based on clinical, histological, and immunological findings, some authors believe that the three aforementioned disorders are different clinical expressions of the same disease [40].

Alezzandrini's syndrome associates unilateral retinal degeneration, ipsilateral vitiligo, poliosis, and possibly hearing abnormalities, and may represent a segmental form of VKH. The initial complaint is usually a gradual loss of visual acuity in one eye. Several years after the ocular insult, ipsilateral facial vitiligo and poliosis of the eyebrows and eyelashes usually develop [46].

Sarcoidosis

Sarcoidosis is a systemic granulomatous disorder of unknown aetiology. It has a predilection for involvement of lung, liver, lymph nodes, skin, and eyes [13]. The association of sarcoidosis and vitiligo is well-known in the literature in which, in the majority of the cases, autoimmune thyroiditis has been associated as well [89]. Vitiligo has been seen in association with recurrent scar sarcoidosis [24], and also in association with scar sarcoidosis, autoimmune thyroiditis, and autoimmune chronic hepatitis [57]. Subcutaneous sarcoidosis associated with vitiligo, pernicious anaemia, and autoimmune thyroiditis has been described [7]. The coexistence of vitiligo and sarcoidosis has been observed in patients with circulating autoantibodies [89].

1.3.8.4 Other Reported Associations

Lichen Planus; Lichen Sclerosus

Lichen planus is a common inflammatory disorder affecting the skin, mucous membranes, nails, and hair. Clinical manifestations include flat-topped, variable sized, quite pruritic, violaceous papules and plaques with characteristic Wickham striae [21]. The anatomical colocalization of vitiligo and lichen planus has been reported [4, 6, 76]. The coexistence of vitiligo, lichen planus, and psoriasis in a single individual has been observed. Although the pathogenesis is unclear, the shared Koebner phenomenon could explain part of the rare coexistence of these three skin diseases [94]. Lichen sclerosus is discussed in Chap. 1.3.12.

Urticaria

Urticaria can occur as a clinical manifestation of immunologic and inflammatory mechanisms, or may be idiopathic. In addition to the skin and mucous membranes, the respiratory and gastrointestinal tracts, and the cardiovascular system may also be involved in any combination [83]. Significantly elevated frequencies of chronic urticaria were observed among a large cohort of vitiligo patients in China [56]. The coexistence of vitiligo, malignant melanoma, lupus erythematosus, and urticaria has also been reported [41].

Ichthyosis

The ichthyoses are a heterogeneous group of disorders characterized by abnormal differentiation (cornification) of the epidermis, resulting in generalized scaling of the skin [27]. Compared with the general population in China, significantly elevated frequencies of ichthyosis of unclassified type were detected among the patients with vitiligo [56].

Malignant Melanoma

Cutaneous malignant melanoma is an increasingly common, enigmatic, and potentially lethal malignancy of melanocytes [53]. The coexistence of malignant melanoma and depigmentation is frequently observed [8, 15, 50], but the type of depigmentation is rarely the same as that found in vitiligo (see Chap. 1.2.1). Some authors accept the coexistence of vitiligoid depigmentation and malignant melanoma as a favourable prognostic criterion [8, 14, 62], whereas others reject this claim [78]. A recent study in interferon alpha treated patients was in favour of a better prognosis [37]. Vitiligo can also develop simultaneously with malignant melanoma, lupus erythematosus, and urticaria [41].

Human Immunodeficiency Virus Disease

Cutaneous manifestations of HIV disease are commonly infectious or neoplastic [30, 33]. Autoimmune disorders such as vitiligo have been reported in patients with HIV infection [5, 30] (Chap. 1.3.9).

Acknowledgement I.M. is supported by grant G.0161.07 of the Research Foundation Flanders.

References

1. Ahmed I, Nasreen S, Bhatti R (2007) Alopecia Areata in Children. J Coll Physicians Surg Pak 17:587–590
2. Alkhateeb A, Fain PR, Thody A, Bennett DC et al (2003) Epidemiology of vitiligo and associated autoimmune diseases in Caucasian probands and their families. Pigment Cell Res 16:208–214
3. Amerio P, Tracanna M, De Remigis P et al (2006) Vitiligo associated with other autoimmune diseases: polyglandular autoimmune syndrome types 3B + C and 4. Clin Exp Dermatol 31:746–749
4. Anstey A, Marks R (1993) Colocalization of lichen planus and vitiligo. Br J Dermatol 128:103–104
5. Antony FC, Marsden RA (2003) Vitiligo in association with human immunodeficiency virus infection. J Eur Acad Dermatol Venereol 20:63–65
6. Baran R, Ortonne JP, Perrin C (1997) Vitiligo associated with a lichen planus border. Dermatology 194:199
7. Barnadas MA, Rodriguez-Arias JM, Alomar A (2000) Subcutaneous sarcoidosis associated with vitiligo, pernicious anaemia and autoimmune thyroiditis. Clin Exp Dermatol 25:55–56
8. Berd D, Mastrangelo MJ, Lattime E et al (1996) Melanoma and vitiligo: immunology's Grecian urn. Cancer Immunol Immunother 42:263–267
9. Betterle C, Greggio NA, Volpato M (1998) Clinical review: autoimmune polyglandular disease type 1. J Clin Endocrinol Metab 83:1049–1055
10. Betterle C, Zanchetta R (2003) Update on autoimmune polyendocrine syndromes (APS). Acta Bio Med 74:9–33
11. Birlea SA, Fain PR, Spritz RA (2008) A Romanian population isolate with high frequency of vitiligo and associated autoimmune diseases. Arch Dermatol 144:310–316
12. Bonifati G, Impara G, Morrone A et al (2006) Simultaneous occurrence of linear scleroderma and homolateral segmental vitiligo. J Eur Acad Dermatol Venereol 20:63–65
13. Braverman IM (2003) Sarcoidosis. In: Freedberg IM, Eisen AZ, Wolff K, Austen KF, Goldsmith LA, Katz SI (eds) Fitzpatrick's dermatology in general medicine. McGraw-Hill, New York
14. Bystryn JC, Rigel D, Friedman RJ et al (1997) Prognostic significance of hypopigmentation in malignant melanoma. Arch Dermatol 123:1053–1056
15. Cavallari V, Cannavo SP, Ussia AF (1996) Vitiligo associated with metastatic malignant melanoma. Int J Dermatol 35:738–740
16. Christophers E, Mrowietz U (2003) Psoriasis. In: Freedberg IM, Eisen AZ, Wolff K, Austen KF, Goldsmith LA, Katz SI (eds) Fitzpatrick's dermatology in general medicine. McGraw-Hill, New York
17. Costner MI, Sontheimer RD (2003) Lupus erythematosus. In: Freedberg IM, Eisen AZ, Wolff K, Austen KF, Goldsmith LA, Katz SI (eds) Fitzpatrick's dermatology in general medicine. McGraw-Hill, New York
18. Cruz MW, Maranhao Filho PA, André C et al (1994) Myasthenia gravis and vitiligo. Muscle Nerve 17: 559–560
19. Cunliffe WJ, Hall R, Newell DJ et al (1968) Vitiligo, thyroid disease and autoimmunity. Br J Dermatol 80: 135–139
20. Damico FM, Kiss S, Young LH (2005) Vogt-Koyanagi-Harada disease. Semin Ophthalmol 20:183–190
21. Daoud MS, Pittelkow MR (2003) Lichen planus. In: Freedberg IM, Eisen AZ, Wolff K, Austen KF, Goldsmith LA, Katz SI (eds) Fitzpatrick's dermatology in general medicine. McGraw-Hill, New York
22. Dawber RPR (1968) Vitiligo in mature-onset diabetes mellitus. Br J Dermatol 20:275–278
23. Dawber RPR (1970) Integumentary associations of pernicious anemia. Br J Dermatol 82:221–223
24. Demirkök SS, Arzuhal N, Devranoglu G et al (2007) Recurrent sarcoidosis on a scar associated with vitiligo. J Dermatol 34:829–833
25. Dhar S, Malakar S, Dhar S (1998) Colocalization of vitiligo and psoriasis in a 9-year-old boy. Pediatr Dermatol 15:242–243

26. Dienstag JL (2008) Chronic hepatitis. In: Fauci AS, Braunwald E, Kasper DL, Hauser SL, Longo DL, Jameson L, Loscalzo J (eds) Harrison's principles of internal medicine. McGraw-Hill, New York
27. DiGiovanna JJ (2003) Ichthyosiform dermatoses. In: Freedberg IM, Eisen AZ, Wolff K, Austen KF, Goldsmith LA, Katz SI (eds) Fitzpatrick's dermatology in general medicine. McGraw-Hill, New York
28. Dogra S, Parsad D, Handa S et al (2005) Late-onset vitiligo: a study of 182 patients. Int J Dermatol 44:193–196
29. Drachman DB (2008) Myasthenia gravis and other diseases of the neuromuscular junction. In: Fauci AS, Braunwald E, Kasper DL, Hauser SL, Longo DL, Jameson L, Loscalzo J (eds) Harrison's principles of internal medicine. McGraw-Hill, New York
30. Duvic M, Rapini R, Hoots WK et al (1987) Human immunodeficiency virus-associated vitiligo: expression of autoimmunity with immunodeficiency? J Am Acad Dermatol 17:656–662
31. Eisenbarth GS, Gottlieb PA (2003) The immunoendocrinopathy syndromes. In: Larsen PR, Kronenberg HM, Melmed S, Polonsky KS (eds) Williams textbook of endocrinology. Saunders, Philadelphia
32. Eisenbarth GS, Polonsky KS, Buse JB (2003) Type 1 diabetes mellitus. In: Larsen PR, Kronenberg HM, Melmed S, Polonsky KS (eds) Williams textbook of endocrinology. Saunders, Philadelphia
33. Fauci AS, Lane HC (2008) Human immunodeficiency virus disease: AIDS and related disorders. In: Fauci AS, Braunwald E, Kasper DL, Hauser SL, Longo DL, Jameson L, Loscalzo J (eds) Harrison's principles of internal medicine. McGraw-Hill, New York
34. Finkelstein E, Amichai B, Metzker A (1995) Coexistence of vitiligo and morphea: a case report and review of the literature. J Dermatol 22:351–353
35. Freyschmidt-Paul P, McElwee K, Hoffman R (2005) Alopecia Areata. In: Hertl M (ed) Autoimmune diseases of the skin. Springer, New York
36. Friedman S, Blumberg RS (2008) Inflammatory bowel disease. In: Fauci AS, Braunwald E, Kasper DL, Hauser SL, Longo DL, Jameson L, Loscalzo J (eds) Harrison's principles of internal medicine. McGraw-Hill, New York
37. Gogas H, Ioannovich J, Dafni U et al (2006) Prognostic significance of autoimmunity during treatment of melanoma with interferon. N Engl J Med 354:709–718
38. Gould IM, Gray RS, Urbaniak SJ et al (1985) Vitiligo in diabetes mellitus. Br J Dermatol 113:153–155
39. Grunnet I, Howitz J (1970) Vitiligo and pernicious anemia. Arch Dermatol 101:82–85
40. Guarnieri F, Aragona P, Vaccaro M (2004) Vogt-Koyanagi-Harada syndrome. In: Lotti T, Hercogovà J (eds) Vitiligo problems and solutions. Marcel Dekker, New York
41. Gül Ü, Kiliç A, Tulunay Ö et al (2007) Vitligo associated with malignant melanoma and lupus erythematosus. J Dermatol 34:142–145
42. Hahn BH (2008) Systemic lupus erythematosus. In: Fauci AS, Braunwald E, Kasper DL, Hauser SL, Longo DL, Jameson L, Loscalzo J (eds) Harrison's principles of internal medicine. McGraw-Hill, New York
43. Harris M (2000) Definition and classification of diabetes mellitus and the new criteria for diagnosis. In: LeRoith D, Taylor SI, Olefsky JM Diabetes mellitus. Lippincott Williams & Wilkins, Philadelphia
44. Hauser SL, Goodin DS (2008) Multiple sclerosis and other demyelinating diseases. In: Fauci AS, Braunwald E, Kasper DL, Hauser SL, Longo DL, Jameson L, Loscalzo J (eds) Harrison's principles of internal medicine. McGraw-Hill, New York
45. Hollar CB, Jorizzo JL (2003) Rheumatoid arthritis. In: Freedberg IM, Eisen AZ, Wolff K, Austen KF, Goldsmith LA, Katz SI (eds) Fitzpatrick's dermatology in general medicine. McGraw-Hill, New York
46. Huggins RH, Janusz CA, Schwartz RA (2006). Vitiligo: a sign of systemic disease. Indian J Dermatol Venereol Leprol 72:68–71
47. Jabbour SA (2003) Cutaneous manifestations of endocrine disorders. Am J Clin Dermatol 4:315–331
48. Jameson JL, Weetman AP (2008) Disorders of the thyroid gland. In: Fauci AS, Braunwald E, Kasper DL, Hauser SL, Longo DL, Jameson L, Loscalzo J (eds) Harrison's principles of internal medicine. McGraw-Hill, New York
49. Klisnick A, Schmidt J, Dupond JL et al (1998) Le vitiligo au cours des syndromes auto-immuns multiples: étude retrospective de 11 observations et revue de la litérature. Rev Méd Interne 19:348–352
50. Kovacs OS (1998) Vitiligo. J Am Acad Dermatol 38:647–648
51. Kubota A, Komiyama A, Tanigawa A et al (1997) Frequency and clinical correlates of vitiligo in myasthenia gravis. J Neurol 244:388–401
52. Laberge G, Mailloux CM, Gowan K et al (2005) Early disease onset and increased risk of other autoimmune diseases in familial generalized vitiligo. Pigment Cell Res 18:300–305
53. Langley RGB, Barnhill RL, Mihm MC Jr et al (2003) Neoplasms: cutnaeous melanoma. In: Freedberg IM, Eisen AZ, Wolff K, Austen KF, Goldsmith LA, Katz SI (eds) Fitzpatrick's dermatology in general medicine. McGraw-Hill, New York
54. Lerner AB (1959) Vitiligo. J Invest Dermatol 32:285–310
55. Lipsky PE (2008) Rheumatoid arthritis. In: Fauci AS, Braunwald E, Kasper DL, Hauser SL, Longo DL, Jameson L, Loscalzo J (eds) Harrison's principles of internal medicine. McGraw-Hill, New York
56. Liu JB, Li M, Yang S et al (2005) Clinical profiles of vitiligo in China: an analysis of 3742 patients. Clin Exp Dermatol 30:327–331
57. Marzano AV, Gasparini LG, Cavicchini S et al (1996) Scar sarcoidosis associated with vitiligo, autoimmune thyroiditis and autoimmune chronic hepatitis. Clin Exp Dermatol 21:466–467
58. Masharani U, German MS (2007) Pancreatic hormones & diabetes mellitus. In: Gardner DG, Shoback D (eds) Basic and clinical endocrinology. McGraw-Hill, New York
59. Neufeld M, Maclaren N, Blizzard R (1980) Autoimmune polyglandular syndromes. Pediatr Ann 9:154–162
60. Nickoloff BJ, Qin JZ, Nestle FO (2007) Immunopathogenesis of psoriasis. Clin Rev Allerg Immunol 33:45–56
61. Nordlund JJ, Hann S-K (2000) The association of vitiligo with disorders of other organ systems. In: Hann S-K, Nordlund JJ (eds) Vitiligo. Blackwell Science. Oxford

62. Nordlund JJ, Kirkwood JM, Forget BM et al (1983) Vitiligo in patients with metastatic melanoma: a good prognostic sign. J Am Acad Dermatol 9:689–691
63. Noseworthy JH, Luchinetti C, Rodriguez M et al (2000) Multiple sclerosis. N Engl J Med 343:938–952
64. Nousari HC, Provost TT (2003) Sjögren syndrome. In: Freedberg IM, Eisen AZ, Wolff K, Austen KF, Goldsmith LA, Katz SI (eds) Fitzpatrick's dermatology in general medicine. McGraw-Hill, New York
65. Ochi Y, DeGroot LJ (1969) Vitiligo in Graves' disease. Ann Intern Med 71:935–940
66. Ongenae K, van Geel N, Naeyaert JM (2003) Evidence for an autoimmune pathogenesis of vitiligo. Pigment Cell Res 16:90–100
67. Padula A, Ciancio G, La Civita L et al (2001) Association between vitiligo and spondyloarthritis. J Rheumatol 28: 313–314
68. Prignano F, Betts CM, Lotti T (2008) Vogt-Koyanagi-Harada disease and vitiligo. Where does the illness begin? J Electron Microsc 57:25–31
69. Prinz JC (2005) Psoriasis vulgaris and arthropathica. In: Hertl M (ed) Autoimmune diseases of the skin. Springer, New York
70. Rahman A, Isenberg DA (2008) Sytemic lupus erythematosus. N Engl J Med 358:929–939
71. Read RW (2002) Vogt-Koyanagi-Harada disease. Ophthalmol Clin North Am 15:333–341
72. Rezaei N, Gavalas NG, Weetman AP et al (2007) Autoimmunity as an aetiological factor in vitiligo. J Eur Acad Dermatol Venereol 21:865–876
73. Rudwaleit M, Baeten D (2006) Ankylosing spondylitis and bowel disease. Best Pract Res Clin Rheumatol 20:451–471
74. Sacher M, Blümel P, Thaler H et al (1990) Chronic active hepatitis associated with vitiligo, nail dystrophy, alopecia and a new variant of LKM antibodies. J Hepatol 10:364–369
75. Sandhu K, Kaur I, Kumar B (2004) Psoriasis and vitiligo. J Am Acad Dermatol 51:149–150
76. Sardana K, Sharma RC, Koranne RV et al (2002) An interesting case of colocalization of segmental lichen planus and vitiligo in a 14-year-old boy. Int J Dermatol 41:508–509
77. Schallreuter KU, Lemke R, Brandt O et al (1994) Vitiligo and other diseases: coexistence or true association? Hamburg study on 321 patients. Dermatology 188:269–275
78. Schallreuter KU, Levening C, Berger J (1991) Vitiligo and cutaneous melanoma. Dermatologica 183:239–241
79. Sehgal VN, Rege VL, Desai SC (1976) Vitiligo and myasthenia gravis. Ind J Dermatol Vener Leprol 42:1–2
80. Sharma VK, Dawn G, Kumar B (1996) Profile of alopecia areata in Northern India. Int J Dermatol 35:22–27
81. Sharma VK, Kumar B, Dawn G (1996) A clinical study of childhood alopecia areata in Chandigarh, India. Pediatr Dermatol 13:372–377
82. Smith DI, Heffernan MP (2008) Vitiligo after the resolution of psoriatic plaques during treatment with adalimumab. J Am Acad Dermatol 58(Suppl):S50–S52
83. Soter NA, Kaplan AP (2003)Urticaria and angioedema. In: Freedberg IM, Eisen AZ, Wolff K, Austen KF, Goldsmith LA, Katz SI (eds) Fitzpatrick's dermatology in general medicine. McGraw-Hill, New York
84. Spritz RA (2007) The genetics of generalized vitiligo and associated autoimmune diseases. Pigment Cell Res 20: 271–278
85. Stenson WF, Korzenik J (2003) Inflammatory bowel disease. In: Yamada T, Alpers DH, Kaplowitz N, laine L, Owyang C, Powell DW (eds) Textbook of gastroenterology. Lippincott Williams & Wilkins, Philadelphia
86. Sukavatcharin S, Tsai JH, Rao NA (2007) Vogt-Koyanagi-Harada disease in Hispanic patients. Int Ophthalmol 27:143–148
87. Tan E, Tay YK, Goh CL et al (2002) The pattern and profile of alopecia areata in Singapore – a study of 219 Asians. Int J Dermatol 41:748–753
88. Tan RS (1974) Ulcerative colitis, myasthenia gravis, atypical lichen planus, alopecia areata, vitiligo. Proc R Soc Med 67:195–196
89. Terunuma A, Watabe A, Kato T et al (2000) Coexistence of vitiligo and sarcoidosis in a patient with circulating autoantibodies. Int J Dermatol 39:551–553
90. Toh B-H, Van Driel IR, Gleeson PA (1997) Pernicious anemia. N Engl J Med 337:1441–1448
91. Topaktas S, Dener S, Kenis M et al (1993) Myasthenia gravis and vitiligo. Muscle Nerve 16:566–567
92. Tsuruta D, Hamada T, Teramae H et al (2001) Inflammatory vitiligo in Vogt-Koyanagi-Harada disease. J Am Acad Dermatol 44:129–131
93. Tugal-Tutkun I, Ozyazgan Y, Akova YA et al (2007) The spectrum of Vogt-Koyanagi-Harada disease in Turkey. Int Ophthalmol 27:117–123
94. Ujiie H, Sawamura D, Shimizu H (2006) Development of lichen planus and psoriasis on lesions of vitiligo vulgaris. Clin Exp Dermatol 31:375–377
95. Varga J (2008) Systemic sclerosis (scleroderma) and related disorders. In: Fauci AS, Braunwald E, Kasper DL, Hauser SL, Longo DL, Jameson L, Loscalzo J (eds) Harrison's principles of internal medicine. McGraw-Hill, New York
96. Willer CJ, Ebers GC (2000) Susceptibility to multiple sclerosis: interplay between genes and environment. Curr Opin Neurol 13:241–247
97. Winchester R (2003) Psoriatic arthritis. In: Freedberg IM, Eisen AZ, Wolff K, Austen KF, Goldsmith LA, Katz SI (eds) Fitzpatrick's dermatology in general medicine. McGraw-Hill. New York
98. Yang P, Ren Y, Li B et al (2007) Clinical characteristics of Vogt-Koyanagi-Harada syndrome in Chinese patients. Ophthalmology 114:606–614
99. Yu BD, Eisen AZ (2003) Scleroderma. In: Freedberg IM, Eisen AZ, Wolff K, Austen KF, Goldsmith LA, Katz SI (eds) Fitzpatrick's dermatology in general medicine. McGraw-Hill, New York
100. Zelissen PMJ, Bast EJEG, Croughs RJM (1995) Associated autoimmunity in Addison's disease. J Autoimmunity 8: 121–130

Vitiligo and Immunodeficiencies 1.3.9

Khaled Ezzedine, Sébastien Lepreux, and Alain Taïeb

Contents

Core Messages

- Combined or T-cell immunodeficiencies, inherited or acquired, are commonly associated with autoimmunity.
- Vitiligo has been reported to be associated with cellular or combined immunodeficiencies, suggesting a possible link to pathomechanisms involved in common nonsegmental vitiligo.
- Although there is some evidence that both cellular and humoral immunity can act together in vitiligo pathogenesis, the associations noted in this chapter suggest that cellular immunity is more implicated in disease causation or progression.

1.3.9.1 General Background

Combined or T-cell immunodeficiencies, inherited or acquired, are commonly associated with autoimmunity [7, 8, 19, 21]. During frank losses of immunocompetence, autoimmune diseases that are predominantly CD8 T-cell driven, predominate. Isolated reports of vitiligo associated with cellular or combined immunodeficiencies, especially during human immunodeficiency virus (HIV) infection, suggest a possible link to pathomechanisms involved in common nonsegmental vitiligo (NSV). However, the initial clinical presentation which leads to a vitiligoid condition can be strikingly different from that of common NSV. Figure 1.3.9.1 shows a patient who presented with a vitiligoid

K. Ezzedine (✉)
Service de Dermatologie, Hôpital St André, CHU de Bordeaux, France
e-mail: khaled.ezzedine@chu-bordeaux.fr

M. Picardo and A. Taïeb (eds.), *Vitiligo*,
DOI 10.1007/978-3-540-69361-1_1.3.9, © Springer-Verlag Berlin Heidelberg 2010

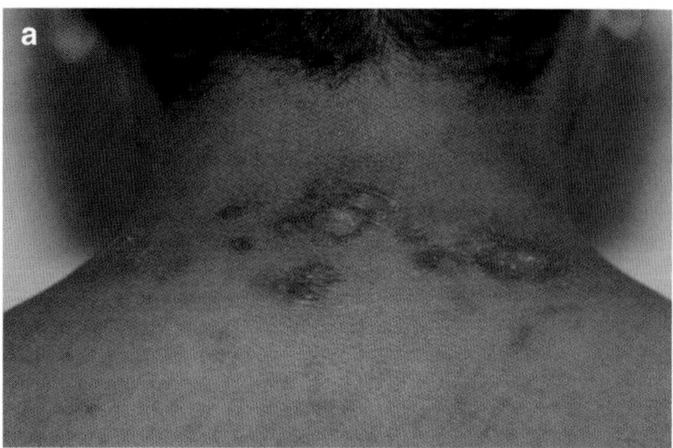

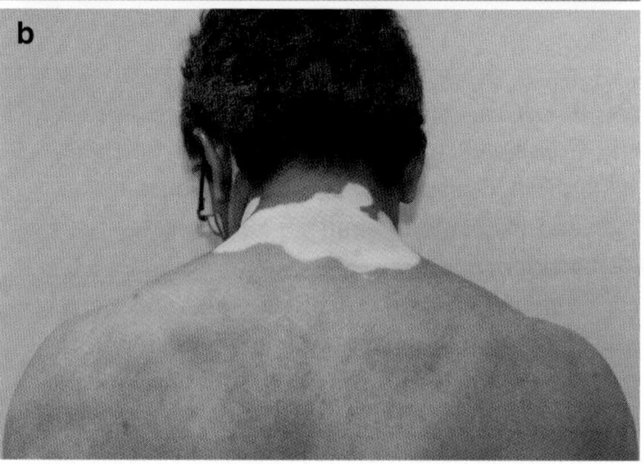

Fig. 1.3.9.1 HIV-associated vitiligo (**a**); lichenoid annular dermatitis evolving into vitiligoid lesions undistinguishable from common vitiligo (**b**)

Table 1.3.9.1 Primary, secondary and other causes of immunodeficiencies associated with vitiligo

Primary immunodeficiencies	Secondary and other causes of immunodeficiencies
Common variable immunodeficiency	Human immunodeficiency virus infection
Ataxia telangiectasia (Chap. 1.3.11)	Idiopathic T-cell lymphocytopenia
Nijmegen breakage syndrome (Chap. 1.3.11)	Protein-losing enteropathy
	Nephrotic syndrome
	Haematological malignancies

condition in the context of HIV infection. In this patient, depigmented lesions followed an actinic lichen planus-like condition on sun-exposed areas before spreading to the rest of the body.

The role of autoreactive T-cells in the pathogenesis of common NSV (Sect. 2.2.7.4) is supported by disease associations with HLA, mostly MHC Class II molecules, which may vary according to populations [20, 29]. In skin biopsies showing inflammatory changes (Chaps. 1.2.2 and 1.3.10), the ratio of CD4 to CD8 T lymphocytes is usually skewed to CD8 predominance. The influence of immunodeficiency in this context is not well-understood. The role of regulatory T-cells (Treg) might be critical to break the tolerance to self-antigens. Even though definitive evidence needs more appropriate techniques, the natural Treg CD4 population seems virtually absent from vitiligo skin [1]. In such an environment, an autoimmune response directed to melanocyte derived self-antigens could expand in genetically predisposed individuals. Although there is limited evidence that both cellular and humoral immunity (Sect. 2.2.7.3) can act together in vitiligo pathogenesis, the associations noted in this chapter suggest that cellular immunity is more implicated in disease causation or progression (Table 1.3.9.1).

Among primary immunodeficiencies, the most severe forms are usually lethal in the first years of life if cell or gene therapy is not available, and a vitiligo phenotype is not clearly established in association. Worthy of mention are ataxia telangiectasia and the Nijmegen breakage syndrome, which lead indirectly to reduced immunoglobulin production and T-cell anomalies (Chap. 1.3.11). Common variable immunodeficiency (CVID) is compatible with longer survival and is the most common, primary inherited immunodeficiency encountered in clinical practice [24].

Besides the commonest cause, HIV infection, other causes of secondary immunodeficiencies which can be associated with vitiligo include chronic undernutrition and other conditions including protein-losing enteropathy [3], nephrotic syndrome [17], and hematological malignancies [23].

Vitiligo could also result from anomalies of the innate immunity [15] (Sect. 2.2.7.2), but so far no inherited or acquired condition fitting a well-established innate immunity disturbance has been detected in this context. Oxidative stress dysregulation has been associated with immunosuppression, as in granulomatous disease [4], and a link to vitiligo is not established. The following sections view the immunodeficiencies with an established, probable, or possible association with vitiligo.

1.3.9.2 Vitiligo and HIV Infection

HIV infection is a common cause of immunodeficiency leading to CD4 T-cell depletion. Clinical consequences associate opportunistic infections and specific manifestations, defining the acquired immunodeficiency syndrome (AIDS) stage [5]. HIV infection is also associated with an early immune dysregulation attributed to a depletion of gut lymphoid tissue. A few cases of vitiligo associated with HIV infection have been reported [2, 6, 10, 13, 14, 24, 28]. The circumstances of occurrence of vitiligo are, however, variable if disease history, immune status, and history of exposition to highly active antiretroviral treatment (HAART) are taken into account (Table 1.3.9.2). Three different situations have been observed, namely (1) vitiligo revealing immunosuppression, (2) improvement of previous vitiligo during the course of immunodeficiency, and (3) modification of vitiligo presentation with antiretroviral treatment and/or immune restoration with inflammatory signs (described as "punctuate advanced erythematous margins"). Concerning vitiligo presenting as a manifestation of HIV infection, all patients reported by Duvic et al. in the 1980s [10] where in an advanced AIDS stage one patient (Patient 2) presented with incident vitiligo revealing severe immunosuppression. The CD4 cell count available for three patients was less than 400 cells/mm^3. There was no information on the viral load.

Concerning the improvement of previous vitiligo, spontaneous repigmentation of vitiligo has been reported in an untreated HIV-positive patient (Patient 6). This patient had seroconverted for HIV infection four years before [14]. He noticed spontaneous and significant repigmentation of vitiligo lesions of 15 years duration. His CD4 cell count was at 390/mm^3 when repigmentation occurred. Besides HIV-induced immune suppression, an alternative hypothesis for improvement is the excessive production of polyclonal antibodies by activated B-lymphocytes [14].

Contrary to the previous scenario, pigmentation has occurred following HAART. A first patient (Patient 7) had generalized vitiligo onset associated with photosensitivity two years after the diagnosis of HIV infection [2]. Two years later, he was commenced on HAART and the skin began to repigment. At vitiligo onset, the CD4 cell count was at 72/mm^3. When vitiligo began to repigment, the CD4 cell count was at 149. Concurrently, the patient developed marked lipodystrophy. The improvement was attributed to the change in CD4 cell count. PUVA therapy was initiated following partial repigmentation and failed to produce any additional benefit. If the associated lipodystrophic effects of antiretroviral drugs noted in this patient are partly attributable to a mitochondrial cytopathy, it is possible to speculate about the role of resulting disturbed oxidative stress responses in vitiligo (Chap. 2.2.6). The role of the immune reconstitution inflammatory syndrome (IRIS) is also discussed in this context – see later [22].

For two other patients of ours (Patients 9 and 10), vitiligo revealed HIV seropositivity. At the time of vitiligo assessment, Patient 9 had a CD4 T-cell count at 4/mm^3 with a viral RNA load at more than 100,000 copies/mL. Patient 10 had a CD4 T-cell count at 5/mm^3 with a viral RNA load at more than 1,000,000 copies/mL. For both of them, the introduction of HAART had no impact on vitiligo, despite a good CD4 and viral RNA load response. Moreover, both UVB TL01 and PUVA failed to improve vitiligo. In Patient 10, biopsies taken from vitiligo lesions at the time of diagnosis demonstrated the persistence of melanocytes with a loss of their cell differentiation markers, that is, protein S100, HMB45 and Melan A and a

Table 1.3.9.2 HIV infection and vitiligo

Patient	References	Delay from HIV infection diagnosis to vitiligo onset	HIV infection status	Vitiligo course
Patient 1	Duvic et al. [10]	3 years	Stage C2	Repigmentation following ribavirin
Patient 2	Duvic et al. [10]	Simultaneous	Stage C2	Unknown
Patient 3	Duvic et al. [10]	14 months	Stage C3	No improvement following ribavirin
Patient 4	Duvic et al. [10]	7 months	Stage C3	Unknown
Patient 5	Duvic et al. [10]	Vitiligo preceeded HIV infection for 14 years	Stage C3	Unknown
Patient 6	Grandhe et al. [14]	Vitiligo preceeded HIV infection for 11 years	Stage B2	Spontaneous repigmentation 2 years following HIV infection diagnosis
Patient 7	Antony et al. [2]	1 year	Stage B3	Repigmentation following HAART
Patient 8	Niamba et al. [24]	Simultaneous	Stage C3	Repigmentation following HAART
Patient 9	Personal data	Simultaneous	Stage B3	No repigmentation following HAART
Patient 10	Personal data	Simultaneous	Stage C3	No repigmentation following HAART

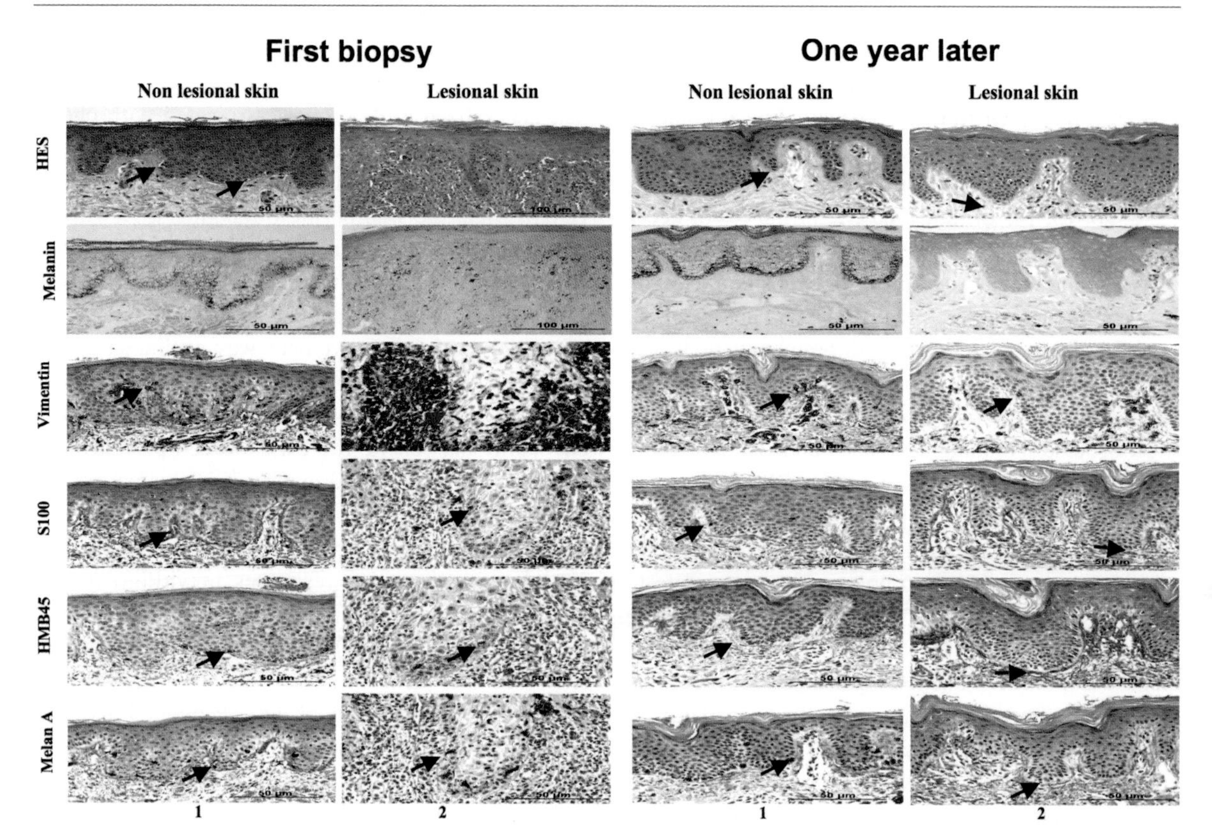

Fig. 1.3.9.2 Histopathologic sequence corresponding to the patient depicted in Fig. 1.3.9.1. Two biopsies were obtained from lesional and non lesional skin. The procedure was repeated after a 1-year interval. First biopsies (*correspond to 13.9.1a*): in non lesional skin (1) melanocytes were present and expressed protein S100, HMB45, Melan A. Melanin was present in basal keratinocytes; in lesional skin (2) perivascular mononuclear infiltrate and lichenoid pattern. Melanocytes were present in the basal layer and immunostaining for proteins S100, HMB45, Melan A was positive. Pigmentary incontinence was observed and there was no melanin within keratinocytes. Second biopsies, 1 year later (*correspond to 13.9.1b*): in non lesional skin (1) no changes; in lesional skin (2) pigmentary incontinence. Melanocytes were still present as shown by vimentin staining although they did not express protein S100, HMB45, Melan A and there was a loss of function (no melanin in keratinocytes)

complete lack of melanin pigment within basal keratinocytes (Fig. 1.3.9.2).

The relation of vitiligo to IRIS which may occur in HIV-infected patients initiating antiretroviral therapy is intriguing. IRIS seems to result from restored immunity to specific antigens either infectious or not [22, 33]. Clinical manifestations attributed to IRIS are the result of an inflammatory response caused by the recovering immune system which does not recognize those residual antigens properly [9]. Theories concerning the pathogenesis of IRIS involve a combination of underlying antigenic burden, the degree of immune restoration and host genetic susceptibility. An increase in memory CD4 T-cells, possibly as a result of the redistribution from peripheral lymphoid tissue, would be associated with an increase in naïve T-cells leading to a combination of both quantitative restoration of immunity, as well as qualitative function and phenotypic expression [18]. In noninfectious causes of IRIS, innate antigens play likely a role of antigenic stimuli [12].

1.3.9.3 Vitiligo and Idiopathic T-Cell Lymphocytopenia

A patient with a long history of vitiligo associated with idiopathic CD4+ T-cell lymphocytopenia (ICTL) has been reported [32]. ICTL is defined as a persistent depletion of peripheral blood CD4 cell count of less than 300 cells/mm^3 in the absence of either HIV infection or other known causes of immunodeficiency. ICTL has a variable clinical spectrum that ranges from patients exhibiting minimal symptoms to those who have died from opportunistic infections [34]. In the Yamauchi et al. case report [32], in addition to the depletion of CD4+ T lymphocytes, the patient presented a depletion of CD8+ T lymphocytes. Vitiligo associated with ICTL could be tentatively explained as a shift in the balance of self- tolerance versus autoimmunity, when T-regulatory cells specific for melanocytic self-antigens like tyrosinase are inactivated because of mutation or deletion [31].

1.3.9.4 Vitiligo and CVID

CVID is the most common primary immunodeficiency. Among patients with CVID, up to 25% present with autoimmune events [26]. The clinical phenotype of the condition is broad and heterogeneous. The autoimmune diseases associated include autoimmune thrombocytopenic purpura and autoimmune hemolytic anemia (frequently), and less commonly, rheumatoid arthritis, vasculitis, and vitiligo [16]. These autoimmune manifestations have suggested a genetic dysregulation provoking autoreactivity during B-cell development, with the production of multiple autoantibodies against various antigenic targets [16]. CVID can be caused by mutation in the TNFRSF13B gene located on chromosome 17, which encodes the transmembrane activator and CAML interactor (TACI). Figure 1.3.9.3 shows a patient with a very early onset of vitiligo (before the age of 1) that was further diagnosed with CVID and various autoimmune disorders, including alopecia areata and Hashimoto's thyroiditis (personal data).

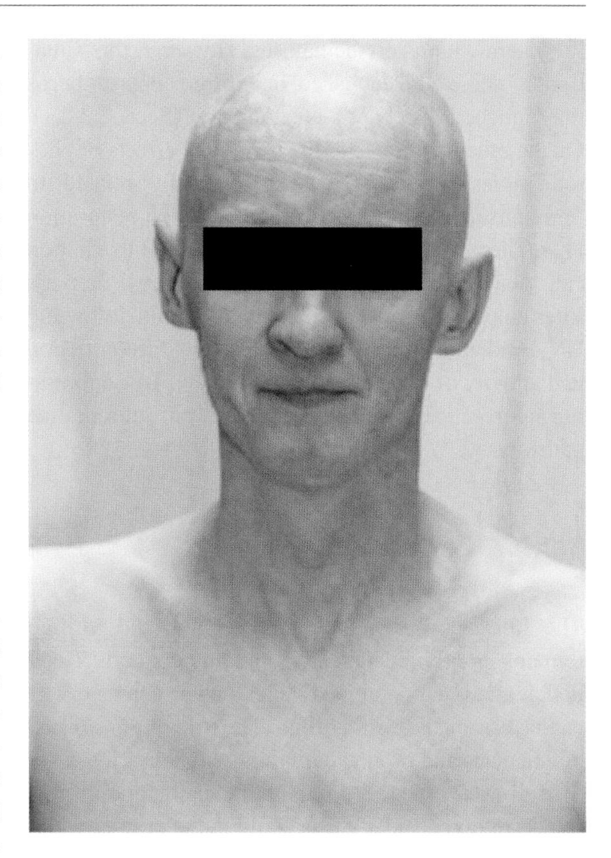

Fig. 1.3.9.3 CVID-associated vitiligo in a 34-year-old man. Lesions began at age 1 and expanded progressively over more than 30% of the body surface. Other clinical manifestations included Hashimoto's thyroiditis, alopecia areata, and recurrent nasopharyngeal and pulmonary infections

1.3.9.5 Complement Deficiencies and Vitiligo

The arguments for a role of the complement cascade are mostly based on in vitro studies and theoretical speculations. However, heterozygous C4 deficiency has been linked to an increased risk for vitiligo [30].

Complement activating antimelanocyte antibodies have been implicated in vitiligo pathogenesis, with complement components directly involved in cell killing (Sect. 2.2.7.3). Along this line, it has been speculated a role in vitiligo pathogenesis for a plasmatic protein called mannose binding lectin (MBL). This protein recognizes mannose terminal residues of microbial glycoproteins or glycolipids, and then activates the classic complement pathway without antibodies, through an associated serin protease [25].

1.3.9.6 Ultraviolet Irradiation and Vitiligo

Several effects of natural and artificial UV irradiation may influence the onset and progression of vitiligo. On the one hand, excessive UV exposure can trigger autoimmune responses in predisposed individuals, as noted in systemic lupus erythematosus. However, this situation is not well-established in vitiligo, but sunburns may induce a Koebner's phenomenon (Sect. 2.2.2.1). On the other hand, one of the beneficial effects of phototherapies for vitiligo was supposedly related to immunosuppression, besides pigmentation restoration [11]. There is now some evidence that UVB irradiation can also induce regulator T-cell (Treg) activity. Interleukin-10 produced in the epidermis following UV irradiation may contribute to the differentiation and activity of the Treg population [31]. By suppressing autoreactive T-cells, Treg may prevent autoimmune pathways [27].

1.3.9.7 Conclusions

The link between immunodeficiencies and vitiligo remains poorly documented and understood. T-cell deficiencies and combined deficiencies such as CVID can be associated with vitiligo. The role of pure B-cell deficiencies and of innate immunity deficiencies is not as well-established. From the clinical standpoint, it is important to consider – especially in the case of widespread vitiligo following a clinically inflammatory phase, an underlying T-cell immunosupression and that triggered by HIV infection should be investigated in priority.

The application to common NSV of this phenomenon is not straightforward. Several reports point to the activation of T-cells found at the margin of progressive vitiligo as being a step amplifying skin melanocyte loss, suggesting an "acceleration phase." This phase may depend on a shift in the balance between immunity and tolerance. In this context, the role of properly functioning Treg, which seem involved in IRIS, needs to be investigated and could be a potential target for the therapy of common vitiligo.

References

1. Abdallah M, Abdel-Nasr MB, Moussa MH et al (2003) Sequential immunohistochemical study of depigmenting and repigmenting minigrafts in vitiligo. Eur J Dermatol 13:548–552
2. Antony FC, Marsden RA (2003) Vitiligo in association with human immunodeficiency virus infection. J Eur Acad Dermatol Venereol 20:63–65
3. Berkowitz DM, Passaro E Jr, Isenberg JI (1977) Hypertrophic protein-losing gastropathy and vitiligo. Report of a second case. Am J Dig Dis 22:554–557
4. Brown KL, Bylund J, MacDonald KL et al (2008) ROS-deficient monocytes have aberrant gene expression that correlates with inflammatory disorders of chronic granulomatous disease. Clin Immunol 129:90–102
5. CDC (1993) 1992 revised classification system for HIV infection and expanded surveillance case definition for AIDS among adolescents and adults. MMWR 41(RR–17): 961–962
6. Cho M, Cohen PR, Duvic M (1995) Vitiligo and alopecia areata in patients with human immunodeficiency virus infection. South Med J 88:489–491
7. Clark R, Kupper TS (2005) Old meets the new: the interaction between innate and adaptive immunity. J Invest Dermatol 125:629–637
8. Dejaco C, Duftner C, Grubeck-Loebenstein B, Schirmer M (2006) Imbalance of regulatory T cells in human autoimmune diseases. Immunology 117:289–300
9. Dhasmana DJ, Dheda K, Ravn P, Wilkinson RJ, Meintjes G (2008) Immune reconstitution inflammatory syndrome in HIV-infected patients receiving antiretroviral therapy: pathogenesis, clinical manifestations and management. Drugs 68:191–208
10. Duvic M, Rapini R, Hoots WK et al (1987) Human immunodeficiency virus-associated vitiligo: expression of autoimmunity with immunodeficiency? J Am Acad Dermatol 17:656–662
11. El-Ghorr AA, Norval M (1997) Biological effects of narrow-band (311 nm TL01) UVB irradiation: a review. J Photochem Photobiol B 38:99–106
12. French MA •(2009) HIV/AIDS: immune reconstitution inflammatory syndrome: a reappraisal. Clin Infect Dis 48: 101–107
13. García-Patos Briones V, Rodríguez Cano L, Capdevila Morell JA, Costells Rodellas A (1994) Vitiligo associated with the acquired immunodeficiency syndrome. Med Clin (Barc) 103:358
14. Grandhe NP, Dogra S, Kumar B (2006) Spontaneous repigmentation of vitiligo in an untreated HIV-positive patient. J Eur Acad Dermatol Venereol 20:234–235
15. Knight AK, Cunningham-Randles C (2006) Inflammatory and autoimmune complications of common variable immune deficiency. Autoimmun Rev 5:156–159
16. Jin Y, Mailloux CM, Gowan K et al (2007) NALP1 in vitiligo-associated multiple autoimmune disease. N Eng J Med 356:1216–1225
17. Kuzmanovska DB, Shahpazova EM, Kocova MJ et al (2001) Autoimmune thyroiditis and vitiligo in a child with minimal change nephrotic syndrome. Pediatr Nephrol 16:1137–1138
18. Lim A, Tan D, Price P et al (2007) Proportions of circulating T cells with a regulatory cell phenotype increase with HIV-associated immune activation and remain high on antiretroviral therapy. AIDS 21:1525–1534
19. Liston A, Enders A, Siggs OM (2008) Unravelling the association of partial T-cell immunodeficiency and immune dysregulation. Nat Rev Immunol 8:545–558
20. Mandelcorn-Monson RL, Shear NH, Yau E et al (2003) Cytotoxic T lymphocyte reactivity to gp100, MelanA/

MART-1, and tyrosinase, in HLA-A2-positive vitiligo patients. J Invest Dermatol 121:550–556
21. Milner JD, Fasth A, Etzioni A (2008) Autoimmunity in severe combined immunodeficiency (SCID): lessons from patients and experimental models. J Clin Immunol S1: S29–S33
22. Murdoch DM, Venter WDF, Van Rie A, Feldman C (2007) Immune reconstitution inflammatory syndrome (IRIS): review of common infectious manifestations and treatment options. AIDS Res Ther 4:9
23. Newman MD, Milgraum S (2008) Leukemia cutis masquerading as vitiligo. Cutis 81:163–165
24. Niamba P, Traoré A, Taieb A (2007) Vitiligo sur peau noire associée au VIH et repigmentation lors du traitement antiretroviral. Ann Dermatol Venereol 134:272–276
25. Onay H, Pehlivan M, Alper S et al (2007) Might there be a link between mannose binding lectin and vitiligo? Eur J Dermatol 17:146–148
26. Park MA, Li JT, Hagan JB et al (2008) Common variable immunodeficiency: a new look at an old disease. Lancet 372:489–502
27. Ponsonby AL, Lucas RM, van der Mei IA (2005) UVR, vitamin D and three autoimmune diseases-multiple sclerosis, type 1 diabetes, rheumatoid arthritis. Photochem Photobiol 81: 1267–1275
28. Tojo N, Yoshimura N, Yoshizawa M et al (1991) Vitiligo and chronic photosensitivity in human immunodeficiency virus infection. Jpn J Med 3:255–259
29. Van den Wijngaard R, Wankowicz-Kalinska A, Le Poole C et al (2000) Local immune response in skin of generalized vitiligo patients. Destruction of melanocytes associated with the prominent presence of CLA+ T cells at the perilesional site. Lab Invest 80:1299–1309
30. Venneker GT, Westerhof W, de Vries IJ et al (1992) Molecular heterogeneity of the fourth component of complement (C4) and its genes in vitiligo. J Invest Dermatol 99:853–858
31. Westerhof W, d'Ischia M (2007) Vitiligo puzzle: the pieces fall in place. Pigment Cell Res 20:345–359
32. Yamauchi PS, Nguyen NQ, Grimes PE (2002) Idiopathic CD4+ T-cell lymphocytopenia associated with vitiligo. J Am Acad Dermatol 46:779–782
33. Zandman-Goddard G, Shoenfeld Y (2002) HIV and autoimmunity. Autoimmunity Rev 1:329–337
34. Zonios DI, Falloon J, Bennett JE, Shaw PA, Chaitt D, Baseler MW et al (2008) Idiopathic CD4+ lymphocytopenia: natural history and prognostic factors. Blood 112:287–294

Inflammatory Vitiligo

1.3.10

Khaled Ezzedine, Julien Seneschal, Ratnam Attili, and Alain Taïeb

Contents

Core Messages

- A particular form of vitiligo reported as 'inflammatory vitiligo with raised borders' has been rarely described, and this condition is associated with lichenoid infiltrates in the margins of progressing lesions.
- Histopathological findings in still-pigmented progressing borders of common NSV and SV may show less-intense, but similar features, suggesting that common vitiligo could be considered as a clinically silent chronic inflammatory skin disorder.
- Clinically inflammatory vitiligo has been sometimes associated with infectious or inflammatory diseases, and may result in this context from various pathomechanisms.

1.3.10.1 Introduction

One of the striking features of vitiligo, when compared with other chronic skin conditions, is the absence of symptoms and signs of inflammation. In contrast with atopic dermatitis, which is characterised locally by a TH2 cytokine imbalance, where erythema and oedema are common during flares, pruritus is not a classic vitiligo symptom. However, since this item has been included in the VETF evaluation form [16], some patients definitely report a mild pruritus preceding flares of depigmentation. Vitiligo also contrasts with psoriasis, which shows a predominant TH1–TH17 cytokine imbalance and where lesions are clinically

K. Ezzedine (✉)
Service de Dermatologie, Hôpital St André,
CHU de Bordeaux, France
e-mail: khaled.ezzedine@chu-bordeaux.fr

M. Picardo and A. Taïeb (eds.), *Vitiligo*,
DOI 10.1007/978-3-540-69361-1_1.3.10, © Springer-Verlag Berlin Heidelberg 2010

erythematous and scaly, associated with pruritus during pears.

However, a particular form of vitiligo defined as 'inflammatory vitiligo with raised borders' has been reported long ago [4, 6, 20]. The clinical presentation of this particular form of vitiligo consists in depigmented patches with an erythematous micro-papular edge. This condition is associated with inflammatory infiltrates in the margin of progressing lesions. In general, clinically inflammatory vitiligo (CIV) may occur in isolation, but has infrequently been associated with various disorders including infectious diseases such as human immunodeficiency virus (HIV) infection [5, 7, 10, 17] lichen sclerosus [19], atopic dermatitis [14], or Vogt–Koyanagi–Harada (VKH) disease [18, 21]. Recent histological studies indicate that common vitiligo can also be micro-inflammatory at the progressing edge of de-pigmented lesions.

1.3.10.2 Isolated Clinically Inflammatory Vitiligo

A few cases of CIV without documented association with other diseases have been reported (Table 1.3.10.1). In all reported cases, progressive lesions with erythema, fine scaling with or without pruritus are common features. An erythematous raised border has been documented. Marginal hyperpigmentation surrounding hypopigmented patches has also been reported. Inflammatory vitiligo is generally considered as a progressive (unstable) form since extensive involvement seems to be the rule. It seems to affect both sexes at any age. Interestingly, in almost half the patients, concomitant inflammatory and non-inflammatory patches are observed. Based on histopathological studies, this unusual clinical manifestation can be interpreted as disease progression due to an inflammatory phase.

1.3.10.3 Clinically Inflammatory Vitiligo Associated with Other Disorders

Inflammatory vitiligo has also been associated with specific disorders (Table 1.3.10.2).

The clinically inflammatory stage of HIV-associated cases is different from CIV with raised borders (Chap. 1.3.9). It consists in nummular lichenoid lesions which result in complete pigment loss. Interestingly, in this setting the histopathological sequence shows a loss of melanocytic antigens preceding the definitive loss of pigment cells. Contrarily, in two cases of vitiligo occurring respectively in the setting of chronic hepatitis associated with hepatitis-C infection [17] and atopic dermatitis [14], clinical features were close to that of inflammatory vitiligo with raised borders. In these case reports, patients presented both inflammatory patches with raised border in recent lesions and noninflammatory older patches. Interestingly, Tsuruta et al. reported a VKH patient with a 1-year history of a superimposed thin inflammatory raised erythema and plaque-type inflammatory erythema occurring in the setting of non-segmental vitiligo (NSV) of 20-year duration. Two particular features were noted: first, within the patches of vitiligo with raised

Table 1.3.10.1 Published cases of clinically inflammatory vitiligo

Reference	Age/gender	Type of vitiligo	Pruritus	Duration of disease	Presence of other non-inflammatory patches	Evolution
[20]	47/F	NSV	Yes	2 months	No	Regressive
[6]	12/M	NSV	No	2 months	Yes	Regressive
[4]	46/M	NSV	Yes	1 year	No	Extensive
[9]	5/M	NSV	?	3 years	Yes	Regressive
	24/M	NSV	?	7 years	Yes	Regressive
[11]	10/M	NSV	?	? recent	?	Regressive
	31/M	?	Yes	?	?	Regressive
[22]	61/M	NSV	Yes	5 months	Yes	Extensive
[2]	77/M	NSV	Yes	?	?	Extensive
[12]	70/M	NSV	?	6 months	No	Stable
	20/M	NSV	?	2 years	No	Stable
[15]	21/M	SV	Yes	2 years	No	Progressive

Table 1.3.10.2 Inflammatory vitiligo in the setting of infectious or inflammatory diseases

Reference	Age/gender of patient	Type of vitiligo	Associated disease	Pruritis	Duration of disease	Presence of other non-inflammatory patches	Evolution
[18]	48/M	NSV	Vogt-Koyanagi-Harada disease	Yes	1 year	Yes	Extensive
[14]	31/M	?	Atopic dermatitis	Yes	5 years	?	Extensive
[17]	38/M	NSV	Hepatitis C virus infection	Yes	2 years	Yes	Extensive
[19]	41/W	SV	Lichen sclerosus	Yes	6 months	No	Extensive

inflammatory borders, erythema was separated from the totally depigmented areas by an incompletely vitiliginous area; second, the presence of plaque-type erythema is uncommonly seen in the so-called inflammatory vitiligo [18]. Weisberg et al. reported the case of a 41-year-old African–American woman who had a 6-month history of inflammatory segmental vitiligo (SV) that started on the left foot, and then spread proximally up her leg to her thigh, and then to her buttocks with an involvement of pubic hairs that turned white. Concomitantly, the patient had severe burning and itching in a clinically whitish pink skin of the vaginal area consistent with the diagnosis of lichen sclerosus [19].

1.3.10.4 Differential Diagnosis

A clinical presentation similar to that noted in HIV infection-associated inflammatory vitiligo may occur in patients with graft versus host disease (GVHD) following allogeneic haematopietic cell transplantation (Prof Beylot-Barry, personal communication, Fig. 1.3.10.1). The depicted GVHD patient presented with erythroderma followed by generalised depigmented macules affecting the whole body, clinically consistent with the diagnosis of vitiligo. In addition, he developed alopecia areata. One hypothesis is that GVHD may have triggered skin-targeted autoimmunity towards melanocytes (epidermal and follicular).

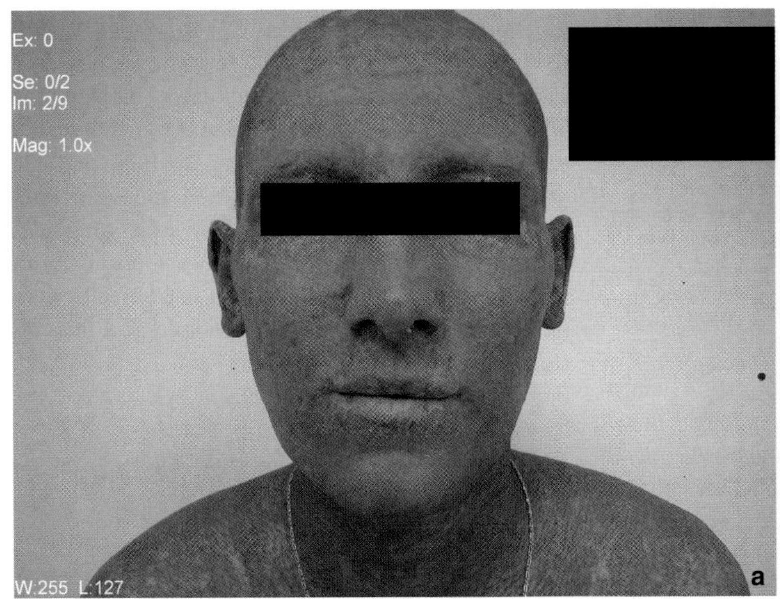

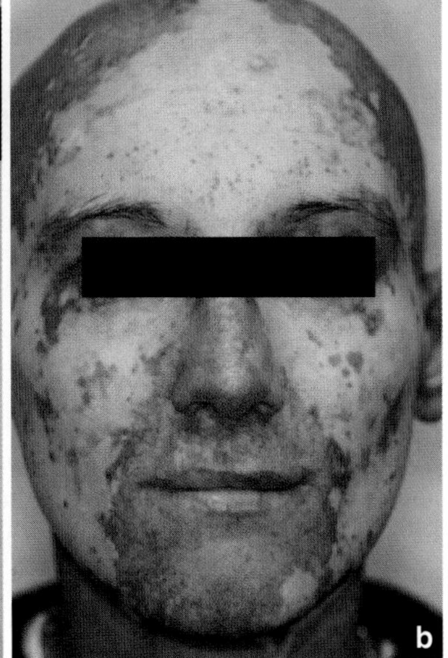

Fig. 1.3.10.1 Graft versus host disease at erythrodermic (**a**) and vitiligoid (**b**) phases

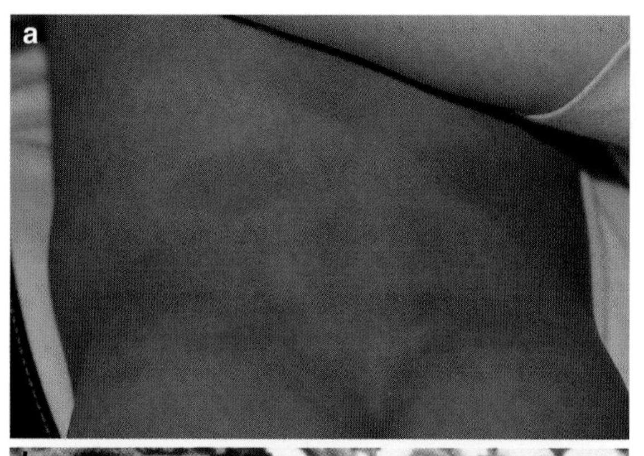

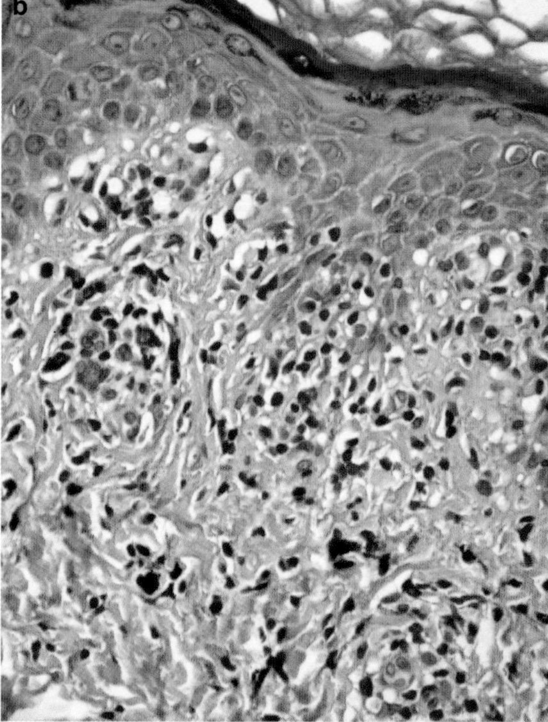

Fig. 1.3.10.2 Annular lichenoid dermatitis of youth (ALDY) (**a**). Histologic analysis of a cutaneous biopsy specimen from the border of one annular patch. Small lymphocytes infiltrate the lower epidermal layers with keratinocytes necrosis, only into the tips of rete ridges. (hematoxylin–eosin, original magnification ×400) (**b**)

A peculiar annular lichenoid dermatitis named '*annular lichenoid dermatitis of youth,*' the clinical appearance of which can initially suggest diagnoses of morphea, mycosis fungoides, or annular erythema, may be mistaken for inflammatory vitiligo [1] (Fig. 1.3.10.2). Lesions consist of persistent asymptomatic erythematous macules and round annular patches with a red-brownish border and central hypopigmentation, mostly distributed on the groin and flanks. Histology reveals a lichenoid dermatitis with massive necrosis/apoptosis of the keratinocytes limited to the tips of rete ridges, in the absence of dermal sclerosis and epidermotropism of atypical lymphocytes. The infiltrate is composed mainly of memory CD4(+) CD30(−) T-cells with few B-cells and macrophages.

1.3.10.5 Histological Features of Clinically Inflammatory Vitiligo versus Common Clinically Non-Inflammatory Vitiligo

Histopathological examination in CIV should include biopsies taken from the inflammatory border. Whatever the case, vitiligo being associated or not with another disease, the perivascular mononuclear infiltrate with an associated lichenoid pattern is almost always present. In addition, degeneration of melanocytes and basal keratinocytes may be observed. Immunohistochemistry may reveal CD8 cells and sparse CD4 cells. Table 1.3.10.3 summarises published reports in case of associated conditions.

Histopathological studies of clinically non-inflammatory NSV that include adjacent pigmented skin have shown inflammatory changes in the pigmented border with decreasing inflammation from the marginal pigmented areas to the fully depigmented vitiligo patches [3, 8, 13, 22]. Our experience based on rapidly progressing paediatric cases biopsied on clinically pigmented skin 0.5 cm of the visible border of depigmentation confirms this finding, and the intensity of lichenoid inflammation and sometimes epidermotropism may simulate mycosis fungoides (Chap. 1.2.2) (Fig. 1.3.10.3). Recent observations suggest that early biopsies taken in clinically non-inflammatory SV show similar pathological features (Fig. 1.3.10.4) [3].

1.3.10.6 Summary and Concluding Remarks

CIV is a rare clinical presentation of common vitiligo, described so far in the context of NSV. When associated with other cutaneous diseases, it may correspond

Table 1.3.10.3 Histological and immunohistochemistry findings in patients with clinically inflammatory vitiligo and associated diseases

Reference	Age/ sex	Duration / progression of disease	Type of vitiligo	Localisation	Histological Findings	Immunohisto-chemistry	Associated disease	Treatment/ evolution
[18]	48/M	20 years/12 months	NSV	Entire body except face	Perivascular mononuclear infiltrate, lichenoid pattern	LFA-1, CD3, CD8 and rare CD4 and CD20	Vogt-Koyanagi-Harada disease	NA
[14]	31/M	5 years/5 years	NSV	Limbs, trunk	Perivascular mononuclear infiltrate, lichenoid pattern	No melano-cytes, CD4 and CD8 in raised border/CD8 outside the border	Atopic dermatitis	Topical corticosteroid/ Inprovement of erythema
[17]	38/M	24 months/24 months	NSV	Buttock, axilla, inguinal, extremities	Degeneration of basal layer, perivascular mononuclear infiltrate	NA	Hepatitis virus C infection	Prednisolone 20 mg daily + PUVA/ Resolution of erythema + partial repigmentation
[19]	41/W	6 months/6 months	SV	Left foot, thigh and buttock + left genital area	Papular dermal sclerosis, perivascular mononuclear infiltrate lichenoid pattern	NA	Lichen sclerosus	Topical corticosteroid/ involvement of erythema + partial repigmentation

* lichen sclerosis NA - not available

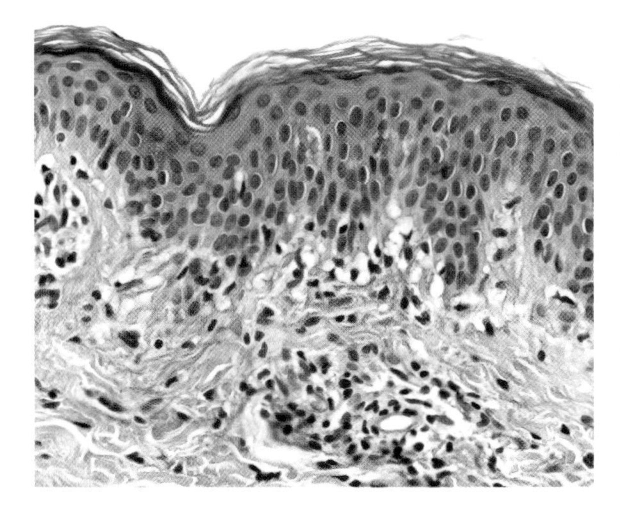

Fig. 1.3.10.3 Microscopic inflammation in common progressive NSV, biopsy taken at the pigmented margin of a clinically non inflammatory lesion (HES ×400). The infiltrate is mostly composed of CD8+ cells (Chap. 1.2.2)

to a particular mode of onset of depigmentation triggered by local inflammation different from that of Sutton's or even Koebner's phenomena, and so far, poorly understood. When associated with HIV infection, it has a clearly distinct clinical presentation for which the outcome of severe cutaneous inflammation leads to a vitiligo-like disorder, difficult to distinguish from common NSV. However, besides such rare observations, consistent histopathological findings in still-pigmented progressing borders in NSV and SV suggest that common vitiligo could be considered as a clinically silent chronic micro-inflammatory skin disorder, clinically visible forms being the emerged tip of an iceberg. This is probably to relate pathophysiologically either to a particular inflammatory cytokine profile of vitiligo, which would provoke minimal inflammation and pruritus, or to the loss of melanocyte as providers of mediators of an inflammatory response which would

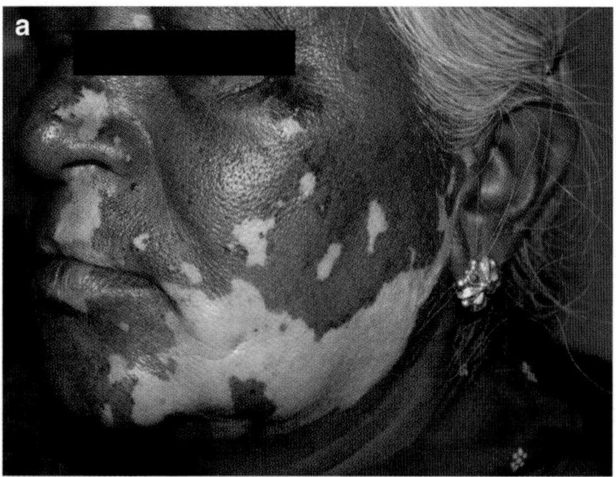

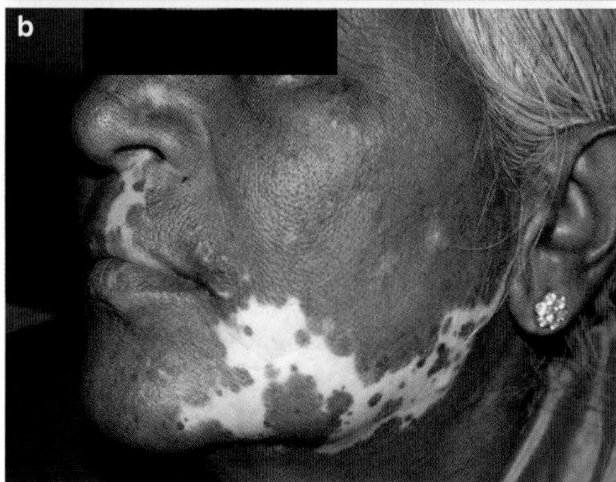

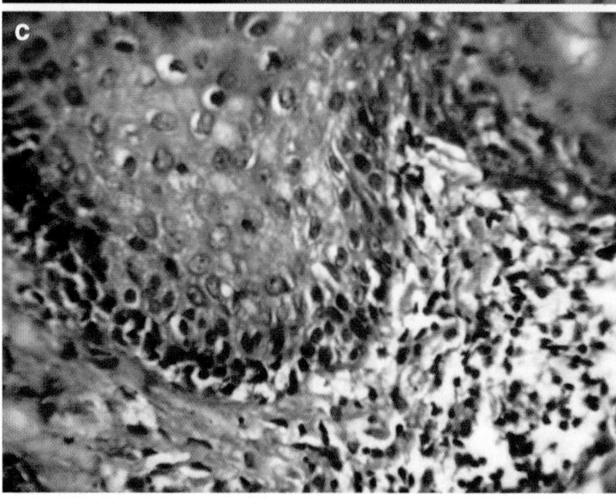

Fig. 1.3.10.4 Segmental vitiligo: recent onset lesion (3 months in (**a**)) and 4 months later (**b**). Microscopic lichenoid inflammation (**c**), on biopsy taken on pigmented skin across the midline (chin), at the margin of the lesion at stage depicted in (**a**)

be lost in vitiligo, or to both phenomena. In this respect, the 'neurone of the skin' view of the melanocyte needs probably to be reinvestigated more closely in terms of production of cytokines and neuropeptides mediating inflammatory skin responses, suggesting a possible melanocyte-mediated neurogenic inflammation being lost in vitiligo.

References

1. Annessi G, Paradisi M, Angelo C et al (2003) Annular lichenoid dermatitis of youth. J Am Acad Dermatol 49: 1029–1036
2. Arata J, Abe-Matsuura Y (1994) Generalized vitiligo preceded by a generalized figurate erythematosquamous eruption. J Dermatol 21:438–441
3. Attili VR, Attili SK (2008) Lichenoid inflammation in vitiligo – a clinical and histopathologic review of 210 cases. Int J Dermatol 47:663–669
4. Buckley W, Lobitz W (1953) Vitiligo with raised inflammatory border. Arch Derm Syphilol 67:316–320
5. Duvic M, Rapini R, Hoots WK et al (1987) Human immunodeficiency virus-associated vitiligo: expression of autoimmunity with immunodeficiency? J Am Acad Dermatol 17:656–662
6. Garb J, Wise F (1948) Vitiligo with raised borders. Arch Derm Syphilol 58:149–153
7. Grandhe NP, Dogra S, Kumar B (2006) Spontaneous repigmentation of vitiligo in an untreated HIV-positive patient. J Eur Acad Dermatol Venereol 20:234–235
8. Hann SK, Park YK, Lee KG et al (1992) Epidermal changes in active vitiligo. J Dermatol 19:217–222
9. Michaëlsson G (1968) Vitiligo with raised borders. Report of two cases. Acta Derm Venereol 48:158–161
10. Niamba P, Traoré A, Taieb A (2007) Vitiligo sur peau noire associée au VIH et repigmentation lors du traitement antirétroviral. Ann Dermatol Venereol 134:272–276
11. Ortonne JP, Baran R, Civatte J (1979) Vitiligo with an inflammatory border. A propos of 2 cases with review of the literature (18 cases). Ann Dermatol Venereol 106:613–615
12. Petit T, Cribier B, Bagot M, Wechsler J (2003) Inflammatory vitiligo-like macules that simulate hypopigmented mycosis fungoides. Eur J Dermatol 17:410–412
13. Sharquie KE, Mehenna SH, Naji AA, Al-Azzawi H (2004) Inflammatory changes in vitiligo: stage I and II depigmentation. Am J Dermatopathol 26:108–112
14. Sugita K, Izu K, Tokura Y (2006) Vitiligo with inflammatory raised borders, associated with atopic dermatitis. Clin Exp Dermatol 31:80–82
15. Swick BL, Walling HW (2008) Depigmented patches with an annular inflammatory border. Clin Exp Dermatol 33:671–672
16. Taïeb A, Picardo M; VETF Members (2007) The definition and assessment of vitiligo: a consensus report of the Vitiligo European Task Force. Pigment Cell Res 20:27–35
17. Tsuboi H, Yonemoto K, Katsuoka K (2006) Vitiligo with inflammatory raised borders with hepatitis C virus infection. J Dermatol 33:577–578

18. Tsuruta D, Hamada T, Teramae H et al (2001) Inflammatory vitiligo in Vogt-Koyanagi-Harada disease. J Am Acad Dermatol 44:129–131
19. Weisberg EL, Le LQ, Cohen JB (2008) A case of simultaneously occurring lichen sclerosus and segmental vitiligo: connecting the underlying autoimmune pathogenesis. Int J Dermatol 47:1053–1055
20. Wise F (1942) Leukoderma with inflammatory borders. Arch Derm Syph 45:218–219
21. Wong SS, Ng SK, Lee HM (1999) Vogt-Koyanagi-Harada disease: extensive vitiligo with prodromal generalized erythroderma. Dermatology 198:65–68
22. Yagi H, Tokura Y, Furukawa Y, Takigawa M (1997) Vitiligo with raised inflammatory borders: involvement of T cell immunity and keratinocytes expressing MHC class II and ICAM-1 molecules. Eur J Dermatol 7: 19–22

Rare Inherited Diseases and Vitiligo 1.3.11

Alain Taïeb and Fanny Morice-Picard

Contents

Core Messages

- Some monogenic disorders such as piebaldism have marked phenotypic overlap with vitiligo, which raise the question of a common pathogenic background.
- Another interesting point of monogenic diseases, which include organ-specific autoimmunity, is the use of the monogenic disease as a model to identify autoantigens of importance in vitiligo.
- Auto-immune polyendocrine syndromes, mitochondrial, and breakage disorders can be associated with a vitiligo phenotype.

1.3.11.1 The Interest of Studying Monogenic Disorders for the Understanding of Common NSV

The candidate gene approach (Chap. 2.2.1) has not been successful for vitiligo. However, the observation of relevant common features in monogenic disorders and complex multigenic ones may offer interesting clues, either for extracting a major genetic component in a complex disorder [1], or for looking at possible common pathogenetic pathways or disease markers. For vitiligo, even though the phenotype is easily recognisable in humans, there are limited human monogenic disease models (Table 1.3.11.1). Unfortunately, the 'vitiligo' phenotype is usually rarely well-documented in genetics publications. We have included only

A. Taïeb (✉)
Service de Dermatologie, Hôpital St André,
CHU de Bordeaux, France
e-mail: alain.taieb@chu-bordeaux.fr

M. Picardo and A. Taïeb (eds.), *Vitiligo*,
DOI 10.1007/978-3-540-69361-1_1.3.11, © Springer-Verlag Berlin Heidelberg 2010

Table 1.3.11.1 Monogenic disorders and NSV

Name OMIM entry Reference if not detailed in text	Inheritance/population	Gene/protein/antigen/	Clinical symptoms/remarks
APS1 (auto-immune polyglandular syndrome, type I)/ APECED (auto-immune polyendocrinopathy-candidiasis-ectodermal dystrophy) 240300	Autosomal recessive (most cases) Finns, Iranian Jews, Sardinians	*AIRE* (auto-immune regulator gene)	Vitiligo not mandatory; auto-immunity to NALP5 frequent in case of hypoparathyroidism
APS 2/ Schmidt syndrome 269200	Autosomal recessive or autosomal dominant with incomplete penetrance?	Unknown Linked to MHC class I on chromosome 6	No straightforward marker, vitiligo not mandatory
Ataxia Telangiectasia 208900	X linked recessive	*ATM*/atm Spontaneous chromosomal instability with multiple rearrangements especially chromosome 7 and 14 Chromosomal hypersensitivity to ionizing radiation and alkylating agents	Skin: Vitiligo Telangiectasias Granulomas Non skin: elevated serum alpha foetoprotein
Nijmegen breakage syndrome 251260	Autosomal recessive Eastern European origin	8q21 *NBS 1*/nibrin Spontaneous chromosomal instability with multiple rearrangements, especially chromosome 7 and 14 Chromosomal hypersensitivity to ionizing radiation and alkylating agents Radioresistant DNA synthesis	Skin: progressive vitiligo, Cafe au lait spots, photosensitivity (?) Non skin: facial dysmorphy, microcephaly ++, mental retardation, immunodeficiency, auto-immune disorders, neoplasias
MELAS 540000	Mitochondrial	Genetically heterogeneous : mutations in MTTL1 (most common) and MTTQ, MTTH, MTTK, MTTS1, MTND1, MTND5, MTND6, and MTTS2	Vitiligo associated in 11% of cases studied with common Mt mutation 3243
Vitiligo-spasticity syndrome/spastic paraplegia with pigmentary abnormalities 270750 [2]	Autosomal recessive Inbred Arab families	Mapped to 1q24-q32	Skin: vitiligo and lentigines of exposed areas (apparent at birth or in infancy); premature greying of body hair, canitia Non skin: microcephaly, thin face, micrognathia, retrognathia Severe spastic paraplegia in childhood +++; mild cognitive impairment
Combined immunodeficiency with auto-immunity and spondylometaphyseal dysplasia 607944 [3]	Autosomal recessive or dominant	Unknown	Skin : Vitiligo and hyperpigmented macules Non skin: Spondylometaphyseal dysplasia + combined humoral and cellular immunodeficiency: recurrent infections (pneumonia, sinusitis, fulminant varicella) auto-immune disorders

monogenic disorders with a confirmed ascertainment of this phenotype, based on personal experience or literature review.

Animal models (see Chap. 2.2.4) may be misleading because the human pigmentation system is the result of a very specific evolutional maturation. The discussion around separate human cutaneous melanocyte compartments (hair follicule and interfollicular glabrous skin) which underlies clinical differences between SV and NSV (Part 2.3) is probably relevant for the analysis of the various vitiligo-related phenotypes in monogenic disorders. Along the same line, the heritability of premature greying genes has been already addressed in some studies [4, 5], but the definitive evidence of a link between the two phenotypes (premature greying and vitiligo) is still missing.

Some clearly distinct monogenic disorders (Part 1.2.1) have marked phenotypic overlap with vitiligo, which raise the question of a common pathogenic background. Piebaldism is an autosomal dominant disorder of melanocyte development characterised by ventral leucoderma/trichia due to *KIT* mutations which impair melanocyte precursor's migration. Given the importance of its encoded protein c-Kit in the maintenance of viable melanocytes (Chap. 2.2.8), the hypothesis of a direct genetic link to common vitiligo had to be raised, but was excluded in genome wide scans. An uncommon variant, the Val620Ala (1859T > C) mutation of the *KIT* gene has been found in the so-called progressive piebaldism leading to progressive loss of pigmentation as well as the progressive appearance of hyperpigmented macules also noted in trichrome vitiligo [6]. Another phenotype, which is clinically related to vitiligo has been observed in a large Canadian family. Vitiligoid patches occurred at adolescence and progressed towards diffuse depigmentation. This 'autosomal dominant vitiligo' (Chap. 2.2.1) is associated with a heterozygous −639G-T transversion identified in a highly conserved area in the promoter region of the *FOXD3* gene [7]. Functional expression studies of the gene variant indicated that it increased transcription in neural crest melanoblast precursors, possibly altering the differentiation profile of the melanoblast lineage.

Another area of interest in monogenic diseases, which include organ specific autoimmunity, is the use of the monogenic disease as a model to identify autoantigens of importance in other disorders, such as vitiligo. APECED/APS1 is a relevant example. The gene causing APECED has been named autoimmune regulator (*AIRE*). It is predominantly expressed in some cells of the immune system and is thought to be involved in transcriptional regulation. Autoantibodies seem to be able to enter target cells and neutralise enzymatic activities. Using sera of a subset of APECED patients with a mixed phenotype of alopecia areata and vitiligo, melanocytes of hair follicle and epidermis were specifically immunostained, but a causal relationship to depigmentation could not be ascertained [8]. However, such an immunoreactivity to hair follicle and epidermal melanocytes has already been shown in common vitiligo [9]. APECED patients have auto-antibodies against several enzymes involved in the biosynthesis of neurotransmitters. Auto-antibodies against aromatic l-amino acid decarboxylase (AADC) in autoimmune polyendocrine syndromes (APS) I patients are associated with the presence of auto-immune hepatitis and vitiligo [10]. Of related interest because of the extensive work already done on the pterin cycle and H_2O_2 accumulation in vitiligo [11], tetrahydrobiopterin-dependent hydroxylases are particularly targeted. They consist of the three highly homologous enzymes: tryptophan hydroxylase (TPH), tyrosine hydroxylase (TH), and phenylalanine hydroxylase (PAH). All three enzymes catalyze the hydroxylation of amino acids and depend on a tetrahydropteridine as a cofactor. They all have central roles in the biosynthesis of the neurotransmitters, serotonin and dopamine (Fig. 1.3.11.1).

TPH catalyzes the hydroxylation of tryptophan into 5-OH tryptophan and is the rate limiting enzyme in the

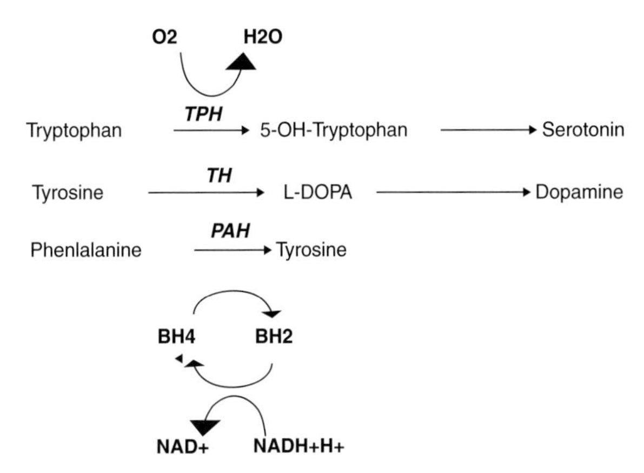

Fig. 1.3.11.1 Pterin-dependent hydroxylases: the role of TPH, TH, and PAH in the biosynthesis of serotonin and dopamine. All three enzymes are dependent on tetrahydrobiopterin (BH_4) as a cofactor. *NAD* nicotineamide adenine dinucleotide. From [12]

synthesis of serotonin. TH is the rate-limiting enzyme in the biosynthesis of catecholamines, where it converts tyrosine into L-DOPA. PAH is mainly expressed in the liver, where it catalyzes the conversion of phenylalanine into tyrosine. Mutations in PAH are the most common defect responsible for phenylketonuria, which is characterised by hair and skin pigment dilution. However, against this hypothesis, the vitiligo phenotype is not clearly associated in APECED patients with one subset of antibodies to tetrahydrobiopterin-dependent hydroxylases [12] and these have not been tested in common NSV or its auto-immune subsets (Sect. 2.2.7.3). Similar to glutamic acid decarboxylase and AADC, which are autoantigens in APECED, as well as enzymes of great importance in the synthesis of GABA, serotonin, and dopamine, it remains possible that the group of tetrahydrobiopterin-dependent hydroxylases have shared immunogenic properties or may play a role in the pathogenesis of vitiligo as in APECED.

1.3.11.2 Discussion of Some Selected Monogenic Disorders

Autoimmune Polyendocrine Syndromes

The APS, also known as polyglandular autoimmune syndromes (PGA), are associated with an increased incidence of NSV. The APS are clinically and genetically heterogeneous, but they reflect multiple endocrine gland insufficiency caused by mainly humoral autoimmunity. APS-1 and APS-2 are the two major APS. The APS-1/APECED gene *AIRE* is on 21q. The gene for Schmidt's syndrome or APS-2 is not yet cloned. According to linkage studies (see Chap. 2.2.1), the *AIRE* gene is not associated with vitiligo coexisting with other autoimmune diseases in European-descent families.[1] This does not exclude some common pathophysiological background (see earlier discussion). The clinical and genetic delineation of so-called APS 3 or 4 being not clear currently (Chap. 1.3.8), these syndromes (which do not include Addison's disease within their clinical spectrum) are not be covered in this chapter.

[1] Recently however, the haplotype AIRE CGCC has been associated with vitiligo in a case-control study [13].

APS1/APECED

Autoimmune polyglandular syndrome Type I or APECED is characterised by the presence of two of the following major clinical symptoms: Addison's disease, hypoparathyroidism, chronic mucocutaneous candidiasis [39]. The spectrum of associated minor clinical diseases includes other autoimmune endocrinopathies (hypergonadotropic hypogonadism, insulin-dependent diabetes mellitus, autoimmune thyroid diseases, and pituitary defects), autoimmune or immune-mediated gastrointestinal diseases (chronic atrophic gastritis, pernicious anaemia, and malabsorption), chronic active hepatitis, skin diseases (vitiligo and alopecia areata), ectodermal dystrophy including enamel and nail defects, keratoconjunctivitis, immunologic defects (cellular and humoral), asplenia, and cholelithiasis. The first manifestations occur commonly in childhood with the three main diseases developing in the first 20 years of life, but other accompanying diseases continue to appear until at least the fifth decade. In a majority of cases, candidiasis is the first clinical manifestation to appear, usually before the age of 5 years, followed by hypoparathyroidism (usually before the age of 10 years), and later by Addison disease (usually before the age of 15 years). Overall, the three main components of APECED occur in chronological order, but they are present together in only about one-third to one-half of the cases. Generally, the earlier the first components appear, the more likely it is that multiple components will develop; conversely, patients who have late manifestations of the disease are likely to have fewer components [14]. Vitiligo is more frequent in the APECED population, but had until recently not been analyzed comprehensively to bring specific and useful information in the vitiligo field. The analysis of autoantibodies in APS-I patients is a useful tool for establishing autoimmune manifestations of the disease as well as providing diagnosis in patients with suspected disease. For Addison's disease, antibodies to 21-hydroxylase– present in adrenal cortex are commonly found. Humoral immunity to steroidogenic P450 cytochromes is also common, and antibody to P450 1A2 might be a hepatic marker autoantigen for patients with APS. For diabetes, autoantibodies against the enzyme AADC of pancreatic beta-cells are good, but low-sensitivity markers. For alopecia areata, autoantibodies against TH correlate with the presence of the phenotype [8]. Antibodies to TPH are associated with intestinal dysfunction. For hypoparathyroidism, the most common autoimmune endocrinopathy of APECED, autoantibodies specific for NACHT leucine-rich repeat protein-5 (NALP5) have been shown in 49% of patients with APS1 and hypoparathyroidism [15]. NALP5 which is predominantly expressed in the cytoplasm of parathyroid chief cells appears to generate tissue-specific autoantibodies. Its role is under close investigation because of a suspected role of the innate immune system in triggering autoinflammation and autoimmunity [16]. NALP proteins, also known as NLRPs, belong to the CATERPILLER protein family involved, like Toll-like receptors, in the recognition of microbial molecules and the subsequent activation of inflammatory and immune responses. Current advances in the function of NALPs support the recently proposed model of a disease continuum bridging autoimmune and autoinflammatory disorders. In APECED, an infectious aetiology was already suggested by Kunin et al. [17], who pointed out that hepatitis had occurred in a number of these cases before the development of endocrinopathy. Since NALP1 variants are associated with autoimmune vitiligo [18], this could be relevant to the vitiligo field. Indeed, the genotype–phenotype correlations for APECED are not well-understood. The only established association between the APECED phenotype and the AIRE genotype is the higher prevalence of candidiasis in the patients with the most common mutation, arg257 to ter, than in those with other mutations. The modifying effect of HLA genes on APECED has been shown for Addison's disease, alopecia areata, and Type I diabetes [19].

Schmidt (APS2) Syndrome

Schmidt syndrome is characterised by the presence of autoimmune Addison disease in association with either autoimmune thyroid disease or Type I diabetes mellitus, or both [14]. Chronic candidiasis is not present. APS2 may occur at any age and in both sexes, but is most common in middle-aged females and is very rare in childhood. Over half of the patients present initially with insulin-dependent diabetes mellitus. The effects on endocrine glands extend to gonadal atrophy and hypoparathyroidism. Alopecia areata, vitiligo, myasthenia gravis, pernicious anaemia, and Graves disease are frequently associated. Schmidt syndrome can be more specifically associated with interstitial myositis [20]. An association of HLA-B8 with APS2 has been found in three generations of a family [21].

Immunodysregulation Polyendocrinopathy and Enteropathy X-Linked (IPEX; 304790)

Interestingly, this more recently described X-linked recessive disease due to mutations in *FOXP3* which block the development of regulatory T-cells has not yet been linked to vitiligo, but induces multiple immune-related diseases, including TH2 predominant (atopic dermatitis) and TH1 predominant (alopecia areata, Type I diabetes) phenotypes. Further observations are necessary to exclude the alteration of this important pathway in the pathogenesis of immune vitiligo.

Mitochondrial Disorders

MELAS (Myopathy, Encephalopathy, Lactic Acidosis, and Stroke-Like Episodes) and MERRF (Myoclonic Epilepsy Associated with Ragged-Red Fibers) Syndromes

These disorder belongs to the mitochondrial encephalomyopathies which include Leigh syndrome (LS; 256000), Kearns-Sayre syndrome (KSS; 530000), and Leber optic atrophy (535000). They are grouped because phenotypes characterised by features of both MERFF (in which myoclonus and ataxia predominate) and MELAS have been described [22]. The most common clinical manifestation in these disorders is encephalomyopathy, because the nerve and muscle cells depend more heavily on aerobic energy production. However, any tissue or organ may be affected.

The most frequent MELAS symptom is episodic sudden headache with vomiting and convulsions, which is found in 80% of cases in patients aged 5–15 years. Progressive encephalopathy and dementia may follow depending on the severity of the phenotype. There is frequently laboratory evidence of myopathy related to abnormal mitochondrial metabolism (elevated resting serum lactate increased with exercise, ragged-red fibres on muscle biopsy, subsarcolemmal pleomorphic mitochondria on electron microscopy); clinically, there is commonly evidence of diabetes, sensorineural hearing/visual loss, and retinal pigment epithelial cell involvement [23]. The common mitochondrial causative mutations are mostly found in the muscle mtDNA molecules, but can be found in other cell type responsible for the main symptoms. The variability of the clinical phenotype is partly due to mutation heteroplasmy, a condition where the normal genome and the mutant variant coexist in mitochondria; for example, the relative amount of mutant mtDNA in muscle correlate with the severity of the clinical presentation in familial observations [22]. An A to G transition at base pair 3,243 in the $tRNA^{Leu(UUR)}$ gene in mtDNA is the most common molecular aetiology of the MELAS syndrome, but other mutations have been described (Table 1.3.11.1).

Vitiligo was associated in 11% of cases of MELAS bearing the common bp 3,243 point mutation [24]. Interestingly, there is no reported evidence of melanocyte loss in this form of vitiligo, but only decreased melanogenesis. Unfortunately, the melanocyte marker used in that study was the S100 protein which stains also Langerhans cells. The mitochondrial mutation was not studied in epidermal or skin cells [24]. In skeletal muscle, the biochemical defect is often segmental [25], suggesting a nonrandom distribution of mutant and wildtype mtDNAs within a muscle cell. A decreased activity of the mitochondrial respiratory chain may lead to a metabolic impairment within the epidermal melanin unit, which is consistent with previous hypotheses concerning the altered redox status in vitiligo [24] (Chap. 2.2.6). For MERRF syndrome, the mutations produces multiple deficiencies in the enzyme complexes of the respiratory chain, most prominently involving NADH-CoQ reductase (complex I) in cytochrome c oxidase (COX) (complex IV), consistent with a defect in translation of all mtDNA-encoded genes. Another possible molecular target is the transport of melanosomes via microtubules, which requires adenosine triphosphate (ATP). This is a possibility since all the mutations in mtDNA affect the respiratory chain and oxidative phosphorylation and may thus lead to a decreased amount of ATP in the cell. Mutations of mtDNA accumulate during normal aging. The most frequent mutation is a deletion, which is increased in photoaged skin. Oxidative stress in turn may play a major role in the generation of large scale mtDNA deletions. Reactive oxygen species (ROS) have been shown to be involved in the generation of aging-associated mtDNA lesions in human cells [26].

Breakage Disorders: Ataxia Telangiectasia and Nijmegen Breakage Syndrome

Ataxia-Telangiectasia (AT)

Patients present in early childhood with progressive cerebellar ataxia and usually later develop conjunctival telangiectases, progressive neurologic degeneration, sinopulmonary infections, and mostly lymphoid malignancies. In general, lymphomas in AT patients tend to be of B-cell origin, whereas the leukemias tend to be of the T-cell type. Oculocutaneous telangiectases typically develop between 3 and 5 years of age (Fig. 1.3.11.2). Vitiligo is a feature already mentioned, [27, 28] but not emphasised in the context of major neurological problems. Elevated levels of alpha-fetoprotein and carcinoembryonic antigen are the most available markers for confirmation of the diagnosis of AT. Specific functional markers however derive from AT cells being abnormally sensitive to killing by ionising radiation, and abnormally resistant to inhibition of DNA synthesis by ionising radiation. Patients with AT and their cultured cells are unusually sensitive to X-ray just as patients and cells with xeroderma pigmentosum are sensitive to ultraviolet. Treatment of malignancy with conventional dosages of radiation can be fatal to AT patients. It has been shown that radiosensitivity is a feature of vitiligo, interpreted as a manifestation of Koebner's phenomenon [29]. The ionising radiation sensitivity of AT cells has been used to identify complementation groups which correspond to intragenic complementation within the *ATM* gene located at 11q22-q23. Ataxia telangiectasia mutated (ATM), the product of the mutated gene, is a 370-kDa protein member of the phospatidyl inositol 3-kinases superfamily. The *ATM* gene itself is not involved in site-specific rearrangements in either lymphocytes or fibroblasts. Lymphocytes show characteristic rearrangements involving the site of the T-cell receptor genes and immunoglobulin heavy chain genes, respectively, on Chromosomes 7 and 14. Increased recombination is a component of genetic instability in AT and may contribute to the cancer risk. The mutations of the *ATM* gene are quite prevalent in the population (2.8%). Interestingly, for a recessive condition, heterozyrous carriers of a mutant AT allele, which are known to be at increased risk of

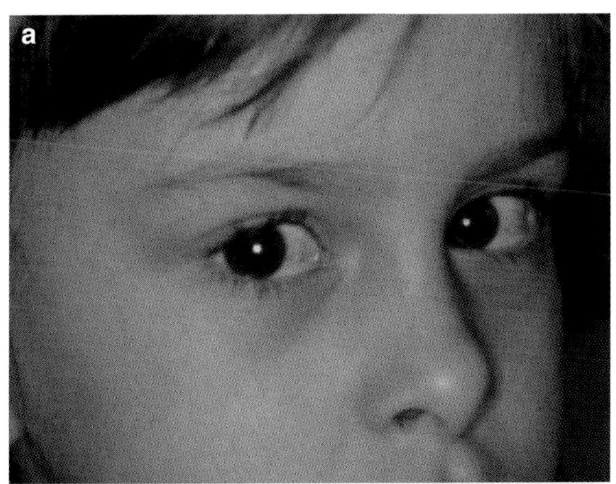

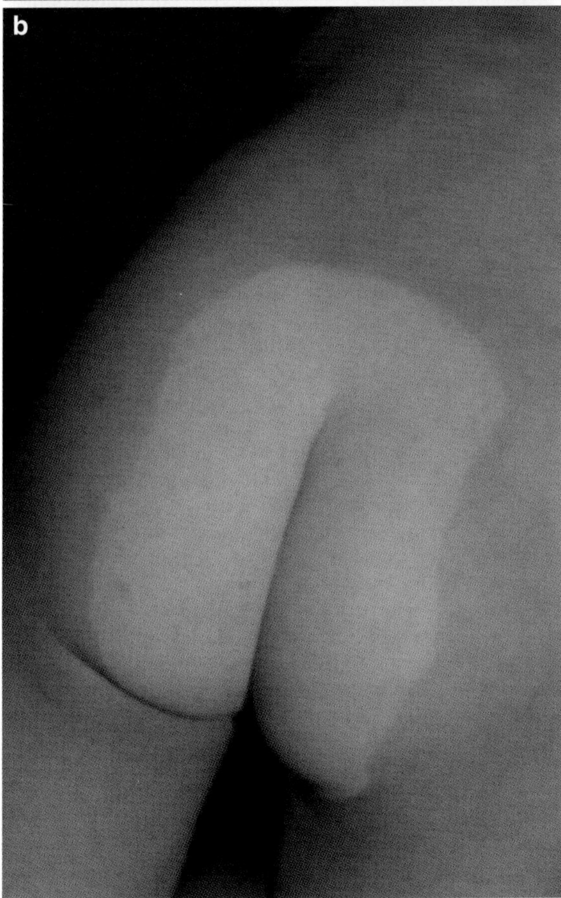

Fig. 1.3.11.2 Ataxia-telangiectasia and vitiligo: equatorial ocular telangiectasias (**a**) and vitiligo starting in the napkin area, indicating a possible role of the Koebner phenomenon (**b**)

cancer, have a phenotype that distinguishes them from normal control individuals in lymphoblastoid cells, with changes in expression level of many genes [30] at baseline and in response to ionising radiation. The phenotypic pleiotropy of AT may result from different tissues expression of ATM targets and/or different complementation by ATM family members whose functions may overlap/replace ATM. Identified targets of ATM protein include ABL, replication protein A, p53, and beta-adaptin. These targets are located in the nucleus and in the cytoplasm. ATM is most likely involved in several distinct signalling pathways. ATM plays an important role in the activation of the tumour suppressor gene product p53. It has been called the p53 booster. In the thymus, p53 is phosphorylated directly by ATM after ionising radiation, probably in the nucleus, leading to transcriptional activation of p21 and consequential cell cycle arrest. In the absence of ATM, this pathway is disrupted, and this defect perhaps results in the immunodeficiency and abnormal cellular responses to ionising radiation seen in patients with AT. In the skin, this may precipitate the apoptotic commitment of melanocytes in AT-related vitiligo [28]. ATM interactions with beta-adaptin in the cytoplasm might mediate axonal transport and vesicle trafficking in the central nervous system and so account for the neuronal dysfunction and eventual neurodegeneration seen in ataxia-telangiectasia. The previous pathways may be interesting for the immune and neural theories of vitiligo, but of considerable interest is that the AT phenotype seems to be a consequence, at least in part, of an inability to respond appropriately to oxidative damage [31] and that the absence of the ATM protein affects stem cells [32]. Ionising radiation oxidises macromolecules and causes tissue damage through the generation of ROS. When compared to normal human fibroblasts, AT dermal fibroblasts exhibit increased sensitivity to *t*-butyl hydroperoxide toxicity. These cells fail to show G1 to G2 phase checkpoint functions or to induce p53 in response to oxidative challenge [33]. ATM-deficient mice are viable and display pleiotropic defects as observed in AT patients including growth retardation, neurologic dysfunction, infertility, thymic atrophy with decreased T-cell population, chromosomal instability, and extreme sensitivity to ionising radiation [34, 35]. Cells derived from ATM-deficient mice grow poorly in culture, are genetically instable, and appear to undergo apoptosis more readily than control cells [36]. In a mouse AT model, premature greying of hair has been observed in heterozygous mice after a sublethal dose of ionising radiation [34, 37]. Furthermore, ATM protein deficiency and telomere dysfunction seem to act together to impair cellular viability and cause adverse effects on stem/progenitor cell reserves [32].

Nijmegen Breakage Syndrome (NBS)

This syndrome has been related to AT as a variant until more recent clear clinical and molecular evidence of a distinct aetiology [38]. The gene product, nibrin, is member of the hMre11/hRad50 protein complex, suggesting that the gene is involved in DNA double strand break repair. The immunological, cytogenetic, and cell biological findings in NBS closely resemble those described in AT (Table 1.3.11.1). For the vitiligo phenotype, it is better documented (14/21 cases studied in the international NBS study group). Telangiectases are rare and *café au lait* spots common. Concerning the nonskin clinical aspects, however, NBS is more similar to Bloom syndrome (BS), which features also severe microcephaly with relatively preserved mental development. Unlike AT, neurological features are rare. NBS and BS lack the increased serum alpha fetoprotein concentrations of AT. BS shows a characteristic cytogenetic feature, the sister chromatid exchanges, which is not seen in NBS and AT.

Studies on the gene product suggest that deficiency in nibrin disrupts a common pathway that functions to sense or repair double stranded DNA breaks. DNA double strand breaks occur often, but seldom lead to aberrations in case of a normal functioning repair mechanism. In the case of dysfunction of repair mechanisms, however, serious aberrations can be expected, particularly in tissues with high proliferative capacity. This does not apply to skin melanocytes, and clues to vitiligo pathogenesis need further specific studies in this disorder.

References

1. Palmer CN, Irvine AD, Terron-Kwiatkowski A et al (2006) Common loss-of-function variants of the epidermal barrier protein filaggrin are a major predisposing factor for atopic dermatitis. Nat Genet 38:441–446
2. Blumen SC, Bevan SN, Abu-Mouch S et al (2003) A locus for complicated hereditary spastic paraplegia maps to chromosome 1q24-q32. Ann Neurol 54:796–803
3. Kulkarni ML, Baskar K, Kulkarni PM (2007) A syndrome of immunodeficiency, autoimmunity, and spondylometaphyseal dysplasia. Am J Med Genet 143A:69–75
4. Halder RM, Grimes PE, Cowan CA et al (1987) Childhood vitiligo. J Am Acad Dermatol 16:948–954
5. Taïeb A, Picardo M; VETF Members (2007) The definition and assessment of vitiligo: a consensus report of the Vitiligo European Task Force. Pigment Cell Res 20:27–35
6. Richards KA, Fukai K, Oiso N et al (2001) A novel KIT mutation results in piebaldism with progressive depigmentation. J Am Acad Dermatol 44:288–292
7. Alkhateeb A, Fain PR, Spritz RA (2005) Candidate functional promoter variant in the FOXD3 melanoblast developmental regulator gene in autosomal dominant vitiligo. J Invest Derm 125:388–391
8. Hedstrand H, Perheentupa J, Ekwall O et al (1999) Antibodies against hair follicles are associated with alopecia totalis in autoimmune polyendocrine syndrome type I. J Invest Dermatol 113:1054–1058
9. Tobin DJ, Bystryn JC (1996) Different populations of melanocytes are present in hair follicles and epidermis. Pigment Cell Res 9:304–310
10. Husebye ES, Gebre-Medhin G, Tuomi TM et al (1997) Autoantibodies against aromatic l-amino acid decarboxylase in autoimmune polyendocrine syndrome type I. J Clin Endocrinol Metab 82:147–150
11. Hasse S, Gibbons NC, Rokos H et al (2004) Perturbed 6-tetrahydrobiopterin recycling via decreased dihydropteridine reductase in vitiligo: more evidence for H2O2 stress. J Invest Dermatol 122:307–313
12. Ekwall O, Hedstrand H, Haavik J et al (2000) Pteridin-dependent hydroxylases as autoantigens in autoimmune polyendocrine syndrome type I. J Clin Endocrinol Metab 85:2944–2950
13. Tazi-Ahnini R, McDonagh AJ, Wengraf DA, Lovewell TR, Vasilopoulos Y, MessengerAG, Cork MJ, Gawkrodger DJ (2008) The autoimmune regulator gene (AIRE) is strongly associated with vitiligo. Br J Dermatol 159:591–596
14. Betterle C, Dal Pra C, Mantero F, Zanchetta R (2002) Autoimmune adrenal insufficiency and autoimmune polyendocrine syndromes: autoantibodies, autoantigens, and their applicability in diagnosis and disease prediction. Endocr Rev 23:327–364
15. Alimohammadi M, Björklund P, Hallgren A et al (2008) Autoimmune polyendocrine syndrome type 1 and NALP5, a parathyroid autoantigen. N Engl J Med 358:1018–1028
16. Taieb A (2007) NALP1 and the inflammasomes: challenging our perception of vitiligo and vitiligo-related autoimmune disorders. Pigment Cell Res 20:260–262
17. Kunin AS, MacKay BR, Burns S et al (1963) The syndrome of hypoparathyroidism and adrenocortical insufficiency, a possible sequel of hepatitis. Am J Med 34:856–865
18. Jin Y, Mailloux CM, Gowan K, Riccardi SL, LaBerge G, Bennett DC, Fain PR, Spritz RA (2007) NALP1 in vitiligo-associated multiple autoimmune disease. N Engl J Med 356:1216–1225
19. Halonen M, Eskelin P, Myhre AG et al (2002) AIRE mutations and human leukocyte antigen genotypes as determinants of the autoimmune polyendocrinopathy-candidiasis-ectodermal dystrophy phenotype. J Clin Endocrinol Metab 87:2568–2574
20. Heuss D, Engelhardt A, Gobel H et al (1995) Myopathological findings in interstitial myositis in type II polyendocrine autoimmune syndrome (Schmidt's syndrome). Neurol Res 17:233–237

21. Eisenbarth GS, Wilson PW, Ward F et al (1978) The polyglandular failure syndrome: disease inheritance, HLA type, and immune function: studies in patients and families. Ann Intern Med 91:528–533
22. Zeviani M, Muntoni F, Savarese N et al (1993) A MERRF/MELAS overlap syndrome associated with a new point mutation in the mitochondrial DNA tRNA(Lys) gene. Eur J Hum Genet 1:80–87
23. Latkany P, Ciulla TA, Cacchillo PF et al (1999) Mitochondrial maculopathy: geographic atrophy of the macula in the MELAS associated A to G 3243 mitochondrial DNA point mutation. Am J Ophthalmol 128:112–114
24. Karvonen SL, Haapasaari KM, Kallioinen M et al (1999) Increased prevalence of vitiligo, but no evidence of premature ageing, in the skin of patients with bp 243 mutation in mitochondrial DNA in the mitochondrial encephalomyopathy, lactic acidosis and stroke-like episodes syndrome (MELAS). Br J Dermatol 140:634–639
25. Matsuoka T, Goto Y, Yoneda M et al (1991) Muscle histopathology in myoclonus epilepsy with ragged-red fibers (MERRF). J Neurol Sci 106:193–198
26. Berneburg M, Grether-Beck S, Kürten V et al (1999) Singlet oxygen mediates the UVA-induced generation of the photoaging-associated mitochondrial common deletion. J Biol Chem 274:15345–15349
27. Cohen LE, Tanner DJ, Schaefer HG et al (1984) Common and uncommon cutaneous findings in patients with ataxia-telangiectasia. J Am Acad Dermatol 10:431–438
28. Taieb A (2000) Intrinsic and extrinsic pathomechanisms in vitiligo. Pigment Cell Res 13:41–47
29. Kim DH, Kim CW, Kim TY (1999) Vitiligo at the site of radiotherapy for malignant thymoma. Acta Derm Venereol 79:497
30. Watts, JA, Morley M, Burdick JT et al (2002) Gene expression phenotype in heterozygous carriers of ataxia telangiectasia. Am J Hum Genet 71:791–800
31. Barlow C, Dennery PA, Shigenaga MK et al (1999) Loss of the ataxia-telangiectasia gene product causes oxidative damage in target organs. Proc Nat Acad Sci 96: 9915–9919
32. Wong KK, Maser RS, Bachoo RM et al (2003) Telomere dysfunction and Atm deficiency compromises organ homeostasis and accelerates ageing. Nature 421:643–648
33. Shackelford RE, Innes CL, Sieber SO et al (2001) The ataxia telangiectasia gene product is required for oxidative stress-induced G1 and G2 checkpoint function in human fibroblasts. J Biol Chem 276:21951–21959
34. Barlow C, Hirotsune S, Paylor R et al (1996) Atm-deficient mice: a paradigm of ataxia telangiectasia. Cell 86: 159–171
35. Elson A, Wang Y, Daugherty CJ et al (1996) Pleiotropic defects in ataxia-telangiectasia protein-deficient mice. Proc Nat Acad Sci 93:13084–13089
36. Jung M, Zhang Y, Lee S, Dritschilo A (1995) Correction of radiation sensitivity in ataxia telangiectasia cells by a truncated I-kappa-B-alpha. Science 268:1619–1621
37. Barlow C, Eckhaus MA, Schaffer AA et al (1999) Atm haploinsufficiency results in increased sensitivity to sublethal doses of ionizing radiation in mice. Nature Genet 21: 359–360
38. The International Nijmegen Breakage Syndrome Study Group (2000) Nijmegen breakage syndrome. Arch Dis Child 82:400–406
39. Eisenbarth GS, Gottlieb PA (2004) Autoimmune polyendocrine syndromes. New Engl J Med 350:2068–2079

Vitiligo in Childhood

1.3.12

Juliette Mazereeuw-Hautier and Alain Taïeb

Contents

J. Mazereeuw-Hautier (✉)
Service de Dermatologie, CHU Purpan, Toulouse, France
e-mail: mazereeuw-hautier.j@chu-toulouse.fr

Core Messages

- Childhood vitiligo differs from adult onset vitiligo in several features, including more segmental forms, higher prevalence of halo nevi, and more common family history for autoimmune diseases.
- The major differential diagnoses are the post-inflammatory hypomelanoses for non-segmental vitiligo (NSV) and nevus depigmentosus for segmental vitiligo (SV).
- Thyroid anomalies are detected in NSV only, and actually
- The course of childhood vitiligo is not well-known, since there is a lack of long-term follow-up studies.

1.3.12.1 Introduction

From the diagnostic standpoint, the diagnosis of vitiligo in childhood is easy in most instances, but specific age-related pitfalls exist (Chaps. 1.2.1 and 1.3.11). From the therapeutic standpoint, early awareness of the diagnosis seems to correlate with a good treatment outcome in this age group, even if there is no evidence available from well-controlled studies to support this statement. Some clinical features of vitiligo of childhood onset, which differ from adult onset cases have been recorded [2, 3, 17, 18]. They are detailed in this chapter.

M. Picardo and A. Taïeb (eds.), *Vitiligo*,
DOI 10.1007/978-3-540-69361-1_1.3.12, © Springer-Verlag Berlin Heidelberg 2010

1.3.12.2 Epidemiology

The exact prevalence of vitiligo in the paediatric age group is unknown, but many studies state that most cases of vitiligo are acquired early in life. Howitz et al. in Denmark found that approximately 25% of the patients noted the onset of vitiligo before the age of 10 years [6]. The mean age of onset varied among different studies from 4 to 8 years [2, 5, 7, 17]. Vitiligo can occur in infants as young as three months. There are reports of congenital vitiligo, although it is not clear whether these reports concern piebaldism (Chap. 1.2.1) or true 'congenital vitiligo' [14]. In fair-skinned individuals, vitiligo patches are usually detected after exposure of the skin to the sun in response increases the contrast with the unaffected skin, which tans normally. In most reported paediatric case series, girls predominate [3, 5, 9, 13, 17], but population-based studies do not confirm a sex bias (Chap. 1.2.1). The higher prevalence noted in India is difficult to explain, but a bias is possible due to the considerable social attention given to the disorder [4].

1.3.12.3 Classification

As in adult onset vitiligo, the most common form of vitiligo in children is the non-segmental type (NSV). Nevertheless, segmental vitiligo (SV) is seen more frequently in children compared to adults [2–4] (Chap. 1.3.2). This difference is difficult to explain for a non-lethal disease, but the recent detection of mixed SV–NSV cases might be part of the explanation (Chap. 1.3.2). The prevalence of SV in childhood varies in the different series from 4.6 to 32.5% [2, 3, 5, 9, 10, 13, 15, 16, 18].

From the series of literature (Table 1.3.12.1), it is difficult to clearly distinguish the clinical characteristics of NSV and SV, since all but one study does not give separate descriptions. These series are descriptive or compare childhood vitiligo to adult vitiligo. The only series comparing NSV to SV found that NSV comprises more plaques, a larger body surface area, more Koebner phenomenon, and more rapid progression of the disease. Furthermore, thyroid anomalies were only seen in NSV [13].

1.3.12.4 Familial Background

Children with vitiligo may have a positive family history for autoimmune diseases. The family incidence found in the different studies varies from 3.3 to 27.3% [2, 7, 9, 15]. Halder et al. demonstrated that a family history was more frequently reported in children with vitiligo, as compared to adults with vitiligo [3]. Pajvani et al. [16] found that children with vitiligo and an extended family history of vitiligo were more likely to have an earlier age of onset of the disease than those with a negative family history. We found a similar percentage of familial autoimmune diseases in children with SV and NSV [13].

Table 1.3.12.1 Reported series of childhood vitiligo

Author	Year	Number of patients	Country	Percentage of segmental form
Halder et al. [3]	1987	82	USA	28
Jaisankar et al. [9]	1992	346	India	21.1
Cho et al. [2]	2000	80	Korea	32.5
Prcic et al. [17]	2002	50	Croatia	ND
Handa and Dogra [5]	2003	625	India	4.6
Kurtev and Dourmishev [11]	2004	61	Bulgaria	ND
Kakourou et al. [10]	2005	54	Greece	14.8
Iacovelli et al. [8]	2005	121	Italy	ND
Al-Mutairi et al. [1]	2005	88	Australia	ND
Hu et al. [7]	2006	541	China	14.4
Pajvani et al. [16]	2006	137	USA	16.1
Prcic et al. [18]	2007	50	Croatia	18
Pagovich et al. [15]	2008	67	USA	20.9
Mazereeuw et al. [13]	2008	114	France	22

Premature diffuse hair greying has also been found to be increased in first- and second-degree relatives of children with vitiligo, as compared to adults with vitiligo and children without vitiligo [3].

1.3.12.5 Clinical Characteristics

SV is characterized by unilateral lesions that usually do not cross the midline, often in a dermatomal distribution (see discussion in Part 2.3). The most common location is the face (Fig. 1.3.12.1a), followed by the trunk, neck, and limbs [13]. The face is also the most common location of NSV (Fig. 1.3.12.1b) [2, 5, 9, 16]—especially around the eyes—and neck, followed by the lower limbs, trunk, neck, and upper limbs. A burning sensation of the eyelids can be a mode of onset of NSV in children of fair complexion with outdoor activities. In NSV, the back is relatively spared and the lesions are confined to the sacrum or a few areas over the spines of the vertebrae. The back is also usually one of the last areas to exhibit vitiligo, in contrast to the abdomen which is a common site for the early appearance of lesions. Involvement of the perineum and in particular, perianal and buttocks skin is a common onset location of NSV in toddlers, suggesting a role of the Koebner's phenomenon triggered by nappies and hygiene care (Fig. 1.3.12.1c).

The skin on the scalp can also be involved in NSV and SV, leading sometimes to a lock of white hair (to differentiate from piebaldism, see Chap. 1.3.6), the prevalence of scalp involvement being comprised between 12.3 and 19.3% [5, 13, 18], and including probably some cases of halo nevi, more difficult to detect in this location. The percentage of scalp involvement in SV and NSV seem to be comparable [13]. Independently of interfollicular skin involvement, premature diffuse greying of the hair is sometimes observed; Jaisankar et al. noted this finding in 4.4% of children [9]. There are a few individuals who develop totally white hair of the scalp, eyebrows, and eyelashes.

Mucosal vitiligo is rare in childhood. The lips, oral mucosa, and gingiva can be involved, but less frequently as in adults. In the literature, the involvement of the mucosa varies from 0 to 13.3% [3, 5, 9].

Concerning repigmentation patterns, a hyperpigmented rim was only observed in the group of children with NSV in our series [13].

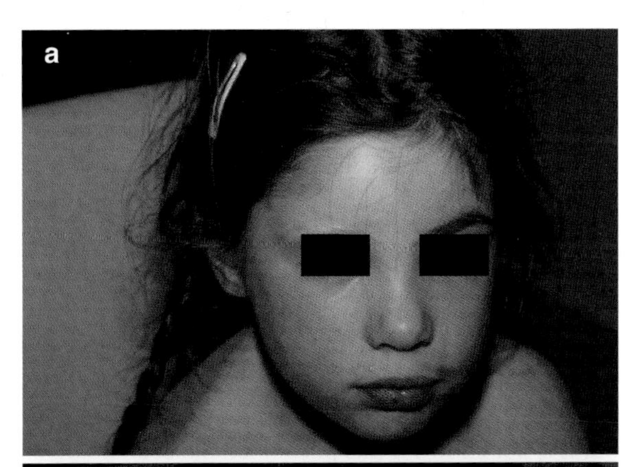

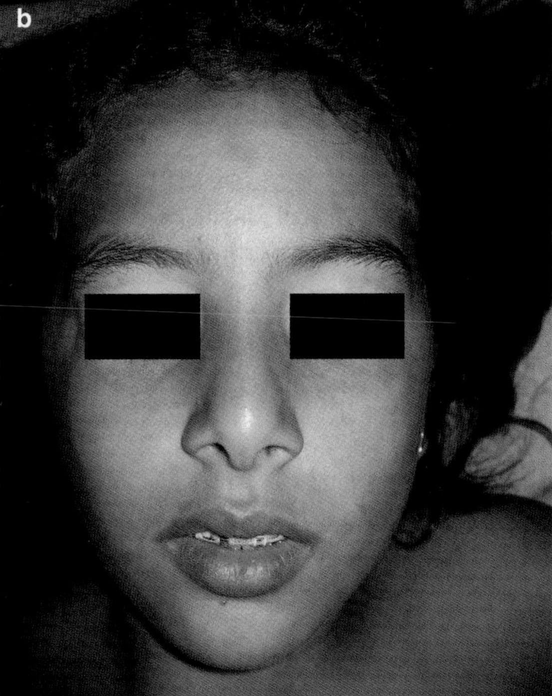

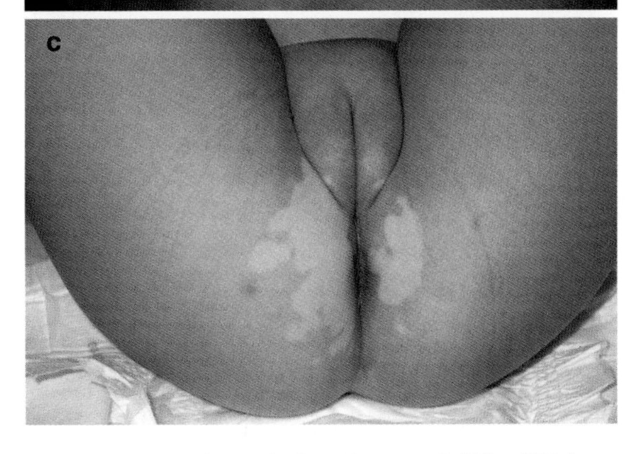

Fig. 1.3.12.1 Location on the face of segmental vitiligo (SV) (courtesy of Dr Plantin) (**a**) or non-segmental vitiligo (NSV) (Courtesy of Dr Lacour) (**b**). NSV located on the buttocks in a toddler (**c**)

1.3.126 Disease Extent and Progression

The extent of lesions varies greatly among patients. Handa and Dogra found that 96.4% of children have less than 20% of the body surface involved , and 89.7% had less than 5% body area involvement [5]. Nevertheless, in general, a larger body surface area is involved in NSV, as compared to SV [13].

Knees, elbows, shins, arms, and hands are often the sites of initial involvement in children. Such locations are frequently scraped and scratched in this age group, but it is not known if this phenomenon occurs more or less commonly than in adults. Only a few studies have recorded the presence of the Koebner's phenomenon in children with vitiligo. Handa and Dogra found it in 11.3% of children [5]. We found that the occurrence of lesions in sites of Koebner's phenomenon was most common in NSV compared to SV (47.2 vs. 24%) [13]. In SV, it was observed only in the involved segment, and was more difficult to assess.

There is a lack of studies on the evolution of childhood vitiligo. In our series, after approximately one year of follow-up, we noted that vitiligo was clinically undetectable in only 5.5% of the patients [13]. As expected, based on the natural history of SV (Chap. 1.3.2), more progression was observed in the group of children with NSV, compared to SV (23.29 vs. 5.56%).

1.3.12.7 Associated Skin Conditions

Vitiligo commonly affects children with a history of atopic dermatitis, but no study has clearly demonstrated this as being a significant statistical association. In this situation, it is necessary to distinguish the depigmented areas of vitiligo from post-inflammatory hypo-pigmentation secondary to eczema (Fig. 1.3.12.2).

Halo nevi can be observed in children with vitiligo and can precede the onset of NSV. However, to which extent halo nevi can be considered as precursors of vitiligo is unknown. Indeed, we often see children with halo naevi who do not go on to develop vitiligo. A precise comparison to the general paediatric population is impossible since the prevalence of halo nevi in not well-known, possibly around 1% (Chap. 1.3.5). Nevertheless, Prcic et al. [18] found that there occurred more halo nevi in children with vitiligo, compared to children without vitiligo (34 vs. 3.3%). In literature, the prevalence of halo nevi in children with vitiligo varies from 2.5 to 34% [2, 5, 7, 18]. It is also unclear whether the prevalence of halo nevi in children with vitiligo is different from that found in adult vitiligo. Prcic found [17, 18]

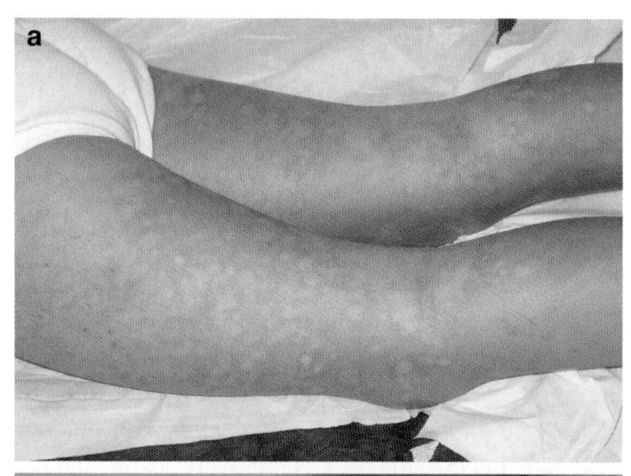

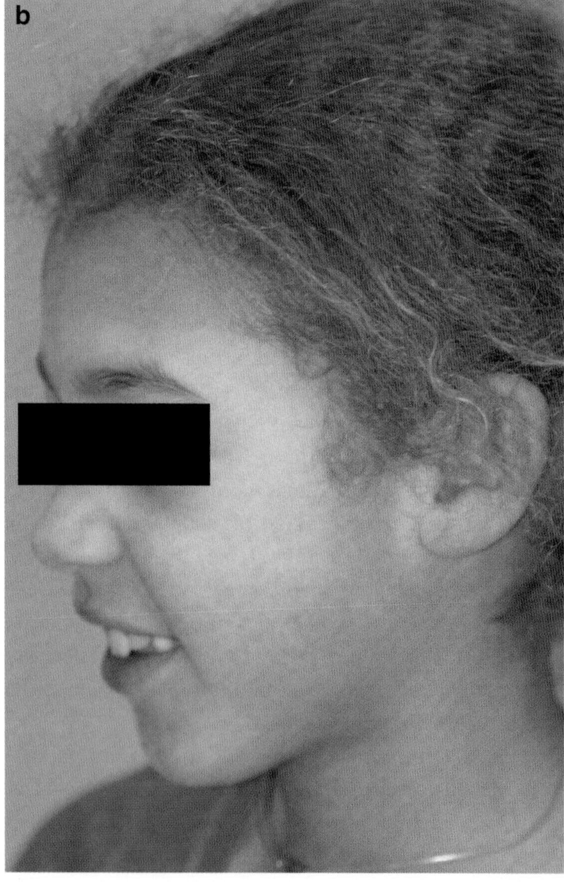

Fig. 1.3.12.2 Post-inflammatory hypopigmentation secondary to eczema. Legs (**a**) with typical flexural dermatitis; Face (**b**), giving a Kabuki mask appearance in a metis girl

that halo nevi were more frequent in childhood vitiligo as compared to adult vitiligo, at variance with data reported earlier by Cho et al. in Korea [2]. We found a non-significant tendency to have more halo nevi in children with NSV versus SV (20 vs. 12%) [13].

1.3.12.8 Associated Autoimmune diseases and Laboratory Investigations

Children suffering from vitiligo are generally healthy. Nevertheless, as in adult vitiligo, other autoimmune diseases can be associated. The most commonly reported autoimmune disease is thyroiditis. The prevalence of thyroid dysfunction in childhood vitiligo is reported with large variations (0–25%) [8, 10, 11, 15, 18]. Some authors have suggested that thyroid dysfunction increases with age. However, Kakourou et al. found no association between thyroid dysfunction and the following parameters: chronological age, age of onset, mean duration, clinical type of non-segmental vitiligo, family history of autoimmunity/thyroid disorder, or sex [10]. In our series, of the 10 affected children out of 114, all but one patient were girls [13]. It seems now established that thyroid anomalies are only more common in NSV [8, 13, 15].

The other autoimmune diseases reported in children with vitiligo include: alopecia areata, diabetes mellitus, Addison's disease isolated or within the autoimmune polyglandular syndrome Type I or APECED (Chap. 1.3.11). The prevalence of these other diseases is very low [2, 5, 7].

Antinuclear antibodies (ANA) may be found in children with vitiligo. This is considered as a marker of the general autoimmune status of the child with vitiligo. Halder et al. [3] found ANAs in 4.8% of children, whereas Cho et al. [2] and Prcic et al. [18] did not find any children testing positively. As for thyroid anomalies, we only found ANA in the group of children with NSV (3.6 vs. 0% in SV) [13].

It seems appropriate to perform a routine initial thyroid screening in NSV which should include anti TPO antibodies and TSH. Kurtev and Dourmishev [11] recommend thyroid function tests to be assessed annually. Other possible laboratory tests suggested include a full blood count, fasting blood sugar, and an autoantibody screen including ANA. However, there is no consensus on this issue in NSV. On the contrary, it makes sense not to test children with SV.

1.3.12.9 Differential Diagnosis

The conditions to exclude are detailed in Chap. 1.2.1. In children, the most common causes of hypomelanoses are the post-inflammatory hypomelanoses for NSV and nevus depigmentosus for SV. Wood's lamp examination is particularly helpful in difficult situations, because the UV-enhancement of vitiligo lesion lacks in both post-inflammatory lesions and nevus depigmentosus. In nevus depigmentosus, there is also a slight ability to tan in the depigmented area and depigmentation is even under normal and Wood's lamp irradiation.

With regard to perineal involvement in NSV, lichen sclerosus is a possible differential diagnosis when the lesions are located to the vulva (Fig. 1.3.12.3a, b). Lichen sclerosus may indeed have a vitiligoid localised variant.

1.3.12.10 Psychological Effects of Childhood Vitiligo

There is a lack of specific studies on psychological effects of vitiligo in children, but reporting negative experiences from childhood vitiligo may influence adult life [12]. Nevertheless, although vitiligo is not a life-threatening disease, it can be a life-altering disease. Children are probably affected differently, depending on the location, extent, and course of the disease, their age, individual capacities, and social environment.

1.3.12.11 Summary of Therapeutic Issues

There are minor differences in the management of childhood versus adult vitiligo due to mostly feasibility and aesthetic demand of treatment (Part 3.12). The major difference comes from the parent's behaviour and coping with the disease, especially in vitiligo families. One important unsolved issue is the possible impact of early aggressive treatment on disease outcome in severe rapidly progressive NSV in children.

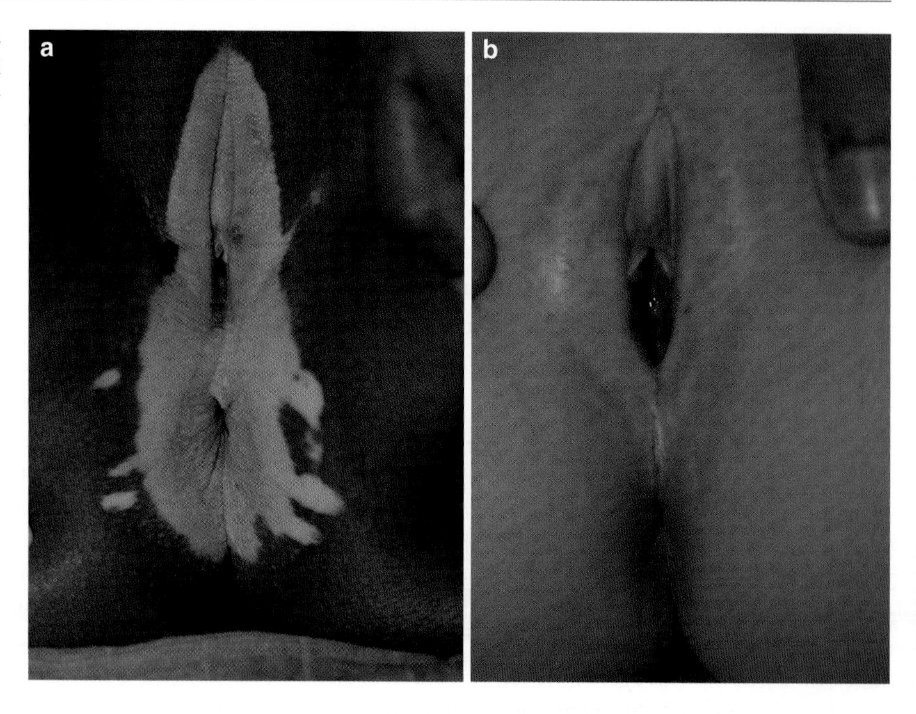

Fig. 1.3.12.3 Vulval hypopigmentation in non segmental vitiligo (**a**) and lichen sclerosus (**b**)

References

1. Al-Mutairi N, Sharma AK, Al-Sheltawy M, Nour-Eldin O (2005) Childhood vitiligo: a prospective hospital-based study. Australas J Dermatol 46:150–153
2. Cho S, Kang HC, Hahn JH (2000) Characteristics of vitiligo in Korean children. Pediatr Dermatol 17:189–193
3. Halder RM, Grimes PE, Cowan CA et al (1987) Childhood vitiligo. J Am Acad Dermatol 16:948–954
4. Handa S, Kaur I (1999) Vitiligo: clinical findings in 1436 patients. J Dermatol 26:653–657
5. Handa S, Dogra S (2003) Epidemiology of childhood vitiligo: a study of 625 patients from north India. Pediatr Dermatol 20:207–210
6. Howitz J, Brodthagen H, Schwartz M, Thomsen K (1997) Prevalence of vitiligo. Epidemiological survey of the Isle of Bornholm, Denmark. Arch Dermatol 113:47–52
7. Hu Z, Liu JB, Ma SS et al (2006) Profile of childhood vitiligo in China: an analysis of 541 patients. Pediatr Dermatol 23:114–116
8. Iacovelli P, Sinagra JL, Vidolin AP et al (2005) Relevance of thyroiditis and of other autoimmune diseases in children with vitiligo. Dermatology 210:26–30
9. Jaisankar TJ, Baruah MC, Garg BR (1992) Vitiligo in children. Int J Dermatol 31:621–623
10. Kakourou T, Kanaka-Gantenbein C, Papadopoulou A et al (2005) Increased prevalence of chronic autoimmune (Hashimoto's) thyroiditis in children and adolescents with vitiligo. J Am Acad Dermatol 53:220–223
11. Kurtev A, Dourmishev AL (2004) Thyroid function and autoimmunity in children and adolescents with vitiligo. J Eur Acad Dermatol Venereol 18:109–111
12. Homan MWL, de Korte J, Grootenhuis MA et al (2008) Impact of childhood vitiligo on adult life. Br J Dermatol1 59:915–920
13. Mazereeuw-Hautier Juliette, Bezio Sophie, Mahe Emmanuel, Bodemer Christine, Eschard Catherine, Viseux Valérie, Labreze Christine, Plantin Patrice, Barbarot Sebastien, Vabres Pierre, Martin Ludovic, Paul Carle, Lacour Jean-Philippe and the Groupe de Recherche Clinique en Dermatologie Pédiatrique (GRCDP). Segmental and non segmental childhood vitiligo have distinct clinical characteristics, a prospective observational study J Am Acad 2009, in press.
14. Nordlund JJ, Lerner AB (1982) Vitiligo. It is important. Arch Dermatol 118:5–8
15. Pagovich OE, Silverberg JI, Freilich E, Silverberg NB (2008) Thyroid anomalies in pediatric patients with vitiligo in New York city. Cutis 81:463–466
16. Pajvani U, Ahmad N, Wiley A et al (2006) Relationship between family medical history and childhood vitiligo. J Am Acad Dermatol 55:238–244
17. Prcic S, Djuran V, Poljacki M (2002) Vitiligo in childhood. Med Pregl 55:475–480
18. Prcic S, Djuran V, Mikov A, Mikov I (2007) Vitiligo in children. Ped Dermatol 24:666

Late-Onset Vitiligo

1.3.13

Davinder Parsad and Dipankar De

Contents

Core Messages

- 6.8% of vitiligo cases in Northern India develop after 50 years of age, and the NSV is proportionally more important when compared to other age groups.
- Late-onset vitiligo is a rather new clinical research topic, and reflects probably a neglected in fair-skinned population and data regarding this issue are lacking.
- The increased average life expectancy in both the developed and developing countries is probably another possible reason for the recent interest in late onset cases of vitiligo.
- Melanocyte, and immune system senescence are possible reasons for late-onset vitiligo.

1.3.13.1 Introduction

Vitiligo appears usually in childhood and early adulthood, and around half the patients have the disease onset before the age of 20 [5].Vitiligo research has largely been restricted to this younger age group or without any age specification. Late onset vitiligo is thus a rather new research issue, but probably also reflects a neglected one in fair-skinned populations and there is not much data regarding this issue. Improved health-care services, nutrition, hygiene, and housing have significantly increased the average life expectancy in the developed and developing countries, another possible reason for the recent interest in late-onset cases of vitiligo.

D. Parsad (✉)
Department of Dermatology,
Postgraduate Institute of Medical Education & Research,
Chandigarh, India
e-mail: parsad@mac.com

M. Picardo and A. Taïeb (eds.), *Vitiligo*,
DOI 10.1007/978-3-540-69361-1_1.3.13, © Springer-Verlag Berlin Heidelberg 2010

1.3.13.2 Definition and Epidemiological Data

Our group has arbitrarily defined late onset vitiligo as vitiligo that appears after the age of 50 years [2], at variance with Huggins et al. who had defined it as vitiligo with onset after the age of 30 years [6]. Our study performed in northern India is so far the only published study on a large number of cases of late onset vitiligo in the literature in English [2], and it is thus difficult to know if our results are relevant to other populations. Of 2,672 patients of vitiligo seen over a period of 10 years, a significant number of patients (6.8%) were late- onset cases [2]. There was no marked sex predominance (87 male, 95 female). The mean age at onset of disease was 55 years, 73.6% between 50 and 60 years, 21.4% between 61 and 70 years, and 4.9% after 70 years. Such skew towards less numerous populations in more aged subgroups is expected as its total population spontaneously decreases. As in other age groups, vitiligo vulgaris was the commonest type observed in 83.5%, followed by focal in 5.5%, segmental in 4.4%, acrofacial in 3%, mucosal in 2.2%, and universal in 0.5%. Mucosal vitiligo, either in pure form or associated with other cutaneous types of vitiligo was observed in 31 (17%) patients. The oral mucosa was involved in 71%, and genital/anal mucosa in 29% of patients. Almost 16% patients had a family history of vitiligo. First-degree relatives were affected in 11.5%, and second-degree relatives in 4.4% of patients [2].

In contrast, Mason and Gawkrodger recently analysed vitiligo presentation in adults in a European population, the study not addressing late onset vitiligo *sensu stricto* [9]. Non-segmental vitiligo was the commonest pattern and none had segmental disease [9]. There are no data on fair-skinned populations about true late onset vitiligo, although it is not rare to find vitiligo as a neglected examination feature in hospitalised elderly patients. In this situation, photoaging features may mask the depigmented background.

1.3.13.3 Significance of Late Onset Vitiligo

Aging is associated with immunosenescence manifested as a decreased immune response to different antigens. However, in the elderly there is a paradoxical increase in the incidence of autoimmune/autoinflammatory diseases. This increase in incidence has been attributed to the age-related accumulation of abnormal proteins which acts as antigens, and to a simultaneous decrease in the efficacy of the scavengers for these altered proteins. Given such a situation of potential, increased incidence of autoimmunity, rare reporting of late-onset vitiligo is surprising, and may rather reflect the neglected status of the disease.

After the age of 25–30 years, there is approximately a 10–20% decrease in the number of melanocytes per decade. This phenomenon occurs both in the photo-exposed and photo-protected skin of a particular individual. However, whether this age-related reduction in number of melanocytes has anything to do with vitiligo in the elderly is not known.

Is there any difference between classical early onset vitiligo and so-called late onset vitiligo? A genetic study suggests that there might be two co-existing modes of inheritance for vitiligo, depending on age of onset [1]. In patients with onset before the age of 30, the so-called early onset vitiligo, the inheritance would follows a dominant mode with incomplete penetrance, while a recessive genotype and exposure to environmental trigger would account for late-onset vitiligo (after 30 years of age). Specific HLA haplotypes are strongly associated with a family history of vitiligo; severity of disease, age of onset, and population geography [6].

In children, segmental vitiligo constitutes about one-fifth of patients [4, 8] (Chap. 1.3.12), while it is rarely observed in late-onset vitiligo (4.4% of total late-onset vitiligo patients had segmental vitiligo in our study) [2]. There is a tendency for a subset of childhood [7] and adult non-segmental vitiligo [6] to be associated with Hashimoto's thyroiditis and sometimes other autoimmune diseases like pernicious anaemia, alopecia areata, diabetes mellitus, Addison's disease [6]. In our study of late-onset cases, a vitiligo subset seems similarly to be a marker of an auto-immune or auto-inflammatory diathesis since associated auto-immune /endocrine disorders were present in 21.4% of our patients [2]. Insulin-dependent diabetes mellitus was found in 27, thyroid diseases in 15, and rheumatoid arthritis in 9, 12 of them having more than one associated disease.

1.3.13.4 Therapeutic Considerations

Grimes has observed that vitiligo in elderly patients is slowly progressive and less responsive to treatment [4]. Topical corticosteroids, calcipotriol, and narrowband ultraviolet B remain the mainstay of treatment. Topical calcineurin inhibitors including tacrolimus and pimecrolimus may be welcome new additions. PUVA and systemic steroids may not be suitable because of either relative contraindications in elderly patients or associated comorbidities. Vitiligo activity was found to be stable in 64.8% of our late onset vitiligo patients. Based on this, a surgical approach may be cautiously considered in patients with limited disease.

How does a disease like vitiligo without perceived symptoms except aesthetic consequences affect a person who has crossed his prime of life? In contrast to the common belief, it has been shown that the elderly people are often anxious to look attractive. As others, elderly patients often come to the physician with the quest that the patches should not progress and should be cured, at least the lesions over the visible parts of the body.

References

1. Arcos-Burgos M, Parodi E, Salgar M et al (2002) Vitiligo: complex segregation and linkage disequilibrium analyses with respect to microsatellite loci spanning the HLA. Hum Genet 110:334–342
2. Dogra S, Parsad D, Handa S et al (2005) Late onset vitiligo: a study of 182 patients. Int J Dermatol 44:193–196
3. Grimes PE (2008) http://aad2008.omnibooksonline.com/data/papers/FRM-526-C.pdf. (Accessed 05 Nov 2008)
4. Halder RM, Grimes PE, Cowan CA et al (1987) Childhood vitiligo. J Am Acad Dermatol 16:948–954
5. Handa S, Dogra S (2003) Epidemiology of childhood vitiligo: a study of 625 patients from North India. Pediatr Dermatol 20:207–210
6. Huggins RH, Schwartz, Janniger CK (2005) Vitiligo. Acta Dermatovenerol Alp Panonica Adriat 14:137–142
7. Iacovelli P, Sinagra JL, Vidolin AP et al (2005) Relevance of thyroiditis and of other autoimmune diseases in children with vitiligo. Dermatology 210:26–30
8. Jaisankar TJ, Baruah MC, Garg BR (1992) Vitiligo in children. Int J Dermatol 31:621–623
9. Mason CP, Gawkrodger DJ (2005) Vitiligo presentation in adults. Clin Exp Dermatol 30:344–345

Evaluation, Assessment and Scoring

1.4

Alain Taïeb and Mauro Picardo

Contents

Core Messages

- When visiting a patient for the first time, it is necessary to take enough time in order to include a comprehensive clinical evaluation. Therefore, a checklist of assessment items is useful to get a comprehensive individual image of the patient and his/her disease. The VETF scoring system combines analysis of extent, stage of disease (staging), and disease progression (spreading).
- Wood's lamp is useful for a combined assessment of staging and spreading in the same selected area.
- Subjective items,which impact the quality of life are also important to take into account for management purposes.

1.4.1 Introduction

The first contact with a patient coming with a previous diagnosis of vitiligo is of major importance. He/she generally says that the previous doctors have not been keen to engage in a conventional doctor–patient relationship about their disease, which they said was benign but out of therapeutic reach. This common approach denies the status of patient and explains the long search for alternative sources of cure. Secondly, it is important to check the accuracy of the diagnosis (Chap. 1.2.1) because in a subset of patients, the depigmenting disease has been labeled loosely "vitiligo" for some reason, and patients are grateful to obtain a diagnosis, if not a cure. This is commonly the case in

A. Taïeb (✉)
Service de Dermatologie, Hôpital St André,
CHU de Bordeaux, France
e-mail: alain.taieb@chu-bordeaux.fr

M. Picardo and A. Taïeb (eds.), *Vitiligo*,
DOI 10.1007/978-3-540-69361-1_1.4, © Springer-Verlag Berlin Heidelberg 2010

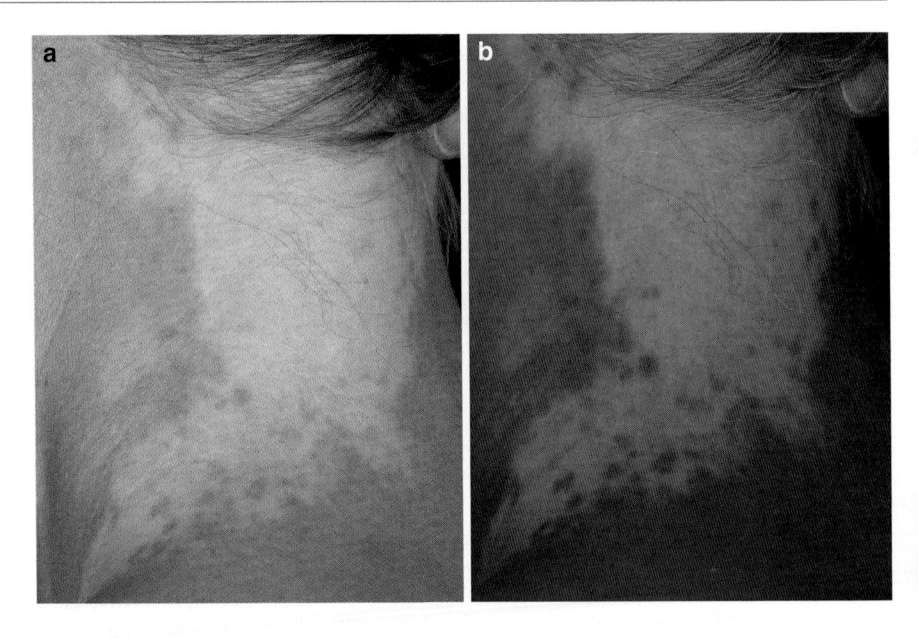

Fig. 1.4.1 Wood's lamp examination: It is best performed in a completely dark room. The examiner should adapt to the darkness for at least 30 s before starting examining the patient. Note the bright reflection of white patches under Wood's lamp (**b**) and details on intermediate pigment tones to compare with (**a**). A magnifying lens incorporated to the lamp is useful to analyse (terminal and vellus) hair pigmentation

children consulting for segmental hypopigmentation (Chap. 1.3.12).

Currently,there is a lack of consensus in methods of assessment of vitiligo, which makes it generally impossible to compare the outcomes of different studies of the same treatment and to perform meta-analyses [17]. In order to address these problems, the Vitiligo European Task Force (VETF) was founded in 2003 during the ESPCR meeting in Ghent with three initial goals: (a) the proposal of a consensus definition of the disease (Chap. 1.2.1); (b) the design of biometric tools to assess disease severity/ stability; and if possible (c) the derivation of a consensus scoring system. The second and third step are still going on, but following the method already used for the SCORAD index in atopic dermatitis [3], an evaluation form was discussed and gradually improved at several meetings which included the input of patient's support groups, and tested in patients with generalized vitiligo in a dozen of academic medical centers. Secondly, the most relevant descriptive variables were selected for severity assessment and prognosis. Following this step, the VETF held a workshop in Rome to test and validate the evaluation sheet on a series of patients [14]. This chapter is written following the remarks and criticism, which have followed the publication of this report [14]. An updated and simplified version has been produced more recently [15].[1]

1.4.2 Step by Step Evaluation

It is necessary to take enough time for a first visit, which includes a comprehensive clinical evaluation. The time required is roughly 20 min (an identical extra time is needed for counseling and replying to questions). For lightly pigmented patients, a dark room is needed for Wood's lamp examination (Fig. 1.4.1). The help of a standardized sheet as available in the appendix page xx allows not to miss some issues and to secure the relevant information for routine clinical work or research, which is summarized in Table 1.4.1. It includes the analysis of the Koebner phenomenon (Sect. 2.2.2.1) which is of particular interest for prevention [6]. A question about vitiligo on genitals is on the VETF checklist because it causes a strong embarrassment to patients. Itch preceding lesions may correspond to the micro-inflammatory nature of the disease. This symptom was mentioned by the patient's support groups. However, since common disorders such as atopic dermatitis may coexist, especially in children, the exact value of this symptom needs more in depth investigations, and questioning should address a topographic relation to the vitiligo patches.

[1] The last version is also available online at http://content.nejm.org/cgi/content/full/360/2/160/DC1.

Table 1.4.1 Evaluation check-list for NSV (adapted from Taieb and Picardo [14])

Subject features	Disease features	Family	Interventions
Phototype (Fitzpatrick's skin type)	Duration	Premature hair graying	Type and duration of previous treatments including opinion of patient on previous treatments (list): useful/not useful
Ethnic origin	Activity, based on patient's opinion (progressive, regressive, stable over the last 6 months)	Vitiligo (detail if needed in family tree)	Current treatment(s)
Age at onset occupation	Previous episodes of repigmentation, and if yes, spontaneous or not (details)		Other diseases and treatments (list)
Stress/anxiety levels	Koebner's phenomenon on scars or following mechanical trauma		
Halo nevus	Itch before flares		
History of autoimmune diseases: if yes, which type	Thyroid disease, if yes detail, including presence of thyroid autoantibodies		
Global QoL assessment (10 cm analog scale) "how does vitiligo currently (last week) affects your everyday life"	Vitiligo on genitals	History of autoimmune diseases (details if needed in family tree)	
	Clinical photographs and if possible UV-light photographs		

Family history of vitiligo and/or premature hair graying is currently a routine part of the assessment, as well as personal and family history of thyroid disease and presence of thyroid autoantibodies and autoimmune diseases. Hair graying is defined as more than 50% of white/grey hairs before the age of 40. This item is generally more difficult to assess in the female ancestry. It is commonly found in pedigrees of vitiligo patients [7, 14], and suggests the role of an aging defect in the pigmentary system (Sect. 2.2.2.10). For the personal/ family history of chronic possible auto-immune/ inflammatory disorders, Table 1.4.2 indicate some hints to ask questions, because the related diseases are not always known by non professionals. Clinical examination should guide the need of specialized investigations (see below). Halo nevi, considered as a marker of cellular autoimmunity (Chap. 1.3.5), are assessed by history and at examination. A global quality of life assessment is recommended ("how does vitiligo currently (last week) affects your everyday life" or "how much have you been bothered by the white spots last week" assessed on a visual analog scale). The impact of stress on disease onset/progression is also taken into account. This form has been designed for NSV, but can serve also for segmental or unclassified forms of the disease.

Table 1.4.2 Assessment of chronic autoimmune/inflammatory diseases

Ask by name	Ask by symptoms or treatments
Alopecia areata	Chronic diarrhea
Psoriasis, psoriatic arthritis	Arthritis
Lupus	Hair loss
Scleroderma	Chronic bowel disease
Lichen planus, lichen sclerosus	Chronic skin disease
Rheumatoid arthritis	
Ulcerative colitis	Insulin
Crohn's disease	Thyroxin
Biermer's (pernicious) anemia	(Hydro)cortisone
Addison's disease	Vitamin B12
Celiac disease	

1.4.3 Associated Disorders and Laboratory Workup

Because of the magnitude of the association of autoimmune thyroid disease with NSV (25% in our recruitment), the assessment of thyroid function (thyrotropin, thyroid hormones) and the presence of antibodies to thyroid peroxidase (TPO)/thyroglobulin should be routinely performed. A specific follow-up is needed when Hashimoto's thyroiditis or Graves disease are diagnosed with referral to an endocrinologist. Associated auto-immune disease frequencies are apparently not similar according to ethnic background [2, 11]. Some data emphasize the need for a larger preventive screening because vitiligo may precede the onset of organ-specific auto-immune diseases [4]. If a personal and/or family history of auto-immune/auto-inflammatory disorders is detected (Table 1.4.2), broadening the etiologic investigations more specifically should be considered. In the context of familial auto-immune syndromes (Chaps. 1.3.8 and 1.3.11) such as autoimmune polyendocrine syndrome (APS), anti 21-hydroxylase autoantibodies (Addison's) and anti-NALP 5 antibodies (hypoparathyroidism) [1] can be looked for. For precursors to diabetes, antibodies can be searched against GAD65 and ICA512. A specialized advice is strongly suggested.

1.4.4 Skin Biopsy

A skin biopsy is usually not requested for common NSV, except when another diagnosis cannot be ruled out (Chap. 1.2.1). However, this point of view needs probably to be challenged. As skin inflammation in vitiligo seems clinically not detectable (Chap. 1.3.10), this simple test might allow a more precise "inflammatory" staging of vitiligo in near future (pending the implementation of less invasive tools). The finding of a micro-inflammatory border in active NSV (Chap. 1.2.2) – and/or of other local markers of disease activity – might become useful to make more appropriate therapeutic decisions.

1.4.5 Scoring

Vitiligo treatments have been previously analyzed using the proportion of treated patients who achieve a specified degree of repigmentation, usually >50% for a "good" response [16], which is much lower than patient's expectations. A quantitative score has been recently proposed [8]. The VETF has partially validated a clinical method of assessment which combines, analysis of the extent (rule of nines), grading of depigmentation, and progression. Figure 1.4.2 details the scoring system. Several problems have been raised during the patient's session [14]. For staging, color of the selected patch is clearly not homogeneous, especially in SV. Staging chosen by investigators generally reflects the worst stage and not the most representative stage. This poses problems when a few white hairs are present in association with skin repigmentation and more than 30% is required for the worst stage in the revised scoring system [15]. Another difficulty was related to the need to magnify the lesions to assess hairs, especially vellus hairs. Wood's lamp equipment for vitiligo assessment should include a magnifying lens. Analysis of spreading (progressive/stable/regressive) was the most difficult item in a blind (nonpatient influenced) test. The view already expressed in textbooks, that concavity vs. convexity as related to the general shape of a patch may predict progression vs. regression, is probably an indication to take with caution. Perifollicular repigmentation may occur with progressing marginal depigmentation. Partial depigmentation in a border of a patch may be interpreted as repigmentation. Surprisingly, at the Rome workshop [14], the investigator's opinion was right in the majority of cases if the patient's opinion was chosen as the gold standard. Overall, it was felt that this item should be graded more accurately using the patient's opinion.

The concordance between investigators using the same assessment grid was measurable, but it seems reasonable to predict that the results can be improved if some interactive training is available. There are probably, as noted for atopic dermatitis workshops [10], intrinsic high and low scorer profiles which can be minimized when the causes of variability between observers have been identified.

1.4.6 Interobserver Variability

As tested at a workshop with patients [14], the interobserver agreement was acceptable for the three items; extent, staging, and spreading, and individual scorer's

a

Area		% Area	Staging* (0-4)	Spreading* (–1 +1)
Head and neck (0-9%)				
Trunk	(0-36%)			
Arms	(0-18%)			
Legs	(0-36%)			
Hands and feet				
Totals	(0-100%)		0-20	(–5 +5)

*largest patch in each area

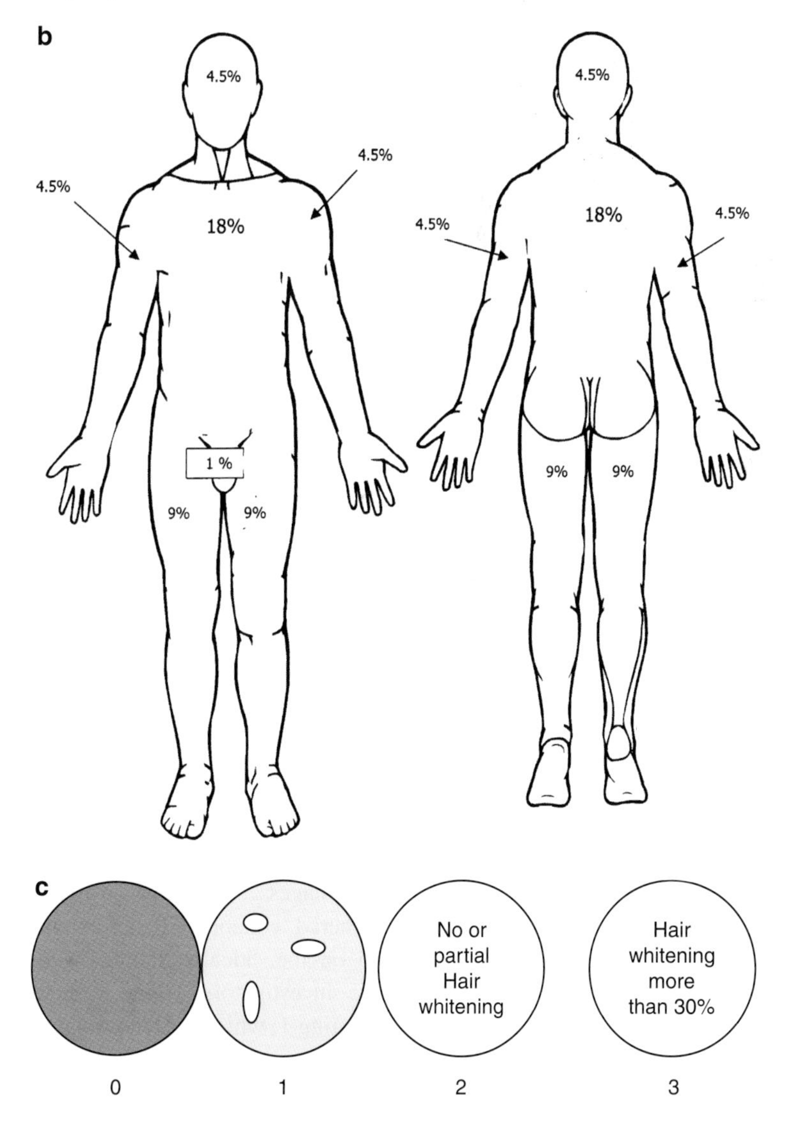

Fig 1.4.2 Scoring system (adapted from Taieb and Picardo [15]). The system assesses three dimensions of the disease (extent, staging, spreading/progression), which are summarized in a table for practical purposes (**a**). To assess *extent* (**b**), it is useful to refer to the patient's palm including digits, which averages 1% of body surface area. We recommend to draw the patches and mark the evaluated patches on figure, if necessary with a higher magnification for detail; if a child is under 5, head and neck totals 18%, legs 13.5% each; no change in other parts. If any, indicate halo nevi on the graph. *Staging* of vitiligo is based on the assumption that repigmentation requires melanocytes to be still present either in the epidermis or in the hair follicle (reservoir) to repigment the skin. Repigmentation patterns reflect this assumption: homogeneous diffuse repigmentation occurs when melanocytes remain in the interfollicular epidermis, and either marginal (from the border of the patch) or perifollicular (melanocyte precursors migrating from the hair follicle). Based on this, four stages (0–3) (**c**) are distinguished from normal pigmentation (0) to complete depigmentation (3), based on the assessment of the largest patch in each territory. Intermediate stages are defined as follows: stage 1 means incomplete depigmentation; stage 2 means complete depigmentation (may include hair whitening in a minority of hairs, less than 30%). If stage 3 persists after attempts of medical therapy, this would indicate a need for surgical treatment. *Spreading* is introduced to include a dynamic dimension, since rapidly progressive vitiligo needs urgent intervention to stabilize the disease. +1 means additional patches in a given area or demonstrated ongoing depigmentation using Wood's lamp in light skin colored patients; 0 means stable disease; –1 means observed ongoing repigmentation

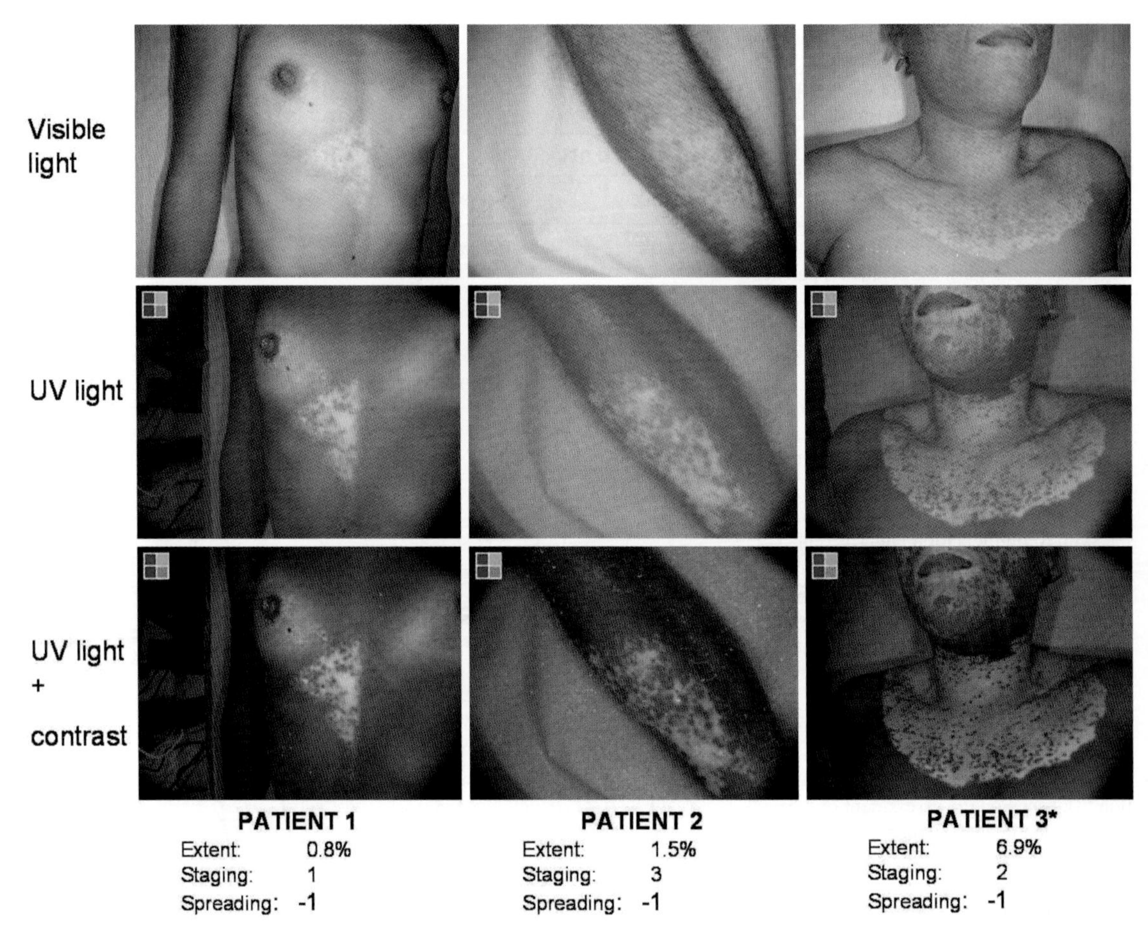

*For patient 3, phototherapy was ongoing, explaining erythema. The area scored comprised only the neck and upper thorax lesion in this patient

Fig. 1.4.3 Examples of assessments in three patients using the "ideal observer" method. This "ideal observer" is a virtual investigator, assigning the reference mean area involved and the modal value (mode) for the staging and spreading variables. The underlying assumption is that the value expressed by the majority of specialists is most likely to be the true value, as there is no gold standard

profiles could be defined. Black and white digitized photographs (Fig. 1.4.3) were chosen as references for the discussion of data. However, it was noted that erythema (in patients treated with phototherapy) will appear gray in digitized black and white images.

To check the validity of the data gathered, concordance analysis was carried out based on the "ideal observer" method (Table 1.4.3). The investigators tended to underestimate staging 2 for staging 1, which resulted in being overestimated (27 indications against 10 in Staging of Table 1.4.3). The frequency of assigning incorrect values decreased with staging: 40% for stage 1, 34% for stage 2, 30% for stages 3 and 4, respectively. Table 1.4.3(spreading) shows a wide dispersion of data for spreading. Based on these data, it seems easier to identify regression (identified in 72.5% of cases) even if one investigator assigned progression instead of regression, than stability and progression. To an extent (Table 1.4.3, extent), the investigators' findings were very close, because 92% of the evaluations were within a range of 1% of the mean value. Denominators were important to make a good judgment, as larger areas such as trunk and legs are more difficult to score. Image analysis may help in different settings (clinical research mostly) to reduce interobserver variability (Sect. 2.2.3.1).

Table 1.4.3 Interobserver variability

	Panelist scores				
	1	2	3	4	Total
Staging					
1	6	1	3	0	10
2	17	46	6	1	70
3	4	7	28	1	40
4	0	0	3	7	10
tot	27	54	40	9	130
	Panelist scores				
	-1	0	1		
Spreading					
−1	29	10	1		40
0	11	27	12		50
1	11	8	21		40
tot	51	45	34		130
	±0.2%	±0.5%	±1%		
Extent					
Head & neck	17 (85%)	19 (95%)	20 (100%)		20
arms	11 (55%)	13 (65%)	20 (100%)		20
Leg	13 (43%)	17 (57%)	29 (97%)		30
trunk	23 (38%)	36 (60%)	51 (85%)		60
	64 (49%)	85 (65%)	120 (92%)		130

The ideal observer method was used for: staging: K¼ 0.499; SE¼ 0.059 ($p < 0.00001$); spreading: K¼ 0.388; SE¼ 0.064 ($p < 0.0001$); and extent. The reference values correspond to the modal (staging and spreading) or mean (extent) score of the 10 panelists

1.4.7 Correlations Between Assessment Variables

Possible correlations between the three main assessment variables (extent, staging, and spreading) have been investigated. Some association was suggested between high staging and limited extent of vitiligo, when assessment was limited to only one patch. Similarly, larger patches tended to be assumed by investigators to spread less than smaller patches, but the difference was not significant here, and a wide variability in spreading assessment makes an interpretation of this item difficult. Stage 2 increased with extent, and stages 3 and 4 were mainly assigned to areas >1% [14].

1.1.8 Subjective Items

Response to disfiguring diseases is affected by basic ego strength [12]. Thus, psychological factors should be taken seriously into account in the global care of vitiligo patients. Quality of life impact (Part 1.5) is generally considered, overall as moderate in vitiligo, but the patient's phototype, cultural background and gender may influence the differences in data reported [9]. The perceived severity of vitiligo is explained mainly by the patients' personality and their psychological features and less significantly by the clinical objective criteria. Perceived severity of the illness and patient's psychological features such as trait-depression are predictors of the patients' QoL. Patients who tend to be more anxious in their daily life and who have a poor self-esteem perceive their illness as more severe even if vitiligo is less extensive [9]. A higher prevalence of alexithymia and depression or anxiety was found in vitiligo patients as compared with the general population [13].

There are now some specific arguments for including a psychological support in the care of vitiligo patients, including consideration of their personality and their difficulty of living with this disease (Part 1.5). Consequently, some factors which are easy to assess such as perceived severity and patient's personality (trait anxiety, trait depression, trait self esteem) could be included in the assessment and management of this chronic disfiguring disease. A simple perceived severity scale is clearly useful in clinical practice as a screening tool. However, the development of a vitiligo specific QoL scale would probably be helpful.

References

1. Alimohammadi M, Björklund P, Hallgren A et al (2008) Autoimmune polyendocrine syndrome type 1 and NALP5, a parathyroid autoantigen. N Engl J Med 358:1018–1028
2. Alkhateeb A, Fain PR, Thody A et al (2003) Epidemiology of vitiligo and associated autoimmune diseases in Caucasian probands and their families. Pigment Cell Res 16:208–214
3. Anonymous (1993) Severity scoring of atopic dermatitis: the SCORAD index. Consensus Report of the European Task Force on Atopic Dermatitis. Dermatology 186:23–31
4. Betterle C, Caretto A, De Zio A et al (1985) Incidence and significance of organ-specific autoimmune disorders (clinical, latent or only autoantibodies) in patients with vitiligo. Dermatologica1 71:419–423
5. Daneshpazhooh M, Mostofizadeh GM, Behjati J et al (2006) Anti-thyroid peroxidase antibody and vitiligo: a controlled study. BMC Dermatol 6:3
6. Gauthier Y (1995) The importance of Koebner's phenomenon in the induction of vitiligo vulgaris lesions. Eur J Dermatol 5:704–708

7. Halder RM, Grimes PE, Cowan CA et al (1987) Childhood vitiligo. J Am Acad Dermatol 16:948–954
8. Hamzavi I, Jain H, McLean D et al (2004) Parametric modeling of narrowband UV-B phototherapy for vitiligo using a novel quantitative tool: the Vitiligo Area Scoring Index. Arch Dermatol 140:677–683
9. Kostopoulou P, Jouary T, Quintard B et al (2009) Objective vs subjective factors in the psychological impact of vitiligo: the experience from a French referral center. Br J Dermatol 161:128–33
10. Kunz B, Oranje AP, Labreze L et al (1997) Clinical validation and guidelines for the SCORAD index: consensus report of the European Task Force on Atopic Dermatitis. Dermatology 195:10–19
11. Liu JB, Li M, Yang S et al (2005) Clinical profiles of vitiligo in China: an analysis of 3742 patients. Clin Exp Dermatol 30:327–331
12. Porter JR, Beuf AH, Lerner A, Nordlund J (1979) Psychological reaction to chronic skin disorders: a study of patients with vitiligo. Gen Hosp psychiatry 1:73–77
13. Sampogna F, Raskovic D, Guerra L et al (2008) Identification of categories at risk for high quality of life impairment in patients with vitiligo. Br J Dermatol 159: 351–359
14. Taïeb A, Picardo M; VETF Members (2007) The definition and assessment of vitiligo: a consensus report of the Vitiligo European Task Force. Pigment Cell Res 20: 27–35
15. Taieb A, Picardo M (2009) Clinical practice. Vitiligo. N Engl J Med 360:160–169
16. Westerhof W, Nieuweboer-Krobotova L (1997) Treatment of vitiligo with UV-B radiation vs topical psoralen plus UV-A. Arch Dermatol 133:1525–1528
17. Whitton ME, Ashcroft DM, Barrett CW, Gonzalez U (2006) Interventions for vitiligo. Cochrane Database Syst Rev (1): CD003263

Quality of Life

1.5

Davinder Parsad

Contents

Core Messages

- Vitiligo can induce considerable psychosocial stress and psychiatric comorbidity. Therefore, it is important to recognize and manage the psychological component not only to improve coping but also to obtain a better treatment response.
- The training in assertiveness and relaxation skills and helping in building self-confidence would have substantial effects on the quality of life as well as treatment outcomes.
- Social and psychological well-being increase when patients with facial disfigurement are helped to develop social skills and to confront their difficulties.
- There is a need for a standardized quality of life evaluation tool, which can quantify the psychosocial stress of the vitiligo subject and could be used for treatment evaluation.

1.5.1 Introduction and Historical Perspective

Vitiligo has currently a major impact on the quality of life of patients, many of whom feel distressed and stigmatized by their condition [2, 8, 9].

The Sanskrit word *kilas* is derived from kil, meaning to throw or to cast away, so kila means that which throws away color (Part 1.1). Since ancient times, patients of vitiligo have suffered the same physical

D. Parsad
Department of Dermatology, Postgraduate Institute of Medical Education & Research, Chandigarh, India
e-mail: parsad@mac.com

M. Picardo and A. Taïeb (eds.), *Vitiligo*,
DOI 10.1007/978-3-540-69361-1_1.5, © Springer-Verlag Berlin Heidelberg 2010

and mental abuses as lepers of that age and they were considered to have "shweta kustha." Vitiligo is particularly disfiguring for people with dark skin, and carries such a social stigma in Indian society that patients are sometimes considered unmarriageable [4]. Society greets vitiligo patients in much the same way as it does any one else who appears to be different. In some parts of India, they are still stared at, or subjected to whispered comments, antagonism, insult or isolation. A woman with vitiligo may face numerous social problems and experience great difficulties in getting married. If vitiligo develops after marriage, it provides a ground for divorce. In some parts of India, the disease is often considered as a punishment by God, presumably caused by unconscious feelings of guilt. Patients are subjected to various dietary restrictions like avoidance of milk, fish, citrus fruits, etc [13].

1.5.2 Modern Psychological Studies

Porter et al. [16, 17] in the late 1970s, brought the psychosocial effects of vitiligo to the attention of dermatologists. A questionnaire survey among 62 vitiligo patients in a hospital-based outpatient setting indicated that two-thirds felt embarrassed by the disease, more than half of them felt ill at ease, a majority of patients felt anxious, concerned, and worried about the disease itself. They also studied the effect of vitiligo on sexual relationship and found that embarrassment during sexual relationship was especially frequent for men with vitiligo [18]. Many patients chose adapted clothing and used large amounts of cosmetics in order to hide the skin lesions. Although 80% perceived their friends and family as supportive, strangers expressed less understanding and patients felt uncomfortable meeting them. Salzer and Schallreuter [19] reported that 75% found their disfigurement moderately or severely intolerable. An important component of stigma and stigmatization is the impossibility of concealing the affected skin parts and consequent visibility. In a recent study, stigmatization experienced by vitiligo patients was found psychologically relevant since patients with visible lesions experienced a higher level of stigmatization [21].

Appearance of skin can condition an individual's self-image, and any pathological alteration can have psychological consequences [20]. Patients often develop negative feelings about their skin, which are reinforced by their experiences over a number of years. Most patients with vitiligo report embarrassment, which can lead to a low self-esteem and social isolation [11]. Facial vitiligo lesions may be particularly embarrassing and the frustration of resistant lesions over exposed part of hands and feet can lead to anger and disillusionment. Particularly in teenagers, mood disturbances including irritability and depression are common. Patients with vitiligo are extensively sensitive to the way others perceive them and they will often withdraw because they anticipate being rejected. Sometimes, strangers and even close friends can make extremely hurtful and humiliating comments. The impact of such factors is profound, subjecting them to emotional distress, interference with their employment, or use of tension-lessening, oblivion-producing substances such as alcohol [5]. Severe depression has been known to lead to suicide attempts [3]. Vitiligo can also result in problems in interpersonal relations and induce depression and frustration. In a study from India, vitiligo has been shown to be associated with high psychiatric morbidity [11]. One fourth of vitiligo patients attending a specialized clinic were found to have psychiatric morbidity, and a diagnosis of adjustment disorder was made in the majority of cases. Psychiatric morbidity was significantly correlated with dysfunction arising out of illness [11].

1.5.3 Socio-Economic and Educational Consequences

Patients often suffer financial loss because they have to take time off work to attend hospital appointments to perform therapies such as phototherapies. Lesions in exposed sites can adversely affect a person's chances of getting a job at interview and so restrict career choices. Vitiligo beginning in childhood can be associated with significant psychological trauma that may have long lasting effects on personal self-esteem [6]. Children with vitiligo usually avoid sport or restrict such activities and often lose vital days in school.

1.5.4 Quality of Life Evaluation

Most of the studies have used Dermatology Life Quality Index (DLQI), a widely validated questionnaire that is easy to use and allows comparison between several skin disorders. However, there is a need for a uniformly acceptable scale which can more specifically quantify psychosocial stress associated with this disease. Moreover, this type of quality life scale shall be useful as a tool to assess treatment effectiveness or to compare treatment outcome.

Although a limited number of studies have paid attention to the psychosocial effects of vitiligo, they point towards an appreciable psychosocial impact on those afflicted. Kent and Al' Abadie [7] found that the stigmatization experience accounted for 39% of the variance in the quality of life of vitiligo patients. On the other hand, self-esteem, a number of symptoms on the distress checklist, race, and general health accounted for only 12% of the variance in quality of life.

There may be a relationship between stress and vitiligo since psychological stress can increase levels of neuroendocrine hormones, leading to a damage of melanocytes in the skin, affect the immune system altering the level of neuropeptides [1]. Recently, increased levels of neuropeptide-Y have been shown in the plasma and skin tissue fluids of patients with vitiligo [22]. Liu et al. [10] studied the occurrence of cutaneous nerve endings and neuropeptides in vitiligo vulgaris, and suggested that emotional trauma and stressful life-events can cause large adrenal secretions and this can result in acute onset of vitiligo.

Because of the possible connection between stress and exacerbation of vitiligo [15], psychological and psychotherapeutic interventions may be helpful. In a study of 150 vitiligo patients, we assessed the nature and extent of the social and psychological difficulties associated with the disease and their impact on treatment outcome by using DLQI [14]. Our study clearly demonstrated that patients with high DLQI scores responded less favorably to a given therapeutic modality.

These results suggest that additional psychological approaches may be particularly helpful in these patients. In a preliminary study by Papadopoulos et al. [12], it has been shown that counseling, can help to improve the body image, self-esteem and quality of life of patients with vitiligo, and may also have a positive effect on course of the disease.

References

1. Al'Abadie MSK, Kent G, Gawkrodger DJ (1994) The relationship between stress and the onset and exacerbation of psoriasis and other skin conditions. Br J Dermatol 130:199–203
2. Bolognia JL, Pawelek JM (1988) Biology of hypopigmentation. J Am Acad Dermatol 19:217–255
3. Cotterill JA, Cunliffe WJ (1997) Suicide in dermatological patients. Br J Dermatol 137(2):246–250
4. Fitzpatrick TB (1993) The scourge of vitiligo. Fitzpatrick's J Clin Dermatol 52:68–69
5. Ginsburg IH (1996) The psychological impact of skin diseases: an overview. Dermatol Clin 14:473–484
6. Hill-Beuf A, Porter JDR (1984) Children coping with impared appearance. Social and psychologic influences. Gen Hosp Psychiatry 6:294–300
7. Kent G, Al' Abadie M (1996) Factors affecting responses on Dermatology Life Quality Index items among vitiligo sufferers. Clin Exp Dermatol 21:330–333
8. Lerner AB (1959) Vitiligo. J Invest Dermatol 32:285–310
9. Lerner AB, Nordlund JJ (1978) Vitiligo. What is it? Is it important? JAMA 239:1183–1187
10. Liu PY, Bondesson L, Löntz W, Johansson O (1996) The occurrence of cutaneous nerve endings and neuropeptides in vitiligo vulgaris: a case-control study. Arch Dermatol Res 288:670–675
11. Mattoo SK, Handa S, Kaur I et al (2002) Psychiatric morbidity in vitiligo: prevalence and correlates in India. J Eur Acad Dermatol Venereol 16:573–578
12. Papadopoulos L, Bor R, Legg C (1999) Coping with the disfiguring effects of vitiligo: a preliminary investigation into the effects of cognitive-behaviour therapy. Br J Med Psych 72:385–396
13. Parsad D, Dogra S, Kanwar AJ (2003) Quality of life in patients with vitiligo. Health Qual Life Outcomes 1:58
14. Parsad D, Pandhi R, Dogra S et al (2003) Dermatology Life Quality Index score in vitiligo and its impact on the treatment outcome. Br J Dermatol 148:373–374
15. Picardi A, Pasquini P, Cattaruzza MS et al (2003) Stressful life events, social support, attachment security and alexithymia in vitiligo. A case-control study. Psychother Psychosom 72:150–158
16. Porter JR, Beuf AH, Nordlund JJ, Lerner AB (1978) Personal responses to vitiligo. Arch Dermatol 114:1348–1385
17. Porter JR, Beuf AH, Nordlund JJ, Lerner AB (1979) Psychological reaction to chronic skin disorders. A study of patients with vitiligo. Gen Hosp Psychiatry 1:73–77
18. Porter J, Beuf A, Lerner A et al (1990) The effect of vitiligo on sexual relationship. J Am Acad Dermatol 22:221–222
19. Salzer B, Schallreuter K (1995) Investigations of the personality structure in patients with vitiligo and a possible association with catecholamine metabolism. Dermatology 190:109–115
20. Savin J (1993) The hidden face of dermatology. Clin Exp Dermatol 18:393–395
21. Schmid-Ott G, Kunsebeck HW, Jecht E, Shimshoni R et al (2007) Stigmatization experience, coping and sense of coherence in vitiligo patients. J Eur Acad Dermatol Venereol 21:456–461
22. Tu C, Zhao D, Lin X (2001) Levels of neuropeptide-Y in the plasma and skin tissue fluids of patients with vitiligo. J Dermatol Sci 27:178–182

Natural History and Prognosis

1.6

Davinder Parsad

Contents

D. Parsad
Department of Dermatology, Postgraduate Institute of Medical Education & Research, Chandigarh, India
e-mail: parsad@mac.com

Core Messages

- The natural course of vitiligo is highly unpredictable, but some evolution patterns can be delineated.
- Disease activity scores are needed especially to precisely define stability.
- The Koebner's phenomenon, tri(multi-)chrome vitiligo, mucosal involvement, leukotrichia, a positive family history, presence of antithyroid antibodies or association with other autoimmune diseases predict a relatively poor prognosis.
- The prognosis of vitiligo might be predicted by the location of the initial lesions, a bad prognosis is expected if the initial sites are the posterior trunk and hands, less progression is expected when the initial sites are the face, upper or lower extremities.
- Leukotrichia means poor prognosis for repigmentation, with the assumption that no melanocytes are available within the depigmented area.
- The repigmentation induced by narrowband UBV therapy in NSV seems more stable than that induced by PUVA. However, ability to retain the pigment depends also on host factors and site of repigmentation (follicular, marginal, and interfollicular).

M. Picardo and A. Taïeb (eds.), *Vitiligo*,
DOI 10.1007/978-3-540-69361-1_1.6, © Springer-Verlag Berlin Heidelberg 2010

1.6.1 Natural Course

The natural course of vitiligo is highly unpredictable. Segmental vitiligo (SV) has a predominantly early onset and spreads rapidly over the affected area, but the activity usually ceases after a short period (Chap. 1.3.2). Non-segmental vitiligo (NSV) spreads progressively, with possible remission periods, over the whole body throughout the life of the patient and is frequently associated with a poor prognosis (Chap. 1.3.1). Further course of the disease varies in different patients and is unpredictable. Some patients may develop just a few more lesions over the next several months or years and there may be only a mild increase in the size of the initial lesions (*slowly progressive vitiligo*). However, in a small proportion of such patients, few episodes of relatively faster activity occur (*slowly progressive with periodic rapid exacerbations*). In another group of patients, the disease may progress at a much faster rate after an initial slow progression (*rapidly progressive vitiligo*). Such patients usually develop several clusters of new lesions and tend to develop an extensive disease within a period of only a few months. On rarer occasions, the spread is extremely rapid from the very beginning, leading to extensive depigmentation within a few days to a few weeks (*explosive vitiligo*). Slowly progressive vitiligo, with a simmering disease ultimately leading to a very extensive involvement, is also not uncommon (*slow but continuously progressive vitiligo*). In a small group of patients, after the initial slow progression, the disease may become static (*stable or static vitiligo*), or there may even be a spontaneous repigmentation of the lesions (*regressive vitiligo*).

1.6.2 Stability vs. Active Disease

Clinically, a vitiligo patient can be in one of the three stages of evolution, namely progressive, stable or regressive. The progressive and regressive stage definition do not pose much problem, with respectively, new lesions or lesions increasing in size; and lesions decreasing in size without appearance of new lesions. It is the definition of stable vitiligo that varies from clinician to clinician. The literature reveals that no consensus regarding the clinical evaluation of disease activity and therapeutic response has been reached to date. Moellman et al. define active disease as "when lesions are enlarging in 6 weeks before examination" [14].Cui et al. define the same as "development of new lesions or extension of old lesions in 3 months before examination"[1]. Uda et al. [21] define active vitiligo as "spread without regression within the last half year." Falabella et al. define stable vitiligo as "a condition that has not progressed for at least 2 years" [10]. No vitiligo activity scoring system that is sensitive to individual variations in the clinical pattern and also to the therapeutic response has been devised till date. An objective criterion, the vitiligo disease activity score (VIDA), was suggested by Njoo et al. [15] to follow the course of lesions, but it has limitations. It is a six-point scale on which the activity of the disease is evaluated by the appearance of new vitiligo lesions or enlargement of preexisting lesions, gauged during a period ranging from less than 6 weeks to 1 year.

1.6.3 Treatment Outcome and Stability of Repigmentation

In order to make a reasonable choice of therapy with the highest probability of success and to predict prognosis for an individual patient, it is important to identify disease characteristics that help predict the outcome of therapy as well as prognosis [16]. Many factors like emotional stress, sunburn and pregnancy have been speculated to be aggravating factors of vitiligo in the existing literature; however, objective clinical findings of progressive vitiligo are yet to be elucidated [4]. Inability to define the outcome and progression of the disease often leads to a disappointed and dissatisfied patient who ends up "shopping" doctors. Since the progression and response to therapy show considerable inter-individual patient variability, the greatest challenge facing the physician is to predict the outcome of the disease at the very outset. What seems to be clear is that leukotrichia means poor prognosis for repigmentation, with the assumption that no melanocytes are available within the depigmented area; even under conditions of full vitiligo stability, no repigmentation will usually occur with medical therapy. In a study of our group, it has been shown that perifollicular repigmentation was more stable than diffuse type of repigmentation [18]. Ability to retain the pigment depends not only on the source of stimulus but also on the site of activation. The repigmentation induced by narrowband UBV therapy was found to be more stable than that induced by PUVA [17].

1.6.4 Clinical Markers of Prognosis

At present, disease activity is mainly assessed by the medical history and extent of the involvement. In the absence of reliable laboratory indicators, clinical parameters can be used [3, 4, 6, 7]. Results of several independent epidemiological studies show that the Koebner phenomenon (KP) occurs in most patients with vitiligo and it has been frequently postulated that the KP may indicate active disease [9, 11] (Sect. 2.2.2.1).

The term trichrome vitiligo (see discussion of the term in Chap. 1.3.1) describes lesions that have a tan zone of varying width between normal and totally depigmented skin, which exhibits an intermediate hue. It has been shown that presence of trichome vitiligo also denotes progressive disease [5]. The presence of mucosal involvement has been shown to be a poor prognostic indicator that portends disease progression [2]. The presence of leukotrichia has been linked to poor response to medical treatment. However, as has been noted by various studies, leukotrichia per se is not related to disease activity [2]. It has been seen that a positive family history is associated with a statistically significant, higher rate of progression of the disease. Similarly the presence of antithyroid antibodies or association with other autoimmune diseases predicts a relatively poor prognosis [8]. According to Hann et al., when the initial sites were the posterior trunk and hands, there was more widespread progression to other body areas, whereas there was less progression when the initial sites were the face, and upper or lower extremities [4]. These results indicate that the prognosis of vitiligo may be predicted by the location of the initial lesions.

1.6.5 Prognosis and Subtypes of Vitiligo

Attempts have been made to classify vitiligo either by distribution of the patches, age of onset or degree of loss of color. According to the VETF nomenclature revision based on Koga [12] and Koga and Tango [13], NSV and SV are the preferred terms to designate the major subtypes [20] (Chap. 1.2.1). The prediction of pattern and final extension in SV is still a matter of debate (Chap. 1.3.2). Clinically, there are five recognized types of NSV: focal, generalized, acrofacial, mucosal and universalis, but whether this subclassification is useful or not, is not established for diagnostic and prognostic purposes; according to the VETF, there is no pressing evidence that these are distinct disorders [20]. Another important question which remains unanswered is, whether or not all types of vitiligo including mucosal vitiligo share a common genetic background, and if in the future, genetic testing may help management decisions. Obviously, the worst prognosis is for subjects with vitiligo universalis (Chap. 1.3.3). In the acrofacial subset, treatment outcome may not be so good because of the involvement of acral and mucosal parts, but usually acrofacial vitiligo does not show widespread or rapid extension [19].

References

1. Cui J, Arita Y, Bystryn JC (1993) Cytolytic antibodies to melanocytes in vitiligo. J Invest Dermatol 100:812–815
2. Dave S, Thappa DM, Dsouza M (2002) Clinical predictors of outcome in vitiligo. Indian J Dermatol Venereol Leprol 68:323–325
3. Halder RM, Grimes PE, Cowan CA (1987) Childhood vitiligo. J Am Acad Dermatol 16:948–954
4. Hann SK, Chun WH, Park YK (1997) Clinical characteristics of progressive vitiligo. Int J Dermatol 36:353–355
5. Hann SK, Kim YS, Yoo JH, Chun YS (2000) Clinical and histopathologic characteristics of trichrome vitiligo J Am Acad Dermatol 42:589–596
6. Hann SK, Lee HJ (1996) Segmental vitiligo: clinical findings in 208 patients. J Am Acad Dermatol 35:671–674
7. Hann SK, Park YK, Whang KC (1986) Clinical study of 174 patients with generalized vitiligo. Kor J Dermatol 24: 798–805
8. Harning R, Cui J, Bystryn JC (1991) Relation between the incidence and level of pigment cell antibodies and disease activity in vitiligo. J Invest Dermatol 97:1078–1080
9. Hatchome N, Kato T, Tagamit H (1990) Therapeutic success of epidermal grafting in generalized vitiligo is limited by the Koebner phenomenon. J Am Acad Dermatol 22:87–91
10. Falabella R, Arrunategui A, Barona MI, Alzate A (1995) The minigrafting test for vitiligo: detection of stable lesions for melanocyte transplantation. J Am Acad Dermatol 32: 228–232
11. Gauthier Y (1995) The importance of Koebner's phenomenon in the induction of vitiligo vulgaris lesions. Eur J Dermatol 5:704–708

12. Koga M (1977) Vitiligo: a new classification and therapy. Br J Dermatol 97:255–261
13. Koga M, Tango T (1988) Clinical feature and course of type A and type B vitiligo. Br J Dermatol 118:223–228
14. Moellmann G, Klein-Angerer S, Scollay DA et al (1982) Extracellular granular material and degeneration of keratinocytes in the normally pigmented epidermis of patients with vitiligo. J Invest Dermatol 79:321–330
15. Njoo MD, Das PK, Bos JD, Westerhof W (1999) Association of the Koebner phenomenon with disease activity and therapeutic responsiveness in vitiligo vulgaris. Arch Dermatol 135:407–413
16. Ortonne JP (2000) Special features of vitiligo. In: Hann SK, Nordlund JJ (eds) Vitiligo. A monograph on the basic and clinical science. Blackwell, Oxford, pp 70–77
17. Parsad D, Kanwar AJ, Kumar B (2006) Psoralen-ultraviolet A vs. narrow-band ultraviolet B phototherapy for the treatment of vitiligo. J Eur Acad Dermatol Venereol 20: 175–177
18. Parsad D, Pandhi R, Dogra S, Kumar B (2004) Clinical study of repigmentation patterns with different treatment modalities and their correlation with speed and stability of repigmentation in 352 vitiliginous patches. J Am Acad Dermatol 50:63–67
19. Parsad D, Pandhi R, Juneja A (2003) Effectiveness of oral Ginkgo biloba in treating limited, slowly spreading vitiligo. Clin Exp Dermatol 28:285–287
20. Taïeb A, Picardo M; VETF Members (2007) The definition and assessment of vitiligo: a consensus report of the Vitiligo European Task Force. Pigment Cell Res 20:27–35
21. Uda H, Takei M, Mishima Y (1984) Immunopathology of vitiligo vulgaris, Sutton's leukoderma and melanoma-associated vitiligo in relation to steroid effects, II: the IgG and C3 deposits in the skin. J Cutan Pathol 11:114–124

Defining the Disease: Editor's Synthesis 1.7

Alain Taïeb and Mauro Picardo

Contents

Core Messages

- The initial evaluation of a vitiligo patient is a step frequently neglected.
- More epidemiological data could delineate heritability vs. environmental factors and natural history.
- Clinical data do not support the concept of a common involvement of non skin melanocytes in non-segmental vitiligo (NSV) (with the exception of the oral and genital mucosae, and vitiligo universalis).
- Major initial determinants are probably situated directly in the skin, and activated in a subset of patients by mechanical stressors (Koebner's phenomenon).
- NSV can be considered in a subset of patients as a marker of an auto-inflammatory/autoimmune diathesis, but other predisposing inherited traits are involved (including premature melanocyte aging).

1.7.1 Clinical Assessment Is Important

The initial evaluation of a vitiligo patient is a step which is frequently neglected, as witnessed by the only recent delineation of a mixed form of vitiligo combining segmental vitiligo (SV) and non-segmental vitiligo (NSV). It should include a thorough history-taking and examination including Wood's lamp in a dark cabinet for all low phototype individuals. Clinical evaluation yields important information for routine management purposes and also for asking relevant clinical research questions. Assessing coping and quality of life issues should not be

A. Taïeb (✉)
Service de Dermatologie, Hôpital St André,
CHU de Bordeaux, France
e-mail: alain.taieb@chu-bordeaux.fr

M. Picardo and A. Taïeb (eds.), *Vitiligo*,
DOI 10.1007/978-3-540-69361-1_1.7, © Springer-Verlag Berlin Heidelberg 2010

forgotten. In particular, the role of stress has been emphasised by some studies and patients' support groups. There is a need to clarify this issue of stress acting as a trigger factor, or more importantly as influencing disease progression, because it could open new perspectives for management. The evaluation of the impact of the disease on non skin melanocytes (eye, ear, central nervous system) is not necessary for common NSV. For vitiligo universalis, there is a frontier with Vogt Koyanagi Harada (VKH) syndrome, which is unclear, and those patients should be evaluated more carefully in this respect.

1.7.2 Epidemiological Studies, Including Twin Studies, Are Needed

The spectrum of clinical manifestations in vitiligo is limited, but simple items such as variations in age at the onset and extension or pattern of the disease, natural course with possible spontaneous repigmentation, suggest a complex and mysterious interplay between the host and the environment. Incidence or prevalence rates seem stable, without major changes observable in a short time span, as noted for some chronic Th1 or Th2 predominant diseases, which have been the basis of the hygiene hypothesis, suggesting that a westernized life style changes our microbial environment and influence the population risk for chronic inflammatory disorders [1]. However informative studies such as repeated twin studies have not been performed in vitiligo as in atopic dermatitis [4]. More epidemiological data are clearly needed in this field to better delineate heritability vs. environmental factors and natural history.

1.7.3 Variable Melanocytic Targets According to Clinical Subtypes

Clinical aspects suggest different cellular targets/territories according to the subtypes of vitiligo,0 as shown for the marked preference of leukotrichia in SV. The mucosal pigmentary system seems to be involved more frequently in patients of dark complexion, but this aspect needs more accurate studies in other populations.

Data gathered in vitiligo studies worldwide do not support the concept of a common involvement of non skin melanocytes (with the exception of the oral and genital mucosae, and vitiligo universalis), contrary to what is found in VKH syndrome which targets ocular, auditory and central nervous system melanocytes [2]. However, as mentioned before, vitiligo universalis may correspond to a variant involving more generalized melanocytic targets. This favours a chronic skin condition with major initial determinants situated directly in the skin, and which might be activated in a subset of patients by mechanical stressors (Koebner's phenomenon, see also Sect. 2.2.2.1).

1.7.4 Lessons from Associated Diseases and Rare Syndromic Cases

History assessment of patients underlines two facts: the association in 20% of cases with a personal or familial history of auto-immune/inflammatory disorders, which define the usually so-called auto-immune vitiligo, where vitiligo is a marker of an auto-immune diathesis; and the association in 11% of patients with a family history of premature hair greying, which witnesses a visible premature senescence of the pigmentary system [5]. Syndromic vitiligo cases occurring in monogenic heritable diseases, underline possible similar pathophysiological pathways which belong either to the immune diathesis, with more emphasis on the humoral responses (APECED), or defects in the oxidative stress and aging process (MIDAS syndrome, ataxia-telangiectasia and Nijmegen breakage syndrome). The immune-mediated destruction of nevus cells (halo-nevus), and the rare clinically inflammatory onset of vitiligo, as well as vitiligo changes in VKH suggest clearly that, inflammation and cell-mediated immunity is important in a subset of patients. Interestingly, the recent emphasis on pruritus in vitiligo may correspond to a far more common micro-inflammatory component which is not clinically symptomatic (Chap. 1.3.10).

1.7.5 Predictive Classifications of Facial Segmental Vitiligo?

For the common and disfiguring facial involvement in SV, there is a current debate about the possible prediction of territories involved after the detection of the initial macule (Chap. 1.3.2). There is clearly a need of establishing a good database including patients of

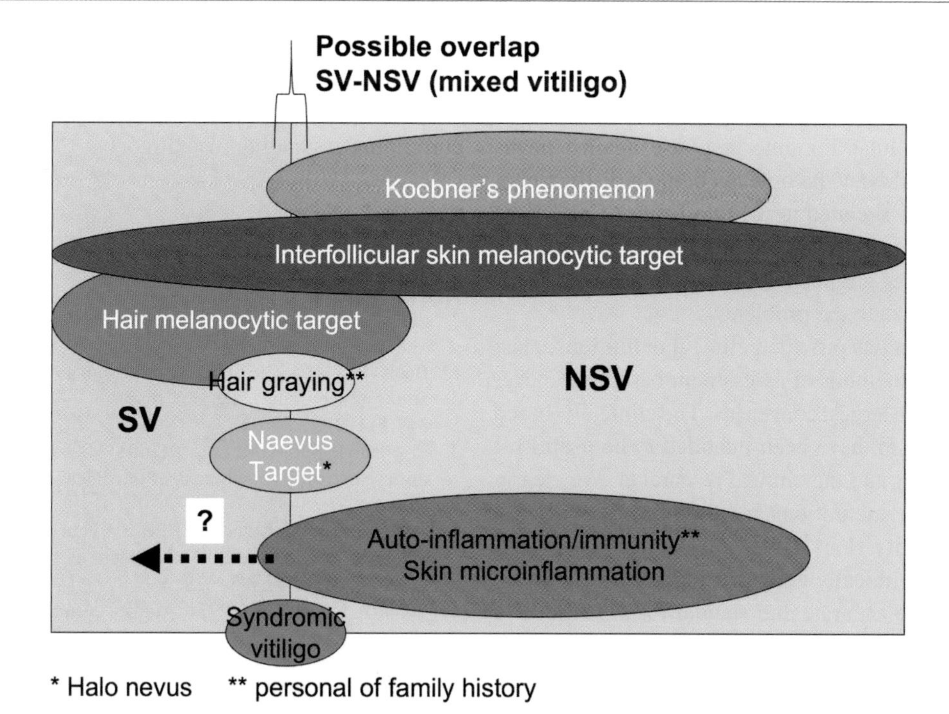

Fig. 1.7.1 Possible research angles of attack based on clinical findings in vitiligo. Segmental vitiligo (SV) and non-segmental vitiligo (NSV) can overlap (mixed vitiligo). The epidermal melanocytic target is equally shared between SV and NSV, but the hair melanocytic target is skewed towards SV. The history of hair greying and the presence of halo nevus seem equally shared between the two main types (to be confirmed by larger studies). The Koebner's phenomenon and personal/familial history of auto-inflammation/immunity seem mostly associated with NSV. Syndromic vitiligo may share some pathways influencing either survival or immune loss of melanocyte with common non syndromic vitiligo (SV and NSV). The question of microinflammation is open for SV

various ethnic backgrounds to settle this important issue. The patterns of extension in NSV, if predictable, deserve also a better attention (Chap. 1.3.1 and Part 1.6). Newer non invasive techniques of assessment such as *in vivo* confocal microscopy should be tested (Sect. 2.2.3.1).

1.7.6 Suboptimal Use of Pathology to Assess and Probably Understand Vitiligo

Pathology has been considered as providing limited information on the nature and course of the disease, because diagnosis was easy to make on clinical grounds. Vitiligo, based on the microscopic examination of long lasting NSV depigmented macules, was synonym with an epidermis without pigment cells. As discussed in Chaps. 1.2.2 and 1.3.10, the study of the progressing but still pigmented edges of NSV lesions (or with pinpoint depigmentation) and of distant normal-looking skin, has provided evidence of microdepigmentation and microinflammation. The systematic pathology assessment of SV for evidence of microinflammation is still lacking (but some evidence has been provided in Chap. 1.3.10 by Attili). This new vision raises several questions: if vitiligo is inflammatory pathologically, why is it usually not clinically? Erythema and pruritus are uncommon findings in the setting of even rapidly progressing vitiligo, suggesting a unique pattern of responsiveness or a so far unexpected active role of melanocytes in symptomatic inflammation in non depigmenting disorders. Is microinflammation primary, as proposed by Aslanian (Chap. 1.2.2), or secondary to some basic imbalance in pigment cell homeostasis? Whatever the answers are, the growing evidence of microinflammation should change concepts around vitiligo, and should have major therapeutic consequences in the future.

1.7.7 Summary

In summary, Figure 1.7.1 proposes a synthetic vision of the vitiligo clinical data. There are several points which need clarification, because these are quite difficult to reconcile with the common conceptions of vitiligo. Is there a common predisposition shared for developing cellular immune response against nevus cells in SV and NSV, halo nevus being a distinct phenomenon [3] associated to the two forms? Can SV be inflammatory as NSV? Some simple clinical and histological studies are needed to provide more accurate angles of attacks to the problem.

This section has provided clinical definitions, classification and methods of assessment based, whenever possible, on the largest consensus. The clinically-based definitions given, have been intended to be useful for the practitioner and the clinical researcher. We clearly need also better definitions and monitoring of disease activity /stability (Part 1.4), which may include evidence of non clinically graded inflammation with skin biopsies. More accurate and standardized methods of assessment need to be developed for specific needs including clinical trials (Sect. 2.2.3.1). A basic science understanding of vitiligo should represent soon an available alternative to this clinically-based approach in the context of developing translational research. The next section will thus review *in vivo* and *in vitro* clinical and experimental data which form the basis of our current understanding of vitiligo.

References

1. Bach JF (2002) The effect of infections on susceptibility to autoimmune and allergic diseases. N Engl J Med 347:911–920
2. Prignano F, Betts CM, Lotti T (2008) Vogt-Koyanagi-Harada disease and vitiligo: where does the illness begin? J Electron Microsc 57:25–31
3. Schallreuter KU, Kothari S, Elwary S et al (2003) Molecular evidence that halo in Sutton's naevus is not vitiligo. Arch Dermatol Res 295:223–228
4. Schultz Larsen F (1993) Atopic dermatitis: a genetic-epidemiologic study in a population-based twin sample. J Am Acad Dermatol 28:719–723
5. Taïeb A, Picardo M; VETF Members (2007) The definition and assessment of vitiligo: a consensus report of the Vitiligo European Task Force. Pigment Cell Res 20:27–35

Section 2

Understanding the Disease

Pathophysiology Overview

2.1

Mauro Picardo and Alain Taïeb

Contents

M. Picardo (✉)
Istituto Dermatologico San Gallicano, via Elio Chianesi, 00144 Roma, Italy
e-mail: picardo@ifo.it

Core Messages

- Vitiligo is a multifactorial disorder and a good angle of attack is still lacking.
- Different intrinsic, metabolic and functional defects appear to affect melanocytes and other cell types.
- The stage at which inflammation and autoimmunity are involved remains unclear.
- The understanding of the role of the melanocyte stem cells will provide a new insight into the therapeutical approaches.

2.1.1 From Where to Start? A Good Hierarchy of Relevant Data Is Needed

The discussion on the pathogenesis of vitiligo has been for decades, a magnet for endless speculation, and this indicates that some aspects of vitiligo are still confused. Several theories have been proposed to explain the disappearance of functioning melanocytes, but even the concept of "disappearance" is a matter of debate [4, 6, 14, 15, 18, 20]. There is, however, a general consensus that common NSV originate from melanocytes loss and not from simple melanogenesis inhibition. Several morphological, functional or metabolic alterations of the melanocytes, apparently not related, have been described. In the 1970–1980s, the immune-mediated, the autocytotoxic, or the neural mechanisms were considered as independent pathways accounting for melanocyte damage (Part 1.1).

M. Picardo and A. Taïeb (eds.), *Vitiligo*,
DOI 10.1007/978-3-540-69361-1_2.1,

Progressively, the researchers come across cell biology, genetics, biochemistry, immunology, and microbiology with new insights. The first step, as delineated in Part. 1, is to start from a firm ground, with clear definitions, and to postpone interpretation after data collection. Ideally, the observation of the disease and its natural history/treatment influences, annotated databases and biological collections for clinical and epidemiological studies, tissue collection for histopathology and other direct in vivo approaches are needed to provide the pathophysiological debate with solid arguments, before going to in vitro experiments and animal models.

2.1.2 Time for a Critical Reappraisal of the Convergence Theory

Starting from the proposal of a convergent theory with the melanocyte at its centre [14], several authors have contributed their point of view attempting to solve the puzzle. However, some clinical issues such as the link of SV to NSV have not been clearly integrated [21]. Furthermore, for defining a good starting point we need to reconcile the convergent theory with clinical and experimental data supporting an underlying generalized intrinsic biochemical defect, independent of melanocyte-specific metabolisms, enlarging thus, the spectrum of possible involved cells [1, 3, 8, 10, 13, 17]. Is this defect primary or secondary? Is it just more pronounced in pigment cells, as the emerging tip of an iceberg?

2.1.3 Melanocyte Loss: Survival Defect, True Destruction, or Multistep Process with Immune Acceleration?

Current theories, based on the newest basic science trends, give indications on putative mechanisms explaining how epidermal melanocytes may actually disappear or become non-functional. Death by cytotoxicity, apoptosis or following detachment have likewise been proposed [2, 5, 12, 22]. The in vivo data from skin biopsies has been so far disappointing as far as tracing this event is concerned ("crime without cadaver" as pointed out by Gauthier). It is possible that biopsies of vitiligo lesions gave little arguments just because early lesions are rarely biopsied. Biopsy material from established lesions contains no or few melanocytes, and it is difficult to capture the essence of the mechanism. Peripheral biopsies of progressive NSV lesions, even in non clinically inflammatory cases, demonstrate predominantly CD8+ T cell lichenoid infiltrates. These are in tune with a possible cell-mediated cytotoxic mechanism of the loss and form the basis of the concept of vitiligo being a microinfammatory skin disease. But the early initiating events of this phase are not known.

2.1.4 The Genetics Angle: Unbiased and Productive?

The genetic approach suggests a wide range of predisposing factors, none being universal, major differences having been detected across population of various ethnic backgrounds (Chap. 2.2.1). Most cases of NSV occur sporadically, but about 15–20% of patients have one or more affected first-degree relatives. Familial aggregation of NSV follows a non-Mendelian pattern suggesting complex polygenic, multifactorial inheritance. Case–control studies have reported the link of NSV with genes (such as *CTLA4, PTPN22, MBL2*, and *IL10*) already associated with autoimmune diseases. Genetic associations between vitiligo and other candidate genes (*GCH1, CAT, COMT, ACE, GPX1, AIRE*) have been suggested, even if the concerned case–control studies are small and require further validation. Targeted family-based association analysis recently proposed NALP1 as a breakthrough susceptibility gene for NSV associated with other autoimmune diseases. NALP1 is central to the innate immune system. The binding of bacterial derivatives or other environmental ligands can induce the assembly of NALP1 within the inflammasome, with subsequent production of active interleukin-1β [19]. This emerging new rationale for an increased skin susceptibility towards hazardous stimuli, stimulating the innate immune system and possibly cutaneous inflammation (Sect. 2.2.7.2) is now being closely scrutinized in vitiligo patients, and gene expression profiling may prove helpful to follow this idea.

2.1.5 Inflammation and Auto-Immunity. The Role of Stress

Arguments for a humoral immune response targeting melanocytes exist in a subset of patients (Sect. 2.2.7.3). Mainly, the CD8+ T cell infiltrate present in progressing vitiligo is probably, if not primarily, the cause of the disease, at least implicated in its acceleration phase clearly noted in some NSV patients, which may lead to vitiligo universalis (Chap. 1.3.3). However, its (probably) melanocytic targets are not yet clearly identified (Sect. 2.2.7.4). Experimental data highlight a link between oxidative stress and immune system activation. Following an external danger stimulus, an oxidative stress can frequently occur inside the cell, also determining the expression and the release of proteins that belong to the heat shock protein family. Melanocytes may produce and release the highest amount of hsp70, thus activating immune responses [12]. The pathogenetic role of the production of the hsp70 has also been tested in mouse through the gene gun vaccination with melanocyte differentiation antigens (TRP1 or gp100) and hsp70. The mouse hsp70 vaccinated early developed depigmentation, testifying for the ability of hsp70 to enhance antigen uptake by dendritic cells [6].

2.1.6 Identifying and Characterizing Skin and Non Skin Cellular Anomalies in Vitiligo

The main target cell of the disease is the epidermal and/or hair follicle melanocyte. There is evidence of the involvement of non skin melanocytes in common NSV (Chap. 1.3.7) [5]. This exceptional involvement seems to be related to an acceleration phase of the disease in extensive/universalis cases and characterizes the very rare Vogt-Koyanagi-Harada syndrome (Chaps. 1.3.3 and 1.3.8). There are marked clinical differences between SV and NSV in hair follicle melanocyte involvement, which may correspond to a different, even if poorly understood, pathogenesis. Vitiligo lesional melanocytes show morphological alterations, including cytosol vacuolization and limited dendrite formation [11] (Sect. 2.2.3.2). Associated with these structural features, several functional alterations have been described. In fact, vitiligo lesional and non-lesional melanocytes are characterized by an altered redox status, possibly due to the compromised activity of the intracellular antioxidants (catalase and glutathione peroxidase, mainly) or increased ROS production (Chap. 2.2.6). The final effect of this condition would be the high susceptiblity to toxic compounds, including melanin derivatives, and to physical trauma [5, 12, 18]. The other major epidermal cell type, the keratinocyte, appears to be involved (Chaps 2.2.5 and 2.2.10). A defective intracellular signal transduction of the TNF-α-activated pathway in vitiligo keratinocytes has been reported, possibly accounting for limited survival and subsequent loss of production of specific melanocyte growth factors [1, 9, 10, 13, 16, 17]. So far, neglected neighbouring cells such as dermal fibrobasts, may control adhesion checkpoints and might actually be involved in vitiligo, through the release of soluble factors [2].

As discussed before, besides the recognition of the major target cell of the disease, it is not completely settled whether or not melanocyte alterations can be the consequence of a more generalized biochemical/biological defect. Peripheral blood mononuclear cells appear to be characterized by metabolic deregulations and oxidative stress, similar to those found in melanocytes and epidermis [3, 8].

2.1.7 The Need for Translational Research

The multifactorial pathogenetic process leading to the functional loss of melanocytes may benefit from the data obtained in animal models (Chap. 2.2.4) [7]. The spontaneous autoimmune vitiligo Smyth line chicken provides, indeed a good chance to study vitiligo at onset and during progression. Smith chicken becomes depigmented after the hatch and the depigmentation can be complete or partial. Smyth chicken vitiligo is associated, as in humans, with uveitis and thyroid disorders. The relevance of the chicken model is also supported by the occurrence of intrinsic melanocyte defects (irregularly shaped melanosome, low catalase activity), genetic background, cell-mediated immune response (CD8+ T cell infiltrate and Th1 cytokines production), and external danger triggers (turkey herpesvirus). Other avian models support the intrinsic melanocyte fragility (Barred Plymouth Rock and White Leghorn chicken breeds). Other animal models have been proposed, including grey horses, the vitiligo mouse and the Sinclair pig, in which vitiligo spontaneously develops.

2.1.8 Conclusions and Scope of this Book Section

Vitiligo is still a poorly understood disease, and its multifactorial basis is indubitably a disadvantage to pick up a relevant item among so many, to begin unfolding the puzzle. The immunological and genetic approaches have provided until now, the most powerful insights. However, they cannot indicate with certainty the initial event causing the immune activation and subsequent amplified melanocyte damage. The following chapters of this section provide a more in depth analysis of the pathomechanisms mentioned in this overview.

References

1. Bondanza S, Maurelli R, Paterna P et al (2007) Keratinocyte cultures from involved skin in vitiligo patients show an impaired in vitro behaviour. Pigment Cell Res 20: 288–300
2. Cario-André M, Pain C, Gauthier Y et al (2006) In vivo and in vitro evidence of dermal fibroblasts influence on human epidermal pigmentation. Pigment Cell Res 19:434–442
3. Dell'Anna ML, Maresca V, Briganti S et al (2001) Mitochondrial impairment in peripheral blood mononuclear cells during the active phase of vitiligo. J Invest Dermatol 117:908–913
4. Dell'Anna ML, Picardo M (2006) A review and a new hypothesis for non-immunological pathogenetic mechanisms in vitiligo. Pigment Cell Res 19:406–411
5. Dell'Anna ML, Ottaviani M, Albanesi V et al (2007) Membrane lipid alterations as a possible basis for melanocyte degeneration in vitiligo. J Invest Dermatol 127(5): 1226–1233
6. Denman CJ, McCracken J, Hariharan V et al (2008) HSP70i accelerates depigmentation in a mouse model of autoimmune vitiligo. J Invest Dermatol 128:2041–2048
7. Erf GF, Trovillion CT, Plumlee BL et al (2008) Smyth line chicken model for autoimmune vitiligo: opportunity to examine events leading to the expression of vitiligo in susceptible individuals. Pigment Cell Res 21:265
8. Giovannelli L, Bellandi S, Pitozzi V et al (2004) Increased oxidative DNA damage in mononuclear leukocytes in vitiligo. Mut Res 556:101–106
9. Imokawa G (2004) Autocrine and paracrine regulation of melanocytes in human skin and in pigmentary disorders. Pigment Cell Res 17:96–110
10. Kim NH, Jeon S, Lee HJ et al (2007) Impaired PI3K/Akt activation-mediated NF-kB inactivation under elevated TNF-alpha is more vulnerable to apoptosis in vitiliginous keratinocytes. J Invest Dermatol 127:2612–2617
11. Kim YC, Kim YJ, Kang HY et al (2008) Histopathologic features in vitiligo. Am J Dermatopathol 30:112–116
12. Kroll TM, Bommiasamy H, Boissy RE et al (2005) 4-tertiary butyl phenol exposure sensitizes human melanocytes to dendritic cell-mediated killing: relevance to vitiligo. J Invest Dermatol 124:798–806
13. Lee YA, Kim NH, Choi WI et al (2005) Less keratinocyte-derived factors related to more keratinocyte apoptosis in depigmented than normally pigmented suction-blistered epidermis may cause passive melanocyte death in vitiligo. J Invest Dermatol 124:976–983
14. Le Poole IC, Das PK, van den Wijngaard RM et al (1993) Review of the etiopathomechanism of vitiligo: a convergence theory. Exp Dermatol 2:145–153
15. Le Poole IC, Wankowicz-Kalinska A, van den Wijngaard RM et al (2004) Autoimmune aspects of depigmentation in vitiligo. J Invest Dermatol Symp Proc 9:68–72
16. Moretti S, Spallanzani A, Amato L et al (2002) New insights into the pathogenesis of vitiligo: imbalance of epidermal cytokines at sites of lesions. Pigment Cell Res 15:87–92
17. Pelle E, Mammone T, Maes D et al (2005) Keratinocytes as a source of reactive oxygen species by transferring hydrogen peroxide to melanocytes. J Invest Dermatol 124: 793–797
18. Schallreuter KU, Bahadoran P, Picardo M et al (2008) Vitiligo pathogenesis: autoimmune disease, genetic defect, excessive reactive oxygen species, calcuim imbalance, or what else. Exp Dermatol 17:139–160
19. Spritz RA (2006) The genetics of generalized vitiligo and associated autoimmune diseases. J Dermatol Sci 41:3–10
20. Taieb A (2000) Intrinsic and extrinsic pathomechanisms in vitiligo. Pigment Cell Res 13:41–47
21. Taieb A, Morice-Picard F, Jouary T et al (2008) Segmental vitiligo as the possibile expression of cutaneous somatic mosaicism: implications for common non-segmental vitiligo. Pigment Cell Mel Res 21:646–652
22. van den Wijngaard RM, Aten J, Scheepmaker A et al (2000) Expression and modulation of apoptosis regulatory molecules in human melanocytes: significance in vitiligo. Br J Dermatol 143:573–581

2.2 Generalized Vitiligo

Genetics

2.2.1

Richard Spritz

Contents

R. Spritz
Human Medical Genetics Program,
University of Colorado Denver, Aurora, Colorado, USA
e-mail: richard.spritz@ucdenver.edu

Core Messages

- Technological and theoretical advances enabled by the human genome project have led to efforts to map and identify specific genes involved in vitiligo susceptibility.
- Specific approaches include the Candidate Gene Approach, the Genome-Wide Approach, and the Gene Expression Approach, each offering specific advantages and disadvantages.
- Generalized vitiligo is epidemiologically associated with a number of autoimmune diseases. This epidemiologic association has a genetic basis, at least in part, as vitiligo patients' close relatives have elevated risk of both vitiligo and other autoimmune diseases, even if those relatives don't have vitiligo.
- Several generalized vitiligo susceptibility genes have now been identified, including loci in the MHC, *PTPN22*, and *NALP1*. The status of other genes, whose involvement has been suggested, remains uncertain.

2.2.1.1 Genetic Epidemiology

Large-scale epidemiological surveys have shown that most cases of generalized vitiligo occur sporadically, though about 15–20% of patients report one or more affected first-degree relatives. Rarely, large multi-generation families segregate generalized vitiligo in patterns that suggest autosomal dominant [3] or autosomal recessive [12] inheritance with incomplete penetrance.

M. Picardo and A. Taïeb (eds.), *Vitiligo*,
DOI 10.1007/978-3-540-69361-1_2.2.1, © Springer-Verlag Berlin Heidelberg 2010

More typically, however, familial aggregation of vitiligo cases occurs in a non-Mendelian pattern that suggests complex polygenic, multifactorial inheritance [4, 10, 17, 23, 24, 36, 54, 60–62, 64, 79].

Strong evidence for genetic factors in the pathogenesis of generalized vitiligo comes from the studies of patients' close relatives. Among Caucasians of European origin, the risk of vitiligo to a patient's siblings is about 6.1% [4], a 16-fold increase over the approximately 0.38% prevalence of generalized vitiligo in at least one Caucasian population [42]. There is a similar risk of generalized vitiligo to patients' other first-degree relatives besides siblings: 7.1% in Caucasians, 6.1% in Indo-Pakistanis, and 4.8% in USA Hispanic/Latinos [4], with lower risks to more distant relatives. Generally similar results have come from studies of Chinese families [79]. Furthermore, in the largest vitiligo twin study to date [4] the concordance for generalized vitiligo in Caucasian monozygotic twins was 23%, more than 60 times the general population risk of 0.38%, and almost four times the 6.1% risk of vitiligo to probands' siblings, thus providing additional strong support for the involvement of genes in conferring risk of generalized vitiligo.

Additional evidence for a genetic component in generalized vitiligo comes from age of onset data: among unselected (mostly sporadic) Caucasian vitiligo patients the mean age of disease onset is 24.2 years [4], but among patients in families with multiple relatives affected by vitiligo the mean age of disease onset is significantly earlier, 21.5 years [54]. Earlier age of disease onset in more "familial" cases and diminishing disease risk with increasing genetic distance from an affected proband are typical characteristics of a polygenic disorder, and formal genetic segregation analyses have suggested that multiple major loci contribute to vitiligo susceptibility in a complex interactive manner [61, 64, 79].

While epidemiologic and twin studies thus indicate that genes play an important role in disease pathogenesis, non-genetic factors must also be very important, perhaps even more important than genes. Identical twins share all of their genes identically, and the limited twin concordance and delayed disease onset indicate that non-genetic, presumably environmental factors must also play a major role in determining the occurrence of vitiligo. Unfortunately, though many different environmental risk factors for generalized vitiligo have been proposed, epidemiologic data, that definitively supporting the involvement of any of these in the pathogenesis of typical cases of generalized vitiligo remain very limited.

Epidemiologic studies have also shown that generalized vitiligo is strongly associated with other autoimmune diseases (Chap. 1.3.8), and that this association apparently has a genetic basis. Generalized vitiligo is a component of the APECED (APS1) and Schmidt (APS2) multiple autoimmune disease syndromes (additional APS categories have been proposed but are not widely accepted), and in several studies vitiligo has been associated with autoimmune thyroid disease [22, 72], pernicious anaemia [25, 35], Addison's disease [88], and perhaps alopecia areata [75, 76]. A survey of over 2,600 unselected Caucasian patients with generalized vitiligo (most with sporadic occurrence of the disease) and their close relatives found significantly increased frequencies of autoimmune thyroid disease, pernicious anaemia, Addison's disease, and systemic lupus erythematosus [4]; overall, about 30% of patients with generalized vitiligo were affected with at least one additional autoimmune disease. These same diseases also occurred at increased frequency in patients' first-degree relatives, regardless of whether or not those relatives had vitiligo themselves [4]. These findings suggest that vitiligo patients and their close relatives have a genetically-determined susceptibility to this specific group of autoimmune diseases. A similar study of families in which multiple individuals had generalized vitiligo found even higher frequencies of these same autoimmune diseases, in both vitiligo patients and their siblings, as well as significantly elevated frequencies of psoriasis, rheumatoid arthritis, and adult-onset autoimmune diabetes mellitus [54]. These families with multiple affected relatives thus exhibited an expanded repertoire of vitiligo-associated autoimmune diseases, indicating that such families segregate even greater genetic susceptibility to these diseases than in typical singleton patients. Generally similar results have come from retrospective studies of vitiligo patients in India [37, 38] and Nigeria [67], although these studies found lower frequencies of some vitiligo-associated autoimmune diseases, most likely due to under-diagnosis in these populations. Together, these studies indicate that pathologic variants in specific genes predispose to a specific subset of autoimmune diseases that includes generalized vitiligo, autoimmune thyroid disease, rheumatoid arthritis, psoriasis, adult-onset autoimmune diabetes mellitus, pernicious anaemia, systemic lupus erythematosus, and Addison's disease. Several of these broad-spectrum autoimmunity genes have now been identified.

2.2.1.2 Identification of Vitiligo Susceptibility Genes

Three different general approaches have been used to identify genes that might mediate susceptibility to vitiligo: the *candidate gene* approach, the *genome-wide* approach, and the *gene expression* approach (summarized in Table 2.2.1.1).

The Candidate Gene Approach

Candidate gene studies typically have tested for the genetic association of specific DNA sequence variants in specific genes, thought to perhaps be involved in susceptibility to vitiligo on the basis of *a priori* biological hypotheses. While the variants tested are usually unlikely to be causal for the disease, they are assumed to

Table 2.2.1.1 Genes and genomic regions suggested for involvement in vitiligo

Chromosome	Gene or locus	Method	Comments
1p36		Linkage	Chinese
1p31.3-p32.2	*AIS1* (*FOXD3*?)	Linkage, positional cloning, sequencing	Rare autosomal dominant; atypical vitiligo phenotype, autoimmunity-associated
1p13	*PTPN22*	Association	Confirmed; associated with many autoimmune disorders
1q31-q32	*IL10*	Association	
2p21	*VIT1* (*FBXO11*)	Expression analysis	No evidence causally involved in vitiligo
2q33	*CTLA4*	Association	Data conflicting
3p21.3	*GPX1*	Association	
3p14.1-p12.3	*MITF*	Linkage	Candidate gene; no evidence for linkage
6p21.3	*TNFA*	Association	No association
6p21.3	MHC (*HLA-DRB1*, *HLA-DRB4*, *HLA-DQB1*)	Association, linkage	Associated with many autoimmune disorders
6p21.3 (within MHC)	*LMP/TAP*	Association	
6p21-p22		Linkage	Chinese
6q24-q25		Linkage	Chinese
6q25.1	*ESR*	Association	
7	*AIS2*	Linkage, association	Autoimmunity-associated; Caucasians
8	*AIS3*	Linkage	Caucasians
10q11.2-q21	*MBL2*	Association	
11p13	*CAT*	Association	Now considered invalid
12q12-q14	*VDR*	Association	
12q13	*MYG1*	Expression analysis	
12q14	*IFNG*	Association	No association with VKH disease
14q12-q13		Linkage	Chinese
14q22.1-q22.2	*GCH1*	Association, sequencing	Now considered invalid
16q24.3	*MC1R*	Association	No association
17p13	*NALP1* (*SLEV1*)	Linkage, association	Confirmed; autoimmunity-associated
17q23	*ACE*	Association	Data conflicting
20q11.2	*ASIP*	Association	No association
21q22.3	*AIRE*	Linkage, sequencing, association	Causes autosomal recessive APECED syndrome (can include vitiligo); no association with typical generalized vitiligo
22q11.2		Linkage	Chinese
22q12	*COMT*	Association	

perhaps be in linkage disequilibrium with true pathological variants. The candidate gene approach obviously is limited to testing hypotheses involving already known biological candidate genes; it cannot discover novel genes or pathways. The most commonly used study design is the case–control analysis, in which allele or genotype frequencies are measured in a collection of ethnically matched unrelated cases vs. ethnically matched unrelated controls. While the case–control study design seems simple, it can be used to study singleton patients who represent the majority of cases, and can detect genetic signals that exert relatively small effects. This study design is unfortunately, highly subject to false-positive errors due to inadequate ethnic matching of cases vs. controls, and population admixture stratification. Retrospective analysis of published "genetic associations" has shown that the great majority represent false-positives, with bias towards publishing apparently positive results [40]. Family-based association study designs are generally more robust as they are not subject to these errors, but require collection and analysis of families and thus are more difficult and more expensive, and so are less often undertaken. Surprisingly, many published genetic association studies of vitiligo fail to include appropriate multiple-testing correction, and so, many of those studies reporting statistically marginal associations unfortunately, are probably spurious. Similarly, most published, negative genetic association studies fail to consider the statistical power of the analysis to detect a genetic effect of a given magnitude, which must be considered in evaluating a negative result.

The earliest genetic studies of vitiligo were case–control genetic association analyses of the major histocompatibility complex (MHC), carried out by typing various MHC markers in patients with various different vitiligo phenotypes vs. controls, from many different populations [6, 7, 30, 31, 59, 68, 71, 81, 85, 87]. In general, these studies have found no consistent association between the occurrence of vitiligo and specific HLA alleles. However, re-analysis of these studies as a group showed that several found association between vitiligo and HLA-DRB4 alleles, and meta-analysis of multiple studies found association of vitiligo with HLA-A2 [59]. Recent studies that utilized more robust family-based association methods found genetic association between generalized vitiligo and HLA-DRB4*0101 and HLA-DQB1*0303 in Dutch patients [87], with HLA-DRB1*03, DRB1*04 and HLA-DRB1*07 alleles in Turkish patients [81], and with alleles of microsatellites located in the MHC in Columbian patients [7]. In Caucasian multiplex generalized vitiligo families, the MHC class II haplotype HLA DRB1A *04-(DQA1*0302)-DQB1*0301 is associated with both increased risk of vitiligo and with relatively early disease onset [28], and in Han Chinese, the MHC haplotype HLA A25-Cw*0602-DQA1*0302 is associated with generalized vitiligo [59]. Genetic association has also been reported between generalized vitiligo and genes of the *LMP/TAP* gene region of the MHC [19], although the significance of this is still unclear. Genetic associations of vitiligo with alleles of MHC loci appear to be the strongest in patients and families with various vitiligo-associated autoimmune diseases, vs. in patients and families with only generalized vitiligo. Many of these non-vitiligo autoimmune diseases are themselves associated with variation in the MHC, and it remains uncertain whether these reported vitiligo-MHC associations are primary or actually are indirect, due to primary genetic association of the MHC with these other diseases. Indeed, genome-wide genetic linkage analyses of vitiligo have shown no apparent linkage signal at the MHC in Caucasians, although a minor linkage signal in this region of chromosome 6 has been reported in Han Chinese [20, 58].

Case–control studies have led to reports of the association of generalized vitiligo with several other candidate genes thought to be involved in autoimmunity, including *CTLA4* [5, 43, 50] *PTPN22* [16], *MBL2* [66], and *IL10* [1]. Variations in *CTLA4* and *PTPN22* have been implicated in a number of other autoimmune diseases, and these genes may, like HLA, function as general autoimmunity susceptibility loci [14, 33, 84]. *MBL2* may participate in the innate immune response to bacterial infections, which, as discussed below, is implicated in the pathogenesis of vitiligo by other studies. *IL10*, encoding interleukin-10, may also be involved in immune response to bacterial pathogens, and has been associated with susceptibility to many other autoimmune diseases. Association of generalized vitiligo with *PTPN22* has been confirmed by multiple case–control and family-based studies [55, 56]; however, family-based studies fail to support association with *CTLA4* [56], suggesting that this reported association may be spurious or driven by concomitant other autoimmune diseases, rather than by vitiligo itself [13].

Genetic associations between vitiligo and a number of other candidate genes have also been reported, almost all based solely on a small number of case–control studies; these, therefore, must be considered as merely suggestive. A reported association between vitiligo/DOPA-responsive dystonia and the GTP-cyclohydrolase (*GCH1*) gene [26] was subsequently shown to probably be spurious [8]. Likewise, reported genetic association between vitiligo and the catalase (*CAT*) gene [18] was not confirmed by subsequent studies [32, 69, 74]. Similarly, reported association between vitiligo and variation in the gene encoding angiotensin converting enzyme (*ACE*) [44] was not confirmed in another study [2], albeit of patients from a different population. Genetic associations have also been reported between vitiligo and the genes encoding the estrogen receptor 1 (*ESR1*) [45], catechol-*O*-methyltransferase (*COMT*) [83], the vitamin D receptor (*VDR*) [11] and glutathione peroxidase (*GPX1*) [74]. These claimed associations have not yet been tested by replication studies, and several are based on marginally significant results and seem to be of uncertain validity.

Several negative studies involving potential vitiligo candidate genes have also been reported. Mutations in the *AIRE* locus result in an autosomal recessive multiple autoimmune disease syndrome (APECED, autoimmune polyendocrine syndrome type 1) that can include generalized vitiligo. However, while *AIRE* is located within a minor vitiligo linkage peak on chromosome 21q, a large, family-based association study of *AIRE* in families with generalized vitiligo and other autoimmune diseases, showed no evidence of association [47]. Likewise, variation in the *TNFA* [86], *MC1R* [63, 80], and *ASIP* [63] genes showed no association with vitiligo. Another negative study specifically tested for genetic linkage to *MITF* in families with multiple cases of vitiligo [82]. Finally, another study found no association of the *IFNG* gene in Japanese patients with Vogt-Koyanagi-Harada disease [41] (Chap. 1.3.7), a multi-system syndrome that includes panuveitis, headache, pleocytosis of the cerebrospinal fluid, alopecia, poliosis, and vitiligo. As noted above, an important consideration in evaluating any such negative study is the magnitude of genetic effect that might have been detectable, given the statistical power of the study.

The Genome-Wide Approach

Whereas candidate gene studies are limited to testing hypotheses based on known biology, the genome-wide approach scans the entire genome to identify genetic markers that flag genomic regions that may contain disease susceptibility genes. Because this approach is based on genomic position rather than known biology, findings are not limited by what is known (or believed), and the genome-wide approach therefore offers the possibility of discovering entirely new disease susceptibility genes, that highlight new pathways to the disease.

There are two quite different genome-wide approaches to disease gene identification. Genetic linkage studies can be used to scan the genome for chromosomal regions that non-randomly co-segregate with vitiligo in families in which multiple relatives are affected by vitiligo. Such "multiplex" families are not common, and the genetic underpinnings of vitiligo in multiplex families may not be the same as in the predominant singleton cases. In general, linkage studies are best able to detect loci that exert major effects but perhaps involve uncommon risk alleles. Genome-wide association studies constitute a powerful new approach, in which allele or genotype frequencies at hundreds of thousands of single-nucleotide polymorphisms (SNPs) distributed across the genome are simultaneously compared in large collections of cases vs. controls. This approach has not yet been applied to study the genetics of vitiligo, although it is likely in the near future.

The first genome-wide findings relevant to vitiligo were indirect, showing genetic linkage between a locus on chromosome 17p13, called *SLEV1*, and systemic lupus erythematosus in multiplex lupus families that included at least one relative with vitiligo [65]. The first genome-wide linkage analysis of vitiligo *per se* investigated a unique, very large, multi-generation family in which generalized vitiligo and other autoimmune diseases were inherited as an apparent autosomal dominant trait with incomplete penetrance. Vitiligo in this family was mapped to a locus termed *AIS1*, located in chromosome segment 1p31.3-p32.2, whereas susceptibility to other autoimmune diseases in the context of a co-inherited *AIS1* mutation was mapped to a region of chromosome 6 that included the MHC [3]. Detailed studies of genes in the *AIS1* region of chromosome 1p in this family identified a promoter variant in *FOXD3*, which encodes an embryonic transcription factor that regulates differentiation and development of neural crest melanoblasts and some mesodermal elements, including pancreatic islet cells. The promoter variant in this family increases transcriptional activity in transfected permissive cells by 50%, and in vivo might thus interfere with melanoblast/melanocyte differentiation or survival, somehow predisposing to vitiligo [5]. Interestingly, the generalized vitiligo phenotype in this family is somewhat unusual, consisting of progressively coalescent skin mottling. Other generalized vitiligo patients do not appear to have mutations of *FOXD3*, and other families do not show linkage to the *AIS1* region of chromosome 1p.

Genome-wide linkage analyses of smaller multiplex families with more typical generalized vitiligo have yielded a number of additional linkage signals that may reflect additional disease susceptibility genes (Table 2.2.1.1). In Caucasians, significant vitiligo linkage signals have been detected on chromosomes 7 (*AIS2*), 8 (*AIS3*), and 17p (*SLEV1*), with indications of additional suggestive linkage signals on several other chromosomes [27, 77]. The chromosome 7 and 17p linkage signals appear to derive primarily from families segregating both vitiligo and other vitiligo-associated autoimmune diseases, whereas the chromosome 8 linkage signal derived from families segregating only vitiligo [77]. In Chinese families with generalized vitiligo, genetic linkage studies have detected an entirely different set of linkage signals [20, 58], particularly on chromosome 4q13-q21, and also included signals at 1p36, 6p21-p22, 6q24-q25, 14q12-q13, and 22q12, none of which align with the linkage signals observed in Caucasian families, suggesting that different genes may be involved in the pathogenesis of vitiligo in different populations around the world. In general, these vitiligo linkage signals do not correspond to the chromosomal locations of most of the various candidate genes that have been suggested for vitiligo, with the possible exception of the MHC in 6p21-p22.

The chromosome 17p vitiligo linkage signal, detected in Caucasian multiplex vitiligo families with various other autoimmune diseases [77], coincided with the location of *SLEV1*, a linkage signal originally detected in multiplex lupus families that included at least one case of vitiligo [65], and subsequently confirmed in other lupus families with various other autoimmune diseases [49]. Taken together, these findings suggested that 17p harbours at least one gene that mediates susceptibility to vitiligo, lupus, and other autoimmune diseases. Targeted family-based association analysis of markers spanning the 17p linkage region, in the same multiplex vitiligo families used for linkage analyses (and subsequently analyzed in a second, independent set of multiplex vitiligo families) identified *NALP1* (also known as *NLRP1*, *CARD7*, *DEFCAP*, and *NAC*) as a major susceptibility gene for generalized vitiligo and the other autoimmune diseases associated with vitiligo [48]. Genetic association between *NALP1* and vitiligo was subsequently confirmed in an independent case–control study of generalized vitiligo patients from northwestern Romania [46]. *NALP1* encodes a key regulator of the innate immune system that is part of the surveillance system of Langerhans cells and T-cells. In response to binding of bacterial "pathogen-associated molecular patterns", possibly including muramyl dipeptide [29], NALP1 is thought to direct the assembly of a "NALP1-inflammasome" that activates the interleukin-1β and perhaps other inflammatory pathways [15], thereby recruiting a subsequent response by the adaptive immune system. NALP1 is also thought to stimulate cellular apoptosis, although these mechanisms remain largely undefined [15, 34]. These findings suggest that bacterial components, sensed by skin-resident immune cells system, might trigger the initiation of vitiligo, and that, drugs that modulate inflammatory or apoptotic pathways may offer new approaches to treatment or even prevention of generalized vitiligo.

The Gene Expression Approach

These analyses have generally attempted to identify genes that are differentially over- or under-expressed in cultured cells or skin tissue from vitiligo patients vs. controls, or from disease tissue vs. normal tissue from patients, either by testing mRNA levels of specific biological candidate genes or by assaying levels of large numbers of mRNAs in the entire cellular transcriptome. Obviously, as melanocytes are deficient in vitiligo lesions, comparison of vitiligo vs. non-vitiligo skin will show differences in the expression of all the genes, specifically expressed in melanocytes. Furthermore, these studies cannot distinguish genes with primary disease-related effects from the many more genes whose expression may be dysregulated on a secondary basis, or that show "differential" expression merely due to individual variation resulting from the outbred genetic background among humans.

Expression of a number of candidate genes, including eight genes of the melanocortin system (*POMC*, *MC1R*, *MC2R*, *MC3R*, *MC4R*, *MC5R*, *ASIP*, *AGRP*), *TYRP1*, and *DCT* have been assayed in lesional vs. non-lesional, skin of vitiligo patients vs. skin of controls, using quantitative reverse transcriptase-PCR [53]. While a number of expression differences for some of these genes were observed comparing lesional skin vs. non-lesional skin, and non-lesional skin of patients vs. skin of controls, it is impossible to say whether any of these are causally related to vitiligo.

VIT1 is a gene located at chromosome 2p21 (previously assigned to 2p16), originally so-named on the basis of its reduced expression in intralesional vitiligo melanocytes [57]. *VIT1*, now officially renamed *FBXO11* appears to encode a widely expressed protein arginine methyltransferase [21], partial deficiency of which is associated with otitis media in both humans [73] and mice [39], and total deficiency of which results in cleft palate, facial clefting, and perinatal lethality in mice [39]. There is no evidence that *FBXO11* is causally involved in vitiligo, and it is likely coincidental that mutations in an adjacent (but oppositely oriented) gene, *MSH6*, have recently been reported in a single patient with early-onset colorectal cancer, systemic lupus erythematosus, and vitiligo [6,70].

MYG1 is a widely expressed gene located at chromosome 12q13, found by differential hybridization to have elevated expression in melanocytes from vitiligo patients [52]. There is no evidence that *MYG1* is causally involved in the pathogenesis of vitiligo.

Expression profiling of 16,000 genes in melanocytes cultured from 5 vitiligo patients vs. 5 controls, showed significant expression differences in 859 genes, many of which are involved in controlling melanocyte development, processing and trafficking of melanogenic enzymes, melanosome biogenesis, cell adhesion, and antigen processing and presentation [78]. These authors concluded that autoimmunity involving melanocytes may be a secondary event in vitiligo patients caused by abnormal melanocyte function, although many of the changes observed may reflect statistical fluctuation due to the very small number of patients and controls studied.

2.2.1.3 Concluding Remarks

Generalized vitiligo is a disease of multifactorial, polygenic origin involving multiple genes and environmental triggers. The risk of vitiligo is about 6–7% to a patient's siblings and other first-degree relatives; this risk declines as the degree of relationship is reduced.

Generalized vitiligo is epidemiologically associated with several other autoimmune diseases, including autoimmune thyroid disease, rheumatoid arthritis, psoriasis, adult-onset autoimmune diabetes mellitus, pernicious anaemia, systemic lupus erythematosus, and Addison's disease. Several genes have been identified that appear to mediate susceptibility to vitiligo and some of these other autoimmune diseases, including loci in the MHC, *PTPN22*, and *NALP1*, and additional vitiligo susceptibility genes will likely be identified in the future. Discovery of vitiligo susceptibility genes will likely provide insights into biological pathways that mediate disease pathogenesis, offering novel interventional targets for treatment and prevention in genetically susceptible individuals.

Acknowledgements This work was supported by grants AR45584, AI46374 and AR056292 from the National Institutes of Health.

References

1. Abanmi A, Al Harthi F, Zouman A et al (2008) Association of interleukin-10 gene promoter polymorphisms in Saudi patients with vitiligo. Dis Markers 24:51–57
2. Akhtar S, Gavalas NG, Gawkrodger DJ et al (2005) An insertion/deletion polymorphism in the gene encoding angiotensin converting enzyme is not associated with generalised vitiligo in an English population. Arch Dermatol Res 297:94–98
3. Alkhateeb A, Stetler GL, Old W et al (2002) Mapping of an autoimmunity susceptibility locus (*AIS1*) to chromosome 1p31.3-p32.2. Hum Mol Genet 11:661–667
4. Alkhateeb A, Fain PR, Thody A et al (2003) Epidemiology of vitiligo and associated autoimmune diseases in Caucasian probands and their relatives. Pigment Cell Res 16:208–214
5. Alkhateeb A, Fain PR, Spritz RA (2005) Candidate functional promoter variant in the *FOXD3* melanoblast developmental regulator gene in autosomal dominant vitiligo. J Invest Dermatol 125:388–391
6. Ando I, Chi HI, Nakagawa H et al (1993) Difference in clinical features and HLA antigens between familial and non-familial vitiligo of non-segmental type. Br J Dermatol 129:408–410
7. Arcos-Burgos M, Parodi E, Salgar M et al (2002) Vitiligo: complex segregation and linkage disequilibrium analyses with respect to microsatellite loci spanning the HLA. Hum Genet 110:334–342
8. Bandyopadhyay D, Lawrence E, Majumder PP et al (2000) Vitiligo is not caused by mutations in GTP-cyclohydrolase I gene. Clin Exp Dermatol 25:152–153
9. Berends MJW, Wu Y, Sijmons RH et al (2002) Molecular and clinical characteristics of MSH6 variants: an analysis of 25 index carriers of a germline variant. Am J Hum Genet 70:26–37

10. Bhatia PS, Mohan L, Pandey ON et al (1992) Genetic nature of vitiligo. J Dermatol Sci 4:180–184
11. Birlea S, Birlea M, Cimponeriu D et al (2006) Autoimmune diseases and vitamin D receptor Apa-I polymorphism are associated with vitiligo in a small inbred Romanian community. Acta Derm Venereol 86:209–214
12. Birlea SA, Fain PR, Spritz RA (2008) A Romanian population isolate with high frequency of vitiligo and associated autoimmune diseases. Arch Dermatol 144:310–316
13. Blomhoff A, Kemp EH, Gawkrodger DJ et al (2005) CTLA4 polymorphisms are associated with vitiligo, in patients with concomitant autoimmune diseases. Pigment Cell Res 18:55–58
14. Brand O, Gough S, Heward J (2005) HLA, CTLA-4, and PTPN22: the shared genetic master-key to autoimmunity? Expert Rev Mol Med 7:1–15
15. Bruey J-M, Bruey-Sedano N, Luciano F et al (2007) Bcl-2 and Bcl-X_L regulate proinflammatory caspase-1 activation by interaction with NALP1. Cell 129:45–56
16. Canton I, Akhtar S, Gavalas NG et al (2005) A single-nucleotide polymorphism in the gene encoding lymphoid protein tyrosine phosphatase (PTPN22) confers susceptibility to generalised vitiligo. Genes Immun 6:584–587
17. Carnevale A, Zavala C, Castillo VD et al (1980) Analisis genetico de 127 families con vitiligo. Rev Invest Clin 32:37–41
18. Casp CB, She JX, McCormack WT (2002) Genetic association of the catalase gene (CAT) with vitiligo susceptibility. Pigment Cell Res 15:62–66
19. Casp CB, She JX, McCormack WT (2003) Genes of the LMP/TAP cluster are associated with the human autoimmune disease vitiligo. Genes Immun 4:492–499
20. Chen JJ, Huang W, Gui JP et al (2005) A novel linkage to generalized vitiligo on 4q13-q21 identified in a genomewide linkage analysis of Chinese families. Am J Hum Genet 76: 1057–1065
21. Cook JR, Lee JH, Yang ZH et al (2006) FBXO11/PRMT9, a new protein arginine methyltransferase, symmetrically dimethylates arginine residues. Biochem Biophys Res Commun 342:472–481
22. Cunliffe WJ, Hall R, Newell DJ et al (1968) Vitiligo, thyroid disease and autoimmunity. Br J Dermatol 80:135–139
23. Das SK, Majumder PP, Chakraborty R et al (1985) Studies on vitiligo. I. Epidemiological profile in Calcutta, India. Genet Epidemiol 2:71–78
24. Das SK, Majumder PP, Majumdar TK et al (1985) Studies on vitiligo. II. Familial aggregation and genetics. Genet Epidemiol 2:255–262
25. Dawber RP (1970) Integumentary associations of pernicious anemia. Br J Dermatol 82:221–223
26. de la Fuente-Fernandez R (1997) Mutations in GTP-cyclohydrolase I gene and vitiligo. Lancet 350:640
27. Fain PR, Gowan K, LaBerge GS et al (2003) A genomewide screen for generalized vitiligo: confirmation of *AIS1* on chromosome 1p31 and evidence for additional susceptibility loci. Am J Hum Genet 72:1560–1564
28. Fain PR, Babu SR, Bennett DC et al (2006) HLA class II haplotype DRB1*04-DQB1*0301 contributes to risk of familial generalized vitiligo and early disease onset. Pigment Cell Res 19:51–57
29. Faustin B, Lartigue L, Bruey J-M et al (2007) Reconstituted NALP1 inflammasome reveals two-step mechanism of caspase-1 activation. Mol Cell 25:713–724
30. Finco O, Cuccia M, Martinetti M et al (1991) Age of onset in vitiligo: relationship with HLA supratypes. Clin Genet 39:48–54
31. Foley LM, Lowe NJ, Misheloff E et al (1983) Association of HLA-DR4 with vitiligo. J Am Acad Dermatol 8:39–40
32. Gavalas NG, Akhtar S, Gawkrodger DJ et al (2006) Analysis of allelic variants in the catalase gene in patients with the skin depigmenting disorder vitiligo. Biochem Biophys Res Commun 345:1586–1591
33. Gough SCL, Walker LSK, Sansom DM (2005) *CLTA4* gene polymorphism and autoimmunity. Immunol Rev 204: 102–115
34. Gregersen P (2007) Modern genetics, ancient defenses, and potential therapies. New Engl J Med 356:1263–1266
35. Grunnet I, Howitz J (1979) Vitiligo and pernicious anemia. Arch Dermatol 101:82–85
36. Hafez M, Sharaf L, El-Nabi SMA (1983) The genetics of vitiligo. Acta Dermatovener 63:249–251
37. Handa S, Kaur I (1999) Vitiligo: clinical findings in 1436 patients. J Dermatol 26:653–657
38. Handa S, Dogra S (2003) Epidemiology of childhood vitiligo: a study of 625 patients from north India. Pediatr Dermatol 20:207–210
39. Hardisty-Hughes RE, Tateossian H, Morse SA (2006) A mutation in the F-box gene, Fbxo11, causes otitis media in the Jeff mouse. Hum Mol Genet 15:3273–3279
40. Hirschhorn JN, Lohmueller K, Byrne E (2002) A comprehensive review of genetic association studies. Genet Med 4:45–61
41. Horie Y, Kitaichi N, Takemoto Y et al (2007) Polymorphism of IFN-γ gene and Vogt-Konanagi-Harada disease. Mol Vis 13:2334–2338
42. Howitz J, Brodthagen H, Schwartz M et al (1977) Prevalence of vitiligo: epidemiological survey of the Isle of Bornholm, Denmark. Arch Dermatol 113:47–52
43. Itirli G, Pehlivan M, Alper S et al (2005) Exon-3 polymorphism of CTLA-4 gene in Turkish patients with vitiligo. J Dermatol Sci 38:225–227
44. Jin SY, Park HH, Li GZ et al (2004) Association of angiotensin converting enzyme gene I/D polymorphism of vitiligo in Korean population. Pigment Cell Res 17:84–86
45. Jin SY, Park HH, Li GZ et al (2004) Association of estrogen receptor 1 intron 1 C/T polymorphism in Korean vitiligo patients. J Dermatol Sci 35:181–186
46. Jin Y, Birlea SA, Fain PR et al (2007) Genetic variations in *NALP1* are associated with generalized vitiligo in a Romanian population. J Invest Dermatol 127:2558–2562
47. Jin Y, Fain PR, Bennett DC et al (2007) Vitiligo-associated multiple autoimmune disease is not associated with genetic variation in *AIRE*. Pigment Cell Res 20:402–404
48. Jin Y, Mailloux CM, Gowan K et al (2007) *NALP1* in vitiligo-associated multiple autoimmune disease. N Engl J Med 356:1216–1225
49. Johansson CM, Zunec R, Garcia MA et al (2004) Chromosome 17p12-q11 harbors susceptibility loci for systemic lupus erythematosus. Hum Genet 115:230–238
50. Kemp EH, Ajjan RA, Waterman EA et al (1999) Analysis of a microsatellite polymorphism of the cytotoxic T-lymphocyte antigen-4 gene in patients with vitiligo. Br J Dermatol 140:73–78
51 .Kingo K, Philips MA, Aunin E et al (2006) MYG1, novel melanocyte related gene, has elevated expression in vitiligo. J Dermatol Sci 44:119–122

52. Kingo K, Aunin E, Karelson M et al (2007) Gene expression analysis of melanocortin system in vitiligo. J Dermatol Sci 48:113–122
53. Laberge G, Mailloux CM, Gowan K et al (2005) Early disease onset and increased risk of other autoimmune diseases in familial generalized vitiligo. Pigment Cell Res 18: 300–305
54. Laberge G, Birlea SA, Fain PR et al (2008) The *PTPN22* -1858C > T (R620W) functional polymorphism is associated with generalized vitiligo in the Romanian population. Pigment Cell Melanoma Res 21:206–208
55. LaBerge GS, Bennett DC, Fain PR et al (2008) PTPN22 is geneticaly associated with not of generalized vitiligo, but CTLA4 is not. J. Invesitg J. Investig Dermatol 128: 1757–1762
56. Le Poole IC, Sarangarajan R, Zhao Y et al (2001) "VIT1", a novel gene associated with vitiligo. Pigment Cell Res 14:475–484
57. Liang Y, Yang S, Zhou Y et al (2007) Evidence for two susceptibility loci on chromosomes 22q12 and 6p21-p22 in Chinese generalized vitiligo families. J Invest Dermatol 127:2552–2557
58. Liu JB, Li M, Chen H et al (2007) Association of vitiligo with HLA-A2: a meta-analysis. J Eur Acad Dermatol Venereol 21:205–213
59. Majumder PP, Das SK, Li CC (1988) A genetical model for vitiligo. Am J Hum Genet 43:119–125
60. Majumder PP, Nordlund JJ, Nath SK (1993) Pattern of familial aggregation of vitiligo. Arch Dermatol 129:994–998
61. Mehta NR, Shah KC, Theodore C, et al (1973) Epidemiological study of vitiligo in Surat area, South Gujarat. Indian J Med Res 61:145–154
62. Na GY, Lee KH, Kim MK et al (2003) Polymorphisms in the melanocortin-1 receptor (MC1R) and agouti signaling protein (ASIP) genes in Korean vitiligo patients. Pigment Cell Res 16:383–387
63. Nath SK, Majumder PP, Nordlund JJ (1994) Genetic epidemiology of vitiligo: multilocus recessivity cross-validated. Am J Hum Genet 55:981–990
64. Nath SK, Kelly JA, Namjou B et al (2001) Evidence for a susceptibility gene, *SLEV1,* on chromosome 17p13 in families with vitiligo-related systemic lupus erythematosus. Am J Hum Genet 69:1401–1406
65. Onay H, Pehlivan M, Alper S et al (2007) Might there be a link between mannose binding lectin and vitiligo? Eur J Dermatol 17:146–148
66. Onunu AN, Kubeyinje EP (2003) Vitiligo in the Nigerian African: a study of 351 patients in Benin City, Nigeria. Int J Dermatol 42:800–802
67. Orecchia G, Perfetti L, Malagoli P et al (1992) Vitiligo is associated with a significant increase in HLA-A30, Cw6 and DQw3 and a decrease in C4AQ0 in northern Italian patients. Dermatology 185:123–127
68. Park HH, Ha E, Uhm YK et al (2006) Association study between catalase gene polymorphisms and the susceptibility to vitiligo in Korean population. Exp Dermatol 15: 377–380
69. Rahner N, Höefler G, Högenauer C et al (2008) Compound heterozygosity for two *MSH6* mutations in a patient with early onset colorectal cancer, vitiligo and systemic lupus erythematosus. Am J Med Genet A 146:1314–1319
70. Schallreuter KU, Levenig C, Kühnl P et al (1993) Histocompatibility antigens in vitiligo: Hamburg study on 102 patients from northern Germany. Dermatology 187: 186–192
71. Schallreuter KU, Lemke R, Brandt O et al (1994) Vitiligo and other diseases: coexistence or true association? Dermatol 188:269–275
72. Segade F, Daly KA, Allred D et al (2006) Association of the FBXO11 gene with chronic otitis media with effusion and recurrent otitis media: the Minnesota COME/ROM family study. Arch Otolaryngol Head Neck Surg 132:729–733
73. Shajil EM, Laddha NC, Chatterjee S et al (2007) Association of catalase T/C exon 9 and glutathione peroxidase 200 polymorphisms in relation to their activities and oxidative stress with vitiligo susceptibility in Gujarat population. Pigment Cell Res 20:405–407
74. Sharma VK, Dawn G, Kumar B (1996) Profile of alopecia areata in Northern India. Int J Dermatol 35:22–27
75. Sharma VK, Kumar V, Dawn G (1996) A clinical study of childhood alopecia areata in Chandigarh. India Pediatr Dermatol 13:372–377
76. Spritz RA, Gowan K, Bennett DC et al (2004) Novel vitiligo susceptibility loci on chromosomes 7 (*AIS2*) and 8 (*AIS3*), confirmation of *SLEV1* on chromosome 17, and their roles in an autoimmune diathesis. Am J Hum Genet 74:188–191
77. Strömberg S, Björklund MG, Asplund A et al (2008) Transcriptional profiling of melanocytes from patients with vitiligo vulgaris. Pigment Cell Melanoma Res 21: 162–171
78. Sun X, Xu A, Wei X et al (2006) Genetic epidemiology of vitiligo: a study of 815 probands and their families from south China. Int J Dermatol 45:1176–1181
79. Széll M, Baltás E, Bodai L et al (2008) The Arg160Trp allele of melanocortin-1 receptor gene might protect against vitiligo. Photochem Photobiol 84:565–571
80. Tastan HB, Akar A, Orkunoglu FE et al (2004) Association of HLA class I antigens and HLA class II alleles with vitiligo in a Turkish population. Pigment Cell Res 17: 181–184
81. Tripathi RK, Flanders DJ, Young TL et al (1999) Microphthalmia-associated transcription factor (MITF) locus lacks linkage to human vitiligo or osteopetrosis: an evaluation. Pigment Cell Res 12:187–192
82. Tursen U, Kaya TI, Erdal ME et al (2002) Association between catechol-*O*-methyltransferase polymorphism and vitiligo. Arch Dermatol Res 294:143–146
83. Vang T, Miletic AV, Bottini N et al (2007) Protein tyrosine phosphatase PTPN22 in human autoimmunity. Autoimmunity 40:453–461
84. Xia Q, Zhou WM, Liang YH et al (2006) MHC haplotypic association in Chinese Han patients with vitiligo. J Eur Acad Dermatol Venereol 20:941–946
85. Yazici AC, Erdal ME, Kaya TI et al (2006) Lack of association of TNF-α-308 promoter polymorphism in patients with vitiligo. Arch Dermatol Res 298:46–49
86. Zamani M, Spaepen M, Sghar SS, et al (2001) Linkage and association of HLA class II genes with vitiligo in a Dutch population. Br J Dermatol 145:90–94
87. Zelissen PM, Bast EJ, Croughs RJ (1995) Associated autoimmunity in Addison's disease. J Autoimmun 8:121–130
88. Ying J, Birlea SA, Fain PR et al (2010) Genomewide association study of generalized vitiligo in Caucasians identifies disease-specific and shared autoimmunity susceptibility loci. New Engl J Med, in press.

Section 2.2.2

Environmental Factors

The Kobner's Phenomenon

2.2.2.1

Yvon Gauthier and Laila Benzekri

Contents

Core Messages

- Fitzpatrick and ourselves described the frequency of vitiligo lesions vulgaris at sites subjected to repeated trauma such as continuous pressure or repeated frictions of various origins.
- In non-segmental vitiligo, the incidence of Kobner's Phenomenon (KP) is very varied according to reports, that is, from 15 to 70%.
- Wounds, scars, burns, abrasion, laser, and other types of surgical abrasion are the more frequent inducing factors of the KP.
- Many mechanisms have been hypothesised: increased release of neuropeptides noxious for melanocytes, detachment and transepidermal elimination, and lower secretion of keratinocyte-derived factors. In normal vitiligo skin, a minor trauma (tape stripping) is possible after 72 h to induce the formation of autophagic vacuoles containing polymelanosomes and the detachment of few melanocytes from the basement membrane.
- In the presence of a history of vitiligo in the family, the onset of permanent depigmentation following repeated scratches in children without vitiligo could indicate a 'vitiligo diathesis.'

Historical Aspects and Definition

Human skin is repeatedly exposed to mechanical trauma and this prominent feature has often been neglected. In 1877, Heinrich Koebner (Fig. 2.2.2.1a) [19] described a phenomenon in psoriasis which bears his name, and has provided an explanation for the sites commonly

Y. Gauthier (✉)
Service de Dermatologie, Hôpital Saint-André.
CHU de Bordeaux, France
e-mail: yvongauthier@free.fr

M. Picardo and A. Taïeb (eds.), *Vitiligo*,
DOI 10.1007/978-3-540-69361-1_2.2.2.1, © Springer-Verlag Berlin Heidelberg 2010

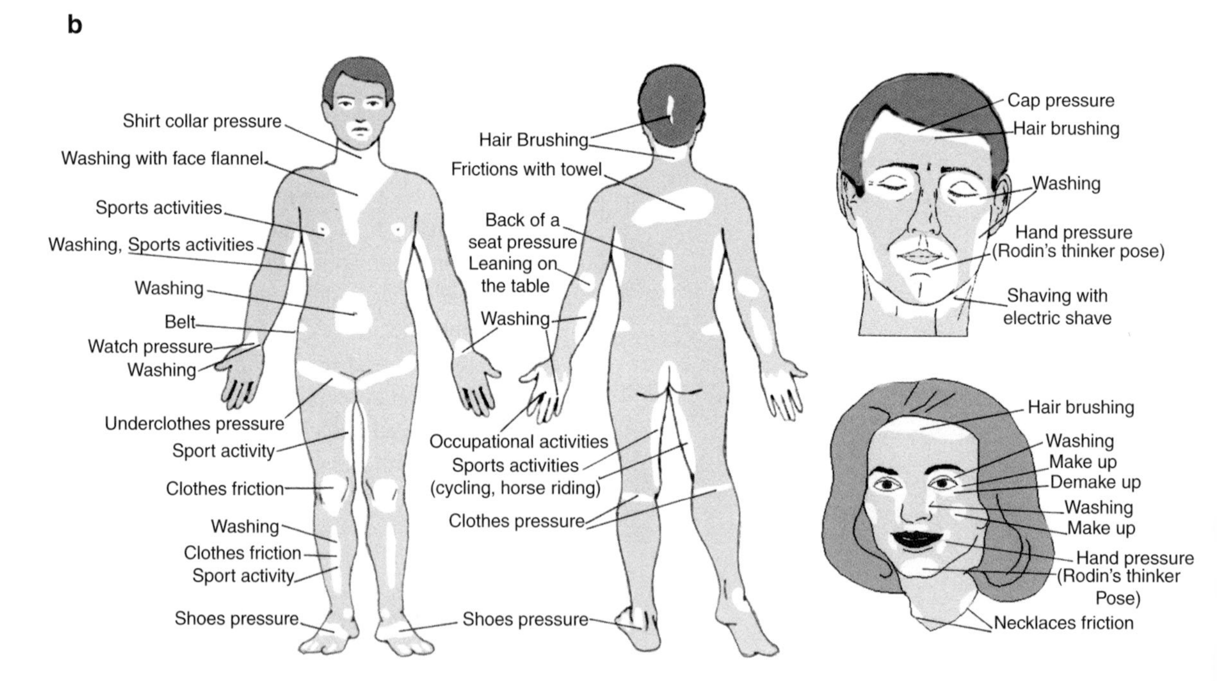

Fig. 2.2.2.1 (**a**) Heinrich Köbner (1838–1904) founder of the dermatology clinic in Breslau (Germany, now Wroclaw, Poland) which was later led by the famous Neisser and Jadassohn. He described his eponymic phenomenon in 1872 in a case of psoriasis; (**b**) the most frequent anatomical sites of vitiligo lesions and the hypothetic factors susceptible to inducc KP

involved in psoriasis. This phenomenon is defined according to Beetley as 'a specific response to an aspecific usually traumatic stimulus [4]'. Koebner's phenomenon (KP) was later found in a number of dermatoses including lichen planus, vitiligo, etc. Kaposi was the first to point out in 1890, that the starting point of some vitiligo patches could be determined by local conditions, and that leukodermas could develop on sites of repeated trauma [17]. Levaï [22, 23] later observed in a number of patients that pressure from a sari tightened at the waist and cutaneous excoriations were frequently involved in the appearance of achromic vitiligo lesions of the trunk and abdomen. Finally, Fitzpatrick described the frequency of vitiligo lesions vulgaris at sites subjected to repeated trauma such as continuous pressure or repeated frictions of various origins [9].

Clinical Features

Aspect

In vitiligo, KP generates the onset of isomorphic depigmented lesions (Fig. 2.2.2.2). Usually, these isomorphic depigmented lesions following scratches are easily recognisable by their artefactual or elongated 'linear' shape [6]. In some cases, the isomorphic response can also take the appearance of trichrome vitiligo [6]. In other cases, the shape of the macules corresponds exactly to the shape of the agent in very intimate contact with the integument (bracelets or watches, fingers pads) [29, 31].

Incidence

In non-segmental vitiligo (NSV), the incidence of KP is very variable according to reports, from 15 up to 70% [28–32]. In the segmental form of vitiligo, KP was neither found by Koga and Tango [20], nor by Njoo et al. [27], but reported by Hann et al. [15] and Barona et al. [3] These discordant findings suggest that a redefinition of KP in vitiligo is needed.

Gopinathan [14] observed depigmentation of clinically non-involved skin in 70% of patients with vitiligo studied after performing scarification procedures. In the same way, Njoo et al. [27] have studied the incidence of the experimentally induced Koebner's phenomenon (Kpe) by performing an epidermo–dermal injury on non-depigmented skin with a 2 mm punch biopsy. The incidence of Kpe was compared to the assessment of Kp by history (KPh). The patients were asked to recall the onset of depigmentation after burns, wounds, or scratches. Kpe occurred in most patients with non-segmental vitiligo and KP-h was less frequently found. This is the likely explanation of discordant data published earlier. The author's conclusion was that the KPh lacks accuracy, because in many cases patients do not recognise or notice the appearance of depigmentation after a skin injury. Moreover, it is also possible that some patients had never had a skin trauma. So, to obtain a realistic assessment of the incidence of KP in vitiligo, experimentally induced KP is probably needed.

Identification of Factors Causing KP

Schematically, several kinds of precipitating factors have to be identified in vitiligo such as:

- The KP following an epidermo–dermal injury including scratches [14, 27, 29] (Fig. 2.2.2.3), wounds, scars, burns, abrasion, laser, and other types of surgical abrasion. This type of KP has been reported in the majority of cases of non-segmental vitiligo.

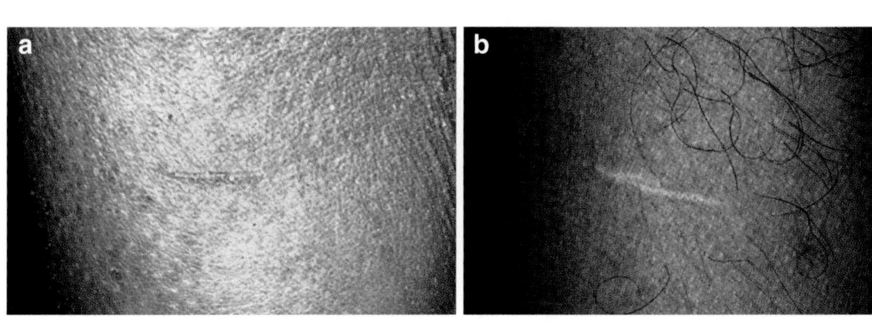

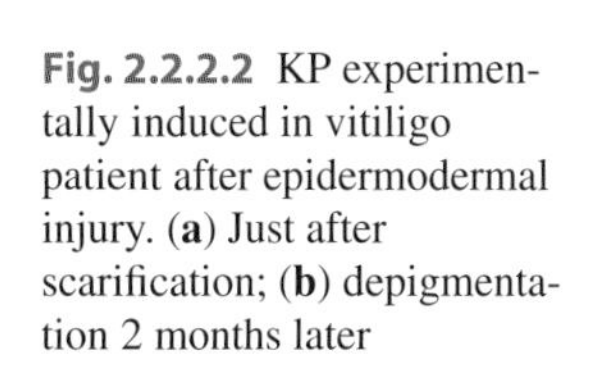

Fig. 2.2.2.2 KP experimentally induced in vitiligo patient after epidermodermal injury. (**a**) Just after scarification; (**b**) depigmentation 2 months later

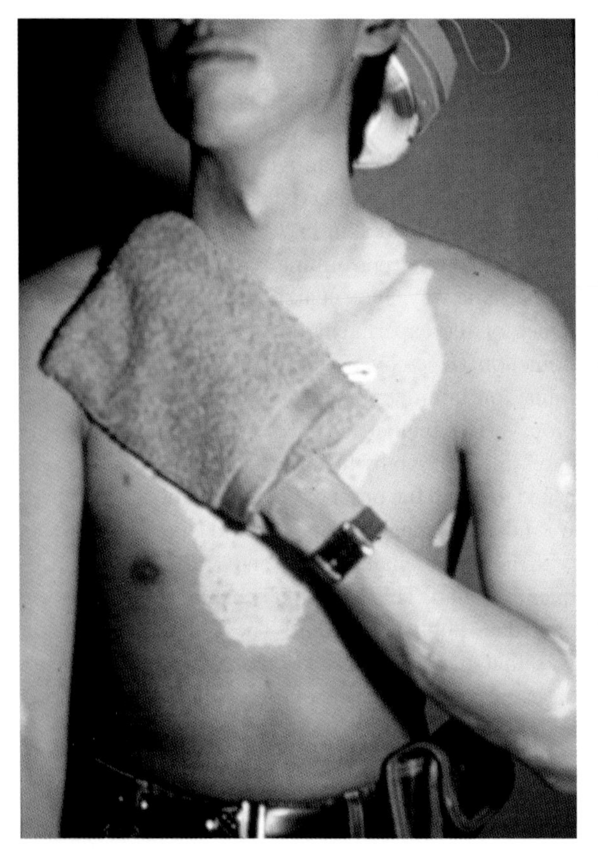

Fig. 2.2.2.3 KP induced by repeated frictions

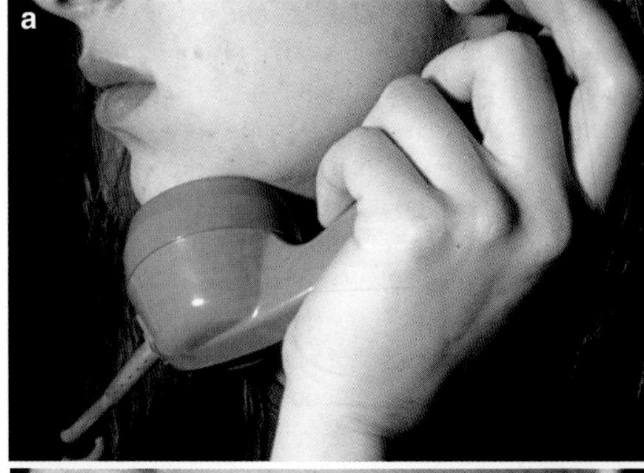

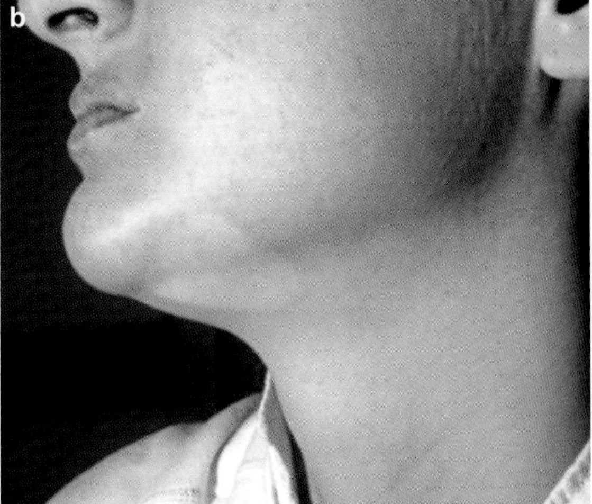

Fig. 2.2.2.4 KP induced by continuous pressures. (**a**) The continuous use of the telephone, (**b**) the consequent KP

- The KP occurring after repeated superficial trauma, such as continued pressure [10, 29] (Fig. 2.2.2.3), frictions [5, 10, 22, 23, 32] (Fig. 2.2.2.4), local pruritus [2, 23], and UV irradiation. This type of KP has been observed only in unstable, progressive non-segmental vitiligo.
- The KP following some cutaneous diseases such as lichen planus, psoriasis [7].

In our experience, at onset of vitiligo, significant epidermo–dermal injury appears to be required to induce koebnerisation, such as cuts, wounds, and surgical wounds. But after a few months or years in progressive NSV, superficial trauma such as constant pressure or repeated frictions could be sufficient to induce depigmentation. KP could then appear when a certain 'threshold' value of external forces is exceeded in a susceptible individual.

Consequently, it has been hypothesised [10, 11] that many lesions of NSV could be related to repeated frictions (occurring during washing, dressing, personal care, sports, and occupational activities) or continuous pressure (from clothing or other items) (Fig. 2.2.2.1b).

Pathomechanisms

The mechanism for onset of the isomorphic response is poorly understood. It is well-established that the clinically 'normal' skin of patients with vitiligo is not normal histologically. The keratinocytes exhibit dystrophic changes with vacuoles, there is an extracellular accumulation of granular material, and autophogosomes can be detected in melanocytes [24]. These observations have suggested that the melanocytes in the normal appearing skin bear sequels of susceptibility to injury. Many hypotheses have been proposed. Microtrauma could lead to increased release of neuropeptides noxious for melanocytes [1] Due to a supposed adhesion abnormality of melanocytes, repeated local mechanical traumas could induce melanocyte loss by means of their detachment and transepidermal elimination [12, 13]. A lower secretion of keratinocyte-derived factors including stem-cell factor (SCF) in vitiliginous keratinocytes could result from keratinocyte apoptosis and might be responsible for passive melanocyte death and might thus explain the KP [21]. More recently, it has been speculated that inflammasome activation via an alternative mechanism involving the stratum corneum chymotryptic enzyme (SCCE or kallikrein 7) could take place. Mechanical traumas might activate this proteolytic enzyme and IL1 beta-driven microinflammation could destabilise the anchoring system of basal melanocytes [18]. In normal skin, after 72 h a minor trauma (tape stripping) is able to induce the formation of autophagic vacuoles containing polymelanosomes and the detachment of few melanocytes from the basement membrane [25]. The KP generated by minor trauma may represent an example of such vulnerability.

The KP in the Vitiligo Clinic

The KP is not exclusively a 'vitiligo associated skin manifestation' as mentioned in most textbooks. The observation and the study of KP on the skin of patients with vitiligo could give valuable indications which are detailed here.

Disease Activity Assessment

It is very difficult to evaluate if vitiligo is active or stable. Many approaches have been reported. According to Moellmann et al. [24] unstable vitiligo is the clinical stage 'when lesions are enlarging over a period of 6 weeks.' Stable vitiligo is defined by Falabella et al. [8] 'as a condition that has not been progressing for at least 2 years.' The use of experimentally induced KP associated with the establishment of a clinical score would be a powerful tool to assess disease activity.

Prediction of Vitiligo in at Risk Individuals

When there is a history of vitiligo in the family, the onset of permanent depigmentation following repeated scratches in children without vitiligo could possibly indicate a 'vitiligo diathesis' [3]. In many cases, depigmentation following scratches which occur frequently in association with halo nevi are preceding the onset of vitiligo. They could be considered as 'warning signs.'

Pathogenesis of Vitiligo

The KP may be utilised to study the early changes occurring in vitiligo lesions. Unfortunately, chronological histological studies of the KP are lacking [14].

Identification of Disease Characteristics that Help Predicting the Outcome of Therapies

Do the patients with a positive KP have a different prognosis than those with a negative KP? According to Njoo et al. [27], experimentally induced KP could not predict the outcome of UV-B (311 nm) phototherapy. This suggests that patients both in active or stable stages of the disease may respond equally well to UV-B therapy.

Before melanocyte transplantation, a minigraft test is frequently performed. It is important to assess the absence of koebnerisation at the donor site. For some authors, koebnerisation at the donor site could be a contraindication to surgical treatment [16]. However, Mulekar reported recently a good repigmentation at recipient areas coexisting with a depigmentation at the donor site. For this author, the degree and depth of skin trauma may be greater at the donor site and could induce more easily the KP locally than on recipient site [26]. In our experience, the most important recommendation is to avoid during at least six months strong mechanical traumas on the grafted area, which may be responsible for the recurrence of the depigmentation.

Prevention of Onset of New Lesions and Improvement of Repigmentation

Before undergoing elective surgical or cosmetic procedures such as facial peels, electrolysis, dermabrasion, laser treatment on the skin of patient with vitiligo, the risk of inducing new lesions must be evaluated. Such patients should be warned about the possibility of developing further vitiligo at treatment sites. Along the same line, the removal of all factors (i.e. strong frictions, continuous pressure) which can lead to KP may prevent the appearance of unsightly achromic lesions on exposed areas of vitiligo patients [10]. Usually, when the mechanical factor is removed, the skin can repigment more easily if the local melanocyte loss is only partial.

Concluding Remarks

The human skin is exposed daily to many environmental factors. Among these, the noxious influence of UV irradiation on the skin has been largely investigated, but the importance of mechanical trauma has been neglected. The KP had to be considered not exclusively as 'an associated skin manifestation' of vitiligo and other cutaneous diseases, but as a fascinating process able to give the physician any information about the pathogenesis and cause of skin disorders.

References

1. Al'Abadie MS, Senior HJ, Bleehen SS et al (1994) Neuropeptide and neuronal marker studies in vitiligo. Br J Dermatol 131:160–165
2. Asboe-Hansen G (1954) Depigmentation itching. Acta Derm Venereol 34:1–3
3. Barona MI, Arrunategui A, Falabella R et al (1995) An epidemiologic case-control study in a population with vitiligo. J Am Acad Dermatol 33:621–625
4. Beetley F (1951) The provocation of cutaneous disease: Koebner's isomorphic phenomenon. Arch Midds Hosp 1:279–287
5. Bleehen SS, Hall-Smith P (1970) Brassiere depigmentation: light and electron microscope studies Br J Dermatol 83:157–160
6. Dupré A, Christol B (1978) Cockade-like vitiligo and linear vitiligo; a variant of Fitzpatrick's trichrome vitiligo. Arch Dermatol Res 262:197–203
7. Eyre RW, Krueger GG (1982) Response to injury of skin involved and uninvolved with psoriasis and its relation to disease activity: Koebner and reverse Koebner reactions. Br J Dermatol 106:153–159
8. Falabella R, Arunategui A, Barona MI et al (1995) The minigrafting test for vitiligo detection of stable lesions for melanocyte transplantation. J Am Acad Dermatol 32: 228–232
9. Fitzpatrick TB (1971) Vitiligo. Dermatology in general Medicine. Mc Graw Hill, New York, pp 1596–1600
10. Gauthier Y (1995) The importance of Koebner's phenomenon in the induction of vitiligo vulgaris lesions. Eur J Dermatol 5:704–708
11. Gauthier Y, Bazeille J-E (1992) Autologous grafting with non cultured melanocytes: a simplified method for treatment of depigmented lesions. J Am Acad Dermatol 26: 191–194
12. Gauthier Y, Cario-André M, Lepreux S et al (2003) Melanocyte detachment after skin friction in non lesional skin of patients with generalized vitiligo. Br J Dermatol 148:95–101
13. Gauthier Y, Cario-André M, Taieb A (2003) A critical appraisal of vitiligo etiologic theories. Is melanocyte loss a melanocytorrhagy? Pigment Cell Res 16:322–327
14. Gopinathan T (1965) A study of the lesion of vitiligo. Arch Dermatol 91:397–404
15. Hann SK, Chun WH, Park YK (1997) Clinical charasteristics of progressive vitiligo. Int J Dermatol 36:353–355
16. Hatchome N, Kato T, Tagami H (1990) Therapeutic success of epidermal grafting in generalized vitiligo is limited by the Koebner's phenomenon. J Am Acad Dermatol 22 : 87–91
17. Kaposi M (1891) Pathologie et traitement des maladies de la peau.Traduction Doyon and Besnier, Masson, France, pp 105–110
18. Jin Y, Mailloux C, Gowan K et al (2007) NALP1 in vitiligo-associated multiple autoimmune disease. N Engl J Med 356:1216–1225
19. Koebner H (1877) Zur Aetiologie der Psoriasis. Vrtljscher Dermatol Syphil 1876 8:559–561
20. Koga M, Tango T (1988) Clinical features and course of type A and type B vitiligo. Br J Dermatol 118:223–228
21. Lee AY, Kim NH, Choo W-J et al (2005) Less keratinocyte-derived factors related to more keratinocyte apoptosis in depigmented than normally pigmented suction-blistered epidermis may cause passive melanocyte death in vitiligo. J Invest Dermatol 124:976–983
22. Levaï M (1958) A study of certain contributory factors in the development of vitiligo in South Indian patients. Arch Dermatol 78:364–371
23. Levaï M (1958) The relation ship of pruritus and local skin conditions to the development of vitiligo. Arch Dermatol 78:372–377
24. Moellmann G, Klein-Angerer S, Scollary DA et al (1982) Extracellular granular material and degeneration of keratinocytes in the normally pigmented epidermis with vitiligo. J Invest Dermatol 79:321–330
25. Mottaz JH, Thorne G, Zelicson As (1971) Response of the epidermal melanocyte to minor trauma. Arch Dermatol 104:611–618
26. Mulekar S (2007) Koebner's phenomenon in vitiligo: not always an indication of surgical failure. Arch Dermatol 143:801–802
27. Njoo MD, Das PK, Bos JD et al (1999) Association of the Koebner's phenomenon with disease activity and therapeutic responsiveness in vitiligo vulgaris. Arch Dermatol 135: 407–413
28. Ormsby OS, Montgomery H (1948) Koebner's phenomenon in dermatology: a study and report of some unusual stigmata of this phenomenon. Lea and Febrigerm Philadelphia, pp 306

29. Ortonne JP (1983) Vitiligo. In: Ortonne JP (ed) Vitiligo and other hypomelanosis of hair and skin. Plenum, New York, pp 163–310
30. Schallreuter KU, Lemke R, Brandt Owesthofen M et al (1994) Vitiligo and other diseases: coexistence or true association? Dermatology 188:268–275
31. Seif El-Nas RH, El-Hefnawi H (1963) Koebner's phenomenon in dermatology: a study and report of some unusual stigmata of this phenomenon. J Egypt Med Assoc 46: 1067–108
32. Sweet RD (1978) Vitiligo as Koebner's phenomenon. Br J Dermatol 99:223–224

Occupational Vitiligo

2.2.2.2

Raymond E. Boissy

Contents

R. E. Boissy
University of Cincinnati College of Medicine,
Cincinnati, OH, USA
e-mail:boissy.raymond@uc.edu

Core Messages

- Certain environmental chemicals may be selectively toxic to melanocytes and thus may be responsible for instigating vitiligo.
- The most common melanotoxic agents are aromatic or aliphatic derivatives of phenols and catechols.
- Vitiligo patients seem more susceptible to these agents, suggesting a genetic predisposition.
- Recent evidence suggests that tyrosinase related protein-1 (Tyrp-1) may mediate the action of tertiary butyl phenol (TBP).
- Tyrp-1 may catalyze the conversion of phenols and catechols into their quinone/indole intermediates that induce oxidative stress thus mediating melanocyte cytotoxicity.

One distinctive form of vitiligo has been classified as occupational or contact vitiligo [7, 8, 22]. This form of vitiligo is unique, in that its onset appears to be correlated with exposure to certain chemicals in the workplace or at home. However, eventually, the cutaneous depigmentation extends from the site of chemical contact and develops into progressive, generalised vitiligo, well beyond the original site [10].

Chemicals Triggering Occupational Vitiligo

There is anecdotal and much experimental evidence demonstrating that certain environmental chemicals

M. Picardo and A. Taïeb (eds.), *Vitiligo*,
DOI 10.1007/978-3-540-69361-1_2.2.2.2, © Springer-Verlag Berlin Heidelberg 2010

may be selectively toxic to melanocytes, and thus may be responsible for instigating vitiligo [10, 24, 36]. Specifically, these environmental toxins are aromatic or aliphatic derivatives of phenols and catechols (i.e. hydroquinone, monobenzyl ether of hydroquinone, 2,4-di-*tert*-butylphenol (DTBP), *p*-*tert*-butylphenol (PTBP), *p*-methyl-catechol, *p*-isopropylcatechol, *p*-choloresorcinol, *p*-cresol, diisopropyl fluorophosphate, and physostingmine). These compounds have been shown to be preferentially toxic to melanocytes both in culture and in vivo [6, 14, 22]. In fact, these compounds have been added to bleaching creams, products used to remove hyperpigmented lesions. Interestingly, these creams are not toxic to all individuals. Even at high dosages, only a subset of humans depigment in response to application. In contrast, patients with extensive vitiligo readily depigment in response to application. This observation suggests that these agents are not simple poisons for melanocytes, but are injurious to only those genetically susceptible (i.e. vitiligo patients).

In addition to phenolic/catecholic derivatives, other chemicals have been shown to precipitate vitiligo [5, 12, 18, 23, 41–43, 47]. These compounds consist of sulfhydryls (β-mercapto-ethylamine hydrochloride, cysteamine dihydrochloride), systemic medications (chloroquine, fluphenazine) and others (mercurials, arsenic, corticosteroids, chloroquine, resorcinol, cp-cresol) (Table 2.2.2.1).

Clinical Observations

Depigmentation

An important anecdotal clue implicating these environmental agents in melanocyte cytotoxicity is seen in cases where a definitive correlative factor is associated with the onset of vitiligo. Depigmentation develops in a significant number of people who work with, or are exposed to phenolic and catecholic derivatives [8]. Cutaneous depigmentation caused by phenolic derivatives was first described by Oliver et al. [35]. They reported that a number of workers who wore rubber gloves in a leather manufacturing company exhibited depigmentation over areas of their hands and forearms covered by the gloves. Subsequent patch test using mono-benzyl ether of hydroquinone (MBEH), a component in the gloves, caused a positive reaction on affected workers only. Additional reports have subsequently been published demonstrating occupational/contact vitiligo developing in people working with rubber [33, 39] and industrial oils [13] containing phenolic antioxidants or plasticisers, phenolic detergents germicides [21], paratertiary-butyphenol containing adhesives [2], and in the general manufacturing of these chemicals [9, 34] (Fig. 2.2.2.5). Studies have shown that greater than 2% of individuals who are exposed to these chemicals rapidly develop a vitiligo-

Table 2.2.2.1 Chemicals that putatively instigate occupational vitiligo

Phenolic/catecholic derivatives	Sulfydryls	Miscellaneous
Butylphenol	Cystamine dihydrocholoride	Arsenic
Butylcatechol	*N*-(2-mercaptoethyl)-dimethylamine hydrochloride	Azaleic acid
Amylphenol	β-Mercaptoethylamine hydrochloride	Benzyl alcohol
Hydroquinone	3-Mercaptopropylamine hydrochloride	Chloroquine
Monobenzyl ether of hydroquinone	Sulfanolic acid	Cinnamic aldehyde
Monomethyl ether of hydroquinone		Corticosteroids
Monoethyl ether of hydroquinone		Diisopropyl fluorophosphates
Phenylphenol		Triethylene-thiophosphoramide
Octylphenol		Physostigmine
Nonylphenol		*Para*-phenylene diamine
Isopropylcatechol		Mercurials
Methylcatechol		Guanonitrofuracin
Butylated hydroxytoluene		Fluphenazine
Butylated hydroxyanisole		
Cresol		

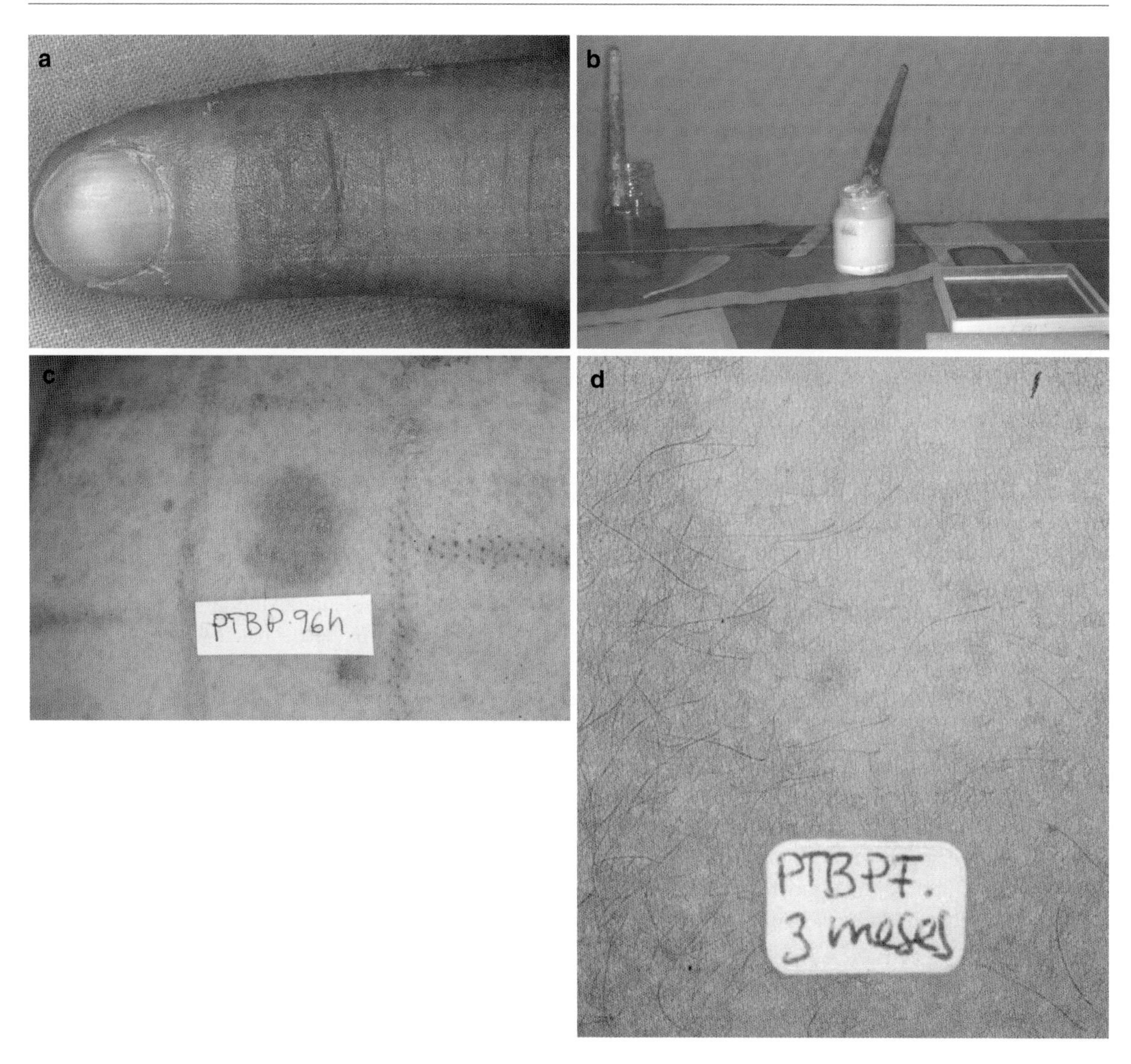

Fig. 2.2.2.5 Local contact depigmentation (Stage I) after use of a resin of *p-tert*-butylphenol in a leather goods worker (**a**). The resin (**b**) is used as an adhesive for making purses or shoes. Depigmentation 3 months after patch testing (**c**, **d**) (courtesy Luis Conde Salazar, Madrid)

like syndrome, while other exposed individuals require years of exposure to develop pigment loss [33]. Exposure to hair dyes containing *para*-phenylene diamine and benzyl alcohol [3, 45], footwear and other clothing articles containing azo dyes [1, 4, 37], or compounded phenol cream [17] have also been correlated with the development of vitiligo in some individuals. All these above observations suggest that there is a genetic variability in the response to these environmental contaminants. In addition, this suggests that melanocytes in vitiligo patients are genetically susceptible to the cytotoxic action of these phenolic/catecholic agents. A list of sites where one of these phenolic/catecholic derivatives, tertiary butyl phenol (TBP), can be found is provided in Table 2.2.2.2, as an example of how widespread these toxic compounds are.

Non-Cutaneous Effects

Exposure to certain phenols and catechols can affect other tissues and induce other diseases in addition to

Table 2.2.2.2 Sources of phenolic/catecholid derivatives

Working area	Home area
Synthetic oils	Varnish and lacquer resins
Rubber antioxidants	Valve plants
Printing ink	Soap antioxidants
Plasticizers	Insecticides
Photographic chemicals	Germicidal detergents
Paints	Disinfectants
Motor oil additives	Deodorants
Latex gloves	
Formaldehyde resins	
Duplicating paper	
De-emulsifiers for oil field use	
Adhesives	

affecting the skin and causing occupational/contact leukoderma. Specifically TBP has produced hepatosplenopathy and diffuse thyroid enlargement among workers [40]. Abnormal liver function tests were also demonstrated among individuals with severe TBP induced leukoderma [19].

Mode of Action of Phenols/Catechols

The mode of action of these compounds on the melanocyte has been suggested. Phenols and catechols are molecules structurally similar to tyrosine, the substrate for tyrosinase that initiates the biochemical pathway for melanin synthesis [22]. It has been proposed that derivatives of these compounds compete with tyrosine for hydroxylation by tyrosinase and interfere with the completion of melanin synthesis [20, 29]. How this subsequently induces cell death/apoptosis is uncertain. However, semiquinone-free radicals generated by the catalytic action of tyrosinase on these phenolic/catecholic derivatives and their induction of cell death through peroxidation of its lipoprotein membranes has been proposed for the melanocyte [14], as well as other model systems [16, 27, 31]. We have recently demonstrated that tyrosinase does not clearly mediate the cytotoxic action of TBP. Melanocytes cultured from individuals varying tenfold in tyrosinase activity levels demonstrate similar dose-dependent sensitivity to TBP [46]. In addition, melanocytes cultured from an individual with oculocutaneous albinism Type 1 expressing null mutations in the gene encoding tyrosinase exhibit similar sensitivities to TBP as normal melanocytes [46]. Recent evidence suggests that tyrosinase related protein-1 (Tyrp-1) may mediate the action of TBP [26]. Transfection of normal Tyrp1 and mutant Tyrp1 into Tyrp-1 deficient melanocytes will elevate or not affect, respectively, the melanocyte's susceptibility to TBP [25]. We propose that Tyrp1 catalyzes the conversion of phenols and catechols into their quinone/indole intermediates that induce oxidative stress thus mediating cytotoxicity.

Quinones and Link with Oxidative Damage

Quinones are a ubiquitous class of compounds that can in turn undergo enzymatic and nonenzymatic redox cycling with their corresponding semiquinone radical and as a result generate superoxide anion radicals [30, 32, 38]. The subsequent enzymatic or spontaneous dismutation of these superoxide anion radicals gives rise to hydrogen peroxide that reacts with trace amounts of iron or other transitional metals to form cellular hydroxyl radicals. The hydroxyl radicals are powerful oxidising agents that are responsible for damage to essential protein, lipid, carbohydrate, and DNA macromolecules. For example, oxidation of cysteine residues in proteins leads to disulfide bond formation that can dramatically alter protein structure and function. Hydroxyl radicals also can catalyze oxidation of lipids, generating lipid hydroperoxides that can lead to formation of lipid peroxide-derived malondialdehyde DNA adducts [28]. Ultimately, the molecular and cellular damage caused by the aberrantly generated plethora of reactive oxygen species (ROS) results in programmed cell death, apoptosis [44]. It has recently been demonstrated that vitiligo melanocytes exhibit (1) more ROS, (2) membrane peroxidation, (3) impaired mitochondrial electron transport chain 1, and (4) more readily induced apoptosis, all characteristics of cells susceptible to death by oxidative stress [11].

Staging

Pertaining to occupational/contact Vitiligo, a recent prospective study was performed on 864 cases of individuals with chemical leukoderma by Ghosh and Mukhopadhyay [15]. This study demonstrated that 31% of affected individuals presented with solitary depigmented macules where multiple patches occurred in 69% of the cases. In addition, patches were limited to the site of exposure in 74% of the cases, whereas patches developed in remote areas in 26% of cases. In regard to therapeutic response, it was demonstrated that repigmentation was more commonly seen in chemical leukoderma cases not associated with pre-existing vitiligo than in individuals who exhibit pre-existing vitiligo. These authors also presented a staging formula for 'chemical leukoderma syndrome' (CLS) [15]. In this formula, Stage I represents chemical leukoderma only at the site of contact, Stage II represents local spread of chemical leukoderma through lymphatics beyond the site of contact, Stage III represents distant spread of chemical leukoderma through hematogenous spread beyond the site of contact, and Stage IV represents distant spread of vitiligo-like patches even after a year of strictly withheld exposure to offending chemical. This staging system seems very appropriate for occupational/contact Vitiligo.

References

1. Bajaj A, Gupta S, Chatterjee A (1996) Footwear depigmentation. Contact Dermatitis 35:117–118
2. Bajaj AK, Gupta SC, Chatterjee AK (1990) Contact depigmentation from free *para*-tertiary-butylphenol in bindi adhesive. Contact Dermatitis 22:99–102
3. Bajaj AK, Gupta SC, Chatterjee AK et al (1996) Hair dye depigmentation. Contact Dermatitis 35:56–57
4. Bajaj AK, Pandey RK, Misra K et al (1998) Contact depigmentation caused by an azo dye in alta. Contact Dermatitis 38(4):189–193
5. Bickley LK, Papa CM (1989) Chronic arsenicism with vitiligo, hyperthyroidism, and cancer. N J Med 86(5):377–380
6. Bleehen SS, Pathak MA, Hori Y et al (1968) Depigmentation of skin with 4-isopropylcatechol, mercaptoamines and other compounds. J Invest Dermatol 50:103–117
7. Boissy RE, Nordlund JJ (1995) Biology of vitiligo. In: Arndt KA, LeBoit PE, Robinson JK et al (eds) Cutaneous medicine and surgery: an integrated program in dermatology. WB Saunders, Philadelphia
8. Boissy RE, Manga P (2004) On the etiology of contact/occupational vitiligo. Pigment Cell Res 17:208–214
9. Chumakov NN, Babanov GP, Smirnov AG (1962) Vitiliginoid dermatoses in workers of phenol-formaldehyde resin works. Bull Dermatol 36:3–8
10. Cummings MP, Nordlund JJ (1995) Chemical leukoderma: fact or fancy. Am J Contact Dermatitis 6:122–127
11. Dell'Anna ML, Ottaviani M, Albanesi V et al (2007) Membrane lipid alterations as a possible basis for melanocyte degeneration in vitiligo. J Invest Dermatol 127(5): 1226–1233
12. Flickinger CW (1976) The benzenediols: catechol, resorcinol and hydroquinone – a review of the industrial toxicology and current industrial exposure limits. Am Industr Hyg Assoc J 37:596–606
13. Gellin GA, Possick PA, Perone VB (1970) Depigmentation from 4-tertiary butyl catechol – an experimental study. J Invest Dermatol 55:190–197
14. Gellin GA, Maibach HI, Misiaszek MH (1979) Detection of environmental depigmenting substances. Contact Dermatitis 5:201–213
15. Ghosh S, Mukhopadhyay S (2008) Chemical leukoderma: clinico-etiological study of 864 cases in perspective of developing country. Br J Dermatol 160:40–47
16. Halliwell B, Chirico S (1993) Lipid peroxidation: its mechanism, measurement, and significance. Am J Clin Nutr 57(Suppl 5):715S–725S
17. Hernandez C, Reddy SG, Le Poole C (2008) Contact leukoderma after application of a compounded phenol cream. Eur J Dermatol
18. Ito Y, Jimbow K, Ito S (1987) Depigmentation of black guinea pig skin by topical application of cysteaminylphenol, cysteinylphenol, and related compounds. J Invest Dermatol 88:77–82
19. James O, Mayes RW, Stevenson CJ (1977) Occupational vitiligo induced by *p-tert*-butylphenol: a systemic disease? Lancet 2:1217–1219
20. Jimbow K, Obata H, Pathak MA et al (1974) Mechanism of depigmentation by hydroquinone. J Invest Dermatol 62: 436–449
21. Kahn G (1970) Depigmentation caused by phenolic detergent germicides. Arch Dermatol 102:177–187
22. Lerner AB (1971) On the etiology of vitiligo and grey hair. Am J Med 51:141–147
23. Levin CY, Maibach H (2001) Exogenous ochronosis. An update on clinical features, causative agents and treatment options. Clin Dermatol 2:213–217
24. Malten KE, Seutter E, Hara I et al (1971) Occupational vitiligo due to paratertiary butylphenol and homologues. Trans St Johns Hosp Dermatol Soc 57:115–134
25. Manga P, Sheyn D, Sarangarajan R et al (2005) The microphthalmia-associated transcription factor and tyrosinase related-protein 1 play a role in melanocyte response to 4-tertiary butyl phenol. J Invest Dermatol
26. Manga P, Sheyn D, Yang F et al (2006) A role for tyrosinase related-protein 1 in 4-*tert*-butylphenol-induced toxicity in melanocytes. Implications for vitiligo. Am J Pathol 169(5): 1652–1662
27. Mans DR, Lafleur MV, Westmijze EJ et al (1992) Reactions of glutathione with the catechol, the ortho-quinone and the semi-quinone free radical of etoposide. Consequences for DNA inactivation. Biochem Pharmacol 43:1761–1768
28. Marnett LJ (1999) Lipid peroxidation-DNA damage by malondialdehyde. Mutat Res 424(1–2):83–95
29. McGuire J, Hinders J (1971) Biochemical basis for depigmentation of skin by phenol germicides. J Invest Dermatol 57:256–261
30. Monks TJ, Hanzlik RP, Cohen GM et al (1992) Contemporary issues in toxicology. Quinone chemistry and toxicity. Toxicol Appl Pharmacol 112:2–16
31. Nakagawa Y, Tayama S, Moore G et al (1993) Cytotoxic effects of biphenyl and hydroxybiphenyls on isolated rat hepatocytes. Biochem Pharmacol 45:1959–1965
32. O'Brien PJ (1991) Molecular mechanisms of quinone cytotoxicity. Chem Biol Interact 80(1):1–41

33. O'Malley MA, Mathias CG, Priddy M et al (1988) Occupational vitiligo due to unsuspected presence of phenolic antioxidant byproducts in commercial bulk rubber. J Occup Med 30(6):512–516
34. Okmura Y, Shirai T (1962) Vitiliginous lesions occurring among workers in a phenol derivative factory. Jpn J Dermatol 7:617–619
35. Oliver EA, Schwartz L, Warren LH (1939) Occupational leukoderma. JAMA 113:927–928
36. Ortonne J-P, Bose SK (1993) Vitiligo: where do we stand? Pigment Cell Res 6:61–72
37. Pandhi RK, Kumar AS (1985) Contact leukoderma due to 'Bindi' and footwear. Dermatologica 170(5):260–262
38. Powis G (1987) Metabolism and reactions of quinoid anticancer agents. Pharmacol Ther 35(1–2):57–162
39. Quevedo WC Jr, Fitzpatrick TB, Szabo G et al (1986) Biology of the melanin pigmentary system. In: Fitzpatrick TB, Eisen AZ, Wolff K (eds) Dermatology in general medicine. McGraw-Hill, New York
40. Rodermund OE, Jorgens H, Muller R et al (1975) [Systemic changes in occupational vitiligo]. Hautarzt 26(6): 312–316
41. Selvaag E (1996) Chloroquine-induced vitiligo. A case report and review of the literature. Acta Derm Venereol 76:166–167
42. Shelley W (1974) *p*-Cresol: cause of ink-induced hair depigmentation in mice. Br J Dermatol 90:169–174
43. Sun CC (1987) Allergic contact dermatitis of the face from contact with nickel and ammoniated mercury in spectacle frames and skin-lightening creams. Contact Dermatitis 17: 306–309
44. Takahashi A, Masuda A, Sun M et al (2004) Oxidative stress-induced apoptosis is associated with alterations in mitochondrial caspase activity and Bcl-2-dependent alterations in mitochondrial pH (pHm). Brain Res Bull 62:497–504
45. Taylor JS, Maibach HI, Fisher AA et al (1993) Contact leukoderma associated with the use of hair colors. Cutis 52(5):273–280
46. Yang F, Sarangarajan R, Le Poole IC et al (2000) The cytotoxicity and apoptosis induced by 4-tertiary butylphenol in human melanocytes is independent of tyrosinase activity. J Invest Dermatol 114:157–164
47. Yusof Z, Pratap RC, Nor M et al (1990) Vogt-Koyanagi-Harada syndrome – a case report. Med J Malaysia 45:70–73

2.2.3 In Vivo Data

Non-Invasive Methods for Vitiligo Evaluation

2.2.3.1

Marco Ardigò, Francesca Muzio, Mauro Picardo, and Valeria Brazzelli

Contents

Core Messages

- The extension and severity of vitiligo guide prognosis and help making therapeutic choices. However, in spite of attempts to standardise clinical judgment, wide variations exist both in assessment rules and interpretation of their use, making intra- and inter-observer variations unavoidable.
- Ultraviolet (UV)-light examination and UV photography remain useful tools for the assessment of Caucasoid patients. Non-invasive instruments that use reflectance spectroscopy provide a convenient and reproducible methodology for the study of vitiligo patients and their follow-up.
- Reflectance confocal microscopy provides microscopical informations *in vivo* about changes in achromic macules both in repigmented areas after treatments and in clinically normal-appearing skin of vitiligo patients.

Introduction

The extension and severity of vitiligo guide diagnosis, prognosis, and therapeutic choices. However, variability in assessment rules and their interpretation make intra- and inter-observer variations unavoidable.

Actually, there is no standardised measure for vitiligo lesions, treatment responses, or for the comparison of different treatment options. If the percentage of repigmentation is clinically estimated, the bias of subjective

M. Ardigò (✉)
San Gallicano Dermatological Institute, IRCCS, Rome, Italy
e-mail: ardigo@ifo.it

M. Picardo and A. Taïeb (eds.), *Vitiligo*,
DOI 10.1007/978-3-540-69361-1_2.2.3.1, © Springer-Verlag Berlin Heidelberg 2010

assessment without objective measurements leads to incorrect evaluations [13, 44, 75].

In order to address these problems, the Vitiligo European Task Force (VETF) was founded in 2003 during the ESPCR meeting in Ghent with three initial goals: (a) the proposal of a consensus definition of the disease; (b) the design of biometric tools to assess disease severity/stability; and if possible (c) the derivation of a consensus scoring system [60].

Defining the Extension of Vitiligo

Objective Methods for the Assessment of Vitiligo Extent

Current methods to assess the involved body surface are: 'the rule of nine', the 'flat hand = 1%' method, the point counting method and the Vitiligo Area Scoring Index (VASI) [27, 60].

'The rule of nine' assumes that the total body surface area comprises 9% for head/neck, each arm, leg, and the four trunk quadrants, leaving 1% for the genitalia. However, the difficulty in assessing surface area involvement is greatest when the denominator is just as large as on the legs, as has been suggested by use of the Lund-Browder chart in vitiligo [30, 32, 70]. The Lund-Browder chart is the most accurate method for estimating burn extent, and could be applied in vitiligo patients too, mainly in the evaluation of children. This chart determines the changes in percentage of various parts of the body that occur during the different stages of development from infancy through childhood. The area of the head makes up a relatively large portion of the total skin area in infants, when compared with adults. This proportion is counterbalanced in infants by the smaller area of the thighs and legs.

In the 'flat hand = 1%' method, a flat hand represents 1% of the total body surface area. Both these methods are very subjective and based on visual assessment.

The point counting method is a simple, practical, and has an accurate technique, which is used in the estimation of the irregularly shaped sectional surface area to obtain the volumes of organs or structures using macroscopic, microscopic, or radiological images [5, 66]. The borders of the lesions are marked with an ordinary ballpoint pen and a piece of paper is immediately placed over the lesion as for each lesion, the copied borders of projection areas are enhanced by redrawing the contours with a pen. In order to estimate the number of points, a transparent sheet that has points (+) printed on it is randomly superimposed on lesion projection area. The numbers of intersections hitting the area of interest are counted. The total area of each lesion is estimated by multiplying the representative area of a point on grid by total number of points counted for the lesion (Fig. 2.2.3.1a). The reliability of the point-counting method has been tested against image analysis [5, 36].

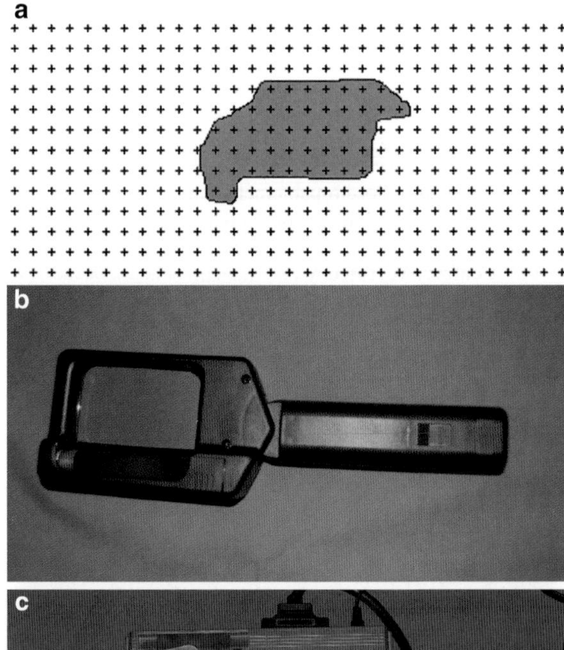

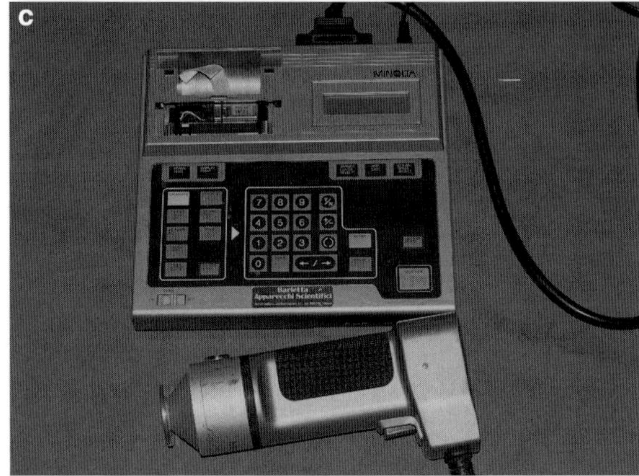

Fig. 2.2.3.1 (**a**) Point counting method. The total area of each lesion is estimated by multiplying the representative area of a point on grid by total number of points counted for the lesion; (**b**) Wood's light. The envelope of the bulb is composed by a deep-*bluish-purple* glass called Wood's glass, a nickel-oxide–doped glass, which blocks almost all visible light above 400 nm. In this instrument a magnifying glass is present; (**c**) Minolta CR-200 Chromameter

The VASI, is a standardised and sensitive method to measure extent and percentage of de/repigmentation, and it is easy to perform [29].

The body of the patient is divided into five separate and mutually exclusive regions: hands, upper extremities (excluding hands), trunk, lower extremities (excluding the feet), and feet. The axillary and inguinal regions are included with the upper and lower extremities, respectively, while the buttocks are included in the lower extremities. The face and neck areas can be assessed separately, but are not included in the overall evaluation. Then, by using the assumption that one-hand unit (the palm plus the volar surface of all the fingers) is equivalent to approximately 1% of the body surface, the physician determines how much of the skin is affected at the baseline in each body region. Depigmentation within each area is estimated to the nearest of 1 of the following percentages: 0, 10, 25, 50, 75, 90, or 100%.

For each body region, the VASI is determined by the product of the area of vitiligo in hand units (which is set at 1% per unit) and the extent of depigmentation within each hand unit-measured patch (possible values of 0, 10, 25, 50, 75, 90 or 100%). The VASI applied to the whole body is then calculated using the following formula that considers the contributions of all body regions (possible range 0–100): VASI = $\Sigma_{\text{(all body size)}}$ (hand units) x (depigmentation).

It could be argued that the VASI has a subjective component, because it involves the physician deciding the amount of pigmentation and the area of involvement.

A more objective technique might involve the use of standardised grids to measure areas marked out on the skin as is done with the ADASI score (atopic dermatitis area and severity index) in Atopic dermatitis, which however has not gained much popularity [6].

Skin Colour Measurements and Monitoring of Lesions

In dermatological practice and clinical research, visual cues are of primary importance for the accurate diagnosis and grading of skin lesions. The natural course of vitiligo can exhibit flares, remissions, and spontaneous repigmentation, its therapy takes months and the clinical follow-up is difficult. For these reasons objective, non-invasive investigations for skin colour measurements and monitoring are needed.

Currently, a wide range of techniques is available to aid the diagnosis and assessment of pigmentary disorders even if few studies so far include vitiligo. Two types of methods can be identified. The first is a macroscopic morphological measurement to study the skin pigmentation as a whole that includes: visual assessment, photography under natural light or ultraviolet (UV), photography with computerised image analysis, tristimulus colorimetry, or spectrophotometry. The second is a non-invasive micromorphological method characterised by an accurate measure of hue and chroma of the substructures of the pigmented lesions and includes confocal laser microscopy [1, 19], a non-invasive in vivo technique, with a resolution close to conventional microscopy.

Macroscopic Morphological Measurements

Photography

Illustrative documentation of dermatological conditions has been an established practice in dermatology throughout the centuries. Photographs can tell us a great deal about the patient's conditions at an instant in time, and serial photographs taken over a period can tell about the progress of the disease or the response to treatment [3, 43]. However, in low-pigment Caucasian patients, conventional photography fails to document adequately pigment changes induced by vitiligo. Thus, all patients with vitiligo of lower phototypes should be examined under both visible and Wood's light (Chap. 1.2.1). Wood's lamp enhances the contrast between skin with little or no pigment and skin with excessive quantities of melanin.

Wood's Light

Wood's lamp was invented in 1903 by a Baltimore physicist, Robert W. Wood (1868–1955) [76]. Wood's light is produced by using a filter opaque to all radiation except that which has a wavelength between 320 and 400 nm, in the UV range, with a peak at 365 nm (Fig. 2.2.3.1b) [4, 45]. The UV radiation penetrates the epidermis, where it is attenuated by melanin, and on entering the dermis it stimulates fluorescence emission by collagen bundles. Part of the emitted fluorescence is directed towards the surface of the skin, but attenuated by haemoglobin in capillaries and epidermal melanin. In particular, fluorescence in tissues occurs when light of shorter wavelengths, in this case 340–400 nm, (as initially emitted by Wood's light) is absorbed, and radiation of longer wavelengths, usually in visible light, is emitted (Fig. 2.2.3.2a) [33]. In vitiligo it has been further suggested that the fluorescence of the amelanotic patches is enhanced by the intraepidermal accumulation of biopterins [54].

The cutaneous lesions that possess an increased concentration of epidermal melanin will appear darker by contrast against the surrounding normal skin and lesions possessing decreased melanin will appear brighter because the UV light is not absorbed in the

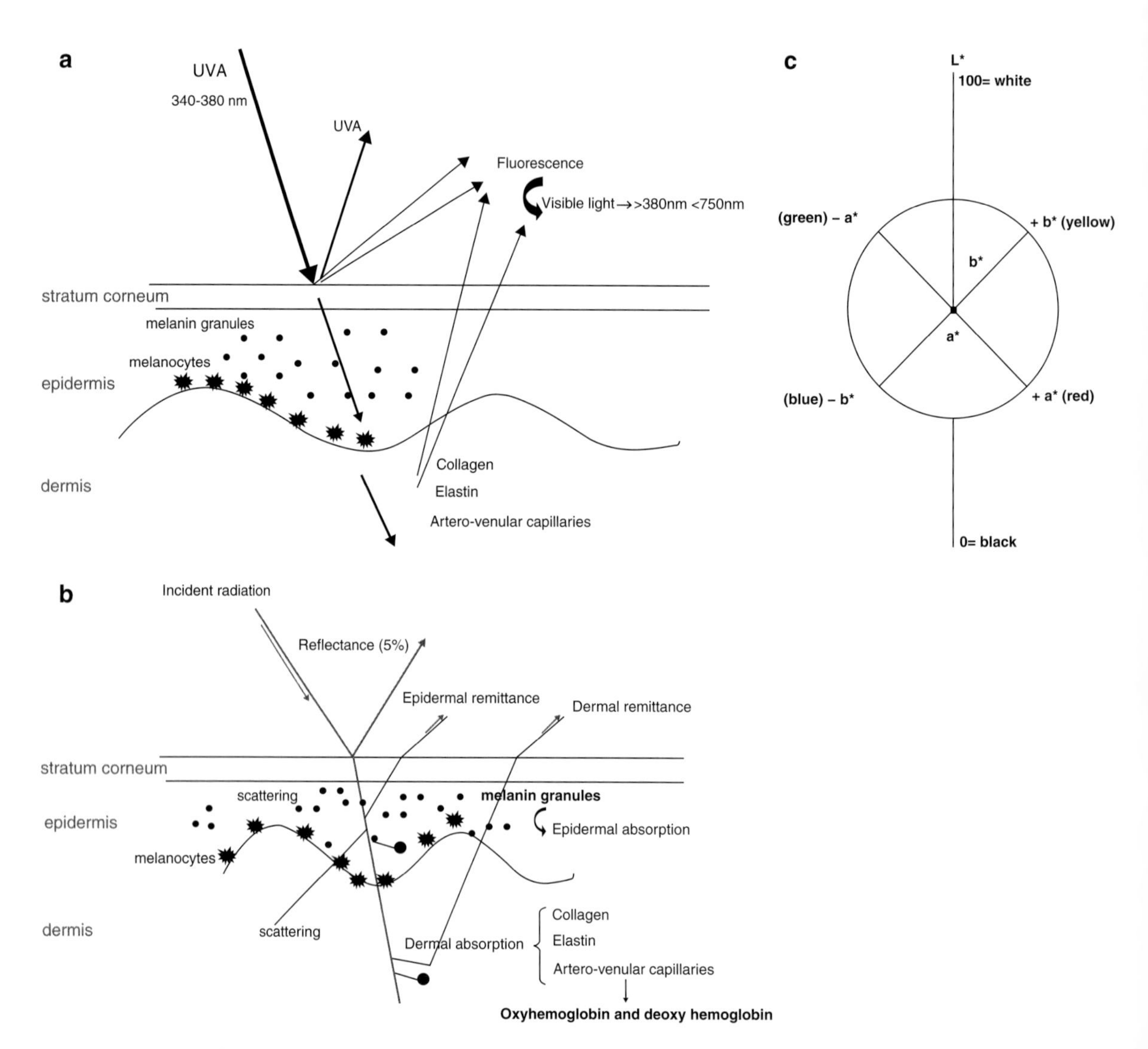

Fig. 2.2.3.2 (**a**) Fluorescence in tissues occurs when light of shorter wavelengths, in this case 340–400 nm emitted by Wood's light is absorbed, and radiation of longer wavelengths, usually in visible light, is emitted. In particular the UVA radiation, that is not absorbed by melanin, enter the dermis where induces the collagen and elastin cross-links to produce fluorescence. The light emitted, even if absorbed by the epidermal melanin, exits the skin and will reach the eye of the observer; (**b**) When light impinges on the skin, about 4–8% of the incident radiation is reflected off the surface. Most of the incident light enters the first layers of the skin and follows a tortuous path until it exits back out of the skin or gets attenuated by skin chromophores (oxyhaemoglobin, deoxy-haemoglobin and melanin granules); (**c**) The vertical dimension L* indicates light intensity or luminance and takes values from 0 (black) to 100 (white); in each horizontal plane, the hue is defined by its coordinates on two perpendicular axes: a* indicates the colour of the object on scale that goes from green (negative values) to red (positive values) and b* indicates the colour of the object on a scale that goes from blue (negative values) to yellow (positive values). Axes a* and b* cross the L* axis at their zero values thus defining the coordinates of a chosen colour considering its hue and purity or saturation

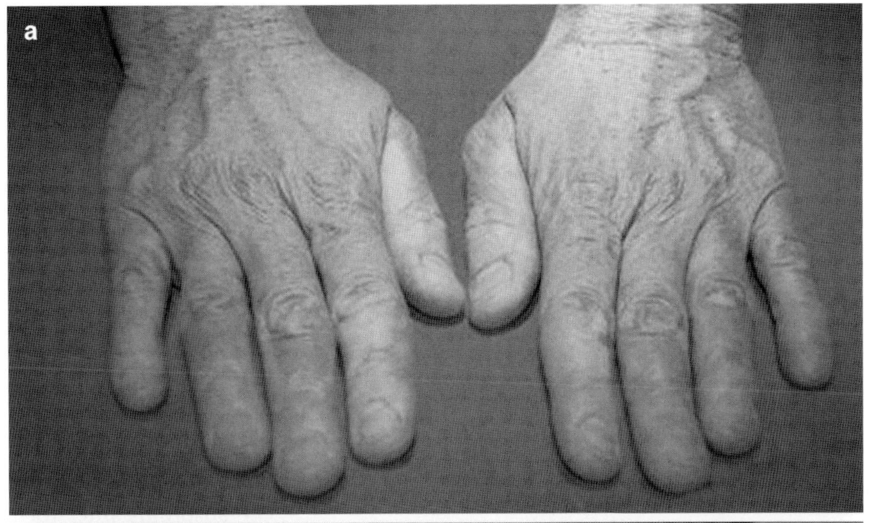

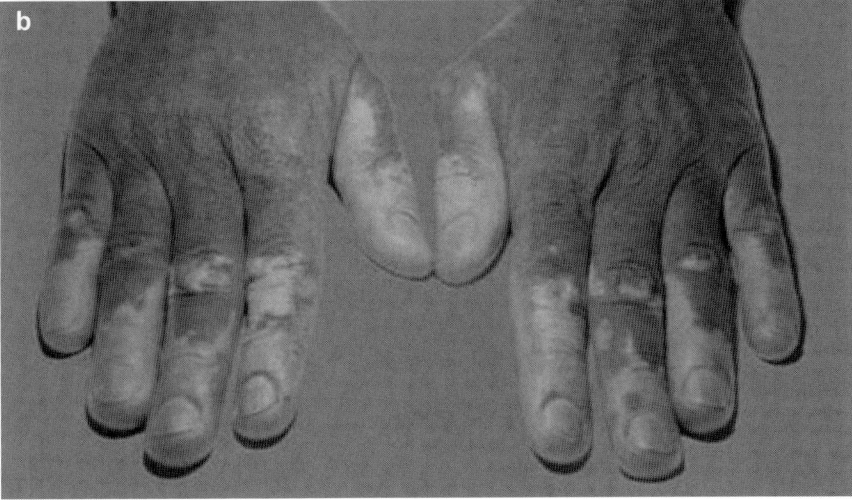

Fig. 2.2.3.3 Depigmented patches on the hands of a patient with vitiligo. (**a**) Standard lighting. (**b**) Wood's light lighting

epidermis, but reflected. On the contrary, variations in dermal pigmentation are less apparent under Wood's lamp than under visible light, because dermal melanins do not affect the amount of light observed [25].

Ideally, the lamp should be allowed to warm up for about 1 min and the examination should be performed in a very dark room, with black occlusive shades or without windows. It is also essential for the examiner to become adapted to the darkness to improve contrast viewing. This procedure enhances variation of epidermal pigmentation not apparent under visible light and improves the assessment of the extent of pigment abnormalities.

The greater the loss of epidermal pigmentation is, the more marked the contrast on Wood's light examination. This explains why vitiligo patches which are completely devoid of melanins are so apparent even in fair-skinned individuals when examined under Wood's light (Fig. 2.2.3.3) [4, 44, 45].

The results of Wood's light examination can be recorded through UV photography (see later).

Visible Light Photography

Conventional visible light photography using standard 35 mm film has long been the favoured method of recording images of the skin both for the purpose of medical record keeping and education as well. Measurements are valid only upon constant lighting (controlled light intensity, diffused light to avoid reflection, unchanged distance between skin and light source, and skin surface orientation). However, storage of a large number of photographic prints can become a problem over time. Under visible light in vitiligo patients anyhow, it is sometimes difficult to distinguish between hypomelanosis and amelanosis on very fair-skin type patients (Type I or II) or in children [28, 58].

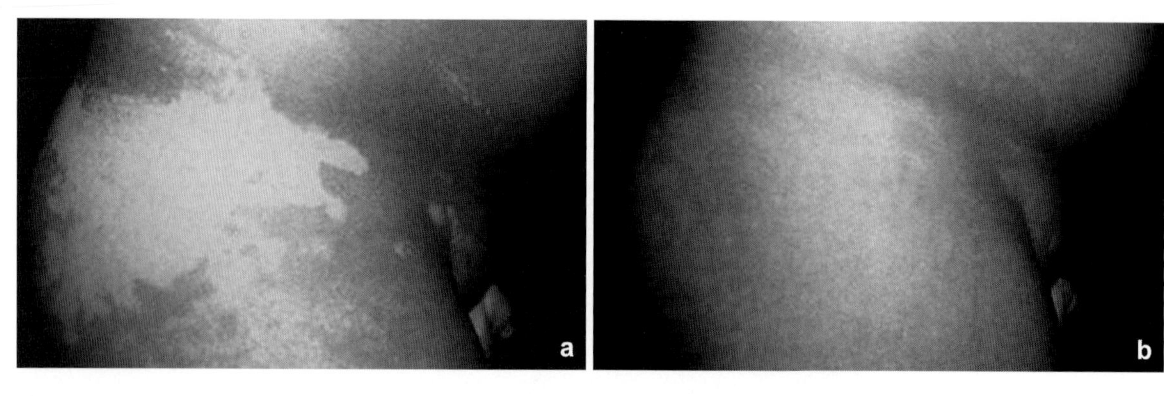

Fig. 2.2.3.4 Photographic documentation of narrow band UVB therapy in a child. (**a**) Before treatment. (**b**) After treatment

Digital Photography

In the last five years, photographic film has become obsolete as digital image sensor pixel counts have increased and compact digital cameras became available at a reasonable price. Digital cameras with resolution capabilities exceeding 1,556 × 1,024 pixels can provide good-quality images of the skin. Digital images are available within seconds and printed out directly using a colour printer, or saved in a tiny memory card or on a hard disk of a computer [3, 53, 68, 67]. Skin diseases such as vitiligo can be monitored through serial imaging to allow more accurate assessment in the progress of the condition or responses to treatment (Fig. 2.2.3.4).

Ultraviolet Light Photography

UV-light photography is based on the principle of UV rays being more selectively absorbed by melanin in the epidermis as compared with visible light. The skin is photographed after illumination with UVA radiation, which causes visible fluorescence (Fig. 2.2.3.2a). The resulting photograph show collagen fluorescence areas as light areas and melanin pigmentation as dark spots; so it displays areas of hyperpigmentation with much greater clarity than visible light photography. There are two ways to use UV radiation to take photographs: UV fluorescence and reflected UV photography [3, 64, 73].

In the case of the UV fluorescence technique, a source of UV radiation filtered with an UV transmission filter – or excitation filter – is aimed at the subject in a completely darkened room. The subject reflects the UV and may also emit a visible fluorescence. The UV radiation is then prevented from entering the lens by an UV absorbing filter (or barrier filter) and fast black-and-white or colour film records any visible fluorescence emitted in the region 400–700 nm [34]. The reflected UV photographic technique records only UV radiation, in the region 320–390 nm, reflected from the subject. All other radiation is prevented from reaching the film. A source of UV is directed at the subject who will then reflect this radiation back into the camera. It is necessary to fit an UV transmission filter over the lens to prevent any visible radiation from impinging on the standard black-and-white film [33].

Reflectance Spectroscopy

Reflectance spectroscopy has been in use for more than 50 years for the objective measurement of skin colour. It is based on different approaches to the analysis of light reflected by the skin [63]. A number of instrumental methods have been used to evaluate skin pigmentation/depigmentation as early as the 1920s and 1930s, when it was recognised that 'the more melanin present, the lower the percentage of light reflected from the surface of the skin and the lower the brilliance' [11, 20, 59].

The colour of the skin depends mainly on its pigment content, on the spectrum of the illuminating light, and on the quality of its surface. When light impinges on the skin, about 4–8% of the incident radiation is reflected off the surface (Fig. 2.2.3.2b) [33].

Most of the incident light enters the first layers of the skin and follows a tortuous path until it exits the skin or gets attenuated by skin chromophores.

The primary chromophores in skin are melanin, oxyhaemoglobin, and deoxy- (or reduced) haemoglobin. Oxyhaemoglobin (oxy-Hb) and deoxy-haemoglobin (deoxy-Hb) absorb specifically between 540 and 575 nm. Melanin heavily absorbs all wavelengths, but demonstrates a monotonic increase towards shorter wavelengths [47].

The constituents of skin that are strong scatterers include collagen and elastin fibres, erythrocytes, sub-cellular organelles (most notably pigmented melanosomes, nuclei, and mitochondria), and cell membranes. Scattering contributes to the shape of the spectrum remitted from the human skin. It is important to note that the depth of the light that reaches into the tissue (penetration depth) depends on both the absorption and scattering characteristics of the tissue. Since both absorption and scattering are wavelength dependent, the penetration depth varies with wavelength as well. Absorption and scattering are the two ways by which white light is transformed into coloured light by interaction with skin [19, 33].

Several objective methods have been developed to measure skin colour. At present the most used methods that determine colour by measuring the intensity of reflected light of specific wavelengths are the following: reflectance tristimulus CIE (Commission Internationale de l'Eclairage), colorimetry, and narrow band reflectance spectrophotometry [24, 61].

Reflectance Tristimulus CIE Colourimetry

Of the various methods used to convert reflectance data into practical colour values, expressed in the L*a*b* colour system (L*:brightness; a*: red–green chromaticity coordinate; b*: yellow–blue chromaticity coordinate), reflectance tristimulus colorimetry is probably the most widely used, because this colour system authorised by the CIE is regarded as the standard for colorimetry in industrial fields [72].

In particular, this system classifies the existing colours in a roughly cylindrical virtual volume placed in a three dimensional space that makes it possible to measure three parameters for each colour: the vertical dimension L* indicates light intensity or luminance and takes values from 0 (as if colours were seen at night: black) to 100 (maximum light: white); in each horizontal plane, the hue is defined by its coordinates on two perpendicular axes: a* indicates the colour of the object on scale that goes from green (negative values) to red (positive values) and b* indicates the colour of the object on a scale that goes from blue (negative values) to yellow (positive values). Axes a* and b* cross the L* axis at their zero values thus defining the coordinates of a chosen colour considering its hue and purity or saturation (Fig. 2.2.3.2c). The three-dimensional space that represents all possible perceivable colours is termed "colour volume" [74].

In other words, the tristimulus analysis converts intensity versus wavelength data (i.e. spectral information) into three numbers that indicate how a colour of an object appears to a human observer, hence the "psycho-photometric" characterisation.

Different hand-held tristimulus reflectance colorimeters are commercially available for the measurement of the skin colour: Micro Color (Dr. Bruno Lange GmbH, Dusseldorf, Germany), the Chromameter CR 200 or CR 300 (Minolta Osaka, Japan) and the Visi-Chroma VC-100 (Biophotonics, Lessines, Belgium), etc. All instruments consist of a measuring sensor placed on the skin, connected to a control box by an optical fiber.

However, the Minolta Chromameter CR200 and 300 have become very popular and are considered more or less standard instruments for colour measurements in the dermato-cosmetic field (Fig. 2.2.3.1c). They are portable instruments with a flexible hand-held probe which can be easily moved. The measured area is 8 mm in diameter. Three consecutive readings (each of them taking a few seconds) are taken at the site of measurement and their mean values are calculated. The instruments are calibrated by a calibration plate (CR-A43) before each measurement.

The evaluation of the intensity of the constitutional skin colour is correlated with the L* and b* components and the colour expressed with these numerical coordinate units is arbitrary (AU). The Chromameter does not give information about the substances that generate the colour because it does not measure the wavelengths of specific chromophores [46].

Clinical applications for colorimetry extend to the determination of erythema and pigmentation or depigmentation, to the studies on skin typology, races, anatomical distribution of pigment and to photoprotection factors, sunscreens or depigmenting agents [48, 55–57]. Other fields of application are the determination of the response to systemic drug treatments such as

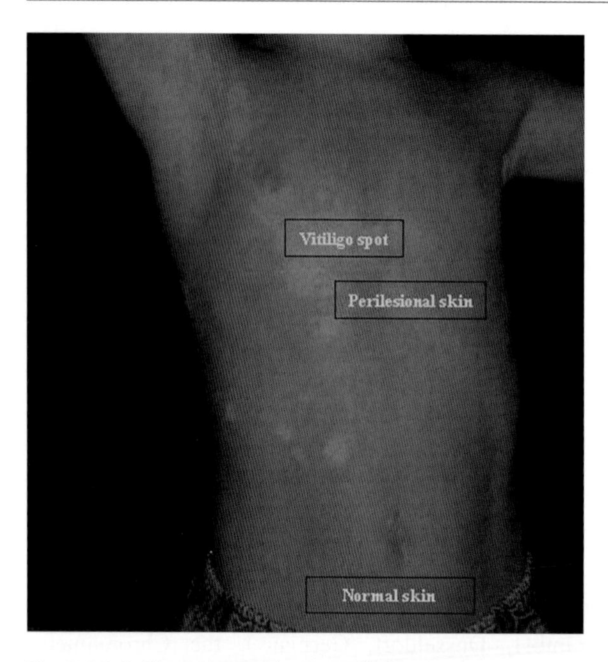

Fig. 2.2.3.5 The lesional skin, the perilesional skin and the normal skin as far as 5 cm from the nearest vitiligo spot in a young patient. Skin pigmentation measurements were performed with a Minolta CR-200 Chromameter

imatinib mesylate which acts as a depigmenting agent and these studies show the gradual lightening of the skin in patients affected by chronic myeloid leukaemia [8, 9].

In vitiligo, the studies employing these methods are few and among these some show the relationship between skin colour change and variations in the amounts of melanin and haemoglobin [26, 62], whereas others analyse the perilesional skin near the vitiligo patch showing that this skin is lighter than the skin far from the vitiligo patch underlining that normal appearing skin in vitiligo is not normal at all (Fig. 2.2.3.5) [10].

Selected Spectral Bands and Narrow-Band Reflectance Spectrophotometry

The aim of reflectance spectrophotometry is to explore the presence of the two main cutaneous chromophores: melanin and haemoglobin [18, 24, 47].

There are many commercially available instruments that measure a 'melanin index' (MI) (or 'pigmentation index') and an 'erythema index' (EI) based on reflectance measurements at selected spectral bands. Some portable optoelectronic such as DermaSpectrometer (Cortex Technology, Hadsund, Denmark), Erythema/Melanin Meter (DiaStron Ltd, Andover, Hampshire, UK), Mexameter (Courage & Khazaka GmbH, Ko¨ ln, Germany), UV Optimise (Matik, Denmark), devices were designed to perform simple reflectance spectrophotometry in specific narrow wavelength spectra. Data are expressed in what is called melanin and erythema indices. The erythema index represents the decimal logarithm of the ratio between the intensity of the reflected red and green lights. The melanin index corresponds to the decimal logarithm of the ratio between the total reflected light and the reflected red light. Each index increases as the skin becomes more erythematous or more pigmented, respectively.

Clinical application of narrow-band reflectance spectrophotometry includes the studies of physiological parameters and anatomical sites, chronobiological skin colour variation, photodamage and photoprotection and the effect of topical and systemic drugs [23, 28, 43].

In addition, several studies have been performed to compare narrow band reflectance spectrophotometry to tristimulus colorimetry [14].

Non-Invasive Micro-Morphological Measurement: In Vivo-Reflectance Confocal Microscopy

The Technology and Its Optical Principles

Confocal microscopy was created in 1957 by Marvin Minsky [38, 39] and subsequently adapted by New and Corcuff for *in vivo* human skin imaging [15–17, 41]. The evolution of the technology included the use of a laser light source [49, 51, 52].

The name 'confocal' derives from 'optically conjugate focal planes.' The machine is constituted mainly by a point source of light (a laser source), a condenser, objective lenses, and a point detector. Point illumination is achieved by focusing a small source of light into the sample. Point detection is achieved by placing a pinhole in front of the detector. The pinhole collects light emanating only from the focus (lines) blocking light from elsewhere (dashed lines) (Fig. 2.2.3.6a). Scanning the illuminated areas of tissue in the focal plane of the objective lens enables optical sectioning corresponding to non-invasive imaging of a very thin

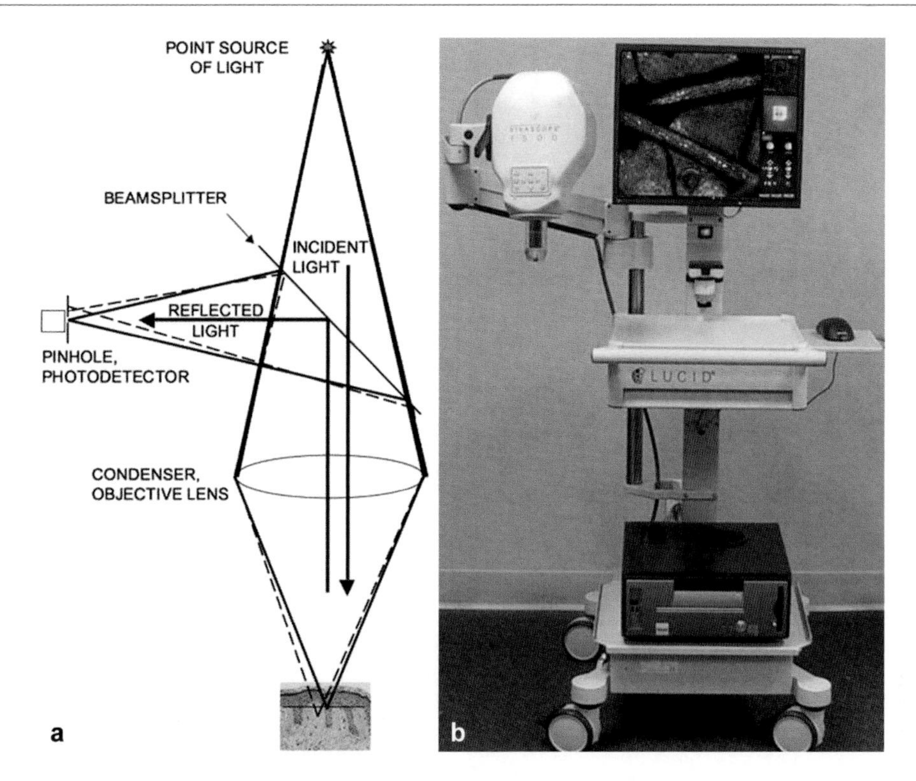

Fig. 2.2.3.6 Reflectance confocal microscope optical function (**a**) and commercial device Vivascope® (**b**)

(<5 µm) section. In grey-scale confocal images, bright (white) structures are composed by components with high refractive index compared with their surroundings (backscattered areas). Backscattering is primarily governed by the structures' refractive index (n) compared to surrounding medium. Highly reflective skin components include melanin ($n = 1.72$) [52, 69, 77], collagen ($n = 1.43$) [71], and keratin ($n = 1.51$) [65]. These components appear bright when surrounded by epidermis ($n = 1.34$) and dermis ($n = 1.41$) [65]. Confocal microscopy allows clinicians to obtain a non-invasive, real time, high-resolution close to histology (lateral resolution of 1 µm).

In standard histology, the microscope illuminates and images a relatively large field of tissue and ultra-structural details are not resolved or visualised. Moreover, thin sections (5 µm) have to be first prepared to enable observation of nuclear, cellular, and ultra-structural details. Differently, confocal microscope illuminates a microscopical area of the tissue from which every single light wavelength is reflected and collected into a detector in order to produce a pixel on a screen. Scanning of the area in two dimensions creates an illuminated plane that produces an image composed by collection of single pixels.

The commercially available confocal reflectance microscope (Vivascope 1500®) (Fig. 2.2.3.6b) uses a laser with wavelength of 830 nm and a 30× objective lens with numerical aperture of 0.9. The laser power is 5–10 mW at skin level with no tissue damage. A metal ring with either a polymer window is attached to the skin. The ring is then connected to the objective lens housing to stabilise the site of imaging. A drop of immersion oil has to be applied to the skin lesion in order to have a refractive index (1.50) sufficiently close to that of the stratum corneum (1.55) and the polymer window (1.52). The water immersion lens requires to be used with water (1.33) or water-gels (1.35) placed between the window and objective lens with a refractive index close to that of the epidermis (1.34) allowing penetration of light through the epidermis and into the dermis.

Reflectance Confocal Microscopy and Vitiligo

Reflectance confocal microscopy used on vitiligo lesional and non-lesional skin provides, non-invasively, microscopical informations useful for disease management [16].

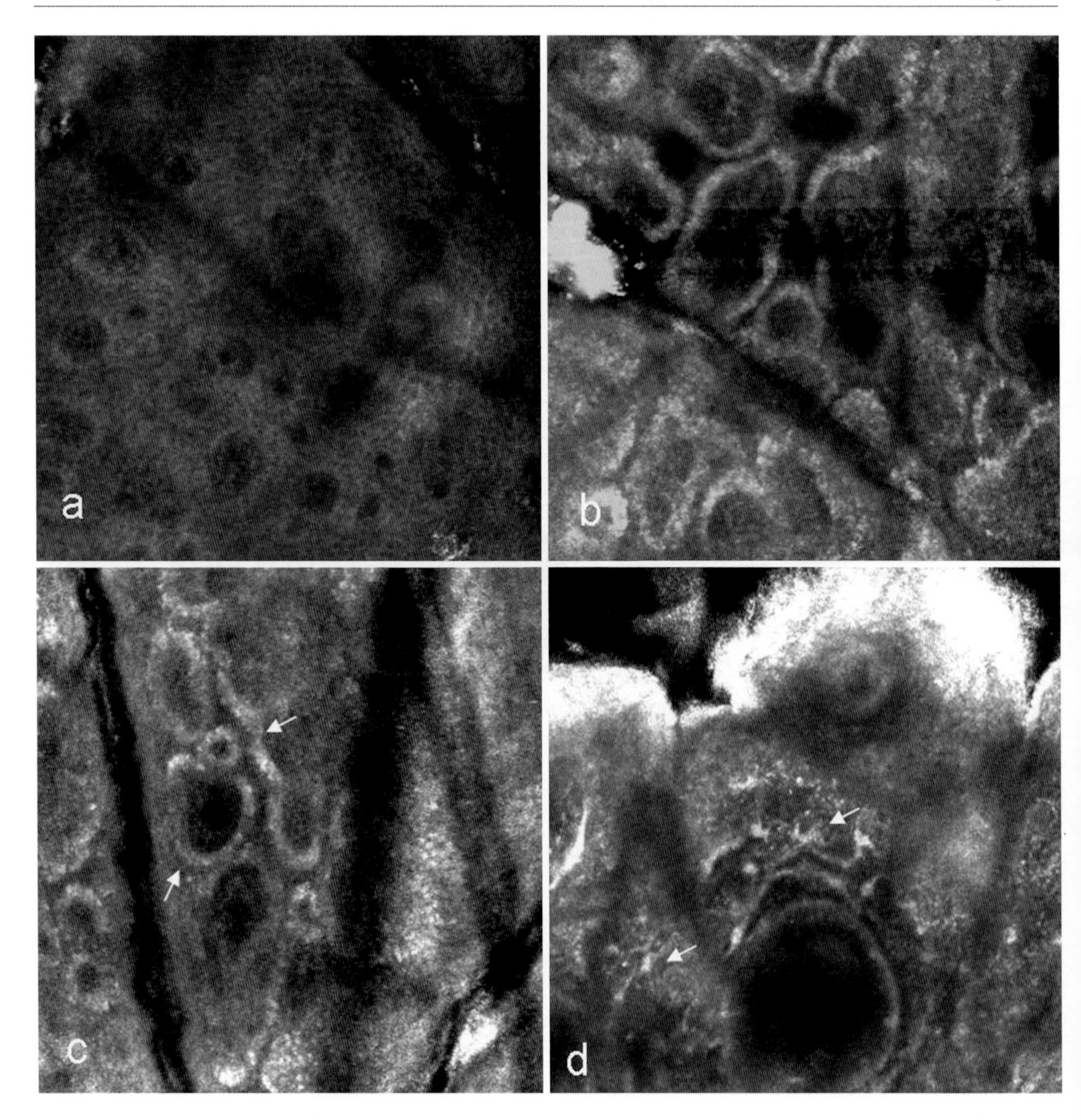

Fig. 2.2.3.7 Reflectance confocal microscopy of dermo–epidermal junction in normal skin (**a**) and vitiligo lesion in which bright papillary rings are absent (**b**). Normal skin of vitiligo patients is characterised by abnormal distribution of brightness pattern around dermal papillae providing "half-ring" features or resembling "scalloped border-like" features (*arrows*) (**c**). Repigmented area after narrow band UVB therapy presenting dendritic melanocytes (*arrows*) (**d**)

Because of confocal microscopy en-face visualisation of the skin layers, in normal skin, melanin presents a higher reflectance index (1.7) in comparison with the total skin (1.4); therefore, melanocytes and pigmented keratinocytes are seen as bright structures on a dark background (Fig. 2.2.3.7a) [13, 31, 35, 63]. Differently, vitiligo lesions show disappearance of the normal brightness at dermo–epidermal junction level with remnant of a 'shadow' of the pre-existing papillary ring (Fig. 2.2.3.7b). Moreover, bright keratinocytes, seen in normal skin, are generally absent in vitiligo lesions above the dermo–epidermal junction [2]. The disappearance of brightness (i.e. pigment) at the dermo–epidermal junction level or above fits perfectly with the progressive loss of melanocytes and the reduction of epidermal pigmentation previously demonstrated with histopathology and histochemistry [22, 35].

Interestingly, non-lesional skin of vitiligo patients shows an abnormal distribution pattern of brightness (i.e. pigment) at the dermo–epidermal junction. The characteristic ring structures are hardly recognisable, bright structures are incompletely distributed around the dermal papillae providing 'half-ring' features or resembling 'scalloped border-like' features (Fig. 2.2.3.7c). It has been speculated that these changes could derive from an initial and progressive disappearing of melanocytes or a congenital defective melanocyte distribution [2].

After UVB-narrow band treatment, repigmented areas show a variable number of activated melanocytes located at the dermo–epidermal junction. Activated melanocytic cells can be seen as bipolar or stellate dendritic structures (Fig. 2.2.3.7d) [2, 37].

In conclusion, reflectance confocal microscopy in vitiligo might be useful for the prognosis and the clinical and therapeutic follow-up. Moreover, considering the findings on non-lesional skin, confocal microscopy might be used to estimate the risk of developing vitiligo in subjects with familial vitiligo or exposed to chemical agents [2, 12].

References

1. Agache P (2004) Skin color measurement in measuring the skin. In: Agache P, Humbert P (eds) Measuring the skin. Springer, Berlin, pp 33–39
2. Ardigò M, Malizewsky I, Dell'anna ML et al (2007) Preliminary evaluation of vitiligo using in vivo reflectance confocal microscopy. J Eur Acad Dermatol Venereol 21:1344–1350
3. Aspres N, Egerton IB, Lim AC et al (2003) Imagin the skin. Australas J Dermatol 44:19–27
4. Asawanonda P, Taylor CR (1999) Wood's light in dermatology. Int J Dermatol 38:801–807
5. Aydin F, Senturk N, Sahin B et al (2007) A practical method for the estimation of vitiligo surface area: a comparison between the point counting and digital planimetry techniques. Eur J Dermatol 17:30–32
6. Bahmer FA (1992) ADASI score: atopic dermatitis area and severity index. Acta Derm Venereol Suppl 176:32–33
7. Bertrand C, Corcuff P (1994) In vivo spatio-temporal visualization of the human skin by real-time confocal microscopy. Scanning 16:150–154
8. Brazzelli V, Prestinari F, Barbagallo T et al (2007) A long-term time course of colorimetric assessment of the effects of imatinib mesylate on skin pigmentation: a study of five patients. J Eur Acad Dermatol Venereol 21:384–387
9. Brazzelli V, Roveda E, Prestinari F et al (2006) Vitiligo-like lesions and diffuse lightening of the skin in a pediatric patient treated with imatinib mesylate: a noninvasive colorimetric assessment. Pediatr Dermatol 23:175–178
10. Brazzelli V, Antonieti M, Muzio F et al (2008) The perilesional skin in vitiloigo: a colorimetric "in vivo" study of 25 patients. Photodermatol Photoimmunol Photomed 24:314–317
11. Brunsting LA, Sheard C (1929) The color of the skin as analyzed by spectrophotometric methods: II. The role of pigmentation. J Clin Invest 7:559–574
12. Boissy RE, Manga P (2004) On the etiology of contact/occupational vitiligo. Pigment Cell Res 17:208–214
13. Chiaverini C, Passeron T, Ortonne JP (2002) Treatment of vitiligo by topical calcipotriol. J Eur Acad Dermatol Venereol 16:137–138
14. Clarys P, Alewaeters K, Lambrecht R et al (2000) Skin color measurements: comparison between three instruments: the Chromameter®, the DermaSpectrometer® and the Mexameter®. Skin Res Technol 6:230–238
15. Corcuff P, Leveque JL (1996) In vivo vision of the human skin with the tandem scanning microscope. Dermatology 186:50–54
16. Corcuff P, Bertrand C, Leveque JL (1993) Morphometry of human epidermis in vivo by real-time confocal microscopy. Arch Dermatol Res 285:475–481
17. Corcuff P, Gonnord G, Pierard GE et al (1996) In vivo confocal microscopy of human skin: a new design for cosmetology and dermatology. Scanning 18:351–355
18. Diffey BL, Oliver RJ, Farr PM (1984) A portable instrument for quantifying erythema induced by ultraviolet radiation. Br J Dermatol 111:663–672
19. Duteil L (2002) Objective methods to assess pigmentation. In: Ortonne JP, Ballotti R (eds) Mechanism of suntanning. Martin Dunitz, London
20. Edwards EA, Duntley SQ (1939) The pigments and color of living human skin. Am J Anatomy 65:1–33
21. Ellengogen R, Jankauskas S, Collini F (1990) Achieving standardized photographs in aesthetic surgery. Plast Reconstr Surg 86:955–962
22. Fain PR, Babu SR, Bennett DC et al (2006) HLA class II haplotype DRB1*04-DQB1*0301 contributes to risk of familial generalized vitiligo and early disease onset. Pigment Cell Res 19:51–57
23. Feather J, Ellis DJ, Leslie G (1988) A portable reflectometer for the rapid quantification of cutaneous haemoglobin and melanin. Phys Med Biol 33:711–722
24. Fullerton A, Fischer T, Lathi A et al (1996) Guidelines for measurement of skin colour and erythema. A report from the Standardization Group of the European Society of Contact Dermatitis. Contact Dermatitis 31:1–10
25. Gilchrest BA, Fitzpatrick TB, Anderson RR et al (1977) Localization of melanin pigmentation in the skin with Wood's lamp. Br J Dermatol 96:245–248
26. Gniadecka M, Wulf HC, Mortensen N et al (1996) Photoprotection in vitiligo and normal skin. Acta Derm Venereol 76:429–432
27. Goldman RJ, Salcido R (2002) More than one way to measure a wound: an overview of tools and techniques. Adv Skin Wound Care 15:236–245
28. Hajizadeh-Saffar M, Feather JW, Dawson JB (1990) An investigation of factors affecting the accuracy of in vivo measurements of skin pigments by reflectance spectrophotometry. Phys Med Biol 35:1301–1315
29. Hamzavi I, Jain H, McLean D et al (2004) Parametric modelling of narrowband UV-B phototherapy for vitiligo using a novel quantitative tool: the Vitiligo Area Scoring Index. Arch Dermatol 140:677–683

30. Hettiaratchy S, Papini R (2004) Initial management of a major burn: II – assessment and resuscitation. BMJ 329:101–103
31. Huzaira M, Rius F, Rajadhyaksha M et al (2001) Topographic variations in normal skin, as viewed by in vivo reflectance confocal microscopy. J Invest Dermatol 116:846–852
32. Knaysi GA, Crikelair GF, Cosman B (1968) The rule of nines; its history and accuracy. Plastic Reconst Surg 41:560–563
33. Kollias N (2007) Skin documentation with multimodal imaging or integrated imaging approaches. In: Wilhelm KP, Elsner P, Berardesca E, Maibach HI (eds) Bioengineering of the skin. Skin imaging and analysis, 2nd edn. Informa Healthcare, New York, pp 221–246
34. Kollias N, Gillies R, Cohen-Goihnab C et al (1997) Fluorescence photography in the evaluation of hyperpigmentation in photodamaged skin. J Am Acad Dermatol 36:226–230
35. LePoole IC, Das PK (1997) Microscopic changes in vitiligo. Clin Dermatol 15:863–873
36. Marrakchi S, Bouassida S, Meziou TJ et al (2007) An objective method for the assessment of vitiligo treatment. Pigment Cell Melanoma Res 21:106–107
37. Middlekamp-Hup MA, Park HY, Lee J et al (2006) Detection of UV-induced and epidermal changes over time using in vivo reflectance confocal microscopy. J Invest Dermatol 126:401–407
38. Minsky M (1961) Microscopy apparatus. US Patent # 3013467. Filed 1957, Awarded 1961
39. Minsky M (1998) Memoir on inventing the confocal scanning microscope. Scanning J 10:123–128
40. Montes LF, Abulafia J, Wilborn WH et al (2003) Value of histopathology in vitiligo. Int J Dermatol 42:57–61
41. New KC, Petroll WM, Boyde A et al (1991) In vivo imaging of human teeth and skin using real-time confocal microscopy. Scanning 13:369–372
42. Neuse WHG, Neumann NJ, Lehmann P et al (1996) The history of photography in dermatology. Arch Dermatol 132:1492–1498
43. Noon JP, Evans CE, Haynes WG et al (1996) A comparison of techniques to assess skin blanching following the topical application of glucocorticoids. Br J Dermatol 134:837–842
44. Ortonne JP, Nordlund J (2006) Mechanisms that cause abnormal skin color. In: Nordlund JJ, Boissy RE, Hearing VJ, King RA, Oetting WS, Ortonne JP (eds) The pigmentary system, 2nd edn. Blackwell, Oxford, pp 489–502
45. Paraskevas LR, Halpern AC, Marghoob AA (2005) Utility of the Wood's light: five cases from a pigmented lesion clinic. Br J Dermatol 152:1039–1044
46. Park SB, Suh DH, Youn JI (1999) A long-term time course of colorimetric evaluation of ultraviolet light-induced skin reactions. Clin Exp Dermatol 24:315–320
47. Pierard GE (1998) EEMCO guidance for the assessment of skin colour. J Eur Acad Dermarol Venereol 10:1–11
48. Queille-Roussel C, Poncet M, Schaefer H (1991) Quantification of skin color changes induced by topical corticosteroidi preparations using Minolta Chroma Meter. Br J Dermatol 124:264–270
49. Rajadhyaksha M, Anderson RR, Webb RH (1999) Video-rate confocal scanning laser microscope for imaging human tissues in vivo. Appl Optics 38:2105–2115
50. Rajadhyaksha M, Gonzalez S, Zavislan JM (2004) Detectability of contrast agents for confocal reflectance imaging os skin and microcirculation, J Biomed Opt 9(2):323–331
51. Rajadhyaksha M, González S, Zavislan JM et al (1999) In vivo confocal scanning laser microscopy of human skin II: advances in instrumentation and comparison to histology. J Invest Dermatol 113:293–303
52. Rajadhyaksha M, Grossman M, Esterowitz D et al (1995) Video-rate confocal scanning laser microscopy for human skin: melanin provides strong contrast. J Invest Dermatol 104:946–952
53. Ratner D, Thomas CO, Bickers D (1999) The uses of digital photography in dermatology. J Am Acad Dermatol 41:749–756
54. Schallreuter KU, Büttner G, Pittelkow MR et al (1994) Cytotoxicity of 6-biopterin to human melanocytes. Biochem Biophys Res Commun 204:43–48
55. Seitz JC, Whitmore CG (1988) Measurement of erythema and tanning responses in human skin using a tristimulus colorimeter. Dermatological 177:70–75
56. Serup J, Agner T (1990) Colorimetric quantification of erythema – a comparison of two colorimeters (Large Micro Color and Minolta Chroma Meter CR-200) with a clinical scoring schema and laser-Doppler flowmetry. Clin Exp Dermatol 15:267–272
57. Shigeaki S, Imura M, Ota M (1985) The relationship of skin color, spectophotometric techniques. J Invest Dermatol 84: 265–267
58. Stalder JF, Le Forestier D (1992) La photographie en pratique dermatologique. Ann Dermatol Venereol 119:695–702
59. Stamatas GN, Zmudzka BZ, Kollias N et al (2004) Noninvasive measurements of skin pigmentation in situ. Pigment Cell Res 17:618–626
60. Taïeb A, Picardo M (2007) The definition and assessment of vitiligo: a consensus report of the Vitiligo European Task Force. Pigment Cell Res 20:27–35
61. Takiwaki H, Overgaard L, Serup J (1994) Comparison of narrow-band reflectance spectrophotometric and tristimulus colorimetric measurements of skin color. Skin Pharmacol 7:217–255
62. Takiwaki H, Miyaoka Y, Kohno H et al (2002) Graphic analysis of the relationship between skin colour change and variations in the amounts of melanin and haemoglobin. Skin Res Technol 8:78–83
63. Taylor SC (2002) Skin of color: biology, structure, function, and implications for dermatologic disease. J Am Acad Dermatol 46:41–62
64. Taylor S, Westerhof W, Im S et al (2006) Noninvasive techniques for the evaluation of skin color. J Am Acad Dermatol 54:282–290
65. Tearney GJ, Brezinski ME, Southern JF et al (1995) Determination of the refractive index of highly scattering human tissue by optical coherence tomography. Opt Lett 20:2258–2260
66. Tuzun Y, Yazici H (1981) A method of measuring skin lesions. Arch Dermatol 117:192
67. Van Geel N, Ongenae K, Vander Haeghen Y et al (2004) Autologous transplantation techniques for vitiligo: how to evaluate treatment outcome. Eur J Dermatol 14:46–51
68. Van Geel N, Vander Haeghen Y, Ongenae K et al (2004) A new digital image analysis system useful for surface assessment of vitiligo lesions in transplantation studies. Eur J Dermatol 14:150–155
69. Vitkin IA, Woolsey J, Wilson BC et al (1994) Optical and thermal characterization of natural (sepia officinalis) melanin. Photochem Photobiol 59:455–462

70. Wachtel TL, Berry CC, Wachtel EE (2000) The inter-rater reliability of estimating the size of burns from various burn area chart drawings. Burns 26:156–170
71. Wang X, Milner TE, Chang MC et al (1996) Group refractive index measurement of dry and hydrated type I collagen films using optical low-coherence reflectometry, J Biomed Opt 1:212–216
72. Weatherall IL, Coombs BD (1992) Skin color measurements in terms of CIELAB color space values. J Invest Dematol 99:468–473
73. Westerhof W (2006) Dermatoscopy. In: Serup J, Jemec GBE, Grove GL (eds) Handbook of non-invasive methods and the skin, 2nd edn. CRC, Boca Raton, pp 113–114
74. Westerhof W (2006) Colorimetry. In: Serup J, Jemec GBE, Grove GL (eds) Handbook of non-invasive methods and the skin, 2nd edn. CRC, Boca Raton, pp 635–651
75. Whitton ME, Ashcroft DM, Barrett CW et al (2006) Interventions for vitiligo. Cochrane Database Syst Rev 1:CD003263
76. Wood RW (1919) Secret communications concerning light rays. J Physiol 5 serie:t IX
77. Yamashita T, Kuwahara T, Gonzales S et al (2005) Non-invasive visualization of melanin and melanocytes by reflectance-mode confocal microscopy. J Invest Dermatol 124: 235–240

Electron Microscopy

2.2.3.2

Yvon Gauthier

Contents

Y. Gauthier
Service de Dermatologie, Hôpital Saint-André.
CHU de Bordeaux, France
e-mail: yvongauthier@free.fr

Core Messages

- The electron microscope has been considered as a useful tool to study the different stages of normal and abnormal melanogenesis.
- Several ultrastructural studies have been performed on vitiligo skin to investigate controversial issues, namely the presence or absence of melanocytes in the epidermis of depigmented skin, and the mechanism of the melanocyte loss.
- Electron microscopy has confirmed the involvement of melanocytes, but has also shown alterations of keratinocytes and Langerhans cells.

Ultrastructural Study of the Epidermis

Electron microscopy has confirmed the involvement of melanocytes, and also that of keratinocytes and Langerhans cells (LC) in NSV. The abnormalities that have been described are contrasted according to the site of the biopsy, namely depigmented lesions, normally pigmented skin, hyperpigmented margins, and the so-called inflammatory raised borders.

Depigmented Macules

Melanocytes

There is a long-standing controversy regarding whether melanocytes in vitiligo macules are lost or still present, but inactivated. An immunohistochemical investigation using a large panel of antibodies suggested that

M. Picardo and A. Taïeb (eds.), *Vitiligo,*
DOI 10.1007/978-3-540-69361-1_2.2.3.2, © Springer-Verlag Berlin Heidelberg 2010

melanocytes are indeed lost in vitiligo lesions [22]. In the majority of ultrastructural studies, this absence has been confirmed. However, it was shown in favour of the theory of biochemical inhibition of melanogenesis that vitiligo skin still contains tyrosinase activity 4–37% relative to that of normal skin [18]. Two studies provide evidence for the fact that melanocytes are still present in the epidermis of patients even after a long duration. Bartosik et al. [1] found a significant amount of melanin in some keratinocytes located in the basal epidermal layers suggesting that rare melanocytes were still remaining. But the possibility that these melanin granules could originate from functioning melanocytes in perilesionnal rather than amelanotic area was envisaged. Tobin et al. [32] considered that a few melanocytes are still present in the depigmented epidermis of patients with vitiligo of long-standing duration because melanocyte could be grown from depigmented epidermal suction blister samples. Moreover, mature Stage IV melanin granules were found in the cytoplasm of basal and suprabasal keratinocytes in the depigmented lesional epidermis. Quite surprisingly in this study, melanocytes could not be demonstrated immunohistochemically. How can we reconcile these observations? It could be hypothesised that vitiligo could result in a progressive and chronic melanocyte loss with rare melanocytes remaining. Another explanation would be that melanin granules could still originate from melanocytes with functional anomalies and loss of usual markers remaining in the durably achromic epidermis.

Niebauer [27] has described dendritic cells lacking cell specific organelles (melanosomes, Birbeck granules)in vitiligo macules. He called these cells 'indeterminate cells (IC).' The significance of these IC is not clear: do IC represent premelanocytes, 'worn out' melanocytes, or indifferenciated stem cells? Several publications have been devoted to IC. According to Mishima et al. [23], two phases can be distinguished in the cytokinetics of vitiligo. In the first stage, the number of melanocytes decreases, the number of IC increases; in the second stage, as the melanocyte population approaches or reaches zero, the population of IC begins to decrease until no melanocytes or IC remain.

According to Bleehen [5], serial EM sections showed that many of the indeterminate dendritic cells were LC, containing only a few characteristic organelles.

Keratinocytes

The keratinocytes of the white macules have usually been referred as normal. However, Breathnach et al. [9] reported mild vacuolisation in basal keratinocytes in the vitiliginous epidermis of one patient.

Langerhans Cells

Melanocytes seem to be replaced mainly by LC that can be easily recognised by their characteristic granules (rod and racquet-shaped organelles) [9]. LC are more numerous in the basal cell layer than in normal control epidermis [3]. Some of these LC contain membrane bound clusters of melanin granules.

Zellickson and Mottaz [33] found increased numbers of LC in vitiliginous skin exposed either to sun or ultraviolet radiation. Perrot et al. [30] have reported older lesions to contain only Langerhans-type dendritic cells with a normal or decreased number of typical granules.

Clinically Normal Appearing Skin Adjacent to Amelanotic Skin

Ultrastructural study of normal appearing skin shows that in addition to the damage of melanocytes, there is some degeneration of neighbouring keratinocytes.

Melanocytes

Many studies have confirmed earlier observations of melanocytes degeneration, including vacuolisation, presence of a dilated endoplasmic reticulum, fatty degeneration, and granular deposits with aggregation of melanosomes. These abnormalities are more or less pronounced in each cell. Melanocytes are usually observed in basal position, sometimes detached from the basal membrane (BM). In rare cases, they are located in suprabasal position [14, 25, 26] and clusters of premelanosomes can be detected in the mid-layers of the epidermis. It was hypothesised that after a more or less long period, their survival is affected and they leave the basal layer and take part to the general movement of the renewing epidermis (Figs. 2.2.3.8 and 2.2.3.9).

However, a small number of melanocytes show no apparent sign of degeneration. Bleehen et al. [5] have found marginal melanocytes to have small, incompletely melanised melanosomes, some showing degenerative changes.

Keratinocytes

Two types of ultrastructural abnormalities have been observed in the epidermis of the patients with NSV as compared to normal controls: deposits of extracellular granular material (EGM), and foci of vacuolar degeneration of keratinocytes located in basal or suprabasal layers [24].

EGM is found in dilated intercellular spaces between the basal layers of keratinocytes and sometimes between basal cells and the basement membrane. The EGM appears to be derived from the cytoplasm of vacuolated keratinocytes (Fig. 2.2.3.8). The aspect of EGM reminds keratinocyte cytoplasm containing free ribosomes. EGM seems to be more abundant in progressive or stable vitiligo.

A *focal vacuolisation of keratinocytes* is reported mostly in the basal layers of the epidermis with preferential degeneration of the parts of the cytoplam in close apposition to a melanocyte. Among the degenerative changes observed in keratinocytes, the following can be listed: a dilatation of the endoplasmic reticulum and a swelling of mitochondria, the formation of cytoplasmic vacuoles with or without limiting membranes, and a frequent loss of desmosomes or hemidesmosomes. Degenerative keratinocyte changes were found reduced in patients repigmenting spontaneously.

Interestingly these two keratinocytes abnormalities: vacuolisation and EGM deposits have been reported previously after skin frictions which could be responsible for the Koebner phenomenon in vitiligo skin. As a result of skin friction, small vacuoles were detected in the oedematous periphery of keratinocytes and the membranes of these cells were often ruptured, allowing their granular contents to spill into the extracellular spaces [17].

Langerhans Cells

The density of the Langerhans cell is quite normal. LC have been observed in their normal mid-epidermal location, but also sometimes in a basal position in the vicinity of melanocytes.

Marginal Hyperpigmented Skin

Melanocytes

There is an increased density of melanocytes and many completely melanised melanosomes are observed in the cytoplasm of keratinocytes [9, 15]. This is probably the explanation of the onset of this hyperpigmented marginal ring frequently observed during phototherapy.

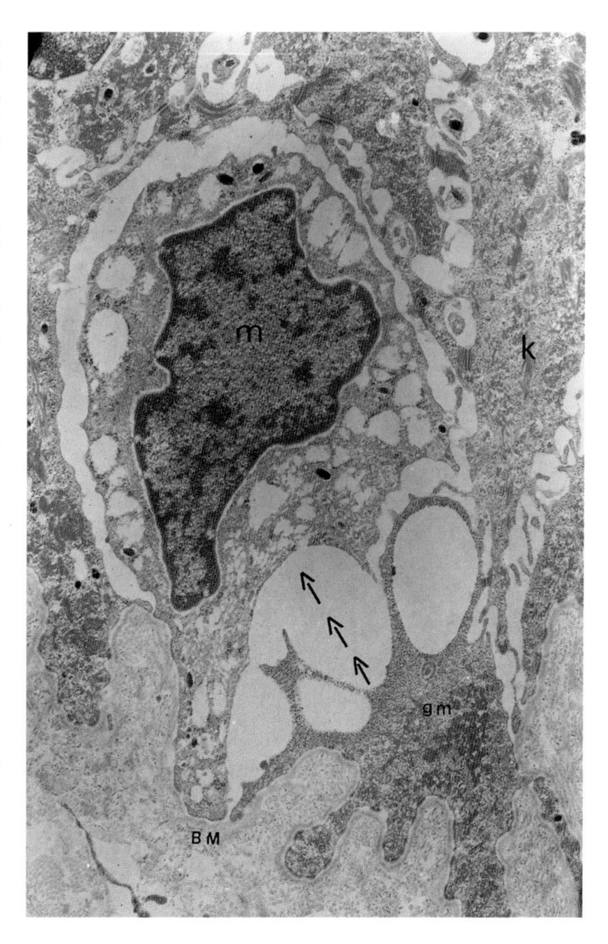

Fig. 2.2.3.8 Melanocyte from the pigmented marginal area (×6,000): Note vacuolated melanocyte (m) detached from the basal membrane (BM), extracellular granular material (gm) derived from the cytoplasm of vacuolated keratinocytes (K)

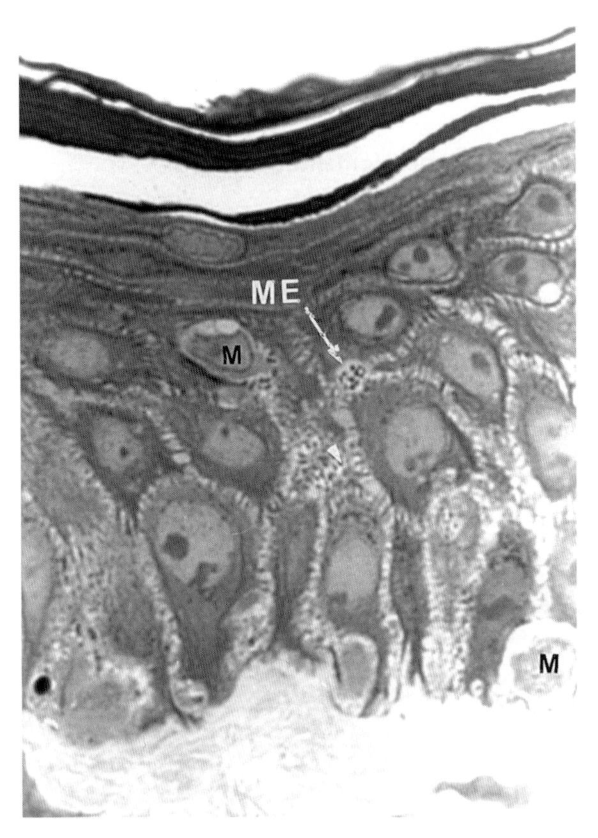

Fig. 2.2.3.9 From the pigmented area: Melanocytes (M) in basal and suprabasal position. Melanin (ME white) deposits in the upper layers of the epidermis (×1,400)

Keratinocytes

The aspect of keratinocytes is usually normally similar to that reported during repigmentation process.

Ultrastructural Studies of Inflammatory Vitiligo

The observation of vitiligo with inflammatory raised borders is unusual (Chaps. 1.2.2 and 1.3.10). It is characterised by slight redness or itching at the borders of early or progressive vitiligo lesions. The cause of this inflammatory reaction with a raised border is not known.

Melanocytes

The melanocytes of the basal layer are markedly decreased in number and contain very few melanosomes in their cytoplasm. Vacuolisation of the cytoplasm or autophagocytosis of melanosomes are not found when the inflammatory changes are limited [19]. On the other hand, in case of severe inflammatory changes a marked dissociation of keratinocytes of the spinous layer has been observed. Some melanocytes with many vacuoles located in their dendrites and cytoplasm were also observed. Lymphocytes were seen in close contact with melanocytes.

Langerhans Cells

Many LC can be observed in basal position and sometimes in close contact with melanocytes (Fig. 2.2.3.10).

Ultrastructural Study of the Basement Membrane

In most instances the basement membrane is normal in vitiliginous skin. Sometimes, a reduplication of the basement membrane in perilesional skin has been reported. Breathnach et al. [9] found multiple replications or layering of basement membrane directly beneath the melanocytes on the pigmented side of the vitiligo margin. Perrot et al. [30] described several layers separated by spaces containing filamentous material and collagen fibers (Fig. 2.2.3.11). Gaps in the epidermal basement membrane have been observed and attributed to a lack of staining caused by the presence of foreign cells [8].

Ultrastructural Study of Cultured Melanocytes from Vitiligo Patients

An inherent melanocyte defect has been suggested after short-term melanocyte culture.

Cultured melanocytes from vitiligo patients have demonstrated abnormalities that consist of dilatation of the rough endoplasmic reticulum (RER), circular RER shapes, and membrane compartmentalisation of melanosomes [6, 7]. These results have led to suggest that cultured melanocytes from most vitiligo patients have an innate defect. These abnormalities were not always simultaneously expressed and were not associated with specific clinical feature of vitiligo.

The selective dilatation of RER was reported in other diseases in which a newly translated glycoprotein is selectively uploaded in the cisternae of the RER. An extensive dilatation of the RER was observed in the cytoplasm of fibroblasts from Ehlers Danlos disease, where defective collagen synthesis occurs. Moreover, hepatocytes from the Z-type mutant form of inherited antitrypsin deficiency also demonstrate extensive dilatation of RER. This dilated RER can be continuously expressed for over one year in culture.

Ultrastructural Study of the Dermis

An increased cellularity of the dermis has been reported in some instances [9]. Mononuclear cells have been found in the dermis and sometimes in the epidermis of the hypomelanotic skin. Rarely, a marked lymphocytic infiltrate in the papillary dermis was observed.

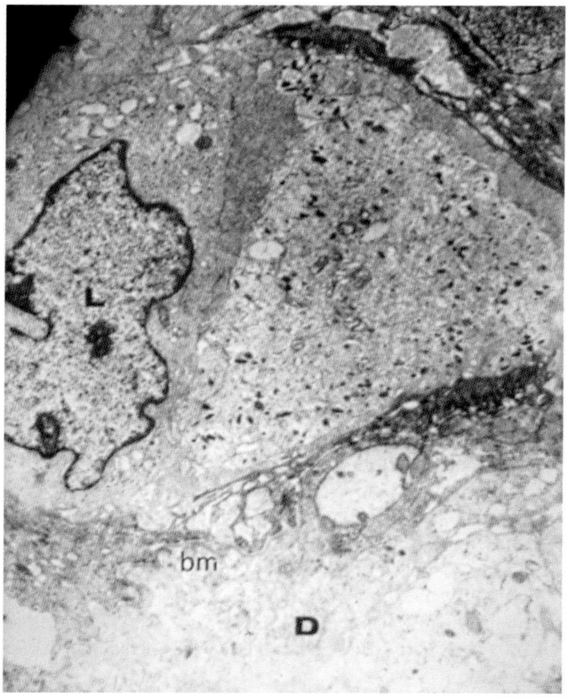

Fig. 2.2.3.10 From the marginal pigmented area: Close contact in the basal layer between a Langerhans cell (L) and a melanocyte (M). *D* dermis (×6,000) (with the collaboration of Surleve-Bazeille)

Melanosomes have been found in macrophages located in the dermis and sometimes in Schwann cells.

Degenerative and regenerative changes of a number of unmyeliminated nerves of the papillary dermis have been described by Breathnach [9], such as swelling of axons and replacement of axonal neurofibrillae by a granular matrix, reduplication of Schwann cell basement membrane, and regenerating axons non-embedded in Schwann cell cytoplam (Fig. 2.2.3.12). Most of these degenerative changes have been found in small and superficial nerves. In some cases, free nerves endings have been observed in the epidermis. Similar biological changes of human cutaneous nerves have been reported after long-term repeated exposure to UVA-solar irradiation [21].

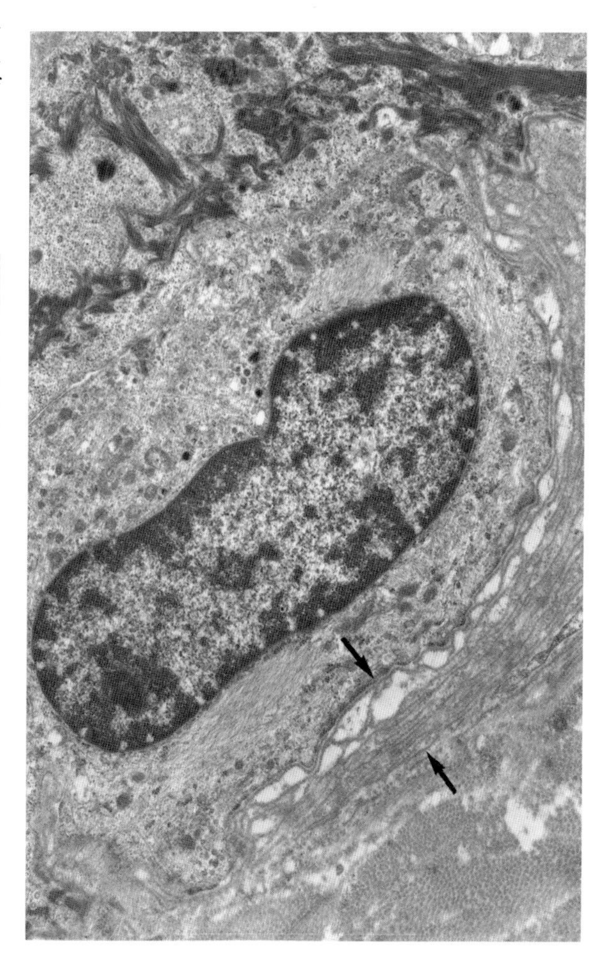

Fig. 2.2.3.11 From the pigmented area: Layering of basement membrane (between the *arrows*) directly beneath the melanocyte located in the margin (10,000) (with the collaboration of Surlève-Bazeille)

Concluding Remarks

Ultrastructural studies on vitiliginous skin have shown that there is a lack of melanocytes in the epidermis. How these melanocytes are lost is still a matter of speculation and has generated several theories. EM observations suggest that the entire epidermal melanin unit is probably affected in vitiligo since alterations and vacuolisation can be observed even in the keratinocytes. The basement membrane generally is normal in vitiliginous skin. An increased cellularity of the dermis has been reported in some instances.In normal appearing skin, melanocytes are usually observed in basal position, sometimes detached from the BM. In rare cases, they are located in suprabasal position.

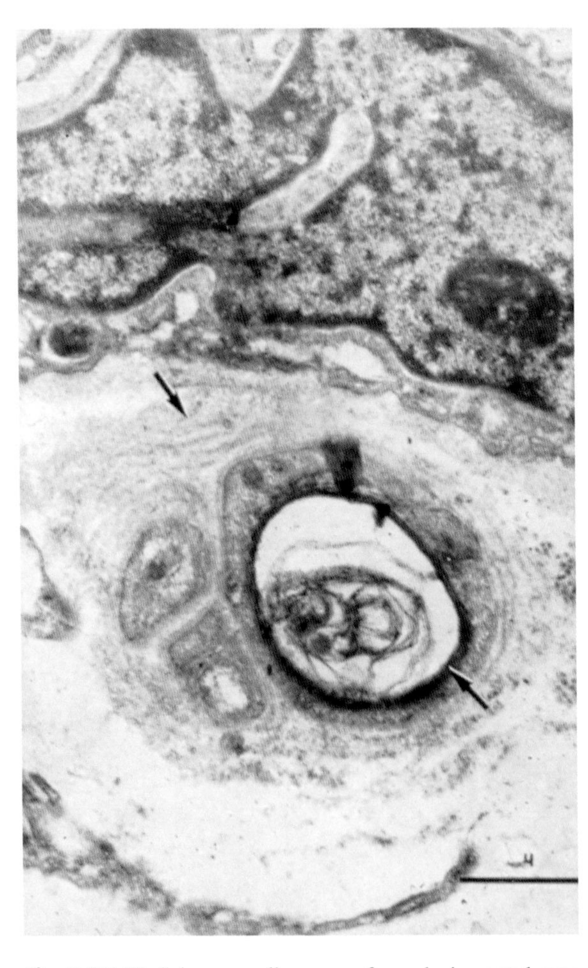

Fig. 2.2.3.12 Schwann cell process from depigmented area (×40,000):Note reduplication of basement membrane(*upper arrow*) and degenerative aspect of Schwann cell (*lower arrow*) (with the collaboration of Surleve-Bazeille)

References

1. Bartosik J, Wulf HC, Kobayasi T (1998) Melanin and melanosomes complexes in long standing stable vitiligo: ultrastructural study. Eur J Dermatol 8: 95–97
2. Bhawan J Bhutani LK (1983) Keratinocyte damage in vitiligo. J Cutan Pathol 10:207–212
3. Birbeck MS, Breathnach AS, Everall JD An electron microscope study of basal melanocytes and high level clear cells (Langerhans cells) in vitiligo. J Invest Dermatol 37:51–64
4. Bleehen SS (1976) The treatment of vitiligo with topical corticosteroids: light and electron microscopic studies. Br J Dermatol 94:43–49
5. Bleehen SS (1979) Histology of vitiligo. Pigment Cell Res 5:54–61
6. Boissy RE, Liu YY, Medrano E et al (1991) Structural aberrations of the rough endoplasmic reticulum and melanosome compartmentalisation in long term cultures of melanocytes from vitiligo patients. J Invest Dermatol 97:395–404
7. Boissy RE (1991) Dilated rough endoplasmic reticulum and premature death in melanocytes cultured from vitiligo mouse. Am J Pathol 138:1511–1525
8. Bose SK, Ortonne JP (1964) Focal gaps in the basement membrane of involved skin of vitiligo: are they normal? J Dermatol 21:152–159
9. Breathnach AS, Bor S, Wyllie LM. Electron microscopy of peripheral nerve terminals and marginal melanocytes in vitiligo. J Invest Dermatol;47:125–140
10. Brown J, Winkelmann RK, Wolff K (1967) Langerhans cells in vitiligo. J Invest Derm 49:386–390
11. Fitzpatrick TB, Parrish JA, Pathak MA (1957) Melanin biosynthesis and the pathophysiology and treatment of vitiligo with Trioxalen. Giornale Italiano di Dermatologia 110:121–130
12. Galadari E. (1993) Ultrastructural study of vitiligo. Int J Dermatol 32:269–271
13. Gauthier Y, Surlève-Bazeille JE (1974) Ultrastructure des fibres nerveuses périphériques dermiques dans le vitiligo. Bull Soc Fr Dermatol Syphil 81:550–554
14. Gauthier Y, Cario-André M, Lepreux S et al (2003) Melanocyte detachment after skin friction in non lesional skin of patient with generalized vitiligo. Br J Dermatol 148:95–101
15. Gonzalves RP, Bechelli LM, Simao T et al (1978) Etude quantitative ultrastructurale de la couche basale de l'epiderme dans les taches de vitiligo et dans la melanodermie périlésionnelle. Ann Dermatol Venereol 205:395–400
16. Hann SK, Park YK, Lee KG (1992) Epidermal changes in active vitiligo. J Dermatol 19:217–222
17. Hunter JA, MC Vittie E, Comaisk JS (1975) Light and electronmicroscopic studies of physical injury to the skin: friction. Br J Dermatol 90:491–498
18. Husain I, Vijayan E, Ramaiah A, Pasricha JS, Madan NC (1982) Demonstration of tyrosinase in the vitiligo skin of human beings by a sensitive fluorometric method as well as by 14C(U)-L-tyrosine incorporation into melanin. J Invest Dermatol 78:243–252
19. Ishi M, Hamada T (1981) Ultrastructural studies of vitiligo with inflammatory raised borders. J Dermatol 8:312–322
20. Kao M (1990) Depletion and repopulation of Langerhans cells in non segmental type of vitiligo. J Dermatol 17: 287–296

21. Kumakiri M, Hashimoto K, Willis I (1978) Biological changes of human cutaneous nerves caused by ultraviolet irradiation: an ultrastructural study. Br J Dermatol 99:65–70
22. Le Poole C, Van den Wijngaard MC, Westerhof W et al (1993) Presence or absence of melanocytes in vitiligo lesions: an immunohistochemical investigation. J Invest Dermatol 100:816–822
23. Mishima Y (1972) Dendritic cell dynamics in progressive depigmentation. Arch Dermatol Res 87:243–267
24. Moellmann G, Klein-Angerer S, Scobly DA (1982) Extracellular granular material and degeneration of keratinocytes in the normally pigmented epidermis of patients with vitiligo. J Invest Dermatol 79:321–330
25. Morahashi M, Hashimoto K, Goodman TF et al (1977) Ultrastructural studies of vitiligo, Vogt-Koyanagi syndrome and incontinentia pigmenti achromians. Arch Dermatol 113:755–766
26. Mottaz JH, Thorne G, Zelickson A (1971) Response of the epidermal melanocytes to minor trauma. Arch Dermatol 104:611–618
27. Niebauer G (1965) On the dendritic cells in vitiligo. Dermatologica 130:317–324
28. Ortonne JP (1983) Vitiligo and other hypomelanoses of hair and skin. In: Ortonne JP, Mosher DB, Fitzpatrick TB (eds) Disorders with circumscribed hypomelanosis. Plenum, New York
29. Ortonne JP, Thivolet J (1980) PUVA induced repigmentation of vitiligo: scanning electron microscopy of hair follicles. J Invest Dermatol 74:40–42
30. Perrot H et al (1974) Etude ultrastructurale du vitiligo. Lyon Med 232:439–446
31. Pinkus H (1959) Vitiligo: what is it? J Invest Dermatol 32:281–284
32. Tobin JD, Swanson N, Pittelkow M et al (2000) Melanocytes are not absent in lesional skin of long duration vitiligo. J Pathol 191:407–416
33. Zelickson AS, Mottaz JH (1968) Epidermal dendritic cells. Arch Dermatol 98:652–658

Animal Models

2.2.4

Gisela F. Erf

Contents

G. F. Erf
Center of Excellence for Poultry Science,
University of Arkansas, Fayetteville, Arkansas, USA
e-mail: gferf@uark.edu

Core Messages

- Chronic autoinflammatory/autoimmune disorders typically are multifactorial in nature, requiring several components such as genetic susceptibility, immune system components, and environmental factors for expression. Appropriate experimental animal models have become an essential tool to delineate and dissect the relative contributions of these components.
- The spontaneous autoimmune vitiligo described in the Smyth-line chicken recapitulates the entire spectrum of clinical and biological manifestations of human vitiligo, providing a unique opportunity to examine processes prior to expression and throughout the progression of the disease.
- Induction of autoimmune vitiligo in the mouse by the administration of plasmids encoding melanocyte differentiation antigens in the context of stress/danger signals provides a defined in vivo model to examine immune recognition of melanocyte proteins.

2.2.4.1 Animal Models of Autoimmune Vitiligo

The Multifactorial Nature of Autoimmune Disorders: Application to Vitiligo

Autoimmunity has been identified as a major aetiological factor in vitiligo, although many other factors

M. Picardo and A. Taïeb (eds.), *Vitiligo*,
DOI 10.1007/978-3-540-69361-1_2.2.4, © Springer-Verlag Berlin Heidelberg 2010

including infections, stress, neural abnormalities, aberrant melanocyte function, and genetic susceptibility have been implicated [46, 47, 59, 63, 67, 70, 77]. This multifactorial nature of vitiligo involving genetic susceptibility and disease precipitating factors in addition to immunopathology is in itself in accordance with the general concepts described for tissue-specific autoimmune disease. In tissue-specific autoimmune diseases such as vitiligo, the genetic susceptibility is frequently associated with an inherent target cell defect that predisposes the target cell to immunorecognition and may include aberrant immune activity at various levels (e.g. dendritic cells/macrophages, T-cells, and B-cells). The autoimmune destruction of cells has been found to be associated with a lack of regulatory function within the immune system, heightened immune activity, and altered responsiveness of immune components and target cells to endogenous and exogenous factors. The role of environmental factors in the development of autoimmune disease is also multifaceted and may include infections by microbes, exposure to chemicals, and a wide array of other stress factors that can provoke an autoimmune response to the target cells [22, 40, 56]. Unfortunately, the relative contributions of these components in organ-specific autoimmune disease cannot be easily delineated and dissected, especially in human patients when they become apparent only after the clinical manifestation of the disease.

Naturally Occurring Animal Models of Vitiligo

In order to understand the initial aetiology and pathogenic events leading to onset and the progression of autoimmune disease, appropriate animal models are required. In this context, experimental animal models that spontaneously develop the autoimmune disease would reflect the situation in humans more closely than experimental models where the autoimmune disease was induced. In biomedical research, the mouse has become the most studied animal model, due in part to its short generation time, its small size, and the extensive availability of genetically defined strains of mice, research reagents, and research procedures. Murine models for spontaneous autoimmune disease are, however, rare, and there is currently no mouse model that develops spontaneous autoimmune vitiligo [22, 40, 84].

For vitiligo research, several animal models of naturally occurring vitiligo were identified and described in a review by Boissy and Lamoreux published in 1988 [4]. These included the Smyth line chicken, the grey horse, the vitiligo mouse, and the Sinclair pig. Of these spontaneous vitiligo models, only the Smyth line chicken continues to be studied extensively and, as outlined later, is currently the only animal model for spontaneous autoimmune vitiligo that recapitulates the entire spectrum of clinical and biological manifestations of the human disease.

Grey Horse

Horses that carry the dominant Grey (G^G) allele (e.g. the Lipizzaner, Arabian horse, Andalusian) are generally born coloured. After birth, they begin to show white hair that is intermixed with their original colour. The amount of white hair increases with age until the coat is completely white at maturity (Fig. 2.2.4.1a–c). This progressive greying and depigmentation of the hair over time is more extensive and happens more quickly in homozygotes (*G/G*) than in heterozygotes (*G/g*). In most cases, pigmentation of the skin and eyes is not affected [4]. In mature grey horses, there is a high incidence of melanoma (>80%) [69, 72]. The melanomas typically are located on the ventral surface of the tail and the perineal region as well as head, neck, and external genitalia. Albeit not malignant, the melanomas can cause intense discomfort to the horses especially in sensitive areas. As melanoma development is rarely observed in other coloured horses, its high incidence in the grey horse suggests an intimate association between melanoma development and processes involved in the grey coat colour generation. Occasionally, areas of progressive vitiligo involving the epidermis as well as the hair can also be observed in the grey horse [4, 33, 57]. These areas of skin and hair depigmentation tend to involve the face, perioral, perianal, and perigenital areas. Pigmentation loss has been associated with the presence of autoantibodies that bind to surface antigens on pigment cells. These similarities between the vitiligo observed in the grey horse and human vitiligo suggest the usefulness of the grey horse as an experimental model for vitiligo [33, 57]. Research on the aetiology and pathogenesis of the vitiligo in the grey horse has however not been pursued extensively, presumably due

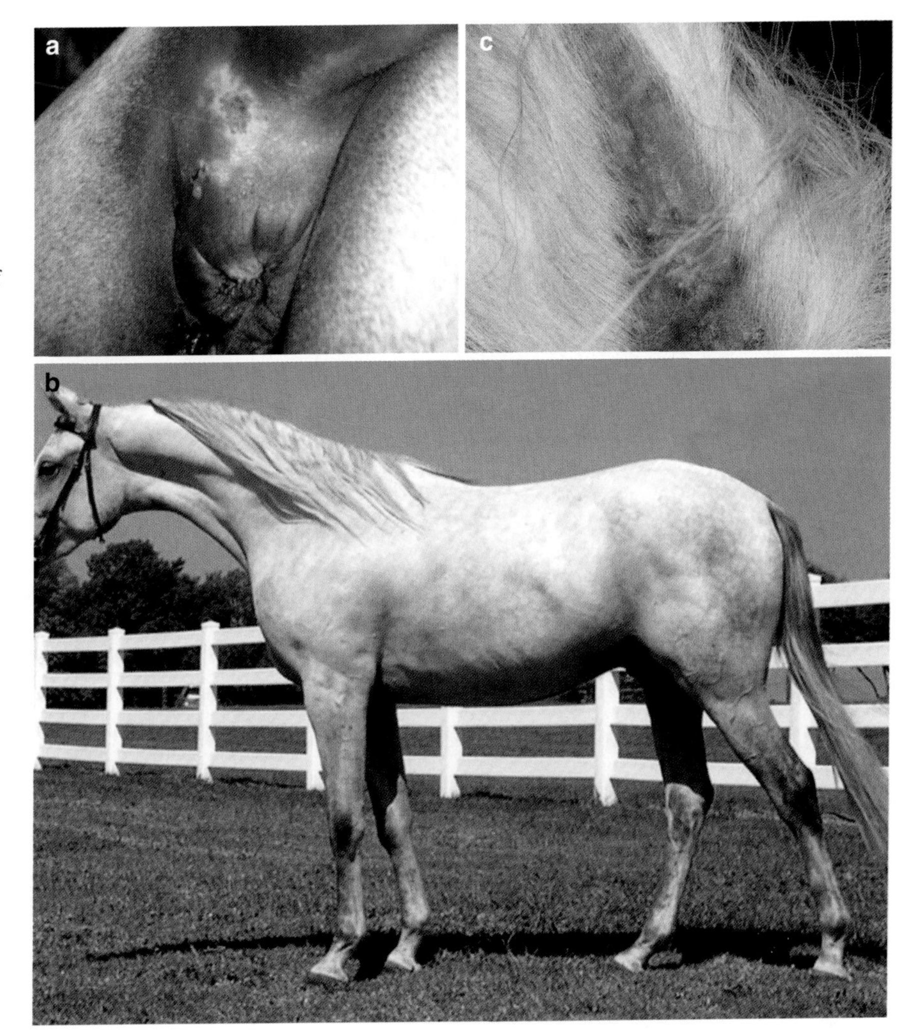

Fig. 2.2.4.1 Grey horse model for spontaneous vitiligo. (**a**) Grey horse with vitiligo on regression site of melanoma (from www.equusdreams.com/GreyHorseMelanomas-AVigil.htlm). (**b**, **c**) Grey horse and grey horse with melanoma at the underside of the tail (courtesy of Dr N Jack, Dept Animal Science, Division of Agriculture, University of Arkansas, Fayetteville, AR)

to the substantial cost of rearing and maintaining research animals of this large size and long generation time. In recent years, the *Grey* gene locus has been mapped to horse chromosome 25 (ECA25) and several genes responsible for different coat colour phenotypes or associated with pigmentation disorders and melanoma have been excluded as candidate genes. Recent comparative linkage mapping for the grey colour gene in horses suggests that the grey phenotype is caused by a mutation in a novel gene [50, 64, 80].

The Sinclair Miniature Swine

The Sinclair miniature swine (Fig. 2.2.4.2a–d) is characterised by spontaneous development and subsequent regression of cutaneous malignant melanoma lesions [36, 37, 52, 53, 58]. The incidence of melanoma is 85% by one year of age. Tumour regression closely follows development and is characterised by a decrease in tumour volume and sequential changes in tumour pigmentation from black to white. The pigmentation changes are not limited to the tumour and are associated with the development of uveitis and vitiligo. There is substantial evidence for immune system involvement in tumour regression and in the development of the generalised vitiligo disorder. Immune mechanisms include components of both cellular and humoral immunity, recognising common antigens shared by normal and malignant swine pigment cells [3, 17, 20, 68]. Autoimmune depigmentation following sensitisation to melanoma antigens has also been reported in other animal models (e.g. mouse melanoma-transfer studies,

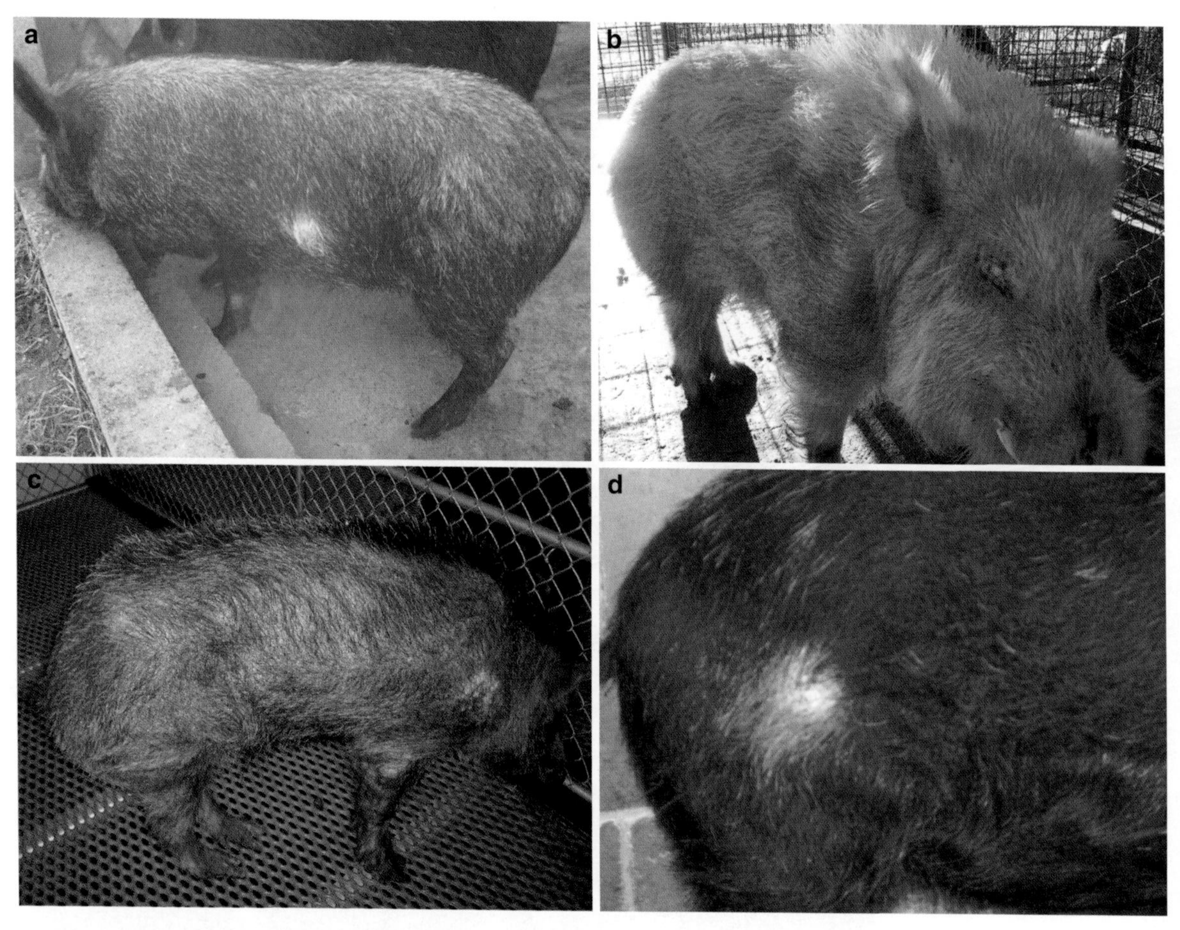

Fig. 2.2.4.2 Sinclair Swine pig. (**a**) Six months old pig with vitiligo at regression site of melanoma on left side. (**b**) Total depigmentation after melanoma; even the cornea is depigmented; this is not an albino pig (courtesy of Dr D. Kraemer and O. Ash, Texas A&M University, Sinclair Swine Research Project, College Station, TX). (**c**) Another vitiligo pig with regressing melanoma, NV. (courtesy of Dr L Gomez-Raya, University of Neveda, Reno, NV). (**d**) Six years old sow with vitiligo at regression site of melanoma on right hip

vaccination with melanocyte antigens) as well as in human patients receiving immunotherapy [38, 44, 45, 54]. Hence, the Sinclair miniature pig appears to be an excellent model for spontaneous melanoma formation, melanoma regression, and vitiligo development. Research on this animal model however continues to be sporadic and focuses primarily on melanoma development and regression [34, 62].

Water Buffalo

Vitiligo is also known to occur in water buffalos (Fig.2.2.4.3). Histological, histochemical, and ultrastructural analyses of biopsies taken from involved and uninvolved skin from two female buffalos revealed cytological aberrations of melanocytes similar to those reported in humans and other vitiligo models [19].

The Vitiligo (C57Bl/J6-vit/vit) Mouse

The vitiligo (C57Bl/J6-vit/vit) mouse developed by Dr. Aaron Lerner was a promising murine model for vitiligo, exhibiting not only progressive loss of cutaneous hair and ocular pigmentation with age, but also a muted response to contact allergens a phenomenon observed in humans with vitiligo [1, 10, 49]. However, the *vit* gene was later mapped to the *mi* (microphthalmia) locus and the mutation was

Fig. 2.2.4.3 Water buffalo with vitiligo (courtesy of Dr F. Roperto, University of Naples Federico II, Naples, Italy)

subsequently designated as $mi^{vit/vit}$ [43]. The human homolog (*MITF;* microphthalmia-associated transcription factor) of the mouse *mi* gene carries no mutations in humans with vitiligo and, while a good candidate gene at the phenotypic level, does not appear to be involved in the pathogenesis of familial human vitiligo [81]. Hence, the $mi^{vit/vit}$ mouse does not appear to reflect mechanisms involved in spontaneous autoimmune vitiligo and was not further pursued as an animal model for this disease.

The Barred Plymouth Rock and White Leghorn Chickens

In 1992, Bowers et al. [13] reported another chicken model for vitiligo that focused specifically on premature melanocyte death due to inherent sensitivity to environmental factors. This model included two pigmentation mutants: the Barred Plymouth Rock (BPR) (Fig. 2.2.4.4) and the White Leghorn (WL) chicken. The feather pigmentation pattern of BPR chickens consists of alternating bands of pigmented and non-pigmented segments of feather pigmentation. The barring gene of BPR chickens was shown to

Fig. 2.2.4.4 Plymouth Rock Hen (from Fowlvisions.com)

inhibit the deposition of melanin in the feathers, beak, shanks, and eye of chickens. The alternating

Table 2.2.4.1 The Smyth line chicken model (formerly the DAM-line of chickens)

Production	Robert Smyth, Poultry Geneticist, University of Massachusetts, Amherst, MA
Maintanance	Gisela F. Erf, Avian Immunologist, Division of Agriculture, University of Arkansas, Fayetteville, AR
Features	Vitiligo-like, autoimmune, post-hatch loss of pigmentation in feathers and eye
Onset	6–14 weeks of age
Incidence	80–95% by 20 weeks (young adult)
Severity	Erratic to complete loss of pigmentation
Unique features	Recapitulates the entire spectrum of clinical and biological manifestations of the human disease
	Availability of MHC-matched ($B^{101/101}$) control lines of chickens, including the parental Brown line, from which the Smyth line was developed (<2% incidence of vitiligo), and the Light Brown Leghorn chicken (vitiligo resistant)
	Easy, repeatable, minimally invasive access to the autoimmune lesion in the feather, prior to and throughout the development of vitiligo
	The strong direct association of herpesvirus of turkey (HVT) administration at hatch and the incidence of vitiligo expression (incidence: >80% with HVT, <20% without HVT) allows for examination of disease precipitating/protective factors in genetically susceptible individuals
	Spontaneous expression of other autoimmune diseases (e.g. autoimmune thyroiditis, alopecia) in the Smyth line chicken provides opportunity to study the kaleidoscope of autoimmune diseases

black and white feather band pattern is caused by autophagocytic degeneration of melanocytes. Melanocytes are not present in the white bands. Feather melanocytes of WL chickens undergo premature cell death in the feather collar prior to deposition of melanin and the feathers are hypomelanocytic. The Wild-type control for the BPR/WL animal model is the Jungle Fowl where feather melanocytes do not undergo cell death until they have grown out with the feather. While both BPR and WL feather melanocytes die prematurely, there is no immunopathology associated with their death. The fact that BPR and WL feather-derived melanocytes live for months in tissue culture suggests that these mutant melanocytes have an inherent sensitivity to toxic factors that are produced and accumulated in the local feather environment. Hence, removal of feather melanocyte and subsequent culture in the presence and absence of toxic factors (melanogenesis-related products, oxidative radicals) with and without anti-oxidants provided an excellent system to study inherent melanocyte sensitivity. Moreover, the fact that BPR melanocytes were more sensitive than Jungle fowl melanocytes, and WL melanocytes were more sensitive than BPR melanocytes, provided additional opportunity to examine the mechanism associated with this sensitivity. The observation of a role of altered antioxidant capacity and sensitivity to oxidative stress in the premature death of BPR and WL melanocytes constitutes an important connection to the altered anti-oxidant capacity observed in human vitiligo and in Smyth line chickens [13–15, 29, 70].

2.2.4.2 The Smyth Line Chicken Animal Model of Spontaneous Autoimmune Vitiligo

The SL chicken is characterised by a spontaneous, vitiligo-like, post-hatch loss of melanin producing pigment cells (melanocytes) in feather and choroidal tissue (Fig. 2.2.4.5, Table 2.2.4.1). Vitiligo occurs in approximately 80–95% of hatch-mates, with about 70% of those affected expressing complete depigmentation in adulthood (>20 weeks of age). There are many similarities between SL and human vitiligo. Both are characterised by autoimmune destruction of melanocytes, usually first seen during adolescence and early adulthood. In both SL chickens and humans, pigmentation loss may be either partial or complete. Remelanisation of amelanotic tissue occurs, although severe pigment loss and remelanisation are more frequent in the chicken. In addition to vitiligo, SL chickens exhibit uveitis, often resulting in blindness (5–15%), and have associated autoimmune diseases such as hypothyroidism (4–8%) and an alopecia areata-like feathering defect (2–3%) [22, 74, 75, 84]. Similarly, in humans it is not uncommon to find thyroidal and other autoimmune diseases associated with vitiligo. Moreover, SL vitiligo, like human vitiligo, is a multifactorial disorder involving a genetic component (manifested in part as an inherent melanocyte defect; e.g. abnormal melanosome membranes), an immune system component (melanocyte-specific cell-mediated immunity), and environmental triggers (e.g. herpesvirus of turkey; HVT) [22, 84].

Fig. 2.2.4.5 Smyth line chicken model for spontaneous autoimmune vitiligo. (**a**) One-day-old Smyth line chick; Smyth line chicks hatch with an intact pigmentary system. (**b**) Hen with normal pigmentation (**c**) Vitiliginous Smyth line females; (**d**) Pictures from left to right: (1) Growing feathers used for visual assessment of vitiligo and collection for down-stream analysis. (2) Cross-section of the growing portion of a non-vitiliginous feather. Melanocytes are aligned with their cell-bodies facing the pulp and their dendrites extending into the barb ridge where they deposit pigment into barbule cells (keratinocytes). (3) Cross-section of the growing portion of a vitiliginous feather. This frozen section was immunohistochemically stained with mouse-anti-chicken CD8 monoclonal antibody to identify cytotototoxic lymphocytes (CD8+; brown cells). Note: CD8+ lymphocytes surround the cell bodies of the melanocytes and can be observed along the melanocyte dendrites in the barb ridge. Pigment transfer to barbule cells has been disrupted. (4) Cross-section of a vitiliginous feather (same sample as in picture 3) immunohistochemically stained for MHC class II expression (brown stain). Note: The observed MHC class II expression reflects very active immunopathology associated with melanocyte loss. In chickens, as in humans, activated T cells, B cells and macrophages express MHC class II and increased MHC class II expression is associated with interferon-γ production. It is not clear from this staining pattern whether melanocytes express MHC class II

Hence, SL chickens offer unique opportunities to examine the interplay between genetic susceptibility, immune system activity, and environmental factors. As pigment cells are located in the feather rather than the skin, the target tissue is easily accessible and can be sampled prior to and throughout the development of SL vitiligo in the same individual. We know of no other tissue-specific autoimmune disease model where the developing lesion can be monitored this easily; especially using a minimally invasive procedure like pulling a few small feathers (Fig. 2.2.4.5). Another advantage of this animal model is that the incidence of SL vitiligo by 20 weeks of age is predictably high (>80%) and low (<20%) with and without HVT-administration at hatch, respectively, providing unique opportunity to examine the influence of environmental factors on the expression of SL vitiligo. Lastly, this animal model includes two MHC (*B* locus)-matched control lines ($B^{101/101}$): the parental Brown Line (BL, <2% incidence of vitiligo) and the Light Brown Leghorn (LBL, similar plumage colours, vitiligo resistant) [78]. The development and characteristics of this animal model have been reviewed extensively [22, 74, 84]. Below, is a summary demonstrating the suitability of this animal model for research on the aetiology, pathology, prevention, and treatment of spontaneous autoimmune vitiligo.

The Genetics Basis of Smyth Line Autoimmune Vitiligo

The SL chickens and control lines of chicken were developed by Dr. J. Robert Smyth, Jr. at the University of Massachusetts, Amherst, MA. The origin, development, and characteristics of these lines of chicken have been reviewed extensively by Smyth [74]. The mutant SL (previously known as the delayed amelanosis (DAM)-chicken) was developed from one female hatched in 1971 from a non-pedigreed mating of the Massachusetts Brown line (BL) [76]. The ability to develop a line of chickens with a predictably high incidence of vitiligo starting with one vitiliginous hen clearly demonstrates the heritability of this disorder. Throughout the years, Dr. Smyth developed various BL and SL sublines. Among the criteria used for the development of sublines was the MHC-haplotype (*B101*, *B102* and *B103)* associated with SL vitiligo. The *B101* and *B102* haplotypes were also identified in LBL chickens. MHC-sublines (SL101, SL102, SL103, BL101, BL102, BL103, and LBL101) were subsequently developed. The characteristics of these MHC-sublines were summarised by Smyth and McNeil [75]. Unfortunately, most of these valuable lines have been destroyed following the closing of the Poultry Farm at the University of Massachusetts and the retirement of Dr. Smyth in 1996. The lines homozygous for the *B101* MHC-haplotype (SL101, BL101, and LBL101) are the only lines remaining and are currently maintained at the University of Arkansas, Fayetteville, AR by the author. Together, these lines constitute the current animal model for autoimmune vitiligo.

The genetic basis of autoimmune vitiligo and line-associated traits has long been described as being under the control of multiple autosomal genes. A more recent molecular characterisation of SL and BL sublines revealed a high level of inbreeding within lines (0.948 for SL101; 0.902 for BL101) and high genetic similarity between SL101 and BL101 lines (similarity index of 0.049 ± 0.006) [79]. Hence, it appears that a limited number of genes are responsible for the SL phenotype. With the availability of the chicken genome sequence and other sophisticated bioinformatics and experimental resources for chicken research, it is now possible to conduct genome-wide expression analysis and high-resolution quantitative trait loci (QTL) mapping in chickens [39]. QTL mapping using the SL model is currently underway in Sweden by Dr. Susanne Kerje and Dr. Olle Kämpe (Department of Medical Science) and Dr. Leif Andersson (Department of Medical Biochemistry and Microbiology) at Uppsala University and will soon shed light on the specific genes responsible for the depigmentation and other abnormalities seen in the SL chicken.

Inherent Melanocyte Defect in Smyth Line Autoimmune Vitiligo

Previous studies by J. Robert Smyth Jr. and co-workers describe the presence of a competent pigment system in SL chicks at hatch (Fig. 2.2.4.5a). The earliest abnormalities within SL melanocytes, prior to visible onset of SL vitiligo, were irregularly shaped melanosomes containing pigmented membrane extensions, hyperactive melanisation, and selective autophagocytosis of melanosomes [6–9]. Similar degenerative processes were also observed *in vitro* in embryo-derived SL melanocytes, including heightened lipid peroxidation and catalase activity [9, 29]. The occurrence of these melanocyte malfunctions *ex vivo* provides strong evidence for an inherent melanocyte defect in SL vitiligo. However, as shown through immunosuppression studies, the inherent melanocyte defect alone is not sufficient to cause SL vitiligo without a functioning immune system, but it appears to play a role in provoking a melanocyte-specific autoimmune response [7, 16, 32, 42, 61]. Morphological and biological melanocyte defects/alterations have also been reported in vitiligo patients and non-SL animal models for vitiligo. In both human and animals, cultured vitiligo melanocytes grow more slowly than normal melanocytes, are more dependent on antioxidants in the medium, and present structural alterations [9, 11, 12, 14, 15, 48, 51, 66, 70].

Immune System Involvement in Smyth Line Autoimmune Vitiligo

Several studies have provided evidence supporting the role of the immune system in the pathology of SL vitiligo. Destruction of melanocytes is preceded by increased levels of circulating inflammatory cells and the infiltration of the feather pulp and barb ridge by macrophages and lymphocytes (Fig. 2.2.4.5d). Prior to onset of vitiligo, the infiltrate consisted primarily of T-helper cells (CD4+), but with onset of vitiligo and active melanocyte destruction, cytotoxic T cells (CD8+) predominated (Fig. 2.2.4.6) [22–24, 73]. Melanocyte death was shown to occur by apoptosis, apparently induced by cytotoxic T cells [82]. Overall, the immunopathology of SL vitiligo described earlier is very similar to observations in affected skin of vitiligo patients [67] and supports the involvement of cell-mediated immune mechanisms in melanocyte destruction. In SL chickens, direct evidence for a role of cell-mediated immunity in the development of SL vitiligo was provided by immunosuppression studies and by in vivo demonstration of anti-melanocyte cell-mediated immunity in vitiliginous but not in non-vitiliginous SL and control chickens [83]. Current examination of feather tissues collected prior to and near onset of SL vitiligo using targeted gene expression analyses by real-time quantitative RT-PCR showed that the onset of SL vitiligo was preceded and accompanied by high expression of the pro-inflammatory cytokines interleukin (IL)-1β, IL-8, IL-12 and

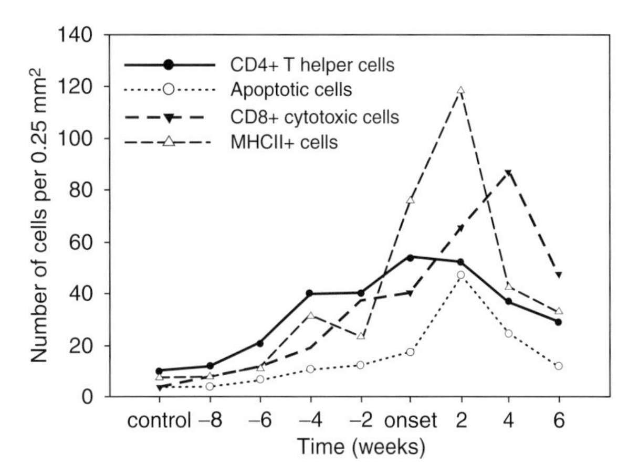

Fig. 2.2.4.6. T cell infiltration, MHC class II expression and melanocyte apoptosis in Smyth line feathers prior to and throughout the development of vitiligo. Three growing feathers were collected from Smyth line chickens every 2 weeks from 6- to 20-weeks of age (10 birds/time point). Tips were snap-frozen for immunohistochemical- and Tunnel-staining. Feather infiltration of CD4+, CD8+, and MHC class II+ cells, as well as the presence of apoptotic cells were examined by bright field microscopy. The presence and severity of vitiligo was scored on a weekly basis. Because age of onset differed, data were expressed with respect to visible onset of vitiligo. Onset refers to the presence of depigmented barbs emerging from the shafts of growing feathers

interferon (IFN)-γ, but low expression levels of IL-4. This further supports the presence of cellular, Th1-mediated immune activity directed against melanocytes [30, 65]. Additionally, vitiliginous SL chickens have melanocyte-specific autoantibodies that have been shown to bind to chicken as well as human melanocytes, and to be specific to tyrosine-related protein-1 and heat-shock proteins (HSP), but their role in the aetiology of SL vitiligo has not been established [2, 30, 71]. Maternal autoantibodies to melanocytes have been shown to be transferred from the hen to her chick via the egg, but disappeared from the chick's circulation within 10 days of hatch and their transfer was not correlated with vitiligo expression [27].

Environmental Factors Involved in the Expression of Smyth Line Autoimmune Vitiligo

In addition to an inherent melanocyte defect and an autoimmune component, we reported the role of environmental factors, specifically vaccination with live turkey herpesvirus (HVT) at hatch, in the expression of

SL vitiligo [25]. Without HVT, the incidence of SL vitiligo is <20%, but with HVT, the incidence is >80% and generally around 95%. HVT is an alpha-herpesvirus commonly used in commercial chicken production as a vaccine to protect chickens from Marek's disease caused by Serotype 1 Marek's disease viruses (MDV-1). HVT is a non-oncogenic Serotype 3 MDV isolated from turkeys that causes only minor inflammatory lesions. And like other MDV, it exhibits strong tropism for feather follicles, where it infects the feather follicle epithelium [18, 35]. Additional studies on the role of HVT in SL vitiligo revealed that killed HVT had no effect on the expression of SL vitiligo. Therefore, the ability of HVT to cause infection must be of importance. Administration of other live virus vaccines at hatch (i.e. Newcastle disease virus, NDV; infectious bronchitis virus; IBV), instead of HVT, did not trigger the expression of SL vitiligo; suggesting that viral infection and associated anti-viral immune activity, as such, are not responsible for triggering expression of SL vitiligo. Unlike HVT, NDV and IBV do not translocate to the feather; hence the presence of HVT where melanocytes are located may be the key to its effect on SL vitiligo expression [22]. Lastly, comparison of HVT-vaccinated and non-HVT-vaccinated SL and parental control BL chicks, revealed heightened cell-mediated immune activity to HVT in SL compared to BL chicks [26].

Based on these observations we hypothesise that the translocation of the HVT-infection to the feather epithelium brings anti-viral immune activity to the feather where the melanocytes are present. The resulting local antiviral cell-mediated immune activity in the melanocytes' environment causes alterations in the already inherently defective melanocytes that result in their recognition by the immune system leading to the development of melanocyte-specific immune activity and autoimmune destruction of melanocytes.

The SL chicken model provides a unique opportunity to study the role of precipitating factors in the expression of vitiligo in susceptible individuals in vivo. Using non-HVT vaccinated chicks expected to have a low incidence of SL vitiligo (<20%) the effects of various immunomodulators, inflammatory agents, microbial products, and other endogenous and exogenous factors on triggering the expression of SL vitiligo in susceptible individuals can be examined. For example, when recombinant chicken IFN-γ was administered subcutaneously or intra-abdominally to non-HVT-vaccinated SL chicks twice per week for the first 6 weeks of their lives, 87.5% of females and 0% of males developed SL vitiligo by 20 weeks of age. The incidence of SL vitiligo in vehicle-injected chicks was 25% in females, 0% in males [28]. This finding was interesting, not only because of the ability of this cytokine to cause expression of SL vitiligo, but also because of the observed gender difference. Gender differences in autoimmune diseases are common. When HVT-vaccinated SL chicks are raised in conventional floor pens on litter, SL vitiligo tends to be expressed earlier in females compared to males, but by 20 weeks of age this gender difference is no longer significant. The gender difference is however significant when HVT-vaccinated SL chicks are raised in cages in isolation [25].

We are currently conducting a pilot study examining the role of local feather inflammation in the expression of SL vitiligo. For this, study, chicks were kept in brooder cages in a chicken house. Inflammatory bacterial cell wall products (lipopolysaccharide, LPS; muramyl dipeptide, MDP) were directly injected into the feather of vitiligo-susceptible non-HVT-vaccinated SL chicks. Histological analysis of injected feathers, collected at different time-points post-LPS, -MDP, and -vehicle control (PBS) injection, is underway. Visual assessment of pigmentation of feather-injected birds revealed that nearly 40% of the birds receiving LPS or MDP had developed a vitiligo-like progressive pigmentation loss [31].

Opportunities Provided by the Smyth Line Chicken Model for Autoimmune Vitiligo

Past and ongoing studies have clearly demonstrated the suitability and usefulness of the Smyth line chicken model as a model for human vitiligo. The spontaneous development of autoimmune vitiligo in this animal model together with the demonstrated similarity in many of the complex features of human vitiligo offer excellent opportunities to address the interrelationship of components involved in this multifactorial disease. Moreover, the predictably high incidence of vitiligo, the easy and repeatable access to the target tissue, and the identification of an environmental trigger of disease expression make this a highly valuable and unique animal model for spontaneous autoimmune vitiligo and tissue-specific autoimmune disease. The vitiligo research community is fortunate to have a spontaneous vitiligo model that can provide important insight into the aetiology and pathogenesis of the disease and is useful for the testing and development of strategies for prevention and treatment of vitiligo. Much appreciation needs to be extended to Dr. J. Robert Smyth Jr. for his excellent scientific insight in recognising the value of the one vitiliginous hen and for his tireless effort, vision, and generosity in the development of the Smyth and control lines of chickens.

2.2.4.3 Experimental Model of Induced Autoimmune Vitiligo: Mouse Model for the Role of Stress

In vitiligo, a wide variety of stress factors have been reported to provoke an autoimmune response to melanocytes ranging from chemicals, overexposure to sunlight (UV), skin injury, infection, and emotional stress [5, 41, 46, 55]. Cells under stress will produce stress proteins such as HSP which serve a protective function during the cell-stress episode. The observation that HSP70 can enhance dendritic cell uptake of antigen and is an ideal vaccine adjuvant for anti-tumour vaccines, together with the implication that HSP70 is involved in precipitating vitiligo in susceptible individuals [46, 60], led to the design of a reproducible in vivo mouse model of autoimmune vitiligo [21]. For this, mice were gene-gun vaccinated with eukaryotic expression plasmids encoding melanocyte differentiation antigens (e.g. TRP-2, gp100) with or without human or mouse-derived HSP70. Inclusion of HSP70, independent of origin, accelerated depigmentation in this model. Interestingly, the progressive depigmentation induced by HSP70 involved areas not directly exposed to the stress (Fig. 2.2.4.7). Depigmentation was accompanied by the induction of prolonged antibody-responses to HSP70 and correlated with T cell-mediated cytotoxicity towards targets loaded with a H-2K$^{(b)}$-restricted TRP-2 peptide. These observations strongly support the role of HSP70 in the observed depigmentation, and provide opportunity to examine the role of cellular stress in immunorecognition of melanocyte proteins. Moreover, this induced autoimmune vitiligo model can be extended to assess the ability of other environmental factors to precipitate or protect from the induction of a melanocyte-specific autoimmune attack.

Fig. 2.2.4.7 Hair depigmentation in C57BL6 mice gene-gun vaccinated with 4 μg of DNA encoding human gp100. (courtesy of Dr. J. Guevara-Patino, University of Chicago and Dr. Caroline Le Poole, Loyola University, Chicago, IL)

2.2.4.4 Concluding Remarks

Vitiligo is a multifactorial disease requiring animal models to delineate and dissect the components and mechanisms involved in this complex disease. Excellent animal models of both spontaneous and induced autoimmune vitiligo are available to the vitiligo research community for in vivo research on mechanisms leading to the expression and progression of vitiligo as well as for the development of treatment and prevention strategies.

Past and current research efforts have clearly established the Smyth line chicken model as an appropriate and suitable animal model of vitiligo. Currently, the Smyth line chicken model is the only animal model of spontaneous autoimmune vitiligo shown to recapitulate the entire spectrum of clinical and biological manifestations of the human disease. The Smyth line chicken model together with the recently developed induced autoimmune vitiligo mouse model provide the necessary tools to delineate and dissect the mechanisms involved in this complex disease. Insight gained from these models can then be translated into strategies for vitiligo prevention, treatment, and management in humans.

References

1. Amornsiripanitch S, Barnes LM, Nordlund JJ et al (1988) Immune studies in the depigmenting C57BL/Ler-vit/vit mice. An apparent isolated loss of contact hypersensitivity. J Immnol 140:3438–3445
2. Austin LM, Boissy RE (1995) Mammalian tyrosinase-related protein-1 is recognized by autoantibodies from vitiliginous Smyth chickens. An avian model for human vitiligo. Am J Pathol 146:1529–1541
3. Berkelhammer J, Ensign BM, Hook RR et al (1982) Growth and spontaneous regression of swine melanoma: relationship of in vitro leukocyte reactivity. JNCI 68:461–468
4. Boissy RE, Lamoreux ML (1988) Animal models of an acquired pigmentary disorder-vitiligo. Prog Clin Biol Res 256:207–218
5. Boissy RE, Manga P (2004) On the etiology of contact/occupational vitiligo. Pigment Cell Res 17:208–214
6. Boissy RE, Smyth JR Jr, Fite KV (1983) Progressive cytologic changes during the development of delayed feather amelanosis and associated choroidal defects in the DAM chicken line. Am J Pathol 111:197–212
7. Boissy RE, Lamont SJ, Smyth JR Jr (1984) Persistence of abnormal melanocytes in immunosuppressed chickens of the autoimmune "DAM" line. Cell Tissue Res 235:663–668
8. Boissy RE, Moellmann G, Smyth JR Jr (1985) Melanogenesis and autophagocytosis of melanin within feather melanocytes of delayed amelanotic (DAM) chickens. Pigment Cell 1:731–739
9. Boissy RE, Moellmann G, Trainer AT et al (1986) Delayed-amelanotic (DAM or Smyth) chicken: melanocyte dysfunction in vivo and in vitro. J Invest Dermatol 86:149–156
10. Boissy RE, Moellmann GE, Lerner AB (1987) Morphology of melanocytes in hair bulbs and eyes of vitiligo mice. Am J Path 127:380–388
11. Boissy RE, Liu YY, Medrano EE et al (1991) Structural aberration of the rough endoplasmic reticulum and melanosome compartmentalization in long-term cultures of melanocytes from vitiligo patients. J Invest Dermatol 97: 395–404
12. Bowers RR, Gatlin JE (1985) A simple method for the establishment of tissue culture melanocytes from regenerating fowl feathers. In Vitro Cell Dev Biol 21:39–44
13. Bowers RR, Harmon J, Prescott S et al (1992) Fowl model for vitiligo: genetic regulation on the fate of the melanocytes. Pigment Cell Res Suppl 2:242–248
14. Bowers RR, Lujan J, Biboso A et al (1994) Premature avian melanocyte death due to low antioxidant levels of protection: fowl model for vitiligo. Pigment Cell Res 7:409–418
15. Bowers RR, Nguyen B, Buckner S et al (1999) Role of antioxidants in the survival of normal and vitiliginous avian melanocytes. Cell Mol Biol 45:1065–1074
16. Boyle ML III, Pardue SL, Smyth JR Jr (1987) Effects of corticosterone on the incidence of amelanosis in Smyth delayed amelanotic line chickens. Poult Sci 66:363–367
17. Burns RP, Tidwell M (1986) Experimental ocular malignant melanoma in Sinclair swine. Curr Eye Res 5:257–262
18. Calnek BW, Witter RL (1991) Marek's disease. In: Calnek BW, Barnes HJ, Beard CW, Reid WM, Yoder HW Jr (eds) Diseases of poultry. Iowa State University Press, Ames, Iowa
19. Cerundolo R, De Caprariis D, Esposito L et al (1993) Vitiligo in two water buffaloes: histological, histochemical, and ultrastructural investigations. Pigment Cell Res 6:23–28
20. Cui J, Chen D, Misfeldt ML et al (1995) Antimelanoma antibodies in swine with spontaneously regressing melanoma. Pigment Cell Res 8:60–63
21. Denman CJ, McCracken J, Hariharan V et al (2008) HSP70i accelerates depigmentation in a mouse model of autoimmune vitiligo. J Invest Dermatol 128:2041–2048
22. Erf GF (2008) Autoimmune diseases of poultry. In: Davison F, Kaspers B, Schat K (eds) Avian immunology. Elsevier, London
23. Erf GF, Smyth JR Jr (1996) Alterations in blood leukocyte populations in Smyth line chickens with autoimmune vitiligo. Poult Sci 75:351–356
24. Erf GF, Trejo-Skalli AV, Smyth JR Jr (1995) T cells in regenerating feathers of Smyth line chickens with vitiligo. Clin Immunol Immunopathol 76:120–126
25. Erf GF, Bersi TK, Wang X et al (2001) Herpesvirus connection in the expression of autoimmune vitiligo in Smyth line chickens. Pigment Cell Res 14:40–46
26. Erf GF, Johnson JC, Parcells MS et al (2001) A role of turkey herpesvirus in autoimmune Smyth line vitiligo. In: Shat KA (ed) Current progress on avian immunology research. American Association of Avian Pathologists, PA, pp 226–231
27. Erf GF, Lockhart BR, Griesse RL et al (2003) Circulating melanocyte-specific autoantibodies and feather-infiltrating lymphocytes in young Smyth line chickens prior to visible onset of vitiligo. Pigment Cell Res 16:420–421
28. Erf GF, Wang X, Bersi TK et al (2003) A role of interferon gamma in autoimmune vitiligo of Smyth line chickens. In: Workshop on Molecular Pathogenesis of Marek's Disease and Avian Immunology, Limassol, Cyprus (published on CD, 5pp)
29. Erf GF, Wijesekera HD, Lockhart BR et al (2005) Antioxidant capacity and oxidative stress in the local environment of feather-melanocytes in vitiliginous Smyth line chickens. Pigment Cell Res 18:69
30. Erf GF, Plumlee BL, Bateman KD et al (2007) Examination of early events in the development of autoimmune vitiligo in the Smyth line chicken model. Pigment Cell Res 20:329
31. Erf GF, Trovillion CT, Plumlee BL et al (2009) Smyth line chicken model for autoimmune vitiligo: opportunity to examine events leading to the expression of vitiligo in susceptible individuals. Pigment Cell and Melanoma Res 21:265

32. Fite KV, Pardue S, Bengston L et al (1986) Effects of cyclosporine in spontaneous, posterior uveitis. Curr Eye Res 5:787–796
33. Gebhart W, Niebauer G (1977) Connections between pigment loss and melanogenesis in gray horses of the Lipizzaner breed. Yale J Biol Med 50:45
34. Greene JF Jr, Morgan CD, Rao A et al (1997) Regression by differentiation in the Sinclair swine model of cutaneous melanoma. Melanoma Res 7(6):471–477
35. Holland MS, Mackenzie CD, Bull RW et al (1998) Latent turkey herpesvirus infection in lymphoid, nervous, and feather tissues of chickens. Avian Dis 42:292–299
36. Hook RR Jr, Aultman MD, Adelstein EH et al (1979) Influence of selective breeding on the incidence of melanoma in Sinclair miniature swine. Int J Cancer 24: 668–672
37. Hook RR Jr, Berkelhammer J, Oxenhandler RW (1982) Melanoma: Sinclair swine melanoma. Am J Pathol 108: 130–133
38. Hurwitz AA, Ji Q (2004) Autoimmune pigmentation following sensitization with melanoma antigens. Methods Mol Med 102:421–427
39. International Chicken Polymorphism Map Consortium (2004) A genetic variation map for chicken with 2.8 million single-nucleotide polymorphisms. Nature 432:717–722
40. Krishnamoorthy G, Holz A, Wekerle H (2007) Experimental models of spontaneous autoimmune disease in the central nervous system. J Mol Med 85:1161–1173
41. Kroll TM, Bommiasamy H, Boissy RE et al (2005) 4-Tertiary butyl phenol exposure sensitizes human melanocytes to dendritic cell-mediated killing, relevance to vitiligo. J Invest Dermatol 124:798–806
42. Lamont SJ, Smyth JR Jr (1981) Effect of bursectomy on development of a spontaneous postnatal amelanosis. Clin Immunol Immunopathol 21:407–411
43. Lamoreux ML, Boissy RE, Womack JE et al (1992) The *vit* gene maps to the mi (microphthalmia) locus of the laboratory mouse. J Hered 83:435–439
44. Lane C, Leitch J, Tan X et al (2004) Vaccination-induced autoimmune vitiligo is a consequence of secondary trauma to the skin. Cancer Res 64:1509–1514
45. Lengagne R, Le Gal F-A, Garcette M et al (2004) Spontaneous vitiligo in an animal model for human melanoma: role of tumor-specific CD8+ T cells. Cancer Res 64: 1496–1501
46. Le Poole IC, Luiten RM (2008) Autoimmune etiology of generalized vitiligo. In: Nickoloff BJ, Nestle FO (eds) Current directions in autoimmunity: dermatologic immunity, vol 10. Karger, Basel, pp 227–243
47. Le Poole IC, Das PK, van den Wijngaard RMJGJ et al (1993) Review of the etiopathomechanism of vitiligo: a convergence theory. Exp Dermatol 2:145–153
48. Le Poole IC, Boissy RE, Sarangarajan R et al (2000) PIG3V, an immortalized human vitiligo melanocyte cell line, expresses dilated endoplasmic reticulum. In Vitro Cell Dev Biol Anim 36:309–319
49. Lerner AB, Shiohara T, Boissy RE et al (1986) A possible mouse model for vitiligo. J Invest Dermatol 87:299–304
50. Locke MM, Penedo MCT, Bricker SJ et al (2002) Linkage of the grey coat colour locus to microsatellites on horse chromosome 25. Anim Genet 33:329–337
51. Medrano EE, Nordlund JJ (1990) Successful culture of adult human melanocytes obtained from normal and vitiligo donors. J Invest Dermatol 95:441–445
52. Millikan LE, Hook RR, Manning PJ (1973) Immunobiology of melanoma. Gross and ultrastructural studies in a new melanoma model: the Sinclair swine. Yale J Biol Med 46: 631–645
53. Misfeldt ML, Grimm DR (1994) Sinclair miniature swine: an animal model of human melanoma. Vet Immunol Immunopathol 43:167–175
54. Nagai H, Hara I, Horikawa T et al (2000) Elimination of CD4+ T cell enhances anti-tumor effect of locally secreted interleukin-12 on B16 mouse melanoma and induces vitiligo-like coat color alteration. J Invest Dermatol 115: 1059–1064
55. Namazi MR (2007) Neurogenic dysregulation, oxidative stress, autoimmunity, and melanocytorrhagy in vitiligo: can they be interconnected? Pigment Cell Res 20:360–363
56. National Institutes of Health Autoimmune Diseases Coordinating Committee (2005) Autoimmune diseases research plan. In: Progress in autoimmune disease. Research NIH, Bethesda, MD
57. Naughton GK, Mahaffey M, Bystryn J-C (1986) Antibodies to surface antigens of pigmented cells in animals with vitiligo. Proc Soc Exp Biol Med 181:423–426
58. Nordlund JJ, Lerner AB (1977) On the causes of melanomas. Am J Pathol 89:443–448
59. Nordlund JJ, Lerner AB (1982) Vitiligo-it is important. Arch Dermatol 118:5–8
60. Overwijk WW, Lee DS, Irvine KR et al (1999) Vaccination with a recombinant vaccinia virus encoding "self" antigen induces autoimmune vitiligo and tumor destruction in mice: requirement for CD4(+) T lymphocytes. Proc Natl Acad Sci U S A 96:2982–2987
61. Pardue SL, Fite KV, Bengston L et al (1987) Enhanced integumental and ocular amelanosis following termination of cyclosporine administration. J Invest Dermatol 88: 758–761
62. Pathak S, Multani AS, McConkey DL et al (2000) Spontaneous regression of cutaneous melanoma in Sinclair swine is associated with defective telomerase activity and extensive telomere erosion. Int J Oncol 17:1219–1224
63. Pawelek J, Korner A, Bergstorm A (1980) New regulation of melanin biosynthesis and autodestruction of melanoma cells. Nature 286:617–619
64. Pielberg G, Mikko S, Sanberg K et al (2005) Comparative linkage mapping of the Grey coat color gene in horses. Anim Genet 36:390–395
65. Plumlee BL, Wang X, Erf GF (2006) Interferon-gamma expression in feathers from vitiliginous Smyth line chickens. J Immunol 176:S283
66. Puri N, Phil M, Mojandar M et al (1987) In vitro growth characteristics of melanocytes obtained from adult normal and vitiligo subjects. J Invest Dermatol 88:434–438
67. Rezaei N, Gavalas NG, Weetman AP et al (2007) Autoimmunity as an aetiological factor in vitiligo J Eur Acad Dermatol Venereol 21:865–876
68. Richerson JT, Burns RP, Misfeldt ML (1989) Association of uveal melanocyte destruction in melanoma-bearing swine with large granular lymphocyte cells. Invest Ophthalmol Vis Sci 30:2455–2460
69. Rodriguez M, Garcia-Barona V, Pena L et al (1997) Grey horse melanotoic condition: a pigmentary disorder. J Equine Vet Sci 17:677–681

70. Schallreuter KU, Wood JM, Berger J (1991) Low catalase levels in the epidermis of patients with vitiligo. J Invest Dermatol 97:1081–1085
71. Searle EA, Austin LM, Boissy YL et al (1993). Smyth chicken melanocyte autoantibodies: cross-species recognition, in vivo binding, and plasma membrane reactivity of the antiserum. Pigment Cell Res 6:145–157
72. Seltenhammer MH, Simhofer H, Scherzer S et al (2003) Equine melanoma in a population of 296 grey Lipizzaner horses. Equine Vet J 35:153–157
73. Shresta S, Smyth JR Jr, Erf GF (1997) Profiles of pulp infiltrating lymphocytes at various times throughout feather regeneration in Smyth line chickens with vitiligo. Autoimmunity 25:193–201
74. Smyth JR Jr (1989) The Smyth chicken: a model for autoimmune amelanosis. Poult Biol 2:1–19
75. Smyth JR Jr, McNeil M (1999) Alopecia areata and universalis in the Smyth chicken model for spontaneous autoimmune vitiligo. J Invest Dermatol Symp Proc 4:211–215
76. Smyth JR Jr, Boissy RE, Fite KV (1981) The DAM chicken: a model for spontaneous postnatal cutaneous and ocular amelanosis. J Hered 72:150–156
77. Spritz RA (2007) The genetics of generalized vitiligo and associated autoimmune diseases. Pigment Cell Res 20: 271–278
78. Sreekumar GP, Erf GF, Smyth JR Jr (1996) 5-Azacytidine treatment induces autoimmune vitiligo in the parental control strains of the Smyth line chicken model for autoimmune vitiligo. Clin Immun Immunopathol 81: 136–144
79. Sreekumar GP, Smyth JR Jr, Ponce de Leon FA (2001) Molecular characterization of the Smyth chicken sublines and their parental controls by RFLP and DNA fingerprint analysis. Poult Sci 80:1–5
80. Swinburne JE, Hopkins A, Binns MM (2002) Assignment of the horse grey coat colour gene to ECA25 using whole genome scanning. Anim Genet 33:338–342
81. Tripathi RK, Flanders DJ, Young TL et al (1999) Microphthalmia-associated transcription factor (MITF) locus lacks linkage to human vitiligo or osteoperosis: an evaluation. Pigment Cell Res 12:187–192
82. Wang X, Erf GF (2004) Apoptosis in feathers of Smyth line chickens with autoimmune vitiligo. J Autoimmun 22: 21–30
83. Wang X, Erf GF (2003) Melanocyte-specific cell mediated immune response in vitiliginous Smyth line chickens. J Autoimmun 21:149–160
84. Wick G, Andersson L, Hala K et al (2006) Avian models with spontaneous autoimmune diseases. Adv Immunol 92:71–117

In Vitro Approaches

2.2.5

Muriel Cario-André and Maria Lucia Dell'Anna

Contents

M. Cario-André (✉)
Inserm U876, Centre de référence des maladies rares de la peau, Université V Segalen Bordeaux 2,
Bordeaux, France
e-mail: muriel.cario-andre@dermatol.u-bordeaux2.fr

Core Messages

- The complexity of cell culture models varies from monolayers for adherent cells or suspensions for nonadherent cells, to cocultures or to epidermal reconstructs (3D model).
- Cocultures and epidermal reconstructs allow the study of paracrine or contact cell–cell effects.
- The tests on nonepidermal nonadherent cells, such as peripheral blood mononuclear cells (PBMC) expand the view, going beyond melanogenesis-associated metabolisms.
- The in vitro studies on hair follicle melanocytes help understand the maturation and differentiation of melanocytes and to unveil the defective steps of growth and migration in melanocyte precursors during the repigmentation.
- New analytic techniques are emerging which can be applied for investigating vitiligo.

2.2.5.1 Cell Isolation and Culture

Melanocytes and keratinocytes can be cultured from both skin and hair follicle. Indeed, epidermal stem cells occur in the basal layer of the epidermis and within hair follicles. Follicular keratinocyte stem cells are situated in the bulge area, whereas those for melanocytes are found in the sub-bulge area.

M. Picardo and A. Taïeb (eds.), *Vitiligo*,
DOI 10.1007/978-3-540-69361-1_2.2.5, © Springer-Verlag Berlin Heidelberg 2010

Isolation of Skin Melanocytes and Keratinocytes

Vitiligo cells are obtained from lesional or nonlesional skin of generalized vitiligo patients. The method for the isolation of primary melanocytes or keratinocytes is similar for both vitiligo and normal skin samples. Split-thickness skin samples are cut in small pieces and trypsinized. Trypsin disrupts the epidermis above the basal layer, and it is neutralized with fetal calf serum or trypsin soybean inhibitor. Epidermis is removed, and the basal layer is scraped to dissociate melanocytes and basal keratinocytes. Cells are seeded at a density of 200,000/cm^2 for melanocytes and 100,000/cm^2 for keratinocytes culture [29]. Usually, the culture media for melanocytes are M2 medium (PromoCell), M254 (Cascade Biologics), or MCDB153 (Sigma), added with specific growth factors. Several different hand-made growth factors' cocktails for culture are used (Tables 2.2.5.1 and 2.2.5.2).

The culture media used for keratinocytes are CellnTechn-07 (Chemicon), MCDB153 (Sigma), or M154 with the appropriate growth factors. Cells at passage 2–3 can be used to reconstruct epidermis or to start cocultures for functional studies.

Isolation and Culture of Skin Fibroblasts

After obtaining keratinocyte and melanocyte suspensions, the scraped skin is rinsed in HBSS and placed epidermal side up in a cell culture treated scarified Petri dish to grow fibroblasts. Fibroblasts are cultured in DMEM supplemented with 10% fetal calf serum and antibiotics and migrate from the dermis to the Petri dish after 15 days of culture. Fibroblasts can be also obtained by treating the dermis with collagenase.

Isolation and Culture of Hair Follicle Melanocytes and Keratinocytes

Scalp specimens are cut into small pieces and the epidermis and upper 1 mm of dermis are carefully removed with a scalpel [21]. Hair follicles are isolated by incubating the tissue in Eagle's minimal essential medium (EMEM) containing dispase and then collagenase. The released hair follicles are washed repeatedly with PBS until hair follicles appear pure by microscopic examination. Single-cell suspensions are then obtained by treatment with trypsin-EDTA [41]. Hair follicle

Table 2.2.5.1 Recapitulative of medium used to culture vitiligo melanocytes

Medium	bFGF (ng/mL)	PMA (nM)	Transferring (µg/mL)	a-toc (µg/mL)	Insulin (µg/mL)	FCS (%)	hCS	BPE (µg/mL)	References
M154	3	16	5		5	0.5	0.18 µg/mL	2	[10]
MCDB153	0.3		5	1	5	5	0.5 µg/mL	30	[29]
MCDB153	0.6	8		1	5	4		13	[36]
MCDB153	0.6	8	5	1	5	5	0.5 µg/mL	30	[4][a]
MCDB153					20	3	1.75 µM	140	[7]

In all culture media: penicillin and streptomycin are 10,000 U/mL and 10,000 ng/mL, respectively). *a-toc* tochoferol; *hCS* hydrocortisone
[a]Also contains 20 µg/mL catalase. Ham F10 contain 1% ultracer, 2 mM Glu, 10 ng/mL PMA, 0.1 nM IBMX

Table 2.2.5.2 Recapitulative of medium used to culture vitiligo keratinocytes

Medium	EGF (ng/mL)	Glutamin	Adenine	CT	Thyronine	Insulin (µg/mL)	FCS	Hydrocortisone	BPE (µg/mL)	References
Celln Tech-07										[7]
MCDB 153	10					5		1.4 µM	70	[7]
DMEM: HamF 12 (2:1)	10	4 mmol/L	0.18 mmol/L	0.1 nM	2 nM	5	10%	0.4 µg/mL	70	[5]

In all culture media: penicillin and streptomycin are 10,000 U/mL and 10,000 ng/mL, respectively). *a-toc* tochoferol; *hCS* hydrocortisone

melanocytes (HFM) cultures are established using media supplemented with either artificial mitogens or natural melanocyte mitogens as follows. Contaminating fibroblasts, when present, are eliminated by treating the cultures with 150 μg/mL geneticin sulfate (G418) [17]. Follicular keratinocytes are separated from the HFM cultures by differential trypsinization. The identity of the isolated cells is confirmed by immunophenotyping with the melanocyte-lineage specific marker NKI/beteb against glycoprotein100 (gp100) [30, 43]. Hair follicle keratinocytes (HFK) are established by preparing single-cell suspensions from isolated hair follicles. At the primary culture stage, follicular melanocytes are first selectively trypsinized from the culture. This step is carried out under microscopic observation. Remaining keratinocytes are then switched to the keratinocyte specific medium, which does not support melanocyte growth.

Isolation and Freezing of Peripheral Blood Mononuclear cells (PBMC)

The Peripheral Blood Mononuclear cells (PBMC) have been analyzed in vitiligo in order to characterize their immunological status, the capability to recognize specific melanocyte antigens, the cytotoxic effects toward melanocyte or melanoma cell lines, redox status and response to DNA damage. The isolation of vitiligo PBMC is performed through the stratification onto Ficoll density gradient, which allows separating mononuclear from polynuclear and red cells. The PBMC localize at interface between serum and Ficoll, whereas the polynucleates, after centrifugation, go to the bottom of the tube. After recovery, PBMC are washed twice with saline solution (NaCl 0.9%) and used as planned. The procedure should be carefully performed (short time between blood withdrawal and PBMC isolation, gently manipulation) in order to avoid any physical stress able to affect vitiligo PBMC independently of in vitro test. Theoretically, vitiligo PBMC may be used even after freezing in DMSO/Serum. However, our experience suggests avoiding freezing when ROS generation or some functional membrane-dependent parameters will be assayed. On the other hands, the cellular pellet, stored at −80°C, can be used for enzymatic assays [8, 9, 11, 14, 33].

2.2.5.2 In Vitro Reconstructed Epidermis

As early as in 1979, Pruniéras et al. [35] have demonstrated that it is possible to obtain a fully differentiated epidermis in vitro by simply raising keratinocytes up to the air–liquid interface. Apparently, the interface with air stimulates synthesis of profillagrin by keratinocytes and thus the appearance of the granular phenotype when keratohyalin granules develop [34]. Epidermis can be reconstructed using various supports: dead de-epidermized dermis (DDD) [35], gel of collagen (Episkin), lattices including fibroblasts and collagen [1], lattices including fibroblasts and collagen-glycosa minoglycan-chitosan [3], human fibrous sheet [26] basal inert substrates such as porous filters [38]. This model has been perfected first by adding melanocytes [20] and secondly Langerhans cells [37], but models with Langerhans cells have not been tested in vitiligo experiments. The reconstructed epidermis with keratinocytes and melanocytes on DDD reproduces the epidermal melanin unit (EMU) and is suitable to study pigmentation [6].

Preparation of Dead De-Epidermized Dermis

Human dermis is obtained from plastic surgery specimen from normal adults, mostly breast reduction specimen. Skin samples are thinned, cut into very small pieces, and incubated at 37°C in Hank's balanced salt solution until epidermis can be removed without excessive scraping. After removal, the dermis are rinsed in 70° ethanol and submitted to two cycles of freezing–thawing and stored in Hank's balanced salt solution at −20°C until use.

Epidermal Reconstruction

Melanocytes and keratinocytes (from normal or nonlesional vitiligo skin) at passages 2 or 3 are seeded in an incubation chamber placed on the epidermal side of DDD at 4 × 10^5 cells/cm^2 at a melanocyte/keratinocyte ratio of 1:20 (5%) for normal melanocytes [6] and

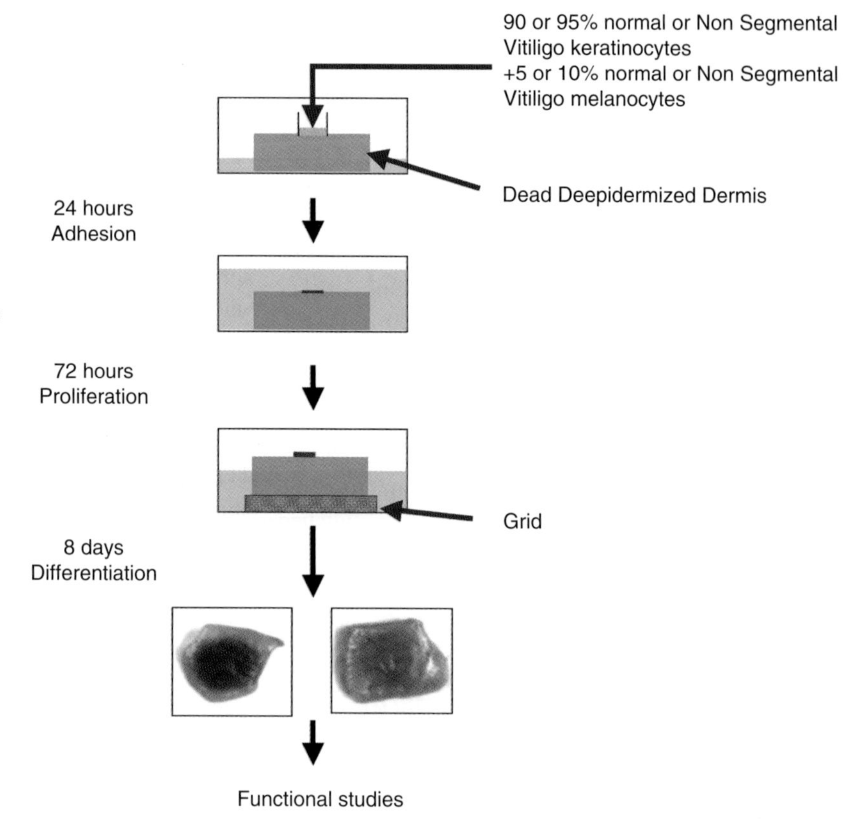

Fig. 2.2.5.1 Schematic summary of the protocol for epidermal reconstructs. Melanocytes and keratinocytes (from normal or nonlesional vitiligo skin) are seeded in an incubation chamber placed on the epidermal side of DDD. The incubation chamber is removed thus and the DDD is immersed for 3 days. DDD are shifted to the air/liquid interface for 8 days before functional studies

1:20 or 1:10 (5–10%) for vitiligo melanocytes since vitiligo melanocytes have a defective adhesion [7]. Twenty-four hours after seeding, the incubation chamber is removed and the DDD is immersed for 3 days. The DDD are shifted to the air/liquid interface for 8 days before functional studies (Fig. 2.2.5.1). The model can be improved by seeding fibroblasts in an incubation chamber placed on the dermal side of DDD 72 h before seeding keratinocytes and melanocytes (Fig. 2.2.5.2).

2.2.5.3 Functional Studies of NSV Cells Using Monolayers: Melanocytes, Keratinocytes, and Fibroblasts

Cultures can be used to characterize the phenotype of vitiligo melanocytes as compared to control pigment cells under various treatments such as UV or pharmacological agents. The cells are usually seeded the day before to obtain 60–70% confluency on the day of treatment.

Various techniques can be used to observe melanocyte behavior. Direct observation by microscopy on fixed material of specifically stained melanocytes (melan-A, DOPA, c-kit, S-100, HMB-45) gives information on shape, dendriticity, and pigmentation. Since nonlesional vitiligo melanocytes are difficult to growth, it is often difficult to have enough cells to allow melanin content determination by spectrophotometry. Thus DOPA staining may be used as alternative method. Vitiligo melanocyte and keratinocyte cultures can be tested in vitro in order to determine their specific or differential susceptibility to noxious stimuli or to physiological growth factors. UVB, cumene hydroperoxide, and *tert*-buthyl-phenol are the most used stimuli [10, 19, 20, 24, 25, 28, 44–46]. Among tested compounds, imatinib mesilate has been reported to inhibit melanogenesis in both normal and vitiligo melanocytes by decreasing the number of highly pigmented melanocytes. Moreover, it inhibits the growth of vitiligo fibroblasts than that of normal fibroblasts more efficiently [6]. Cell proliferation and mortality can be assessed through MTT test, manual cell count, annexinV/propidium staining, DNA ladder test, or caspase

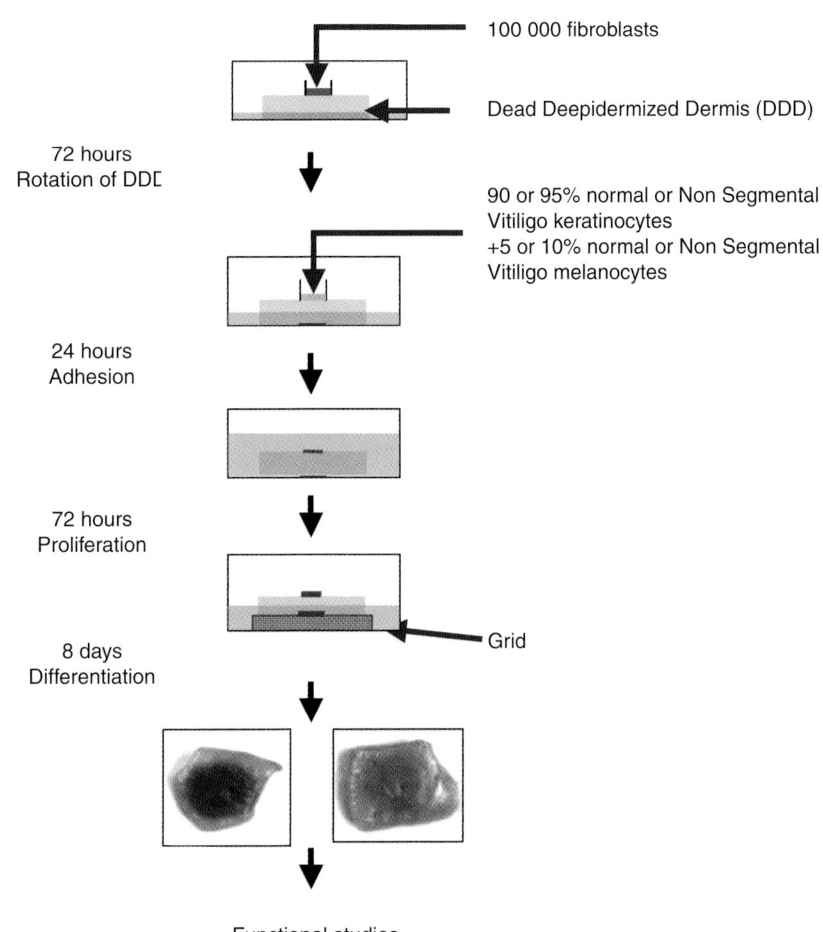

Fig. 2.2.5.2 Schematic representation of epidermal reconstruction using a first step of dermal colonization with fibroblasts

activation. In addition, the morphology and mode of melanosome transfer in vitro can be studied using atomic force microscopy, which allows to estimate and quantize the measure and the distribution of the dendrites including internal melanosomes distribution and arrangement [46]. Senescence has been studied on keratinocytes from involved and uninvolved vitiligo skin. Culture on 3T3 feeder-layer induces the formation of colonies of keratinocytes which vary in size according to the state of cell proliferation, differentiation, or senescence. Lesional NSV keratinocytes are characterized by a lower proliferative potential, as indicated by a shorter in vitro life span. Moreover, the expression of p16, PCNA, p53, and p63 markers differs between lesional and nonlesional cells. Lesional keratinocytes show a lower level of the senescence marker p16 and a higher level of the melanocyte growth factor SCF [42].

2.2.5.4 Functional Studies Using Reconstructed Epidermis

Monolayer cultures and coculture [13] are useful to study vitiligo melanocytes and keratinocytes direct (cell–cell contact) or indirect interaction (soluble factors/culture insert), but they do not reproduce the tridimensional interactions of cells of the EMU. Epidermal reconstructs, which reproduce the EMU and basal membrane attachment, are thus a handful "in vivo-like" model. Indeed, reconstructions of different levels of complexity can be prepared (reconstructs with keratinocytes alone, keratinocytes and melanocytes, keratinocytes, melanocytes and fibroblasts), including chimeric reconstructs (normal vs. pathological cells). The analysis of the behavior of melanoctes and keratinocytes is done in a more physiological environment than that of monolayer cultures

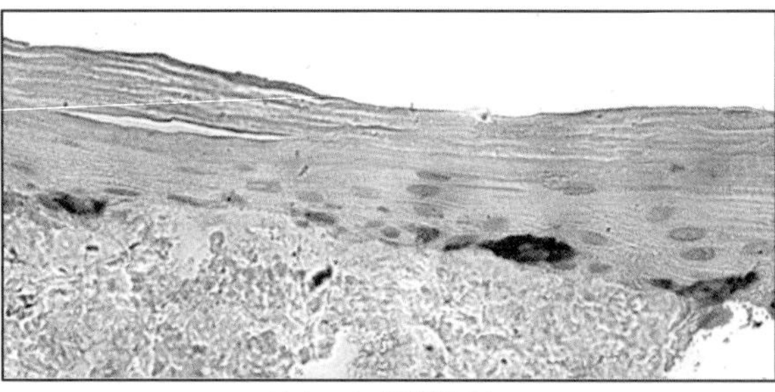

Untreated REKnMn

Serum-treated REKnMn

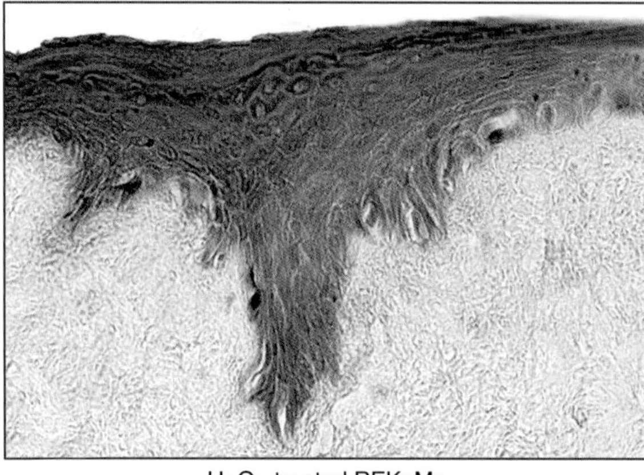

H_2O_2 treated REKnMn

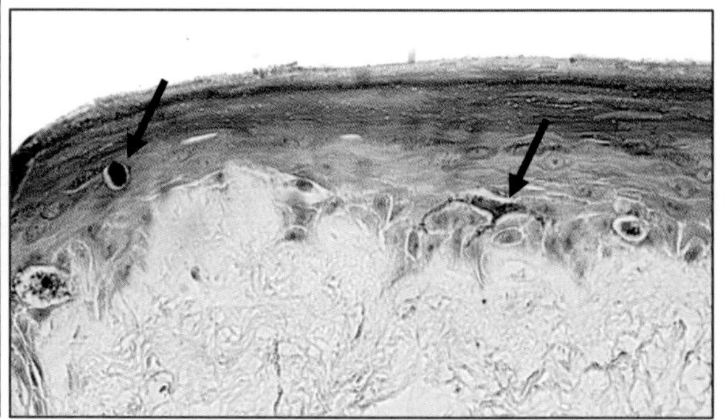

H_2O_2 treated REKvMv

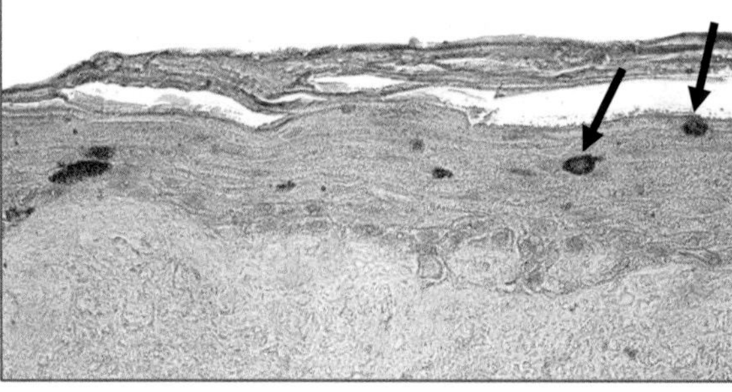

H_2O_2 treated REKnMn

H_2O_2 treated REKvMv

Fig. 2.2.5.3 Reconstructed epidermis were made with normal keratinocytes and melanocytes REKnMn (**a–c**, **e**) or with NSV uninvolved keratinocytes and melanocytes REKvMn (**d**, **f**). REKnMn were treated with serum from NSV patients (**b**) or with H_2O_2 during 3 h (**c**) or 24 h (**e**). REKvMn were treated with H_2O_2 (**d**, **f**). Reconstructed epidermis embedded in paraffin, were stained with hematoxylin-eosin to assess the general morphology of the epidermis. Melanin was visualized using Fontana-Masson (**c**, **d**) stain. Sections have been analyzed for Melan-A (**a**, **b**, **e**, **f**) or cadherin expression. Melan-A staining will allows to count melanocytes in basal, spinous, granular layers and stratum corneum. In order to evaluate melanocyte detachment the percentage of melanocytes located in spinous and granular layers and lower stratum corneum was calculated. Serum can induce melanocytes detachment (*arrow*) in normal or vitiligo reconstructs whereas H_2O_2 induce variable effects according to reconstructions

and cocultures, and allows to test conveniently compounds which are suspected to be implicated in vitiligo etiology or susceptible to improve the attachment and survival of melanocytes upon the basal layer. An example of epidermal reconstructs tested with epinephrin, norepinephrin, dopamine, hydrogen peroxide, or vitiligo sera is illustrated in Fig. 2.2.5.3.

Our main results using this model [7] were the following: reconstructs made with melanocytes from nonlesional generalized vitiligo skin have a significantly reduced number of basal layer melanocytes, whereas the presence of vitiligo keratinocytes enhances this effect. Vitiligo sera may induce melanocyte detachment independently of the disease activity or extent. Hydrogen peroxide induces melanocyte detachment in reconstructs containing vitiligo melanocytes and normal keratinocytes, but not in normal controls. Finally, epinephrine, but not norepinephrine, allows melanocyte detachment. Epidermal reconstructs are useful to address several research questions such as: Is the melanocyte primarily affected? Is the cellular environment important (keratinocytes, fibroblasts)? Are other (soluble) factors implicated in the development of vitiligo? However, this model has some limitation since we have not yet been able to introduce, for instance, Langerhans cells or other immune cells to study their implication in vitiligo etiology; the study of topical molecules to improve vitiligo treatment is less easy than that of soluble factors, and that long-term studies (more than 3 weeks) are not possible, since there is no renewal of the basal layer.

2.2.5.5 New Analytic Techniques

Fluorescence-Based Assays

Fluorochrome-conjugated monoclonal or polyclonal antibodies are widely applied to visualize and quantify the expression of some surface or internal melanocyte markers [16, 23, 31, 37]. The first data were derived by immuno histochemistry methods on frozen and paraffin-embedded sections, followed by fluorescence and confocal microscopy on slide-cultured cells (Table 2.2.5.3). Phenotypic characterization of cultured vitiligo melanocytes with respect to control cells was carried out (Fig. 2.2.5.4). A reduced expression of c-Kit and ET-1 receptor, of tyrosinase and MITF were found in vitiligo melanocytes with a pattern progressively varying from the edge of the white spot to the nonlesional area [23]. The flow cytometric approach permits a quali-quantitative multiparametric and time-dependent analysis. Membrane and intracellular staining together with the analysis of the physical and time parameters permits the structural and functional characterization of the melanocytes and their sorting for further cultures. The current flow cytometers can analyze up to 16 parameters. Beside the antibody-based approach, the cytomic has been used to detect in vitiligo melanocytes the intracellular ROS production (DCFH-DA or dhRho123 staining), the membrane lipo-peroxidation [9] ($BODIPY^{581/591}$ staining), the content and the transmembrane cardiolipin distribution (NAO fluorescence

Table 2.2.5.3 List of antigens against which antibodies exist for detection on paraffin embedded sections. Most part of these antibodies needs heat antigen retrieval techniques

Antigen	Cells	Protein features	Differentiation featuring	Staining pattern	Sensitivity
MART-1	Melanocytes, melanoma cells, Nevi	Plasma membrane protein	Epidermal melanocytes	cytoplasmic	Low/good
S-100	Melanocytes, Langerhans cells, melanoma cells, Nevi, astrocytes, Schwann's cells, ependymomas, astrogliomas	calcium binding protein	Epidermal melanocytes	cytoplasmic and nuclear	Very good
gp-100	Melanocytes, melanoma cells, Nevi, clear cells sarcomas	Neuraminidase sensitive oligosaccharide side chain of a glycoconjugate present in immature melanosomes	Hair follicle melanocytes, epidermal melanocytes	cytoplasmic	Low
c-Kit	Melanocytes, hematopoietic stem cells, mastocytes	Transmembrane tyrosine kinase receptor for SCF	Hair follicle and epidermal melanocytes	Cytoplasmic and nuclear	
Tyrosinase	Melanocytes, melanoma cells, Nevi	Melanosomal membrane protein	Epidermal melanocytes	Cytoplasmic	Good
TRP-1	Melanocytes, melanoma cells, Nevi	Melanosomal membrane protein (melanosomes stage III and IV)	Epidermal melanocytes	Cytoplasmic	
Vimentin	Melanocytes, Langerhans cells, fibroblasts, endothelial cells	Subunit protein of the intermediate filaments	Hair follicle and epidermal melanocytres	Cytoplasmic	
MITF-M	Melanocytes	Transcription factor in melanogenetic pathway	Hair follicle (cyt) and epidermal melanocytes (nuclear)	Cytoplasmatic and nuclear	Excellent

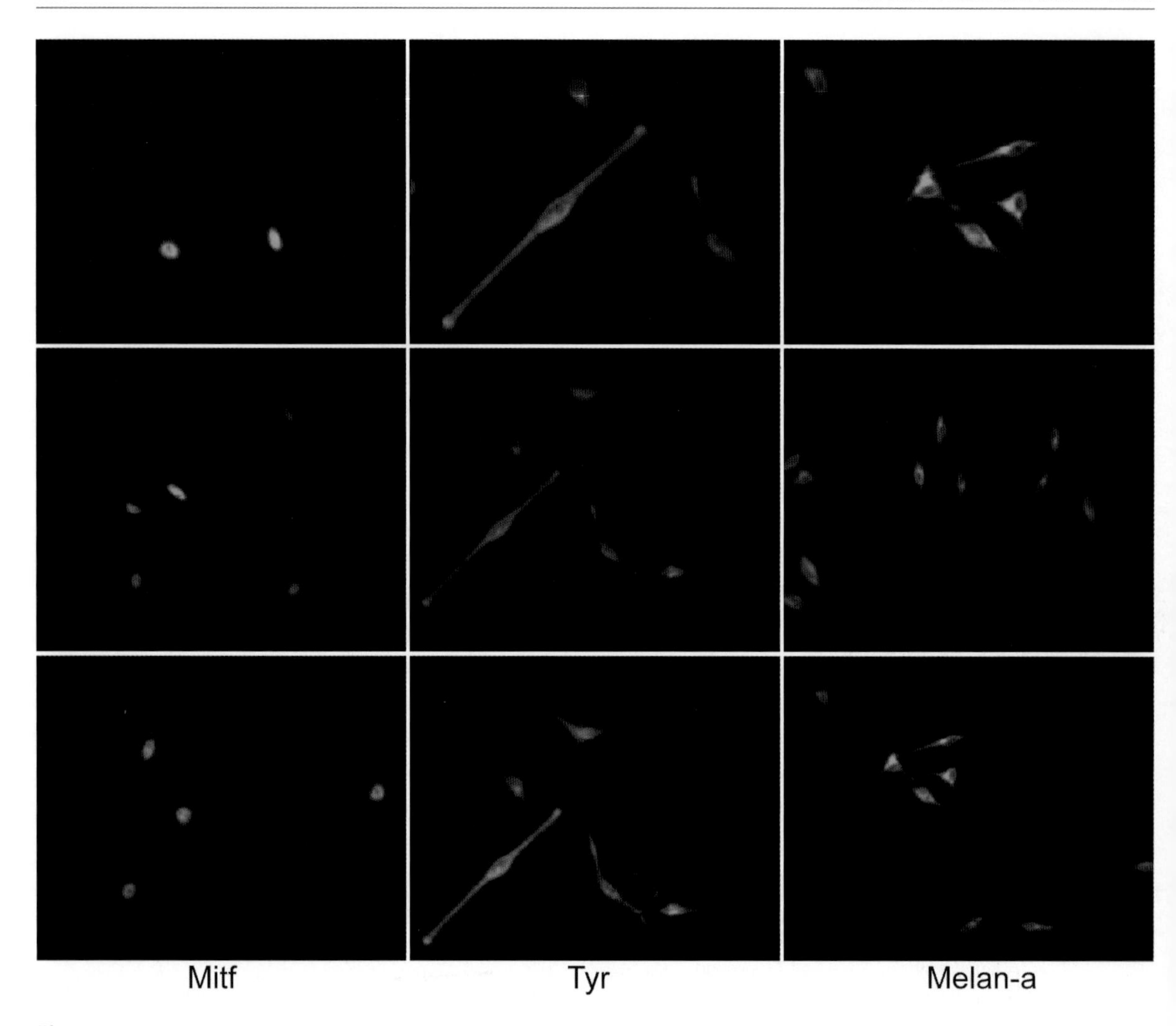

Fig. 2.2.5.4 Human normal epidermal melanocytes in vitro cultured were stained with monoclonal antibodies against MITF, tyrosinase, or Melan-A. The melanocytes were fixed and stained with the different antibodies, conjugates with Alexa Fluor 488. The picture is representative of a normal culture of human epidermal melanocytes (40×)

pattern) [10] (Fig. 2.2.5.5). The flow cytometer may be also used for fluorescence resonance energy transfer (FRET) analysis. Cells are stained with nonyl acridine orange (NAO) (donor) and Mitotracker orange, a dye with a high affinity for mitochondrion proteins and voltage sensible (acceptor). The mitochondrial mass and the polarization state of the inner mitochondrial membrane can be monitored using FRET index based on NAO FL1 (green fluorescence) decreased and FL2 (red fluorescence) increase from single to double-labeled tubes [12]. The limitation associated with flow cytometry is that it is sample consuming. Several new approaches, based on "fluo world," are now available. Most of these innovative technologies have not yet been applied to the study of vitiligo, but they open promising perspectives. The laser scanning cytometer (LSC) permits slide-based cytometry (SBC) and the hyperchromatic approach. Its potential employ in vitiligo study arises from its intrinsic features: nonconsumptive (unlike the flow cytometer), iterative restaining, differential photobleaching (fluorochromes differentiated on the basis of their specific photostability), and photoactivation (for nano-particles or photo-caged dyes). A single cell can be reanalyzed, whereas the information gained per specimen is only limited by the number of available antibodies and sterical hindrance [40]. The LSC when combined with fine-needle sampling (FNS) may be used to monitor the cell structural and functional modifications subsequent to in vitro treatment, where FNS further reduces the amount of sample

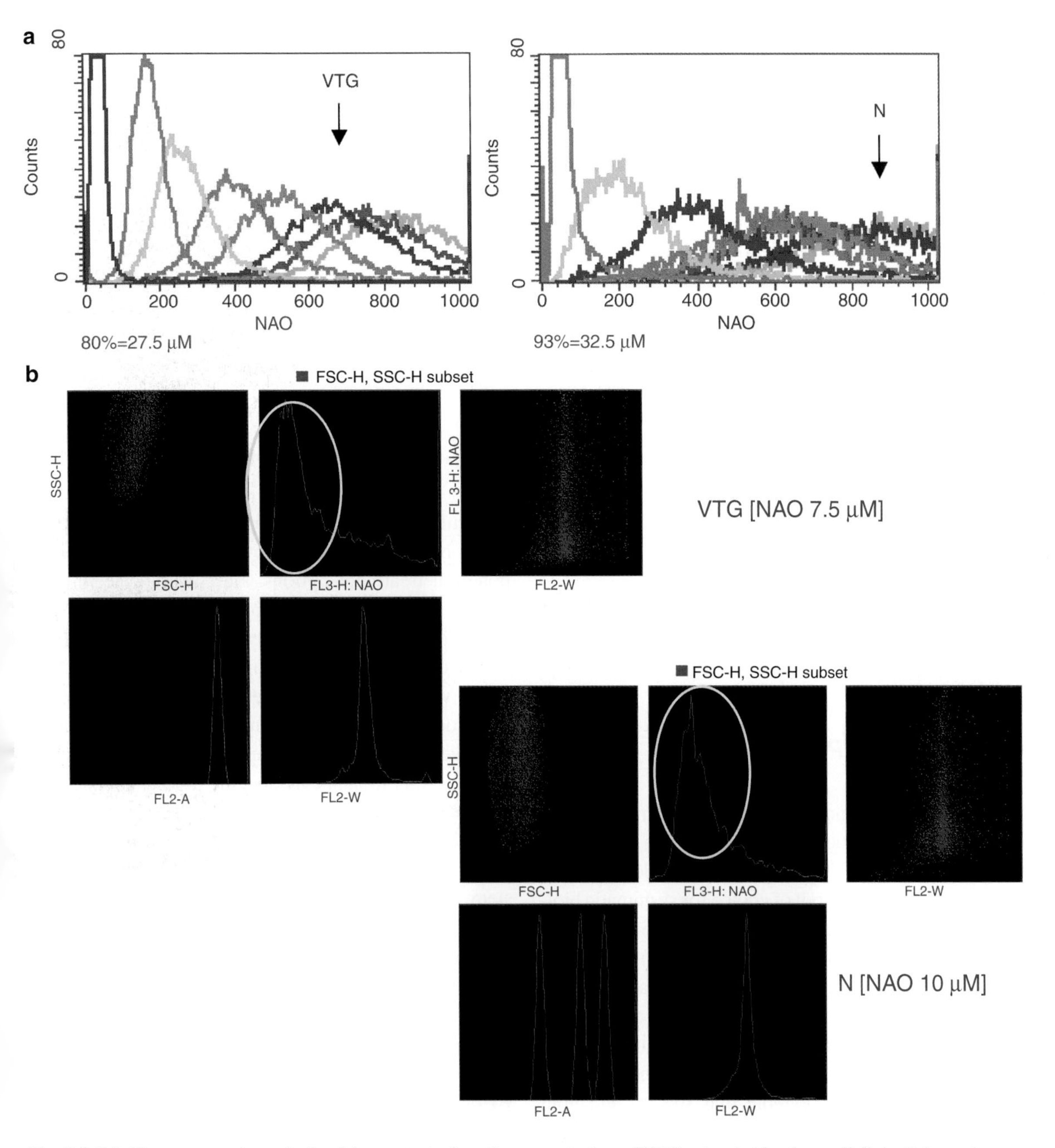

Fig. 2.2.5.5 Flow cytometric analysis of the content of cardiolipin in the mitochondrial membrane. PBMC (**a**) from vitiligo and healthy subjects were fixed and stained with nonyl acridine orange (NAO), which provides a measure of the membrane content of nonoxidised cardiolipin. The *arrows* indicate the concentration of NAO saturated by the cardiolipin. Primary human melanocytes (**b**) from vitiligo and healthy subjects were also stained with NAO. The NAO concentration saturated by cardiolipin is dependent on cardiolipin content

needed for the analysis. The SBC is a significant advance research in the measurement of short-lived processes in adherent cells and small samples, two crucial features of vitiligo samples. New light microscope architecture is provided by iMIC, which integrates time-lapse studies, FRET measurements, laser microdissection, and slitscan confocal measurements with incoherent illumination (340–680 nm).

Proteomic

Recently, mass spectrometry in conjunction with free-flow electrophoresis of sucrose density gradient has allowed the identification of early stage melanosome proteins [32]. Two-dimensional differential image-gel electrophoresis (2D-DIGE) and liquid chromatography-tandem mass spectrometry (LC-MS/MS) allow the analysis and identification of the proteic components of the organelles of melanocytes with melanosomes at different maturation stages [22]. Calreticulin, a soluble Ca^{++}-binding chaperone protein, is involved together with calnexin in the folding of newly synthesized proteins and glycoproteins (including tyrosinase) and for quality control pathways in endoplasmic reticulum. The LC-MS/MS analysis has highlighted that calreticulin expression is dependent on the maturation stage of melanocytes. Even if the proteomic assay have been so far carried out on murine healthy melanocytes, the scenario possibly designed by this approach may be crucial for the maturation process leading from nonpigmented melanocytes to pigmented melanocytes in vitiligo.

Metabolomic/Lipidomic

Metabolomic, as a global study of small molecules in biological fluids, tissues and cells, is gaining recognition for being a very sensitive and amplified readout of physiology, and therefore a tool of choice for systematic biology studies in complex organisms. Serum lipidomics (global profiling of lipid molecular species) can be utilized to predict the expression of inflammatory genes. Lipidomics, defined as the large-scale study of pathways and networks of cellular lipids, is an emerging and rapidly expanding research field. It is often necessary to evaluate a wide range of molecular species and lipid classes to gain insights into pathophysiology. In the future, research in lipidomics will expand to include interactions of lipids with lipids, proteins, and other cellular components. Mass Spectroscopy, Nuclear Magnetic Resonance, and Fluorescence Spectroscopy have played a crucial role in lipid characterization, identification, and quantitation [18]. The study of vitiligo pathogenesis should actually gain a relevant burst from the metabolomic approach, considering that most of the current studies underline the leading role of cellular and metabolic networks in melanocyte function impairment.

Transcriptomic

Recently, oligonucleotide-based microarrays have been used to explore the pattern of gene expression of vitiligo melanocytes. Data were first filtered and normalized to remove up to 9800 probes. Interestingly, the most up-regulated genes are related to the network of endosome and lysosome organelles. The next step was to analyze the various clusters of the differentially expressed genes. This approach can represent the basis for further in-depth analyses to better clarify the complex vitiligo pathomechanisms [39].

2.2.5.6 Concluding Remarks

Cell culture models are useful to investigate the differences between normal and vitiligo cells, and to test potential treatment options. New analytic techniques using a limited amount of biological material are very promising. Beyond the study of vitiligo pathomechanisms or therapies using unmodified patient's cells, the in vitro approach can be also adapted to generate melanocyte-specific silencing or overexpression of putative target genes or, a step further, to design animal models using melanocyte-specific expression of modified genes. The study of various issues pertaining to vitiligo including the role and regulation of transcription factors, organelle genesis, intracellular transport, stem cell maintenance, and senescence can be envisaged [15]. This step will allow the development of vitiligo models using normal cells bypassing the difficulty to culture vitiligo cells, which is currently the limiting factor for studying vitiligo in vitro.

References

1. Bell E, Sher S, Hull B et al (1983) The reconstitution of living skin. J Invest Dermatol 81:2s–10s
2. Bessou S, Surlève-Bazeille JE, Sorbier E, Taïeb A (1995) Ex vivo reconstruction of the epidermis with melanocytes and the influence of UVB. Pigment Cell Res 8:241–249
3. Black AF, Bouez C, Perrier E et al (2005) Optimization and characterization of an engineered human skin equivalent. Tissue Eng 11:723–733
4. Boissy RE, Liu YY, Medrano EE et al (1991) Structural aberration of the rough endoplasmic reticulum and melanosome compartmentalization in long-term cultures of melanocytes from vitiligo patients. J Invest Dermatol 97:395–404

5. Bondanza S, Maurelli, R, Paterna P et al (2007) Keratinocytes cultures from involved skin in vitiligo patients show an impaired in vitro behaviour. Pigment Cell Res 20:288–300
6. Cario-André M, Bessou S, Gontier E et al (1999) The reconstructed epidermis with melanocytes: a new tool to study pigmentation and photoprotection. Cell Mol Biol 45:931–942; Review (Erratum in: Cell Mol Biol (2000); 446:489)
7. Cario-André M, Pain C, Gaythier Y et al (2007) The melanocythorragic hypothesis of vitiligo tested on pigmented, stressed, reconstructed epidermis. Pigment Cell Res 20:385–393
8. Dell'Anna ML, Maresca V, Briganti S et al (2001) Mitochondrial impairment in peripheral blood mononuclear cells during the active phase of vitiligo. J Invest Dermatol 117:908–913
9. Dell'Anna ML, Urbanelli S, Mastrofrancesco A et al (2003) Alterations of mitochondria in peripheral blood mononulcear cells of vitiligo patients. Pigment Cell Res 16: 553–559
10. Dell'Anna ML, Ottaviani M, Albanesi V et al (2007) Membrane lipid alterations as a possible basis for melanocyte degeneration in vitiligo. J Invest Dermatol 127: 1226–1233
11. Donmez-altuntas H, Sut Z, Ferahbas A et al (2008) Increased micronucleus frequency in phytohaemagglutinin-stimulated blood cells of patients with vitiligo. JEADV 22:162–167
12. Dykens JA, Fleck B, Ghosh S et al (2002) High-throughput assessment of mitochondrial membrane potential in situ using fluorescence resonance energy transfer. Mitochondrion 1:461–473
13. Eves PC, Beck AJ, Shard AG et al (2005) A chemically defined surface for the co-culture of melanocytes and keratinocytes. Biomaterials 26:7068–7081
14. Giovannelli L, Bellandi S, Pitozzi V et al (2004) Increased oxidative DNA damage in mononuclear leukocytes in vitiligo. Mut Res 556:101–106
15. Goding CR (2007) Melanocytes: the new black. Pigment Cell Res 39:275–279
16. Graham A, Westerhof W, Thody AJ (1999) The expression of a-MSH by melanocytes is reduced in vitiligo. Ann NY Acad Sci 885:470–473
17. Halaban R, Alfano FD (1984) Selective elimination of fibroblasts from cultures of normal human melanocytes. In vitro 20:447–450
18. Han X (2007) Potential mechanisms contributing to sulfatide depletion at the earliest clinically recognizable stage of Alzheimer's disease: a tale of shotgun lipidomics. J Neurochem 103:171–179
19. Ivanova K, van der Wijngaard R, Gerzer R et al (2005) Nonlesional vitiliginous melanocytes are not characterized by an increased proness to nitric oxide-induced apoptosis. Exp Dermatol 14:445–453
20. Jimbow K, Chen H, Park JS et al (2001) Increased sensitivity of melanocytes to oxidative stress an dabnormal expression of tyrosinase-related protein in vitiligo. Br J Dermatol 144:55–65.
21. Kauser S, Thody AJ, Schallreuter KU et al (2004) β-Endorphin as a regulator of human hair follicle melanocyte biology. J Invest Dermatol 123:184–195
22. Kawase A, Kushimoto T, Kawa Y et al (2008) Proteomic analysis of immature murine melanocytes at different stages of maturation: a crucial role for calreticulin. J Dermatol Sci 49:43–52
23. Kitamura R, Tsukamoto K, Harada K et al (2004) Mechanisms underlying the dysfunction of melanocytes in vitiligo epidermis: role of SCF/KIT protein interactions and the downstream effector, MITF-M. J Pathol 202:463–475
24. Kroll TM, Bommiasamy H, Boissy RE et al (2005) 4-tertiary butyl phenol exposure sensitizes human melanocytes to dendritic cell-mediated killing: relevance to vitiligo. J Invest Dermatol 124:798–806
25. Lee AY, Kim NH, Choi WI et al (2005) Less keratinocyte-derived factors related to more keratinocyte apoptosis in depigmented than normally pigmented suction-blistered epidermis may cause passive melanocyte death in vitiligo. J Invest Dermatol 124:976–983
26. Lee DY, Lee JH, Yang JM et al (2006) A new dermal equivalent: the use of dermal fibroblast culture alone without exogenous materials. J Dermatol Sci 43:95–104
27. Le Poole IC, Stennett LS, Bonish BK et al (2003) Expansion of vitiligo lesions is associated with reduced epidermal Cdw60 expression and increased expression of HLA-DR in perilesional skin. Br J Dermatol 149:739–748
28. Maresca V, Roccella M, Roccella F et al (1997) Increased sensitivity to peroxidative agents as a possible pathogenetic factor of melanocyte damage in vitiligo. J Invest Dermatol 109:310–313
29. Medrano EE, Nordlund JJ (1990) Successful culture of adult human melanocytes obtained from normal and vitiligo donors. J Invest Dermatol 95:441–445
30. Na GY, Paek SH, Park BC et al (2006) Isolation and characterization of outer root sheath melanocytes of human hair follicles. Br J Dermatol 155:902–909
31. Norris A, Todd C, Graham A et al (1996) The expression of the c-kit receptor by epidermal melanocytes maybe reduced in vitiligo. Br J Dermatol 134:299–306
32. Ouvry-Patat SA, Torres MP, Quek HH et al (2008) Free-flow electrophoresis for top-down proteomics by Fourier transform ion cycloton resonance mass spectrometry. Proteomics 8:2798–2808
33. Palermo B, Campanelli R, Garbelli S et al (2001) Specific cytotoxic T lymphocyte responses against Melan-A/MART1, tyrosinase and gp100 in vitiligo by the use of major histocompatibility complex/peptide tetramers: the role of cellular immunity in the aetiopathogenesis of vitiligo. J Invest Dermatol 117:326–332
34. Poumay Y, Coquette A (2007) Modelling the human epidermis in vitro: tools for the basic and applied research. Arch Dermatol Res 298:361–369
35. Prunieras M, Regnier M, Schlotterer M (1979) [New procedure for culturing human epidermal cells on allogenic or xenogenic skin: preparation of recombined grafts]. Ann Chir Plast 24:357–362
36. Richmond B, Huizing M, Knapp J et al (2005) Melanocytes derived from patients with Hermansky-Pudlak syndrome types 1, 2 and 3 have distinct defects in cargo trafficking. J Invest Dermatol 124:420–427
37. Régnier M, Staquet MJ, Schmitt D et al (1997) Integration of Langerhans cells into a pigmented reconstructed human epidermis. J Invest Dermatol 109:510–512
38. Rosdy M, Bjorklund MG, Asplud A et al (1990) Terminal epidermal differentiation of human keratinocytes grown in

chemically defined medium on inert filter substrates at the hair-liquid interface. J Invest Dermatol 95:409–414
39. Stromberg S, Bjorklund MG, Asplud A et al (2008) Transcriptional profiling of melanocytes from patients with vitiligo vulgaris. Pigment Cell Melanoma Res 21:162–171
40. Tellez CS, Davis DW, Prieto VG et al (2007) Quantitative analysis of melanocytic tissue array reveals inverse correlation between activator protein-2 alpha and protease-activated receptor-1 expression during melanoma progression. J Invest Dermatol 127:387–393
41. Tobin DJ, Colen SR, Bystryn JC (1995) Isolation and long-term culture of human hair-follicle melanocytes. J Invest Dermatol 104:86–89
42. Van den Wijngaard RMJGJ, Aten J, Scheepmaker A et al (2000) Expression and modulation of apoptosis regulatory molecules in human melanocytes: significance in vitiligo. Br J Dermatol 143:573–581
43. Vennegor C, Hageman P, Van Nouhuijs H et al (1988) A monoclonal antibody specific for cells of the melanocyte lineage. Am J Pathol 130:179–192
44. Yang F, Boissy RE (1999) Effects of 4-tertiary butylphenol on the tyrosinase activity in human melanocytes. Pigment Cell Res 12:237–245
45. Yang F, Sarangarajan R, Le Poole IC et al (2000) The cytotoxicity and apoptosis induced by 4-tertiary butylphenol in human melanocytes are independent of tyrosinase activity. J Invest Dermatol 114:157–164
46. Zhang RZ, Zhu WY, Xia MY et al (2004) Morphology of cultured human epidermal melanocytes observed by atomic force microscopy. Pigment Cell Res 17:62–65

Oxidative Stress

2.2.6

Mauro Picardo and Maria Lucia Dell'Anna

Contents

Core Messages

- The studies of the metabolic deregulations leading to toxic damage of the melanocytes appear to be relevant.
- An oxidative stress process is associated with melanocyte degeneration.
- In vitiligo patients, systemic oxidative stress can be detected, suggesting the involvement of metabolisms not exclusively related to the melanogenic process.

2.2.6.1 General Aspects

In contrast to the generally easy clinical diagnosis of vitiligo, the cellular mechanisms leading to the appearance of the white lesions is still uncertain. Besides the immunological approach, supported by in vivo and in vitro data underlying the presence of anti-melanocyte specific antibodies and T cells (CD8, mainly), several evidences indicate the occurrence of an oxidative stress as a pathogenetic factor. The early hypotheses, defined as autocytotoxic and neurogenic, have suggested that biochemical alterations leading to the intra or extracellular generation of free radicals and other toxic intermediates can induce melanocyte degeneration. Melanocytes are intrinsically exposed to high level of toxic compounds because during the melanin synthesis, potentially toxic intermediates are produced. In addition, melanocytes, due to their anatomical localization, are specifically exposed to UV irradiation and physical toxic agents. In vitiligo, melanocytes may be thus the theater of the loss of the redox balance because

M. Picardo (✉)
Istituto Dermatologico San Gallicano, via Elio Chianesi, 00144 Roma, Italy
e-mail: picardo@ifo.it

M. Picardo and A. Taïeb (eds.), *Vitiligo*,
DOI 10.1007/978-3-540-69361-1_2.2.6,

of the increased generation of free radical and toxic metabolites, or because of a defective antioxidant pattern [3, 9, 16, 22, 41, 44].

The generation of oxidative stress has been associated with the alteration of catecholamines metabolism and melanin synthesis [6, 14, 32, 36, 38]. Moreover, the dysregulation of some metabolic pathways, even not strictly related to the melanogenetic activity, has been considere.

2.2.6.2 Catecholamine and Biopterin Metabolisms

The attention to the potential pathogenetic role of the catecholamins comes from clinical data, which suggest that psychophysical stress (relative death, school tests, work's problems, etc.) can be associated with the onset or the worsening of the clinical manifestation of vitiligo, and by the morphological demonstration of the presence of nerve ending close to melanocytes. Different authors have reported increased plasma level of dopamine (DA) and norepinephrine (NE), as well of the urinary metabolites homovanilic acid (HVA) and vanylmandelic acid (VMA), in early and active phase of the disease [6, 32]. It has been demonstrated that both melanocytes and keratinocytes are involved in the synthesis/recycling of catecholamine and present functional β2-adreceptors on the cell membrane [5, 12]. Two possible mechanisms have been suggested to explain the melanocyte damage: a direct one due to the toxicity of the quinone and semiquinone moieties and oxyradicals generated by the oxidation of the catecholamines, and an indirect mechanism related to the vasoconstriction associated with NE release and the subsequent production of free radicals [5]. The decreased activity of catechol *O* methyl transferase (COMT) in the skin of vitiligo patients could explain the increased local oxidation of the neuromediators [45].

Moreover, an alteration of the epidermal and systemic metabolic synthetic pathway of biopterins, cofactors required for the catecholamine, as well as serotonin and melatonin, possibly due to a genetic polymorphism, has been reported. The alteration could lead to the accumulation of the intermediated product 7-tetrahydrobiopterin, capable of inhibiting melanin synthesis and toxic for melanocytes. The altered recycling of the biopterins generate an increased amount of hydrogen peroxide, which in turn can produce the oxidation of different proteins with enzymatic activity, including catalase [36–39, 41]. The increased production of NE during the deregulated catecholamine pathway may take part in the melanocyte damage. The synthesis of biopterins is based on a multistep process, where several enzymes work in a reciprocal dependent manner. Most of the involved enzymes are characterized by the presence of methionine and tryptophan in active or binding sites. Both the aminoacides have been reported to be susceptible to deactivation by oxidation. The functional effect of their oxidative-mediated deactivation should be the shift of the synthetic pathway toward the production of $7BH_4$ and L-phenilalanine. The build-up of L-phenylalanine and $7BH_4$ could be responsible for the increased H_2O_2 generation because of the short circuit of biopterins recycling. The biopterin pathway is interconnected with the thioredoxin one. The block of the tyrosine synthesis has been suggested to cause depigmentation through the accumulation of 7-tetrahydrobiopterin within the epidermis and catechol in the serum. The accumulation of biopterins inhibits the production of tyrosine by phenylalanine. In addition, the degree of phenylalanine uptake, depending in turn on the calcium homeostasis, which is altered in vitiligo, could further affect the melanogenic pathway (Fig. 2.2.6.1). Vitiligo melanocytes are thus deprived of the essential precursor for melanin synthesis and are exposed to the toxic intermediates. Finally, hydrogen peroxide could be responsible for the alteration of β2-adrenoceptor-activated pathway and of the enzyme acetylcholinesterase, which once inactivated, participates in maintaining or promoting the oxidative stress [12].

2.2.6.3 Cellular Alterations Related to the Oxidative Stress

In vitiligo, some morphofunctial alterations suggested ROS-mediated damage. Initial studies on cultured vitiligo melanocytes reported that these cells present longer latency period and low replication rate, and require catalase supplementation to support cell growth [2, 28, 35]. Several morphological features have been subsequently described, including melanosome compartmentalization, dilated RER, impaired cell–cell contacts, and lipid vacuoles [2, 23, 24, 29, 33]. Cultured vitiligo melanocytes

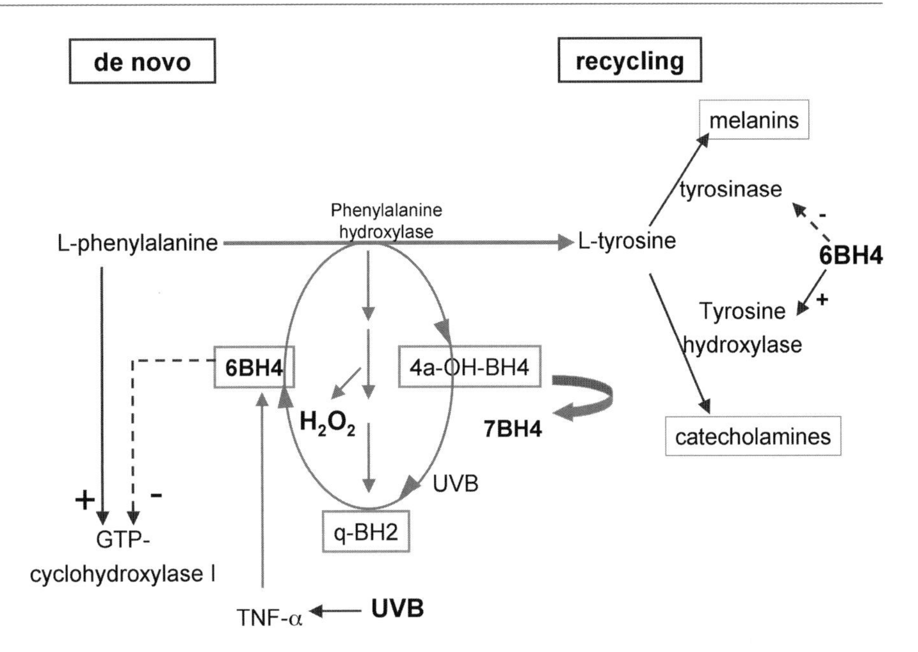

Fig. 2.2.6.1 The scheme represents the biopterin metabolic pathway and network

show increased ROS level, and low amount as well as decreased activity of the catalase, the principal enzyme involved in H_2O_2 removal. In vitro vitiligo melanocytes are more susceptible to the toxic effect of pro-oxidant stimuli such as cumene hydroperoxide and UV [10, 16, 19, 27]. Inside the cells, the level of the free radicals increases when the production is boosted or the enzymatic and nonenzymatic detoxifying systems are defectives. In vitiligo skin, in vivo millimolar H_2O_2 level has been reported by the Fourier Transform Raman spectroscopy [38, 40].

The generation of ROS is a physiological cellular process. Several metabolic pathways (monoamino oxidase A, NADPH oxidase, biopterin synthesis, nitric oxide synthase, estro-progestin metabolism, and xanthine oxidase) physiologically produce H_2O_2. The xanthine oxidase enzyme is actually a combination of couple of enzymes, because it is transcribed as dehydrogenase and immediately converted to the oxidase form, through a proteolytic or oxidative pathway. The conversion of the dehydrogenase from into the oxidase one has been implicated in oxidative-stress associated diseases.

The ROS are generated inside the cells by the Electron Transport Chain during the mitochondrial energy production [8, 9]. The ROS production cannot simply be considered a side effect of different metabolic pathways, as at micromolar concentration, ROS are crucial mediators of the intracellular signal transduction. H_2O_2 production is actually regulated by a fine integrated network of antioxidant and detoxifying enzymes provided by catalase, thioredoxin reductase/thioredoxin, glutathione peroxidase, and reductase. The enzyme responsible for the removal of H_2O_2 is catalase, primarily placed in peroxysome, the site of major H_2O_2 production. The antioxidant activity of catalase is complemented by the seleno-enzyme glutathione peroxidase (GPx), even though the latter is located in a different cellular site and has different chemical features and substrate specificities. The expression and the activity of GPx have been reported lower in melanocytes. While catalase is active only toward H_2O_2, GPx reduces many different peroxide compounds, including organic hydroperoxides. Superoxide dismutases (SODs) catalyze the dismutation of O_2^- to H_2O_2 and O_2. The various SODs isoenzymes are crucial for the H_2O_2 formation in both intracellular and extracellular spaces. Schematically, inside the cells, SOD is present in two forms, one with a prevalent cytosolic (Cu/ZnSOD) localization and the second one principally abundant in the mitochondrial matrix (MnSOD). In the extracellular fluids, it is present as an additional isoenzyme containing Cu and Zn (EC-SOD) [9].

The hyperproduction of ROS may lead to the inactivation of the catalase, through the oxidation of porphyrin ring, methionin, and tryptofan residues, producing the reduction of the catalase activity. Through the same mechanism, even the activity of the thioredoxin reductase, an enzyme involved in protein repair, as well as of methionine sulfoxide reductase A, has been reported to be compromised. Biopterins could be oxidized by H_2O_2, affecting thus the tyrosine-dependent pathways. At the same time, ROS affect propiomelanocortin cleavage giving rise to a reduced amount of α-MSH and β-endorphin. Recently, the ROS-mediated reduced activity of MITF, the transcription factor controlling the melanin synthesis pathway, has been proposed [17]. MITF is usually phosphorylated and activated by receptor-associated kinases allowing the subsequent production of the melanogenetic enzymes. To bind M-Box and then promote the transcription of tyrosinase, TRP-1, and DCT, MITF needs the phosphorylation in Ser^{73} and Ser^{49}. However, the oxidation of MITF can compromise the melanogenetic pathway, as well as the cell-cycle progression ($p16^{INK4A}$-dependent) and cell survival (Bcl-2-dependent). Finally, the oxidative damage of MITF may affect even its ability to antagonize TNF-α activity and to bind $6BH_4$, further contributing to the oxidative stress.

The role of an oxidative stress is suggested even by the mechanism proposed to explain the chemical vitiligo, occurring in subjects professionally exposed to phenol-derived substances. A TRP-1 dependent pathway has been suggested as possible source of toxic quinones and ROS at least in in vitro models. Following the exposure to phenol compounds, the assembly may be affected. The compromised assembly of the TRP-1 in a multiproteic complex may affect the stability of the TRP-1 itself with subsequent production of the melanin toxic intermediates [3, 46]. Moreover, the chemical compounds may allow the increased membrane expression of hsp70, favoring the activation of the dendritic cells and of the immune system [19].

An altered intracellular membrane-dependent signal transduction may take place in vitiligo melanocytes, possibly affecting the response to growth factors and dangerous stimuli. A marked membrane lipoperoxidation, together with loss and dislocation of the cardiolipin across the mitochondrial membrane has been described in cultured vitiligo melanocytes [10], which could be the cause of the increased mitochondrial ROS production and for the possible reduced response to the specific growth factors.

The oxidative-mediated damage has been suggested to be involved even in the alteration of other epidermal cells, compromising their survival/proliferative potential, and also production of melanocyte-specific growth factors. Lesional keratinocytes, in fact, can present vacuolar degeneration, which has been associated with the increased exposure to H_2O_2, and produce an inadequate amount of the specific melanocyte growth factors, including membrane-bound SCF, which could lead to melanocyte apoptosis [20, 21] (Chap. 2.2.8). Moreover, keratinocytes themselves, following the exposure to prooxidant stimuli, can produce and release high amount of proinflammatory cytokines, such as IL-6, IL-1α, and TNF-α leading to lymphocytes recruitment [30, 31] (Chap. 2.2.8 and Sect. 2.2.7.4). However, TNF-α was reported to directly interfere with some mitochondrial activities through the production of peroxides (including hydrogen peroxide) leading to a mitochondria-dependent cell death or, at least, to the activation of the inflammatory genes, through the nuclear translocation of NF-κB [9, 41].

2.2.6.4 The Possible Genetic Background

The expression of the genes of the FOXD3 pathway (SLUG, Wnt-2, SOX9, SOX10), as well of calreticulin, GGA1, and MATP (the latest two genes, codifying for proteins involved in the trafficking of melanosome proteins, regulate the post-Golgi vesicle-mediated transport) has been found differently modulated in vitiligo [43]. On this basis, the accumulation of toxic melanin intermediates may be due to an impaired melanosome transport, inner coating of the melanosome membranes, or to defective expression of proteins regulating melanogenic enzymes synthesis. A catalase polymorphism (T/C SNP), leading to an incorrect subunit assembly, was reported in Caucasian population [4]. A reduced expression of VIT1 (22q11) could account for an altered G/T mismatch repair, calcium homeostasis, and protein degradation [25]. In acrofacial vitiligo, COMT polymorphism (G/A, val/met) could be responsible for a thermo-sensibility of the enzyme with a subsequent high production of quinones [45]. However, the true value of the different genetic studies has to be further confirmed, because they are frequently carried out on a limited number of subjects and without validation by other authors.

2.2.6.5 The Systemic Oxidative Stress

Some biochemical pathways, altered in the vitiligo epidermis, have been reported to be deregulated even in other nonepidermal related compartments. The alteration of the catecholamine biosynthesis was observed in truth even in blood cells, associated with higher plasma level of catechols [7, 33]. In red blood cells and peripheral blood mononuclear cells (PBMC) a defective activity of catalase and GPx, associated with an increased production of lipid peroxidation by-products, has been described [1, 7, 18]. Melanocytes and PBMC, show altered antioxidants pattern, with low activity of catalase and glutathione peroxidase, high activity of SOD and xanthine oxidase, associated with the presence of index of lipid peroxidation. The alteration of the antioxidant apparatus in peripheral blood cells has been suggested to be due to the H_2O_2 produced inside the epidermis and systemically distributed by the blood [39, 42]. However, the ROS production could take place inside the nonepidermal cells, independently on melanocyte-specific metabolisms [7, 8]. In fact, the increased ROS production in PBMC has been reported to be associated with some mitochondrial alterations, such as the loss of the transmembrane potential, the cardiolipin loss, or dislocation [8, 10]. Moreover, the increased ROS production can be inhibited by drugs acting on the mitochondrial transition pores, such as cyclosporin A [7]. In addition, the concentration of the shuttle enzyme MDH, physiologically ensuring the adequate level of NADH inside the mitochondria, was reported to be defective [8]. The last alteration could be associated with the impairment of the mitochondrial activity leading to the ROS production. Finally, the DNA oxidative damage in PBMC has been described indicating that the site of production may be intracellular [13].

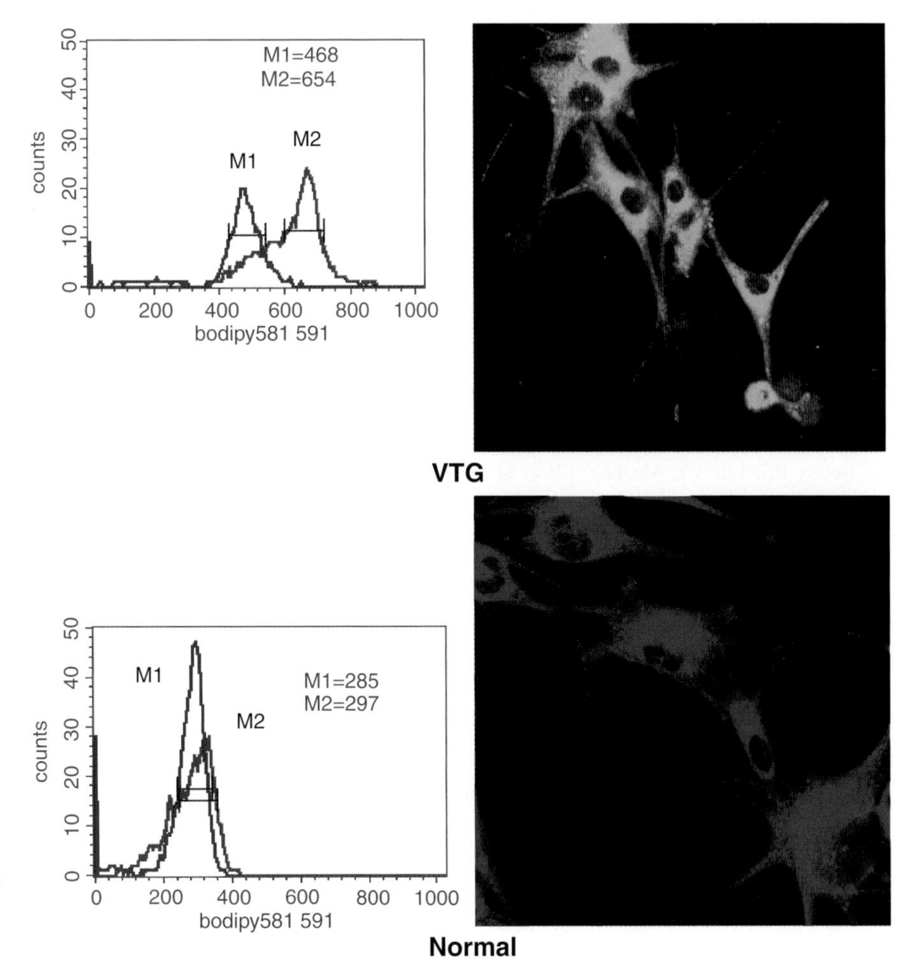

Fig. 2.2.6.2 Fluorescence-based assay indicates membrane lipid peroxidation in vitiligo cells, as showed by the green shift of the Bodipy$^{581/591}$

2.2.6.6 Concluding Remarks

The occurrence of the oxidative stress seems to be correlated with vitiligo onset and progression, indicating that the alterations of different metabolisms leading to oxidative damage of the cells may occur. However, whether the cellular oxidative stress is a primary or a secondary event has not been defined. The oxidative stress can affect the membrane organization and the cellular adhesion, agents with a subsequent anomalous exposure, and release of proteins on the membrane (i.e., hsp70) leading to an increased susceptibility to toxic. The membrane lipids are easy target of the ROS and an altered composition of lipids may enhance the propagation of the peroxidative damage (Fig. 2.2.6.2). The lipid-dependent membrane structure and function may be thus compromised by the oxidative damage, affecting the expression/release at mitochondrial and cellular levels of proteins. The released mitochondrial proteins will transduce to the nucleus, potentially death signals. At the cellular level, the over-expression of hsp protein members may mediate and enhance the antigen presentation and immune activation [11, 34, 41]. The oxidative-stress related alteration and damage has been considered as a target of the therapy in patients with active disease.

References

1. Agrawal D, Shajil EM, Marfatia YS et al (2004) Study of the antioxidant status of vitiligo patients of different age groups in Baroda. Pigment Cell Res 17:289–294
2. Boissy RE, Liu YY, Medrano EE et al (1991) Structural aberration of the rough endoplasmic reticulum and melanosome compartmentalization in long-term cultures of melanocytes from vitiligo patients. J Invest Dermatol 97: 395–404
3. Boissy RE, Manga P (2004) On the etiology of contact/occupational vitiligo. Pigment Cell Res 17:208–214
4. Casp CB, She JX, McCormack WT (2001) Genetic association of the catalase gene (CAT) with vitiligo susceptibility. Pigment Cell Res 15:62–66
5. Chavan B, Gillbro JM, Rokos H et al (2006) GTP cyclohydrolase feedback regulatory protein controls cofactor 6-tetrahydrobiopterin synthesis in the cytosol and in the nucleus of epidermal keratinocytes and melanocytes. J Invest Dermatol 126:2481–2489
6. Cucchi ML, Frattini P, Santagostino G et al (2000) Higher plasma catecholamine and metabolite levels in the early phase of nonsegmental vitiligo. Pigment Cell Res 13:28–32
7. Dell'Anna ML, Maresca V, Briganti S et al (2001) Mitochondrial impairment in peripheral blood mononuclear cells during the active phase of vitiligo. J Invest Dermatol 117:908–913
8. Dell'Anna ML, Urbanelli S, Mastrofrancesco A et al (2003) Alterations of mitochondria in peripheral blood mononulcear cells of vitiligo patients. Pigment Cell Res 16:553–559
9. Dell'Anna ML, Camera E, Picardo M (2005) Free radicals. In: Bos JD (ed) Skin immune system, 3rd edn. CRC press, Boca Raton, pp 287–313
10. Dell'Anna ML, Ottaviani M, Albanesi V et al (2007) Membrane lipid alternations as a possible basis for melanocyte degeneration in vitiligo. J Invest Dermatol 127: 1226–1233
11. Dell Anna HL, Picardo M (2006). A review and a new hypothesis for non-immunological pathogenetic mechanisms in vitiligo. Pigment Cell Res 19:406–411
12. Gillbro JM, Marles LK, Hibberts NA et al (2004) Autocrine catecholamine biosynthesis and the b2-adrenoceptor signal promote pigmentation in human epidermal melanocytes. J Invest Dermatol 123:346–353
13. Giovannelli L, Bellandi S, Pitozzi V et al (2004) Increased oxidative DNA damage in mononuclear leukocytes in vitiligo. Mut Res 556:101–106
14. Hann SK (1999) A role of the nervous system in the pathogenesis of segmental vitiligo. Pigment Cell Res 7:26
15. Hasse S, Gibbons NCJ, Rokos H et al (2004) Perturbed 6-tetrahydrobiopterin recycling via decreased dihydropteridine reductase in vitiligo: more evidence for H2O2 stress. J Invest Dermatol 122:307–313
16. Jimbow K, Chen H, Park JS et al (2001) Increased sensitivity of melanocytes to oxidative stress and abnormal expression of tyrosinase-related protein in vitiligo. Br J Dermatol 144:55–65
17. Jimenez-Cervantes C, Martinez-Esparza M, Perez C et al (2001) Inhibition of melanogenesis in response to oxidative stress: transient downregulation of melanocyte differentiation markers and possible involvement of microphtalmia transcription factor. J Cell Sci 114:2335–2344
18. Koca R, Armutcu F, Altinyazar HC et al (2004) Oxidant-antioxidant enzymes and lipid peroxidation in generalized vitiligo. Exp Dermatol 29:406–409
19. Kroll TM, Bommiasamy H, Boissy RE et al (2005) 4-tertiary butyl phenol exposure sensitizes human melanocytes to dendritic cell-mediated killing: relevance to vitiligo. J Invest Dermatol 124:798–806
20. Lee AY, Youm YH, Kim NH et al (2004) Keratinocytes in the depigmented epidermis of vitiligo are more vulnerable to trauma (suction) than keratinocytes in the normally pigmented epidermis, resulting in their apoptosis. Br J Dermatol 151:995–1003
21. Lee YA, Kim NH, Choi WI et al (2005) Less keratinocyte-derived factors related to more keratinocyte apoptosis in depigmented than normally pigmented suction-blistered epidermis may cause passive melanocyte death in vitiligo. J Invest Dermatol 124:976–983
22. Le Poole IC, Das PK, van den Wijngaard RM et al (1993) Review of the etiopathomechanism of vitiligo: a convergence theory. Exp Dermatol 2:145–153
23. Le Poole IC, Das PK (1997) Microscopic changes in vitiligo. Clin Dermatol 15:863–873
24. Le Poole IC, van den Wijingaard RMJGJ, Westerhof W et al (1997) Tenascin is overexpressed in vitiligo lesional skin and inhibits melanocyte adhesion. Br J Dermatol 137:171–178

25. Le Poole IC, Sarangarajan R, Zhao Y et al (2001) "VIT1", a novel gene associated with vitiligo. Pigment Cell Res 14: 475–484
26. Le Poole IC, Wankowicz-Kalinska A, van den Wijngaard RM et al (2004) Autoimmune aspects of depigmentation in vitiligo. J Invest Dermatol Symp Proc 9:68–72
27. Maresca V, Roccella M, Roccella F et al (1997) Increased sensitivity to peroxidative agents as possible pathogenic factor of melanocyte damage in vitiligo. J Invest Dermatol 109:310–313
28. Medrano EE, Nordlund JJ (1990) Successful culture of adult human melanocytes obtained from normal and vitiligo donors. J Invest Dermatol 95:441–445
29. Montes LF, Abulafia J, Wilborn WH et al (2003) Value of histopathology in vitiligo. Int J Dermatol 42:57–61
30. Moretti S, Spallanzani A, Amato L et al (2002) New insights into the pathogenesis of vitiligo: imbalance of epidermal cytokines at sites of lesions. Pigment Cell Res 15:87–92
31. Moretti S, Spallanzani A, Amato L et al (2002) Vitiligo and epidermal microenvironment: possible involvement of keratinocytes-derived cytokines. Arch Dermatol 138:273–274
32. Morrone A, Picardo M, De Luca C et al (1992) Catecholamines and vitiligo. Pigment Cell Res 5:65–69
33. Panucio AL, Vignale R (2003) Ultrastructure studies in stable vitiligo. Am J Dermopathol 25:16–20
34. Picardo M (2009). Lipid-mediated signalling and melanocyte function. Pigment Cell Mel Res 22:152–153
35. Puri N, Mojamdar M, Ramaiah A (1987) In vitro growth characteristics of melanocytes obtained from adult normal and vitiligo subjects. J Invest Dermatol 88:434–438
36. Rokos H, Beazley WD, Schallreuter KU (2002) Oxidative stress in vitiligo: photo-oxidation of pterins produces H_2O_2 and pterin-6-carboxylic acid. Biochem Biophys Res Commun 292:805–811
37. Schallreuter KU, Wood JM, Pittelkow MR et al (1996) Increased monoamine oxidase A activity in the epidermis of patients with vitiligo. Arch Dermatol Res 288: 14–18
38. Schallreuter KU, Moore J, Wood JM et al (1999) In vivo and in vitro evidence for hydrogen peroxide (H_2O_2) accumulation in the epidermis of patients with vitiligo and its successful removal by a UVB-activated pseudocatalase. J Invest Dermatol Symp Proc 4:91–96
39. Schallreuter KU, Moore J, Wood JM et al (2001) Epidermal H_2O_2 accumulation alters tetrahydrobiopterin (6BH4) recycling in vitiligo: identification of a general mechanism in regulation of all 6BH4-dependent processes? J Invest Dermatol 116:167–174
40. Schallreuter KU, Wood JM, Berger J (2001) Low catalase levels in the epidermis of patients with vitiligo. J Invest Dermatol 97:1081–1085
41. Schallreuter KU, Baharodan P, Picardo M et al (2008) Vitiligo pathgenesis autoimmune disease, generic defect, excessive reactive oxygen species, calcium imbalnac, or what else? Exp Dermatol 17:139–160
42. Schallreuter KU, Chinchiarelli G, Cemeli E et al (2006). Estrogens can contribute to hydrogen peroxide generation and quinone-mediated DNA damage in peripheral blood lymphocytes from patients with vitiligo. J Invest Dermatol 126:1036–1042
43. Stromberg S, Bjorklund MG, Asplund A et al (2008) Transcriptional profiling of melanocytes from patients with vitiligo vulgaris. Pigment Cell Mel Res 21:162–171
44. Taieb A (2000) Intrinsic and extrinsic pathomechanisms in vitiligo. Pigment Cell Res 13:41–47
45. Tursen U, Kaya TI, Derici MEEE et al (2002) Association between catechol-*O*-methyltransferase polymorphism and vitiligo. Arch Dermatol Res 294:143–146
46. Yang F, Boissy RE (1999) Effects of 4-tertiary butylphenol on the tyrosinase activity in human melanocytes. Pigment Cell Res 12:237–245

Immune/Inflammatory Aspects 2.2.7

Contents

2.2.7.1 Introduction

Alain Taïeb

In the intermingling of genetic and environmental influences delineated in the earlier chapters, immune responses have a key role in the pathogenesis of NSV. The bulk of the work has until recently concentrated on adaptive immunity, through antibody (Sect. 2.2.7.3) or cell related (Sect. 2.2.7.4) pathomechanisms, but a clear evidence to the assumption that melanocytes are directly killed by autoantibodies or cytotoxic cells is still lacking in common NSV. Indeed, inflammation remains a low-key feature in vitiligo, with regard to the astoundingly "silent" loss of pigment cells. However, progressive depigmentation in vitiligo is consistently associated with cellular infiltrates in at least a margin of actively depigmenting skin. This identifiable, mostly CD8+ cellular component is clearly associated with an acceleration phase of the disease, if not directly causative. On the other hand, the pathogenic role of autoantibodies in the disease remains obscure, but as pointed out by Kemp et al., (2.2.7.3) identifying their target antigens could provide information for the development of diagnostic tests for vitiligo and could indicate important T-cell responses in patients.

The initial steps leading to an activation of an autoimmune/inflammatory response are probably backed by the immune constitution of the individual, but they are overall poorly understood. Recently, it was suggested that either physical trauma (the Koebner's phenomenon) [1] or chemical molecules (such as phenolic compounds which trigger occupational vitiligo) [3] could play a role and link "danger" responses to an inappropriate activation of the innate immunity which encompasses many cell types, mediators, and receptors.

A. Taïeb
Service de Dermatologi,
Hôpital St André, CHU de Bordeaux,
France
e-mail: alain.taieb@chu-bordeaux.fr

M. Picardo and A. Taïeb (eds.), *Vitiligo*,
DOI 10.1007/978-3-540-69361-1_2.2.7, © Springer-Verlag Berlin Heidelberg 2010

This inappropriate activation would affect the homeostasis of the epidermis and affect intrinsically fragile melanocytes. As indicated by Peroni and Girolomoni (Sect. 2.2.7.2) keratinocytes-derived stimuli, dendritic cells, proteins involved in the activation of the inflammasome complex, which is present in keratinocytes, and complement are potential major actors. A recent breakthrough in this perspective was the finding that single-nucleotide polymorphisms in the inflammasome NALP1 gene appear to independently contribute to risk of generalized vitiligo and other autoimmune diseases in Caucasians [2] (Sect. 2.2.7.2).

References

1. Gauthier Y, Cario-André M, Taieb A (2003) A critical appraisal of vitiligo etiologic theories. Is melanocyte loss a melanocytorrhagy? Pigment Cell Res 16:322–327
2. Jin Y, Mailloux CM, Gowan K et al (2007) NALP1 in vitiligo-associated multiple autoimmune disease. N Engl J Med 356:1216–1225
3. Manga P, Sheyn D, Yang F et al (2006) A role for tyrosinase related-protein 1 in 4-*tert*butylphenol-induced toxicity in melanocytes. Implications for vitiligo. Am J Pathol 169: 1652–1662

2.2.7.2 The Role of Innate Immunity in Vitiligo

Anna Peroni and Giampiero Girolomoni

Contents

Core Messages

> Innate immune system is distinct, but not separate from the adaptive immune system. Particularly, many elements of the innate immunity exert a crucial role in determining the characteristics of the subsequent adaptive immune response.
> Innate immunity encompasses many cell types, mediators, and receptors.
> The role of innate immunity in vitiligo pathogenesis is far from being clarified, but many findings support a contribution. A major involvement to date has been suggested for keratinocytes-derived stimuli, dendritic cells-mediated direct killing, and alterations of proteins involved in the activation of the inflammasome and complement.

The Innate Immune System: An Overview

For many decades, mechanisms of adaptive immunity have been the main focus of immunological research, but in more recent years the importance of the innate

A. Peroni (✉)
Dermatological Clinic, University of Verona, Verona, Italy
e-mail: anna.peroni@univr.it

immune system has been rediscovered. The innate immune system is the most ancient and highly conserved, being present in barely all metazoan organisms [54]. It has traditionally been interpreted as independent from adaptive immune system and characterized by a stereotyped response to infective stimuli, useful to control the pathogens while the adaptive immune response is forming. Innate and adaptive immunities show fundamental differences. Innate immunity is characterized by a limited number of germline-encoded receptors that can recognize some highly conserved microbial structures and promote a stereotyped response. On the contrary, adaptive immunity can arm specific response against a virtually unlimited series of antigens through site-specific somatic recombination; moreover only adaptive immunity features immunological memory that allows a more rapid and efficient reaction in case of a second encounter with the same antigen. The two systems are, however, strictly intermingled: cells of the innate immune system participate as effectors in adaptive immunity responses, and they can also determine the type of adaptive response, principally by the action of dendritic cells (DC) [20]; another link between the two systems is represented by the natural killer (NK)-T cell, a subset of T-cells that express NK as well as T-cell receptors [57].

The innate immune system is formed by physical barriers (epithelial layers of skin and mucosa), cells (skin, pulmonary and gut epithelial cells, mast cells, NK cells, DC, granulocytes, and monocytes-macrophages), and soluble molecules (complement, cytokines and chemokines, antimicrobial peptides) [20]. An important milestone in the understanding of the functioning of innate immunity is the discovery of germline-encoded pattern recognition receptors (PRRs) and their primary role in the identification of various alarm signals, from outside (conserved molecular patterns shared by many bacteria or viruses) or inside (endogenous ligands generated under abnormal conditions or as indirect result of infection) [38]. The best characterized of the PRRs are Toll-like receptors (TLRs), a family of transmembrane proteins expressed mostly on phagocytic cells, with different members being able to recognize distinct microbial structures and leading to the activation of NF-κB transcription factor [4, 44]. More recently, other cytosolic PRRs have been characterized, called NOD-like receptors (NLRs) [34]; their role is still an area of intensive research, but some of them (Nod1 and Nod2) share structural and functional characters with TLRs [71], while others (NLR family apoptosis inhibitory proteins, NAIPS, and NACHT-, LRR-, and Pyrin domain-containing proteins, NALPS) control activation of the inflammasome [52], that leads even to rapid cell death in some cell types [22]. A third family of PRRs is the RIG-I-like receptors (RLRs) that recognize viruses [69]. These three families of PRRs cooperate in innate immunity [23]. Soluble PRRs have been identified too, like mannose binding lectin (MBL) that is involved in opsonization and complement activation [26]. Finally, a new intriguing research matter concerns activation of innate immune system not only by infectious agents, but also by sterile tissue damage [8].

The Skin Innate Immune System

The skin innate immune system is constituted by many different cellular and soluble elements that are summarized in Table 2.2.7.1.

Cellular Components

The stratum corneum, a nonviable layer of anucleated terminally differentiated keratinocytes with extracellular lipid lamellae, is the primary protective shield of the skin [58]. Keratinocytes are actively involved in immune defenses. They can be activated via TLR (they express TLR1, 2 and 5, while conflicting reports exist on TLR4) and NF-κB [53] as well as by proinflammatory cytokines and release cytokines, chemokines, and antimicrobial peptides [6, 20]. These substances act by direct killing of the pathogen, initiation of inflammation via endothelial cell activation and chemoattraction

Table 2.2.7.1 Principal elements of the skin immune system

Cells	Soluble molecules
Keratinocytes	Antimicrobial peptides
Melanocytes	Complement
Langerhans cells	Chemokines
Dermal dendritic cells	Cytokines
Monocytes	
Mast cells	
Endothelial cells	
NK cells	
Granulocytes	

of inflammatory cells, and also by activating DC [20]. Phagocytes (monocytes-macrophages and granulocytes) and DC engulf the pathogens, and then kill them in the first case, while processing for antigen presentation to T- lymphocytes in the second case. Infiltrating neutrophils exert different effects including release of antimicrobial peptides, direct killing of microbes, release of cytokines and proteases necessary for activation of many proinflammatory mediators. Cytokines and growth factors initiate also a reparative response that limits tissue damage and favor the return to a homeostatic condition. Epidermal growth-factor receptor plays a very important role in these processes [62]. The DC, including epidermal Langerhans cells (LC) and dermal DC are the major antigen-presenting cells and are fundamental to determine the action of T cells: whether or not to respond, how to respond (Th1, Th2 or T regulatory (Treg cells), and where to respond; moreover they have a crucial role in the peripheral tolerance by directly rendering unresponsive auto-reactive T-cells or inducing the formation of suppressive Treg cells or conversely overruling the action of established T reg cells [20]. Epidermal LC are strategically located to detect and transmit, vertically and horizontally, surface danger signals [56]; despite the fact that LC were described 140 years ago, their in vivo function still remains elusive. More recently, different dermal DC subpopulations have been characterized according to phenotype and function [56]. Among these, plasmacytoid DC (pDCs) are a major source of IFN-α and are actively involved in antiviral responses, and differ from conventional DC as they uniquely express TLR7 and 9, and not other TLRs [16]. Plasmacytoid DC have also been associated with the pathogenesis of a group of autoimmune diseases (including systemic lupus erythematosus and psoriasis) characterized by the presence of anti self-nucleic acid and self-nuclear components antibodies [16, 45]. Accordingly, with the "Toll hypothesis," autoimmune diseases may originate from imperfect discrimination of microbe from self through receptors that detect microbial DNA and RNA [28].

Crucial for the recognition of self/ nonself are the NK cells, which not only utilize the "missing self" concept typical for innate immunity, but also screen the cell surfaces for the presence of receptors of the adaptive immune system (e.g. HLA-E), which are in this case utilized as self markers. This aspect is particularly interesting because, if it is true that autoimmunity is strictly dependent from adaptive immune system [35], interactions between innate and adaptive immune systems are crucial for the process: innate elements are responsible for labeling a particular antigen as "dangerous" and passing this information to the adaptive immune response, that subsequently may initiate a misdirect immune response, for example against melanocytes as in vitiligo.

Another important cell type involved both in innate and acquired skin immune system is the mast cell [27, 65]. Traditionally considered principally involved in T_H2- and IgE-associated immune responses, now mast cell is known to be implicated in a variety of inflammatory and immunological processes, including modulation of T cells and DC functions, antigen presentation to T cells through MHC class I and II molecules [19, 27], selective recruitment of T cells and NK cells [24]. Finally in recent years, mast cells have been implicated in models of several cutaneous or systemic autoimmune diseases [19, 65] and in T reg cell-mediated allograft tolerance [50].

Melanocytes may exert some immunological functions [67]. They are capable of producing cytokines [5, 70], can phagocytize, process, and present antigens on MHC class II, as "amateur" antigen-presenting cells. Even melanization, which is their main recognized task, could be seen as part of innate immune system with the production and distribution of pigment, which is essential for protection from ultraviolet damage.

Soluble Components

Antimicrobial peptides are predominantly small cationic polypeptides with the ability of directly killing a broad spectrum of bacteria, fungi, and viruses; many of them have also other functions. Some of these peptides are normally expressed in healthy skin, such as lysozyme, dermicidin, Rnase7, Psoriasin (S100A7), while others are up-regulated under specific conditions, like human β-defensin-2 and human β-defensin-3; finally cathelicidin LL-37 is both constitutionally expressed and up-regulated during inflammatory and immune responses [14, 68]. Interestingly enough, many other peptides that have been ascribed alternate functions in the skin also demonstrate antimicrobial activity, such as α-MSH, substance P and chemokines (CXCL9, CXCL10, CXCL11) [14]. In addition, antimicrobial peptides may be involved in the regulation of melanin synthesis by acting on the melanocortin 1 receptor [15]. Complement is a system of plasmatic proteins usually not present within the skin that can be activated by microorganisms and recruited during inflammatory processes. Complement can be activated by different mechanisms: in the classical pathway C1 recognizes IgM, IgG1, or IgG3 bound on microbial or cellular surface, in the alternative pathway complement proteins bind directly structures on microbial surface, finally, in the last discovered lectin pathway, the plasmatic protein MBL recognizes mannose terminal residues of microbial glycoproteins or glycolipids and then activates the classical pathway, without antibodies, through an associated serin-protease [63, 77]. The final result of these activation pathways is the formation of the membrane attack complex (MAC) (C5b, C6-9) that leads to cell lysis. Some complement fractions (C3b, C4b) and MBL acts as opsonine as well, thus favoring microbial engulfment by neutrophils and macrophage. Finally some complement fractions are chemoattractant for neutrophils (C3a, C5a) or directly activate mast cells (C3a, C4a, C5a) [21]. Chemokines and cytokines are very numerous and have a plenty of functions, whose description override the purposes of this text. All cell types of innate immunity are major sources of cytokines and chemokines [1].

The Innate Immune System in Vitiligo

Keratinocytes

Keratinocytes form functional and structural units with melanocytes, and exert several effects on the latter. Indeed, structural [11, 61] and functional [12, 66] abnormalities in vitiligo keratinocytes have been reported. Functional alterations of vitiliginous keratinocytes concerns principally adrenergic metabolism (Chap. 2.2.6), and that represents one of the cross-talk systems between epidermal cells [29]. Another way by which keratinocytes can influence adjacent melanocytes is the production of several soluble and membrane factors (Chap. 2.2.7). Among these, endothelin 1 (ET-1) produced by keratinocytes has paracrine effects on melanocytes influencing their survival and pigmentation, and is thought to play a role in the skin response to UV radiation. It has been shown that ET-1 gene (EDN1) polymorphisms contribute to the increased susceptibility to localized (focal and segmental) vitiligo in Korean population [39]. Particularly haplotype frequencies of EDN1 polymorphisms differed significantly between vitiligo patients and healthy controls, whereas the genotype distributions and allele frequencies did not [39]. Futhermore, keratinocytes in depigmented skin are more susceptible to TNF-α mediated apoptosis through impaired kit and NF-κB activation compared with normally pigmented epidermis [40, 46]; stem cell factor (SCF) decrease followed keratinocytes apoptosis and SCF deprivation from the culture medium determined melanocytes apoptosis [47]. These observations lead to the hypothesis that a minor trauma could lead to melanocytes depletion primarily by inducing keratinocytes apoptosis and subsequent reduction of SCF [47], therefore suggesting a possible mechanism for Koebner phenomenon in vitiligo (Sect. 2.2.2.1). Moreover vitiligo keratinocytes have a shorter life span in vitro than cells from uninvolved skin, and all of them do not support melanocytes properly in culture [12]. Finally, the presence of anti-keratinocyte antibodies has been reported, directed toward keratinocytes cytoplasmic components, but they seem to be more likely a consequence of cellular damage rather than a causative factor for vitiligo [79].

Dendritic Cells

The LC density in vitiligo has been variably reported as decreased, normal, or increased [61], being different in the various zones of trichrome type [32]. Recently, ultrastructural abnormalities have also been described [61] (see also Sect. 2.2.3.2). There are also some evidences of functional impairment of LC in vitiliginous skin that results in relative absence of contact dermatitis to monobenzyl ether of hydroquinone [73]. But, exact correlation of these alterations with vitiligo pathogenesis remains to be clarified. A new role for DC in vitiligo has emerged from the demonstration of their capability to kill tumor cells leaving surrounding healthy cells untouched [49]. DC-mediated killing depends upon the expression of TNF family members on DC surface coupled with the expression of appropriate receptors on target cells. Healthy cells do not express the latter and are therefore protected [49]. More recently DC-mediated killing of stressed melanocytes has been demonstrated in vitiligo experimental models and in vivo [42]. Particularly, the process has been related to increased release by vitiligo melanocytes of heat-shock protein (HSP) 70 in the extracellular milieu, where this can induce an immune response against the very cell from which it is derived, and to enhanced expression of TNF-related apoptosis inducing ligand (TRAIL) receptors on stressed melanocytes that renders them more susceptible to DC direct killing [42].

Macrophages

The presence of macrophages has been proved particularly in perilesional skin [48, 74], and even in noninflammatory stable vitiligo of long duration [61]. Their functional role in vitiligo may depend on H_2O_2 production via NADPH oxidase activity [67]. Further evidence for active involvement of macrophages in vitiligo pathogenesis is linked to their expression of immunoglobulins receptors. It has been shown in a mouse model that macrophages expressing common γ chain of the activating FcγRs can mediate vitiligo both in presence and absence of complement C3 fraction [72].

Natural Killer Cells

Only few studies evaluated NK cells in vitiligo patients, mostly of peripheral blood and with variable results. Particularly, percentage of peripheral blood NK has been detected to be significantly higher in patients with vitiligo and negative autoantibody test [31], but lower [51] or not statistically different [2] from healthy controls in nonsegmental vitiligo patients. Finally, Korsunskaya et al. found a reduction of the relative number of NK CD16+ cells, but with unchanged absolute number in the peripheral blood of patients with vitiligo [41]. Only Abdel-Naser et al. did not find significant number of CD16+ or CD56+ cells in vitiliginous skin and marginal skin in patients with generalized vitiligo [3].

Pattern Recognition Receptors

NACHT leucine-rich protein 1 (NALP1) is one of the proteins involved in the activation of the inflammasome, a multiprotein cytosolic complex that receives signals from both TLRs and NLRs pathways and leads to the activation of the proinflammatory cytokines IL-1β and IL-18 [9]. Particularly, NALP1 determines activation of caspase-1, which processes pro-IL-1 to its mature form. Oligomerization of NALP1 and activation of Caspase-1 occur in a two-step mechanism, requiring microbial product, muramyl-dipeptide, a component of peptidoglycan, followed by ribonucleoside triphosphates [25]. The NALP1 is localized mainly in the nucleus and is widely expressed, but is at highest levels in LC and T-cells [43]. Alterations of the inflammasomes are involved in the pathogenesis of some chronic inflammatory diseases [9]. A role for inflammasome in vitiligo has been suggested only recently. The gene encoding for NALP1 has been identified on 17p13 [36], a chromosomal region, yet related to vitiligo [55], and at least two common NALP1 variants appear to independently contribute to risk of generalized vitiligo and other autoimmune diseases in Caucasians [36]. Further, the same single-nucleotide polymorphisms (SNPs) in the NALP1 gene (rs6502867 and rs2670660) showed independent effect on disease risk in another study [37]. The exact mechanism by which NALP1 influences vitiligo pathogenesis remains to be clarified.

Moreover, in very recent years a role in vitiligo pathogenesis has been proposed also for MBL [60], a soluble PRR also active in complement activation. The MBL deficiency has been associated with other autoimmune diseases, such as systemic and cutaneous lupus erythematosus and dermatomyiositis. Reduction of MBL alleles or levels, particularly if associated to an impaired calcium intake by skin cells, may cause an inadequate clearance of apoptotic cells, thus leading to continuous stimulation of the immune system and antibody production. Moreover MBL deficiency could also result in increased predisposition to attack from viruses.

Antimicrobial Peptides

There is no evidence to date of direct involvement of defensins or cathelicidins in vitiligo pathogenesis. Cathelicidn LL-37 has been implicated in the pathogenesis of autoimmune responses. In fact, it has been demonstrated that LL-37 may protect plasmid DNA and mediate its cellular uptake and subsequent eukaryotic expression [64]. This suggested the hypothesis that LL-37 could lead to cellular uptake of bacterial genes at sites of microbial infection and potentially to autoimmune phenomena [13]. LL-37 may combine with

DNA released by injured eukaryotic cells, and these complexes can activate plasmocytoid DC to release IFN, in turn a potent activator of myeloid DC and thus of autoimmune phenomena [45]. This aspect becomes particularly intriguing in light of the direct correlation between progression of vitiligo lesions and the presence of cytomegalovirus DNA [30]. Finally LL-37 has been demonstrated to induce reactive oxygen species generation and IL-8 production by peripheral blood neutrophils of healthy donors [81]. Both IL-8 [80] and oxidative stress have been implicated in vitiligo pathogenesis.

Complement

Complement-activating antimelanocyte antibodies have been implicated in vitiligo pathogenesis, with complement components directly involved in cell killing, but other mechanisms have been hypothesized. Some alterations in C4 functions (heterozygous C4 deficiency) and gene expression (C4B*Q0 null allele) have been related to increased risk for vitiligo [75]. As membrane regulators of complement activation, membrane cofactor protein (MCP o CD46), decay accelerating factor (DAF o CD55), and CD59 protect cells from elimination by autologous complement. Particularly, MCP and DAF inhibit, via different mechanisms, the formation of C3/C5 convertases of the classical and alternative pathways, while CD59 interferes with the assembly of the cytolytic MAC [7]. Skin melanocytes express functionally active MCP and DAF [74, 76] and the contribution of DAF in protecting them against complement attack is much more than that of MCP [76]. The expression of CD59 on melanocytes has been reported by some authors [74], but not by others [76]. Immunohistochemical studies have shown that *in situ* expression levels of DAF and MCP are lower in lesional and perilesional epidermis compared with nonlesional epidermis [74].

Concluding Remarks

The exact role of innate immunity in vitiligo pathogenesis is far from being clear, but many findings support its possible contribution. A major involvement to date has been suggested for keratinocytes-derived stimuli, DC-mediated direct killing, alterations of proteins involved in the activation of the inflammasome and in complement pathway, but many other aspects should be further investigated. Keratinocytes can cooperate in the pathogenesis through ET-1 gene polymorphisms and dysfunctions of the SCF/c-kit/MITF-M signalling pathway [10, 59]. On the latter the fibroblast-derived secreted antagonist of the Wnt pathway dickkopf 1 (DKK1) also acts [17, 18, 78]. DC-mediated killing of stressed melanocytes has been demonstrated in vitiligo experimental models and in vivo. The process has been related to increased release by vitiligo melanocytes of HSP-70 in the extracellular milieu and to enhanced expression of TRAIL receptors on stressed melanocytes that renders them more susceptible to killing. Single-nucleotide polymorphisms in the NALP1, one of the proteins involved in the activation of the inflammasome gene, appear to independently contribute to the risk of generalized vitiligo as well as other autoimmune diseases.

References

1. Abbas AK, Lichtman AH (2005) Innate immunity. In: Abbas AK, Lichtman AH (eds) Cellular and molecular Immunology, 5th edn. Elsevier Italia, Milano
2. Abdel-Naser MB, Ludwig WD, Gollnick H et al (1992) Nonsegmantal vitiligo: decrease of the CD45RA+ T-cell subset and evidence for peripheral T-cell activation. Int J Dermatol 31:321–326
3. Abdel-Naser MB, Krüger-Krasagakes S, Krasagakis K et al (1994) Further evidence for involvement of both cell mediated and humoral immunity in generalized vitiligo. Pigment Cell Res 7:1–8
4. Akira S, Takeda K, Kaisho T (2001) Toll-like receptors: critical proteins linking innate and acquired immunity. Nat Immunol 1:135–145
5. Al Badri AMT, Todd PM, Garioch J et al (1993) An immunohistological study of cutaneous lymphocytes in vitiligo. J Pathol 170:149–155
6. Albanesi C, Scarponi C, Giustizieri ML et al (2005) Keratinocytes in skin inflammation. Curr Drug Targets Inflamm Allergy 4:329–334
7. Asghar SS, Pasch MC (1998) Complement as a promiscuous signal transduction device. Lab Invest 78:1203–1225
8. Barton GM (2008) A calculated response: control of inflammation by the innate immune response system. J Clin Invest 118:413–420
9. Becker CE, O'Neill LA (2007) Inflammasomes in inflammatory disorders: the role of TLRs and their interactions with NLRs. Semin Immunopathol 29:239–248
10. Beuret L, Flori E, Denoyelle C et al (2007) Up-regulation of MET expression by alpha-melanocyte-stimulating hormone and MITF allows hepatocyte growth factor to protect

melanocytes and melanoma cells from apoptosis. J Biol Chem 282:14140–14147
11. Bhawan J, Bhutani LK (1983) Keratinocyte damage in vitiligo. J Cutan Pathol 10:207–212
12. Bondanza S, Maurelli R, Paterna P et al (2007) Keratinocyte cultures from involved skin in vitiligo patients show an impaired in vitro behaviour. Pigment Cell Res 20:288–300
13. Borregard N, Theilgaard-Mönch K, Cowland JB et al (2005) Neutrophils and keratinocytes in innate immunity-cooperative actions to provide antimicrobial defence at the right time and place. J Leukoc Biol 77:439–443
14. Braff MH, Bardan A, Nizet V et al (2005) Cutaneous defence mechanisms by antimicrobial peptides. J Invest Dermatol 125:9–13
15. Candille SI, Kaelin CB, Cattanach BM et al (2007) A beta-defensin mutation causes black coat colour in domestic dogs. Science 318:1418–1423
16. Cao W, Liu YJ (2007) Innate immune functions of plasmocytoid dendritic cells. Curr Opin Immunol 19:24–30
17. Chang HY, Chi JT, Dudoit S et al (2002) Diversity, topographic differentiation, and positional memory in human fibroblasts. Proc Natl Acad Sci USA 99:12877–12882
18. Chang HY (2007) Patterning skin pigmentation via dickkopf. J Invest Dermatol 127:994–995
19. Christy AL, Brown MA (2007) The Multitasking Mast Cell: positive and negative roles in the rogression of autoimmunity. J Immunol 179:2673–2679
20. Clark R, Kupper T (2005) Old meets the new: the interaction between innate and adaptive immunity. J Invest Dermatol 125:629–637
21. Cole DS, Morgan BP (2003) Beyond the lysis: how complement influences cell fate. Clin Sci 104:455–466
22. Cookson BT, Brennan MA (2001) Proinflammatory programmed cell death. Trends Microbiol 9:113–114
23. Creagh EM, O'Neill LAJ (2006) TLRs, NLRs and RLRs: a trinity of pathogen sensors that co-operate in innate immunity. Trends Immunol 27:352–357
24. Dawicki W, Marshall JS (2007) New and emerging roles for mast cells in host defence. Curr Opin Immunol 19:31–38
25. Faustin B, Lartigue L, Bruey JM et al (2007) Reconstituted NALP1 inflammasome reveals two-step mechanism of caspase-1 activation. Mol Cell 25:713–724
26. Fraser IP, Koziel H, Ezekowitz RAB (1998) The serum mannose-binding protein and the macrophage mannose receptor are pattern recognition molecules that link innate and adaptive immunity. Semin Immunol 10:363–372
27. Galli SJ, Nakae S, Tsai M (2005) Mast cells in the development of adaptive immune responses. Nat Immunol 6: 135–142
28. Goodnow CC (2006) Immunology: discriminating microbe from self suffers a double toll. Science 312:1606–1608
29. Grando SA, Pittelow MR, Schallreuter KU (2006) Adrenergic and cholinergic control in the biology of epidermis: physiological and clinical significance. J Invest Dermatol 126: 1948–1965
30. Grimes PE, Sevall JS, Vodjani A (1996) Cytomegalovirus DNA identified in skin biopsy specimens of patients with vitiligo. J Am Acad Dermatol 35:21–26
31. Hann SK, Park YK, Chung KY et al (1993) Peripheral blood lymphocyte imbalance in Koreans with active vitiligo. Int J Dermatol 32:286–289
32. Hann SK, Kim YS, Yoo JH et al (2000) Clinical and histopathologic characteristics of trichrome vitiligo. J Am Acad Dermatol 42:589–596
33. Imokawa G (2004) Autocrine and paracrine regulation of melanocytes in human skin and in pigmentary disorders. Pigment Cell Res 17:96–110
34. Inohara N, Chamaillard M, Mc Donald C et al (2005) NOD-LRR proteins: role in host-microbial interactions and inflammatory diseases. Ann Rev Biochem 74:355–383
35. Janeway CA Jr, Medzhitov R (2002) Innate immune recognition. Annu Rev Immunol 20:197–216
36. Jin Y, Mailloux CM, Gowan K et al (2007) NALP1 in vitiligo-associated multiple autoimmune disease. N Engl J Med 356:1216–1225
37. Jin Y, Birlea SA, Fain PR et al (2007) Genetic variations in NALP1 are associated with generalized vitiligo in a romanian population. J Invest Dermatol 127:2558–2562
38. Kabelitz D, Medzhitov R (2007) Innate immunity – cross talk with adaptive immunity through pattern recognition receptors and cytochines. Curr Opin Immunol 19:1–3
39. Kim HJ, Choi CP, Uhm YK et al (2007) The association between endothelin-1 gene polymorphisms and susceptibility to vitiligo in a Korean population. Exp Dermatol 16: 561–566
40. Kim NH, Jeon S, Lee HJ et al (2007) Impaired PI3K/Akt activation-mediated NF-kB inactivation under elevated TNF-alpha is more vulnerable to apoptosis in vitiliginous keratinocytes. J Invest Dermatol 127:2612–2617
41. Korsunskaya IM, Suvorova KN, Dvoryankova EV (2003) Modern aspects of vitiligo pathogenesis. Dokl Biol Sci 388: 38–40
42. Kroll TM, Bommiasamy H, Boissy RE et al (2005) 4-Tertiary butyl phenol exposure sensitises human melanocytes to dendritic cell-mediated killing: relevance to vitiligo. J Invest Dermatol 124:798–806
43. Kummer JA, Broekhuizen R, Everett H et al (2007) Inflammasome components NALP1 and 3 show distinct but separate expression profiles in human tissues suggesting a site-specific role in the inflammatory response. J Histochem Cytochem 55:443–452
44. Kwai T, Akira S (2007) Antiviral signaling through pattern recognition receptors. J Biochem 141:137–145
45. Lande R, Gregorio J, Facchinetti V et al (2007) Plasmacytoid dendritic cells sense self-DNA coupled with antimicrobial peptide. Nature 449:564–569
46. Lee AY, Youm YH, Kim NH et al (2004) Keratinocytes in the depigmented epidermis of vitiligo are more vulnerable to trauma (suction) than keratinocytes in the normally pigmented epidermis, resulting in their apoptosis. Br J Dermatol 151:995–1003
47. Lee AY, Kim NH, Choi WI et al (2005) Less keratinocyte-derived factors related to more keratinocyte apoaptosis in depigmented than normally pigmented suction-blistered epidermis may cause passive melanocyte death in vitiligo. J Invest Dermatol 124:976–983
48. Le Poole IC, van den Wijngaard RM, Westerhoff W et al (1996) Presence of T cells and macrophages in inflammatory vitiligo skin parallels melanocyte disappearance. Am J Pathol 148:1219–1228
49. Lu G, Janjic BM, Janjic J et al (2002) Innate direct anticancer effector function of human immature dendritic cells. II.

Role of TNF, lymphotoxin-alpha(1)beta(2), Fas ligand, and TNF-related apoptosis-inducing ligand. J Immunol 15:1831–1839
50. Lu LF, Lind EF, Gondek DC et al (2006) Mast cells are essential intermediaries in regulatory T-cell tolerance. Nature 442:987–988
51. Mahmoud F, Abul H, Haines D et al (2002) Decreased total numbers of peripheral blood lymphocytes with elevated percentages of CD4+CD45RO+ and CD4+CD25+ of T-helper cells in non-segmantal vitiligo. J Dermatol 29:68–73
52. Mariathasan S, Monack DM (2007) Inflammasome adaptors and sensors: intracellular regulators of infection and inflammation. Nat Rev Immunol 7:31–40
53. Mc Inturff JE, Modlin RL, Kim J (2005) The role of toll-like receptors in the pathogenesis and treatment of dermatological disease. J Invest Dermatol 125:1–8
54. Medzhitov R (2007) Recognition of microorganisms and activation of the immune response. Nature 449:819–826
55. Nath SK, Kelly JA, Namjou B et al (2001) Evidence for a susceptibility gene, SLEV1, on chromosome 17p13 in families with vitiligo-related systemic lupus erythematosus. Am J Hum Genet 69:1401–1406
56. Nestle FO, Nickoloff BJ (2007) Deepening our understanding of immune sentinels in the skin. J Clin Invest 117: 2382–2385
57. Nickoloff BJ, Wrone-Smith T, Bonish B et al (1999) Response of murine and normal human skin to injection of allogenic blood-derived psoriatic immunocytes: detection of T cells expressing receptors typically present on natural killer cells including CD94, CD158 and CD161. Arch Dermatol 135:546–552
58. Nickoloff BJ, Denning M (2001) Sensing and killing bacteria by the skin: innate immune defense system: good and bad news. J Invest Dermatol 117:170
59. Norris A, Todd C, Graham A et al (1996) The expression of the c-kit receptor by epidermal melanocytes may be rediced in vitiligo. Br J Dermatol 134:299–306
60. Onay H, Pehlivan M, Alper S et al (2007) Might there be a link between mannose binding lectin and vitiligo? Eur J Dermatol 17:146–148
61. Panucio AL, Vignale R (2003) Ultrastructural studies in stable vitiligo. Am J Dermatopathol 25:16–20
62. Pastore S, Mascia F, Mariani V et al (2007) The epidermal growth factor receptor system in skin repair and inflammation. J Invest Dermatol 127:660–667
63. Petersen SV, Thiel S, Jensenius JC (2001) The mannan-binding lectin pathway of complement activation: biology and disease association. Mol Immunol 38:133–149
64. Sandgren S, Witturp A, Chenf F et al (2004) The human antimicrobial peptide LL-37 transfers extracellular DNA plasmid to the nuclear compartment of mammalian cells via lipid rafts and proteoglycan-dependent endocytosis. J Biol Chem 279:17951–17956
65. Sayed BA, Christy A, Qirion MR et al (2008) The master switch: the role of mast cells in autoimmunity and tolerance. Annu Rev Immunol 26:705–739
66. Schallreuter KU, Pittelkow MP (1988) Defective calcium uptake in keratinocytes cell cultures from vitiliginous skin. Arch Derm Res 280:137–139
67. Schallreuter KU, Bahadoran P, Picardo M et al (2008) Vitiligo pathogenesis: autoimmune disease, genetic defect, excessive reactive oxygen species, calcuim imbalance, or what else. Exp Dermatol 17:139–160
68. Schröder JM, Harder J (2006) Antimicrobial skin peptides and proteins. Cell Mol Life Sci 63:469–486
69. Seth RB, Sun L, Chen ZJ (2006) Antiviral innate immunity pathways. Cell Res 16:141–147
70. Smith N, Le Poole I, van den Wijngaard R et al (1993) Expression of different immunological markers by cultured human melanocytes. Arch Dermatol Res 285:356–365
71. Strober W, Murray PJ, Kitani A et al (2006) Signalling pathways and molecular interactions of NOD1 and NOD2. Nat Rev Immunol 6:9–20
72. Trcka J, Moroi Y, Clynes RA et al (2002) Redundant and alternative roles for activating Fc receptors and complement in an antibody-dependent model of autoimmune vitiligo. Immunity 16:861–868
73. Uheara M, Miyauchi H, Tanaka S (1984) Diminished contact sensitivity response in vitiliginous skin. Arch Dermatol 120:195–198
74. van den Wijngaard R, Asghar SS, Pijnenborg A et al (2002) Aberrant expression of complement regulatory proteins, membrane cofactor protein and decay accelerating factor, in the involved epidermis of patients with vitiligo. Br J Dermatol 146:80–87
75. Venneker GT, Westerhof W, de Vries IJ et al (1992) Molecular heterogeneity of the fourth component of complement (C4) and its genes in vitiligo. J Invest Dermatol 99: 853–858
76. Venneker GT, Vodegel RM, Okada N et al (1998) Relative contributions of decay accelerating factor (DAF), membrane cofactor protein (MCP) and CD59 in the protection of melanocytes from homologous complement. Immunobiology 198:476–484
77. Walport MJ (2001) Complement. N Engl J Med 344:1058–1066; 1141–1144
78. Yamaguchi Y, Itami S, Watabe H et al (2004) Mesenchymal-epithelial interactions in the skin: increased expression of dickkopf1 by palmoplantar fibroblasts inhibits melanocyte growth and differentiation. J Cell Biol 26:275–285
79. Yu HS, Kao CH, Yu CL (1993) Coexistence and relationship of antikeratinocyte and antimelanocyte antibodies in patients with non-segmental-type vitiligo. J Invest Dermatol 100:823–828
80. Yu HS, Chang KL, Yu CL et al (1997) Alterations in IL-6, IL-8, GM-CSF, TNF-alpha, and IFN-gamma release by peripheral mononuclear cells in patients with active vitiligo. J Invest Dermatol 108:527–529
81. Zheng Y, Niyonsaba F, Ushio H et al (2007) Cathelicidin LL-37 indices the generation of reactive oxygen species and release of human a-defensins from neutrophils. Br J Dermatol 157:1124–1131

2.2.7.3 Humoral Immunity

E. Helen Kemp, Anthony P. Weetman,
and David J. Gawkrodger

Contents

Core Messages

- The frequent association of vitiligo with autoimmune diseases, together with studies demonstrating that vitiligo patients can have autoantibodies and autoreactive T-lymphocytes against pigment cells support the theory that there is an autoimmune involvement in the etiology of the disease.
- The pathogenic role of autoantibodies in the disease remains obscure, although identifying their target antigens could provide information for the development of diagnostic tests for vitiligo and could indicate important T-cell responses in patients.

E. H. Kemp (✉)
School of Medicine and Biomedical Sciences,
University of Sheffield, Sheffield,
UK
e-mail: e.h.kemp@sheffield.ac.uk

Melanocyte Antibodies in Vitiligo

Correlations of Melanocyte Antibodies with Vitiligo

Antibodies to pigment cells have been demonstrated in the sera of vitiligo patients in several studies using a variety of research techniques, including immunoprecipitation, immunoblotting, and immunofluoresence (Table 2.2.7.2) [13, 15, 17, 24, 25, 46, 47, 58]. Data from these reports also suggest that melanocyte antibodies can be detected in sera from healthy subjects, but their prevalence is significantly lower (Table 2.2.7.2) [13, 15, 17, 24, 25, 46, 47, 58]. As well as circulating antibodies, antibody deposits have been noted in the basement membrane zones of depigmented areas in individuals with the disease [64]. Melanocyte antibodies have been largely characterized as IgG [13, 15, 17, 24, 25, 46, 47, 58, 64] and belonging to subclasses IgG1, IgG2, and IgG3 [68], although reports have also suggested that vitiligo is closely correlated with levels of IgA melanocyte antibodies [2].

Correlations have been described between the incidence and level of melanocyte antibodies and disease activity in vitiligo: 8/10 patients with active vitiligo, 0/14 with inactive disease, and 0/19 controls were found to have circulating pigment cell antibodies [26]. In addition, the presence of melanocyte antibodies is related to the extent of disease being detected in 50% of patients with minimal vitiligo (<2% of skin area involved) compared with 93% of patients with greater depigmentation (5–10% of skin area involved) [48]. Furthermore, immunofluorescence has indicated that the binding of vitiligo patient antibody to cultured melanocytes increases with both disease extent and activity [71]. The level and activity of melanocyte antibodies also appears to decrease in vitiligo patients, who have repigmentation following treatment with immunosuppressive corticosteroids [22, 57].

Although the described observations provide evidence of the presence of melanocyte antibodies in individuals with vitiligo and correlations with disease status, they do not define the role of the humoral immune response in the development of depigmentation. This question has been addressed, in part, by research aimed at identifying the targets of vitiligo-associated melanocyte antibodies, and this is discussed in the following section.

Table 2.2.7.2 Melanocyte antibodies in vitilgo

Number of patients with melanocyte antibodies (%)	Number of controls with melanocyte antibodies (%)	System of detection	References[a]
17/19 (89)	4/20 (20)	Immunoblotting of melanoma cell extracts	Rocha et al. [58]
44/48 (92)	14/35 (40)	Immunoblotting of melanoma cell extracts	Hann et al. [24]
17/55 (30.9)	0/60 (0)	Indirect immunofluoresence on melanoma cells	Farrokhi et al. [17]
24/29 (83)	2/28 (7)	Immunoprecipitation of melanoma cell extracts	Cui et al. [15]
18/23 (78)	3/22 (14)	Immunoprecipitation of melanoma cell extracts	Cui et al. [13]
50/61 (82)	Not tested	Immunoprecipitation of melanocyte extracts	Naughton et al. [46]
12/12 (100)	0/12 (0)	Immunoprecipitation of melanocyte extracts	Naughton et al. [47]
26/28 (93)	16/26 (62)	Immunoblotting of melanocyte extracts	Hann et al. [25]

[a]All studies show a p value of <0.0001 comparing patients and controls except Hann et al. [25] where $p = 0.0082$, and Naughton et al. [46] where the p value was not calculated

Targets of Melanocyte Antibodies in Vitiligo

Initial immunoprecipitation studies using melanoma cell extracts revealed that antibodies in vitiligo patients were most commonly directed against pigment cell antigens with molecular weights of 35, 40–45, 75, 90, and 150 kDa [13]. Several of the proteins (40–45, 75 and 150 kDa) appeared to be common tissue antigens, while others (35 and 90 kDa) were preferentially expressed on pigment cells [13]. In immunoblotting studies with melanocyte extracts, antigens of 45, 65, and 110 kDa have been identified [25, 57], while vitiligo-associated antibodies have been demonstrated to recognize melanoma cell proteins of 68, 70, 88, 90, 110, and 165 kDa [24, 58]. Differences in the detected antigens could be determined by several factors, including the method employed [13, 24, 25, 57, 58], the origin of the cell extract used [13, 24, 25, 57, 58], and the patient cohort under study. Notably, significant variations between the prevalence of antibodies to different pigment cell antigens and the clinical type of vitiligo have been reported [25]. Although the identity of these particular antigens has not been confirmed, several specific autoantigens in vitiligo have been reported, and these are summarized in Table 2.2.7.3.

Tyrosinase, a melanogenic enzyme was first described as an autoantigen recognized by antibodies in 61% of vitiligo patients in immunoblotting experiments [61]. Likewise, in an enzyme-linked immunosorbent assay (ELISA) with mushroom tyrosinase, 39% of vitiligo patients were considered positive for tyrosinase antibodies [4]. In contrast, using a sensitive radiobinding assay, tyrosinase antibodies were detected in only 10.9% of individuals with vitiligo [31], and in a further study, tyrosinase antibodies could not be demonstrated in vitiligo patients [69]. Similar contrasts have been found in relation to antibodies against tyrosinase-related protein (TRP)-2. A frequency of 5.9% has been reported in one cohort of vitiligo patients [32], whereas 67% of individuals with the disease were positive for antibody reactivity to TRP-2 in another study [53]. The differences in detection rates of antibodies to these particular melanocyte proteins may be related to the vitiligo patient group studied or to the type of assay used for the analysis (Table 2.2.7.3). Interestingly, although TRP-1 has been demonstrated as a major autoantigen in vitiligo Smyth chickens [3], antibodies to TRP-1 have only been identified in a minority (5.9%) of patients with the disease [34]. Again, these dissimilar results could be a reflection of the different assays used to measure antibody levels.

The melanosomal matrix protein gp100 (Pmel17) has been reported as a humoral autoantigen in only 5.9% of vitiligo patients [33]. Interestingly, the melanosomal protein MelanA/MART1 appears not to be a target of autoantibodies in vitiligo [66], although it is frequently recognized by cytotoxic T cells in individuals with the disease [52]. Recently, the melanocyte transcription factors SOX9 and SOX10 were characterized as vitiligo-associated autoantigens in patients with autoimmune polyendocrine syndrome type 1 (APS1), although the frequency of SOX10 and SOX9 antibodies was only 1.1 and 3.2%, respectively, in patients with isolated vitiligo [27]. The technique of phage-display of peptides encoded by melanoctye cDNA has been used to identify the melanin-concentrating hormone receptor (MCHR1) as a target of antibodies in vitiligo patients [37]. Of the vitiligo patients

Table 2.2.7.3 Melanocyte antigens in patients with vitiligo

Antigen	Number of patients with antibodies (%)	Number of controls with antibodies (%)	System of detection	Reference
MCHR1	9/55 (16.4)	0/28 (0)	Radiobinding assay with recombinant human MCHR1	Kemp et al. [37]
MelanA/MART1	0/51 (0)	0/20 (0)	Radiobinding assay and immunoblotting with recombinant human MelanA/MART1	Waterman et al. [66]
Pmel17	3/53 (5.9)	0/20 (0)	Radiobinding assay with recombinant human Pmel17	Kemp et al. [33]
SOX10	3/93 (3.2)	0/65 (0)	Radioimmunoassay assay with recombinant human SOX10	Hedstrand et al. [27]
SOX9	1/93 (1.1)	0/65 (0)	Radioimmunoassay assay with recombinant human SOX9	Hedstrand et al. [27]
Tyrosinase	16/26 (61)	0/31 (0)	Immunoblotting with recombinant human tyrosinase	Song et al. [61]
Tyrosinase	7/18 (39)	0/12 (0)	ELISA with mushroom tyrosinase	Baharav et al. [4]
Tyrosinase	5/46 (10.9)	0/20 (0)	Radiobinding assay with recombinant human tyrosinase	Kemp et al. [31]
Tyrosinase	0/54 (0)	0/29 (0)	ELISA with human tyrosinase and immunopreciptation DOPA stain assay	Xie et al. [69]
TRP-1	3/53 (5.9)	0/20 (0)	Radiobinding assay with recombinant human TRP-1	Kemp et al. [34]
TRP-2	3/53 (5.9)	0/20 (0)	Radiobinding assay with recombinant human TRP-2	Kemp et al. [32]
TRP-2	10/15 (67)	0/21 (0)	Immunoblotting with recombinant human TRP-2	Okamoto et al. [53]
TRP-2	20/30 (67)	1/35 (2)	ELISA with recombinant human TRP-2	Okamoto et al. [53]
35 kDa protein	1/23 (4)	0/22 (0)	Immunoprecipitation of melanoma cell extracts	Cui et al. [13]
40–45 kDa protein	17/23 (74)	3/22 (14)	Immunoprecipitation of melanoma cell extracts	Cui et al. [13]
75 kDa protein	13/23 (57)	2/22 (9)	Immunoprecipitation of melanoma cell extracts	Cui et al. [13]
90 kDa protein	8/23 (35)	0/22 (0)	Immunoprecipitation of melanoma cell extracts	Cui et al. [13]
150 kDa protein	1/23 (4)	0/22 (0)	Immunoprecipitation of melanoma cell extracts	Cui et al. [13]
65 kDa protein	8/18 (44)	1/14 (7)	Immunoblotting of melanocyte extracts	Park et al. [57]
68 kDa protein	7/19 (37)	0/20 (0)	Immunoblotting of melanoma cell extracts	Rocha et al. [58]
90 kDa protein	5/19 (26)	0/20 (0)	Immunoblotting of melanoma cell extracts	Rocha et al. [58]
165 kDa protein	2/19 (11)	0/20 (0)	Immunoblotting of melanoma cell extracts	Rocha et al. [58]
70 kDa protein	32/44 (73)	14/35 (40)	Immunoblotting of melanoma cell extracts	Hann et al. [24]
88 kDa protein	26/44 (60)	4/35 (11)	Immunoblotting of melanoma cell extracts	Hann et al. [24]
110 kDa protein	26/44 (60)	4/35 (11)	Immunoblotting of melanoma cell extracts	Hann et al. [24]
45 kDa protein	13/28 (46)	5/26 (19)	Immunoblotting of melanocyte extracts	Hann et al. [25]
65 kDa protein	7/28 (25)	0/26 (0)	Immunoblotting of melanocyte extracts	Hann et al. [25]
110 kDa protein	9/28 (31)	0/26 (0)	Immunoblotting of melanocyte extracts	Hann et al. [25]

included in the study, 9/55 (16.4%) had antibodies to the receptor, making this antigen the most common, yet identified in patients with vitiligo.

Several epitopes recognized by antibodies in vitiligo patients have been reported (Table 2.2.7.4). For tyrosinase, Pmel17 and MCHR1, antibody binding sites have been defined by the use of deletion constructs combined with radiobinding assays [20, 35, 36]. This methodology is useful to detect linear/continuous epitopes, but is not effective for the identification of conformational/discontinuous antigenic domains, which are usually associated with disease activity [44]. Peptides that react with antibodies in vitiligo patient sera have been revealed by phage-display analysis [30]. Such amino acid sequences may mimic vitiligo-related self-antigens and may represent conformational epitopes. However, the antigenic targets which these peptides might represent have yet to be reported.

Table 2.2.7.4 Epitopes targeted by antibodies in vitiligo

Antigen (amino acid residues)	Epitope	Number of patients with reactivity (%)	Reference
Tyrosinase (240–254)	DEAKCDICTDEYMGG	1/5 (20)	Kemp et al. [35]
Tyrosinase (289–294)	CNGTPE	1/5 (20)	Kemp et al. [35]
Tyrosinase (295–300)	GPLRRN	1/5 (20)	Kemp et al. [35]
Tyrosinase (435–447)	NGDFFISSKDLGYD	1/5 (20)	Kemp et al. [35]
Tyrosinase (461–479)	QDYIKSYLEQASRIWSWLL	1/5 (20)	Kemp et al. [35]
Pmel17 (326–341)	QVPTTEVVGTTPGQAP	4/23 (17)	Kemp et al. [36]
Pmel17 (634–644)	VPQLPHSSSHW	1/23 (40)	Kemp et al. [36]
MCHR1 (1–138)	MLCPSKTDGSGHSGRIHQETHGEGKRDKI SNSEGRENGGRGFQMNGGSLEAEHASRMS VLRAKPMSNSQRLLLLSPGSPPRTGSISY INIIMPSVFGTICLLGIIGNSTVIFAVVK KSKLHWCNNVPDIFIINLSVVD	7/9 (78)	Kemp et al. [20]
MCHR1 (139–298)	LLFLLGMPFMIHQLMGNGVWHFGETMCTL ITAMDANSQFTSTYILTAMAIDRYLATVH PISSTKFRKPSVATLVICLLWALSFISIT PVWLYARLIPFPGGAVGCGIRLPNPDTDL YWFTLYQFFLAFALPFVVITAAYV	5/9 (56)	Kemp et al. [20]
Random peptide	SHMPLANQYQWA	33/65 (51)	Jadali et al. [30]
Random peptide	NHVQWEQFWDS	50/65 (77)	Jadali et al. [30]

So far, although a wide variety of antigens have been identified as targets for vitiligo-associated antibodies, none has yet been unequivocally recognized as one of the major autoantigens in the disease. In addition, no antibody with an identified specificity has been isolated from vitiligo lesions.

The Origin of Melanocyte Antibodies in Vitiligo

How pigment cell antibodies arise in vitiligo has not yet been elucidated and several mechanisms could account for their presence. Antibodies might result from a genetic dysregulation of the immune system at the B- or T-cell level resulting in a lack of tolerance to melanocyte antigens and the subsequent appearance of anti-pigment cell autoantibodies [11, 59]. Particularly, this may be the case with respect to individuals with *AIRE* gene mutations who develop vitiligo as part of their APS1 [45].

Alternatively, antigens released from pigment cells disrupted by cellular immune reactions [38, 39, 52, 56] or nonimmune processes [16, 38, 60] could initiate humoral immune responses that target melanocytes. Indeed, antibodies to TRP-2 have been identified in melanoma patients during immunotherapy with this pigment cell-specific antigen [53], indicating that exposure to melanogenic proteins can stimulate B-cell reactivity. Interestingly, TRP-1 has been found to be expressed transiently on the surface of melanocytes, and therefore accessible as an antibody target [63] and other autoantigens, including the MCHR1, are cell-surface receptors so are exposed to possible antibody interaction without the need for cellular damage [13, 17, 21, 37, 57]. Conceivably, a humoral immune response to pigment cells could also occur in vitiligo patients if melanocytes expose antigens that are similar to either an infecting agent or to other cells that are themselves the primary target of antibodies. Indeed, several vitiligo autoantigens appear to be expressed on cells other than melanocytes [13, 37].

In summary, it has not been determined how melanocyte antibodies arise and this process may be patient-specific and/or antigen-dependent. Whether or not melanocytes are the targets of a primary or secondary humoral immune response, a major question remains concerning the contribution pigment cell antibodies make to the development of depigmentation in vitiligo. This is discussed in the following section.

Pathogenic Effects of Melanocyte Antibodies

With respect to pathogenic effects, vitiligo-associated antibodies are able to destroy melanocytes in vitro by complement-mediated damage and antibody-dependent cellular cytotoxicity (ADCC) [51]. Complement-mediated cytolysis of melanocytes by vitiligo patient antibodies appears to be cell selective and more common in individuals with active disease [14]. In addition, antibodies displaying ADCC against the MCHR1 have been detected in vitiligo patients [21] and IgG purified from vitiligo patients can destroy melanoma cells both in vitro and in vivo [18]. Passive immunization of nude mice grafted with human skin has also indicated that IgG from vitiligo patients can induce melanocyte destruction [19]. Furthermore, IgG melanocyte antibodies from individuals with vitiligo can induce human leukocyte antigen (HLA)-DR and intercellular adhesion molecule (ICAM)-1 expression on and interleukin-8 release from melanocytes [70]. Such changes that may enhance the antigen-presenting activity of pigment cells allowing antigen-specific immune effector cell attack resulting in melanocyte destruction.

Antibodies against MCHR1 have been shown to block the function of the receptor in a heterologous cell line [21]. Stimulation of MCHR1 in cultured melanocytes with melanin-concentrating hormone (MCH) can down regulate the actions of α-melanocyte-stimulating hormone, including the production of melanin, suggesting that the MCH/MCHR1 signaling pathway has a role with the melanocortins in regulating melanocyte function [28]. Any adverse effects of MCHR1 antibodies upon the functioning of the receptor in pigment cells could potentially disrupt normal melanocyte behavior, a feature that could precede the clinical manifestation of vitiligo. However, this has not yet been reported and is still the object of study. More recent work has found that 69% (9/13) of vitiligo patient sera tested induced melanocyte detachment in a reconstructed epidermis model, although this was unrelated to either the extent or the activity of the disease [10]. Further studies are needed to confirm that this serum effect is antibody mediated and, if so, that the antibody activity is specific to vitiligo patient sera.

Overall, the exact contribution that melanocyte antibodies have to the development of vitiligo remains unresolved. It is likely that in most cases antibodies to melanocytes are not the primary pathogenic agent of melanocyte destruction in vitiligo, but arise from an immune response to pigment cells damaged by cellular immune reactions [38, 39, 52, 56] or nonimmune mechanisms [16, 38, 60]. The presence of melanocyte antibodies might then further aggravate melanocyte loss if they have cytotoxic reactivity or adversely affect the function of pigment cells [14, 18, 19, 21, 51]. In the case of antibodies against common cellular antigens, the selective destruction of pigment cells in vitiligo might occur because they are intrinsically more sensitive to immune-mediated injury than other skin cells, for example, keratinocytes or fibroblasts [50].

Nevertheless, even if antibodies to pigment cells are not an agent of the disease, the most valuable contribution that further studies of melanocyte antibody reactivity can make is their help in identifying relevant target antigens. These could provide for the development of diagnostic and prognostic tests which are not yet available for vitiligo. Of particular interest would be profiling the antibody responses in individuals with different clinical types of vitiligo, as this might provide useful information in disease classification. Previously, significant variations between the prevalence of antibodies to different pigment cell antigens and vitiligo subtype have been reported [25]. Autoimmunity is also a feature more closely associated with generalized (nonsegmental) vitiligo compared with the segmental form of the disease [62]. Melanocyte antibodies could serve as markers for important T-cell responses to pigment cells [55] in patients with the disease, a further reason to analyze melanocyte antibody reactivities in vitiligo. Moreover, pathogenic melanocyte antibodies may be of use in the therapy of melanoma if melanoma cells express targeted antigens.

Melanocyte Antibodies in Melanoma-Associated Hypopigmentation

Approximately 10% of patients with metastatic malignant melanoma spontaneously develop vitiligo-like depigmentation [9, 49]. In some cases, this phenomenon has been associated with an improved prognosis [9, 49]. Such observations have led to the postulation that the appearance of vitiligo-like hypopigmentation in some melanoma patients results from immune responses to pigment cell antigens that are shared by normal melanocytes and melanoma cells [8]. Indeed, antibody responses to pigment cell antigens have been observed in some melanoma patients [12, 29]. These include reactivity to tyrosinase, TRP-1, TRP-2 and Pmel17 [29], as well as the 40–45, 75 and 90 kDa antigens [12], that are recognized by antibodies in vitiligo patients. Specific studies concerning antibody reactivity in melanoma patients with associated hypopigmentation have documented responses to TRP-2 [53], melanoma cells [41, 43] and tyrosinase [42, 43]. However, the pathogenic effects of these identified antibody responses, if any, have not been exactly determined and it may be that T-cell immunity is responsible for pigment cell destruction in melanoma patients.

Immunological approaches to melanoma treatment have also resulted in the development of vitiligo-like lesions in some patients. For example, melanoma patients immunized with a polyvalent melanoma cell vaccine have been shown to develop both depigmented lesions and an antibody response to TRP-2 [53], with patients having a favorable survival outcome. Again antibody reactivity has not been shown to be responsible for the loss of pigment cells, and T cells may be the primary pathogenic agent. Such studies do suggest that immune responses to common antigens present on melanoma cells and normal melanocytes can occur after active, specific immunotherapy and can result in depigmentation and tumor rejection.

The immune responses that cause depigmentation in melanoma may have some bearing on understanding the pathogenesis of vitiligo. Although melanoma-associated hypopigmentation and autoimmune vitiligo have a similar clinical appearance, and both can be associated with abnormal immune reactivities to pigment cell antigens, it has not yet been established if they have an identical etiology [41]. As some investigators have shown, different immune mechanisms can be responsible for depigmentation in melanoma hypopigmentation depending on the target antigen [6, 54].

Other Antibodies in Vitiligo

Circulating organ-specific autoantibodies, particularly to the thyroid, adrenal glands, and gastric parietal cells, are commonly detected in the sera of vitiligo patients (Table 2.2.7.5) [5, 7, 40, 72]. Moreover, anti-nuclear antibody and IgM-rheumatoid factor have been detected at a significant frequency in vitiligo patients (Table 2.2.7.5) [17]. Anti-keratinocyte intracellular antibodies that correlate with disease extent and activity have also been detected in vitiligo patients [58]. High antibody titers against the benzene ring structure have been demonstrated in vitiligo patients [67], suggesting that exposure to such compounds might play a part in the induction of aberrant immunological responses in individuals with chemically-induced vitiligo.

Interaction of Humoral and Cellular Immune Responses in Vitiligo

There is evidence that in some cases both cellular and humoral immunity can act together in vitiligo pathogenesis. Firstly, in patients with recent onset vitiligo, both T and B cells were found to be simultaneously increased [23]. Secondly, a concurrent elevation in leukocyte inhibition factor released by activated T lymphocytes and circulating IgG was seen in vitiligo patients, prompting the suggestion that a T cell-mediated B cell-activation had occurred [65]. Finally, both

Table 2.2.7.5 Other antibodies detected in vitiligo patients

Antibody reactivity	Number of patients with antibodies (%)	Reference
Gastric parietal cells	11/65 (17)	Zauli et al. [72]
Gastric parietal cells	6/20 (30)	Mandry et al. [40]
Gastric parietal cells	17/80 (21)	Brostoff et al. [7]
Gastric parietal cells	13/96 (13.7)	Betterle et al. [5]
Thyroid cytoplasm	22/80 (28)	Brostoff et al. [7]
Thyroid peroxidase	10/20 (50)	Mandry et al. [40]
Thyroid peroxidase	19/96 (20)	Betterle et al. [5]
Thyroglobulin	8/20 (40)	Mandry et al. [40]
Thyroglobulin	7/80 (9)	Brostoff et al. [7]
Adrenal gland	3/80 (4)	Brostoff et al. [7]
Pancreatic islet cells	7/96 (7.2)	Betterle et al. [5]
Anti-nuclear antibody	4/55 (7.3)	Farrokhi et al. [17]
IgM-rheumatoid factor	6/55(10.8)	Farrokhi et al. [17]

autoantibodies and autoreactive T lymphocytes were seen to reduce the number of pigment cells in in vitro experiments in a dose-dependent manner [1].

Concluding Remarks

Autoantibodies to pigment cell antigens are common in patients with vitiligo and several targets have been reported, but the exact involvement of antibodies in the development of the disease has not been determined. Identifying relevant antibody targets in vitiligo could provide tools for the development of diagnostic tests. Antigens recognized by vitiligo antibodies could serve as markers for important T-cell responses in patients with the disease.

References

1. Abdel Naser MB, Kruger-Krasagakes S, Krasagakis K et al (1994) Further evidence for involvement of both cell mediated and humoral immunity in generalised vitiligo. Pigment Cell Res 7:1–8
2. Aronson PJ, Hashimoto K (1987) Association of IgA anti-melanoma antibodies in the sera of vitiligo patients with active disease. J Invest Dermatol 88:475
3. Austin LM, Boissy RE (1995) Mammalian tyrosinase-related protein-1 is recognised by autoantibodies from vitiliginous Smyth chickens. Am J Pathol 146:1529–1541
4. Baharav E, Merimski O, Shoenfeld Y et al (1996) Tyrosinase as an autoantigen in patients with vitiligo. Clin Exp Immunol 105:84–88
5. Betterle C, Del Prete GF, Peserico A et al (1976) Autoantibodies in vitiligo. Arch Dermatol 112:1328
6. Bowne WB, Srinivasan R, Wolchok JD et al (1999) Coupling and uncoupling of tumor immunity and autoimmunity. J Exp Med 190:1717–1722
7. Brostoff J, Bor S, Feiwel M (1969) Autoantibodies in patients with vitiligo. Lancet 2:177–178
8. Bystryn J-C (1987) Immune mechanisms in vitiligo. Clin Dermatol 15:853–861
9. Bystryn J-C, Rigel D, Friedman RJ et al (1987) The prognostic significance of vitiligo in patients with melanoma. Arch Dermatol 123:1053–1055
10. Cario-Andre M, Pain C, Gauthier Y et al (2007) The melanocytorrhagic hypothesis of vitiligo tested on pigmented, stressed, reconstructed epidermis. Pigment Cell Res 20:385–393
11. Cooke A, Fehervari Z (2007) Central and peripheral tolerance. In: Wiersinga WM, Drexhage HA, Weetman AP, Butz S (eds) The thyroid and autoimmunity. Georg Thieme, Stuttgart, pp 1–11
12. Cui J, Bystryn J-C (1995) Melanoma and vitiligo are associated with antibody responses to similar antigens on pigment cells. Arch Dermatol 131:314–318
13. Cui J, Harning R, Henn M, Bystryn J-C (1992) Identification of pigment cell antigens defined by vitiligo antibodies. J Invest Dermatol 98:162–165
14. Cui J, Arita, Y, Bystryn JC (1993) Cytolytic antibodies to melanocytes in vitiligo. J Invest Dermatol 100:812–815
15. Cui J, Arita Y, Bystryn J-C (1995) Characterisation of vitiligo antigens. Pigment Cell Res 8:53–59
16. Dell'Anna ML, Picardo M (2006) A review and a new hypothesis for non-immunological pathogenetic mechanisms in vitiligo. Pigment Cell Res 19:406–411
17. Farrokhi S, Farsangi-Hojjat M, Noohpisheh MK et al (2005) Assessment of the immune system in 55 Iranian patients with vitiligo. J Eur Acad Dermatol Venereol 19:706–711
18. Fishman P, Azizi E, Shoenfeld Y et al (1993) Vitiligo autoantibodies are effective against melanoma. Cancer 72: 2365–2369
19. Gilhar A, Zelickson B, Ulman Y et al (1995) In vivo destruction of melanocytes by the IgG fraction of serum from patients with vitiligo. J Invest Dermatol 105:683–686
20. Gottumukkala RVSRK, Waterman EA, Herd LM et al (2003) Autoantibodies in vitiligo patients recognise multiple domains of the melanin-concentrating hormone receptor. J Invest Dermatol 121:765–770
21. Gottumukkala RVSRK, Gavalas NG, Akhtar S et al (2006) Function blocking autoantibodies to the melanin-concentrating hormone receptor in vitiligo patients. Lab Invest 86: 781–789
22. Hann SK, Kim HI, Im S et al (1993) The change of melanocyte cytotoxicity after systemic steroid treatment in vitiligo patients. J Dermatol Sci 6:201–205
23. Hann SK, Park YK, Chung KY et al (1993) Peripheral lymphocyte imbalance in Koreans with active vitiligo. Int J Dermatol 32:286–289
24. Hann SK, Koo SW, Kim JB et al (1996) Detection of antibodies to human melanoma cells in vitiligo and alopecia areata by Western blot analysis. J Dermatol 23:100–103
25. Hann SK, Shin HK, Park SH et al (1996) Detection of antibodies to melanocytes in vitiligo by western blotting. Yonsei Med J 37:365–370
26. Harning R, Cui J, Bystryn J-C (1991) Relation between the incidence and level of pigment cell antibodies and disease activity in vitiligo. J Invest Dermatol 97:1078–1080
27. Hedstrand H, Ekwall O, Olsson MJ et al (2001) The transcription factors SOX9 and SOX10 are melanocyte autoantigens related to vitiligo in autoimmune polyendocrine syndrome type 1. J Biol Chem 276:35390–35395
28. Hoogduijn MJ, Ancans J, Suzuki I et al (2002) Melanin-concentrating hormone and its receptor are expressed and functional in human skin. Biochem Biophys Res Commun 296:698–701
29. Huang SKS, Okamoto T, Morton DL et al (1998) Antibody responses to melanoma/melanocyte autoantigens in melanoma patients. J Invest Dermatol 111:662–667
30. Jadali Z, Eslami B, Sanati MH et al (2005) Identification of peptides specific for antibodies in vitiligo using a phage library. Clin Exp Dermatol 30:694–701
31. Kemp EH, Gawkrodger DJ, MacNeil S et al (1997) Detection of tyrosinase autoantibodies in vitiligo patients using

^{35}S-labelled recombinant human tyrosinase in a radioimmunoassay. J Invest Dermatol 109:69–73
32. Kemp EH, Gawkrodger DJ, Watson PF et al (1997) Immunoprecipitation of melanogenic enzyme autoantigens with vitiligo sera: evidence for cross-reactive autoantibodies to tyrosinase and tyrosinase-related protein-2 (TRP-2). Clin Exp Immunol 109:495–500
33. Kemp EH, Gawkrodger DJ, Watson PF et al (1998) Autoantibodies to human melanocyte-specific protein Pmel17 in the sera of vitiligo patients: a sensitive and quantitative radioimmunoassay (RIA). Clin Exp Immunol 114:333–338
34. Kemp EH, Waterman EA, Gawkrodger DJ et al (1998) Autoantibodies to tyrosinase-related protein-1 detected in the sera of vitiligo patients using a quantitative radiobinding assay. Br J Dermatol 139:798–805
35. Kemp EH, Waterman EA, Gawkrodger DJ et al (1999) Identification of epitopes on tyrosinase which are recognised by autoantibodies from patients with vitiligo. J Invest Dermatol 113:267–271
36. Kemp EH, Waterman EA, Gawkrodger DJ et al (2001) Molecular mapping of epitopes on melanocyte-specific protein Pmel17 which are recognised by autoantibodies in patients with vitiligo. Clin Exp Immunol 124:509–515
37. Kemp EH, Waterman EA, Hawes BE et al (2002) The melanin-concentrating hormone receptor 1, a novel target of autoantibody responses in vitiligo. J Clin Invest 109:923–930
38. Kroll TM, Bommiasamy H, Boissy RE et al (2005) 4-Tertiary butyl phenol exposure sensitizes human melanocytes to dendritic cell-mediated killing: relevance to vitiligo. J Invest Dermatol 124:798–806
39. Lang KS, Caroli CC, Muhm D et al (2001) HLA-A2 restricted, melanocyte-specific CD8+ T lymphocytes detected in vitiligo patients are related to disease activity and are predominantly directed against MelanA/MART1. J Invest Dermatol 116:891–897
40. Mandry RC, Ortiz LJ, Lugo-Somolinos A et al (1996) Organ-specific autoantibodies in vitiligo patients and their relatives. Int J Dermatol 35:18–21
41. Merimsky O, Shoenfeld Y, Yecheskel G et al (1994) Vitiligo- and melanoma-associated hypopigmentation: a similar appearance but a different mechanism. Cancer Immunol Immunother 38:411–416
42. Merimsky O, Baharav E, Shoenfeld Y et al (1996) Anti-tyrosinase antibodies in malignant melanoma. Cancer Immunol Immunother 42:297–302
43. Merimsky O, Shoenfeld Y, Baharav E et al (1996) Melanoma-associated hypopigmentation: where are the antibodies? Am J Clin Oncol 19:613–618
44. Morgenthaler NG, Hodak K, Seissler J et al (1999) Direct binding of thyrotropin receptor autoantibody to in vitro translated thyrotropin receptor: a comparison to radioreceptor assay and thyroid stimulating bioassay. Thyroid 9: 467–475
45. Nagamine K, Peterson P, Scott HS et al (1997) Positional cloning of the APECED gene. Nat Genet 17:393–398
46. Naughton GK, Eisinger M, Bystryn J-C (1983) Antibodies to normal human melanocytes in vitiligo. J Exp Med 158: 246–251
47. Naughton GK, Eisenger M, Bystryn J-C (1983) Detection of antibodies to melanocytes in vitiligo by specific immunoprecipitation. J Invest Dermatol 81:540–542
48. Naughton GK, Reggiardo MD, Bystryn J-C (1986) Correlation between vitiligo antibodies and extent of depigmentation in vitiligo. J Am Acad Dermatol 15:978–981
49. Nordlund JJ, Kirkwood JM, Forget BM et al (1983) Vitiligo in patients with metastatic melanoma: a good prognostic sign. J Am Acad Dermatol 9:689–696
50. Norris DA, Capin L, Muglia, JJ et al (1988) Enhanced susceptibility of melanocytes to different immunologic effector mechanisms in vitro: potential mechanisms for post-inflammatory hypopigmentation and vitiligo. Pigment Cell Res 1(Suppl):113–123
51. Norris DA, Kissinger RM, Naughton GK et al (1988) Evidence for immunologic mechanisms in human vitiligo: patients' sera induce damage to human melanocyes in vitro by complement-mediated damage and antibody-dependent cellular cytotoxicity. J Invest Dermatol 90:783–789
52. Ogg GS, Dunbar PR, Romero P et al (1998) High frequency of skin-homing melanocyte-specific cytotoxic T lymphocytes in autoimmune vitiligo. J Exp Med 6:1203–1208
53. Okamoto T, Irie RF, Fujii S et al (1998) Anti-tyrosinase-related protein-2 immune response in vitiligo and melanoma patients receiving active-specific immunotherapy. J Invest Dermatol 111:1034–1039
54. Overwijk WW, Lee DS, Surman DR et al (1999) Vaccination with recombinant vaccinia virus encoding a "self" antigen induces autoimmune vitiligo and tumor cell destruction in mice: requirement for CD4^{+} lymphocytes. Proc Natl Acad Sci USA 96:2982–2987
55. Oyarbide-Valencia K, van den Boorn JG, Denman CJ et al (2006) Therapeutic implications of autoimmune vitiligo cells. Autoimmunity Rev 5:486–492
56. Palermo B, Campanelli R, Garbelli S et al (2001) Specific cytotoxic T lymphocyte responses against MelanA/MART1, tyrosinase and gp100 in vitiligo by the use of major histocompatibility complex/peptide tetramers: the role of cellular immunity in the etiopathogenesis of vitiligo J Invest Dermatol 117:326–332
57. Park YK, Kim NS, Hann SK et al (1996) Identification of autoantibody to melanocytes and characterisation of vitiligo antigen in vitiligo patients. J Dermatol Sci 11:111–120
58. Rocha IM, Oliveira LJ, De Castro LC et al (2002) Recognition of melanoma cell antigens with antibodies present from patients with vitiligo. Int J Dermatol 39:840–843
59. Secarz E, Rjaa-Gabaglia C (2007) Etiology of autoimmmune disease: how T cells escape self-tolerance. Methods Mol Biol 380:271–284
60. Schallreuter KU, Chavan B, Rokos H et al (2005) Decreased phenylalanine uptake and turnover in patients with vitiligo. Mol Genet Metabol 86:S27–S33
61. Song YH, Connor E, Li Y et al (1994) The role of tyrosinase in autoimmune vitiligo. Lancet 344:1049–1052
62. Taieb A (2000) Intrinsic and extrinsic pathomechanisms in vitiligo. Pigment Cell Res 13(Suppl 8):41–47
63. Takechi Y, Hara I, Naftzger C et al (1996) A melanosomal membrane protein is a cell surface target for melanoma therapy. Clin Cancer Res 2:1837–1842
64. Uda H, Takei M, Mishima Y et al (1984) Immunopathology of vitiligo vulgaris, Sutton's leukoderma and melanoma-associated vitiligo in relation to steroid effects. II. The IgG and C3 deposits in the skin. J Cut Pathol 11:114–124

65. Uz-Zaman T, Begum S, Waheed MA (1992) In vitro assessment of T lymphocyte functioning in vitiligo. Acta Derm Venereol 72:266–267
66. Waterman EA, Kemp EH, Gawkrodger DJ et al (2002) Autoantibodies in vitiligo patients are not directed to the melanocyte differentiation antigen MelanA/MART1. Clin Exp Immunol 129:527–523
67. Wojdani A, Grimes PE, Loeb LJ et al (1992) Detection of antibenzene ring antibodies in patients with vitiligo. J Invest Dermatol 98:644
68. Xie P, Geohegan WD, Jordan RE (1991) Vitiligo autoantibodies. Studies of subclass distribution and complement activation. J Invest Dermatol 96:627
69. Xie Z, Chen DL, Jiao D et al (1999) Vitiligo antibodies are not directed to tyrosinase. Arch Dermatol 135:417–422
70. Yi YL, Yu CH, Yu HS (2000) IgG anti-melanocyte antibodies purified from patients with active vitiligo induce HLA-DR and intercellular adhesion molecule-1 expression and an increase in interleukin-8 release by melanocytes. J Invest Dermatol 115:969–973
71. Yu HS, Kao CH, Yu CL (1993) Coexistence and relationship of antikeratinocyte and antimelanocyte antibodies in patients with non-segmental-type vitiligo. J Invest Dermatol 100:823–828
72. Zauli D, Tosti A, Biasco G et al (1986) Prevalence of autoimmune atrophic gastritis in vitiligo. Digestion 34:169–172

2.2.7.4 Cell-Mediated Immunity

I. Caroline Le Poole and David A. Norris

Contents

Core Messages

- Progressive depigmentation in vitiligo is associated with consistent observations of immune infiltrates in a narrow lesional margin of actively depigmenting skin, resolving when depigmentation comes to a halt.
- Melanocyte destruction in vitiligo is primarily a cell-mediated process controlled by antigen-specific T-lymphocytes, which mobilize other immunological effector mechanisms as well.
- T-cells infiltrating vitiligo skin are reactive, at least in part, with the same antigens recognized by T-cells infiltrating melanoma tumors.
- Innate immune responses are likely important in the initiation phase of the autoimmune response in vitiligo.

I. C. Le Poole (✉)
Loyola University Chicago, Maywood, Illinois
e-mail: ilepool@lumc.edu

Immune Infiltrates in Vitiligo Skin

Immune infiltrates in the skin of vitiligo patients with active disease were initially considered a rare phenomenon observed only in a subset of approximately 0.5% of patients with overt inflammatory vitiligo [1]. Among this patient group, investigators reported the presence of inflammatory infiltrates well before the development and availability of antibodies [2–6]. Immunostaining of cryopreserved skin sections has facilitated the characterization of inflammatory cells found in the skin of both inflammatory and generalized vitiligo patients [1, 7]. A reappraisal of inflammatory infiltrates in inflammatory and generalized vitiligo skin has since confirmed that both CD4+ and CD8+ T-cells are particularly abundant among the frontier of the progressing lesion, directing attention primarily toward the normally pigmented skin.

Infiltrates in vitiligo skin were also found to be replete with macrophages and these macrophages can be heavily pigmented, suggesting that damaged and dying melanocytes are actively cleared from the skin [8]. This is accompanied by an increased abundance of circulating CD36+ monocytes in vitiligo [9]. As monocytes can give rise to CD68+ macrophages as well as to dendritic cells, a rich source of cell populations is available that is capable of presenting specific antigens to T cells in draining lymph nodes [10]. Moreover, macrophages may engage in antibody-dependent cellular cytotoxicity [11], another potential mechanism of melanocyte death in vitiligo.

Recent findings support an abundance of CD11c+ dendritic cells in close proximity to remaining melanocytes in perilesional skin [12]. Antigen presenting cells capable of processing and presenting melanocyte-specific antigens to T-cells in skin-draining lymph nodes are especially meaningful, as melanocytes found at the margin of depigmenting lesions have repeatedly been found to express MHC Class II molecules, suggesting an active role for the target cells in antigen presentation and participating in their own demise [13]. Such findings are congruent with MHC Class II expression among target cells in autoimmune conditions.

B-cells were not detected among infiltrating cells, in contrast to their presence reported among inflammatory infiltrates in lesional psoriatic skin [14]. The latter autoimmune/inflammatory disease otherwise displays some striking similarities with vitiligo, yet the target cells are different: keratinocytes versus melanocytes. The absence of B cells on-site in vitiligo perilesional skin by no means negates the importance of a humoral component to autoimmune reactivity in vitiligo, as described in Sect. 2.2.7.3.

T Cells and Depigmentation

The role of T-cells in vitiligo pathogenesis is also supported by reports of disease associations with HLA in different populations [15–17]. Among HLA associations, most attention has been focused on MHC Class II molecules [18–21], yet associations with MHC Class I have also been reported [22–24]. Associations with MHC Class I molecules specifically suggest a role of cytotoxic T-cells in disease pathogenesis.

Given the narrow margin of disease activity in actively depigmenting skin of most patients, and the fact that infiltrates dissipate in patients with stable disease, it is not surprising that consistent involvement of T-cells in depigmentation was long overlooked. After reports of immune infiltrates in inflammatory vitiligo skin, immune infiltrates were found in generalized vitiligo as well, and infiltrating T-cells are consistently observed in skin biopsies spanning the border of actively depigmenting skin of both inflammatory and generalized vitiligo skin [1, 7]. Whether depigmenting skin in segmental vitiligo displays similar immune infiltrates, remains to be investigated.

In both inflammatory and generalized vitiligo, the ratio of CD4 to CD8 T-cells decreases compared to unaffected control skin, supporting the involvement of a cytotoxic response to melanocytes. Indeed a prevalence of CD8+ T-cells is suggestive of a Type 1 cytokine environment. Evidence for Type 1 cytokine patterns in vitiligo was initially derived from the Smyth line chicken, an exemplary spontaneous model of human autoimmune vitiligo [25] (see Chap. 2.2.4). In humans, indirect evidence comes from upregulated expression of genes downstream from IFN-γ, including MHC Class II expression, expression of ICAM-1 and expression of ganglioside GD3 in perilesional skin biopsies [13]. Elevated expression of the triad of primary cytokines TNF-α, IL-1 and IL-6 has been associated with vitiligo, indicating that proinflammatory conditions

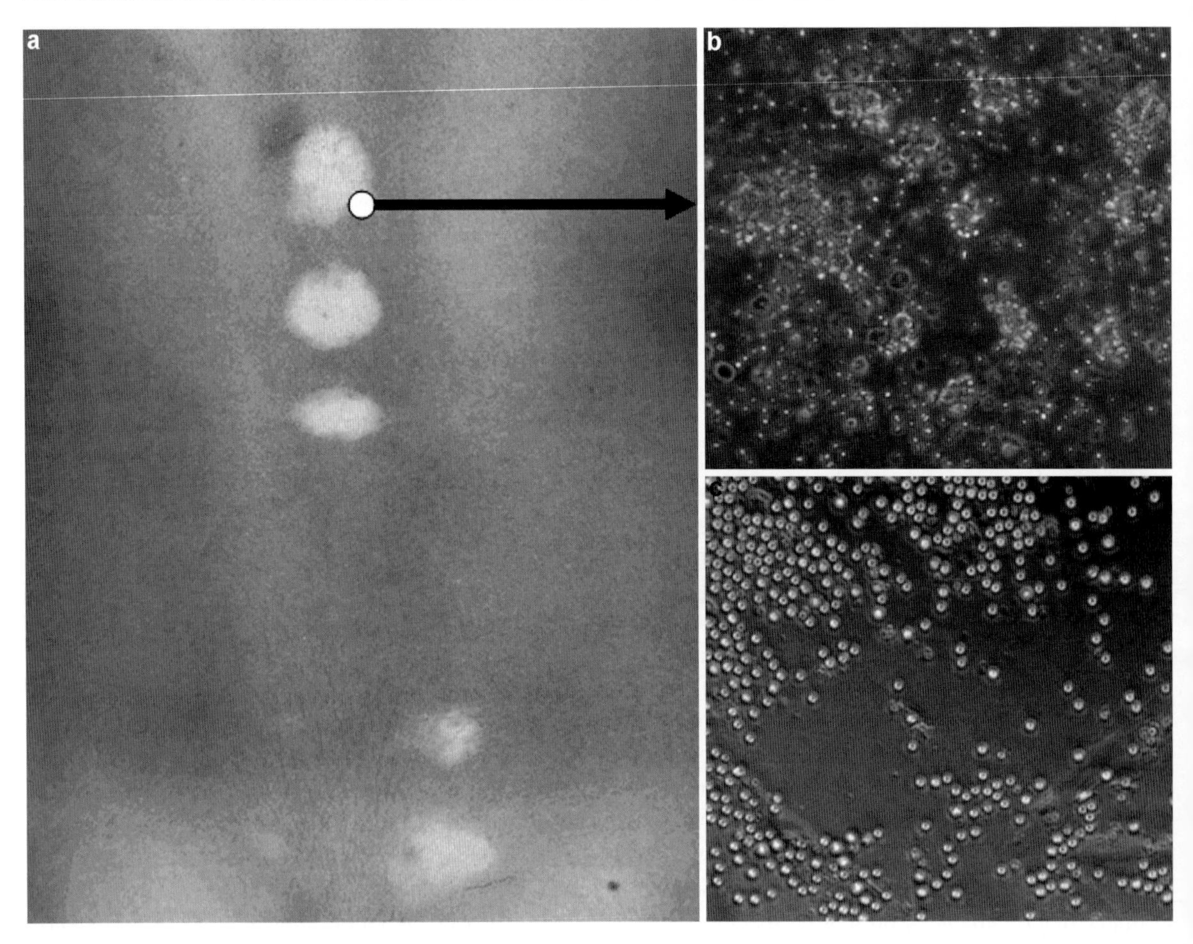

Fig. 2.2.7.1 T cells derived from depigmenting vitiligo skin react with melanocytes. (**a**) Potential biopsy site to isolate T cells emigrating from a vitiligo skin biopsy to be cultured in presence of IL-2 and CD3/CD28 coated beads and recombined with HLA-matched (**b**) melanocytes and (**c**) fibroblasts in vitro. Note clustering and apparent activation of T cells in presence of melanocytes but not fibroblasts

are met in depigmenting skin. Interestingly, GD3 has been shown to inhibit IL-1 expression by macrophages, suggesting that macrophages are not a primary source of IL-1 in vitiligo skin [26]. Besides infiltrating inflammatory cells, epidermal cells including melanocytes and keratinocytes (see also the following section), are important contributors to the cytokine milieu as shown by cytokine synthesis and secretion in vitro [27–32]. Ongoing inflammation in patients with active disease is further evidenced by elevated cytokine levels in the circulation [33].

Also supporting a cytotoxic T-cell mediated response to melanocytes is the fact that CD8+ T-cells are frequently found juxtaposed to remaining melanocytes in the epidermis [33]. The relevance of cellular infiltrates to progressive depigmentation was further supported by the observation that successful repigmentation of lesional skin by mini-grafting was associated with reduced expression of markers of cell-mediated immunity, particularly CD8 and LFA-1 expression, in removed skin specimens [34].

Whereas histological findings strongly suggest a causative role for T-cells in depigmentation, true evidence awaited the isolation and characterization of vitiligo derived T-cells in vitro (Fig 2.2.7.1). Initially such evidence was provided by analysis of circulating T- cells drawn from peripheral blood. Early tetramer analysis revealed that melanocyte differentiation antigen MART-1, which triggers T-cells that infiltrate melanoma tumors, was likewise being recognized by vitiligo patient-derived T-cells [35]. MART-1 reactive T cells were found to be increasingly abundant in patients with active disease [36]. Among T cells reactive with a single antigen, the possibility exists that different T-cell receptors are responsible for antigen recognition. It is also possible that clonal T-cell undergo rapid expansion in response to their cognitive antigen, in which case one would expect preponderance of a particular T-cell clone [37]. In reality, precise analysis of T-cells cloned from peripheral blood has since revealed the prevalence of a limited set of T-cell receptor genes, with particular preference for a single α-subunit among T-cell clones simultaneously expressing a variety of TCR-β subunits. These data suggest that the α-subunit is the primary determinant of antigen specificity, whereas the -β subunit likely fine tunes the TCR affinity of CD8+ T cells for MART-1 [37].

Unraveling a potential role for cytotoxic T cells in vitiligo next involved in vitro cultures of skin-infiltrating T cells isolated from perilesional biopsies of patients with active disease [38]. Bulk cultured T-cell retained similar CD4/CD8 ratios over prolonged culture periods, indicating their true representation of T-cell reactivity in vivo. These bulk cultures displayed a tendency towards Type 1 cytokine profiles, and pooled clones derived from these cultures displayed cytotoxicity towards autologous melanocytes [39]. More recently, perilesional vitiligo T-cells were shown to cause apoptosis of melanocytes in nonlesional skin explants [40]. These experiments provided direct evidence for a major etiologic role of cytotoxic T cells in progressive vitiligo, as suggested earlier by analysis of T cells from depigmenting skin of melanoma patients.

Besides reactivity to MART-1, a melanosomal antigen first identified as a target molecule for T-cells, reactivity of skin-derived T-cells to gp100 has since been described [41]. Moreover, tyrosinase-reactive T-cells were described among circulating T-cells in vitiligo patients [42, 43]. It thus appears that melanocyte differentiation antigens are immunogenic targets in both melanoma and vitiligo. As vitiligo-derived T- cells appear to be recognize their targets with increased avidity compared to melanoma-derived T-cells, melanoma treatment may be derived from TCR that are cloned from T- cells infiltrating vitiligo skin [35, 41, 44]. Likewise, immune escape mechanisms observed for melanoma tumors may offer clues for future treatment of vitiligo patients beyond the use of immunosuppressive ointments or PUVA as treatments of choice for vitiligo patients [45, 46].

In summary, there is accumulating evidence for a role of cytotoxic T cells in clinically inflammatory and generalized vitiligo, representing the vast majority of vitiligo cases.

Homing of Immune Reactive T-Cells

For any circulating leukocyte to home to the skin, cells in the new host environment must facilitate such migration by secreting chemokines which can be recognized by the chemokine receptors expressed by circulating leukocyte subpopulations. The process of leukocyte entry into the tissue is complex, involving attachment, rolling and extravasations of leukocytes under the control of sequential pairs of receptors and ligands [47, 48]. The CLA is expressed on skin homing lymphocytes, interacting with P- or E-selectin on the endothelial cell surface to facilitate extravasation. During the skin homing process, this interaction is preferentially followed by CCL17 (chemokine) induced CCR4 (chemokine receptor) engagement on T-cells. Firm adhesion is mediated by LFA-1/ICAM-1 interactions [49].

TCR engagement possibly lends antigen specificity to T-cell extravasation [49]. Another intriguing point is that the abundance of T-cells in the skin is in part determined by emigration through the lymphatics, involving CCR7 expression by lymphocytes [49]. Clearly skin homing is further defined by T-cell subset-specific molecular interactions. The majority of skin infiltrating T-cells in vitiligo express CLA, supporting their intended travel to the skin to contribute to depigmentation [7].

Upon arrival into the skin, the expression of (co) stimulatory receptors on effecter and target cells is required for effector responses [50]. For autoimmune disease in general, the CTLA-4/B7 pathway has received particular attention, and CTLA-4 knockout mice transgenic for a melanocyte-reactive T-cell receptor developed significantly accelerated depigmentation [51]. Initial reports suggested that particular CTLA-4 polymorphisms may be associated with vitiligo [52, 53]. More recently, an extensive study on 126 families with vitiligo did not confirm association of the most common polymorphisms with disease [54]. ICAM-1 constitutes another molecule of interest: in interactions with CD4+ helper T-cells, antibodies to ICAM-1 and LFA-1 differentially inhibit T-cell proliferation in separate combinations of effectors and target cells. ICAM-1 expression is a consistent feature in perilesional skin of vitiligo and cultured vitiligo melanocytes [55], rendering melanocytes increasingly vulnerable to T-cell mediated responses, and possibly to NK cells as well [56].

It should be noted that differential expression of cytokines is focal and may be determined primarily by the microenvironment. In this respect, Vitamins A and D were shown to suppress CLA expression by human T-cells [57]. Pathways defining the skin homing pattern and the synapse that defines the interaction between target and effecter cells thus lend themselves to therapeutic modulation [58, 59], offering potential means to treat autoimmune vitiligo.

Vitiligo versus Melanoma T-Cell Responses

An interest in cytotoxic T-cell responses to melanocyte differentiation antigens was first developed in melanoma research. A breakthrough came with the identification of MART-1/MelanA as a target antigen for T-cells, named after its recognition by T-cells and found to be expressed by cells of the melanocyte lineage [60, 61]. Subsequent melanocyte-associated antigens identified by reactivity of melanoma-infiltrating T-cells included tyrosinase, gp100, and TRP-2 [62]. The first direct evidence of cytotoxic T-cell responses causing depigmentation of the skin came from melanoma patients [63].

Interestingly, the majority of T-cells infiltrating melanoma tumors were reactive with MART-1 and gp100 [64], which are the antigens identified for T-cells infiltrating the skin of vitiligo patients. In a more general sense, the melanosome appears to be a particularly immunogenic compartment within melanocytic cells [65, 66]. Possible exceptions to the rule that melanocyte antigens originate in the melanosome appear to the melanocortin receptor-1 (MC1R) and Rab38, although both traffic through the melanosome [67–70] (see also Chap. 2.2.9).

An association between mutations in the MC1R and vitiligo has been proposed [71]. The MC1R receptor for α-MSH and agouti signaling protein mediates cellular responses opposing those of the melanin concentrating hormone receptor MCHR1, binding melanin concentrating hormone and targeted by the humoral response in vitiligo [72, 73] (see Sect. 2.2.7.3). MCHR1 engagement on PBMC can repress CD3-stimulated T-cell proliferation, thus serving an immunomodulatory function in its own right [74]. Melanocyte-restricted expression of MCHR1 within the skin again renders the pigment cell a preferred target of the autoimmune response in vitiligo patients [75]. T-cells reactive with MC1R were found in skin depigmenting secondary to the development of melanoma, as well as in melanoma tumors [69]. Susceptibility to melanoma is associated with mutations in this receptor, associated with red hair and fair skin, but such mutations are systemic in nature and therefore unlikely to be immunogenic per se. If mutations arise within antigenic epitopes it is however possible that the altered sequence will have an altered affinity for MHC molecules, or for reactive TCRs.

It will be of great interest to determine whether additional antigens are targeted by T-cells in vitiligo that have yet to be identified in melanoma, and whether such antigens may offer therapeutic opportunities when included in anti-melanoma vaccines. In the case of melanoma, a portion of the T-cells is reactive with progression antigens not shared with normal melanocytes, and it is possible that melanocytes similarly express immunogenic molecules that are lost during malignant transformation. T-cells from melanoma tumors exhibit reduced reactivity to their targets when compared to T-cells from vitiligo skin [76], prompting the hypothesis that the environment is more permissive for immune responses in vitiligo skin than in melanoma tumors. This could mean that processes such as epitope spreading are more likely to occur in vitiligo, and responses to secondary immunogens may thus be more easily detectable in vitiligo skin. In part, increased avidity among vitiligo T-cells may be explained by T-cell tuning, allowing T-cells with higher than otherwise permissible TCR affinity to exit primary lymphoid organs, and migrate to the periphery before encountering antigen under higher avidity conditions in the periphery and contributing to autoimmune disease [77].

Remarkably, avidity differences between tumor and skin-derived T-cells have been noted even when comparing depigmenting skin and the tumor environment within the same melanoma patient, where tumor-derived T-cells expressed relatively more inhibitory CD94/NKG2 receptors than leukoderma-derived T-cells [78]. Similar differences may contribute to differences in avidity noted when comparing antigenic peptide concentrations that evoke half-maximum IFN-γ secretion among responsive autoimmune vitiligo skin-derived T cells and melanoma-derived T-cells [41].

IFN-γ secreted by activated T cells within the tumor environment can repress expression of target antigens MART-1 and gp100 within the tumor environment, affecting subsequent T-cell recognition of melanoma cells (but not normal melanocytes) [79]. By contrast, the epidermal cytokine environment in vitiligo may be conducive to ongoing immune responses [29, 80].

The potential "superiority" of vitiligo patient-derived over melanoma-derived T-cells supports their therapeutic potential for melanoma therapy [41, 76, 81].

CD4+ T Cells and Vitiligo Progression

The tumor environment in melanoma is characterized by multiple factors that prevent immunologic cytotoxicity; this is not the case in vitiligo skin. Although dissimilarities between melanoma cells and melanocytes are a factor, for example, T-cell derived IFN-γ suppresses target antigen expression in melanoma cells, but not in melanocytes as described above [79]. a large part of the difference can be attributed to the types of CD4 T-cells infiltrating tissue in either situation.

As a helper T-cell population, CD4+ T cells are required to initially activate an immune response [82]. Offering help in the maturation and activation process of T- and B-cells, the cytokine profiles of CD4+ T cells infiltrating the tissue will skew the response toward either humoral or cell-mediated immunity. In the effector phase of the immune response, inflammatory cytokines are increasingly important to maintain effector responses [83]. The cytokine patterns detectable in vitiligo skin suggest that cell-mediated responses are favored, while in melanoma tumors, the cytokine profiles are not uniform and frequently include immunosuppressive cytokines [44, 84].

As in healthy skin, part of the CD4+ population in melanoma consisted of natural regulatory T-cells (Treg) in place to keep responses to self antigens in check [85]. Although definitive evidence awaits verification, it appears that the natural Treg population is virtually absent from vitiligo skin, regardless of the location of biopsies in nonlesional, peri-lesional, or lesional vitiligo skin [86]. In that environment, an autoimmune response directed to melanocyte-derived self-antigens can expand in an unbridled fashion, and even leave some room for affinity maturation of the T-cells population involved. Interestingly, the reported abundance of GD3 expression in vitiligo skin may in part be a compensatory mechanism, given the fact that GD3 can indirectly inactivate T cells [87].

Melanocytes as Antigen-Presenting Cells

The dendritic morphology of melanocytes is shared with the macrophage, dendritic cell, and Langerhans cell populations. The distribution pattern and strategic position of melanocytes within the skin, with dendrite tips that appear to be in touch with one another to form a network, are reminiscent of Langerhans cells that play an essential role in immunosurveillance of the skin [88]. In the absence of melanocytes, contact sensitivity responses are impaired as noted in lesional vitiligo skin [89].

A very essential difference between melanocytes and the other dendritic cell populations of the skin was found in the very limited ability of melanocytes to migrate [90]. This is essential as it will undermine attempts of the melanocyte to repopulate areas of depigmentation. Indeed, when vitiligo progression comes to a halt, there appears to be little incentive to replenish the epidermis in the absence of strong stimuli such as UV, frequently resulting in stable lesions that can almost be considered the "scars" of the depigmentation process.

The functional similarities between melanocytes and other dendritic skin cell populations are remarkable. Both melanocytes and macrophages are phagocytic, and melanocytes can express markers otherwise found on macrophages—a property previously described for astrocytes as well [1, 91, 92]. In melanocytes, endosomes that have captured extracellular contents appear to fuse within the cell with melanosomes and not so much with lysosomes. This may be critical, since endosome—melanosome fusion can trigger processing of the contents of the fusion product, including the very antigens shown to be immunogenic in melanoma and vitiligo.

HLA-DM catalyzes release of the invariant chain to make place for antigenic peptides to be presented in the context of MHC Class II [93]. In fact, it has been shown that melanosomal targeting sequences are essential for trafficking antigens toward the endosomal compartment and subsequent presentation of derivative antigenic peptides in the context of MHC class II [94]. Within the endosomal/lysosomal/melanosomal compartment, this preferential sorting process is further influenced by HLA-DO [95]. Since melanosome resident proteins including MART-1 and gp100 are already on location, these may be presented in the context of MHC Class II in the infrequent event that a cell of melanocyte origin expresses MHC Class II. Such expression has been reported in both vitiligo and melanoma [65]. In the case of melanoma, CD4+ T-cells reactive with melanosomal antigens have been reported to infiltrate the tumor tissue [94]. Thus, at least part of the CD4+ T-cell population infiltrating vitiligo skin is likely to be reactive with melanosomal antigens as well. The other issue that remains is whether melanocytes are capable of presenting antigens in the context of MHC Class II. Evidence for antigen presentation in the context of two different variants of HLA-DR was provided using T-cell clones reactive with mycobacterial hsp65 and HLA-matched melanocyte cultures [96]. Stimulated with IFN-γ, HLA-DR + melanocytes were capable of presenting pulsed peptides as well as processing mycobacterial homogenate for presentation to hsp65-reactive CD4+ T-cells, resulting in both proliferative and cytotoxic responses to the target cells. Whether CD4+ T-cells infiltrating vitiligo skin will likewise exhibit cytotoxic responses toward melanocytes, remains to be determined. Meanwhile, a neoantigen expressed on melanocytes was targeted and contributed to autoimmune depigmentation in TCR transgenic mice, where the majority of CD4+ T-cells carry a class II restricted TCR that respond to the same neoantigen [97].

Dendritic Cell Mediated Killing

Dendritic cells, including the Langerhans cell subset have been implemented in processing and presentation of antigens. Dendritic cells that have phagocytosed antigens of interest will migrate to draining lymph nodes under the influence of cytokines and chemokines. In the epidermis, Langerhans cells are in direct contact with melanocytes and in lesional skin of vitiligo, the Langerhans cells can oddly position themselves along the basement membrane, as if to replace lost melanocytes [98]. Although this phenomenon has not been convincingly explained, it is possible that Langerhans cells need melanocyte-derived cytokines IL-1 and TNF-α to pass the basal membrane and are subsequently trapped in the epidermis of lesional skin [39].

Based on other models of autoimmunity, one must consider models for cell damage as key components for T-cell sensitization and cytotoxic effect in vitiligo. In order to submit antigens for processing and presentation, there must be a danger signal sent from the melanocyte. Either it has undergone necrosis, spilling its contents into the extracellular milieu or it is at least in crisis. Apoptosis is the other cytotoxic endpoint, and might be induced by multiple toxic factors in the epidermis as a precursor to vitiligo and T-cell sensitization. Apoptotic cells are generally considered less immunogenic, as this process neatly cleans up the site without molecular spills [99]. Such cells in crisis can express receptors of the TNF family that can signal incoming dendritic cells to become cytotoxic toward such targets. Initially it was shown that dendritic cells can be cytotoxic toward a wide array of tumor cells [100], distinguishing them from normal tissue cells by virtue of membrane expression of members of the TNF receptor family [101]. This mechanism was later expanded to normal tissue cells in crisis, which has likely implications for vitiligo [12]. The cytotoxic capacity of dendritic cells was recently demonstrated for Langerhans cells as well [102].

Additional Players in Cellular Immunity

Patients experience itching of perilesional skin of expanding lesions (unpublished observation). Although little is known about the presence or absence of IgE in perilesional skin vitiligo skin, or subsequent mast-cell activation and histamine release, it is likely that mast cells contribute to an inflammatory environment and carry, at least in part, responsibility for the observed itch. The RAST tests indicated that a higher percentage of vitiligo patients exhibit elevated circulating IgE over controls [103] and MSH was reported to stimulate the release of histamine by mast cells in vitro [104]. A very limited number of reports address colocalization of atopic dermatitis and vitiligo, which would support a role for mast cells in itchy skin during active depigmentation [105–107]. The ligand for c-kit has been implicated in hypopigmentation [108], however, only a single report addresses the presence of c-kit + mast cells in vitiligo skin, and no special reference was made to actively depigmenting areas [109]. Further investigations are needed to support the potential involvement of mast cell degranulation in depigmenting skin.

Another open question remains, whether the NK and NKT cell subsets play a role in vitiligo pathogenesis. A report claiming increased numbers of circulating CD56+ lymphocytes does suggest that NK cells may be involved [110]. No intrinsic functional differences were found for NK cells from patients versus controls in cytotoxic assays toward melanocytes [75]. It remains possible that an increased number of NK cells—if homing to the skin—will be activated within the skin microenvironment and contribute to melanocyte cell death. For NKT cells, the opposite was reported: a reduced number of NKT cells were found in blood samples from vitiligo patients as compared to controls [111]. As the numbers extravasating to the skin have not been evaluated, the role of both NK and NKT cells in vitiligo remains an open question, yet recent genetic data pertaining to NALP-1 certainly implicate innate immune responses in vitiligo [112]. NALP-1 inflammasomes sense metabolic and microbial stress, leading to conversion of proinflammatory IL-1β into its active form [113], and NALP-1 overexpression is associated with apoptosis [114]. Given the importance of innate immune responses to prime subsequent, more refined specific immunity and vice versa [115–117], such finding provide clues for the genetic basis for aberrant immune responses to self in vitiligo.

Heat Shock Proteins in Immune Activation

Even with genetic predisposition toward autoimmune disease, a trigger is needed to set off the response. This is reflected in the fact, that vitiligo lesions are not constant or present from birth. Instead, it appears that specific conditions must be met to overcome immune tolerance and allow for immune responses to self. The common distribution pattern of vitiligo lesions in body orifices and pressure points of the skin suggested that mechanical stress may be a precipitating factor for vitiligo. This is further supported by frequent reports of skin wounds preceding the development of disease, as well as a significant number of patients reporting a Koebner phenomenon associated with development of new lesions [118]. Other forms of stress include, but are not limited to excessive UV exposure and prolonged contact with bleaching phenols. Emotional trauma frequently precedes disease progression (unpublished observation). In vaccination-induced autoimmune vitiligo in mice, trauma and inflammation were shown to trigger depigmenting effector responses [82, 83, 119].

On a cellular level, stress will suppress protein synthesis, and translation is limited to stress proteins that chaperone intracellular proteins and protect the cell from undergoing apoptosis until the danger has subsided. Where the stress response proves inadequate to maintain cell viability, secondary necrosis may follow, implicated in induction of autoimmune responses [120]. Indeed the mode of cell death affects immunogenicity [99, 121]. Stress proteins themselves, in particular HSP70, have been implicated in immune activation, demonstrating enhanced uptake and processing of antigens by DC (Fig 2.2.7.2). Stress proteins have been implicated in successful antimelanoma responses [122] and are currently in clinical trials as tumor vaccine adjuvants. It was recently shown, that HSP70 accelerated depigmentation in a mouse model of vitiligo [123].

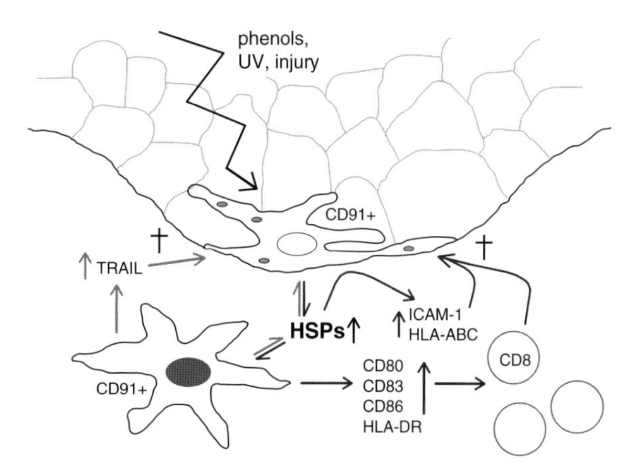

Fig. 2.2.7.2 Stress-induced autoimmune vitiligo. In response to wounding or burning of the skin, exposure to bleaching phenols or excessive sun-exposure, melanocytes under stress express and secrete heat shock proteins (HSPs; in particular: HSP70). HSPs chaperone melanocyte-specific peptides and induce a cytotoxic phenotype by dendritic cells, and enhance antigen uptake, processing and presentation to T cells in draining lymph nodes. Recruited T cells subsequently encounter target cells with elevated expression of MHC class I and ICAM-1, resulting in further cytotoxicity towards melanocytes

HSP70-induced autoimmune responses appear to be mediated by induction of IL-6 [124]. The stress protein can be secreted by live cells in membrane-associated form, creating havoc even in the absence of apparent cell death [125]. Although stress per se is not always preventable, the consequences may be, and such intervention can likely halt disease progression in vitiligo patients with active disease.

Concluding Remarks

In progressive vitiligo, melanocytes are eliminated by a cell-mediated immune response. Alternative mechanisms of melanocyte damage may also contribute to cell death such as autoantibodies, autocytotoxicity, biochemical or oxidant damage, and neural toxicity. Several components of this cytotoxic mechanism, that is, stress-induced DC activation, T-cell recruitment and trafficking, melanocyte physiology and regulatory T-cell responses are potential targets for treatment.

References

1. Le Poole IC, van den Wijngaard RMJGJ, Westerhof W, Das PK (1996) Presence of T cells and macrophages in inflammatory vitiligo parallels melanocyte disappearance. Am J Pathol 148:1219–1228
2. Buckley WR, Lobitz WC Jr (1953) Vitiligo with a raised inflammatory border. AMA Arch Derm Syphilol 67: 316–320
3. Garb J, Wise F (1948) Vitiligo with raised borders. Arch Dermatol Syph 58:149–153
4. Ishii M, Hamada T (1981) Ultrastructural studies of vitiligo with inflammatory raised borders. J Dermatol 8:313–322
5. Michaëlsson G (1968) Vitiligo with raised borders: report of two cases. Acta Dermatol Venereol 48:158–161
6. Wätzig V (1974) Vitiligo with inflammatory marginal dam. Dermatol Monatsschr 160:409–413
7. Van den Wijngaard R, Wankowicz-Kalinska A, Le Poole C et al (2000) Local immune response in skin of generalized vitiligo patients. Destruction of melanocytes associated with the prominent presence of CLA + T cells at the perilesional site. Lab Invest 80:1299–1309
8. Hann SK, Park YK, Lee KG et al (1992) Epidermal changes in active vitiligo. J Dermatol 19:217–222
9. Baumer FE, Frisch W, Milbradt R, Holzmann H et al (1990) Increased expression of the OKM5 antigen in blood monocytes in vitiligo. Z Hautkr 65:917–919
10. Cao T, Ueno H, Glaser C et al (2007) Both Langerhans cells and interstitial DC cross-present melanoma antigens and efficiently activate antigen-specific CTL. Eur J Immunol 37:2657–2667
11. Norris DA, Kissinger RM, Naughton GM, Bystryn JC (1988) Evidence for immunologic mechanisms in human vitiligo: patient's sera induce damage to human melanocytes in vitro by complement mediated damage and antibody-dependent cellular cytotoxicity. J Invest Dermatol 90:783–789
12. Kroll TM, Bommaiasamy H, Boissy RE et al (2005) 4-Tertiary butyl phenol exposure sensitizes human melanocytes to dendritic cell mediated killing: relevance to vitiligo. J Invest Dermatol 124:798–806
13. Le Poole IC, Stennett LS, Bonish BK et al (2003) Expansion of vitiligo lesions is associated with reduced epidermal CDw60 expression and increased expression of HLA-DR in perilesional skin. Br J Dermatol 149:739–748
14. De Boer OJ, van der Loos CM, Hamerlinck F et al (1994) Reappraisal of in situ immunophenotypic analysis of psoriasis skin: interaction of activated HLA-DR + immunocompetent cells and endothelial cells is a major feature of psoriatic lesions. Arch Dermatol Res 286:87–96
15. Arcos-Bugos M, Parodi E, Salgar M et al (2002) Vitiligo: complex segregation and linkage disequilibrium analyses with respect to microsatellite loci spanning the HLA. Hum Genet 110:334–342
16. Taştan HB, Akar A, Orkuynoğlu FE et al (2004) Association of HLA class I antigens and HLA class II alleles with vitiligo in a Turkish population. Pigment Cell Res 17:181–184
17. Xia Q, Zhou WM, Liang YH et al (2006) MHC haplotypic association in Chinese Han patients with vitiligo. J Eur Acad Dermatol Venereol 20:941–946
18. Buc M, Fazekasová H, Cechová E et al (1998) Occurance rates of HLA-DRB1, HLA-DQB1, and HLA-DPB1 alleles in patients suffering from vitiligo. Eur J Dermatol 8:13–15
19. Fain PR, Babu SR, Bennett DC, Spritz RA (2006) HLA class II haplotype DRB1*04-DRB1*0301 contributes to risk of familial generalized vitiligo and early disease onset. Pigment Cell Res 19:51–57
20. Orozco-Topete R, Córdova-López J, Yamamoto-Furusho JK et al (2005) HLA-DRB1*04 is associated with the genetic susceptibility to develop vitiligo in Mexican patients with autoimmune thyroid disease. J Am Acad Dermatol 52: 182–183
21. Zamani M, Spaepen M, Ashgar SS et al (2001) Linkage and association of HLA class II genes with vitiligo in a Dutch population. Br J Dermatol 145:90–94
22. Abanmi A, Al Harti F, Al Baqami R et al (2006) Association of HLA loci alleles and antigens in Saudi patients with vitiligo. Arch Dermatol 298:347–352
23. Liu JB, Li M, Chen H et al (2007) Association of vitiligo with HLA-A2: a meta-analysis. J Eur Acad Dermatol Venereol 21:205–213
24. Wang J, Zhao YM, Wang Y et al (2007) Association of HLA lass I and II alleles with generalized vitiligo in Chines Hans in North China. Shonghua Yi Xue Yi Chuan Xue Za Zhi 24:221–223
25. Wick G, Andersson L, Hala K et al (2006) Avian models with spontaneous autoimmune diseases. Adv Immunol 92:71–117
26. Hoon DS, Jung T, Naungayan J et al (1989) Modulation of human macrophage function by gangliosides. Immunol Lett 20:269–275

27. Krüger-Krasagakes S, Krasagakis K, Garbe C, Diamantstein T (1995) Production of cytokines by human melanoma cells and melanocytes. Recent Results Cancer Res 139: 155–168
28. Li YL, Yu CL, Yu HS (2000) IgG anti-melanocyte antibodies purified from patients with active vitiligo induce HLA-DR and intercellular adhesion molecule-1 expression and an increase in interleukin-8 release by melanocytes. J Invest Dermatol 115:969–973
29. Moretti S, Spallanzani A, Amato L et al (2002) Vitiligo and epidermal microenvironment: posible involvement of keratinocyte-derived cytokines. Arch Dermatol 138:273–274
30. Smit N, Le Poole I, van den Wijngaard R et al (1993) Expression of different immunological markers by cultured human melanocytes. Arch Dermatol Res 285:356–365
31. Swope VB, Abdel-Malek Z, Kassem LM, Nordlund JJ (1991) Interleukins 1 alpha and 6 and tumor necrosis factor-alpha are paracrine inhibitors of human melanocyte proliferation and melanogenesis. J Invest Dermatol 96:180–185
32. Swope VB, Sauder DN, McKenzie RC et al (1994) Synthesis of interleukin-1 alpha and beta by normal human melanocytes. J Invest Dermatol 102:749–753
33. Tu CX, Gu JS, Lin XR (2003) Increased interleukin-6 and granulocyte-macrophage colony stimulating factor levels in the sera of patients with non-segmental vitiligo. J Dermatol Sci 31:73–78
34. Abdallah M, Abdel-Nasr MB, Moussa MH et al (2003) Sequential immunohistochemical study of depigmenting and repigmenting minigrafts in vitiligo. Eur J Dermatol 13:548–552
35. Ogg GS, Dunbar P, Romero P et al (1998) High frequency of skin-homing melanocyte-specific cytotoxic T lymphocytes in autoimmune vitiligo. J Exp Med 188:1203–1208
36. Lang KS, Caroll CC, Muhm A et al (2001) HLA-A2 restricted, melanocyte specific CD8(+) T lymphocytes detected in vitiligo patients are related to disease activity and are predominantly directed against MART-1. J Invest Dermatol 116:891–897
37. Mantovani S, Garbelli S, Palermo B et al (2003) Molecular and functional bases of self-antigen recognition in long-term persisten melanocyte-specific CD8 + T cells in one vitiligo patient. J Invest Dermatol 121:308–314
38. Wańkowicz-Kalińska A, van den Wijngaard RM, Tigges BJ et al (2003) Immunopoloarization of CD4 + and CD8 + T cells in type-1-like is associated with melanocyte loss in human vitiligo. Lab Invest 83:683–695
39. Wang B, Amerio P, Sauder DN (1999) Role of cytokines in epidermal Langerhans cell migration. J Leukoc Biol 66:33–39
40. van den Boorn JG, Konijnenberg D, Dellemijn TAM, van der Veen JPW, Bos JD, Melief CJ, Vyth-Dreese FA, Luiten RM (2007) Cytotoxic perilesional T cells cause in situ melanocyte apoptosis in vitiligo vulgaris skin. Pigment Cell Res 20:328; abstract
41. Oyarbide-Valencia K, van den Boorn JG, Denman CJ et al (2006) Therapeutic implication of autoimmune vitiligo T cells. Autoimmun Rev 5:486–493
42. Mandelcorn-Monson RL, Shear NH, Yau E (2003) Cytotoxic T lymphocyte reactivity to gp100, MelanA/MART-1, and tyrosinase, in HLA-A2-positive vitiligo patients. J Invest Dermatol 121:550–556
43. Palermo B, Campanelli R, Garbelli S et al (2001) Specific cytotoxic T lymphocyte responses against Melan-A/MART1, tyrosinase and gp100 in vitiligo by the use of major histocompatibility complex/peptide tetramers: the role of cellular immunity in the etiopathogenesis of vitiligo. J Invest Dermatol 117:326–332
44. Paglia D, Oran A, Lu C et al (1995) Expression of leukemia inhibitory factor and interleukin-11 by human melanoma cell lines: LIF, IL-6, and IL-11 are not co-regulated. Interferon Cytokine Res 15:455–460
45. Lepe V, Moncada B, Castanedo-Cazares JP (2003) A double-blind randomized tial of 01% tacrolimus vs 0.05% clobetasol for the treatment of childhood vitiligo. Arch Dermatol 139:581–585
46. Wu CS, Lan CC, Wang LF et al (2007) Effects of psoralens plus ultraviolet A irradiation on cultures epidermal cells in vitro and patients with vitiligo in vivo. Br J Dermatol 156: 122–129
47. Luo BH, Carman Cv, Springer TA (2007) Structural basis of integrin regulation and signaling Ann Rev Immunol 25: 619–647
48. Ma Q, Shimaoka M, Lu C et al (2002) Activation-induced conformnational changes in the I domain region of lymphocyte function-asociated antigen. J Biol Chem 277: 10638–10641
49. Mirellli FM, Cannella L, Dazzi F, Mirenda V (2008) The highway code of T cell trafficking. J Pathol 214:179–189
50. Norris DA (1990) Cytokine modulation of adhesion molecules in the regulation of immunologic cytotoxicity of epidermal targets. J Invest Dermatol 95:111S–120S
51. Gattinoni L, Ranganathan A, Surman DR et al (2006) CTLA-4 dysfunction of self/tumor-reactive CD8+ T-cell dependent. Blood 108:3818–3823
52. Itirli G, Pehlivan M, Alper S et al (2005) Exon-3 polymophism of CTLA-4 gene in Turkish patients with vitiligo. J Dermatol Sci 38:225–222
53. Kemp EH, Ajjan RA, Waterman EA et al (1999) Analysis of a microsattelite polymorphism of the cytotoxic T-lymphocyte antigen-4 gene in patients with vitiligo. Br J Dermatol 140:73–78
54. Laberge GS, Bennett DC, Fain PR, Spritz RA (2008) PTPN22 is genetically associated with risk of generalized vitiligo, but CTLA4 is not. J Invest Dermatol 128: 1757–1762
55. Hedley SJ, Metcalfe R, Gawkrodger DJ et al (1998) Vitiligo melanocytes in long-term culture show normal constitutive and cytokine-induced expression of intercellular adhesion molecule-1 and major hitocompatibility complex calss I and class II molecules. Br J Dermatol 139:965–973
56. Abdool K, Cretney E, Brooks AD et al (2006) NK cells use NKG2D to recognize a mouse renal cancer (Renca), yet require intercellular adhesion molecule-1 expression on the tumor cells for optimal performing-dependent effector function. J Immunol 177:2575–2583
57. Yamanaka K, Dimitroff CJ, Fuhlbrigge RC et al (2008) Vitamins A and D are potent inhibitors of cutaneous lymphocyte-associated antigen expression. J Allergy Clin Immunol 121:148–157
58. Rychli J, Nehe B (2006) Therapeutic strategies in autoimmune diseases by interfering with leukocyte endothelium interaction. Curr Pharm Des 12:3799–3806

59. Zollner TM, Assudullah K, Schön MP (2007) Targeting leukocyte trafficking to inflamed skin: still an attractive therapeutic approach? Exp Dermatol 16:1–12
60. Coulie PG, Brichard V, van Pel et al (1994) A new gene coding for a differentiation antigen recognized by autologous cytotlytic T lymphocytes on HLA-A2 melanomas. J Exp Med 180:35–42
61. Kawakami Y, Eliyahu S, Delgado CH et al (1994) Cloning of the gene coding for a shared human melanoma antigen recognized by autologous T cells infiltrating into tumor. Proc Natl Acad Sci USA 91:3515–3519
62. Kawakami Y, Suzuki Y, Shofuda T et al (2000) T cell immune responses against melanoma and melanocytes in cancer and autoimmunity. Pigment Cell Res 13S:163–169
63. Yee C, Thompson JA, Roche P et al (2000) Melanocyte destruction after antigen-specific immunotherapy of melanoma: direct evidence of T cell-mediated vitiligo. J Exp Med 192:1637–1644
64. Benlalaem H, Labarrière N, Linard B et al (2001) Comprehensive analysis of the frequency of recognition of melanoma-associated antigen (MAA) by CD8 melanoma infiltrating lymphocytes (TIL): implications for immunotherapy. Eur J Immunol 31:2007–2015
65. Das PK, van den Wijngaard RM, Wankowicz-Kalinska A, Le Poole IC (2001) A symbiotic concept of autoimmunity and tumout immunity: lessons from vitiligo. Trends Immunol 22:130–136
66. Sakai C, Kawakami Y, Law LW et al (1997) Melanosomal proteins as melanoma-specific immune targets. Melanoma Res 7:83–95
67. Osanai K, Takahashi K, Nakamura K et al (2005) Expression and characterization of Rab38, a new member of the Rab small G protein family. Biol Chem 386:143–153
68. Walton SM, Gerlinger M, de la Rosa O et al (2006) Spantaneous CD8 T cell responses against the melanocyte differentiation antigen RAB38 NY-MEL-1 in melanoma patients. J Immunol 177:8212–8218
69. Wankowicz-Kalinska A, Maillard RB, Olson K et al (2006) Accumulation of low-avidity anti-melanocortin receptor 1 (anti-MC1R) CD8 + T Cells in the lesional skin of a patient with melanoma-related depigmentation. Melanoma Res 16:165–174
70. Wasmeier C, Romao M, Plowright L et al (2006) Rab38 and Rab 32 control post-Golgi trafficking of melanogenic proteins. J Cell Biol 175:271–281
71. Széll M, Baltás E, Bodai L et al (2008) The arg160Trp allele of melanocortin-1 receptor gene might protect against vitiligo. Photchem Photobiol 84:565–571
72. Hoogduijn MJ, Ancans J, Suzuki I et al (2002) Melanin-concentrating hormone and its receptor are expressed and functional in human skin. Biochem Biophys Res Commun 296:698–701
73. Kemp EH, Waterman EA, Hawes BE et al (2002) The melanin-concentrating hormone receptor 1, a novel target of autoantibody responses in vitiligo. J Clin Invest 109: 923–930
74. Verlaet M, Adamantidis A, Coumans B et al (2002) Human immune cells express ppMHC mRNA and functional MCHR1 receptor. FEBS Lett 527:205–210
75. Durham-Pierre DG, Walters CS, Halder RM et al (1995) Natural killer cell and lymphokine-activated killer cell activity against melanocyte in vitiligo. J Am Acad 33:26–30
76. Palermo B, Garbelli S, Mantovani S et al (2001) Qualitative difference between the cytotoxic T lymphocyte responses to melanocyte antigens in melanoma and vitiligo. Eur J Immunol 35:3153–3162
77. Van den Boorn JG, Le Poole IC, Luiten RM. (2006) T-cell avidity and tuning: the flexibility connection between tolerance and autoimmunity. Int Rev Immunol 35:235–258
78. Pedersen LØ, Vetter CS, Mingari MC et al (2002) Differential expression of inhibitory or activating CD94/NKG2 subtypes on MART-1-reactive T cells in vitiligo versus melanoma: a case report. J Invest Dermatol 118:595–599
79. Le Poole IC, Riker AI, Quevedo et al (2002) Interferon-gamma reduces melanosomal antigen expresión and recognition of melaoma cells by cytotoxic T cells. Am J Pathol 160:521–528
80. Moretti S, Spallanzani A, Amato L et al (2002) New insights into the pathogenesis of vitiligo: imbalance of epidermal cytokines at sites of lesions. Pigment Cell Res 15:87–92
81. Palermo B, Garbelli S, Mantovani S, Giachino C (2005) Transfer of efficient anti-melanocyte T cells from vitiligo donors to melanoma patients as a novel immunotherapeutical strategy. J Autoimmune Dis 31:2–7
82. Steitz J, Brück J, Lenz J et al (2005) Peripheral CD8 + T cell tolerance against melanocytic self-antigens in the skin is regulated in two steps by CD4 + T cells and local inflammation: implications for the pathophysiology of vitiligo. J Invest Dermatol 124:144–150
83. Steitz J, Wenzel J, Gaffal E, Tüting T (2004) Initiation and regulation of CD8 + T cells recognizing melanocytic antigens in the epidermis: implcations for the pathophysiology of vitiligo. Eur J Cel Biol 83:797–803
84. Armstrong CA, Tara DC, Hart CE et al (1992) Heterogeneity of cytokine production by human malignant melanoma cells. Exp Dermatol 1:27–45
85. Baumgartner J, Wilson C, Palmer R et al (2007) Melanoma induces immunosuppression by up-regulating FOXP3(+) regulatory cells. J Surg Res 141:72–77
86. Le Poole IC, Denman C, Martin AE, Wainwright D, Qin J, Hernandez C, Overbeck A (2006) Functional regulatory T cells are present in peripheral blood but absent from skin of vitiligo patients. J Invest Dermatol 126:155
87. Hoon DS, Irie RF, Cochran AJ (1988) Gancliosides from human melanoma immunomodulate response of T cells to interleukin-2. Cell Immunol 111:410–419
88. Breathnatch AS (1963) A new concept of the relation between Langerhans cells and the melanocyte. J Invest Dermatol 40:279–281
89. Hatchome N, Aiba S, Kato T et al (1987) Possible functional impairment of Langerhans' cells in vitiliginous skin. Reduced ability to elicit dinitrochlorobenzene contact sensitivity reaction and decreased stimulatory effect in the allogeneic mixed skin cell lymphocyte culture reaction. Arch Dermatol 123:51–54
90. Le Poole IC, van den Wijngaard RMJGJ, Westerhof W et al (1994) Organotypic culture of human skin to study melanocyte migration. Pigment Cell Res 7:33–43
91. Le Poole IC, van den Wijngaard RMJGJ, Westerhof W et al (1993) Phagocytosis by normal human melanocytes in vitro. Exp Cell Res 205:388–395
92. Leenstra S, Das PK, Troost D et al (1995) Human malignant astrocytes expres macrophage phenotype. J Neuroimmunol 56:17–25

93. Ferrari G, Knight AM, Watts, Pieters J (1997) Distinct intracellular compartments involved in invariant chain degradation and antigenic peptide loading of major histocompatibility complex (MHC) class II molecules. J Cell Biol 139:1433–1446
94. Lepage S, Lapointe (2006) Melanosomal targeting sequences from gp100 are essential for MHC class II-restricted endogenous epitope presentation and mobilization to endosomal compartments. Cancer Res 66:2423–2432
95. Van Lith M, van Ham M, Griekspoor A et al (2001) Regulation of MHC class II antigen presentation by sorting of recycling HLA-DM/DO and class II within the multivesicular body. J Immunol 167:884–892
96. Le Poole IC, Mutis T, van den Wijngaard RM et al (1993) A novel, antigen presenting function of melanocytes and is possible relationship to hypopigmentary disorders. J Immunol 151:7284–7292
97. Lambe T, Leung JC, Bouriez-Jones T et al (2006) CD4 T cell dependent autoimmunity against a melanocyte neoantigen induces spontaneous vitiligo and depends upon Fas-Fas ligand interaction. J Immunol 177:3055–3062
98. Westerhof W, Groot I, Krieg SR et al (1986) Langerhans' cell population studies with OKT6 and HLA-DR monoclonal antibodies in vitligo patients treated with oral phenylalanine loading and UVA irradiation. Acta Derm Venereol 66:259–262
99. Ullrich E, Bonmort M, Mignot G et al (2007) Tumor stress, cell death and the ensuing immune response. Cell Death Diff 15:21–28
100. Janjic BM, Pimenov A, Whiteside TL et al (2002) Innate direct anticancer effector function of human immature dendritic cells. I. Involvement of an apoptotosis-inducing pathway. J Immunol 15:1823–1830
101. Lu G, Janjic BM, Janjic C et al (2002) Innate direct anticancer effector function of human immature dendritic cells. II. Role of TNF, lymphotoxin-elapha(1)beta(2), Fas ligand, and TNF-related apoptosis-inducing ligand. J Immunol 168:1831–1838
102. Le Poole IC, Elmasri WM, Denman CJ et al (2007) Langerhans cells and dendritic cells are cytotoxic toards HPV16 E6 and E7 expressing target cells. Cancer Immunol Immunother 57:789–797
103. Perfetti L, Cespa M, Nume A, Orecchia G (1991) Prevalence of vitiligo. A preliminary report. Dermatologica 182: 218–220
104. Grützkau A, Henz BM, Kirchhof L et al (2000) Alpha-melanocyte stimulating hormone acts as a selective inducer of secretory functions in human mast cells. Biochem Biophys Res Commun 278:14–19
105. Ichimiya M, Ohmura A, Muto M (1998) Numerous hypopigmented patches associated with atopic dermatitis. J Dermatol 25:759–761
106. Nader-Djalal N, Ansarin K (1996) Hypopigmented skin lesions associated with atotpic dermatitis in asthma. J Asthma 33:231–238
107. Sugita K, Izu K, Tokura Y (2006) Vitiligo with inflammatory raised borders, associated with atopic dermatitis. Clin Exp Dermatol 31:80–82
108. Wehrle-Aller B (2003) The role of Kit-ligand in melanocyte development and epidermal homeastasis. Pigment Cell Res 16:287–296
109. Norris A, Todd C, Graham A et al (1996) The expression of the c-kit receptor by epidermal melanocytes may be reduced in vitiligo. Br J Dermatol 134:299–306
110. Salmasi JM, Khartonova NI, Kazimirsky et al (2003) Characterization of lymphocyte surface markers in patients with vitiligo. Russ J Immunol 8:47–52
111. Mahmoud F, Abul H, Haines D et al (2002) Decreased total numbers of peripheral blood lymphocytes with elevated percentages of CD4 + CD45RO + and CD4 + CD25 + of T-helper cells in non-segmental vitiligo. J Dermatol 29: 68–73
112. Jin Y, Mailloux CM, Gowan K et al (2007) NALP1 in vitiligo-associated multiple autoimmune disease. N Engl J Med 356:1216–1225
113. Church LD, Cook GP, McDermott MF (2008) Primer: inflammasomes and interleukin 1 beta in inflammatory disorders. Nat Clin Pract Rheumatol 4:34–42
114. Liu F, Lo CF, Ning X et al (2004) Expression of NALP1 in cerbellar granule neurons stimulate apoptosis. Cell Signal 16:1013–1021
115. D'Hooge E, Buttiglieri S, Bisignano G et al (2007) Apoptotic renal cell carcinoma cells are better inducers of cross-presenting activity than their necrotic counterpart. Int J Immunopathol Pharmacol 20:707–717
116. Reschner A, Hubert P, Delvene P et al (2008) Innate lymphocyte and dendritic cells cross-talk: a key factor in the regulation of the immune response. Clin Exp Immunol 152:219–226
117. Winter H, van den Engel NK, Rüttinger D et al (2007) Therapeutic T cells induce tumor-directed chemotaxis of innate immune cells through tumor-specifc secetion of chemokines and stimulation of B16BL6 melanoma to secrete cytokines. J Transl Med 24:56
118. Handra S, Dogra S (2003) Epidemiology of childhood vitiligo: a study of 625 patients from north India. Pediatr Dermatol 20:207–210
119. Lane C, Leitch J, Tan X et al (2004) Vaccination-induced autoimmune vitiligo is a consequence of secondary trauma to the skin. Cancer Res 64:1509–1514
120. Silva MT, do Vale A, Dos Santos NM (2008) Secondary necrosis in multicellular animals: an outcome of apoptosis with pathogenic implications. Apoptosis 13:463–482
121. Tesniere A, Panaretakis T, Kepp O et al (2008) Molecular characteristics of immunogenic cancer cell death. Cell Death Diff 15:3–12
122. Sanchez-Perez L, Kottke T, Daniels GA (2006) Killing of normal melanocytes, combined with heat shock protein 70 and CD40L expression, cures large established melanomas. J Immunol 177:4168–4177
123. Denman CJ, McCracken J, Hariharan V et al (2008) HSP70i accelerates depigmentation in a mouse model of autoimmune vitiligo. J Invest Dermatol 128:2041–2048
124. Kottke T, Sanchez-Perez L, Diaz RM et al (2007) Induction of hsp70-mediated Th17 autoimmunity can be exploited as immunotherapy for metastatic prostate cancer. Cancer Res 67:11970–11979
125. Vega VL, Rodriguez-Silva M, Frey T (2008) Hsp70 translocates into the plasma membrane after stress and is released into the extracellular environment in a membrane-associated form that activates-associated macrophages. J Immunol 180:4299–4307

Cytokines and Growth Factors

2.2.8

Genji Imokawa and Silvia Moretti

Contents

S. Moretti (✉)
Department of Dermatological Sciences,
University of Florence, Florence, Italy
e-mail: silvia.moretti@unifi.it

Core Messages

- An imbalance of keratinocyte-derived cytokines and dysregulations of cytokine/receptor interactions both capable of affecting melanocyte activity and survival has been demonstrated in vitiligo epidermis.
- Melanogenic growth factors such as stem-cell factor (SCF) and endothelin (ET)-1, as well as inflammatory cytokines with anti-melanogenic properties, such as IL-1, IL-6, and tumor necrosis factor (TNF)-α, are implicated.
- The over-expression of TNF-α may facilitate the apoptosis of keratinocytes, leading to a decrease in the production of ET-1 and SCF, thus enhancing melanocyte disappearance.
- At the junction of lesional vitiligo epidermis, melanocytes remain and express tyrosinase, S100α and ET-B receptor (ETBR), but not c-kit or MITF-M.
- At the center of lesional vitiligo epidermis, there is a complete loss of melanocytes expressing c-kit, S100α, ETBR, and/or tyrosinase.
- This deterioration in the expression of c-kit by melanocytes and its downstream effectors, including MITF-M, may be associated with the dysfunction and/or loss of melanocytes in vitiligo epidermis.
- The cytokine imbalance in vitiligo epidermis may be also related to an impaired keratinocyte senescence process.

M. Picardo and A. Taïeb (eds.), *Vitiligo*,
DOI 10.1007/978-3-540-69361-1_2.2.8,

2.2.8.1 Introduction: Melanocytic Homeostasis and Cytokines/Growth Factors

Imokawa G, Moretti S, Picardo M, and Taïeb A

Contents

The human epidermal melanin unit represents the symbiotic relationship in which one melanocyte transports pigment-containing melanosomes through its dendrites to approximately 36 associated keratinocytes [5, 6]. In response to several stimuli, melanogenesis is regulated via specific melanogenic cytokines or chemokines produced/released from keratinocytes coordinated with the expression of their respective receptors on melanocytes. Moreover, recent data suggest that even fibroblast-derived growth factors can contribute to the regulation of melanocyte activities and skin pigmentation, suggesting the existence of a dermal–epidermal melanin unit (Table 2.2.8.1).

So far, there has been no evidence to suggest that under nonstimulated, in vivo conditions, the levels of keratinocyte-derived melanogenic cytokines play a pivotal role in maintaining genetically determined constitutive skin color. However, in human skin substitutes, in the presence of similar melanocyte numbers, significantly more melanin is produced when dark rather than light keratinocytes present are associated with a higher expression of endothelin (ET)-1 or stem cell factor (SCF) mRNA, suggesting that that skin color might be regulated not only by melanocytes, but even by keratinocytes, through the production of melanogenic cytokines [15]. In this connection, pivotal roles for SCF and ET-1 have been demonstrated in several hyperpigmentary disorders, including UVB-induced pigmentation [7, 9, 11, 12]. Furthermore, as suggested by monogenic disorders and mouse models, the complete loss of epidermal pigmentation is generally mediated via point mutations of melanogenic ligand-specific receptors, such as c-kit [4] or ET-B receptor (ETBR), as well as of melanogenic cytokines, such as ET-3 or SCF [8, 13, 14].

Recently, more attention has been given to fibroblast-derived factors. Palmoplantar fibroblasts express high levels of the inhibitor of the Wnt/β-catenin signaling pathway dickkopf1 (DKK1), which decreases the growth and differentiation of melanocytes and accounts

Table 2.2.8.1 Mealanogenic cytokines

Factor	Cells	Receptor	Pathway	Effect
α-MSH	Keratinocytes	MC1-R[g]	MITF	↑TRP-1[h], TRP-2
ACTH	Keratinocytes	MC1-R	MITF	↑TRP-1, TRP-2
ET-1[a]	Keratinocytes	ETBR		Proliferation, melanogenesis, dendriticity UVB/A-induced
GM-CSF[b]	Keratinocytes	GM-CSFR		Proliferation, melanogenesis, dendriticity UVB/A-induced
SCF[c]	Keratinocytes and fibroblasts	c-Kit	STAT1/3/5, MAP, SCF	↑TRP-1, TRP-2
BFGF[d]	Keratinocytes and fibroblasts	b-FGFR	STAT1/3/5, MAP, SCF	↑TRP-1, TRP-2
KGF[e]	Keratinocytes and fibroblasts	KGFR	STAT1/3/5, MAP, SCF	↑TRP-1, TRP-2
HGF[f]	Keratinocytes and fibroblasts	c-MET	STAT1/3/5, MAP, SCF	↑TRP-1, TRP-2

[a]Endothelin 1
[b]Granulocyte macrophage colony stimulating factor
[c]Stem cell factor
[d]Basic fibroblast growth factor
[e]Keratinocyte growth factor
[f]Hepatocyte growth factor
[g]Melanocortin receptor 1
[h]Tyrosinase related protein

for the hypopigmentation of palms and soles. Fibroblast-derived SCF and hepatocyte growth factor (HGF) are mitogens for human melanocytes in vitro and in vivo and can influence melanocyte proliferation and melanin distribution in human reconstructed skin [3]. The keratinocyte growth factor (KGF), promotes the phagocytosis of melanosomes by keratinocytes in coculture [1, 2] and, in combination with IL-1α, increases melanin deposition both in the basal layer and in the whole epidermis. In vivo an increased expression of SCF, KGF and HGF has been demonstrated in lentigo senilis [10].

This series of data has generated the hypothesis that paracrine cytokines and their receptors may be involved in the melanocyte dysfunction or loss that occurs in vitiligo. This section reviews the current evidence accumulated in this field. There is still a debate about the status of some modifications of secreted peptides which is reflected in the content of the following section. That may reflect different experimental settings. We would like to emphasize the fact that the stage of the disease is far from being standardized for in vivo studies. The level of inflammation produced by exogenous cells such as lymphocytes which are present at the limits of the lesions is difficult to dissect in nonsegmental vitiligo (NSV). Unfortunately, studies of this kind have not been done in SV, which is not known to be associated with an inflammatory/autoimmune component.

References

1. Cardinali G, Ceccarelli S, Kovacs D et al (2005) Keratinocyte growth factor promotes melanosome transfer to keratinocytes. J Invest Dermatol 125:1190–1199
2. Cardinali G, Bolasco G, Aspite N et al (2008) Melanosome transfer promoted by keratinocyte growth factor in light and dark skin-derived keratinocytes. J Invest Dermatol 128:558–567
3. Cario-André M, Pain C, Gauthier Y et al (2006) In vivo and in vitro evidence of dermal fibroblasts influence on human epidermal pigmentation. Pigment Cell Res 19:434–442
4. Ezoe K, Holmes SA, Ho L et al (1995) Novel mutations and deletions of the KIT (steel factor receptor) gene in human piebaldism. Am J Hum Genet 56:58–66
5. Fitzpatrick TB, Breatnach AS (1963) The epidermal melanin unit system. Dermatol Wochenschr 147:481–489
6. Fitzpatrick TB, Szabo G, Seji M et al (1979) Biology of melanin pigmentary system. In: Fitzpatrick TB, Eisen A, Wolff K, Freedberg I, Austen K (eds) Dermatology in general Medicine. McGraw-Hill, New York
7. Funasaka Y, Boulton T, Cobb M et al (1992) Kit-kinase induces a cascade of protein tyrosine phosphorylation in normal human melanocytes in response to mast cell growth factor and stimulates mitogen-activated protein kinase but is down-regulated in melanomas. Mol Biol Cell 3:197–209
8. Giebel LB, Spritz RA (1991) Mutation of the KIT (mast/stem-cell growth factor receptor) protooncogene in human piebaldism. Proc Natl Acad Sci USA 88:8696–8699
9. Gordon PR, Mansur CP, Gilchrest BA (1989) Regulation of human melanocyte growth, dendricity, and melanization by keratinocytes derived factors. J Invest Dermatol 92:565–572
10. Hattori H, Kawashima M, Ichikawa Y et al (2004) The epidermal stem cell factor is over-expressed in lentigo senilis: implication for the mechanism of hyperpigmentation. J Invest Dermatol 122:1256–1265
11. Imokawa G, Kobayashi T, Miyagishi M et al (1997) The role of endothelin-1 in epidermal hyperpigmentation and signalling mechanisms of mitogenesis and melanogenesis. Pigment Cell Res 10:218–228
12. Moretti S, Massi D, Baroni G et al (2005) Imbalance of cytokine transcripts in non segmental vitiligo. Pigment Cell Res 18(suppl 1):72
13. Norris A, Todd C, Graham A et al (1996) The expression of the c-kit receptor by epidermal melanocytes may be reduced in vitiligo. Br J Dermatol 134:299–306
14. Sakamoto A, Yanagisawa M, Sakurai T et al (1991) Cloning and functional expression of human cDNA for the ETB endothelin receptor. Biochem Biophys Res Commun 178:656–663
15. Valyi-Nagy IT, Murphy GF, Mancianti ML et al (1990) Phenotypes and interactions of human melanocytes and keratinocytes in an epidermal reconstruction model. Lab Invest 62:314–324

2.2.8.2
An Overview of Epidermal Cytokines and Growth Factors in Vitiligo

Silvia Moretti

Contents

The Role of Melanogenic Cytokines

An alteration of the expression of keratinocyte-derived cytokines in vitiligo epidermis has been suggested. But literature does not appear univocal, since some melanocyte-stimulating cytokines have been described as decreased [2, 11, 15–17] or increased [8] in lesional epidermis (Fig. 2.2.8.1). This is possibly due to the different detection techniques or controls used, or to a different recruitment of vitiligo patients. However, a critical evaluation of the findings reported suggests that SCF at both protein and transcript levels is reduced in depigmented lesions, in vivo [11, 15–17] possibly due to the fact that keratinocytes in the depigmented lesions seem to be more prone to apoptosis as compared with those of the normally pigmented counterpart [10], and incapable of producing adequate amounts of SCF for melanocyte survival [11]. In vitro functional studies have shown that apoptosis of cultured normal human keratinocytes is associated with a concentration-dependent decreased production of SCF mRNA and protein, and that deprivation of SCF or keratinocyte feeder in the culture medium induces a marked decrease in melanocytes resulting of keratinocyte apoptosis [11]. Further support to the role of keratinocyte-derived SCF in regulating melanocyte activity and possible implication in vitiligo is also given by additional in vitro data, showing that proliferation of cultured melanocytes is enhanced by tacrolimus-treated keratinocyte supernatant, whose SCF concentration increases dose-dependency with tacrolimus treatment [9].

Other melanogenic cytokines have been less extensively studied, but a substantial decrease of GM-CSF and bFGF protein levels [11, 15, 16], and a reduced expression of ET-1 transcript [17] have been shown in the depigmented epidermis compared with the normally pigmented epidermis in vitiligo patients.

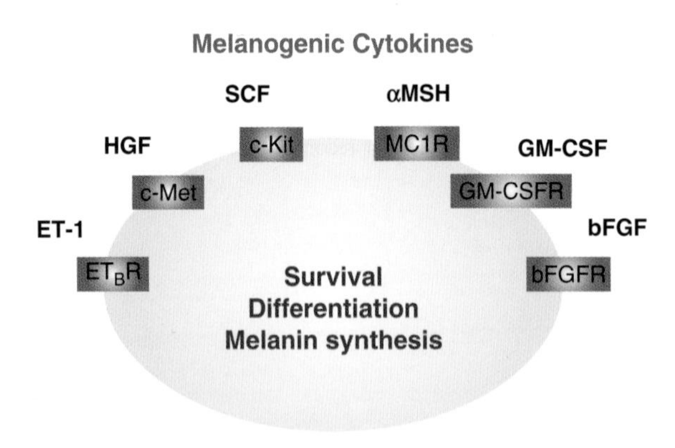

Fig. 2.2.8.1 Cytokines produced by keratinocytes and capable of affecting melanocytes in the epidermal microenvironment. Basic fibroblastic growth factor (bFGF), stem cell factor (SCF), endothelin (ET), hepatocyte growth factor (HGF), α melanocyte-stimulating harmone (αMSH), and granulocyte-monocyte colony stimulating factor (GM-CSF) are melanogenic cytokines

The Role of Inflammatory Cytokines

Human keratinocytes synthesize the inflammatory cytokines IL-1α, IL-6 and tumor necrosis factor (TNF)-α, which also have noninflammatory effects on melanocytes. These cytokines are paracrine inhibitors of human melanocyte proliferation and melanogenesis, eliciting a dose-dependent decrease in tyrosinase activity in vitro [19] and in vivo in a mouse model [12].

In the affected skin of vitiligo patients, a higher expression of these cytokines has been demonstrated at both protein and transcript levels [1, 4, 7, 15–17]. However, the exact role of these cytokines in the pathogenesis of the disease is not fully understood. Besides their anti-melanogenic properties, IL-1α, IL-6, and TNF-α can induce the expression of adhesion molecules, in particular ICAM-1, on melanocyte membranes, thus promoting lymphocyte recruitment that can play a role in melanocyte loss [14]; TNF-α can induce the production of peroxides, including hydrogen peroxide, leading to a mitochondria-dependent cell death [3, 5]. In addition, TNF-α can activate the downstream signaling molecule nuclear factor (NF)-κB [6], allowing its nuclear translocation and the expression of a number of target genes that protect the cell from cell death [13]. In TNF-α-treated normal human keratinocytes, the inhibition of NF-κB activation results in keratinocyte apoptosis [7]. In vitiligo epidermis, besides the high TNF-α levels, an impaired NF-κB activation has been described suggesting that NF-κB dysfunction may promote TNF-α-dependent apoptosis in keratinocytes leading to impaired melanocyte survival. The significant reduction of the expression of TNF-α in lesions undergoing repigmentation after topical tacrolimus treatment suggests its involvement in the pathogenesis of vitiligo [4].

Other inflammation-related cytokines, such as IL-10 and IFN-γ have been demonstrated to be increased in vitiligo skin [4], possibly related to the presence of an ongoing inflammatory process.

Summary: How the Epidermal Cytokine Network May Be Impaired in Vitiligo Skin

The altered expressions of some paracrine melanogenic cytokines support a biological basis for the impaired growth and activity of the pigment cells. The increased production of TNF-α could stimulate apoptotic phenomena in some keratinocytes [7], thus reducing the secretion of melanogenic cytokines such as SCF [11] leading as a consequence to a decreased melanocyte survival. Interestingly, keratinocytes from involved vitiligo skin demonstrate a shorter life span in vitro than those from uninvolved areas, do not maintain melanocytes in culture at a physiological ratio, and exhibit an altered expression of senescence markers suggesting an impaired senescence process and an attempt to regulate it [2]. Since senescent cells (fibroblasts and keratinocytes) produce more paracrine growth factors for melanocytes [18], and lesional vitiligo keratinocytes show an initial delayed senescence in culture, the reduced amount of keratinocyte-derived melanogenic cytokine SCF, may be supported by the postponed senescence process [2]. A different phase of life span of vitiligo keratinocytes evaluated could also explain discrepancies concerning SCF production in vitiligo epidermis reported in literature [2, 8, 11, 15, 16].

In conclusion, vitiligo epidermis exhibits an impaired expression of keratinocyte-derived cytokines and growth factors capable of affecting melanocytes survival and activities, and such an imbalance likely plays a role in melanocyte disappearance in vitiligo [3].

References

1. Birol A, Kisa U, Kara F et al (2006) Increased tumor necrosis factor alpha (TNF-α) and interleukin 1 alpha (IL-1α) levels in the lesional skin of patients with nonsegmental vitiligo. Int J Dermatol 45:992–993
2. Bondanza S, Maurelli R, Paterna P et al (2007) Keratinocyte cultures from involved skin in vitiligo patients show an impaired in vitro behaviour. Pigment Cell Res 20:288–300
3. Dell'Anna ML, Picardo M (2006) A review and a new hypothesis for non immunological pathogenetic mechanism in vitiligo. Pigment Cell Res 19:406–411
4. Grimes PE, Morris R, Avaniss-Aghajani E et al (2004) Topical tacrolimus therapy for vitiligo: therapeutical responses and skin messenger RNA expression of proinflammatory cytokines. J Am Acad Dermatol 51:52–61
5. Haycock JW, Rowe SJ, Cartledge S et al (2000) α-Melanocyte stimulating hormone reduces impact of proinflammatory cytokine and peroxide-generated oxidative stress on keratinocytes and melanoma cell lines. J Biol Chem 275:15629–15636
6. Hsu H, Shu HB, Pan MG et al (1996) TRADD-TRAF2 and TRADD-FADD interactions define two distinct TNF receptor 1 signal transduction pathways. Cell 84:299–308
7. Kim NH, Jeon S, Lee HJ, Lee AY (2007) Impaired PI3K/Akt activation-mediated NF-kB inactivation under elevated TNF-α is more vulnerable to apoptosis in vitiliginous keratinocytes. J Invest Dermatol 127:2612–2617
8. Kitamura R, Tsukamoto K, Harada K et al (2004) Mechanisms underlying the dysfunction of melanocytes in vitiligo epidermis: role of SCF/KIT protein interactions and its downstream effector, MITF-M. J Pathol 202:463–475
9. Lan CCE, Chen GS, Chiou MH et al (2005) FK506 promotes melanocyte and melanoblst growth and creates a favourable milieu for cell migration via keratinocytes: possible mechanisms of how tacrolimus ointment induces repigmentation in patients with vitiligo. Br J Dermatol 153:498–505
10. Lee AY, Youm YH, Kim NH et al (2004) Keratinocytes in the depigmented epidermis of vitiligo are more vulnerable to trauma (suction) the keratinocytes in the normally pigmented epidermis, resulting in their apoptosis. Br J Dermatol 151:995–1003
11. Lee AY, Kim NH, Choi WI et al (2005) Less keratinocyte-derived factors related to more keratinocyte apoptosis in depigmented than normally pigmented suction-blisterd epidermis may cause passive melanocyte death in vitiligo. J Invest Dermatol 124:976–983
12. Martinez-Esparza M, Jimenez-Cervantes C, Solano F et al (1998) Mechanisms of melanogenesis inhibition by tumor necrosis factor-α in B16/F10 mouse melanoma cells. Eur J Biochem 255:139–146
13. May MJ, Ghosh S (1998) Signal transduction through NF-kappa B. Immunol Today 19:80–88

14. Morelli JG, Norris DA (1993) Influence of inflammatory mediators and cytokines on human melanocytes function. J Invest Dermatol 100(suppl):191S–195S
15. Moretti S, Spallanzani A, Amato L et al (2002) Vitiligo and epidermal microenvironment: possibile involvement of keratinocyte-derived cytokines. Arch Dermatol 138:273–274
16. Moretti S, Spallanzani A, Amato L et al (2002) New insights into the pathogenesis of vitiligo: imbalance of epidermal cytokines at sites of lesions. Pigment Cell Res 15:87–92
17. Moretti S, Massi D, Baroni G et al (2005) Imbalance of cytokine transcripts in non segmental vitiligo. Pigment Cell Res 18(suppl 1):72
18. Okazaki M, Yoshimura K, Uchida G et al (2005) Correlation between age and the secretions of melanocyte-stimulating cytokines in cultures keratinocytes and fibroblasts. Brit J Dermatol 153(suppl 2):23–29
19. Swope VB, Abdel-Malek Z, Kassem LM et al (1991) Interleukin 1α and 6 and tumor necrosis factor -α are paracrine inhibitors of human melanocyte proliferation and melanogenesis. J Invest Dermatol 96:180–185

2.2.8.3 In Vivo Studies of Melanogenic Cytokines and Receptors in Vitiligo

Imokawa G

Contents

This section is based on the author's published work on the study of six cases of NSV using the suction blister technique to harvest epidermis selectively on top of depigmented and nondepigmented skin [10] (Fig. 2.2.8.2).

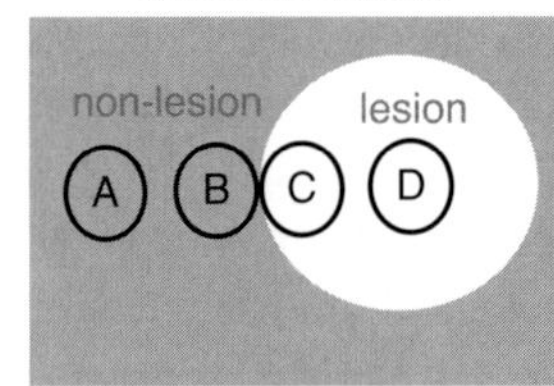

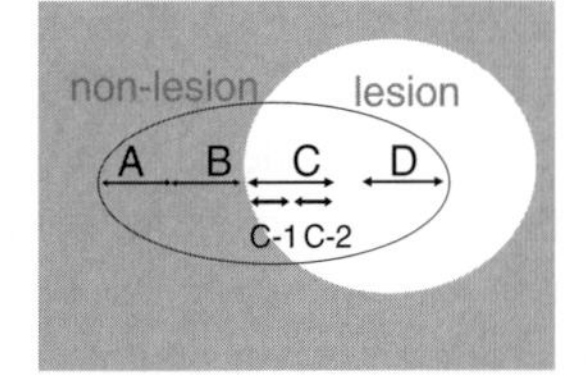

Fig. 2.2.8.2 Skin sites used by suction blister technique and immunohistochemistry. The epidermal sheets were obtained from vitiligo patients using the suction blister technique. Total tissue protein was obtained from extracts of the epidermal sheets. Skin biopsy specimens were used for immunohistochemistry. Melanin deposit was detected by Fontana-Masson staining to determine the border of the lesion and the numbers of melanocytes positively stained with anti-tyrosinase, S-100α, ET_B receptor and KIT protein antibody were counted and expressed as per 200 basal cell in the border of vitiligo lesion and at the center of the vitiligo lesion

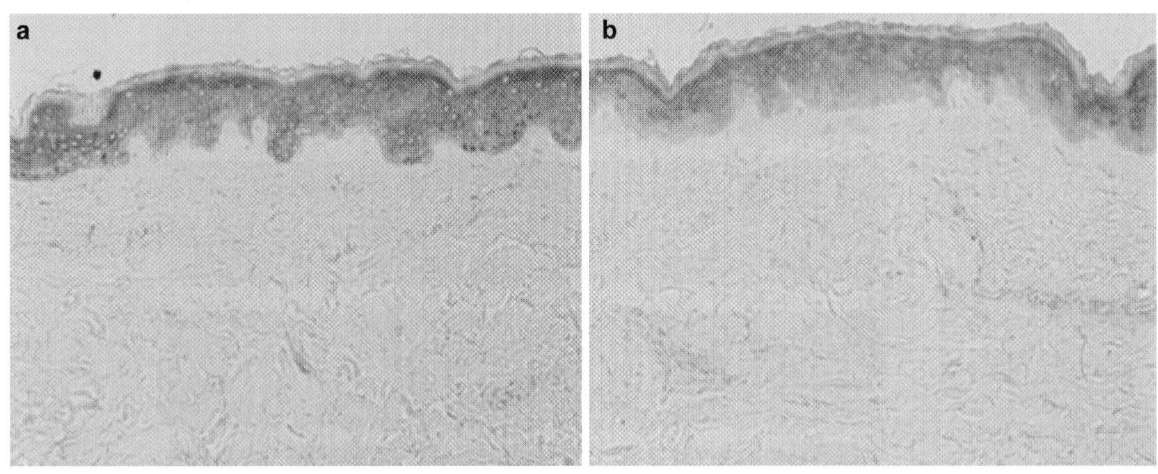

Fig. 2.2.8.3 Immunohistochemistry with antibodies to SCF in vitiligo skin. (**a**) Nonlesional skin (×180), (**b**) the center of the lesional skin (×180)

Melanogenic Cytokines SCF and ET-1 in Lesional NSV Epidermis

Immunohistochemistry using an antibody to SCF (Fig. 2.2.8.3) or ET-1 revealed a similar level of SCF or ET-1-immunopositive staining between the lesional and nonlesional epidermis, indicating that there may be no abnormality in the expression level of melanogenic cytokines in the vitiligo epidermis. However, RT-PCR analysis of mRNAs encoding SCF, ET-1 or other melanogenic cytokines, such as granulocyte macrophage colony stimulating factor (GM-CSF) revealed that gene expression of SCF in vitiligo lesional epidermis was significantly increased compared with nonlesional controls (Fig. 2.2.8.4a). The gene expression of ET-1 in the lesional epidermis was also increased. Further, 3 out of the 6 cases showed a slight expression of GM-CSF in the lesional epidermis. In contrast, there were no or similar levels of expression for other melanogenic cytokines, HGF or basic fibroblast growth factor (bFGF) mRNAs respectively, in the nonlesional and in the lesional vitiligo epidermis. Consistent with immunohistochemistry for SCF, western blotting of proteins extracted from the epidermal sheet reveals that membrane-bound SCF was markedly increased in the lesional vitiligo epidermis (Fig. 2.2.8.4b). In contrast, soluble SCF was not detectable in either type of epidermis. Thus, all data related to melanogenic cytokine expression suggests that there is rather an increased expression of ET-1 and SCF transcripts in lesional vitiligo epidermis compared with nonlesional epidermis, and that SCF protein expression is increased in lesional epidermis.

The Disruption of SCF-c-kit Interaction on Melanocyte Membranes as an Early Event in the Vitiligo Depigmenting Process

The earlier findings led us to determine whether the SCF receptor (c-kit) or ETBR were altered in expression or in their ability to respond to their ligands. This could result in a deficiency for inducing melanogenesis or maintaining melanocyte function/survival in vitiligo skin.

Immunostaining with antibodies to tyrosinase or S100α demonstrated melanocytes in the nonlesional epidermis and at the edge of the lesional epidermis (Fig. 2.2.8.4c, d). In the center of the lesion, there was a complete loss of melanocytes. Immunostaining with an antibody to the ETBR revealed ETBR-positive melanocytes in the nonlesional epidermis and at the edge of the lesional epidermis (Fig. 2.2.8.4e). Although c-kit positive melanocytes were present in the nonlesional epidermis, the number of c-kit positive melanocytes was remarkably decreased at the edge of the lesional epidermis compared with nonlesional epidermis (Fig. 2.2.8.4f). At the center of the lesion, there was a complete loss of melanocytes expressing ETBR or c-kit. Confocal microscopy

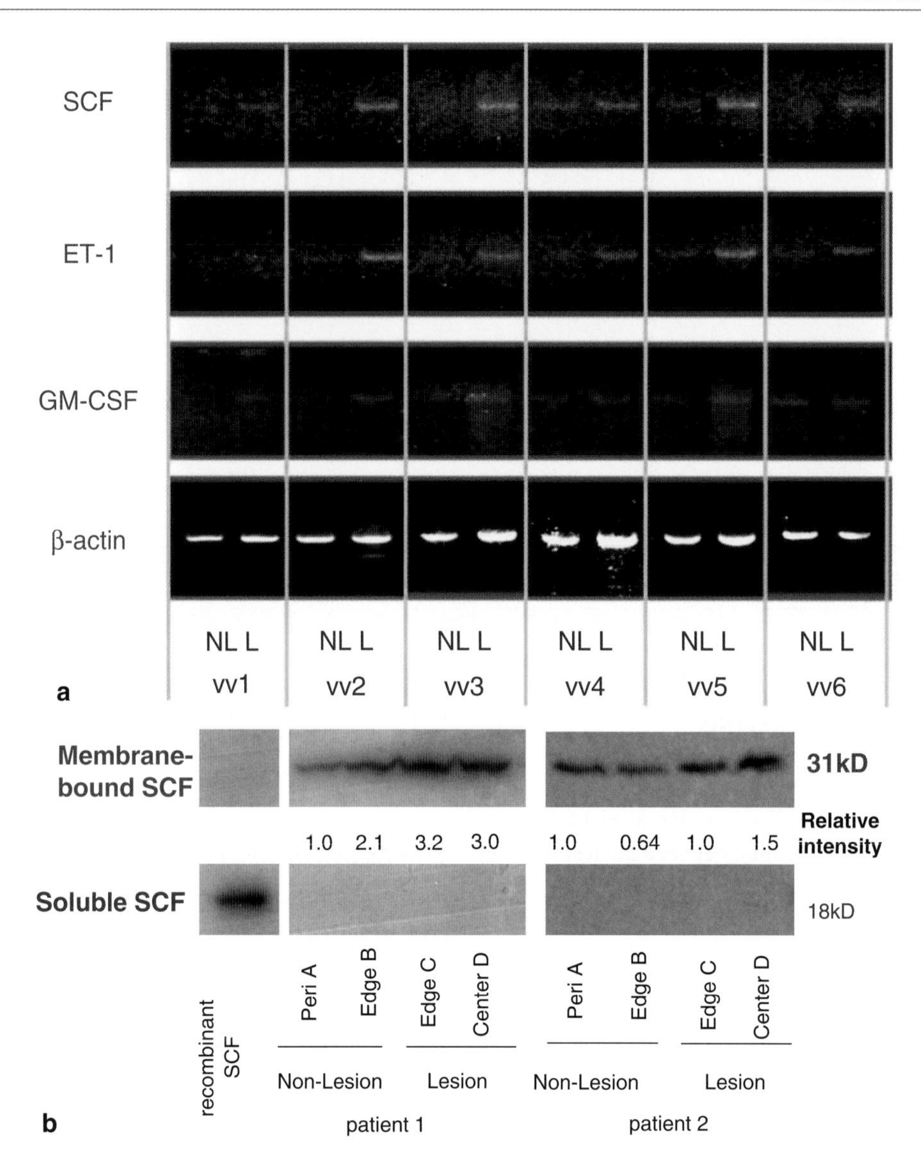

Fig. 2.2.8.4 (**a**) RT-PCR analysis of mRNAs encoding SCF, ET-1 and GM-CSF in the epidermis of vitiligo lesions and in the perilesional skin. *NL* nonlesional epidermis; *L* lesional epidermis of vitiligo. (**b**) Western blotting of SCF protein in the epidermis of vitiligo lesions and in nonlesional skin. SCF is markedly increased in the lesional vitiligo epidermis, including the edge of the lesion (*Edge* C) and the center of the lesion (*Center* D) compared with the nonlesional vitiligo epidermis (Peri A+ *Edge* B).

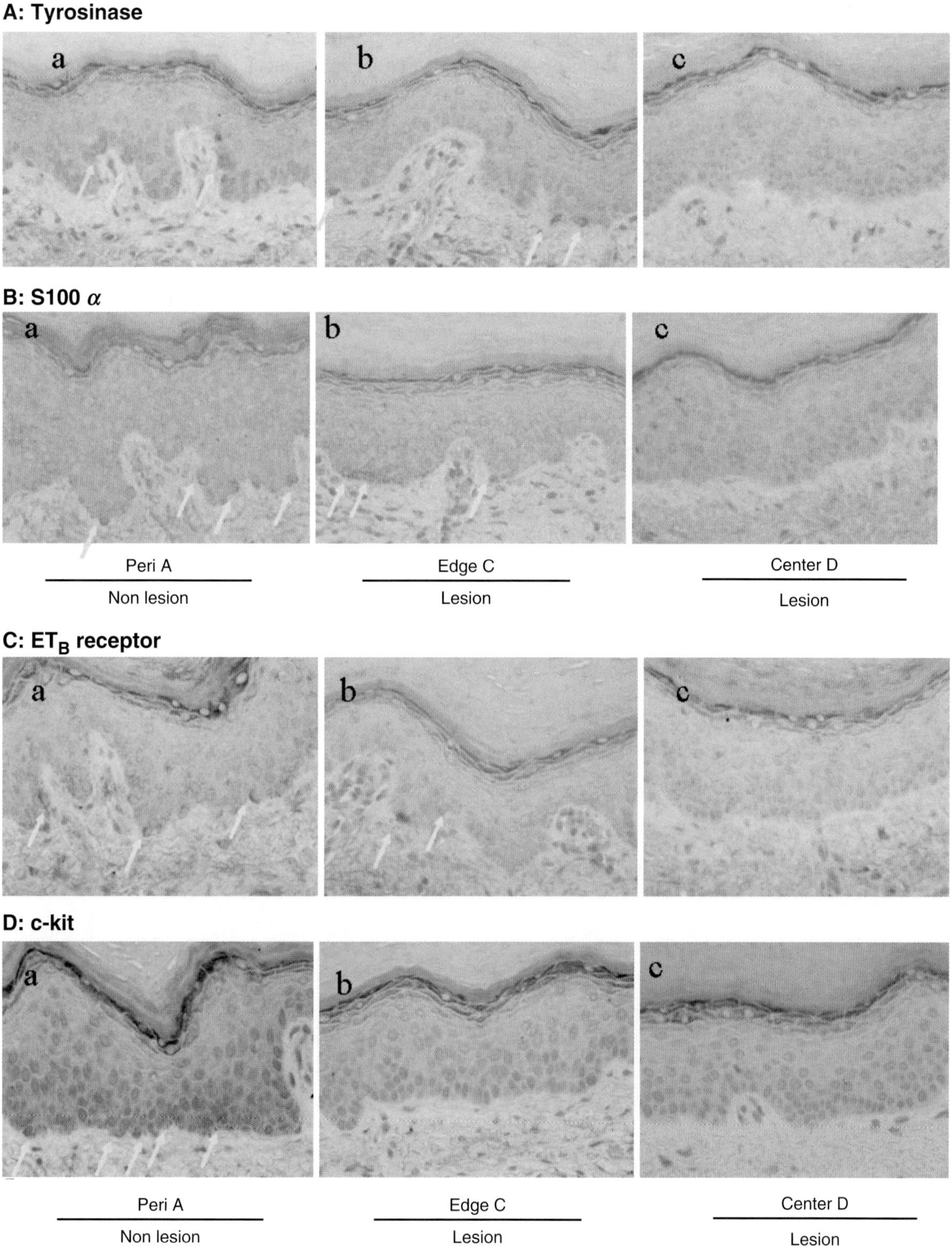

A: Tyrosinase

a b c

B: S100 α

a b c

Peri A — Non lesion | Edge C — Lesion | Center D — Lesion

C: ET_B receptor

a b c

D: c-kit

a b c

Peri A — Non lesion | Edge C — Lesion | Center D — Lesion

Fig. 2.2.8.4 (**c**) Immunohistochemistry with antibodies to tyrosinase (A), S100α (B), ETBR (C) and KIT protein (D) in vitiligo skin. (a) nonlesional skin (×180), (b) the edge of lesional skin (×180), (c) the center of the lesional skin (×180). Immunostaining with antibodies to tyrosinase, S100α and ETBR demonstrated immuno-positive melanocytes in the nonlesional epidermis and at the edge of the lesional epidermis, as indicated by the arrows (A, B and C). Although KIT protein positive melanocytes exist in the nonlesional epidermis, as indicated by the arrows, the number of KIT protein positive melanocytes is significantly decreased at the edge of the lesional epidermis compared with the nonlesional epidermis (D). In the center of the lesion, there is a complete loss of melanocytes expressing tyrosinase, S100α, ETBR and KIT protein

following double immunostaining (Fig. 2.2.8.5a) demonstrated that melanocytes expressing tyrosinase in the nonlesional epidermis were c-kit positive, whereas tyrosinase positive melanocytes were c-kit negative at the edge of the lesion. At the center of the lesional epidermis, there was a complete loss of melanocytes expressing tyrosinase or c-kit. Thus, in summary, although a complete loss of immunoreactive melanocytes was found at the center of vitiligo lesional epidermis, there were almost a normal number of melanocytes expressing tyrosinase, ETBR, and S100α protein at the edge of the lesional nonpigmented epidermis compared with the nonlesional pigmented epidermis. In contrast, very few melanocytes still expressed c-kit at the same edge, suggesting an early pathogenic event.

Quantitation of the number of melanocytes expressing melanogenic factors (Fig. 2.2.8.5b) confirmed that the number of cells expressing c-kit was significantly decreased at the edge of the lesional epidermis compared with the nonlesional epidermis, while there was only a slight decrease in the number of S100α, ETBR, and tyrosinase immuno-positive cells at the same edge.

Thus, among the melanocyte-associated molecules tested, a selective deficiency in c-kit expression occurs in melanocytes localized at the edge of vitiligo lesional epidermis. The disruptions of SCF/c-kit interactions are probably due to a loss or a deficiency of SCF receptor function on melanocytes, despite the increased production of its ligand, SCF (probably due to a feedback mechanism resulting from c-kit deficiency). Our suggestion is that this disruption may lead to abolish the melanogenic potential of melanocytes.

Implication of MITF-M, the Melanocytic Master Transcription Factor

MITF consists of at least seven isoforms with distinct amino-termini, referred to as MITF-M, MITF-H, MITF-A, MITF-B, MITF-C, MITF-D, and MITF-E [1–3, 14, 17, 18]. The MITF-M, isoform, is exclusively expressed in melanocytes and in melanoma cells [1, 2], and regulates the transcription of genes leading to melanocyte differentiation. MITF expression is regulated downstream of the SCF/c-kit linkage [5, 7] as well as the ET-1/ETBR linkage [9]. MITF is phosphorylated by an activated MAP kinase [6], undergoes ubiquitination [19, 20] and is translocated to the nucleus to activate the transcription of several genes, including tyrosinase, TRP-1, ETBR [13, 15] as well as the melanosomal matrix protein Pmel 17.

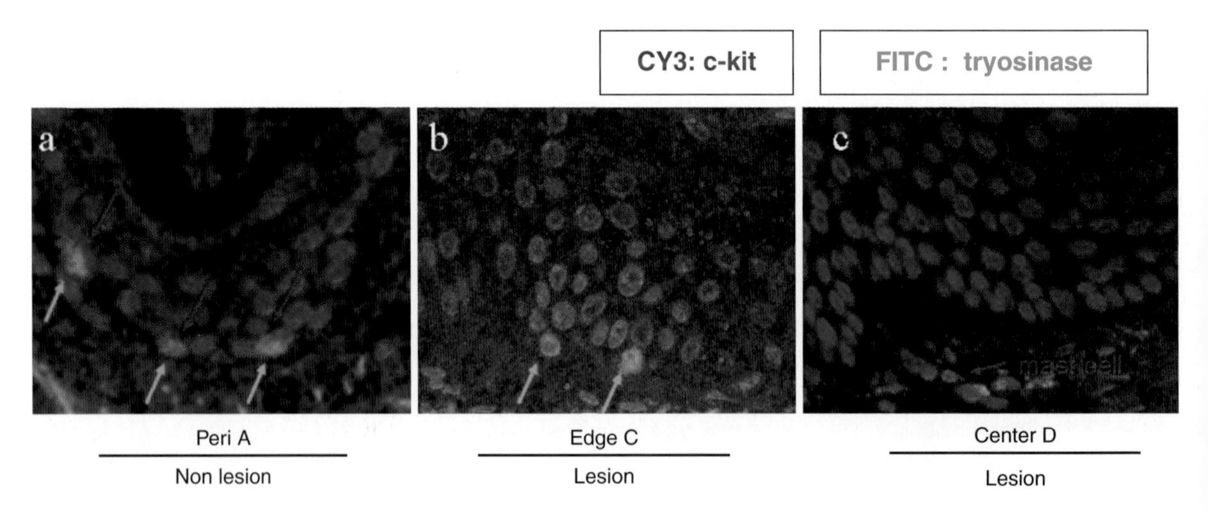

Fig. 2.2.8.5 (**a**) Immunohistochemistry by double immunofluorescence using antibodies to tyrosinase and KIT protein. (**a**) nonlesional skin (×180), (**b**) the edge of lesional skin (×180), (**c**) the center of the lesional skin (×180). *Green* and *red* fluorescences represent tyrosinase and KIT protein, respectively. Nuclei were counterstained by DAPI (*blue fluorescence*). Melanocytes expressing tyrosinase in the nonlesional epidermis are KIT protein positive, as indicated by the *green* and *red arrows*, whereas the tyrosinase positive melanocytes are KIT protein negative at the edge of the lesion

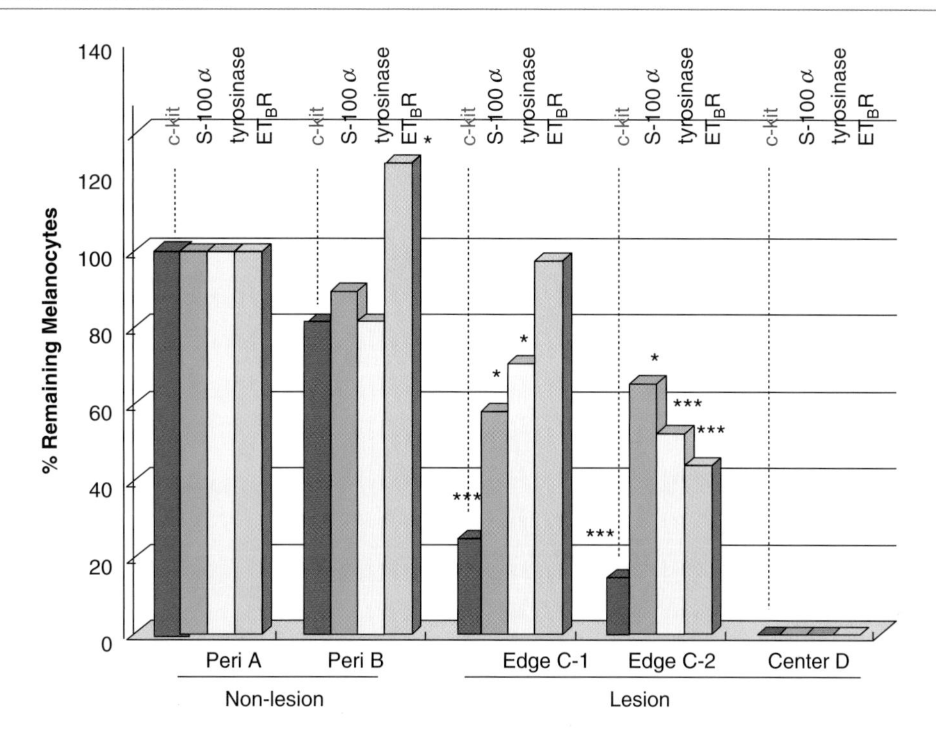

Fig. 2.2.8.5 (**b**) Quantitative analysis of melanocytes immuno-positive for tyrosinase (A) S100α (B), ETBR (C) and KIT protein (D) over 200 basal cells. Quantitation of immuno-positive cells in the epidermis revealed slight or moderate decreases in the number of S100α, tyrosinase and/or ETBR positive cells at the edges (edge C-1 or edge C-2, respectively, as explained in Fig. 2.2.8.2) of lesional epidermis. In contrast, the number of cells expressing c-kit was significantly decreased at the edges (edge C-1 and edge C-2) of the lesional epidermis compared with the nonlesional area. It appears a decrease in the number of melanocytes expressing c-kit, S100α and tyrosinase but not ETBR even at the edge (*Edge* B) of the nonlesional epidermis. Melanin deposit was detected by Fontana-Masson staining to determine the border of the lesion and the numbers of melanocytes positively stained with anti-tyrosinase, S-100α, ET_B receptor and KIT protein antibody were counted and expressed as per 200 basal cells at the border of vitiligo lesion and at the center of the vitiligo lesion. The number of immuno-positive melanocytes is expressed as a percentage of the number found in the nonlesional epidermis (*Peri A*). *: $p < 0.05$, ***: $p < 0.005$ compared with *Peri A*. The number of S100α, tyrosinase and ETBR positive cells in the epidermis reveals slight or moderate decreases at the edges (*edge C-1* or *edge C-2*, respectively) of the lesional epidermis. The number of cells expressing KIT protein is significantly decreased at the edges (*edge C-1* and *edge C-2*) of the lesional epidermis compared with the nonlesional area

There is a genetic cooperation between MITF and Bcl2 in human melanocytes [12]. Thus, a deficiency or loss of MITF expression in melanocytes is closely associated with their apoptosis as a result of the down-regulation of the suppressive apoptotic molecule Bcl2. Further, c-kit and MITF show complex interactions, in that MITF is needed for the maintenance of KIT expression in melanoblasts and KIT signaling modulates MITF activity and stability in melanocytes [6]. Based on this evidence, it was of great interest to determine whether MITF-M expression was altered in melanocytes located at the border between lesional and nonlesional skin.

The expression pattern of MITF-M shown on Fig. 2.2.8.6a seems consistent with the expression pattern of c-kit (Fig. 2.2.8.6b). While the faint decrease in the expression of c-kit protein is nearly comparable with that of MITF-M protein at the edge of the nonlesional skin, there is a definite decrease in the expression both of c-kit and of MITF-M in the lesion. Western blotting (Fig. 2.2.8.7) demonstrates that expression of the M-isoform is decreased at the edges of the vitiligo lesion. Even at the center of the lesion, there is a slight expression of MITF-M protein, suggesting that melanocytes might still exist there, but do not express tyrosinase, S100α or ETBR.

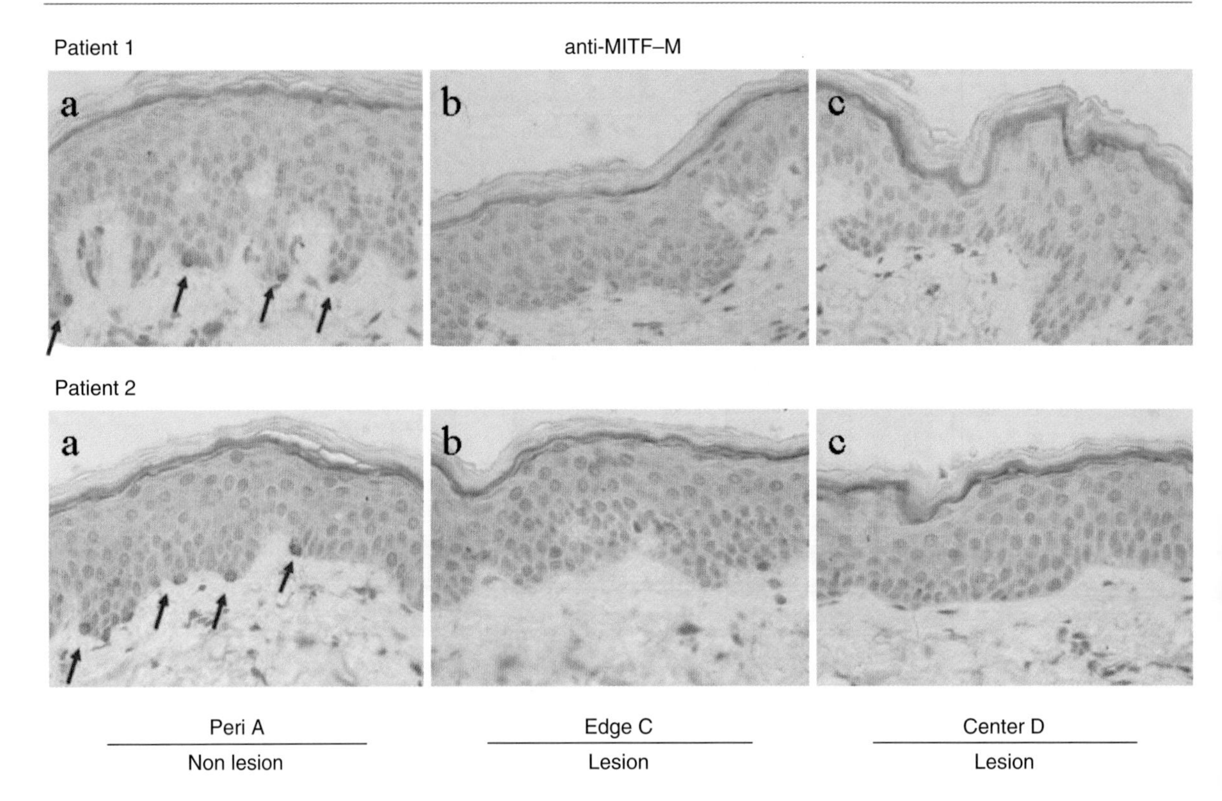

Fig. 2.2.8.6 Immunohistochemistry with an antibody to MITF-M in the epidermis of a vitiligo lesion. A: patient I, B: patient II, a: nonlesional skin (×180), b: the edge of lesional skin (×180), C : the center of the lesional skin (×180). MITF-M positive cells indicated by the arrows. The number of MITF-M positive melanocytes is decreased at the edge of the lesional epidermis compared with the non-lesional epidermis. Immunohistochemistry revealed also that the number of cells expressing MITF-M was significantly decreased at the edge of the lesional epidermis (Edge C) compared with the nonlesional skin (Peri A) and that MITF-M protein expression is markedly decreased in the lesional (Edge C and Center D) as well as in a case at the edge of the nonlesional epidermis (Edge B) compared with the nonlesional skin (Peri A)

Concluding Remarks: Possible Molecular Mechanisms in Vitiligo Melanocytes Dysfunction

The observed deficiency of c-kit expression in melanocytes localized at the edge of the vitiligo lesion may result from the diminished expression of MITF-M. Since a deficiency in MITF has been closely linked to the susceptibility of melanocytes to undergo apoptosis mediated via Bcl2 [12], our observations may indicate that melanocytes localized in vitiligo lesional skin as well as at the edge of the nonlesional skin have a high predisposition toward apoptosis. In addition, c-kit expressed on melanocyte membranes plays an essential role in attracting SCF (which is produced in a membrane-bound form by epidermal keratinocytes), and has a trafficking role by which melanocytes can remain within the epidermis. Thus, the c-kit deficiency observed in vitiligo melanocytes may indicate that vitiligo melanocytes are less able to maintain their interactions with keratinocytes within the epidermis, thus possibly allowing them to detach from the epidermis [4].

In conclusion, the expression patterns observed for c-kit, MITF-M or tyrosinase in melanocytes located at the border between the lesional and the nonlesional vitiligo epidermis indicates that during the sequential process leading to the dysfunction or loss of melanocytes, the initial event may result from a deficiency in MITF-M function within melanocytes in nonlesional epidermis. Such a deficiency would lead to the decreased expression of its target molecule, tyrosinase, as well as to a decrease in c-kit expression. The diminished signaling via the SCF/c-kit linkage and the distinct decrease in tyrosinase expression may underlie the complete loss of melanocyte function at the border between the lesional and nonlesional vitiligo skin. Further, the probable increased susceptibility of melanocytes to apoptosis via

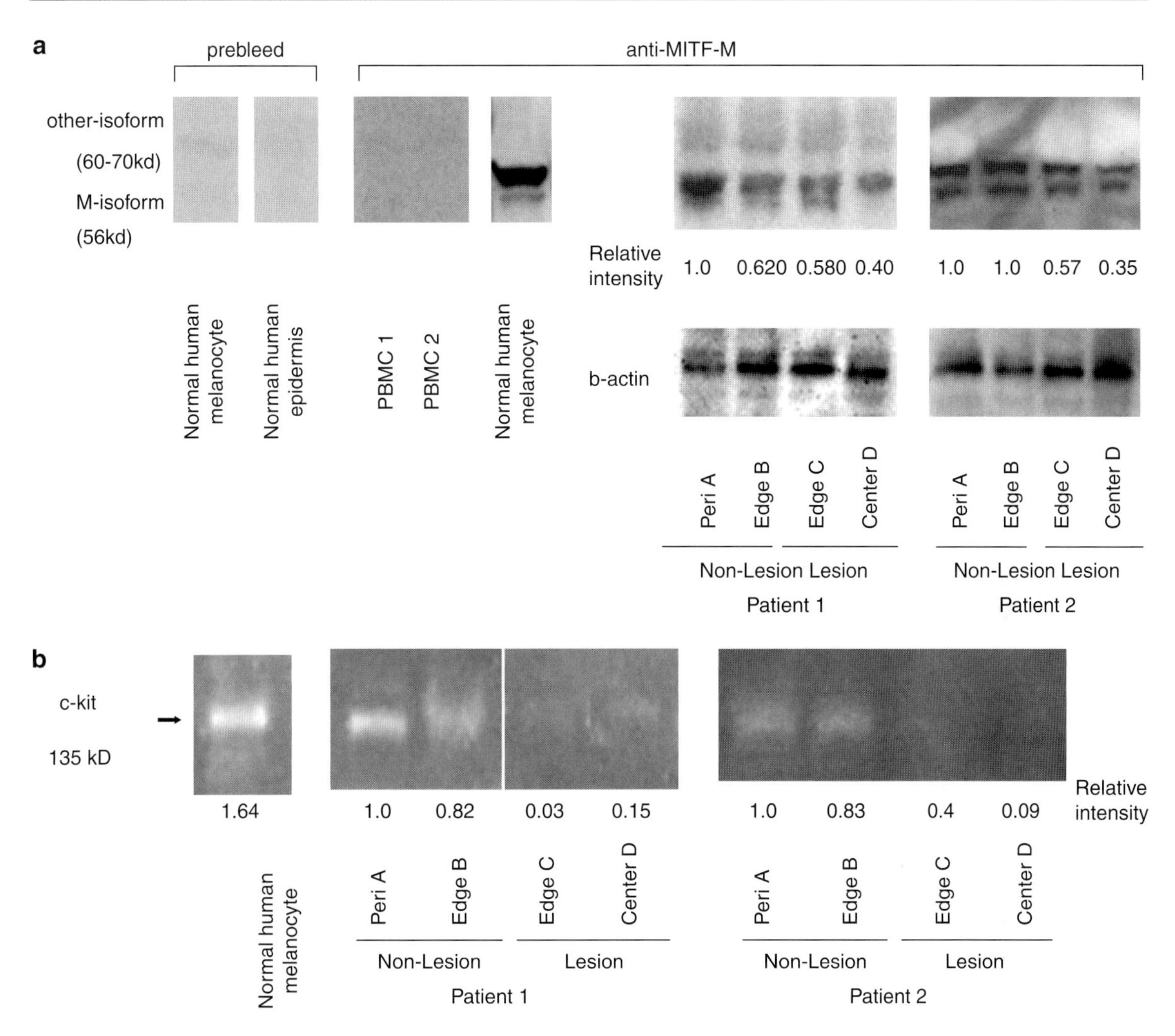

Fig. 2.2.8.7 A: Western blotting of MITF-M protein in the epidermis of lesional vitiligo and in nonlesional skin. Expression of the M-isoform is decreased at the edges of the vitiligo lesion (*Edge C*) and in a case, of the nonlesional epidermis (*Edge B*) compared with the nonlesional epidermis (*Peri A*). Even at the center of the lesion, there is a slight expression of MITF-M protein, suggesting that melanocytes might still exist there but did not express tyrosinase, S100α or ETBR. Methods: the extracts of separated epidermis from vitiligo lesions (10 μg protein/lane) were subjected to 10% SDS-polyacrylamide gel electrophoresis, followed by electroblotting and then immunostaining with antibodies to MITF-M as noted. Protein staining with Coomassie blue was concomitantly performed according to standard procedures and the loading was confirmed to be consistent in all lanes of each specific gel. **B**: Western blotting of KIT protein in the epidermis of vitiligo lesions and in nonlesional skin. The extracts of separated epidermis from vitiligo lesions (10 μg protein/lane) were subjected to 10% SDS-polyacrylamide gel electrophoresis, followed by electroblotting and then immunostaining with antibodies to KIT protein as noted. Protein staining with Coomassie blue was concomitantly performed according to standard procedures and the loading was confirmed to be consistent in all lanes of each specific gel. KIT protein expression is markedly decreased at the edges (*Edge C*) as well as at the center (*Center D*) of the vitiligo lesion compared with the nonlesional epidermis (*Peri A* and *Edge B*)

MITF/Bcl2 interaction [12] may also be involved in the loss of melanocytes in lesional epidermis. Although biological mechanisms involved in the MITF-M deficiency observed in the vitiligo lesional melanocytes remain unknown, recent evidence that oxidative stress such as H_2O_2 generation leads to the down-regulation of MITF expression in cultured human melanocytes [8] suggests that the auto-cytotoxic mechanism [11, 16] by superoxides may underlie the MITF-M deficiency in vitiligo skin. Thus, we propose the hypothesis that the decreased expression of c-kit and its downstream events, including the expression of MITF-M, in melanocytes may be responsible for the dysfunction and/or the loss of melanocytes in vitiligo epidermis.

References

1. Amae S, Fuse N, Yasumoto K et al (1998) Identification of a novel isoform of microphthalmia-associated transcription factor that is enriched in retinal pigment epithelium. Biochem Biophys Res Commun 247:710–715
2. Fuse N, Yasumoto K, Suzuki H et al (1996) Identification of a melanocyte-type promoter of the microphthalmia-associated transcription factor gene. Biochem Biophys Res Commun 219:702–707
3. Fuse N, Yasumoto K, Takeda K et al (1999) Molecular cloning of cDNA encoding a novel microphthalmia-associated transcription factor isoform with a distinct amino-terminus. J Biochem 126:1043–1051
4. Gauthier Y, Cario-André M, Taieb A (2003) A critical appraisal of vitiligo etiologic theories. Is melanocyte loss a melanocytorrhagy? Pigment Cell Res 16:322–332
5. Hachiya A, Kobayashi A, Ohuchi A et al (2001) The paracrine role of stem cell factor/c-kit signaling in the activation of human melanocytes in ultraviolet B-induced pigmentation. J Invest Dermatol 116:578–586
6. Hemesath TJ, Price ER, Takemoto C et al (1998) MAP kinase links the transcription factor Microphthalmia to c-Kit signalling in melanocytes. Nature 391:298–301
7. Hou L, Panthier JJ, Arnheiter H (2000) Signaling and transcriptional regulation in the neural crest-derived melanocyte lineage: interactions between KIT and MITF. Development. 127:5379–5389
8. Jimenez-Cervantes C, Martinez-Esparza M, Perez C et al (2001) Inhibition of melanogenesis in response to oxidative stress: transient downregulation of melanocyte differentiation markers and possible involvement of microphthalmia transcription factor. J Cell Sci 114:2335–2344
9. Jin SK, Nishimura KE, Akasaka E et al (2009) Epistatic connections between MITF and endothelin signaling in Waardenburg syndrome and other pigmentary disorders. FASEB J
10. Kitamura R, Tsukamoto K, Harada K et al (2004) Mechanisms underlying the dysfunction of melanocytes in vitiligo epidermis: role of SCF/KIT protein interactions and its downstream effector, MITF-M. J Pathol 202: 463–475
11. Maresca V, Roccella M, Roccella F et al (1997) Increased sensitivity to peroxidative agents as a possible pathogenic factor of melanocyte damage in vitiligo. J Invest Dermatol 109:310–313
12. McGill GG, Horstmann M, Widlund HR et al (2002) Bcl2 regulation by the melanocyte master regulator Mitf modulates lineage survival and melanoma cell viability. Cell 109:707–718
13. Mochii M, Mazaki Y, Mizuno N et al (1998) Role of Mitf in differentiation and transdifferentiation of chicken pigmented epithelial cell. Dev Biol 193:47–62
14. Oboki K, Morii E, Kataoka TR et al (2002) Isoforms of mi transcription factor preferentially expressed in cultured mast cells of mice. Biochem Biophys Res Commun 290: 1250–1254
15. Sakamoto A, Yanagisawa M, Sakurai T et al (1991) Cloning and functional expression of human cDNA for the ETB endothelin receptor. Biochem Biophys Res Commun 178: 656–663
16. Schallreuter KU, Wood JM, Berger J (1991) Low catalase levels in the epidermis of patients with vitiligo. J Invest Dermatol 97:1081–1085
17. Takeda K, Yasumoto K, Kawaguchi N et al (2002) Mitf-D, a newly identified isoform, expressed in the retinal pigment epithelium and monocyte-lineage cells affected by Mitf mutations. Biochim Biophys Acta 1574:15–23
18. Udono T, Yasumoto K, Takeda K et al (2000) Structural organization of the human microphthalmia-associated transcription factor gene containing four alternative promoters. Biochim Biophys Acta 1491:205–219
19. Wu M, Hemesath TJ, Takemoto CM et al (2000) c-Kit triggers dual phosphorylations, which couple activation and degradation of the essential melanocyte factor Mi. Genes Dev 14:301–312
20. Xu W, Gong L, Haddad MM et al (2000) Regulation of microphthalmia-associated transcription factor MITF protein levels by association with the ubiquitin conjugating enzyme hUBC9. Exp Cell Res 255:135–143

Proopiomelanocortin and Related Hormones

2.2.9

Markus Böhm

Contents

M. Böhm
Department of Dermatology, University of Münster, Münster, Germany
e-mail: bohmm@uni-muenster.de

Core Messages

- The cutaneous proopiomelanocortin (POMC) system is an important regulatory system, which controls pigmentation, inflammation, and the stress response of the skin.
- Based on their potent melanotropic effects, selected melanocortin peptides derived from POMC have been tested in the past for their efficacy in vitiligo.
- To investigate the potential role of the POMC system in the pathogenesis of vitiligo, a limited number of experimental studies have been performed until now. These studies aimed at (*a*) determining the peripheral blood levels of circulating POMC peptides, (*b*) examining possible genetic defects of distinct POMC components, and (*c*) investigating the in situ expression of various components of the POMC system.

2.2.9.1 General Aspects of the Cutaneous Proopiomelanocortin System

Proopiomelanocortin (POMC) is a ~31 kDa prohormone and the precursor for two biologically active families of peptides, the melanocortins and endorphins [10]. The term *melanocortin* emphasizes the originally discovered bioactivity of these peptides, that is, their stimulatory (melanotropic) effect on pigment cells and their steroidogenic (corticotropic) effect on adrenocortical cells. *Endorphins*, on the other hand, are classified into

M. Picardo and A. Taïeb (eds.), *Vitiligo*,
DOI 10.1007/978-3-540-69361-1_2.2.9, © Springer-Verlag Berlin Heidelberg 2010

the family of endogenous opioids, which are well-known for their pain-relieving effect in the central nervous system.

The *melanocortins* include adrenocorticotropin (ACTH) and the various melanocyte-stimulating hormones (α–, β– and γ–MSH). All melanocortin peptides share a common central peptide motif, His^6–Phe^7–Arg^8–Trp^9, which is required for their melanotropic activity [15]. Of note, the full sequence of α-MSH is contained within the amino acids 1–13 of the N-terminal portion of ACTH. In contrast to ACTH, α-MSH requires N-acetylation and C-amidation to obtain its full biologic activity [10]. The biological action of all melanocortin peptides is mediated via melanocortin receptors (MC-Rs), which belong to the superfamily of G-protein-coupled receptors (GPCRs) with seven transmembrane domains. Five MC-R subtypes, MC-1-5R, have been cloned and biochemically characterized [9]. They differ in their relative binding affinity to each melanocortin. Regarding MC-1R, both α-MSH and ACTH bind with similar affinity. After the binding of a high-affinity ligand, all MC-Rs activate Gs followed by increased adenylate cyclase activity and rise of intracellular cAMP. The two naturally occurring antagonists of melanocortin peptides are agouti-signaling protein (ASIP) and agouti-related protein (AGRP). While ASIP blocks ligand binding to MC-1R and MC-4R, AGRP is antagonistic to MC-3R and MC-4R [39].

The second prototype of a biologically active POMC-derived peptide is β-endorphin (β-ED). Like other endogenous opioids, β-ED does not interact with MC-Rs, but binds to the opioid receptors (ORs). Three OR types, the μ-OR (MOR), the δ-OR (DOR) and the κ-RO (KOR) are distinguished [19]. β-ED binds to both MORs and DORs. Other endogenous opioids such as enkephalin bind DORs over MORs while endomorphin is a selective MOR agonist and dynorphin is a selective KOR agonist. Importantly, enkephalin and dynorphin are not POMC-derived peptide, but are synthesized from other preproenkephalin and preprodynorphin, respectively. Of note, ligand binding to ORs are mostly *inactivating* on adenylate cyclase due to activating $G_{i/o}$ and $G_{z,}$ although stimulatory effects have also been documented [13]. Interestingly, there is evidence that β-ED has also melanotropic activity in vitro, on epidermal and follicular melanocytes [17, 18].

The biosynthesis of melanocortins and endorphins is mediated via prohormone convertases (PCs). These enzymes belong to the evolutionary conserved family of serine proteases of the subtilisin/kexin type. PC1 and PC2 are considered the main POMC-processing enzymes, although other members of the PC family, for example, PACE4 and furin convertase are also capable of processing POMC [29]. PC1 generates ACTH and β-lipotropic hormone whereas PC2 is critical for the production of α-MSH and β-ED. In the human system, there is currently no evidence of a biological function of β-lipotropic hormone.

Although POMC, POMC-derived peptides, and PCs were long considered to be restricted to the neuroendocrine system, it is now established that the skin expresses a functional analogon of the hypothalamic–pituitary–adrenal axis. Accordingly, all components of an autonomous functional POMC system including POMC, POMC-derived peptides PCs, and the upstream POMC regulator corticotropin releasing hormone are detectable in the majority of skin cell types in vitro and in situ [32]. In analogy to the induction of the classical hypothalamic–pituitary–adrenal axis by stress, various forms of *cutaneous* stressors, for example, ultraviolet (UV) light or proinflammatory cytokines induce the cutaneous POMC system.

Another important aspect in the cutaneous POMC system is the expression of MC-Rs outside the pigmentary system [6]. For example, MC-1R has been detected in epidermal keratinocytes, fibroblasts, microvascular endothelial cells, and other cutaneous cells types. Moreover, MC-1R is not the only MC-R subtype detectable in these nonpigmentary skin cells since human dermal papilla cells, and most recently, epidermal melanocytes were express MC-4R concomitantly with MC-1R [4, 35]. The extracutaneous effects of α-MSH are thus pleiotropic and not only include induction of melanocyte proliferation and melanogenesis [1] but also regulation of keratinocyte proliferation and differentiation [31], collagen metabolism [3, 21] and cytoprotection [5]. In addition, α-MSH has multiple modulatory effects on cells of the immune system [7], which in synergy with its immunomodulatory effects on resident skin cells, support a role of the cutaneous POMC system as a regulator of inflammatory, and immune responses of the skin.

2.2.9.2 Melanocortin Peptides: Potential Therapeutic Agents in Vitiligo?

Based on their pioneering discovery of α-MSH as a melanotropic hormone in man [22], it is conceivable that Lerner and colleagues proceeded to test the potential efficacy of melanocortins in patients with vitiligo. In a pilot "experiment" performed on a 29-year-old man with extensive and progressive vitiligo, racemized α-MSH (0.33 mg) was administered intradermally for ten consecutive days on three separate vitiliginous areas on the right arm [23]. This patient had previously undergone surgery to assess the role of the sympathetic nervous system on human pigmentation. However, no change was seen in the vitiliginous or normal areas after α-MSH treatment. Furthermore, microscopic examination of DOPA-stained skin sheets from the injected sites and from normal skin did not reveal any significant increase in the number of melanocytes. Fifteen months later, the same patient received α-MSH intramuscularly in a dose-escalating scheme starting from 5 to 10 mg twice weekly for up to 23 weeks. In contrast to nonlesional skin, which became darker (and accordingly displayed an increased amount of melanin upon microscopic examination), no changes were observed in the vitiliginous areas.

In subsequent studies, ACTH or the truncated ACTH peptide, tetracosactid, was tested. In general, these studies were small retrospective case studies in which the peptides were injected either intramuscularly or intradermally. In the first of these studies, 27 patients with vitiligo and no response to previous photochemotherapy with psoralen were treated intramuscularly with 25–40 IU of "long-acting" ACTH twice weekly [11]. Ten to twelve injections were given followed by a break of 2–4 weeks with repetition of such courses for up to four times. After 6 months, 16 patients showed a repigmentation of 80%, 6 patients a repigmentation of 50%, and 4 cases an improvement of 20% or less. One patient did not respond. No side effects were reported. In another case study, Hernandez-Pérez [14] treated 20 patients with nonsegmental vitiligo. The patients suffered from vitiligo for more than 5 years with involvement of more than 60% of body surface. The ACTH gel (40 mg) was given again intramuscularly twice weekly for 5 weeks. After a 5-week break, another course of ACTH treatment was performed and repigmentation was classified as follows: "excellent" (80% and more clearing),"good" (50–80% clearing), "regular" (20–50%) and "poor" (less than 20%). In 6 patients, the response was excellent, among one case with 100% repigmentation and four cases with at least 80% improvement. In 14 patients the ACTH response was classified as poor. Unfortunately, depigmentation recurred rapidly in the responding patients after cessation of the second ACTH course with no further therapeutic effect of additional ACTH treatment. Like in the first ACTH case series, no side effects were observed.

In the final case series, tetracosactid, consisting of the first 24 amino acids of ACTH was injected intralesionally into 9 patients with vitiligo [2]. All the patients had local vitiligo for at least 6 months, and had not responded to previous treatment with photochemotherapy with psoralen and topical Class IV corticosteroids. Controls consisted of two vitiligo patients with no treatment. Prior to and under therapy with the injected tetracosactid, the vitiligo lesions were measured using curvimeter for circular and semicircular lesions. Irregular lesions were documented by transparent graph paper. Tetracosactid (1 mg in 1 mL) was injected intradermally into each lesion once weekly for two months. A statistically significant improvement of the injected lesions ($p < 0.001$) was observed and consisted in a wave of pigmentation starting from the periphery of the injected site, often leaving a central depigmented area. The treatment response was most apparent in the first months of therapy. Interestingly, the response of the vitiligo lesions to intradermal tetracosactid was variable in different body sites. No further outcome data were provided by the authors.

2.2.9.3 POMC-Derived Peptides and Related Peptides in Peripheral Blood of Patients with Vitiligo

Three studies have addressed the peripheral blood levels of POMC-derived peptides and related peptides in patients with vitiligo. Mozzanica et al. [25] studied the plasma levels and circadian rhythm of α-MSH, β-ED and met-enkephalin in vitiligo patients. Using radioimmunoassay (RIA), these authors studied the plasma levels of the above peptides. Twelve of the patients had "more widespread" vitiligo, 2 patients had localized vitiligo. The duration of the disease of the vitiligo patients ranged from 2 to 30 years. 8 out of the 14 vitiligo patients had active vitiligo. 12 healthy individuals served as controls. No alteration in the pg/ml levels of α-MSH was found in patients with vitiligo versus healthy controls. However, plasma levels of β-ED were higher in stable vitiligo than in healthy individuals. Moreover, the peripheral blood levels of met-enkephalin were higher in vitiligo patients, especially in those with active disease. Interestingly, circadian rhythm of both β-ED and met-enkephalin was lost in patients with vitiligo. In another more recent study 40 patients with various forms of vitiligo (18 patients with focal, 11 patients with generalized, and 11 patients with segmental vitiligo) and 23 healthy controls were evaluated for their β-ED plasma levels [8]. Twenty-three of the vitiligo patients had progressive disease, 17 cases had stable disease. The overall plasma levels of β-ED in the tested vitiligo patients as detected by RIA were significantly higher than in normal controls. However, no correlation was detected between the relative β-ED plasma level and the progressiveness of the vitiligo.

Recently, Pichler et al. determined the plasma levels of both α-MSH and ACTH in 40 patients with non-segmental vitiligo [27]. Twenty-one patients had active and 19 patients had stable disease. Sixteen of the patients had a history of an additional autoimmune disease such as Hashimoto's thyroiditis or Graves' disease. In addition to the above, melanocortin peptides serum cortisol levels were determined to further detect possible alterations within the classical HPA axis. α-MSH was detected by RIA while ACTH and cortisol were determined by immunometric assays. Compared with 40 matched normal individuals, vitiligo patients displayed a significantly *lower* median α-MSH plasma level but a *higher* ACTH plasma level. No difference was found in the levels of morning serum cortisol. The data on the plasma levels α-MSH in vitiligo are thus in contrast to the previously published findings [25].

2.2.9.4 Genetic Abnormalities of POMC Components in Vitiligo

Two studies have addressed if patients with vitiligo have genetic defects of distinct components of the POMC system [26, 36]. The so-far conducted studies focused on the genes for MC-1R and ASIP, both master regulators of murine coat color [24, 28]. Moreover, MC-1R is highly polymorphic and distinct loss of function alleles has been associated with red hair and pale skin [38]. However, patients with red hair and pale skin are typically not affected by visible vitiligo as they have little epidermal eumelanin and do not respond to UV with tanning.

In order to test if patients with vitiligo exhibit an altered frequency of MC-1R and ASIP alleles, Na et al. performed a single nucleotide polymorphisms (SNP) analysis of 114 Korean patients and 111 normal controls [26]. Using direct sequencing of the MC-1R coding sequence, the frequency of the Arg67Gln (G200A), Val92Met (G274A), Ile120Thr (T359C), Arg160Arg (C478A) and Gln163Arg (A488G) SNP was determined in both cohorts. Although some variations were detected none of the associations were statistical significant indicating that both MC-1R and ASIP SNP do not play a pathogenetic role in vitiligo. In another study, SNP of the MC-1R and ASIP genes were investigated in 97 Hungarian Caucasian individuals and 59 healthy controls [36]. Patient characteristics included age, gender, duration of vitiligo (the majority suffering from the disease for more than 10 years), family history and presence of other autoimmune diseases. In addition, the possible connection between vitiligo and pigmentation was analyzed for vitiligo patients and controls by grouping both cohorts according to skin, hair and eye color. The full-length *MC-1R* gene was sequenced after PCR amplification. The 3′ untranslated region (3′ UTR) SNP of ASIP was detected by restriction fragment length polymorphism. No differences in the allele frequency of the ASIP SNP could be found. As the Arg160Trp allele of MC-1R however showed a significantly higher frequency among fair-skinned individuals than in dark-skinned individuals, both vitiligo and controls were regrouped. The frequency of MC-1R was then analyzed separately in fair skinned and dark skinned subgroups of vitiligo patients and controls. The C478T SNP of MC-1R showed a significant difference in allele frequency in fair-skinned vitiligo patients and in fair-skinned healthy controls ($p = 0.0262$, odds ration 3.6) with a higher allele frequency in the control group. The authors suggested that the Arg160Trp amino acid change is protective against vitiligo. Although the results of this study are interesting, the overall number of sequenced patients is too low. Moreover, the data of the latter study are confounded by stratification, for which additional testing (Bonferroni's correction for multiple testing) would be needed.

2.2.9.5 Expression of the Cutaneous POMC System in Vitiligo

Thody and coworkers were the first who reported the presence of α-MSH in mammalian skin [37]. They found ng/g amounts of immunoreactive α-MSH in human, rat, gerbil and mouse skin as determined by α-MSH RIA. The highest concentration of α-MSH was detected in human skin, especially within the epidermis. Using suction blister tops, the immunoreactive amounts of α-MSH in the epidermis were also measured in normally appearing and in lesional skin of a limited number of vitiligo patients ($n = 4$). The amounts α-MSH in lesional and nonlesional skin of vitiligo patients was similar but the overall concentration (96.7 ± 12.9 and 86.6 ± 10.1 pg/blister top) appeared less than in normal epidermis (148.6 ± 11.5 pg/blister top; $n = 2$). Subsequent immunohistochemical studies on skin biopsies (lesional, perilesional and nonlesional) from vitiligo patients ($n = 6$) and control subjects ($n = 6$) further supported a reduction in the level of α-MSH in lesional and perilesional skin, especially within the melanocytes [12]. Accordingly, the number of α-MSH-positive *melanocytes* detected by double immunostaining with the melanocyte antibody Mel-5 was significantly reduced in nonlesional and perilesional skin of vitiligo patients compared with controls. Similar to α-MSH there was a reduction in the number of melanocytes staining for PC1 and PC2 in vitiligo skin (both in nonlesional and perilesional skin) compared with control skin. In contrast, epidermal ACTH immunostaining did not differ in its intensity and localization in vitiliginous and control skin. These data suggested that reduced α-MSH amounts within melanocytes of normally appearing and perilesional skin of patients with vitiligo could contribute to melanocyte damage, e. g. via increased oxidative stress. As can be expected, melanocytes reappearing in the repigmented skin of vitiligo patients after successful therapy again display α-MSH immunoreactivity [16].

In another study involving full-thick skin biopsies from 31 vitiligo patients and 24 healthy subjects, cutaneous expression of several components of the POMC systems was further assessed by quantitative real-time RT-PCR [20]. The patients (mean age: 30 years; mean disease duration: 19.2 years) included 25 cases with generalized vitiligo, four cases with localized non-segmental vitiligo, and 1 patient with universal vitiligo. POMC mRNA expression was 1.9-fold lower in lesional skin than in control skin. In contrast, the relative mRNA amounts of both MC-1R and also MC-4R were found to be slightly elevated in nonlesional skin of vitiligo compared with lesional skin and control skin. Regarding the increased amounts of MC-1R and MC-4R mRNA in nonlesional skin, compensatory upregulation by systemic influences was suggested [20]. RNA analysis of the melanogenesis-specific enzymes tyrosinase, tyrosinase-related protein 1 (TRP1) and dopachrome tautomerase showed a positive correlation between the MC-1R and TRP1 mRNA levels in skin of healthy controls. However, no correlation was found between the MC-1R mRNA expression levels and the expression of tyrosinase, TRP1 and dopachrome tautomerase in vitiligo. In addition, no differences were detected in the mRNA amounts of ASIP and AGRP in vitiligo skin and in controls.

Recently, two reports further advanced our knowledge on the cutaneous POMC system in vitiligo. Spencer et al. [33] reevaluated the in situ expression of α-MSH, β-MSH, β-ED and ACTH in skin of patients with vitiligo before and after UVB-activated pseudocatalase PC-KUS treatment ($n = 6$) and in skin of healthy controls ($n = 6$). In lesional skin of vitiligo patients, epidermal immunoreactivity for both α-MSH and β-ED immunoreactivity was significantly reduced. The reduced immunoreactivity of β-ED in vitiliginous skin is surprising since the peripheral blood *and* tissue fluids of this POMC-derived peptides were found to be elevated in patients with vitiligo as noted earlier [8]. In contrast to previous work by the Thody et al. [12], moreover, the latter authors detected *increased* epidermal ACTH immunoreactivity in vitiliginous skin compared with control skin. Immunoreactivity for β-MSH was similar in vitiligo skin and in control skin. Since it was previously shown that lesional skin of vitiligo patients accumulates H_2O_2 in the 10^{-3} M range [30], the possibility of oxidation and thus structural alteration leading to reduced antibody epitope recognition was tested by incubating α-MSH, β-MSH, β-ED and ACTH in vitro with H_2O_2. As shown by dot blot analysis with POMC-peptide antibodies, H_2O_2 appeared to oxidize α-MSH, β-MSH, β-ED and ACTH with β-ED being the most sensitive peptide. Functional in vitro studies could in fact demonstrate a significantly reduced in vitro melanotropic activity of β-ED exposed to H_2O_2. Based on FT-Raman spectroscopy and computer simulation the authors further suggested that Met within ACTH, α-MSH and β-ED, is sensitive towards oxidation by H_2O_2 resulting in Met sulfoxide. In support of this concept, treatment with pseudocatalase PC-KUS led to an increased immunoreactivity of both α-MSH and β-MSH in repigmented skin of patients with vitiligo [33].

In addition to POMC-derived peptides, it was recently shown that the Ca^{2+}-dependent furin convertase, a member of the PC family of POMC-processing enzymes, could be a target for oxidative stress-induced damage in vitiligo [34]. This enzyme is expressed in both normal human melanocytes and keratinocytes in culture at RNA and protein level. Furin expression as assessed by immunohistochemistry was significantly reduced in both lesional and nonlesional skin of untreated patients with progressive vitiligo compared with healthy controls (n = 10 vitiligo patients and controls). In contrast to α-MSH, β-MSH, β-ED and ACTH, antibody antigen recognition was not affected by exposure of furin protein to millimolar amounts of H_2O_2. However, exposure of furin protein to H_2O_2 *in vitro* resulted in a 55% reduction in Ca^{2+} binding as detected by binding of $^{45}Ca^{2+}$ binding assays. Computer simulation and molecular modeling predicted that this H_2O_2-induced decrease in Ca^{2+} binding is due to loss of one Ca^{2+} from the binding site 1 of the catalytic site of furin convertase [34].

2.2.9.6 Concluding Remarks

Several lines of evidence indicate that the POMC is affected in vitiligo (Table 2.2.9.1). These alterations

Table 2.2.9.1 Reported alterations of the POMC system and its possible pathophysiological role in vitiligo

POMC component	Detected alteration	Pathophysiologic consequence	Reference
β-ED	Plasma level ↑ Circadian rhythm lost	Indicator of stress?	Mozzanica et al. [25]
β-ED	Plasma level ↑	"	Caixia et al. [26]
α-MSH	Plasma level	Reduced melanogenesis, cytoprotection, altered immune function?	Pichler et al. [27]
MC1R	Allele frequency of Arg160Trp SNP ↓*	Protection against vitiligo?	Széll et al. [29]
α-MSH	Epidermal amount in lesional and nonlesional skin ↓**	Reduced melanogenesis, cytoprotection, altered immune function?	Thody et al. [33]
α-MSH	Immunoreactivity in melanocytes of perilesional skin ↓	"	Graham et al. [34]
PC1, PC2	Immunoreactivity in melanocytes of nonlesional and perilesional skin ↓	"	Graham et al. [34]
POMC	Cutaneous mRNA amount in lesional skin ↓	"	Kingo et al. [36]
MC-1R, MC-4R	Cutaneous mRNA amount in nonlesional skin ↑	Compensatory response?	Kingo et al. [36]
α-MSH, β-ED	Epidermal immunoreactivity in lesional skin ↓	Reduced melanogenesis, cytoprotection, altered immune function?	Spencer et al. [37]
β-ED	Amount in suction blister fluid of lesional skin ↑	Indicator of cutaneous of cutaneous stress?	Caixia et al. [26]
ACTH	Epidermal immunoreactivity in vitiligo skin ↑	Compensatory response?	Spencer et al. [37]
FC	Epidermal immunoreactivity in lesional and nonlesional skin and non-lesional skin ↓	Reduced melanogenesis, cytoprotection, altered immune function?	Spencer et al. [39]

*Not confirmed by others [28]
**Only n = 2 samples analyzed

include the peripheral blood levels of certain POMC-derived peptides, the in situ expression of POMC and selected POMC-derived peptides, expression of furin convertase, and the expression of MC-Rs in skin of patients with vitiligo. However, it remains to be clarified if these alterations are cause or consequence of vitiligo. At present, our knowledge on the cutaneous POMC system in vitiligo is still very fragmentary due to insufficient numbers of examined vitiligo patients and controls, lack of "hard" quantitative protein data in many of the studies, and lack of more functionally oriented research. If melanocortin peptides are of therapeutic value alone or in combination with any other current vitiligo treatment remains to be determined in future state-of the art clinical trials.

References

1. Abdel-Malek Z, Swope VB, Suzuki I et al (1995) Mitogenic and melanogenic stimulation of normal human melanocytes by melanotropic peptides. Proc Natl Acad Sci USA 92: 1789–1793
2. Al-Omari H, al-Sugair S, Ab Hussain A (1989) Intralesional injection of tetracosactid in the treatment of localized vitiligo. Int J Dermatol 28:682
3. Böhm M, Raghunath M, Sunderkotter C et al (2004) Collagen metabolism is a novel target of the neuropeptide alpha-melanocyte-stimulating hormone. J Biol Chem 279: 6959–6966
4. Böhm M, Eickelmann M, Li Z et al (2005) Detection of functionally active melanocortin receptors and evidence for an immunoregulatory activity of alpha-melanocyte-stimulating hormone in human dermal papilla cells. Endocrinology 146:4635–4646
5. Böhm M, Wolff I, Scholzen TE et al (2005) Alpha-melanocyte-stimulating hormone protects from ultraviolet radiation-induced apoptosis and DNA damage. J Biol Chem 280:5795–5802
6. Böhm M, Luger TA, Tobin DJ, Garcia-Borron JC (2006) Melanocortin receptor ligands: new horizons for skin biology and clinical dermatology. J Invest Dermatol 126:1966–1975
7. Brzoska T, Luger TA, Maaser C et al (2008) Alpha-melanocyte-stimulating hormone and related tripeptides: biochemistry, antiinflammatory and protective effects in vitro and in vivo, and future perspectives for the treatment of immune-mediated inflammatory diseases. Endocr Rev 29: 581–602
8. Caixia T, Daming Z, Xiran L (2001) Levels of beta-endorphin in the plasma and skin tissue fluids of patients with vitiligo. J Dermatol Sci 26:62–66
9. Cone RD, Lu D, Koppula S et al (1996) The melanocortin receptors: agonists, antagonists, and the hormonal control of pigmentation. Recent Prog Horm Res 51:287–318
10. Eipper BA, Mains RE (1980) Structure and biosynthesis of pro-adrenocorticotropin/endorphin and related peptides. Endocr Rev 1:1–27
11. Gokhale BB, Gokhale TB (1976) Corticotropin and vitiligo (preliminary observations). Br J Derrmatol 95:329
12. Graham A, Westerhof W, Thody AJ (1999) The expression of alpha-MSH by melanocytes is reduced in vitiligo. Ann NY Acad Sci 885:470–473
13. Harrison C, Smart D, Lambert DG (1998) Stimulatory effects of opioids. Br J Anaesth 81:20–28
14. Hernández-Pérez E (1979) Vitiligo treated with ACTH. Int J Dermatol 18:578–579
15. Hruby VJ, Wilkes BC, Hadley ME et al (1987) Alpha-melanotropin: the minimal active sequence in the frog skin bioassay. J Med Chem 30:2126–2130
16. Ichimiya M (1999) Immunohistochemical study of ACTH and alpha-MSH in vitiligo patients successfully treated with a sex steroid-thyroid hormone mixture. J Dermatol 26: 502–506
17. Kauser S, Schallreuter KU, Thody AJ et al (2003) Regulation of human epidermal melanocyte biology by beta-endorphin. J Invest Dermatol 120:1073–1080
18. Kauser S, Thody AJ, Schallreuter KU et al (2004) Beta-endorphin as a regulator of human hair follicle melanocyte biology. J Invest Dermatol 123:184–195
19. Kieffer BL, Evans CJ (2009) Opioid receptors: From binding sites to visible molecules in vivo. Neuropharmacology, 56(Suppl 1):205–212
20. Kingo K, Aunin E, Karelson M et al (2007) Gene expression analysis of melanocortin system in vitiligo. J Dermatol Sci 48:113–122
21. Kokot A, Sindrilaru A, Schiller M et al (2009) α-Melanocyte-stimulating hormone suppresses bleomycin-induced collagen synthesis and reduces tissue fibrosis in a mouse model of scleroderma. Arthritis Rheum 60:592–603
22. Lerner AB, McGuire JS (1961) The effect of alpha- and beta melanocyte stimulating hormone on the skin colour of man. Nature 189:176–79
23. Lerner AB, Snell RS, Chanco-Turner ML, McGuire JS (1966)Vitiligo and sympathectomy. The effect of sympathectomy and alpha-melanocyte stimulating hormone. Arch Dermatol 94:269–278
24. Miller MW, Duhl DM, Vrieling H et al (1993) Cloning of the mouse agouti gene predicts a secreted protein ubiquitously expressed in mice carrying the lethal yellow mutation. Genes Dev 7:454–467
25. Mozzanica N, Villa ML, Foppa S et al (1992) Plasma alpha-melanocyte-stimulating hormone, beta-endorphin, met-enkephalin, and natural killer cell activity in vitiligo. J Am Acad Dermatol 26: 693–700
26. Na GY, Lee KH, Kim MK et al (2003) Polymorphisms in the melanocortin-1 receptor (MC1R) and agouti signaling protein (ASIP) genes in Korean vitiligo patients. Pigment Cell Res 16:383–387
27. Pichler R, Sfetsos K, Badics B et al (2006) Vitiligo patients present lower plasma levels of alpha-melanotropin immunoreactivities. Neuropeptides 40:177–183
28. Robbins LS, Nadeau JH, Johnson KR et al (1993) Pigmentation phenotypes of variant extension locus alleles result from point mutations that alter MSH receptor function. Cell 72:827–834
29. Seidah NG, Benjannet S, Hamelin J et al (1999) The subtilisin/kexin family of precursor convertases. Emphasis on PC1, PC2/7B2, POMC and the novel enzyme SKI-1. Ann N Y Acad Sci 885:57–74

30. Schallreuter KU, Moore J, Wood JM et al (1999) In vivo and in vitro evidence for hydrogen peroxide (H_2O_2) accumulation in the epidermis of patients with vitiligo and its successful removal by a UVB-activated pseudocatalase. J Investig Dermatol Symp Proc 4:91–96
31. Slominski A, Paus R, Wortsman J (1991) Can some melanotropins modulate keratinocyte proliferation? J Invest Dermatol 97:747
32. Slominski A, Wortsman J, Luger T et al (2000) Corticotropin releasing hormone and proopiomelanocortin involvement in the cutaneous response to stress. Physiol Rev 80: 979–1020
33. Spencer JD, Gibbons NC, Rokos H et al (2006) Oxidative stress via hydrogen peroxide affects proopiomelanocortin peptides directly in the epidermis of patients with vitiligo. J Invest Dermatol 127:411–420
34. Spencer JD, Gibbons NC, Böhm M, Schallreuter KU (2008) The Ca2+-binding capacity of epidermal furin is disrupted by H_2O_2-mediated oxidation in vitiligo. Endocrinology 149:1638–1645
35. Spencer JD, Schallreuter KU (2009) Regulation of pigmentation in human epidermal melanocytes by functional high affinity β-MSH/MC4-R signalling. Endocrinology 150: 1250–1258
36. Széll M, Baltás E, Bodai L et al (2008) The Arg160Trp allele of melanocortin-1 receptor gene might protect against vitiligo. Photochem Photobiol 84:565–571
37. Thody AJ, Ridley K, Penny RJ et al (1983) MSH peptides are present in mammalian skin. Peptides 4:813–816
38. Valverde P, Healy E, Jackson I et al (1995) Variants of the melanocyte-stimulating hormone receptor gene are associated with red hair and fair skin in humans. Nat Genet 11: 328–330
39. Voisey J, van Daal A (2002) Agouti: from mouse to man, from skin to fat. Pigment Cell Res 15:10–18

Other Hypotheses

2.2.10

Mauro Picardo

Contents

Core Messages

- An accelerated senescence process may involve follicular melanocytes .
- Increased propensity of melanocytes to detach from the basal membrane may indicate another possible intrinsic cellular defect.
- According to the oxidative point of view, an alteration of the membrane lipid synthesis can deliver highly toxic intermediates and determine aberrant intracellular signalling.

2.2.10.1 Introduction

Besides the most widely acknowledged pathomechanisms, several possible causes for melanocyte disappearance have been hypothesized. The main alternative and innovative hypotheses are the aging-mediated or the detachement-associated melanocyte loss, the viral damage, and the mitochondrial energetic defects.

2.2.10.2 The Early Melanocyte Aging Theory

Until now most of our attention has been focused on resident epidermal melanocytes. However, exchanges between hair and epidermal compartments have been demonstrated for melanocytes. The hair follicle niche melanoblast, studied mostly in the mouse,

M. Picardo
Istituto Dermatologico San Gallicano, via Elio Chianesi
00144 Roma, Italy
e-mail: picardo@ifo.it

M. Picardo and A. Taïeb (eds.), *Vitiligo*,
DOI 10.1007/978-3-540-69361-1_2.2.10,

differentiates and migrates to the bulb whilst maintaining a reserve of indifferentiated cells inside the bulge. The niche allows thus the continuous production of new melanocytes avoiding cell exhaustion and senescence. A minority of melanocyte stem cells mature into differentiated and migrating melanocytes whereas the bulk remains quiescent. The exhaustion of the quiescent stock will affect the next melanocyte production. The ability to respond to specific growth factors (ie SCF, and the c-Kit/SCF axis appears to be broken in vitiligo) may determine the continuous production of differentiated and functioning epidermis-committed melanocytes [1] (see chapter 2.2.8). Until now the molecular and genetic background for melanocyte senescence has been investigated mainly on melanoma development. In particular, the pattern of expression of the retinoblastoma (RB) family of proteins as well as the activation of some kinases (ERK2) involved in stress signals, has been reported as featuring the senescent melanocyte status in contrast to the malignant one. The in vitro culture experiments may provide a model to test the physiological value of different growth factors and their role in determining the cell ability to divide or senesce. Vitiligo may be considered as a model for the study of stem cells and senescence process. According to this hypothesis, vitiligo melanocytes may present defects in proliferative and differentiation potential at the bulge level. It may be considered a localized senescence process through a mechanism similar to the "free radical-mediated graying", associated with a progressive melanocyte loss from the hair bulb. In accordance with this hypothesis, around one third 35% of cases of vitiligo are associated with early hair graying affecting relatives or more rarely the patient. The clinical value of a possible accelerated senescence process is relevant in terms of therapy. The presence or not of a functioning follicular reservoir might be the crucial factor for the success of an autologous graft.

2.2.10.3 The Melanocyte Detachment ("Melanocytorrhagic") Theory

Frequently, the onset of vtiligo spot takes place at the site of mild physical pressure or excoriation, and the histological evaluation indicates the increased amount of tenascin in papillary dermis (see Sect. 2.2.2.1). On the basis of the above clinical observation, an in vivo model of melanocyte detachment has been defined [5]. After the application of a calibrated physical stress, the progressive detachment and loss of epidermal melanocytes was shown with non dendritic cells found in the upper epidermal layers, without (histological) evidence of frank death. Even if the cause of tenascin extracellular deposition is still unknown, it has been suggested to inhibit melanocyte adhesion to fibronectin allowing their detachment. Among possible causes of tenascin production, a cytokine-mediated induction has been proposed. Melanocyte adhesion is a phenomenon mediated by the interaction of several classes of proteins and by the shape of the melanocyte, able to contact basement membrane sites through its dendrites. The adhesion between melanocyte and keratinocyte is mainly due to cadherin-bcatenin complexes, whereas the adhesion to the basal membrane is mediated by the Ca-independent a6b1integrins. On the other hand, dendrites promote the anchoring and adhesion events through Ca-dependent integrins. Moreover, the dendrites shape is dependent on cadherin binding properties. The primary or secondary nature of the loose attachment of vitiligo melanocytes is a matter of debate. In vitiligo epidermis, oxygen radicals and the melanin toxic derivatives could indeed be involved in the compromised multiproteic junctions as well in abnormal dendrite morphology. The in vitro addition of H_2O_2 has been reported to affect the dendrites retraction, the first step of melanocyte detachement. Moreover, keratinocyte oxidative damage may contribute to tenascin production and release in the extracellular space.

2.2.10.4 The Viral Hypotheses

Some sporadic and still controversial data have been considered as supporting a possible viral infection triggering the onset of vitiligo. Epidemiological data have shown the increased prevalence of HTLV1 (human T cell lymphotropic virus type 1) in patients with vitiligo (0.7% versus 0.22% in control blood donors) [9]. HTLV1 is the etiological agent of adult T cell lymphoma/leukemia, tropical spastic paraparesis, and uveitis. A trigger role for cytomegalovus has been also proposed [6], on the basis of the presence in skin biopsies of CMV DNA in 38% of the patients. The highlighted characteristics of the virus in favour of vitiligo pathogenesis included easy contamination of hair follicles and nerve endings.

2.2.10.5 Deficient Clearance of Apoptotic Fragments

Other data, to consider with caution, suggest that a deficient clearance of apoptotic fragments by mannose-binding lectin (MBL) may be involved (chapter 2.2.7.2). MBL is a calcium-dependent lectin involved in the early phase of the innate immunity, before the production of antibodies. It activates the complement, promotes opsonophagocytosis, modulates the inflammatory response, and induces apoptosis. Inadequate clearance of apoptotic bodies has been associated with some autoimmune diseases through the perpetuation of immune activation [8]. The above suggestion may fit the recently proposed hypothesis of the inflammasome (see chapter 2.2.1). Cellular debris or apoptotic fragments could activate NALP1, within the inflammasome complex, activating the IL1 pathway and alerting the immune system. In vitiligo patients, an altered calcium metabolism together with the polymorphism of MBL may lead to decreased level of MBL determining a compromised clearance of apoptotic fragments.

2.2.10.6 The membrane Lipid Defect Mechanism

According to the oxidative-mediated damage, a subtle link between ROS generation and membrane structure has been also put forward. On the basis of this hypothesis melanocytes, as well as other cell types, may present biochemical defects, probably due to a predisposing genetic background, affecting the structure and functionality of the membranes. A compromised membrane could render cells differentially sensitive to external and internal agents (UV, cytokines, catechols, melanin intermediates, growth factors withdrawal), usually ineffective on normal cell activity and survival. The impaired arrangement of lipids, involving fatty acids and cholesterol, may affect the transmembrane location of proteins with enzymatic or receptor activities. A possible effect of this membrane defect could be a defective activity of the proteins of the mitochondrial electron transport chain, which is strictly associated with membrane lipids, thus impairing the ATP production and affecting cell survival. The final result could depend on the intensity or duration of the stimuli, i.e. a mild aggression leading to a reduction of ATP production would impair adhesion; a greater stimulus acting as pro-apoptotic agent would affect the mitochondrial cell survival checkpoints; finally, a major stress causing cell death would be associated with an inflammatory infiltrate.

Finally, the altered expression and release of transmembrane proteins could be the basis for the exposure of antigens considered as non self by the immune system, which would deliver in turn an inappropriate immune response [2, 3, 4, 7]. In addition, hsp70 production related to membrane damage could improve immune recognition and the ensuing immune system activation.

References

1. Arck PC, Overall R, Spatz K et al (2006) Towards a free radical theory of graying: melanocyte apoptosis in the aging human hair follicle is an indicator of oxidative stress induced tissue damage. FASEB J 20:1567–1569
2. Broquet AH, Thoma G, Masliah J et al (2003) Expression of the molecular chaperone Hsp70 in detergent-resistant microdomains correlates with its membrane delivery and release. J Biol Chem 278:21601–21606
3. Dell'anna ML, Ottaviani M, Albanesi V et al (2007) Membrane lipid alterations as a possible basis for melanocyte degeneration in vitiligo. J Invest Dermatol 127: 1226–1233
4. D'Silva P, Liu Q, Walter W et al (2004) Regulated interactions ofmtHsp70 with Tim44 at the translocon in the mitochondrial inner membrane. Nat Struct Mol Biol 11: 1084–1091
5. Gauthier Y, Cario André M, Taieb A (2003) A critical appraisal of vitiligo etiologic theories: is melanocyte loss a melanocitorrhagy? Pigment Cell Res 16:322–332
6. Grimes PE, Sevall JS, Vojdani A (1996) Cytomegalovirus DNA identified in skin biopsy specimens of patients with vitiligo. J Am Acad Dermatol 35:21–26
7. Kroll TM, Bommiasamy H, Boissy RE et al (2005) 4-tertiary butyl phenol exposure sensitizes human melanocytes to dendritic cell-mediated killing: relevance to vitiligo. J Invest Dermatol 124:798–806
8. Onay H, Pehlivan M, Alper S et al (2007) Might there be a link between mannose binding lectin and vitiligo? Eur J Dermatol 17:146–148
9. Nobre V, Guedes AC, Proietti FA (2007) Increased prevalence of human T cell lymphotropic virus type 1 in patients attending a Brazilian dermatology clinic. Intervirology 50:316–318

Segmental Vitiligo: A Model to Understand Vitiligo?

2.3

Seung-Kyung Hann, Alain Taïeb, Yvon Gauthier, Laïla Benzekri, Hsin-Su Yu, Cheng-Che Eric Lan, and Ching-Shuang Wu

Core Messages

- The hair follicle melanocytic compartment is more commonly affected in segmental vitiligo (SV)
- Distribution in SV parallels that of other acquired pigmentation disorders such as nevus spilus, pointing out to some underlying developmental defect
- Prognosis can usually be made easily based on initial distribution
- The association of SV to nonsegmental vitiligo (NSV) indicates a continuum between the two subsets with shared predisposing genetic factors, including genes operating specifically in the skin
- Some pedigrees associating SV and NSV further suggest a mechanism of loss of heterozygosity for a dominant gene, which is controlling part of the cutaneous phenotype
- The role of the nervous system in the pathogenesis of SV is probable but is poorly established because neural dysfunction is not usually associated with hypopigmentation
- In vitro studies indicate how phototherapies induce repigmentation, and this knowledge is applicable to SV
- Future directions for the treatment of SV should take neurocutaneous interactions into consideration

Introduction

Alain Taïeb

Segmental vitiligo (SV) has been opposed to generalized, nonsegmental vitiligo (NSV). However, recent evidence showing that the two forms are not mutually exclusive challenge classical vitiligo pathophysiological schemes. There is a growing consensus that SV may provide clues to understand basic mechanisms underlying NSV and repigmentation mechanisms. This section reviews some of the current concepts applicable to SV derived from clinical observations and experimental data. Some opinions given thereafter by leading experts in the field are difficult to reconcile, and we have thus left this section in a format that leaves the debate open.

S.-K. Hann (✉)
Korea Institute of Vitiligo Research, Drs. Woo & Hann's skin clinic, Yongsan-Gu, Seoul, South Korea
e-mail: skhann@paran.com

M. Picardo and A. Taïeb (eds.), *Vitiligo*,
DOI 10.1007/978-3-540-69361-1_2.3, © Springer-Verlag Berlin Heidelberg 2010

2.3.1 Particular Clinical Characteristics of Segmental Vitiligo

Seung-Kyung Hann

Contents

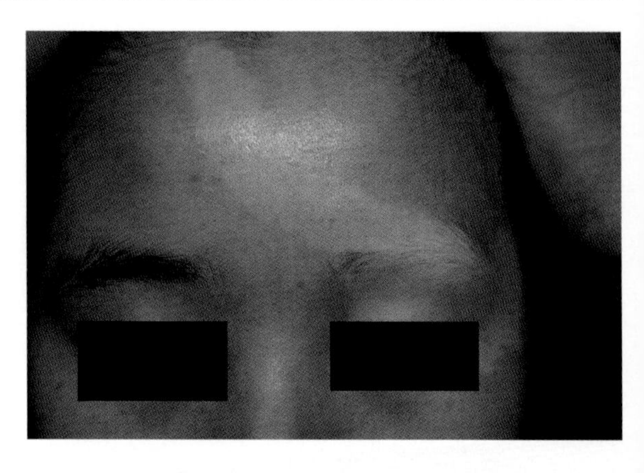

Fig. 2.3.1 The lesion which starts from the right side of the forehead and crosses the midline of the face and spreads down to the eyeball, nose and cheek of the left side of the face

In 1977, Koga performed a sweat-secretion stimulation test using physiostigmine and accordingly reclassified vitiligo into nonsegmental type (Type A), and segmental type (Type B) [6]. He proposed that the nonsegmental type results from immunologic mechanisms, while segmental type results from dysfunction of the sympathetic nervous system in the affected skin (see below, Gauthier and Benzekri). Following are personal observations, which suggest alternative hypotheses.

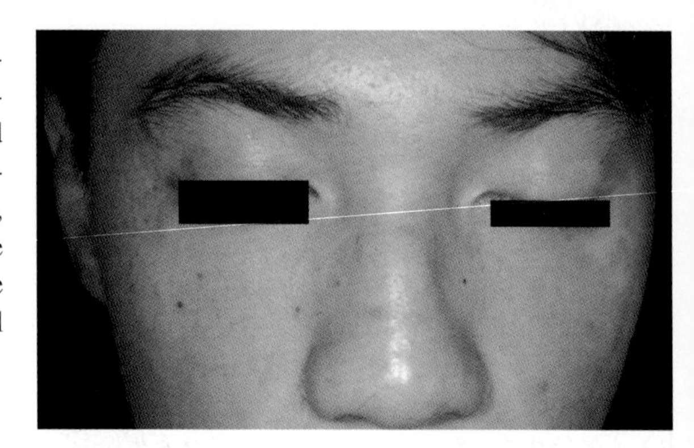

Fig. 2.3.2 Segmental vitiligo and nevus spilus coexist symmetrically

Clinical Observations and Hypotheses for SV Patterning

The key aspect of SV cases is that the lesions do not cross the midline (with exceptions as featured on Fig. 2.3.1), and are distributed along a unilateral dermatome, thus enabling usually a reliable prognosis [3, 4, 7]. However, the actual distribution of the depigmentation does not always correspond very well to a true dermatome, as is seen in other cutaneous disorders like herpes zoster (see the discussion by Gauthier and Benzekri and Chapter 1.3.2). It has been suggested that SV might follow Blaschko's lines (BL) (see later, Taïeb) or acupuncture lines [2]. Another possibility is that the distribution of SV follows a so-far unknown pathway corresponding to a group of identical clonal cells.

Differing characteristics of melanocyte survival and melanogenesis may influence the location/course of the lesions. In Fig. 2.3.2, the 14-year-old male patient had vitiligo around the left eye while nevus spilus occurred in a symmetrical fashion. Both conditions appeared almost simultaneously several months before, suggesting a variant twin-spotting phenomenon [1]. The same cutaneous pattern may also occur in different diseases – SV (Fig. 2.3.3a) and nevus spilus (Fig. 2.3.3b). These findings suggest that clones of melanocytes with different functional characteristics exist in acquired pigmentary disorders. Melanocytes in nevus spilus can produce more melanin than is customary, and may have a normal lifespan. The type of melanocytes found in SV of the same location can be easily lost by an unknown mechanism. The exact nature of this phenomenon has not been fully elucidated, but might reflect the embryonic migration pathway of a melanocyte colony with an inherited defect, which is not readily observable at

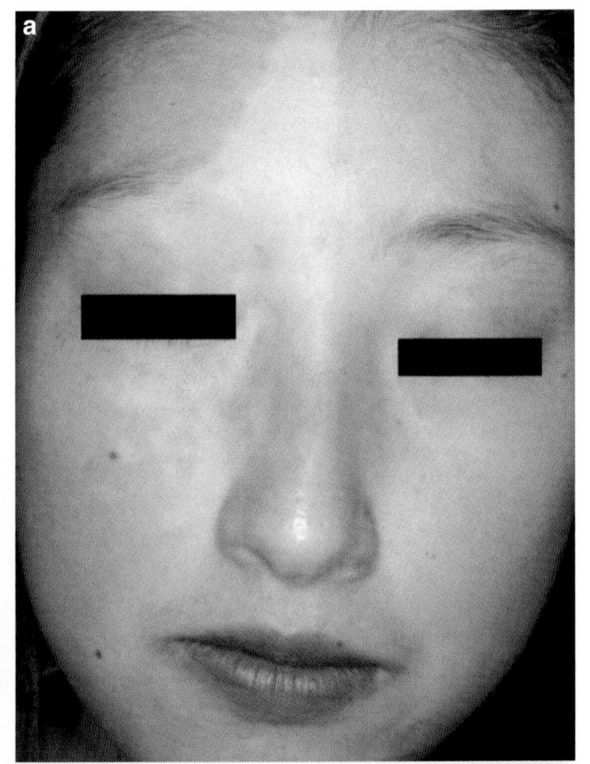

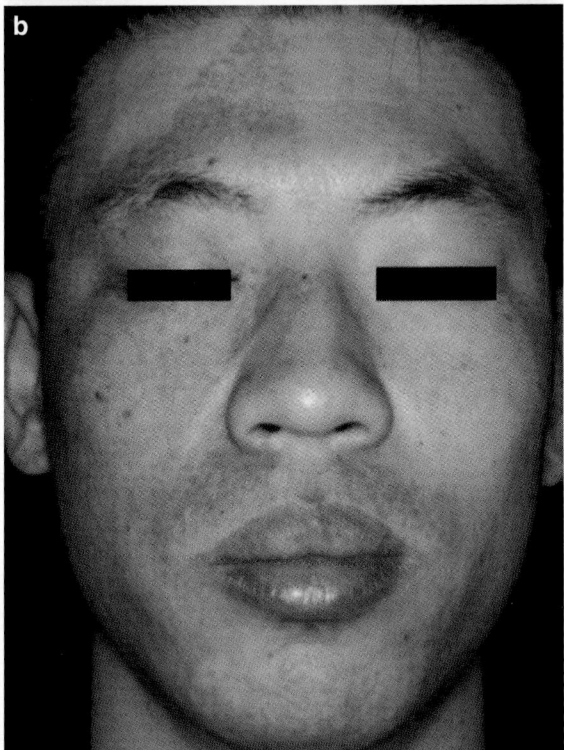

Fig. 2.3.3 Peculiar distribution of segmental vitiligo crossing trigeminal dermatomes (**a**), and nevus spilus of similar distribution (**b**)

birth or in the first months of life, thus suggesting an associated environmental trigger.

Preferential Follicular Reservoir Involvement in SV

Melanocytes of the hair bulb are responsible for the hair color. They transfer their melanosomes to the surrounding hair keratinocytes. In the hair follicles, melanin granules are mainly situated in the cortex, where their long axes parallel the hair surface (Chap. 1.3.6). Indeed, it is striking that the follicular compartment of the melanocyte organ may be spared while vitiligo affects the epidermal compartment. This dissociated behavior of epidermal and follicular melanocytes is very common in NSV. Body leukotrichia varies from 10% to more than 60% in the literature for all types of vitiligo. Poliosis occurs in both nonsegmental and SV. The occurrence of leukotrichia on eyebrow, scalp and eyelash has been estimated at 48.6% in a group of 101 patients with SV [3]. If cases of body involvement of SV (for which this item is less accurately documented) had been excluded, the incidence of leukotrichia-associated SV would be even higher. It is assumed that SV affects both epidermal and follicular melanocytes in its early phase of disease as compared to NSV (Fig. 2.3.4). Thus, the presence of leukotrichia should be accorded special but still not well- understood

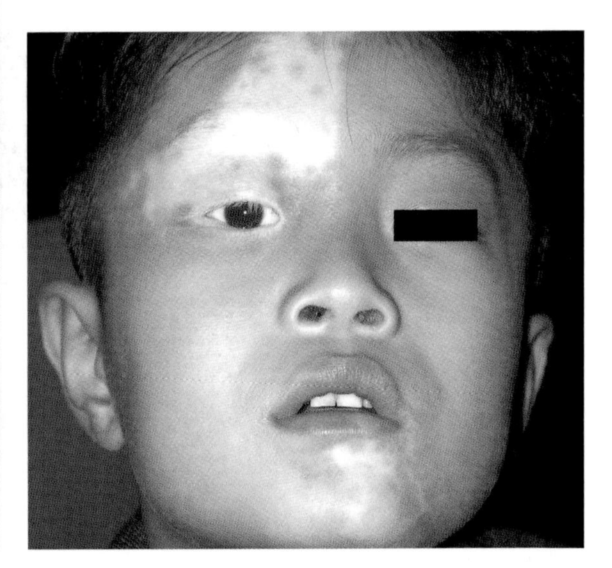

Fig. 2.3.4 Leukotrichia of eyebrow in SV demonstrates common and simultaneous involvement of non follicular and follicular melanocytic compartment

significance. Repigmentation of hair is difficult to treat with UV and other medical treatments. Transplantation of melanocytes is required for rapid and complete repigmentation [5] (Chap 1.3.6 and Parts 3.7).

References

1. Baba M, Akcali C, Seçkin D, Happle R (2002) Segmental lentiginosis with ipsilateral nevus depigmentosus: another example of twin spotting? Eur J Dermatol 12:319–321
2. Bolognia JL, Orlow SJ, Glick SA (1994) Lines of Blaschko. J Am Acad Dermatol 31:157–190
3. Hann SK, Lee HJ (1996) Segmental vitiligo: clinical findings in 208 patients. J Am Acad Dermatol 35:671–674
4. Hann SK, Park YK, Chun WH (1997) Clinical features of vitiligo. Clinic Dermatol 15:891–897
5. Hann SK, Im S, Park YK, Hur W (1992) Repigmentation of leukotrichia by epidermal grafting and systemic psoralen plus UV-A. Arch Dermatol 128:998–999
6. Koga M (1977) Vitiligo: a new classification and therapy. Br J Dermatol 97:255–261
7. Koga M, Tango T (1988) Clinical features and courses of type A and type B vitiligo. Br J Dermatol 118:223–228

2.3.2 The Concept of Mosaicism Applied to SV

Alain Taïeb

Contents

A clear association exists between NSV and a personal or familial autoimmune diathesis, which targets in particular the thyroid gland in around 25% of patients [24]. Assuming that vitiligo is itself a primary organ-specific autoimmune disorder, with the cutaneous melanocytic system being the target of a systemic immune response, is frequently the next reasoning step. However, observations derived from SV suggest that a primary cutaneous cause should be considered with more attention [3, 26].

The Lines of Blaschko and Segmental Vitiligo

The lines described by Alfred Blaschko in 1901 (BL) were based on drawings of cutaneous nevoid lesions [2]. They have been revisited extensively by Happle [5] and considered to follow the dorso-ventral development of cellular components of the skin. When visible, BL reflect an underlying mosaicism demonstrated first in monogenic keratin disorders [17]. BL have been postulated to correspond to a migration pattern restricted to cells of ectodermal or neuroectodermal origin (like melanocytes) [26], but several observations of purely dermal diseases following this pattern mitigate this view [14], and I would rather favour the idea that BL reflect the development of the entire skin. The two types of lines (thin or large, see Fig. 2.3.5a, b) may somewhat correspond to the underlying cellular origin of somatic mosaicism, dermal disorders tending to form broad bands [25]. Other types of patterns which do not

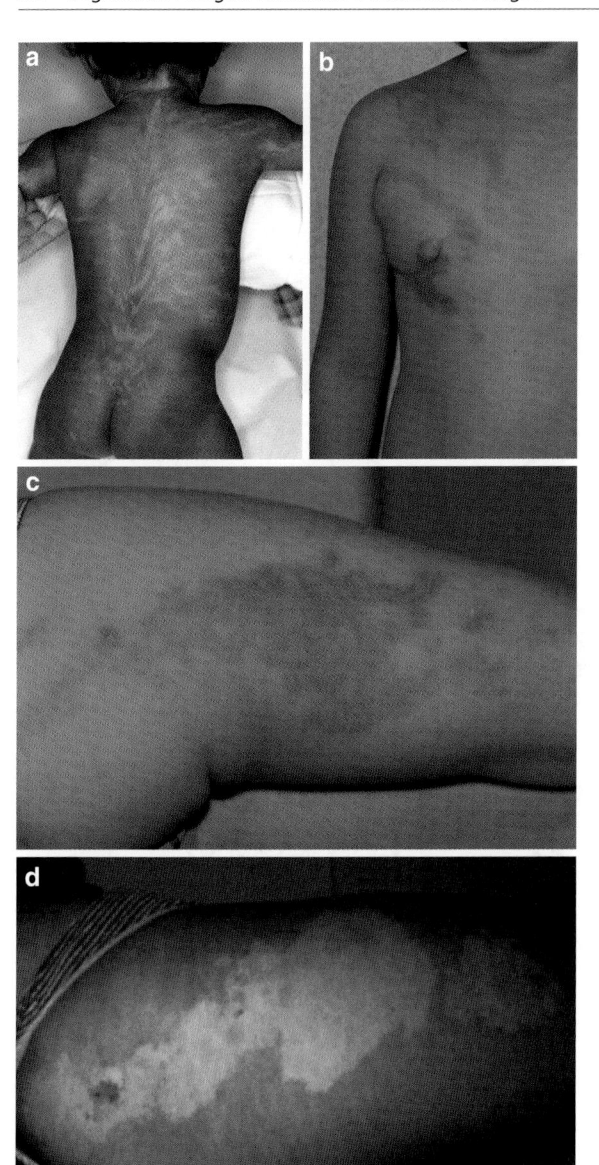

Fig. 2.3.5 Blaschko's bands vs. Blaschko's lines. Hypomelanosis of Ito (**a**) demonstrates nicely the thin blaschkolinear hypopigmented streaks originally described by Alfred Blaschko (courtesy of Dr Odile Enjolras); large band Blaschkolinear granuloma annulare (**b**) (reported by [14]); (**c**) shows same patient as in (**b**), note the difficulty to relate the pattern to Blaschko's bands (compare to **d**); Segmental vitiligo of the thigh considered as following a dermatomal distribution (**d**)

fit BL (checkerboard, phylloid, garment-type) are also observed in the dermatology clinic [8], and may concern the melanocytic system or other types of cells.

Some blaschkolinear distribution patterns in SV are strikingling similar to that of epidermal nevi [23], which are the established expression of somatic mosaicism for basal keratins [18] and other epidermally expressed genes [4]. Mosaicism may sometimes also involve germline cell mutations, a fact which explains the coexistence of segmental and generalized patterns in some rare pedigrees [16]. The dermatomal distribution is usually considered as reflecting better the distribution of SV (Chap. 1.3.2 and the contribution by Gauthier and Benzekri in this chapter), but cases intersecting dermatomes without "filling" their theoretical territory of distribution are difficult to relate to this etiologic background. The sympathetic anomalies noted in favour of the neurogenic theory of SV [27] may also be a confounding factor related to the absence of melanocytes which have been considered as "neurons of the skin" [13] and can release several neuromediators. The large blaschkolinear pattern, when isolated on a limb can mimick a dermatomal distribution (Fig. 2.3.5c, d). On the face, the lines of Blaschko have been drawn more recently [7], and some cases of SV fit clearly better Blaschko's bands than dermatomal territories [23, 25] (Fig. 2.3.6). This being stated, the hypothesis of a cutaneous mosaicism in SV based on distribution analysis does not rule out other triggering factors, especially neurogenic ones (see next section by Gauthier and Benzekri). It does not also exclude the possibility of an autoinflammatory component extending locally beyond the area demarcated by the developmental lines, but there is currently limited histological evidence of an inflammatory infiltrate in SV [11] (Chap. 1.3.10 showing a case of histologically inflammatory SV).

The Association of SV and NSV

Following the report of a pediatric case associating SV and NSV [3], other cases have been reported [15, 20–21, 25] (Fig. 2.3.7). This phenomenon may have been neglected previously due to NSV masking SV, which indeed was the case for the princeps case unveiled by UVB narrow band phototherapy [3]. This could be the case when patients are examined in adulthood after a disease of several years duration. In patients with mostly fair skin coming for SV, one should look systematically at minor symmetrical sites of involvement of the overall integument, using Wood's lamp in a completely dark room.

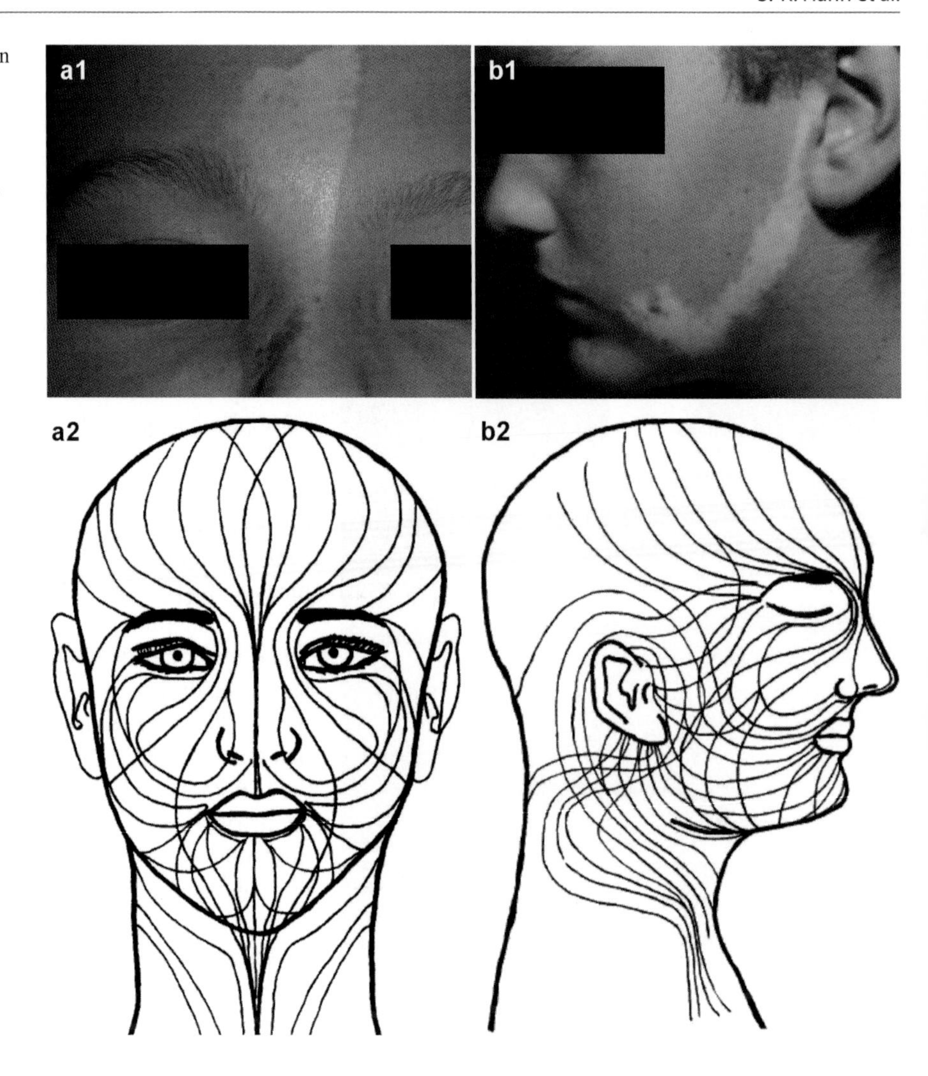

Fig. 2.3.6 Facial distribution of some cases of SV fits better Blaschko's lines than dermatomes. Two examples with the corresponding summary of lines by Happle and Assim (from [16], with permission)

The usually more severe and earlier presentation of SV in the context of associated NSV favours a dosing effet of a common predisposing gene and a loss of heterozygosity (LOH) mechanism like in monogenic disorders caused by cutaneous somatic mosaicism [19]. This corresponds to type II mosaicism according to Happle, as opposed to type I which results from only one mutation, causing an identifiable phenotype if dominant [6]. This sequence (early SV, late NSV) may also reflect the role of a first cutaneous gene defect causing SV triggering a generalised immune response against cutaneous melanocytes supported by another immune-related gene defect. However, pedigrees showing the presence of cases with established SV in families with previous cases of NSV [25] reinforce the possibility of a mechanism close to type II mosaicism already demonstrated in a monogenic skin disorder using laser capture of mutated cells and DNA sequencing [19].

Implications for Research of the Concept of Somatic Mosaicism Applied to Vitiligo

Until recently, the popular target theory for skin involvement in vitiligo (the skin pigmentary system as the target of autoimmunity) has underemphasized information coming from the skin itself. If the somatic mosaicism theory holds true for SV, it seems reasonable to speculate that differences in gene expression in involved and non involved SV skin or genetic skin

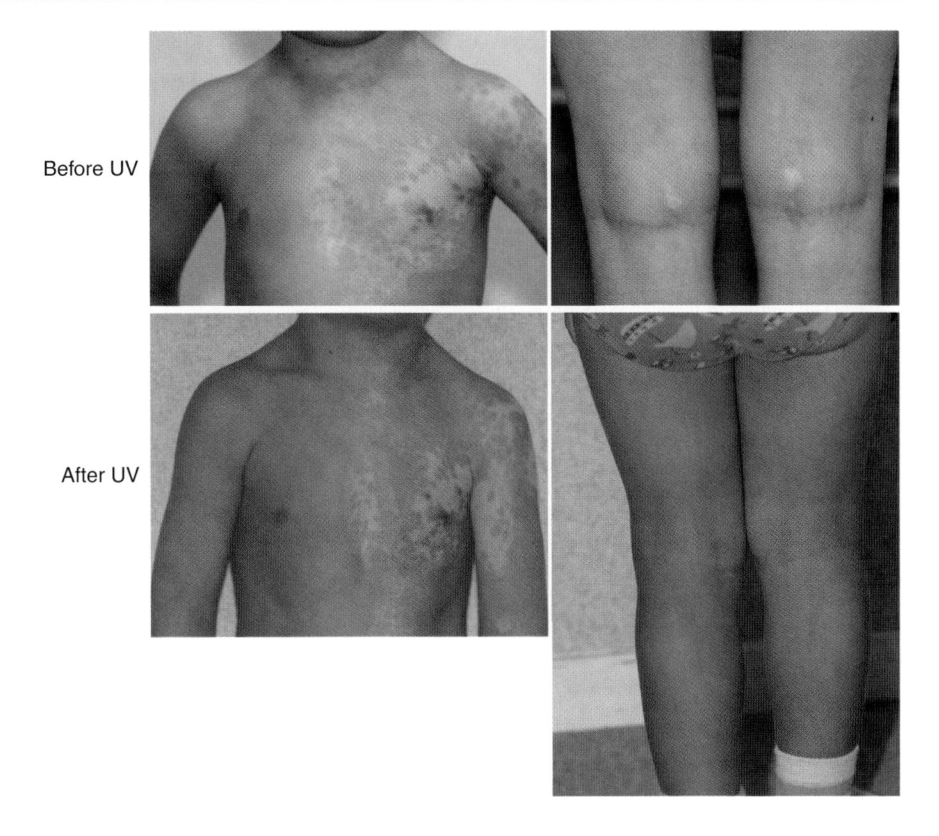

Fig. 2.3.7 Coexisting SV and NSV: better efficacy of NB UVB on nonsegmental involvement

differences in NSV associated with SV should provide relevant information on the cause of the disease. Recent data from melanocyte cultures in NSV have shown abnormal expression profiles of genes involved in melanocyte development, intracellular processing and trafficking of tyrosinase gene family proteins, packing and transportation of melanosomes, cell adhesion and antigen processing and presentation [22]. Focusing on one gene or subset of genes might be a difficult choice. As compared to NSV, SV would probably be more informative in terms of skin causative genes, because the background of inflammation/autoimmunity would be probably more limited: autoimmunity manifested by associated autoimmune disorders is rare in SV [9]. Autoantibodies to melanocytic antigens are also not common [10]. The only shared "immune" phenotype is the presence of halo nevi [1], which seem nearly as frequent in SV and NSV (6.4 vs. 8.6%), and quite high as compared to controls without vitiligo – less than 1% [12] (Chap. 1.3.5). This finding, if confirmed, may suggest that some SV patients have some counteracting protective mechanisms respective to the development of NSV, and which prevent bad take of autologous grafts in SV.

References

1. Barona MI, Arrunategui A, Falabella R, Alzate A (1995) An epidemiologic case-control study in population with vitiligo. J Am Acad Dermatol 33:621–625
2. Blaschko A (1901) Die Nervenverteilung in der Haut in ihre Beziehung zu den Erkrankungen der Haut. Beilage zu den Verhandlungen der Deutschen Dermatologischen Gesellschaft VII Congress, Breslau
3. Gauthier Y, Cario André M, Taïeb A (2003) A critical appraisal of vitiligo etiologic theories. Is melanocyte loss a melanocytorrhagy? Pigment Cell Res 16: 322–332
4. Hafner C, van Oers JM, Vogt T et al (2006) Mosaicism of activating FGFR3 mutations in human skin causes epidermal nevi. J Clin Invest 116:2201–2207
5. Happle R (1985) Lyonization and the lines of Blaschko. Hum Genet 70:200–206
6. Happle R (2001) Segmental type 2 manifestation of autosome dominant skin diseases. Development of a new formal genetic concept. Hautarzt 52:283–287
7. Happle R, Assim A (2001) The lines of Blaschko on the head and neck. J Am Acad Dermatol 44:612–615
8. Happle R (2004) [Patterns on the skin. New aspects of their embryologic and genetic causes, German] Hautarzt 55:960–961; 964–968
9. Iacovelli P, Sinagra JL, Vidolin AP et al (2005) Relevance of thyroiditis and of other autoimmune diseases in children with vitiligo. Dermatology 210:26–30

10. Kemp EH, Gawkrodger DJ, MacNeil S et al (1997) Detection of tyrosinase autoantibodies in patients with vitiligo using 35S-labeled recombinant human tyrosinase in a radioimmunoassay. J Invest Dermatol 109:69–73
11. Kim YC, Kim YJ, Kang HY et al (2008) Histopathologic features in vitiligo. Am J Dermatopathol 30:112–116
12. Larsson PA, Liden S (1980) Prevalence of skin diseases among adolescents 12–16 years of age. Acta Derm Venereol 60:415–423
13. Moellmann G, McGuire J, Lerner AB (1973) Ultrastructure and cell biology of pigment cells. Intracellular dynamics and the fine structure of melanocytes with special reference to the effects of MSH and cyclic AMP on microtubules and 10-nm filaments. Yale J Biol Med 46: 337–360
14. Morice-Picard F, Boralevi F, Lepreux S et al (2007) Severe linear form of granuloma annulare along Blaschko's lines preceding the onset of a classical form of granuloma annulare in a child. Br J Dermatol 157:1056–1058
15. Mulekar SV, Al Issa A, Asaad M et al (2006) Mixed vitiligo. J Cutan Med Surg 10:104–107
16. Nazzaro V, Ermacora E, Santucci B, Caputo R (1990) Epidermolytic hyperkeratosis: generalized form in children from parents with systematized linear form. Br J Dermatol 122:417–422
17. Paller AS, Syder AJ, Chan YM et al (1994) Genetic and clinical mosaicism in a type of epidermal nevus. N Engl J Med 331:1408–1415
18. Paller RA (2004) Piecing together the puzzle of cutaneous mosaicism. J Clin Invest 114:1407–1409
19. Poblete-Gutiérrez P, Wiederholt T, König A et al (2004) Allelic loss underlies type 2 segmental Hailey-Hailey disease, providing molecular confirmation of a novel genetic concept. J Clin Invest 114:1467–1474
20. Schallreuter KU, Krüger C, Rokos H et al (2007) Basic research confirms coexistence of acquired Blaschkolinear vitiligo and acrofacial vitiligo. Arch Dermatol Res 299:225–230
21. Schallreuter KU, Krüger C, Würfel BA et al (2008) From basic research to the bedside: efficacy of topical treatment with pseudocatalase PC-KUS in 71 children with vitiligo. Int J Dermatol 47:743–753
22. Strömberg S, Björklund MG, Asplund A et al (2008) Transcriptional profiling of melanocytes from patients with vitiligo vulgaris. Pigment Cell Melanoma Res 21: 162–171
23. Taïeb A (2000) Intrinsic and extrinsic pathomechanisms in vitiligo. Pigment Cell Res 13:41–47
24. Taïeb A, Picardo M; VETF Members (2007) The definition and assessment of vitiligo: a consensus report of the Vitiligo European Task Force. Pigment Cell Res 20:27–35
25. Taïeb A, Morice-Picard F, Jouary T et al (2008) Segmental vitiligo as the possible expression of cutaneous somatic mosaicism: implications for common non-segmental vitiligo. Pigment Cell Melanoma Res 21:646–652
26. Taibjee SM, Bennett DC, Moss C (2004) Abnormal pigmentation in hypomelanosis of Ito and pigmentary mosaicism: the role of pigmentary genes. Br J Dermatol 151: 269–282
27. Wu CS, Yu HS, Chang HR et al (2000) Cutaneous blood flow and adrenoceptor response increase in segmental-type vitiligo lesions. J Dermatol Sci 23:53–62

2.3.3 The Neurogenic Hypothesis in Segmental Vitiligo

Yvon Gauthier and Laila Benzekri

Contents

An abnormal effect of neurohormones and neuropeptides has been speculated to explain the chronic melanocyte loss in vitiligo, especially in SV due to its distribution, which can fit dermatomes. The neural/neurogenic theory is supported by clinical, histological, ultrastructural, experimental, pathophysiological, and biochemical arguments. Those will be reviewed with emphasis on SV.

Clinical Arguments

Segmental Distribution (Chap. 1.3.2)

In our personal series, 16% of cases localized leukoderma follows partially or completely one or more dermatomes as described by Grant and Haerer [16, 25, 27, 28] (Fig. 2.3.8). The distribution is quite similar to that commonly observed in herpes zoster. The most frequent distribution pattern corresponds to the trigeminal nerve dermatome [17, 18, 25, 27, 34].

Mixed Distribution Overlapping Several Dermatomes

In 64% of cases depigmented macules do not follow a true dermatomal pattern. The depigmentation involves either partially each dermatome, or overlaps two or three dermatomes especially when SV is located on the face

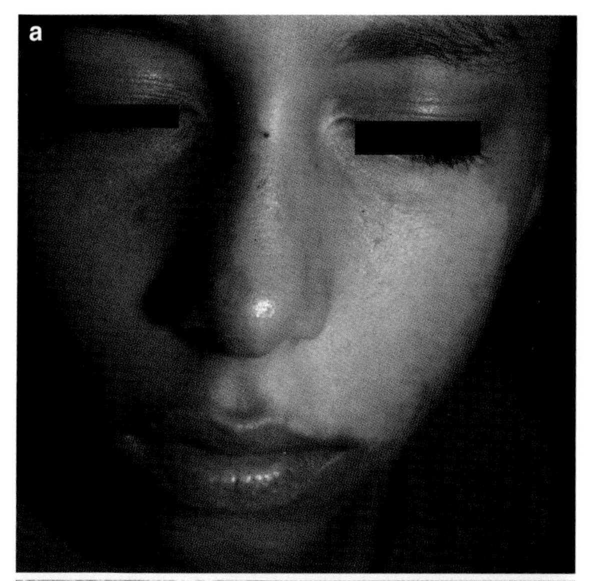

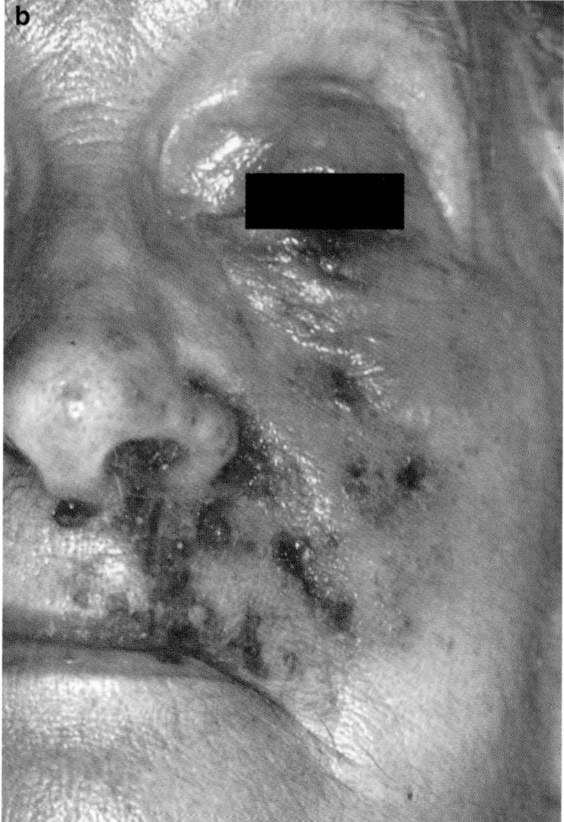

Fig. 2.3.8 Similar segmental distribution of SV and herpes zoster. SV corresponding strictly to V2 dermatome (**a**), Herpes zoster with the same distribution (used with permission of JJ Morand) (**b**)

(Fig. 2.3.9a). We feel that the existence of a distribution which does not fit exactly the dermatomes is not a sufficient argument against the neural hypothesis. One argument is that in well-established cases of herpes zoster of the face, in which a dermatomal distribution is undoubtely implicated, the same distribution than that of mixed SV can be observed [12] (Fig. 2.3.9b).

Localized Vitiligo or Leukodermas Following Nerve Damage

Many clinical observations have been reported in which skin became white in areas corresponding to local neurologic damage. Unilateral vitiligo and poliosis located on the area innervated by both trigeminal and upper cervical nerves was reported in a child with viral encephalitis [26]. Ipsilateral macules of vitiligo at the level of the upper arm, chest, buttocks, sacrum was observed in a 31 years old patient following trauma of the right brachial plexus [7]. After right cervical sympathectomy, a woman developed grayer hair on the left than on the right side [19, 20, 21]. Localized depigmentations similar to SV have been reported in patients with a spinal cord tumor, or following nerve injury [8, 36].

Histological and Ultrastructural Arguments

Abnormalities of the peripheral nervous system around macules of vitiligo have been considered to support the neural hypothesis [3, 9, 31]. However, most studies have been performed on skin samples of patients with NSV.

Electron microscopy has established that direct contact occur between intraepidermal nerve endings and melanocytes [5]. Dystrophic changes in nerve trunks and nerve endings, as well as degenerative and regenerative changes in terminal cutaneous nerves have been found in vitiligo [5, 10]. Using antibodies against neuropeptide Y and calcitonin gene-related peptide (CGRP), several authors have been able to show an increased expression of these peptides in

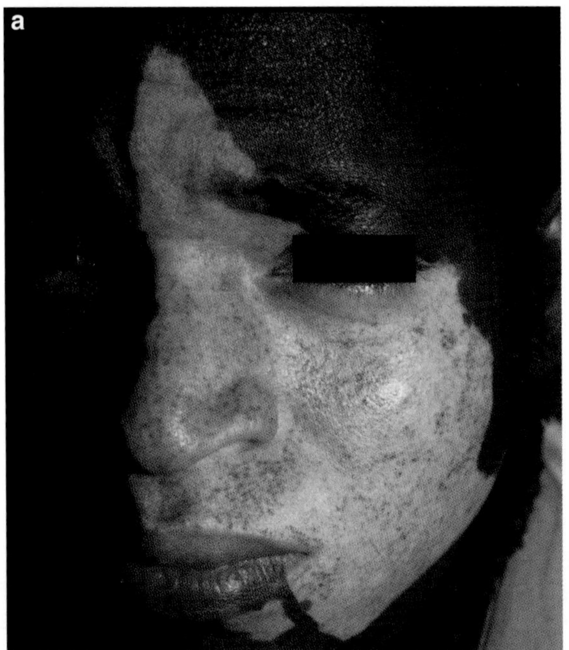

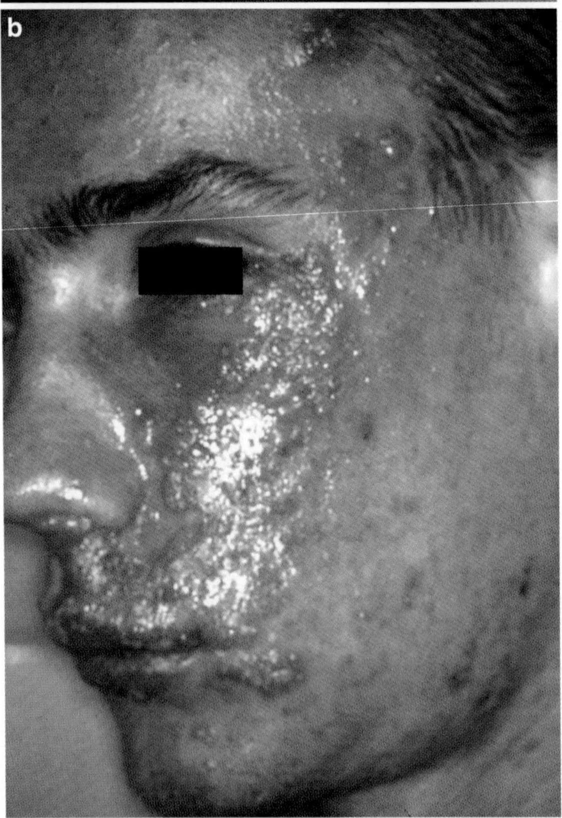

Fig. 2.3.9 Similar mixed distribution of SV and herpes zoster overlapping several dermatomes. SV overlapping V1, V2, V3 dermatomes (**a**), and Herpes zoster with the same distribution (used with permission of JJ Morand) (**b**)

lesional and perilesional skin and frequently an increased number of CGRP positive nerve fibers in involved skin [1, 2].

Experimental Findings in Animals

Abnormalities in neuropeptides or other neurochemical mediators secreted by nerve endings could harm nearby melanocytes. The existence of a dermal adrenergic innervation to melanophores of teleost fish has been demonstrated. Electrical stimulation of nerve fibers of skin causes lightening in the European turbot [15]. Application of epinephrine similarly leads to aggregation of melanin within melanophores in some fishes accompanied by lightening of skin color. Sympathetic denervation has been followed by the development of melanin pigment aggregation.

Embryologically, melanocytes have the same origin as nerve cells. They can react to many compounds that affect the nervous system. In frogs, acetylcholine, norepinephrine, epinephrine and melatonin lighten dermal melanocytes [23].

In guinea pig skin [9], the spread of grafted melanocytes in denervated skin is statistically better that in nonoperated control sites. This could suggest that peripheral nerves could play a role in the regulation of melanocyte function especially for migration. A single subcutaneous injection of epinephrine into rats causes local depigmentation of the hairs, but the interpretation of this phenomenon is not straightforward, since it seems likely that vasoconstriction can induce ischemic depigmentation.

Pathophysiological Arguments

In patients with vitiligo, local abnormalities related to possible neural-mediated aberrations have been reported. However, data are discordant. An increased adrenergic nerve-ending activity in vitiligo macules was suggested by Lerner [7, 22, 23]. Other authors came to the conclusion that there was a dominant cholinergic influence in vitiliginous skin compared to normal skin [6, 11].

Koga's classification [17, 18] proposes that SV could result from the dysfunction of sympathetic nerves in the affected area. On the basis of physostigmine-induced sweating response, a decreased responsiveness of the segmental type was interpreted to be due to an increased cholinergic influence [13, 14, 29]. In SV lesions, the cutaneous blood flow was shown to be increased almost threefold compared to control normal skin and a similar anomaly was shown to a lesser extent in NSV. A significant increase in cutaneous alpha and beta-adrenoreceptor response was found in the segmental type suggesting that a dysfunction of the sympathetic nerves could exist [4, 30, 32, 33, 35, 37]. As noted before, the catecholamine stimulation of adrenoreceptors may cause severe vasoconstriction. It is thus possible that hypoxia-induced oxidative stress at the epidermal level could lead to partial or total depigmentation of one or several dermatomes. However no changes in plasma catecholamine level or adrenoreceptor densities on blood cells have been so far reported [24].

References

1. Al'Abadie M, Gawkrodger DJ, Senior HJ et al (1992) Neuro-ultrastructural and neuropeptides studies in vitiligo. Clin Exp Dermatol (Abstract) 15:824
2. Al'Abadie MS, Senior HJ et al (1994) Neuropeptide and neuronal marker studies in vitiligo. Br J Dermatol 131:160–165
3. Arnozan X, Lenoir L (1922) Vitiligo avec troubles nerveux sensitifs et sympathiques: l'origine sympathique du vitiligo. Bull Soc Fr Dermatol Syphil 12: 124–140
4. Bamshad J (1964) COMT in skin. J Invest Dermatol 43: 111–113
5. Breatnach AS, Bors S, Willie L et al (1966) Electron microscopy of peripheral nerve terminals and marginal melanocytes in vitiligo. J Invest Dermatol 47:125–140
6. Chanco-Turner ML, Lerner AB (1965) Physiologic changes in vitiligo. Arch Dermatol 91:390–396
7. Costea V (1961) Leucoderma patches in the course of traumatic paralysis of the brachial plexus in a subject with insular cavities. Act Dermat Venerol 2:161–166
8. Ferriol L (1905) Vitiligo et tumeurs neurologiques de la moelle. Rev Neurol Paris 13:282–286
9. Fabian G (1951) The spread of black pigment of the denervated skin of the guinea pig. Acta Biol Acad Sci Hung 4:471–479
10. Gauthier Y, Surlève-Bazeille JE (1974) Ultrastructure des fibres nerveuses peripheriques dermiques dans le vitiligo. Bull Soc Fr Dermatol Syphil 81:550–554
11. Gauthier Y, Surleve-Bazeille JE, Gauthier O (1976) Bilan de l'activité cholinesterasique dans le vitiligo: son interêt physiopathologique. Bull Soc Fr Dermatol Syphil 81:321
12. Gauthier Y, Taïeb A (2006) Proposal for a new classification of segmental vitiligo of the face. Pigment Cell Res (Abstract) 19:515
13. Iyengar B (1989) Modulation of melanocyte activity by acetylcholine. Acta Anat 136:139–141
14. Jacklin NH (1965) Depigmentation of the eyelids in eserine allergy. Am J Ophtalmol 59:890–902
15. Jakobowith D, Laties A (1953) Direct adrenergic innervation of a teleost melanophore. Anat Rec 162: 501–504
16. Klippel M, Weil HP (1922) Radicular distribution of vitiligo and nevi. Presse Med 30:388–390
17. Koga M (1977) Vitiligo: a new classification and therapy. Br J Dermatol 97:255–261
18. Koga M, Tango T (1988) Clinical features and course of type A and type B vitiligo. Br J Dermatol 118:223–228
19. Lerner AB (1959) Vitiligo. J Invest Dermatol 39:285–310
20. Lerner AB (1966) Sympathectomy and gray hair. Arch Dermatol 93:235–236
21. Lerner AB (1966) Vitiligo and sympathectomy. Arch Dermatol 94:269–278
22. Lerner AB (1971) Neural control of pigment cells. In: Kawanamura T, Fitzpatrick TB, Seiji M (eds) Biologie of normal and abnormal melanocytes. Tokyo University Park Press, Tokyo, pp 3–16
23. Mac Guire J (1970) Adrenergic control of melanocytes. Arch Dermatol 101:173–180
24. Morrone A, Picardo M, De Luca C (1992) Catecholamines and vitiligo. Pigment Cell Res 5:65–69
25. Musumed V (1951) Sulla distribuzione radicolare della vitiligine. Minerva Dermatol 26:77–79
26. Nelhaus G (1970) Acquired unilateral vitiligo and poliosis of the head and subacute encephalitis with partial recovery. Neurology 20:965–974
27. Ortonne JP, Mosher DB, Fitzpatrick TB (1983) Vitiligo and other hypomelanosis of hair and skin. Plenum, New York, pp 250–258
28. Scholtz JR, Williamson C (1951) Vitiligo in apparent neural distribution. Arch Dermatol Syphil 64:366–368
29. Schallreuter KU, Elwary SM, Gibbons NC et al (2004) Activation/desactivation of acetylcholinesterase by H_2O_2 more evidence for oxydative stress in vitiligo. Biochem Biophys Res Commun 315:502–508
30. Schallreuter KU, Wood JM, Pittelkow M et al (1993) Increased in vitro expression of beta 2 adrenoreceptors in differentiating lesional keratinocytes of vitiligo patients. Arch Dermatol Res 285:216–220
31. Selemy (1955) Vitiligo es psoriasis azonos oldali syringomyelia. Borg Vener Szen 9:94–96
32. Sinan Bit L (1999) Sympathetic skin response in psoriasis and vitiligo. J Auton Nerv Syst 77:68–71
33. Stajano C, Martirena H (1926) Bilateral peripheral distribution of vitiligo according to nerve segments. Ac Fac Med Montevideo 11:563–581
34. Touraine A, Brizard A (1935) La topographie radiculaire du vitiligo. Bull Soc Fr Dermotol Syphil 42:505–515
35. Touraine A, Picquart A (1937) Maladie de Recklinghausen à pigmentation systématisée, vitiligo, pelade. Bull Soc Fr Dermtol Syphil 44:81–85
36. Tremiteria S (1927) Vitiligo from spinal anesthesia. Rinazcenca Medica 4:107–108
37. Wu CS, Yu HS et al (2000) Cutaneous blood flow and adrenoreceptor response increase in segmental type vitiligo lesions. J Dermatol Sci 23:53–62

2.3.4 Segmental Vitiligo: A Model for Understanding the Recapitulation of Repigmentation

Hsin-Su Yu, Cheng-Che Eric Lan, and Ching-Shuang Wu

Contents

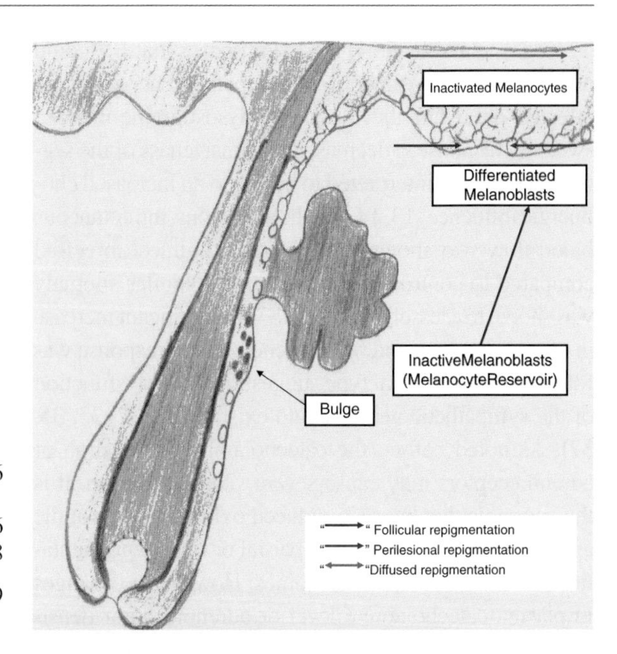

Fig. 2.3.10 Repigmentation scheme of vitiligo. The process of follicular pigmentation is shown in *blue* color. This repigmentation route involves activation, proliferation, migration, and functional development of immature melanoblasts residing on the outer root sheath of the hair follicle. Diffuse (*green*) and perilesional (*red*) patterns of repigmentation require activation of inactive melanocytes and migration of perilesional melanocytes respectively. All three forms of pigmentation pattern may be observed in repigmenting vitiliginous skin

Repigmentation of Segmental Vitiligo and Summary of Repigmentation Schemes

Previous studies on vitiligo have focused on NSV, and very few studies have addressed specifically SV [5, 7, 8, 16, 17, 33]. In order to gain further insights on how to treat this disorder more effectively, a careful analysis of how different modalities induce vitiligo repigmentation would be invaluable. SV is regarded as a stable disease once the initial melanocyte loss has ceased. Since the active destruction of melanocyte is no longer present, the recovery scheme of SV focuses on (1) the activation, migration, and functional development of melanoblasts, (2) the proliferation and migration of functional melanocytes, and (3) the impact of epidermal keratinocytes on pigment cells [41]. Through in vitro experimental results, much insight has been gained regarding the molecular mechanisms involved in vitiligo repigmentation scheme.

The pattern for vitiligo repigmentation may take two forms: the follicular pattern and the diffuse pattern (Part 3.1). The melanoblasts in the outer root sheath (ORS) of the hair follicle serve as the source for follicular repigmentation [32]. In the follicular pattern, recovery from vitiligo is initiated by the activation and proliferation of immature melanoblasts, followed by their upward migration onto the nearby epidermis to form the follicular pigment islands [4]. During this process, the melanoblasts mature morphologically and functionally, and subsequently begin normal transfer of melanin to keratinocytes [44]. On the other hand, perilesional or diffuse patterns of repigmentation involve probably the migration of perilesional melanocytes towards the vitiligo macules or activation of inactive melanocytes within vitiligo lesions respectively [3, 30]. Figure 2.3.10 summarizes the different repigmentation processes described above.

Recapitulation of Vitiligo Repigmentation In Vitro and Clinical Relevance In Vivo

Cells Involved in SV Repigmentation

Epidermal melanocytes and follicular melanocytes express different antigens [35] providing further evidence of intrinsic differences between segmental and generalized vitiligo (see previous section by Hann). Regardless of the initial damaging process, once the disease becomes stable, the most important cells involved in SV repigmentation besides melanoblasts and melanocytes are keratinocytes, which play a crucial role in modulating epidermal pigmentation. Through their growth factors/mitogens, keratinocytes can modulate the biological behavior of neighboring melanocytes. Stem cell factor, basic fibroblast growth factor, endothelins, and nerve growth factor are among the documented factors that participate in this delicate relationship [9, 22, 25, 38, 43] (Chap. 2.2.8). Since these cells are readily available for experimentation, significant insights regarding SV repigmentation can be obtained from comparing in vitro results with in vivo observations.

In Vitro Repigmentation Models

- Comparing the biologic impact of irradiation sources on repigmentation

It is known that PUVA treatment is more effective for treating NSV as compared to SV [34]. The proposed molecular mechanisms responsible for PUVA's impact on vitiligo repigmentation include (1) depletion of vitiligo associated melanocyte antigens [13], and (2) depletion of Langerhans cells in vitiligo skin [12]. However, these immunomodulating effects are probably more important for active NSV and play limited role in SV as active melanocyte loss is not a feature of established SV lesions [14]. The biologic impact of PUVA on epidermal cells include (1) limited stimulatory effect on keratinocytes, (2) increased matrix metalloproteinase (MMP)-2 activity, an important modulator for pigment cell motility, from melanocytes [40], and (3) increase tyrosinase activity of MC [13]. In addition, the photosensitizer 8-methoxypsoralen by itself can stimulate MMP-2 expression from melanoblasts [20]. These biologic effects imparted by PUVA on epidermal cells provide a reasonable explanation for the therapeutic effect of PUVA on SV. Moreover, it is also conceivable that active NSV responds more favorably to PUVA treatment as compared to SV since PUVA also have significant immune regulatory effects in addition to its biologic effects. A schematic diagram summarizing the effects of PUVA is shown in Fig. 2.3.11. Westerhof and Nieuweboer-Krobotova [37] reported that vitiligo patients receiving PUVA and NBUVB experienced a repigmentation rate of 46 and 67%, respectively, after 4 months of phototherapy. The results from in vitro experiments may provide an explanation for this clinical phenomenon. NBUVB is known to have immune modulatory effects including depleting skin infiltrating T cells and arresting maturation of epidermal dendritic cells [26]. In terms of biological effects, NBUVB irradiation affects several cellular targets (1) it causes a direct stimulation of melanocyte locomotion via increased expressions of phosphorylated focal adhesion kinase, (2) it increases MMP-2 activity from melanocytes, (3) it causes indirect stimulation of melanocyte proliferation by increasing the production of basic fibroblast growth factor and endothelin from keratinocytes [39] and (4) it causes a direct stimulation of melanoblasts in terms of melanin formation and cell migration (unpublished data). A summary of these in vitro findings is shown in Fig. 2.3.11.

- Application to SV therapy

Comparing the in vitro effects of NBUVB and PUVA on epidermal cells, it is clear that both modalities have significant immunomodulatory effects, but NBUVB has more direct biological effects on epidermal cells in terms of supporting pigment cell migration, proliferation, and development. Therefore, according to the in vitro repigmentation model, NBUVB may provide better therapeutic efficacy for inducing repigmentation in SV [1], as compared to PUVA therapy. Further head to head clinical trials should be performed to validate this hypothesis.

- Correlating in vitro and in vivo effects of Red visible light and tacrolimus

More recently, helium–neon (He–Ne) laser, which emits within visible red light, has been shown as a novel phototherapeutic modality for treating SV [45]. He–Ne laser promotes vitiligo repigmentation through (1) direct stimulation of melanocyte and melanoblast migration [19, 45], (2) direct stimulation of melanoblasts to undergo functional development [19], (3) increased secretion of nerve growth factor and basic fibroblast growth factor from keratinocytes, and (4) indirect stimulation of melanocyte growth via keratinocytes [45] (Figs. 2.3.11 and 2.3.12). Tacrolimus ointment is a promising topical agent for treating SV. In vitro studies using epidermal cells have shown that tacrolimus (1) directly stimulates keratinocytes to release

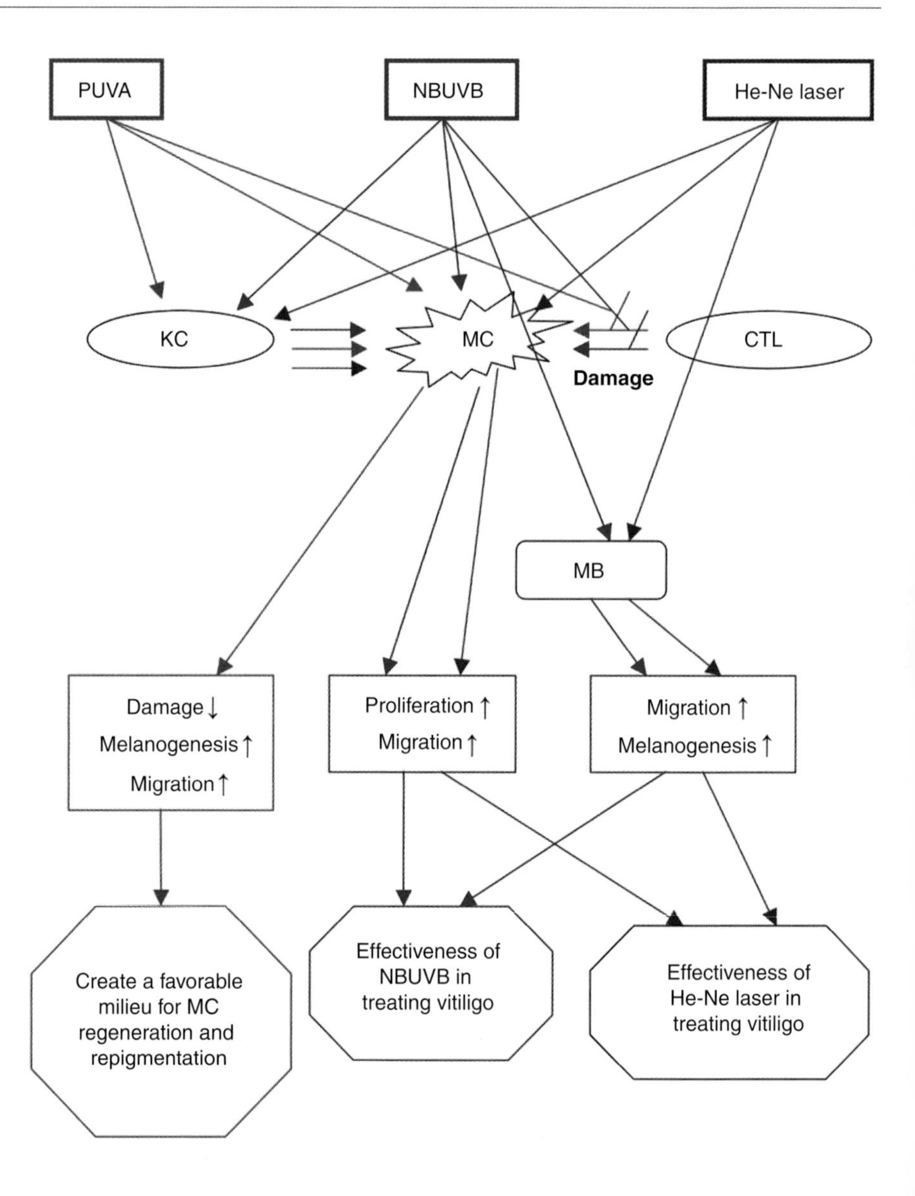

Fig. 2.3.11 Schematic diagram summarizing the mechanisms involved in PUVA, NBUVB and laser red light induced vitiligo repigmentation. PUVA treatment has significant immune modulatory effects on both melanocytes and cytotoxic immune cells. It also directly promotes melanocyte melanin formation and migrations but has limited biologic effects on keratinocytes. Besides immune suppressive effects, NBUVB treatment has significant biologic effects on pigment cells and keratinocytes. More specifically, NBUVB directly or vicariously via keratinocyte increases the proliferation and locomotion of melanocytes. In addition, NBUVB also stimulates melanogenesis and migration of immature melanoblasts. The Ne–He laser red light has also direct effects on melanoblasts. *NBUVB* narrow-band UVB; *KC* keratinocyte; *MC* melanocyte; *CTL* cytotoxic T lymphocyte; *MB* melanoblast

stem cell factor, (2) induces keratinocytes to promote the growth of both melanoblasts and melanocytes, (3) induces keratinocytes to upregulate MMP activities [18], and (4) directly induces pigmentation and migration of melanocytes [10]. Therefore, besides immunomodulatory effects of tacrolimus on immune cells, the direct biological effects of tacrolimus on epidermal cells probably account for its therapeutic effects on SV. Of interest is the effect of different therapies on the cutaneous nervous system. Currently, in vitro data regarding this topic is lacking. However, it has been shown that He–Ne laser irradiation leads to improvement in nerve injury [15, 28] and topical tacrolimus induces neuropeptide release [31] and may lead to neuronal cell differentiation [11]. These reports suggest that neural modulatory effects may have a previously unrecognized role in the treatment of SV. Further studies on how neural modulatory effects are included into the repigmentation scheme may improve treatment regimen for SV.

Therapeutic Perspectives

Although new strides have been made on treatment options for SV, the underlying molecular mechanisms inducing repigmentation must be elucidated for

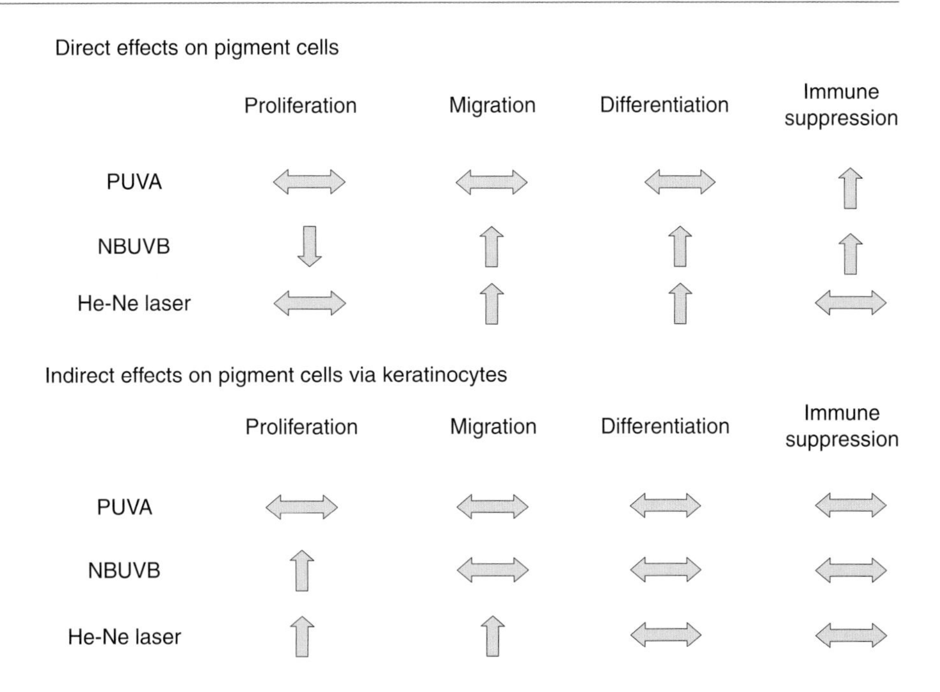

Fig. 2.3.12 Effects of light irradiation on pigment cells directly or indirectly via keratinocytes

optimizing treatment outcomes. Another issue, which requires further attention, is the involvement of neurocutaneous interactions in the pathogenesis of vitiligo (see also previous section by Gauthier and Benzekri in this chapter). Previous reports have indicated that swelling and dystrophic changes, as well as some regeneration, were noted in nerves of vitiligo skin [2]. Intriguingly, it has also been suggested that the stage of the disease may correlate with the condition of the neural tissues. More specifically, the initial events of vitiligo may cause axonal damage, but later stages of disease may favor nerve regeneration [36, 43]. Since recent evidences indicate that skin homeostasis may be regulated by the peripheral neuropeptidergic nerve fibers [27], further work focusing on modulation of neurocutaneous unit may provide a key to more effective vitiligo treatment.

SV is a perfect model for studing repigmentation since it is a relative stable disease with limited active melanocyte loss. Moreover, the cells involved in the repigmentation scheme of SV are readily available for in vitro experimentation. With recent advances in stem cell research, which holds great potential in clinical applications for regenerative medicine, the use of melanocyte stem cells [24] may evolve from in vitro experimentations to in vivo reality as suggested by Yonetani et al. [42].

References

1. Anbar TS, Westerhof W, Abdel-Rahman AT et al (2006) Evaluation of the effects of NBUVB in both segmental and non-segmental vitiligo affecting different body sites. Photodermatol Photoimmunol Photomed 22:157–163
2. Breathnach AS, Bor S, Wyllie L (1966) Electron microscopy of peripheral nerve terminals and marginal melanocytes in vitiligo. J Invest Dermatol 47:125–140
3. Bartosik J, Wulf HC, Kobayasi T (1998) Melanin and melanosome complexes in long standing stable vitiligo-an ultrastructural study. Eur J Dermatol 8:95–97
4. Cui J, Shen LY, Wang GC (1991) Role of hair follicles in the repigmentation of vitiligo. J Invest Dermatol 97:410–416
5. El Mofty AM, El Mofty M (1980) Vitiligo: a symptom complex. Int J Dermatol 19:238–247
6. Hann SK, Lee HJ (1996) Segmental vitiligo: clinical findings in 208 patients. J Am Acad Dermatol 35:671–674
7. Howitz J, Brodthagen H, Schwartz M et al (1977) Prevalence of vitiligo. Epidemiological survey on the Isle of Bornholm, Denmark. Arch Dermatol 113:47–52
8. Halaban R, Hebert DN, Fisher DE (2003) Biology of melanocytes. In: Freedberg IM, Eisen AZ Wolff K Austen KF, Goldsmith LA, Katz, SI (eds) Dermatology in general medicine, 6th edn. McGraw-Hill, New York, pp 127–148.
9. Kang HY, Choi YM (2006) FK506 increases pigmentation and migration of human melanocytes. Br J Dermatol 155: 1037–1040
10. Kano Y, Nohno T, Hasegawa T et al (2002) Immunosuppressant FK506 induces sustained activation of MAP kinase and promotes neurite outgrowth in PC12 mutant cells incapable of differentiating. Cell Struct Funct 27:393–398

11. Kao CH, Yu HS (1990) Depletion and repopulation of Langerhans cells in nonsegmental type vitiligo. J Dermatol 17:287–296
12. Kao CH, Yu HS (1992) Comparison of the effect of 8-methoxypsoralen (8-MOP) plus UVA (PUVA) on human melanocytes in vitiligo vulgaris and in vitro. J Invest Dermatol 98: 734–740
13. Khalid M, Mutjaba G (1998) Response of segmental vitiligo to 0.05% clobetasol propionate cream. Int J Dermatol 37:705–708
14. Khullar SM, Brodin P, Barkvoll P et al (1996) Preliminary study of low-level laser for treatment of longstanding sensory alteration in the inferior alveolar nerve. J Oral Maxillofac Surg 54:2–7
15. Koga A (1977) Vitiligo: a new classification and therapy. Br J Dermatol 97:255–261
16. Koga A, Tango T (1988) Clinical features and course of type A and type B vitiligo. Br J Dermatol 118:223–228
17. Lan CC, Chen GS, Chiou MH et al (2005) FK506 promotes melanocyte and melanoblast growth and creates a favourable milieu for cell migration via keratinocytes: possible mechanisms of how tacrolimus ointment induces repigmentation in patients with vitiligo. Br J Dermatol 153:498–505
18. Lan CC, Wu CS, Chiou MH et al (2006) Low-energy helium-neon laser induces locomotion of the immature melanoblasts and promotes melanogenesis of the more differentiated melanoblasts: recapitulation of vitiligo repigmentation in vitro. J Invest Dermatol 126:2119–2126
19. Lei TC, Virador V, Yasumoto K et al (2002) Stimulation of melanoblast pigmentaion by 8-methoxypsoralen: the involvement of microphthalmia-associated transcription factor, the proetin kinase A signal pathway, and proteasome-mediated degradation. J Invest Dermatol 119: 1341–1349
20 Mori H (1964) Neurohistological studies on vitiligo vulgaris skin with special reference to the vegetative nerves. Jpn J Dermatol 47:411–431
21. Nishimura EK, Jordan SA, Oshima H et al (2002) Dominant role of the niche in melanocyte stem-cell fate determination. Nature 416:854–860
22. Nozaki T (1976) Histochemical and electron microscopic studies of peripheral nerve terminals and clinical studies in vitiligo vulgaris. Jpn J Dermatol 86:1–19
23. Ozawa M, Ferenczi K, Kikuchi T et al (1999) 312-nanometer ultraviolet B light (narrow-band UVB) induces apoptosis of T cells within psoriatic lesions. J Exp Med 189: 711–718
24. Pavlovic S, Daniltchenko M Tobin DJ et al (2008) Further exploring the brain-skin connection: stress worsens dermatitis via substance P-dependent neurogenic inflammation in mice. J Invest Dermatol 128:434–446
25. Rochkind S, Rousso M, Nissan M et al (1989) Systemic effects of low-power laser irradiation on the peripheral and central nervous system, cutaneous wounds, and burns. Lasers Surg Med 9:174–182
26. Silverberg NB, Lin P, Travis L et al (2004) Tacrolimus ointment promotes repigmentation of vitiligo in children: a review of 57 cases. J Am Acad Dermatol 51:760–766
27. Ständer S, Ständer H, Seeliger S et al (2007) Topical pimecrolimus and tacrolimus transiently induce neuropeptide release and mast cell degranulation in murine skin. Br J Dermatol 156:1020–1026
28. Staricco RG (1962) Activation of the amelanotic melanocytes in the outer root sheath of the hair follicle following ultra violet rays exposure. J Invest Dermatol 39:163–164
29. Taïeb A (2000) Intrinsic and extrinsic pathomechanisms in vitiligo. Pigment Cell Res 13:41–47
30. Tallab T, Joharji H, Bahamdan K et al (2005) Response of vitiligo to PUVA therapy in Saudi patients. Int J Dermatol 44:556–558
31. Tobin DJ, Bystryn JC (1996) Different populations of melanocytes are present in hair follicles and epidermis. Pigment Cell Res 9:304–310
32. Warren MA, Bleehen SS (1995) Morphologic observations on the dermal nerves in vitiligo: an ultrastructural study. Int J Dermatol 34:837–840
33. Westerhof W, Nieuweboer-Krobotova L (1997) Treatment of vitiligo with UV-B radiation vs topical psoralen plus UV-A. Arch Dermatol 133:1525–1528
34. Wu CS, Yu HS, Chang HR et al (2000) Cutaneous blood flow and adrenoceptor response increase in segmentals, type vitiligo lesions. J Dermatol Sci 23:53–62
35. Wu CS, Yu CL, Wu CS et al (2004) Narrow-band ultraviolet-B stimulates proliferation and migration of cultured melanocytes. Exp Dermatol 13:755–763
36. Wu CS, Lan CC, Wang LF et al (2007) Effects of psoralen plus ultraviolet A irradiation on cultured epidermal cells in vitro and patients with vitiligo in vivo. Br J Dermatol 156: 122–129
37. Yarr M, Gilchrest BA (1991) Human melanocyte growth and differentiation: a decade of new data. J Invest Dermatol 97: 611–617
38. Yonetani S, Moriyama M, Nishigori C et al (2008) In vitro expansion of immature melanoblasts and their ability to repopulate melanocyte stem cells in the hair follicule. J Invest Dermatol 128:408–420
39. Yokota N (1967) Histochemical studies on vitiligo vulgaris. Jpn J Dermatol 77:109
40. Yu HS (2002) Melanocyte destruction and repigmentation in vitiligo: a model for nerve cell damage and regrowth. J Biomed Sci 9:564–573
41. Yu HS, Wu CS, Yu CL et al (2003) Helium neon laser irradiation stimulates migration and proliferation in melanocytes and induces repigmentation in segmental type vitiligo. J Invest Dermatol 120:56–64

Editor's Synthesis

2.4

Mauro Picardo and Alain Taïeb

Contents

M. Picardo (✉)
Istituto Dermatologico San Gallicano,
via Elio Chianesi, 00144 Roma,
Italy
e-mail: picardo@ifo.it

Core Messages

- Some mechanisms of melanocyte loss have been pooe In Viro
- Pathomechanisms should be envisioned within an enlarged dermal–epidermal melanin unit (including nerve endings)
- Melanocyte Stemnes, innate immunity Somatic Mosaicism (for SV), Membrane Lipids are newamas in the puzzle of utiligo pathophysiology

The current status of our understanding of vitiligo pathomechanisms has been summarized in Table 2.4.1. We have tried to divide it into major fields and levels of evidence. Clearly, in vivo evidence in human disease comes first. In vitro work which relies on cultures should be viewed more cautiously because of artefacts caused by culture conditions, even though appropriate controls are used. Some animal models are probably relevant, but definitive evidence is lacking.

Concerning clinical, epidemiological, and pathological data reviewed in Part 1, which should have provided a solid basis for the understanding of the disease, we have already noticed intriguing weaknesses in the vitiligo field. Indeed, the cooperation among research groups has been developing only recently and variable definitions of type or stage of the disease (Chap. 1.2.1) have led to difficulties in the interpretation of some in vivo data, such as the importance of skin inflammation in the disease. There has also been a rampant temptation (conscious or not) to erase some aspects which were not considered to be consistent with the pathophysiological point of view defended by individual authors.

M. Picardo and A. Taïeb (eds.), *Vitiligo*,
DOI 10.1007/978-3-540-69361-1_2.4, © Springer-Verlag Berlin Heidelberg 2010

Table 2.4.1 Summary of some aspects of our current understanding of vitiligo

Field	Level of evidence		
	Good	Moderate	Limited
Cell biology	Loss of pigment cells in epidermal/follicular compartments, melanocyte structural and functional anomalies, including friction-induced detachment in vivo Keratinocyte anomalies (structural and in culture) reduced keratinocyte-derived melanocyte growth factors (e.g. SCF, a MSH, ET1)	Loss of melanocyte adhesion Melanocyte senescence and stem cell compartment involved Fibroblast functional defects PBMC defects Mitochondrial impairment Defective activity of receptors for growth factors MITF-M deficiency Propopiomelanocortin system alterations (primary or secondary)	Involvement of non skin melanocytes? (not completely settled) Apoptosis of melanocytes TNF-induced disease? Stratum corneum activation of inflammasome or other melanocyte targeting pathways to explain Koebner's phenomenon
Genetics	Polygenic background NALP 1 genetic variants in NSV associated with auto-immune diseases	Causative inflammasome functional alteration Other causative gene variants or gene-modifying variants in subsets of patients (AIRE, CMH haplotypes, oxidative stress regulators...)	Several candidate gene associations (approaches not validated/replicated) Mitochondrial genome defects (not fully tested) Abnormal expression of genes of the FOXD3 pathway (to confirm) Protective MC1-R alleles Predisposing skin-related genetic anomaly (applies mostly to segmental vitiligo)
Biochemistry	Catechols metabolism anomalies Antioxidant systems anomalies Biopterin pathway anomalies	Membrane-associated defects, including mitochondria defects Role of environmental toxic compounds (phenol, catechols, etc.)? (more epidemiologic studies needed)	Melanogenesis defect exclusively
Immunology	Humoral (serum) and cellular (in situ) immune responses to pigment cell proteins T cell lichenoid infiltrates in progressive NSV, mostly CD8+	Relevant cellular targets of immune responses not clearly established Autoimmunity to melanin catabolites Viral infection (direct) Innate immunity defect to bacterial infections (NALP1?)	Contagious disease (but role of microorganisms not ruled out, see role of herpes virus in the Smyth chicken model of vitiligo (Chap. 2.2.4) Injury activation of the skin Innate immune system Deficient clearance of apoptotic fragments

In the following sections we highlight some gaps in our understanding, underline possible unifying scenarii, and propose, based on Part 2 review of data, a hierarchy of possible events occurring upstream of melanocyte loss in the vitiligo puzzle (Fig. 2.4.1).

2.4.1 Do All Supposed Mechanisms of Melanocyte Loss Occur In Vivo?

Table 2.4.1 indicates a wide range of possible events that may provoke melanocyte loss. Interestingly, some popular theories, such as the apoptotic demise of melanocytes, are not yet substantiated. In non-segmental vitiligo, there is now ample evidence that melanocytes of non-clinically affected areas can be morphologically or functionally altered in favour of either an intrinsic abnormality, which could be the basis of the Koebner phenomenon, or of a systemic low-key immune disturbance targeting the cutaneous melanocytic system, or both. This point is crucial for a better understanding and will need more in vivo studies using skin biopsies to look at early changes in the disease. The experimental provocation of Koebner's phenomenon, as indicated in Sect. 2.2.2.1, may be a valuable tool to detect the chronology of events in vitiligo, and especially a link between trauma and inflammation.

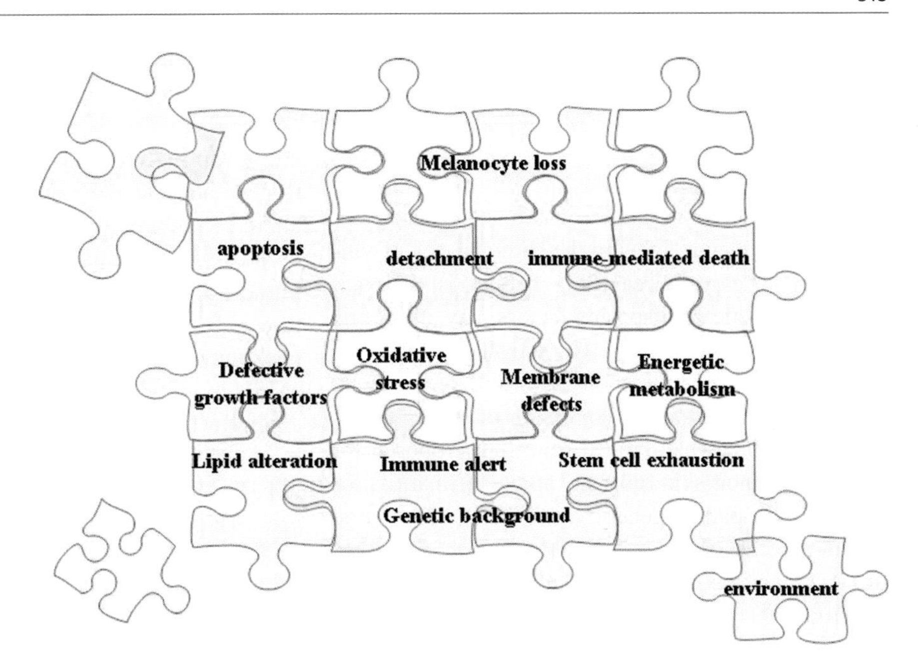

Fig. 2.4.1 A semi-organized puzzle of pathogentic factors

2.4.2 An Enlarged Vision of the Skin Melanogenic Unit May Apply to Vitiligo

While melanocytes are the undisputed target of the disease, other skin cells affect their function and survival. The conventional skin pigmentation unit was the epidermal melanin unit of Fitzpatrick and Breathnach [2]. A dermal–epidermal melanin unit was more recently defended [1]. Based on current knowledge of vitiligo pathophysiology, the skin melanogenic unit should include, beside melanocytes, keratinocytes, fibroblasts, and possibly cutaneous nerve endings of the considered unit. Langerhans cells could also play a role as pivotal cells of the innate immune system, being envisioned as maintaining the immune surveillance of the pigmentary system in case of local danger. Progress in the knowledge of the epidermal and dermal network governing the release of specific cytokines and growth factors opens new perspectives in the pathogenesis and therapy of vitiligo, if the loss of pigment cells can be first stopped.

2.4.3 Melanocyte Stemness and Vitiligo

Until now, the attention has been mostly centred on resident epidermal melanocytes. However, exchanges between hair follicles and interfollicular compartments have been well -demonstrated for melanocytes (Chaps. 1.3.6 and 2.2.10). The melanoblasts of the hair follicle niche, studied mostly in the mouse, differentiate and migrate to the hair bulb while maintaining a reserve of undifferentiated cells inside the bulge [14]. The niche thus allows the continuous production of new melanocytes avoiding cell exhaustion and senescence. Vitiligo may represent a localized senescence process through a mechanism similar to the 'free radical-mediated greying,' [8, 9], associated with a progressive melanocyte loss from the hair bulb. The presence, or not, of a functioning follicular reservoir may also be the critical factor for a successful melanocyte graft besides loss through the Koebner's phenomenon. Usually, a minority of melanocyte stem cells matures into differentiated and migrating melanocytes, whereas the bulk of them remain quiescent. The exhaustion of the quiescent stock will affect the next melanocyte production. The ability to respond to specific growth factors, such as SCF, may determine the continuous production of differentiated and functioning epidermis-committed melanocytes. On this subject, Part 2 has provided several arguments indicating that the c-Kit/SCF axis appears to be broken in vitiligo (Chap. 2.2.8). Vitiligo may be considered as a possible model for the study of stemness and senescence processes.

2.4.4 The Somatic Mosaicism Hypothesis for SV and Deductions for NSV

The recently demonstrated association of SV and NSV has provided new insights in the pathogenesis of vitiligo [13]. The main reasoning behind is that still unknown genetic predisposing factors may affect first the skin pigmentary system. The activation of the skin immune/inflammatory responses would come second leading to a more severe expression of disease. This scenario is somewhat derived from that proposed for another common skin disorder, atopic dermatitis, for which a skin barrier genetic dysfunction comes first and may produce direct inflammation [11], and engages in a facultative second step the immune system into an '"allergic"' Th2 dependent pathway in a subset of patients. This two-phase model is consistent with the common presentation of NSV as an isolated chronic cutaneous disorder in the majority of patients. Since friction/pressure-prone areas are frequently initially affected, it would be tempting to speculate that the primary skin anomaly affects the upper layers of the epidermis to explain the Koebner's phenomenon; it would explain nicely the subsequent epidermal activation of innate immunity-based mechanisms via stratum corneum activation of inactive Il1beta precursors [12]. Pressure may also affect directly basal epidermal layer biology if an abnormal adhesion of melanocytes comes first to explain their detachment [3]. Also, as envisaged earlier, there is a possible genetic cause limiting survival and self-renewal of epidermal and follicular melanocytes, the latter being hit earlier by the disease in SV as compared to NSV [13]. The clear association of NSV to familial hair greying also favours a genetic background affecting melanocyte survival. The expression of several genes influencing such pathways is indeed modified in NSV [10].

2.4.5 Membrane Lipids as Possible Culprits

The lipidomics approach to cellular signalling indicates that lipids may be considered as 'signalling molecules' between extracellular events and cell adaptive response [4]. Alterations involving lipid metabolism may lead to modifications of the structure of the membranes, affecting the sensitivity to pro-oxidant agents, modulating the intracellular redox status (Chap. 2.2.6), and favouring the release of haptened lipids, able in turn to activate the immune system. Moreover, the altered membrane may participate to the propagation of the oxidative damage, responsible for a high level of intracellular ROS, able also in turn to affect the structure and activity of the POMC-derived peptides. This phenomenon might be the basis for the reduction of aMSH in vitiligo epidermis. The responses to the specific growth factors, as well as the processes underlying adhesion and survival events, could also be impaired by the defective metabolism of lipid messengers. The topical application of sphingolipids-mediated restoration of Mitf expression and repigmentation in a mouse model of hair greying is in accordance with this view [8, 9].

2.4.6 The Innate Immunity Hypothesis

Besides the possible exposure of lipid–protein complexes with antigenic properties, apoptosis or other mechanisms of cellular alteration may activate the immune surveillance through the release of membrane fragments. Based on the finding of NALP variants in immune NSV [6], it has been hypothesized that such cellular debris or apoptotic fragments could activate the NALP1 inflammasome complex, activating in turn the IL1 pathway and alerting the immune system. In addition, hsp70 production related to membrane damage (Sect. 2.2.7.4) has been widely associated to the activation of antigen presenting cells. The immune-mediated damage, strongly suggested by a wide range of clinical and experimental data (Chap. 1.2.2, 1.3.10 and 2.2.7), could be thus the most prominent modality of loss of genetically compromised melanocytes best demonstrated during flares or acceleration phases of the disease.

2.4.7 Concluding Remarks

Looking at the most recent experimental approaches, Le Poole et al.'s convergence theory [7] can tentatively be updated, and featured as a pyramid where melanocyte loss is the tip of an enormous puzzle, with a predisposing genetic background at the basement level (Fig. 2.4.1).

A variable genetic background (Level 1) could, under the influence of variable and still poorly understood environmental triggering factors (arrows), activate several pathogenic pathways (Level 2), occuring through different mechanisms (Level 3), and, according to the intensity of the stimuli, may concur to melanocyte loss (Level 4).

For unravelling the mysteries of the basement level 1, the genetic basis of vitiligo, besides the conventional approaches on blood DNA, SV considered as a common skin genetic mosaic revealed postnatally would be a good model. SV is a candidate disease that may be explored as a proof of principle for a new gene discovery strategy to treat multigenic disorders with organ specificity [5, 13]. For a good angle of attack of the puzzle-like pyramid, we still lack a definitive argument with a clinical or basic scientific perspective. The following section will deal with those uncertainties for the currently limited available therapeutic options.

References

1. Cario-André M, Pain C, Gauthier Y et al (2006) In vivo and in vitro evidence of dermal fibroblasts influence on human epidermal pigmentation. Pigment Cell Res 19:434–442
2. Fitzpatrick TB, Breathnach AS (1963) The epidermal melanin unit system. Dermatol Wochenschr 147:481–489
3. Gauthier Y (1995) The importance of Koebner's phenomenon in the induction of vitiligo vulgaris lesions. Eur J Dermatol 5:704–708
4. Hannun YA, Obeid LM (2008) Principles of bioactive lipid signalling:lessons from sphingolipids. Nat Rev Mol Cell Biol 9:139–150
5. Happle R (2007) Superimposed segmental manifestation of polygenic skin disorders. J Am Acad Dermatol 57:690–699
6. Jin Y, Mailloux CM, Gowan K et al (2007) NALP1 in vitiligo-associated multiple autoimmune disease. N Engl J Med 356:1216–1225
7. Le Poole IC, Das PK, van den Wijngaard RM et al (1993) Review of the etiopathomechanism of vitiligo: a convergence theory. Exp Dermatol 2:145–153
8. Picardo M (2009) Lipid-mediated signalling and melanocyte function. Pigment Cell Mel Res 22:152–153
9. Saha B, Singh SK, Mallik S et al (2009) Sphingolipid-mediated restoration of Mitf expression and repigmentation in vivo in a mouse model of hair graying. Pigment Cell Mel Res 22:205–218
10. Strömberg S, Björklund MG, Asplund A et al (2008) Transcriptional profiling of melanocytes from patients with vitiligo vulgaris. Pigment Cell Melanoma Res 21:162–171
11. Taieb A (1999) Hypothesis: from epidermal barrier dysfunction to atopic disorders. Contact Dermatitis 41:177–180
12. Taieb A (2007) NALP1 and the inflammasomes: challenging our perception of vitiligo and vitiligo-related autoimmune disorders. Pigment Cell Res 20:260–262
13. Taïeb A, Morice-Picard F, Jouary T et al (2008) Segmental vitiligo as the possible expression of cutaneous somatic mosaicism: implications for common non-segmental vitiligo. Pigment Cell Melanoma Res 21:646–652
14. Yonetani S, Moriyama M, Nishigori C et al (2008) In vitro expansion of immature melanoblasts and their ability to repopulate melanocyte stem cells in the hair follicle. J Invest Dermatol 128:408–420

Section III

Therapy

Management Overview

3.1

Alain Taïeb and Mauro Picardo

Contents

3.1.1 Management-Oriented Evaluation

Before discussing management with the patient, a thorough assessment is needed. An assessment form has been produced by the Vitiligo European task force (VETF) (see Part 1.4) which summarizes major history taking and examination items. It has been recently updated [http://content.nejm.org/cgi/content/full/360/2/160/DC1] [1]. Skin color and ability to tan (phototype), disease duration, and disease activity are important decision management items, as well as the patient psychological profile and ability of coping with the disease. In some nonsegmental vitiligo (NSV) patients, an "acceleration phase" occurs with a rapid disease progression in a few weeks/months which needs a more urgent intervention, usually minipulse therapy (Chap. 3.5.1). Other useful clinical management items include previous episodes of repigmentation; type, duration, and usefulness of previous treatments. The analysis of the Koebner phenomenon (Sect. 2.2.2.1) is of particular interest for prevention, and needs a review with the patients according their daily habits (hygiene, clothing) and occupations. We use a slide projection in our clinic to show some striking examples. A question about vitiligo on genitals is included because it causes a strong embarassment to patients, as first reported to us by the patient's support groups. A global quality of life assessement is recommended ("how does vitiligo currently affects your everyday life") assessed on a visual analog scale, which is a coarse but useful indicator of coping with the disease (Part 3.13). This is important because we know that the patient's personality and perceived severity of the disease are predictors of quality of life impairment and will guide our management options [2].

A. Taïeb (✉)
Service de Dermatologie, Hôpital St André, CHU de Bordeaux, France
e-mail: alain.taieb@chu-bordeaux.fr

M. Picardo and A. Taïeb (eds.), *Vitiligo*,
DOI 10.1007/978-3-540-69361-1_3.1, © Springer-Verlag Berlin Heidelberg 2010

Because of the frequent association of NSV with autoimmune thyroid disease, especially Hashimoto's thyroiditis, it is recommended to measure on an annual basis the thyrotropin level in patients with antibodies to thyroid peroxidise, which may precede overt thyroiditis. Associated auto-immune diseases frequencies are apparently not similar according to ethnic background, and appropriate management is not settled. Any suggestive manifestations of organ-specific autoimmune diseases should prompt appropriate investigations [3], and a specialist advice can be helpful. The index of suspicion should be raised when a personal and/or family history of auto-immune/auto-inflammatory disorders is obtained.

3.1.2 Treatment Overview

For stopping depigmentation, besides UV therapies, systemic steroids have been evaluated only in open studies and seem to arrest disease progression [4, 8]. Commonly used repigmentation therapies for vitiligo that are supported by data from randomized trials (RCT) include UV light (whole body irradiation or UV targeted to lesions), and topical agents (corticosteroids, calcineurin inhibitors, calcipotriol). There are promising emerging treatments such as prostaglandin E2 [36] (Chap. 3.6). Camouflaging or depigmenting (in widespread/ disfiguring disease) are the other current options.

UV Treatments (Part 3.3)

The currently preferred treatment in adults and compliant children with NSV is narrow-band UVB (NB-UVB), which delivers peak emission at 311 nm [10]. The color match of repigmented skin is excellent, but the response rate remains low based on patients' expectations. With twice weekly NB-UVB treatments for one year, 48–63% of patients with NSV repigment 75% of the affected areas [11]. At least three months of treatment is warranted before identifying a patient as a nonresponder, and approximately nine months of treatment is usually required to achieve the maximal repigmentation. Unfortunately, access to this treatment is frequently limited in some areas due to low availability or long distances to treatment centers. The optimal dose may differ at different sites, and an option is shielding the area with the lower MED after reaching its optimal dose, while continuing to expose higher MED areas until optimal dosing continues [11]. There is no apparent relationship between the degree of initial depigmentation and the response to NB-UVB treatment [10], but the duration of disease is inversely correlated with the degree of treatment-induced repigmentation [11]. The best results are achieved on the face, followed by the trunk and limbs. The poorest outcomes have been noted for hands and feet lesions that at best show a moderate response. Relapses are common at all sites; around 60–70% of patients resume depigmentation in areas repigmented by treatment within one year whatever the regimen, PUVA or NB-UVB [12].

Responses of segmental vitiligo to narrow band UVB are at best limited when the same NB-UVB therapy is applied for six months [11]. The use of targeted high fluency UVB (excimer laser or monochromatic excimer lamp, both at 308 nm) which may reach deeper targets such as amelanotic melanocytes of the hair follicle, and also avoid irradiation of uninvolved skin, may improve outcomes. Preliminary results seem promising in cases of NSV involving limited areas [2, 30]. Red light Helium–Neon laser photherapy has also been reported as promoting repigmentation in SV [31].

Topical Therapies and Combined Therapies (Parts 3.2 and 3.8)

Topical therapies are not suitable for widespread NSV, but may be effective in cases with more localized disease (including SV). Combined treatments are frequently considered when phototherapy alone does not show efficacy after three months, or in an attempt to accelerate response and reduce cumulative UV exposure. As compared with PUVA which determines a predominant perifollicular pattern of repigmentation, topical corticosteroids (and topical calcineurin inhibitors – TCI – such as tacrolimus and pimecrolimus) exhibit a diffuse type, which is faster, but less stable [28]. Based on a meta-analysis, Class 3 (potent) topical corticosteroids achieve more than 75% repigmentation in 56% of patients [17]. The TCI are nowadays preferred for face and neck lesions, because they cause no skin atrophy. The efficacy of TCI is enhanced by occlusive therapy [9], exposure to UV radiation delivered by high fluency UVB devices [30], but not clearly by conventional NB-UVB [23]. The usefulness of these agents as sole therapy in sun-protected sites such as genitals and nipples requires further study. The concerns about the risk of cutaneous or even extracutaneous cancer provoked by topical calcineurin inhibitors have been disproportionate [7]. Calcineurin inhibitors can promote repigmentation without immunosuppression [20]. The combined use of UV and calcineurin inhibitors has not yet been fully approved until the availabililty of more long-term data. Other sources of light such as the helium–neon laser (red light) are proposed to limit potential dangers of combination therapies using UV and calcineurin inhibitors [21]. It is unclear whether topical corticosteroid can also increase the efficacy of UVB as already demonstrated for fluticasone plus UVA [32]. Topical calcipotriene RCT studies indicate limited or no effect in isolation and a possible minor enhancer effect on repigmentation in combination with UV or topical corticosteroids [14].

Surgery (Part 3.7)

Surgical methods such as minigrafting, which consists of transplanting punch grafts from an autologous donor site [16] – currently less used, when other options are possible – or cellular transplantation using autologous epidermal cell suspensions containing melanocytes [18] or ultrathin epidermal grafts [26], or a combination of cellular transplantation and ultrathin grafting are used in cases of focal/segmental vitiligo, if medical approaches fail. UV irradiation is frequently associated [33]. Patients with NSV are considered to be good candidates for surgical techniques depending on their availability and cost, if their disease is stable (over the preceding 1–2 years), and has a limited extent (2–3% of body surface area). Contrary to SV, in which the grafted cells come from disease-free areas, the survival of living, but potentially abnormal transplanted melanocytes is less predictable in NSV. Koebnerization limits efficacy (hands in particular). Based on an RCT, after a strict preoperative selection for disease stability in NSV, cellular transplantation plus UV results in repigmentation of at least 70% of the treated area [33].

Camouflage and Depigmentation (Parts 3.9 and 3.11)

Topical camouflaging products mask esthetic skin disfigurement on a transient, semi-permanent, or permanent basis (tattoos). Benefits can be obtained by the skilled use of corrective cosmetics. For dihydroxyacetone (DHA), the most used self-tanner, the higher the concentration, the better is the response observed, particularly in darker phototypes [5]. Chemical or laser depigmentation can definitely be a choice in a small subset of carefully selected patients, but results are variable both in efficacy and duration [25]

Other Therapies (Parts 3.4 and 3.6)

The use of topically applied antioxidants has not yet been supported by sufficient level of evidence. Reports investigating the efficacy of natural health products (chinese herbs, plant extracts, vitamins) for vitiligo exist, but are of poor methodological quality and contain significant reporting flaws [review in 34]. A RCT supports a moderate adjunctive effect of Polypodium leukotomos to NB-UVB [24], and another that of systemic antioxidants in addition to NB-UVB [15]. Ginkgo biloba as a monotherapy [29] warrants further investigation.

Counseling (Parts 3.10, 3.12–3.14)

The involvement of a psychologist or psychiatrist may be helpful in patients who experience difficulty in coping with the diagnosis. Sunscreens are needed in case of a real risk of sunburn on nonphotoprotected skin, but not on a routine basis, because moderate sun exposure (heliotherapy) is a good substitute to UV therapies. Photoadaptation exists in depigmented vitiligo skin [6, 35]. Repeated frictions for applying sunblockers without real sunburn risk may be more detrimental than beneficial.

3.1.3 Evidence-Based Guidelines

General guidelines for adults and children have been elaborated by the British Association of Dermatologists [19] and guidelines for surgery by the IADVL Dermatosurgery Task Force [27]. A treatment algorithm has also been proposed based on evidence-based medicine principles [22].

Summary Messages

- In NSV, initial assessment should focus on possible aggravation by friction/pressure and the possibility of associated automimmune diseases, in particular thyroid disease.
- Attention is warranted to the psychological impact of the condition, and consideration given to referral for psychological support.
- Patients should be informed that (1) vitiligo is a chronic/relapsing disorder, (2) repigmentation is a slow process, and (3) reactivation of the disease in different body regions or the reappearance of lesions in treated ones may occur.
- The first target of therapy is to stop progression of the disease, which is currently best achieved with more widespread disease using narrow band UVB.
- In rapidly progressing disease or patients failing to stabilize under NB-UVB therapy, systemic corticosteroids or immunosuppressants need to be more thoroughly evaluated.
- Narrow band UVB is also the most useful repigmenting regimen.
- Potent topical corticoteroid or topical calcineurin inhibitor therapy may be used first line for localized disease; on the face, calcineurin inhibitors are currently preferred because of potential side effects of prolonged application of steroids, although their long-term safety requires further study.
- Offering training to camouflage techniques is particularly helpful in dark-skinned patients for facial/hands lesions.
- Cellular transplantation or grafting is an option in specialized centers for patients who have stable and limited lesions that are refractory to other therapy.

References

1. Taïeb A, Picardo M; VETF Members (2007) The definition and assessment of vitiligo: a consensus report of the Vitiligo European Task Force. Pigment Cell Res 20:27–35
2. Kostopoulou P, Jouary T, Quintard B, Marques S, Boutchnei S et al (2009) Objective vs subjective factors in the psychological impact of vitiligo: the experience from a French referral center. Br J Dermatol 161:128–133
3. Betterle C, Caretto A, De Zio A et al (1985) Incidence and significance of organ-specific autoimmune disorders (clinical, latent or only autoantibodies) in patients with vitiligo. Dermatologica 171:419–423
4. Radakovic-Fijan S, Fürnsinn-Friedl AM, Hönigsmann H et al (2001) Oral dexamethasone pulse treatment for vitiligo. J Am Acad Dermatol 44:814–817
5. Rajatanavin N, Suwanachote S, Kulkollakarn S (2008) Dihydroxyacetone: a safe camouflaging option in vitiligo. Int J Dermatol 47:402–406
6. Rivard J, Hexsel C, Owen M, Strickland FM et al (2007) Photoadaptation of vitiliginous skin to targeted ultraviolet B phototherapy. Photodermatol Photoimmunol Photomed. 23: 258–260
7. Rustin MH (2007) The safety of tacrolimus ointment for the treatment of atopic dermatitis: a review. Br J Dermatol 157: 861–873
8. Seiter S, Ugurel S, Tilgen W et al (2000) Use of high-dose methylprednisolone pulse therapy in patients with progressive and stable vitiligo. Int J Dermatol 39:624–627
9. Hartmann A, Brocker EB, Hamm H (2008). Occlusive t nhances efficay of tacrolimus 0.1% oinement in adult patients with vitiligo: results of a placebo-controlled 12 month prospective study. Acta Derm venereol 88 : 474–479
10. Westerhof W, Nieuweboer-Krobotova L (1997) Treatment of vitiligo with UV-B radiation vs topical psoralen plus UV-A. Arch Dermatol 133:1525–1528
11. Anbar TS, Westerhof W, Abdel-Rahman AT et al (2006) Evaluation of the effects of NB-UVB in both segmental and non-segmental vitiligo affecting different body sites. Photodermatol Photoimmunol Photomed 20(22):157–163
12. Yones SS, Palmer RA, Garibaldinos TM et al (2007) Randomized double-blind trial of treatment of vitiligo: efficacy of psoralen-UV-A therapy vs narrowband-UV-B therapy. Arch Dermatol 143:578–584
13. Casacci M, Thomas P, Pacifico A et al (2007) Comparison between 308-nm monochromatic excimer light and narrowband UVB phototherapy (311–313 nm) in the treatment of vitiligo – a multicentre controlled study. J Eur Acad Dermatol Venereol 2:956–963
14. Chiavérini C, Passeron T, Ortonne JP (2002) Treatment of vitiligo by topical calcipotriol. J Eur Acad Dermatol Venereol 16:137–138
15. Dell'Anna ML, Mastrofrancesco A, Sala R et al (2007) Antioxidants and narrow band-UVB in the treatment of vitiligo: a double-blind placebo controlled trial. Clin Exp Dermatol 32:631–636
16. Falabella R (1988) Treatment of localized vitiligo by autologous minigrafting. Arch Dermatol 124:1649–1655
17. Forschner T, Buchholtz S, Stockfleth E (2007) Current state of vitiligo therapy – evidence-based analysis of the literature. J Dtsch Dermatol Ges 5:467–475

18. Gauthier Y, Surleve-Bazeille JE (1992) Autologous grafting with noncultured melanocytes: a simplified method for treatment of depigmented lesions. J Am Acad Dermatol 26(2 Pt 1): 191–194
19. Gawkrodger DJ, Ormerod AD, Shaw L, Mauri-Sole I et al (2008) Guideline for the diagnosis and management of vitiligo. Br J Dermatol 159:1051–1076
20. Lan CC, Chen GS, Chiou MH et al (2005) FK506 promotes melanocyte and melanoblast growth and creates a favourable milieu for cell migration via keratinocytes: possible mechanisms of how tacrolimus ointment induces repigmentation in patients with vitiligo. Br J Dermatol 153:498–505
21. Lan C, Wu C, Chen G et al (2009) Helium-neon laser and topical tacrolimus combination therapy: novel treatment option for vitiligo without additional photocarcinogenic risks. J Eur Acad Dermatol Venereol 23:344–345
22. Lim HW, Hexsel CL (2007) Vitiligo: to treat or not to treat. Arch Dermatol 143:643–646
23. Mehrabi D, Pandya AG (2006) A randomized, placebo-controlled, double-blind trial comparing narrowband UV-B plus 0.1% tacrolimus ointment with narrowband UV-B plus placebo in the treatment of generalized vitiligo. Arch Dermatol 142:927–929
24. Middelkamp-Hup MA, Bos JD, Rius-Diaz F et al (2007) Treatment of vitiligo vulgaris with narrow-band UVB and oral *Polypodium leucotomos* extract: a randomized double-blind placebo-controlled study. J Eur Acad Dermatol Venereol 21:942–950
25. Njoo MD, Vodegel RM, Westerhof W (2000) Depigmentation therapy in vitiligo universalis with topical 4-methoxyphenol and the Q-switched ruby laser. J Am Acad Dermatol 42(5 Pt 1): 760–769
26. Olsson MJ, Juhlin L (1997) Epidermal sheet grafts for repigmentation of vitiligo and piebaldism, with a review of surgical techniques. Acta Derm Venereol 77:463–466
27. Parsad D, Gupta S (2008) IADVL Dermatosurgery Task Force. Standard guidelines of care for vitiligo surgery. Indian J Dermatol Venereol Leprol 74(Suppl):S37–S45
28. Parsad D, Pandhi R, Dogra S et al (2004) Clinical study of repigmentation patterns with different treatment modalities and their correlation with speed and stability of repigmentation in 352 vitiliginous patches. J Am Acad Dermatol 50: 63–67
29. Parsad D, Pandhi R, Juneja A (2003) Effectiveness of oral Ginkgo biloba in treating limited, slowly spreading vitiligo. Clin Exp Dermatol 28:285–287
30. Passeron T, Ostovari N, Zakaria W et al (2004) Topical tacrolimus and the 308-nm excimer laser: a synergistic combination for the treatment of vitiligo. Arch Dermatol 140: 1065–1069
31. Yu HS, Wu CS, Yu CL et al (2003) Helium-neon laser irradiation stimulates migration and proliferation in melanocytes and induces repigmentation in segmental-type vitiligo. J Invest Dermatol 120:56–64
32. Westerhof W, Nieuweboer-Krobotova L, Mulder PG et al (1999) Left-right comparison study of the combination of fluticasone propionate and UV-A vs. either fluticasone propionate or UV-A alone for the long-term treatment of vitiligo. Arch Dermatol 135:1061–1066
33. Van Geel N, Ongenae K, De Mil M (2004) Double-blind placebo-controlled study of autologous transplanted epidermal cell suspensions for repigmenting vitiligo. Arch Dermatol 140:1203–1208
34. Szczurko O, Boon HS (2008) A systematic review of natural health product treatment for vitiligo. BMC Dermatol 8(1):2
35. Caron-Schreinemachers AL, Kingswijk MM, Bos JD et al (2005) UVB 311 nm tolerance of vitiligo skin increases with skin phototype. Acta Derm Venereol 85(1):24–26
36. Kapoor R, Phiske MM, Jerajani HR (2009). Evaluation of safety and efficacy of topical prostaglandin E2 in treatment of vitiligo. Br J Dermatol.160:861–863

Topical Therapies 3.2

Topical Corticosteroids

3.2.1

J. P. Wietze van der Veen, Bas S. Wind, and Alain Taïeb

Contents

J.P.W. van der Veen (✉)
Netherlands Institute for Pigment Disorders
and Department of Dermatology, Academic Medical Centre,
University of Amsterdam, Amsterdam, The Netherlands
e-mail: j.p.vanderveen@amc.uva.nl

3.2.1.1 Introduction

Currently, topical glucocorticosteroids (TCS) or topical calcineurin inhibitors (TCI) are the usual first-line treatments of vitiligo. The TCS have been used widely as topical and sometimes intralesional therapy in vitiligo since their introduction in dermatology in the 1950s. However, based on current criteria for clinical trials, the studies are of poor quality and since 1977, no randomized controlled trial (RCT) has been published.

3.2.1.2 Mode of Action and Rationale for Use in Vitiligo

The classical, genomic mechanism of glucocorticoid (GC) action can be divided into transrepression, which is responsible for a large number of desirable antiinflammatory and immunomodulating effects, and transactivation, which is associated with frequently occurring side effects as well as with some immunosuppressive activities [17].

After cell-membrane passage, GC interacts with the cytosolic glucocorticoid receptor (cGCR). Chaperones and co-chaperones are transported with the receptor into the nucleus where they modulate the action of the receptor. Transactivation is largely mediated through binding of the receptor as homodimer to specific sequences in the promoter or enhancer regions of GC-responsive genes. Dimerization of the GR is a prerequisite for the activation of gene expression. Other mechanisms of activation occur possibly through the activation function domains 1 and 2 of the receptor [17].

Transrepression mechanisms are not fully understood, but the receptor can inhibit the activity of other

M. Picardo and A. Taïeb (eds.), *Vitiligo*,
DOI 10.1007/978-3-540-69361-1_3.2.1, © Springer-Verlag Berlin Heidelberg 2010

transcription factors such as activator protein-1 (AP-1), nuclear factor-κB (NF-κB), and interferon regulatory factor-3 (IRF-3), which regulate the expression of proinflammatory genes. The negative regulation of those transcription factors by the GR has become a paradigm for the antiinflammatory and immunosuppressive action of GCs. Some of the immunosuppressive and antiinflammatory activities of GCs are also mediated by MAP kinase phosphatase-1 [17].

TCS have well-known antiinflammatory effects of GC, which may apply to the inflammatory phase and is well -demonstrated at the margin of vitiligo lesions. TCS, when used in combination with tretinoin and hydroquinone in the treatment of hypermelanoses, suppress the biosynthetic and secretory functions of melanocytes, and thus melanin production. Used alone, they are also depigmenting agents, and the loss of pigment or delay in pigmentation induced may help the migration of cells from the follicular compartment during repigmentation.

3.2.1.3 Studies

Most patients included in the published studies had NSV. The use of a highly potent (clobetasol) or potent (betamethasone) topical steroid can repigment NSV, but only in a small proportion of cases in the limited published series [3, 8]. Clayton [3] found 15–25% repigmentation in 10/23 subjects and >75% in 2/23 (the other 11 showed no response); Kandil [8] found 90–100% repigmentation in 6/23 subjects and 25–90% in 3 (with six showing "beginning" repigmentation). Clayton [3] found all steroid users had skin atrophy with clobetasol, a highly potent topical steroid (used for 8 weeks), whilst Kandil [8] noted hypertrichosis in two subjects and acne in three subjects, related to four months use of the potent topical steroid, betamethasone.

Topical fluticasone alone or combined with UV-A has been studied in 135 adults [19]. Fluticasone (potent TCS) used alone for nine months induced mean repigmentation of only 9% (compared to UV-A alone of 8%), whereas the combination of fluticasone and UV-A induced mean repigmentation of 31%; no steroid atrophy was noted in steroid users.

In a meta-analysis on nonsurgical therapies in vitiligo [4, 15] of data extracted of three RCT on potent [2, 8, 10] and two RCTs on very potent corticosteroids [2, 3], 75% repigmentation was found in one third of the patients treated with potent corticosteroids, while no significant differences were found between very potent corticosteroids and their placebo. In a larger nonrandomized patient series, however, potent and very potent TCS appeared to be equally effective (more than 75% repigmentation) in about 55% of the patients. As in all topical therapies, corticosteroids achieve the best results on sun-exposed areas like face and neck [15, 19]. Other indications for successful treatment are dark skin [11] and lesions of short duration [18] (Fig. 3.2.1.1).

In children, the topical use of the highly potent steroid clobetasol dipropionate was noted to induce better repigmentation than PUVAsol alone, finding >50% repigmentation in 15/22 (vs. 4/23 for PUVAsol) used for six months, but six steroid users developed skin atrophy [9]. Another study in 20 children over an 8-week period compared topical clobetasol and tacrolimus and found 41% repigmentation for clobetasol versus. 49% repigmentation for tacrolimus [4, 14].

Westerhof et al. [19] demonstrated a three times higher repigmentation rate in vitiligo patients treated with fluticasone propionate (FP) combined with UVA as compared to patients treated with either UVA or FP alone. A synergic effect of UVB or calcipotriol plus TCS is possible based on clinical observations, but not proven [12]. The 308 excimer laser was recently shown to be synergic with hydrocortisone 17-butyrate [16].

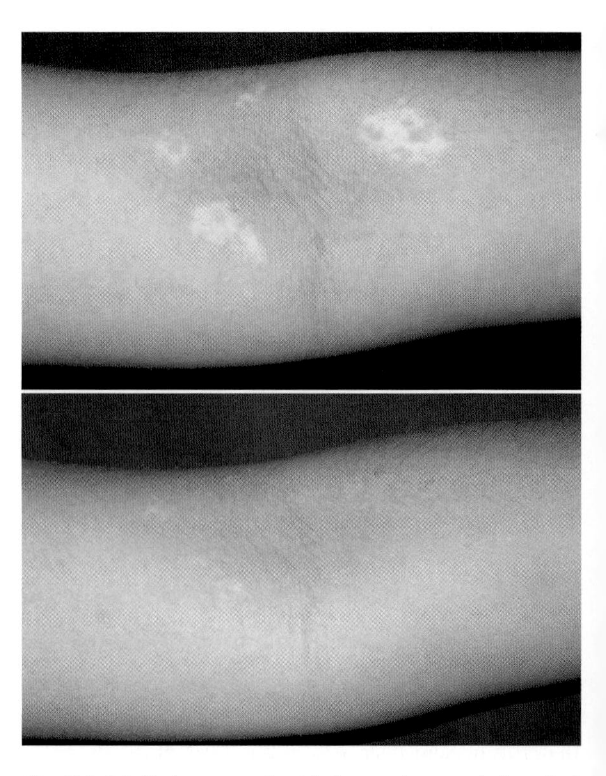

Fig. 3.2.1.1 Patient treated with betamethasone: before (*up*) and after (*down*) the therapy

3.2.1.4 Application Scheme

There is a general agreement to use TCS, whatever their indication, on the basis of once daily application, to limit side effects [5]. Optimal duration for vitiligo is not established, as well as the use of a continuous versus discontinuous regimen [6]. The amount used should be monitored (number of tubes/month is the most manageable like in atopic dermatitis). Potent TCS should be used first. When no results are seen after 3three months of sincere trial, we suggest to try superpotent TCS if no side effects have been noted, or to stop the medication.

3.2.1.5 Side Effects

Side effects like atrophy have been described in 14% of cases with very potent TCS as compared to 2% with potent TCS [17]. Systemic absorption is a concern on thin skin. Children with head and/or neck affected areas are eight times more likely to have an abnormal cortisol levels compared with children affected in other body areas [13]. For local therapy of vitiligo, potent TCS should therefore be preferred over very potent TCS that have more side effects. The lower classes of local corticosteroids have no demonstrated benefit.

A major improvement in the therapeutic index of GCs has been made with the development of compounds that display only local activity. This limited activity is due to the instability of the compounds that markedly reduces undesired systemic effects. For example, mometasone furoate, methylprednisolone aceponate, and budesonide locally inhibit proinflammatory cytokines and chemokines very effectively with negligible systemic effects.

Selective modulation of GR action is a promising concept for separation of anti-inflammatory effects from side effects, which has led to the design of qualitatively new drugs, such as selective glucocorticoid receptor agonists (SEGRAs). These innovative steroidal or nonsteroidal molecules induce transrepression, while transactivation processes are less affected [17].

3.2.1.6 Safety Issues

In children, especially with head and neck lesions, clinically relevant systemic absorption should be considered [13], and there is a general trend to prefer TCI in this location.

Summary Messages

- Topical corticosteroids (TCS) are commonly used as a first line treatment in children and adults with recent onset of segmental or not segmental vitiligo
- Treatment with a potent TCS should be considered for a trial period of no more than 3 months, with appropriate photographic monitoring.
- TCS may stabilize active progressing lesions and role of TCS in promoting repigmentation is possible
- There is no consensus on the best mode of prescription; potent topical corticosteroids are a first choice and should be given not more than once daily to avoid skin atrophy; continuous vs discontinuous schemes need to be tested.
- Topical corticosteroids could be employed when phototherapy is not possible (small children, photosensitivity, aggravation by sunlight etc), limited lesions, progressive lesions during the first months of phototherapy.
- Combination of TCS with UVA and 308 nm excimer laser light is synergic.
- TCS are clearly not suitable for widespread lesions and a considerable repigmentation can hardly be expected without the addition of sun exposure or phototherapy.
- Skin atrophy is a common side effect, which needs to be monitored, and suggest to prefer a discontinuous regimen. Topical calcineurin inhibitors (TCI) are now preferred on the face
- The choice for either potent TCS or TCI should be made based on cost/benefit ratio, side-effects, expected adherence to therapy
- TCS could be a useful adjunctive treatment after skin grafts, and can increase UVA and possibly UVB repigmentation efficacy.

References

1. Barman KD, Khaitan BK, Verma KK (2004) A comparative study of punch grafting followed by topical corticosteroid versus punch grafting followed by PUVA therapy in stable vitiligo. Dermatol Surg 30:49–53
2. Bleehen SS (1976) The treatment of vitiligo with topical steroids. Light and electronmicroscopic studies. Br J Dermatol 94(S12):43–50

3. Clayton R (1977) A double-blind trial of 0.05% clobetasol propionate in the treatment of vitiligo. Br J Dermatol 96: 71–73
4. Coskun B, Saral Y, Turgut D (2005) Topical 0.05% clobetasol propionate versus 1% pimecrolimus ointment in vitiligo. Eur J Dermatol 15:88–91
5. Forschner T, Buchholtz S, Stockfleth E (2007) Current state of vitiligo therapy-evidence-based analysis of the literature. J Dtsch Dermatol Ges 5:467–476
6. Gawkrodger DJ, Ormerod AD, Shaw L et al (2008) Guideline for the diagnosis and management of vitiligo. Br J Dermatol 159:1051–1076
7. Geraldez CB, Gutierrez GT (1987) A clinical trial of clobetasol propionate in Filipino vitiligo patients. Clin Ther 9:474–482
8. Kandil E (1974) Treatment of vitiligo with 0.1% betamethasone 17-valerate in isopropyl alcohol-a double-blind trial. Br J Dermatol 91:457–460
9. Khalid M, Mujtaba G, Haroon TS (1995) Comparison of 0.05% clobetasol propionate cream and topical PUVAsol in childhood vitiligo. Int J Dermatol 34:203–205
10. Koopmans-van Dorp B, Goedhart-van Dijk B, Neering H et al (1973) Treatment of vitiligo by local application of betamethasone 17-valerate in a dimethyl sulfoxide cream base. Dermatologica 146:310–314
11. Kumari J (1984) Vitiligo treated with topical clobetasol propionate. Arch Dermatol 120:631–635
12. Kumaran MS, Kaur I, Kumar B (2006) Effect of topical calcipotriol, betamethasone dipropionate and their combination in the treatment of localized vitiligo. J Eur Acad Dermatol Venereol 20:269–273
13. Kwinter J, Pelletier J, Khambalia A et al (2007) High-potency steroid use in children with vitiligo: a retrospective study. J Am Acad Dermatol 56:236–241
14. Lepe V, Moncada B, Castanedo-Cazares JP et al (2003) A double-blind randomized trial of 0.1% tacrolimus vs 0.05% clobetasol for the treatment of childhood vitiligo. Arch Dermatol 139:581–585
15. Njoo MD, Spuls PI, Bos JD et al (1998) Nonsurgical repigmentation therapies in vitiligo. meta-analysis of the literature. Arch Dermatol 134:1532–1540
16. Sassi F, Cazzaniga S, Tessari G et al (2008) Randomized controlled trial comparing the effectiveness of 308-nm excimer laser alone or in combination with topical hydrocortisone 17-butyrate cream in the treatment of vitiligo of the face and neck. Br J Dermatol 159:1186–1191
17. Schake H, Rehwinkel H, Asadullah K, Cato AC (2006) Insight into the molecular mechanisms of glucocorticoid receptor action promotes identification of novel ligands with an improved therapeutic index. Exp Dermatol 15:565–573
18. Schaffer JV, Bolognia JL (2003) The treatment of hypopigmentation in children. Clin Dermatol 21:296–310
19. Westerhof W, Nieuweboer-Krobotova L, Mulder PGH et al (1999) Left-right comparison study of the combination of fluticasone propionate and UV-A vs either fluticasone propionate or UV-A alone for the long-term treatment of vitiligo. Arch Dermatol 135:1061–1066

Calcineurin Inhibitors

3.2.2

N. van Geel, B. Boone, I. Mollet, and J. Lambert

Contents

N. van Geel (✉)
Department of Dermatology,
Ghent University Hospital, Ghent, Belgium
e-mail: nanny.vangeel@ugent.be

3.2.2.1 Introduction

The topical calcineurin inhibitors – tacrolimus ointment 0.1 and 0.03% (FK 506, Protopic®, Astellas) and pimecrolimus cream 1% (SDZ ASM 981, Elidel®, Novartis) – have been specifically developed for the treatment of inflammatory skin diseases, and approved for the short-term and intermittent long-term treatment of atopic dermatitis in several countries (including the United States of America and European Union nations) [12, 19, 28, 32, 34, 38]. In contrast to topical steroid, they do not have the risk of local side effects, such as skin atrophy, telangiectasia, and glaucoma after prolonged use. Therefore, they are preferentially used in areas more susceptible to these side effects, such as head and neck region, flexures and genital area [23, 35]. Furthermore, because of the limited percutaneous penetration, significant systemic absorption has not been reported following normal use [1]. However, they are currently much more expensive than topical steroids.

Since 2002, the beneficial effect of topical immunomodulators (TIMs) has been reported for patients with vitiligo. How they interfere in vitiligo genesis is still a matter of debate.

3.2.2.2 Mode of Action and Rationale for Use in Vitiligo

Tacrolimus and pimecrolimus are topical ascomycin immunomodulating macrolactams, and act as calcineurin inhibitors, affecting the activation/maturation of T-cells, and subsequently inhibit the production of various Th1 and Th2 type of cytokines (IL-2, IL-3,

M. Picardo and A. Taïeb (eds.), *Vitiligo*,
DOI 10.1007/978-3-540-69361-3_2.2,

IL-4, IL-5, IL-10, GM-CSF, TNF-α and IFN-γ). This has led to speculation that this mechanism may interfere with the autoimmune/inflammatory mediated loss of melanocytes in vitiligo lesions. The inhibition of TNF-α production was suspected by some authors to be especially important in vitiligo, as TNF-α can inhibit melanocyte proliferation and melanogenesis, and it can induce ICAM-1 expression on melanocytes, through which T-lymphocyte induced destruction of melanocytes [13, 16, 20]. However, considering that the lymphocytic infiltrate in perilesional skin is not present in all vitiligo patients, other mechanisms may play a role in TIM-induced repigmentation. Recently, in vitro evidence of a direct interaction between tacrolimus and keratinocytes has been obtained, creating a favorable milieu for melanocytic growth and migration by inducing the release of stem-cell factor (SCF) and enhancing matrix metalloproteinase (MMP)-9 activity. Furthermore, Kang et al. [16] demonstrated that FK506 could even directly stimulate melanogenesis and migration of cultured human melanocytes.

3.2.2.3 Studies

The first clinical trial regarding the efficacy of TIMs was carried out in six patients with generalized vitiligo. Five out of six patients achieved >50% repigmentation of their treated areas using topical tacrolimus for 1–5 months [29].

Many reports have followed this initial trial, investigating mainly tacrolimus, and since 2003, also pimecrolimus (Tables 3.2.2.1 and 3.2.2.2). Unfortunately, only few randomized trials have been published, some of them in combination with UVB (Table 3.2.2.3) and only one study compared pimecrolimus and tacrolimus in vitiligo (Table 3.2.2.4).

3.2.2.4 Tacrolimus and Pimecrolimus Monotherapy

One randomized, double-blind, left–right comparative trial has been conducted and showed that tacrolimus is effective similarly to clobetasol propionate 0.05%. Comparable results were achieved in a pilot study comparing clobetasol propionate 0.05 and 1% pimecrolimus for 2 months in 10 patients with symmetrical lesions [21].

A double-blind vehicle-controlled study, evaluating the efficacy and safety of pimecrolimus cream, 1% predominantly on extremities of 20 vitiligo patients does not show efficacy on the actively treated site, possibly because all the treated lesions were localized on the extremities with 9 out of 20 (45%) on the back of the hands [9].

3.2.2.5 Combination Therapy

Two studies demonstrated that addition of topical tacrolimus to excimer laser therapy could improve laser efficacy [18, 27]. In another study, narrow-band-UVB phototherapy was combined with tacrolimus ointment in 110 vitiligo patients [7]. The authors observed more than 50% of repigmentation in 42% of the lesions. The repigmentation rate appeared strictly related to the site: an improvement of >50% was obtained for lesions located on the face (83%), followed by the limbs (68%) and trunk (53.5%) compared to lesions on both extremities and genital areas, where responses were more disappointing. In a randomized, placebo-controlled, double-blind study of Mehrabi et al. [24], narrow-band-UVB therapy with and without tacrolimus ointment was compared in eight patients. They do not demonstrate an additional effect of tacrolimus. However, it has to be mentioned that this study was underpowered ($n = 8$) to detect significant differences. Besides, the evaluated lesions were all located on non-facial areas.

3.2.2.6 Comparative Studies Between Tacrolimus and Pimecrolimus

Only one case report describes a head-to-head comparison between topical tacrolimus and pimecrolimus. Both agents induced repigmentation of pretibial lesions when used under occlusion overnight. Tacrolimus was slightly more effective (88% repigmentation) than pimecrolimus (73% repigmentation) [25].

Table 3.2.2.1 Case reports and clinical studies on tacrolimus in the treatment of vitiligo

Reference	Study design	*n*	Study duration	Treatment regimen	Results
[29]	Open pilot study	6	4–6 months	0.1% Tacrolimus	Repigmentation in 5/6 patients. More than 25% repigmentation in 3/6 in UV-protected areas
[2]	Open pilot study	12	8 months	0.1% Tacrolimus twice daily	Good to excellent repigmentation in 50% of patients
[17]	Open pilot study	25 children	12 weeks	0.03% Tacrolimus twice daily	Complete repigmentation in 57.9% of patients, best results in face and hear bearing sites
[32]	Retrospective review	57 children	3 months	0.03 or 0.1% tacrolimus once or twice daily	At least partial repigmentation in 84% of patients, best results in head and neck region
[13]	Open prospective study	23	24 weeks	0.1% Tacrolimus twice daily	Varying levels in 89% of patients, best result head and neck region
[21]	Randomized, double-blind, comparative trial	20 children	2 months	0.1% Tacrolimus vs. 0.05% clobetasol propionate	Tacrolimus almost as effective as clobetasol propionate on several body locations
[37]	Case report	3	2–4 months	0.1% Tacrolimus twice daily	Complete repigmentation in 100% of patients
[36]	Open pilot study	15	1.5–9.5 months	0.1% Tacrolimus twice daily (+sunlight exposition in some patients)	At least partial repigmentation in 87% of patients on several body locations
[14]	Open pilot study	6	1–5 months	0.03 or 0.1% tacrolimus twice daily	Moderate to excellent repigmentation in 83% of patients
[33]	Case report	1	18 months	0.1% Tacrolimus twice daily	90% Repigmentation in face and scalp region

Table 3.2.2.2 Case reports and clinical studies on pimecrolimus in the treatment of vitiligo

Reference	Study design	*n*	Study duration	Treatment regimen	Results
[5]	Case report	1	5 months	1% Pimecrolimus vs. calcipotriol cream	Significant improvement with both, but mainly with pimecrolimus of facial lesions
[35]	Case report	2 Children	3–4 months	1% Pimecrolimus twice daily	Almost complete repigmentation of eyelid and genital lesions
[23]	Retrospective study	8	11 months	1% Pimecrolimus twice daily	Mean percentage repigmentation in face 72.5%
[6]	Open prospective study	26	6 months	1% Pimecrolimus twice daily	Repigmentation in 57.7% of lesions. Mean repigmentation of 62% head and neck region
[30]	Open prospective study	30	12 weeks	1% Pimecrolimus twice daily	Repigmentation in 57.7% of lesions. Best results face and truck (mean repigmentation 31 and 36% respectively)
[31]	Open prospective study	19	6 months	1% Pimecrolimus once daily	>25% repigmentation in 68% of patients
[9]	Randomized, placebo-controlled, double-blind trial	20	6 months	1% Pimecrolimus twice daily vs. Placebo	No significant difference and effect on extremities/hands
[8]	Comparative prospective, non blind trial	10	2 months	1% Pimecrolimus vs. 0.05% clobetasol propionate twice daily	Comparable rate of repigmentation in non facial areas
[22]	Case report	1	5 months	1% Pimecrolimus twice daily	Percentage of repigmentation: >90%

Table 3.2.2.3 Topical calcineurin inhibitors in combination with UVB

Reference	Study design	*n*	Study duration	Treatment regimen	Results
[11]	Open prospective study	110	16 weeks	0.03–0.1% Tacrolimus once daily + UVB	Repigmentation of >50% in 42% of lesions, best results in face (83%)
[24]	Randomized, placebo-controlled, double-blind trial	8	12 weeks	0.1% Tacrolimus + UVB vs. placebo + UVB	No statistically significant difference in non facial areas
[26]	Comparative prospective, non blind, pilot study	9	12 weeks	0.1% Tacrolimus twice daily + UVB vs. 0.1% tacrolimus	No repigmentation with tacrolimus monotherapy on UV protected areas
[27]	Comparative, prospective, randomized, intra-individual study	14	12 weeks	0.1% Tacrolimus twice daily + excimer laser vs. excimer laser	Combination therapy is superior to excimer laser monotherapy in UV-resistant areas
[18]	Prospective, double-blind, placebo-controlled study	8	10 weeks	0.1% Tacrolimus twice daily + excimer laser vs. placebo + excimer laser	Significantly greater degree of repigmentation with combination therapy on elbows and knees
[7]	Case report	1	4 months	0.1% Tacrolimus twice daily + UV-B	Repigmentation 95% on face

Table 3.2.2.4 Comparative study between tacrolimus and pimecrolimus

Reference	Study design	*n*	Study duration	Treatment regimen	Results
[15]	Case report	1	18 months	1% Pimecrolimus vs. 0.1% tacrolimus under overnight occlusion	Tacrolimus site 88% and pimecrolimus 73% site repigmentation

3.2.2.7 General Outcome

TIMs seem to affect repigmentation differently according to the anatomical location of the lesions. Most studies show beneficial results mainly in the head and neck region. This may be explained by the greatest density of hair follicles in these areas, and thus of melanocyte reservoir. Besides, the influence of a reduced epidermal thickness might be of importance as this facilitates penetration of large molecules into the skin. Moreover, the head and neck region are sun-exposed areas and UV therapy is a well-known treatment option of vitiligo. Probably, UV-light exposure during TIM treatment seems to play an important or even synergistic role, as demonstrated by Ostovari et al. [26]. Sardana et al. reported that there is sufficient evidence that tacrolimus monotherapy is useful to produce repigmentation, as illustrated by their two cases [29]. However, it is still not known whether monotherapy with tacrolimus ointment requires longer periods to induce repigmentation as compared to combined treatment with UVB. A correlation between the repigmentation rate and patients' age or duration of the disease has not been consistently demonstrated. According to the literature, the achieved results in children using TIMs seem to be comparable to the results in adult patients [17, 21, 32, 35]. Generally, therapeutic options in childhood vitiligo patients are more limited, and TIMs are potential therapeutic alternative with good benefit/ risk ratio.

3.2.2.8 Application Scheme

Data about the most effective treatment scheme using TIM in vitiligo are still missing. In the available vitiligo studies, the application frequency varies between once and twice daily (Fig. 3.2.2.1 and 3.2.2.2). Duration of treatment mentioned in the studies ranged from 10 weeks up to 18 months. Information about the minimal or ideal treatment period in vitiligo, as well as the usefulness of long-term intermittent use, is not available. Furthermore, long-term results about the course of disease after treatment interruption are missing, or are incomplete. Interestingly, the additional overnight occlusion of 0.1% tacrolimus with polyurethane and

Fig. 3.2.2.1 Before (*left*), after (*right*) tacrolimus treatment for non segmental vitilgo

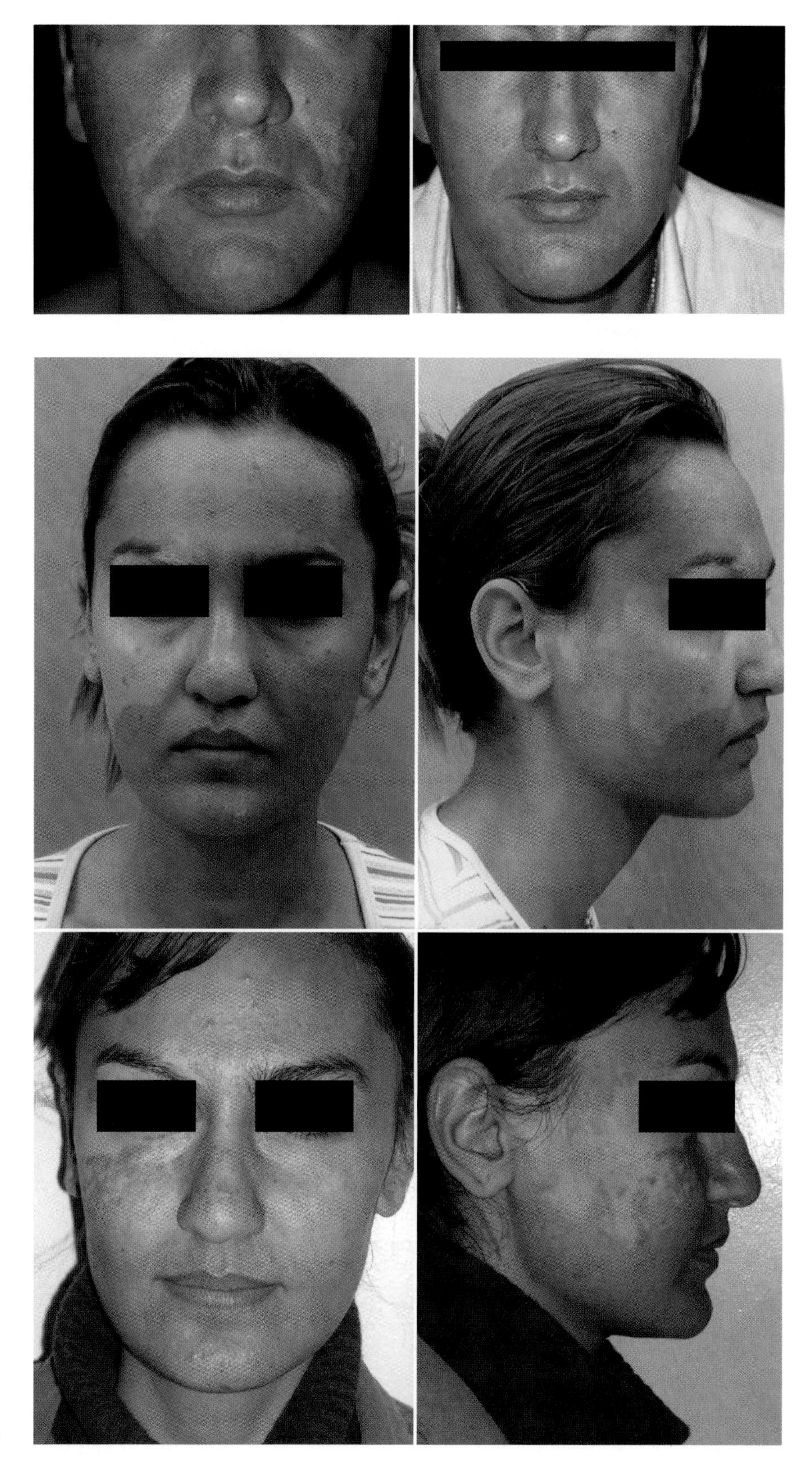

Fig. 3.2.2.2 Before (*top*) and after 6 months of once daily application of 0.1% tacrolimus ointment (*bottom*)

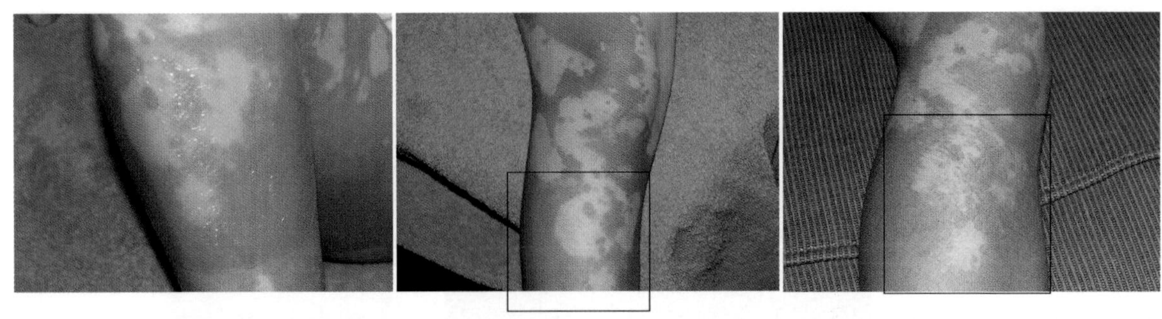

Fig. 3.2.2.3 Before and after tacrolimus under occlusion (hydrocolloid dressing) for 7 weeks. Area of application of hydrocoloid dressing shown in square

hydrocolloid foils might lead to a good repigmentation on the extremities after a previous response failure when used without occlusion. It was mentioned that the hydrocolloid dressings might be more suitable for improving transcutaneous penetration compared to polyurethane dressings, as they produce a much stronger water-binding capacity in the stratum corneum (Fig. 3.2.2.3).

3.2.2.9 Side Effects

The most common reported side effects for TIMs within the first days of treatment are local application reactions such as burning sensation, pruritus, and erythema.

However, the incidence of this side effect is lower in vitiligo patients than in the published atopic dermatitis studies. This is probably due to the presence of an intact epidermis in vitiligo patients versus the excoriated skin of an atopic dermatitis patient. Treatment of vitiligo and other dermatosis with TIMs has not been associated with a statistically significant increase of skin infections or systemic infections [4].

A tacrolimus induced hyperpigmentation in a vitiligo patch in the infraorbital area has been reported and might be related to sun exposure. This hyperpigmentation was temporary, with reappearance of depigmentation within one month of discontinuing topical tacrolimus application [10]. Less frequently reported side effects in the face are acne, hypertrichosis, and rosaceiform eruptions [3].

3.2.2.10 Safety Issues

The use of TIMs, especially in combination with phototherapy, leads to concern with respect to photocarcinogenesis. Currently, as a precaution, vitiligo patients should be advised to use an adequate protection against UV exposure, as long-term carcinogenicity studies using topical tacrolimus with or without the combination of UV exposure are not available yet. The combination therapy of TIM and UV should therefore only be performed in controlled and preferably experimental settings [4].

Since January 2006, a black box warning for tacrolimus and pimecrolimus was announced by the FDA, because of concerns of potential safety risks including skin cancer and lymphoma. However, there is up till now no evidence of a causal relationship between the sporadically reported lymphomas and the use of a topical calcineurin inhibitor, but conclusive safety data will only be available after many years. It should be noted that most studies with TIMs are conducted in patients with atopic dermatitis, in which the barrier function of the skin is disturbed. In vitiligo, the barrier function of the skin is normal which results in a lesser degree of penetration of TIMs. So far, the use of TIMs has not been reported to be associated with significant systemic immunosuppression or increased risk for skin cancer and other malignancies in clinical vitiligo trials [4].

Summary Messages

- Since 2002 topical calcineurin inhibitors (TIM) have been reported as beneficial for patients with vitiligo and have now an important role in the treatment of vitiligo, particularly in areas where the use of potent corticosteroids is contraindicated.
- Besides demonstrated effects on the activation of T cells, which may apply to the microinflammatory phase of vitiligo, in vitro studies demonstrate an influence of TIMs on melanocyte growth and migration via keratinocyte release of growth factors

- The efficacy of TIMs appears to be superior on sun-exposed areas.
- Occlusive dressing application allow repigmentation without UV exposure
- As long as specific safety data are missing and that the costs of these products remain high, it might be advisable to limit the use of TIMs in vitiligo to selected test areas first.
- More RCT studies are essential to confirm the benefits and to assess long term safety of monotherapy or combination treatment strategies.

References

1. Allen A, Siegfried E, Silverman R et al (2001) Significant absorption of topical tacrolimus in 3 patients with Netherton syndrome. Arch Dermatol 137:747–750
2. Almeida P, Borrego L, Rodriguez-Lopez J et al (2005) [Vitiligo. Treatment of 12 cases with topical tacrolimus]. Actas Dermosifiliogr 96:159–163
3. Bakos L, Bakos RM (2007) Focal acne during topical tacrolimus therapy for vitiligo. Arch Dermatol 143:1223–1224
4. Bieber T, Cork M, Ellis C et al (2005) Consensus statement on the safety profile of topical calcineurin inhibitors. Dermatology 211:77–78
5. Bilac DB, Ermertcan AT, Sahin MT, Ozturkcan S (2009) Two therapeutic challenges: facial vitiligo successfully treated with 1% pimecrolimus cream and 0.005% calcipotriol cream. J Eur Acad Dermatol Venereol 23:72–73
6. Boone B, Ongenae K, Van Geel N et al (2007) Topical pimecrolimus in the treatment of vitiligo. Eur J Dermatol 17: 55–61
7. Castanedo-Cazares JP, Lepe V, Moncada B (2003) Repigmentation of chronic vitiligo lesions by following tacrolimus plus ultraviolet-B-narrow-band. Photodermatol Photoimmunol Photomed 19:35–36
8. Coskun B, Saral Y, Turgut D (2005) Topical 0.05% clobetasol propionate versus 1% pimecrolimus ointment in vitiligo. Eur J Dermatol 15:88–91
9. Dawid M, Veensalu M, Grassberger M, Wolff K (2006) Efficacy and safety of pimecrolimus cream 1% in adult patients with vitiligo: results of a randomized, double-blind, vehicle-controlled study. J Dtsch Dermatol Ges 4: 942–946
10. De D, Kanwar AJ (2008) Tacrolimus-induced hyperpigmentation in a patch of vitiligo. Skinmed 7:93–94
11. Fai D, Cassano N, Vena GA (2007) Narrow-band UVB phototherapy combined with tacrolimus ointment in vitiligo: a review of 110 patients. J Eur Acad Dermatol Venereol 21: 916–920
12. Fleischer AB, Ling M, Eichenfield L et al (2002) Tacrolimus ointment for the treatment of atopic dermatitis is not associated with an increase in cutaneous infections. J Am Acad Dermatol 47:562–570
13. Grimes PE, Morris R, Avaniss-Aghajani E et al (2004) Topical tacrolimus therapy for vitiligo: therapeutic responses and skin messenger RNA expression of proinflammatory cytokines. J Am Acad Dermatol 51:52–61
14. Grimes PE, Soriano T, Dytoc MT (2002) Topical tacrolimus for repigmentation of vitiligo. J Am Acad Dermatol 47: 789–791
15. Hartmann A, Brocker EB, Hamm H (2008) Occlusive t nhances efficay of tacrolimus 0.1% oinement in adult patients with vitiligo: results of a placebo-controlled 12 month prospective study. Acta Derm venereol 88:474–479
16. Kang HY, Choi YM (2006) FK506 increases pigmentation and migration of human melanocytes. Br J Dermatol 155: 1037–1040
17. Kanwar AJ, Dogra S, Parsad D (2004) Topical tacrolimus for treatment of childhood vitiligo in Asians. Clin Exp Dermatol 29:589–592
18. Kawalek AZ, Spencer JM, Phelps RG (2004) Combined excimer laser and topical tacrolimus for the treatment of vitiligo: a pilot study. Dermatol Surg 30:130–135
19. Kostovic K, Pasic A (2005) New treatment modalities for vitiligo: focus on topical immunomodulators. Drugs 65: 447–459
20. Lan CC, Chen GS, Chiou MH et al (2005) FK506 promotes melanocyte and melanoblast growth and creates a favourable milieu for cell migration via keratinocytes: possible mechanisms of how tacrolimus ointment induces repigmentation in patients with vitiligo. Br J Dermatol 153:498–505
21. Lepe V, Moncada B, Castanedo-Cazares JP et al (2003) A double-blind randomized trial of 0.1% tacrolimus vs 0.05% clobetasol for the treatment of childhood vitiligo. Arch Dermatol 139:581–585
22. Mayoral FA, Gonzalez C, Shah NS, Arciniegas C (2003) Repigmentation of vitiligo with pimecrolimus cream: a case report. Dermatology 207:322–323
23. Mayoral FA, Vega JM, Stavisky H et al (2007) Retrospective analysis of pimecrolimus cream 1% for treatment of facial vitiligo. J Drugs Dermatol 6:517–521
24. Mehrabi D, Pandya AG (2006) A randomized, placebo-controlled, double-blind trial comparing narrowband UV-B Plus 0.1% tacrolimus ointment with narrowband UV-B plus placebo in the treatment of generalized vitiligo. Arch Dermatol 142:927–929
25. Ormerod AD (2005) Topical tacrolimus and pimecrolimus and the risk of cancer: how much cause for concern? Br J Dermatol 153:701–705
26. Ostovari N, Passeron T, Lacour JP, Ortonne JP (2006) Lack of efficacy of tacrolimus in the treatment of vitiligo in the absence of UV-B exposure. Arch Dermatol 142:252–253
27. Passeron T, Ostovari N, Zakaria W et al (2004) Topical tacrolimus and the 308-nm excimer laser: a synergistic combination for the treatment of vitiligo. Arch Dermatol 140:1065–1069
28. Qureshi AA, Fischer MA (2006) Topical calcineurin inhibitors for atopic dermatitis: balancing clinical benefit and possible risks. Arch Dermatol 142:633–637
29. Sardana K, Bhushan P, Kumar Garg V (2007) Effect of tacrolimus on vitiligo in absence of UV radiation exposure. Arch Dermatol 143:119–120
30. Seirafi H, Farnaghi F, Firooz A et al (2007) Pimecrolimus cream in repigmentation of vitiligo. Dermatology 214: 253–259

31. Sendur N, Karaman G, Sanic N, Savk E (2006) Topical pimecrolimus: a new horizon for vitiligo treatment? J Dermatolog Treat 17:338–342
32. Silverberg NB, Lin P, Travis L et al (2004) Tacrolimus ointment promotes repigmentation of vitiligo in children: a review of 57 cases. J Am Acad Dermatol 51:760–766
33. Smith DA, Tofte SJ, Hanifin JM (2002) Repigmentation of vitiligo with topical tacrolimus. Dermatology 205: 301–303
34. Soter NA, Fleischer AB, Webster GF et al (2001) Tacrolimus ointment for the treatment of atopic dermatitis in adult patients: part II, safety. J Am Acad Dermatol 44:S39–S46
35. Souza Leite RM, Craveiro Leite AA (2007) Two therapeutic challenges: periocular and genital vitiligo in children successfully treated with pimecrolimus cream. Int J Dermatol 46:986–989
36. Tanghetti EA () Tacrolimus ointment 0.1% produces repigmentation in patients with vitiligo: results of a prospective patient series. Cutis 71:158–162
37. Travis LB, Weinberg JM, Silverberg NB (2003) Successful treatment of vitiligo with 0.1% tacrolimus ointment. Arch Dermatol 139:571–574
38. Wolff K (2005) Pimecrolimus 1% cream for the treatment of atopic dermatitis. Skin Therapy Lett 10:1–6

Vitamin D Analogues

3.2.3

Mauro Picardo

Contents

M. Picardo
Istituto Dermatologico San Gallicano, via Elio Chianesi, 00144 Roma, Italy
e-mail: picardo@ifo.it

3.2.3.1 Rationale for use

The idea that Vitamin D analogues may be considered an option for the treatment of the vitiligo arises from the casual observation of the occurrence of cutaneous perilesional hyperpigmentation after their application on psoriatic plaques together with phototherapy [6, 14, 23, 27, 28]. Starting from this clinical evidence, several in vitro and in vivo data support the therapeutic role of calcipotriol and tacalcitol for vitiligo (Table 3.2.3.1). Skin cells, including keratinocytes, melanocytes, and dermal fibroblasts possess Vitamin D receptors (VDR), the activation of which induces several genes associated with proliferation and differentiation, and with the inflammatory reaction. Exposure of human melanocytes to Vitamin D promotes tyrosinase activity and melanogenesis, together with the up-regulation of c-Kit as well as morphological modifications, mainly increases the number and length of the dendrites. With regard to keratinocytes, Vitamin D produces differentiation, decreased proliferation, and the release of the proinflammatory cytokines IL-8 and IL-6, and increased production of IL-10. In vitiligo skin, both melanocytes and keratinocytes have been shown to possess an alteration of the Ca^{++} uptake and a decreased intracellular concentration of the ion, which could compromise the melanogenesis by increasing the level of reduced thioredoxin, which inhibits tyrosinase [6].

Vitamin D_3 analogues have been reported to interfere with intracellular Ca^{++} fluxes, cause G1 block, induce c-Kit upregulation, and inhibit TNF-α, IL-8, and RANTES production. The functional consequences of these activities affect cell survival, proliferation and differentiation processes, and melanogenesis. Moreover, Vitamin D analogues are believed to target the immune

M. Picardo and A. Taïeb (eds.), *Vitiligo*,
DOI 10.1007/978-3-540-69361-3_2.3, © Springer-Verlag Berlin Heidelberg 2010

Table 3.2.3.1 Summary of the studies

Vitamin D analogue	UV	Study	Administration	Duration	Effectiveness	Reference
Calcipotriol	No	Children (18)	2/day	4–6 months	Yes	[10]
Tacalcitol	Sun	Case report children	1/day	1 month	Yes	[2]
Calcipotriol	PUVA	Adults (26)	2/day	3–9 months	Yes	[3]
Tacalcitol	NB-UV\]B	Randomized assessor-blind adults (32)	2/week UV ± 1/day cream	6 months	Yes	[20]
Tacalcitol	308 nm MEL	Single-blind adults (38)	1/week MEL ± 2/day cream	12 sessions	Yes	[22]
Tacalcitol	Sun	Double-blind randomized placebo-controlled (80 adults)	1/day	16 weeks	No	[25]
Calcipotriol	NB-UVB	Comparative prospective, 40 adults	3/week UV ± 2/day cream	30 sessions	No	[4]
Calcipotriol	No	Left–right open (24 adults)	1/day	3–6 months	No	[8]

response interfering with the activated T cells and inhibiting the expression of several cytokine genes, such as those encoding TNF-α and IFN-λ.

3.2.3.2 Studies

Vitamin D_3 analogues have been mainly proposed as supporting therapy during PUVA, NB-UVB, or excimer (308 nm) phototherapy [11, 21, 24].

Initially, open studies reported repigmentation of vitiligo lesions using calcipotriol, a synthetic analogue of calcitriol [$1,25(OH)_2D_3$], and PUVAsol or solar exposure. Subsequent studies reported contradictory results. In fact, the combination of calcipotriol and PUVA was reported as more effective than PUVA alone (70 vs. 52% of the patients obtained mild–moderate repigmentation), in particular in initiating repigmentation [7, 9, 16], also suggesting the up-regulation of c-kit expression as a possible melanocyte-mediated mechanism of action. However, a right/left comparative open study did not show that the addition of calcipotriol significantly increased the response rate to PUVA [5].

Some studies suggest the improvement of the NB-UVB phototherapy when calcipotriol or tacalcitol was locally applied (Fig. 3.2.3.1) before or after the UVB exposure [13, 17, 20], providing earliest pigmentation with lower total UVB dosage, thus reducing the cost and the duration of the treatments. In contrast to this, among the trials that evaluated the effect of calcipotriol plus NB-UVB, two single-blinded left–right clinical studies and one RCT did not show any improvement in the treatment outcome of adding topical calcipotriol [1].

Topical tacalcitol [$1a24(OH)_2D_3$], another synthetic analogue of vitamin D was tested, in a left–right RCT in combination with bi-weekly NB-UVB phototherapy. It improved the extent of pigmentation and increased the rate of response [20] compared to phototherapy alone. Another RCT using tacalcitol in combination with monochromatic excimer light (308 nm MEL) confirmed that tacalcitol improved the efficacy of phototherapy, achieving earlier pigmentation with lower dosage (26 vs. 6% of patients obtained excellent repigmentation, whereas the overall positive response was observed in 72 vs. 60% of the patients) [22]. However the true effectiveness of the combinatory treatment is still considered to be controversial because other studies failed to report any improvement of the NB-UVB, BB-UVB, or 308-nm excimer-induced pigmentation when phototherapy was associated with Vitamin D3 analogues [12, 15].

Moreover, a vehicle-controlled RCT performed on 80 patients reported that the combination of tacalcitol with heliotherapy has no additional advantage compared to the heliotherapy alone [25].

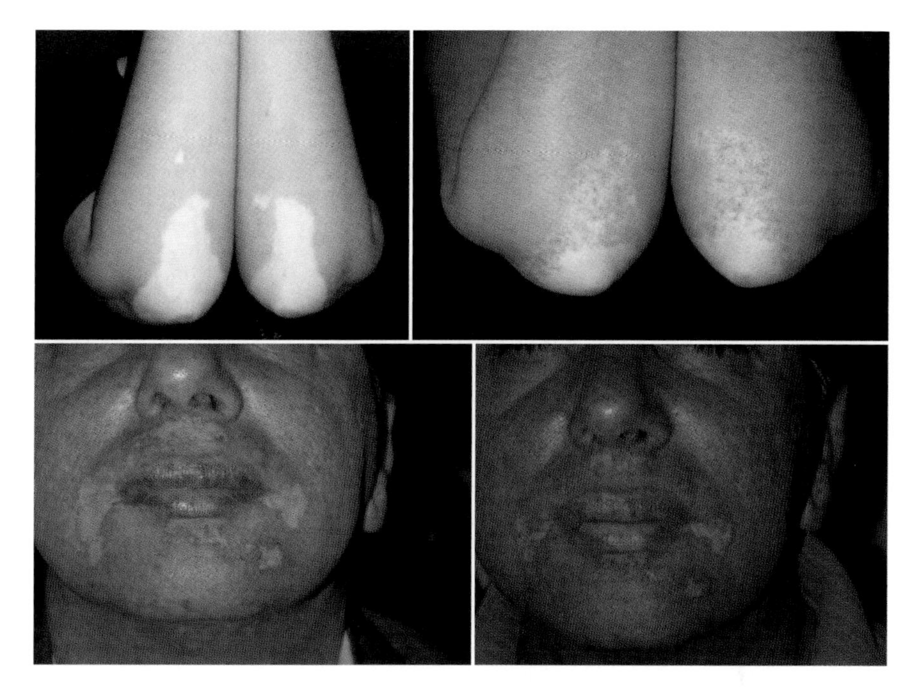

Fig. 3.2.3.1 Two NSJ patients, with different localization of the lesions, treated (*left before treatment, right after 6 months of therapy*) with tacalcitol plus NB-UVB

However, the published trials are mainly based on small numbers of patients and different phototherapy protocols were implemented, characterized by different temporal sequences and modality of Vitamin D_3 analogue application, and variable/different period of treatment–washout before starting the tested therapy, so that a comparison of the results is not easy. In any case, most of the studies reported best results when lesions were localized on the face or on the trunk.

Few studies or isolated case reports described the effectiveness of calcipotriol or tacalcitol in children. A study reported that 78% of the treated patients are responders, among which 21% reported complete repigmentation, whereas 29% showed partial, lower than 80% showed resolution [2, 10].

Calcipotriol has been evaluated alongside corticosteroids (betamethasone dipropionate) in randomized trials involving some pediatric subjects also. Good results (80% of sucess) were obtained with this combinatory therapy in terms of onset, degree, and stability of repigmentation, even if no excellent repigmentation was observed [18, 26]. The most frequent side effects in all the studies was mild to moderate erythema or itching, which is the typical symptom associated with the application of Vitamin D analogues [19].

Summary Messages

- In vitro, vitamin D is capable of interfering with pigmentation
- If any, the therapeutic effect of Vitamin D analogues seems to be obtained in combinatory treatments with phototherapy or topical steroids.
- Vitamin D analogues cause skin irritancy

References

1. Ada S, Sahin S, Boztepe G et al (2005) No additional effect of topical calcipotriol on narrow-band UVB phototherapy in patients with generalized vitiligo. Photodermatol Photoimmunol Photomed 21:79–83
2. Amano H, Abe M, Ishikawa O (2008) First case report of topical tacalcitol for vitiligo repigmentation. Pediatr Dermatol 25:262–264
3. Ameen M, Exarchou V, Chu AC (2001) Topical calcipotriol as monotherapy and in combination with psoralen plus ultraviolet A in the treatment of vitiligo. Br J Dermatol 145:476–479
4. Arca E, Tastan HB, Erbil AH et al (2006) Narrow-band ultraviolet B as monotherapy and in combination with

topical calcipotriol in the treatment of vitiligo. J Dermatol 33:338–343
5. Baysal V, Vildirim M, Erel A, Kesici D (2003) Is the combination of calcipotriol and PUVA effective in vitiligo? JEADV 17:299–302
6. Birlea SA, Costin GE, Norris DA (2008) Cellular and molecular mechanisms involved in the action of vitamin D analogs targeting vitiligo depigmentation. Curr Drug Targets 9:345–359
7. Cherif F, Azaiz MI, Ben Hamida A et al (2003) Calcipotriol and PUVA as treatment for vitiligo. Dermatol Ondine J 9:4
8. Chiaverini C, Passeron T, Ortonne JP (2002) Treatment of vitiligo by topical calcipotriol. JEADV 16:137–138
9. Ermis O, Alpsoy E, Cetin L, Yilmaz E (2001) Is the efficacy of psoralen plus ultraviolet A therapy for vitiligo enhanced by concurrent topical calcipotriol? A placebo-controlled double-blind study. Br J Dermatol 145:472–475
10. Gargoom AM, Duweb GA, Elzorghany AH et al (2004) Calcipotriol in the treatment of childood vitiligo. Int J Clin Pharmacol Res 24:11–14
11. Gawkrodger DJ, Ormerod AD, Shaw L et al (2008) Guideline for the diagnosis and management of vitiligo. Br J Dermatol 159:1051–1076
12. Goldinger SM, Dummer R, Schmid P et al (2007) Combination of 308-nm xenon chloride excimer laser and topical calcipotriol in vitiligo. JEADV 21:504–508
13. Goktas EO, Aydin F, Senturk N et al (2006) Combination of narrow band UVB and topical calcipotriol for the treatment of vitiligo. JEADV 20:553–557
14. Guilhou JJ (1998) The therapeutic effects of vitamin D3 and its analogues in psoriasis. Exp Opin Invest Drugs 7:77–84
15. Hartmann A, Lurz C, Hamm H et al (2005) Narrow-band UVB311 nm vs. broad-band UVB therapy in combination with topical calcipotriol vs. placebo in vitiligo. Int J Dermatol 44:736–742
16. Katayama I, Ashida M, Maeda A et al (2003) Open trial of topical tacalcitol [1alpha24$(OH)_2D_3$] and solar irradiation for vitiligo vulgaris: upregulation of cKit mRNA by cultured melanocytes. Eur J Dermatol 13:372–376
17. Kullavanijaya P, Lim HW (2004) Topical calcipotriene and narrowband ultraviolet B in the treatment of vitiligo. Photoderm Photoimmunol Photomed 20:248–251
18. Kumaran MS, Kaur I, Kumar B (2006) Effect of topical calcipotriol, betamethasone dipropionate and their combinattion in the treatment of localized vitiligo. JEADV 20: 269–273
19. Leone G, Pacifico A (2005) Profile of clinical efficacy and safety of topical tacalcitol. Acta Biomed 76:13–19
20. Leone G, Pacifico A, Iacovelli P et al (2006) Tacalcitol and narrow-band phototherapy in patients with vitiligo. Clin Exp Dermatol 31:200–205
21. Lotti T, Buggiani G, Troiano M et al (2008) Targeted and combination treatments for vitiligo. Comparative evaluation of different current modalities in 458 subjects. Dermatol Ther 21(1):s20–s26
22. Lu-yan T, Wen-wen F, Lei-hong X et al (2006) Topical tacalcitol and 308-nm monochromatic excimer light: a synergistic combination for the treatment of vitiligo. Photoderm Photoimmunol Photomed 22:310–314
23. Ortonne JP, Kaufmann R, Lecha M, Goodfield M (2004) Efficacy of treatment with calcipotriol/betamethasone dipropionate followed by calcipotriol alone compared with tacalcitol for the treatment of psoriasis vulgaris: a randomised, double-blind trial. Dermatology 209:308–313
24. Parsad D, Saini R, Nagpal R (1999) Calcipotriol in vitiligo: a preliminary study. Pediatr Dermatol 16:317–320
25. Rodriguez-Martìn M, Garcia Bustinduy M, Saez rodriguez M, Noda Cabrera A (2009) Randomized, double-blind clinical trial to evaluate the efficacy of topical tacalcitol and sunlight exposure in the treatment of adult nonsegmental vitiligo. Br J Dermatol 160:409–414
26. Travis LB, Silverberg NB (2004) Calcipotriene and corticosteroids combination therapy for vitiligo. Pediatr Dermatol 21:495–498
27. Takahashi H, Ibe M, Kinouchi M et al (2003) Similarly potent action of 1,25-dihydroxyvitamin D_3 and its analogues, tacalcitol, calcipotriol, and maxacalcitol on normal human keratinocyte proliferation and differentiation. J Dermatol Sci 31:21–28
28. Watabe H, Soma Y, Kawa Y et al (2002) Differentiation of murine melanocyte precursore induced by 1,25 dihydroxyvitamin D3 is associated with the stimulation of endothelin B receptor expression. J Invest Dermatol 119: 583–589

Phototherapies 3.3

PUVA and Related Treatments

3.3.1

Agustin Alomar

Contents

3.3.1.1 Introduction

Photochemotherapy (PUVA) combines the use of psoralens with long-length (320–400 nm) ultraviolet radiation (UVA), producing a beneficial effect that cannot be reached by the two components separately. Psoralens can be administered orally or topically (solutions, creams, or bath) with posterior exposition to UVA.

Historically, the concept for this therapy as a repigmenting agent dates back to ancient texts more than 3000 years ago, when topical applications and ingestion of extracts from the plants *Psoralea corylifolia Linnaeus* in India and *Ammi majus Linnaeus* in Egypt were used to treat leukoderma. In 1948, El Mofty demonstrated the efficacy of the extract of the *Ammi majus* plant, 8-mehtoxypsoralen (8-MOP or metoxsalen), used orally and topically and in combination with sunlight or ultraviolet lamps in the treatment of vitiligo [37]. In 1974, high-intensity long-wave (320–400 nm) ultraviolet phototherapy units were introduced, reporting its efficacy in combination with oral 8-MOP in psoriasis, creating the concept of photochemotherapy [26] and its acronym PUVA (posralen + UVA) [35]. Finally, in Scandinavia, the psoralen bath plus UV-A therapy was introduced, which was very effective in some indications such as lichen planus or psoriasis [12].

The main indications of PUVA therapy are psoriasis, atopic dermatitis, vitiligo, cutaneous T-cell lymphomas, and photodermatitis. It has also been employed in other dermatitis characterized by different degrees of cutaneous inflammation and resistant to other treatments with variable results [8, 20].

A. Alomar
Department of Dermatology, Respiratory Medicine, Hospital de la Santa Creu I Sant Pau, Barcelona, Spain
e-mail: aalomar@santpau.cat

M. Picardo and A. Taïeb (eds.), *Vitiligo*,
DOI 10.1007/978-3-540-69361-1_3.3.1,

3.3.1.2 Psoralens

Psoralens belong to the furocoumarin group of compounds that are derived from the fusion of the furan ring with coumarin. These are found in a large number of plants, and there are also several synthetic psoralen compounds. 8-MOP is a plant origin, but it is also available as a synthetic drug. It is used primarily for oral and topical PUVA. The synthetic compound 4,5′,8-trimethylpsoralen (TMP, trioxsalen) is less phototoxic after oral administration, but more phototoxic when administered with bathwater. 5-methoxypsoralen [25] (5-MOP, bergapten) is only used in some European countries and is less erythemogenic and not associated with gastrointestinal intolerance [25, 38, 43].

Without UV radiation, psoralens intercalate in DNA double strand. After the absorption of UVA photons, the formation of 3,4- or 4′,5′-cyclobutane monoadducts with pyrimidine bases of native DNA occurs. Some psoralens, such as 8-MOP, 5-MOP, and TMP, can absorb a second photon and this reaction leads to the formation of a bifunctional adduct that inhibits DNA replication and causes cell-cycle arrest. It is generally assumed that this effect may be the therapeutic mechanism in psoriasis [4, 9, 14, 17].

However, DNA cross-linking does not appear to be a prerequisite for all the therapeutic effects of PUVA, and the successful treatment of other skin diseases is unlikely to be directly due to this molecular reaction. Psoralens also interact with RNA, proteins, and other cellular components and indirectly modify proteins and lipids via oxygen-mediated reactions or by generating free radicals. Although not proven, it is possible that these could be the therapeutic mechanisms of non-hyperproliferative diseases. Psoralens also stimulate melanogenesis. This involves the photoconjugation of psoralens to DNA in melanocytes followed by mitosis and subsequent proliferation of melanocytes, increased formation and melanization of melanosomes, increased transfer of melanosomes to keratynocites, and activation and increased synthesis of tyrosinase via stimulation of cAMP activity[1, 4, 9, 17, 46].

Regarding pharmacokinetics of psoralens, there are several steps between the ingestion and the arrival at the site of action of the molecules. These include disintegration and dissolution of the drug, absorption, first-pass effect, and blood transportation and tissue distribution. There are several features of the pharmacology of psoralens that are very important in therapy. First, the only concentration level that is of importance is the level at the target site in the skin since it is there that an interaction with UVA radiation will yield therapeutic benefit. Direct measurement of the phototoxic response of skin is the only means available for assessing the cutaneous content of psoralens. Second, the available psoralens are all poorly soluble in water, and this is a limiting factor in their absorption from the gastrointestinal tract. Liquid preparation induces earlier, higher, and more reproducible peak plasma levels than crystalline preparations. Third, there is a significant, but saturable, first-pass effect in the liver. This means that a proportion of any dose of psoralen is metabolized by the liver after the absorption from the gut and never reaches skin. However, since this effect can be saturated, as the dose is raised, the proportion of active compound reaching the skin rises. Finally, there are very large inter-individuals and smaller, but still significant, intra-individual variations in the absorption of psoralens that reflect in terms of peak levels in blood and time of peak levels after administration. The metabolism and distribution of individual psoralens show some differences. 8-MOP gives a blood level that is about four times higher than that resulting from an equal dose of 5-MOP and a greater proportion of 5-MOP is bound to protein. In contrast, TMP usually does not produce measurable serum levels or cutaneous phototoxicity after therapeutic doses due to poor absorption and rapid metabolism by the liver [10, 42].

Determinations of action spectra in vivo have shown that psoralen photosensitization occurs with wavelenghts >320 nm. Conventional therapeutic UVA fluorescent tubes and broad-spectrum metal halide lamps, which are filtered in the UVB and the UVC range, cover the psoralen action spectrum well. The typical fluorescent UVA lamp used for PUVA therapy peaks at 352 nm and emits approximately 0.5% in the UVB range. Major advantages of mercury halide units are the stability of output and their high irradiance, enabling short-treatment times. UVA doses are given in J/cm2 and measured with a photometer with a maximum sensitivity at 350–360 nm. Irradiance must be relatively uniform so that the dose does not vary at different anatomic sites. The UVB emission should be kept low to avoid erythematogenic UVB doses before sufficient UVA is absorbed to produce the psoralen photosensitivity reaction [7].

PUVA treatment produces an inflammatory response that manifests as delayed phototoxic erythema. The reaction depends on the dose of the drug, the dose of UVA, and the skin type. PUVA erythema differs from sunburn or UVB since it appears later and lasts longer (lasting up to more than 1 week). PUVA-induced-erythema usually appears after 36–48 h, but in some patients may be delayed until 72 or 96 h after exposure. The more intense the erythema, the later it will appear and reach a maximum. It also has a steep dose-response curve. Thus, the dose required to produce blistering is only few multiples of the dose that produces 1+ erythema. In contrast, the curve for UVB-induced erythema is much less steep, and for UVC-induced erythema, is almost flat. Overdoses of PUVA are frequently followed by edema, intense pruritus, and sometimes by a peculiar stinging sensation in the affected skin area. At this time, erythema is the only available parameter that allows an assessment of the PUVA reaction and thus represents an important factor for determining UVA dose adjustments [21, 22].

The second important effect of PUVA is pigmentation, and it may develop without clinically evident erythema, especially when 5-MOP or TMP is used. This is particularly important in vitiligo and for the preventive treatment of photodermatoses. In normal skin, PUVA pigmentation peaks about 7 days after exposure and may last from several weeks to months. A few PUVA exposures result in a much deeper tan than that produced by multiple exposures to solar irradiation [21, 22].

3.3.1.3 Oral Photochemotherapy

Oral photochemotherapy is used in adult patients with generalized vitiligo. With this modality of treatment, some degree of repigmentation is acquired in 70–80% of patients, but a complete repigmentation is only obtained in about 20% of patients [24]. Relapse occurs to some degree in 75% of patients in about 1 or 2 years after the cessation of therapy [40]. In the guidelines for the treatment of vitiligo performed recently by Njoo et al. [30], PUVA is considered a second-line therapy for generalized vitiligo in adults, although in the past it was considered the gold standard treatment. In their meta-analysis they concluded that narrow-band UVB produced better mean success rates (63 vs. 51%, not statistically significant) with fewer rates of adverse effects and better color match of repigmented skin. Some recent studies confirm these results [5, 36, 47].

This therapeutic modality is not recommended for children under 10–12 years old because of the increased risk of retinal toxicity. Other circumstances in which PUVA should be reconsidered are in patients with history of skin cancer (melanoma or no melanoma), premalignant skin lesions, cataracts, alteration of liver function, Skin Type I, pregnancy, obesity (increased risk of erythema), concomitant immunosupressive therapy, or patients associated with phototoxic/photoallergic treatments [8, 16, 39].

A dosage of 8-MOP of 0.6–0.8 mg/kg is administered orally 1–3 h before exposure, depending on the absorption characteristics of the particular drug brand, and for TMP is 0.6 mg/kg. These two are the photosensitizers most frequently used, and because of weaker phototoxicity, TMP is preferred for treatment with sunlight as radiation source. For 5-MOP, the usual dosage is 1.2–1.8 mg/Kg.

Psoralen dosage, interval between psoralen intake and UVA exposure, and the type and time of intake of food and the drug should be consistent, and the UVA dosage should be varied according to the patient's sensitivity. The initial UVA doses are selected either by skin typing [16, 29] or determining the minimal phototoxicity dose (MPD) [19]. MPD is defined as the UVA dose that produces Grade 1 erythema at 48–72 h in a patient who has ingested the requisite dose of psoralen at the appropriate time interval before exposure on previously non-sun-exposed skin. This dose usually ranges with 8-MOP between 0.5 and 5 J/cm^2, and the initial dose should be the 70% of this dose. In vitiligo, a conservative initial dose is approximately 0.5 J/cm^2 and increments of 0.5 J/cm^2 are given for each subsequent treatment until asymptomatic mild erythema is observed in the vitiligo lesions. The maximum dose per session should not exceed 10 J/cm^2 [11, 44].

Treatments should be given at least twice a week, but not more than three times with at least 1 day, or preferably 48 h, between treatments. A broad-spectrum sunscreen should be applied after therapy to decrease the risk of developing phototoxic reaction. Ultraviolet light exposure should be minimized for the ensuing 24–48 h. In addition, protective UVA sunglasses must be worn for at least 18–24 h after ingesting oral psoralen preparations to avoid ocular damage, specifically cataracts. Male genital areas respond poorly to treatment and have an increased susceptibility to skin cancer; these areas should be shielded from UVA exposure or limited to every third session [6, 8, 14, 16, 39, 43].

Missing therapy sessions is frequent in daily practice and can lead to a loss of efficacy and an increase of adverse effects. When only one session is skipped, the previous dose should be maintained. A reduction of 25 or 50% of the last dose, respectively, should be performed if 1 or 2–3 weeks of treatment are missed. If more than 3 weeks of treatment are missed the treatment should be reinitiated [8].

Patients should be motivated to continue PUVA therapy for at least 6 months before being considered recalcitrant to this treatment, and 12–24 months of continuous therapy may be necessary to acquire maximal repigmentation. Responsiveness is defined as development of multiple perifollicular macules of repigmentation or contraction in size of small lesions (Fig. 3.3.1.1). Complete repigmented areas can remain stable during decades, but if therapy is stopped, partial repigmentation may reverse [14, 33].

Therapeutic responses vary considerably between patients depending on motivation, skin type, severity, number of treatments, age, and location of lesions. Darker-skin types show maximal responses to PUVA and maximal repigmentation occur in those patients achieving erythema Grade 2. Blistering reactions do not enhance repigmentation and may associate with koebnerization of vitiligous lesions. The best response is obtained on the face and neck. The trunk and upper and lower extremities have an intermediate response. Minimal repigmentation occurs on the distal extremities, particularly dorsal hands, feet, and lips. Compared with adults, children have an enhanced response to PUVA, especially in lesions of recent appearance [14, 23, 34].

The main limitations of PUVA are the short- and long-term secondary effects. An acute adverse reaction is the phototoxic reactions that appear in 10–15% of patients and acquire its maximum intensity after 48–72 h of treatment. Its clinical spectrum varies between erythema, blisters, and superficial skin necrosis. When erythema

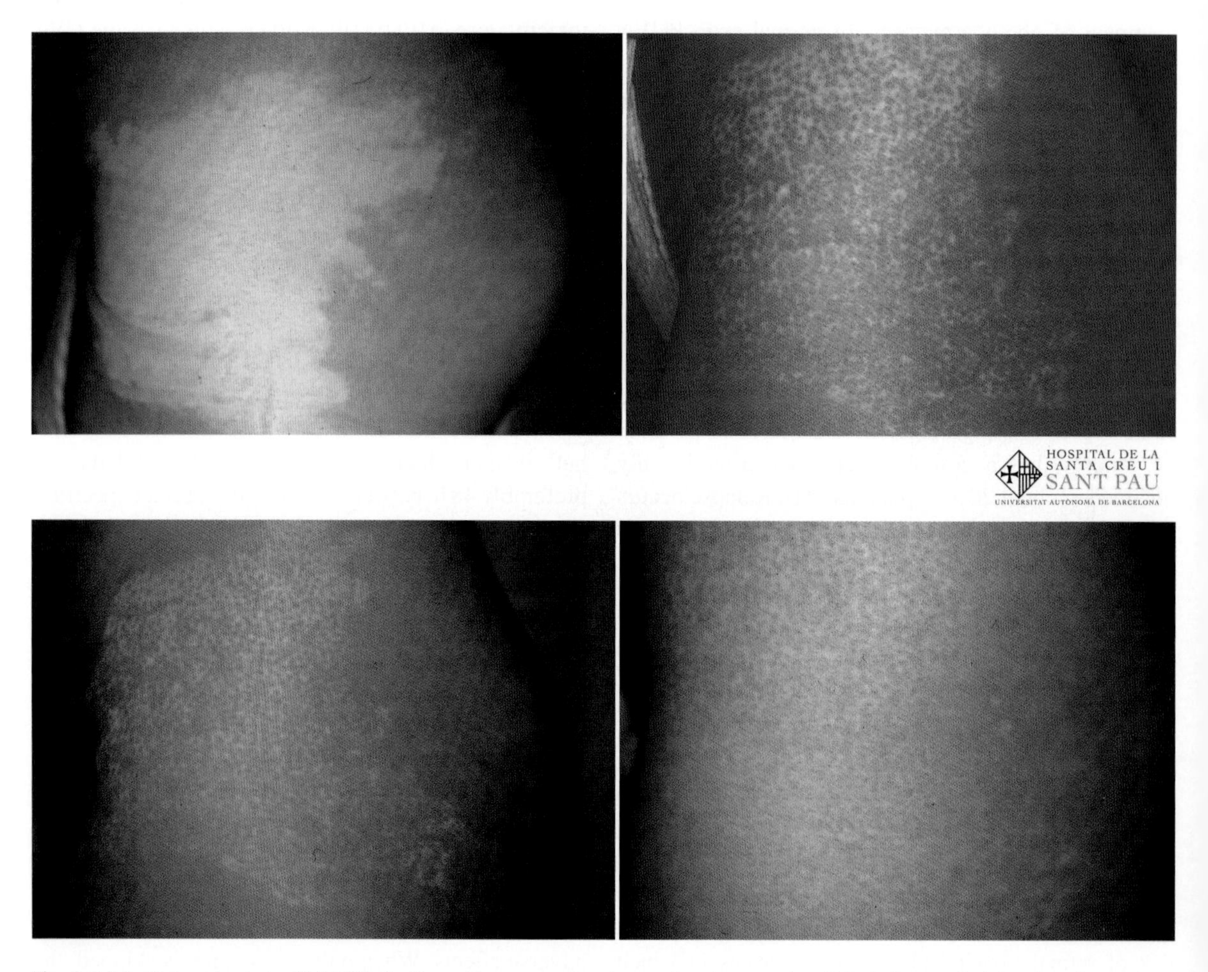

Fig. 3.3.1.1 Patient treated with PUVA for 8 months, progresive stages of repigmentation (*left, before,* and *right,* after therapy)

appears, previous UVA dose should be maintained and if erythema becomes symptomatic (pruritus or pain), one or two sessions should be suspended [28]. In experimental studies, the administration of *Polypodium leucotomus* (7.5 mg/Kg) has shown efficacy in reducing the phototoxicity and hyperpigmentation associated to PUVA therapy [27] (Fig. 3.3.1.2). Other acute secondary effects are pruritus, hyperpigmentation, oral intolerance to 8-MOP, and hepatotoxicity [13]. The most important long-term secondary effect is carcinogenesis. Incidence of squamous cell carcinoma is higher in patients with cumulative doses greater than 1000–1500 J/cm^2. Incidence of basocellular carcinoma and melanoma also rise with cumulative treatments. The main prevention strategies are limiting the number of annual sessions, restricting maintenance treatments, and performing an annual clinical revision of patients that exceed the number of recommended sessions. Other chronic adverse events are cutaneous actinic damage (actinic keratosis, lentigines, and wrinkles) and ocular phototoxicity [29, 41].

3.3.1.4 Topical Photochemotherapy

Topical photochemotherapy is a good therapeutic approach for localized vitiligo in adults and children over two years. In the guidelines performed by Njoo et al [30], it is considered a second-line therapy that should be used if no response is obtained after six months of Class 3 topical corticosteroids combined with UV-A radiation.

A thin coat of 8-MOP cream or ointment at very low concentrations (0.01 or 0.1%) should be applied 30 min before UVA exposure. Psoralen solutions should be avoided since they can produce hyperpigmented lines; to reduce the appearance of this adverse event, sunscreen can be applied around the vitiligo patches. The initial UVA dose is 0.12–0.25 J/cm^2 and is increased weekly 0.12–0.25 J/cm^2 according to the patient's skin phototype, and unless marked erythema appears. One or two weekly sessions are recommended, and sunlight treatments are not usually prescribed for the risk of precipitating phototoxic

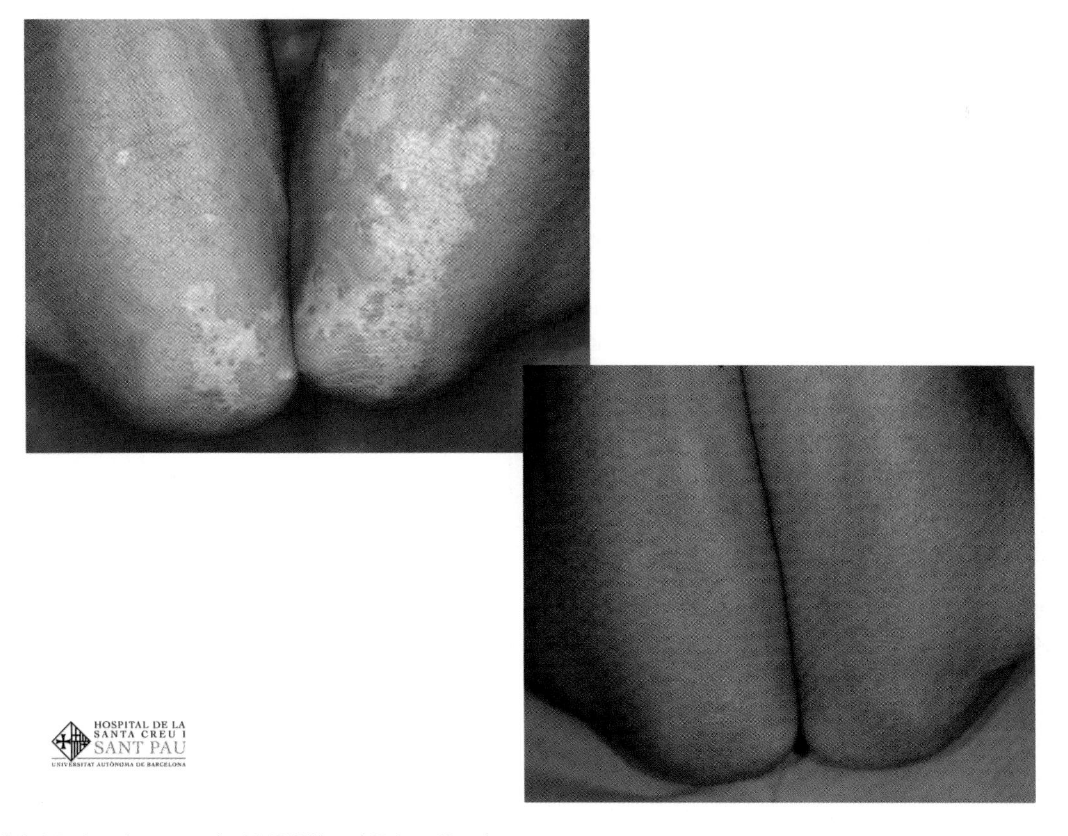

Fig. 3.3.1.2 A patient treated with PUVA and *Polypodium leucotomus*

reactions. For this reason, lesions should be washed out after treatments and a broad-spectrum sunscreen should be applied [15, 18].

The advantages of topical therapy are the need for fewer treatments and considerably smaller cumulative UVA doses, usually achieving lower plasmatic levels and consequently inducing less systemic and ocular phototoxicity. The main disadvantages are severe blistering reactions, perilesional hyperpigmentation, and its lack of effectiveness in limiting the progression of actively spreading vitiligo [45].

3.3.1.5 Khellin + UV

Another photochemotherapeutic regimen for vitiligo consists of Khellin as the photosensitizer, a furanochromone extracted of the plant *Amni visnaga* (5,8 dimetoxi-2-metil-4,5-furo-6,7-chromone), and UVA irradiation (KUVA). The greater advantage is its lack of phototoxicity that makes it safe to use as a home treatment or treatment with natural sunlight, even with a daily regimen. It is also less mutagenic than psoralens, and it promotes less darkening of normal skin.

Khellin can be administered orally at a dose of 100 mg 2 h before treatment. The efficacy rate of this treatment can be compared to PUVA, but approximately 30% of patients present an increase of liver transaminases, which is why this treatment is used in exceptional circumstances. Khellin can also be applied topically in a thin coat being formulated in moisturizing cream or carbopol gel in a concentration between 3 and 5%. It should be applied 30 min before sun exposure during 10–15 min in a daily regimen. It also can be performed with UVA exposure, initiating with a dose of 0.25–0.5 J/cm^2 and increasing 0.25 each session until reaching a dose of 2–3 J/cm^2, or narrow band UVB. Nevertheless, since the peaks of absorption of Khellin conjugates show bands at 256 and 290 nm, UVA radiation should not be the first choice radiation [2, 3, 31, 32].

Topical "KUVA-sun" is very useful in children and in good responding areas, especially in sunny countries where there are several months during which low doses of natural sunlight may be received; however, its efficacy in comparison with oral PUVA or other therapeutic modalities has not been established.

Summary messages

- Photochemotherapy (PUVA) is based on the interaction of long length ultraviolet radiation and psoralens that promotes repetitive and controlled phototoxic reactions.
- There are three type of psoralens (8-mehtoxypsoralen, 4,5', 8-trimethylpsoralen and 5-methoxypsoralen) that can be used topically and orally.
- KUVA is another phothochemotherapy used in vitiligo that uses khellin as a photosensitizer.
- Oral photochemotherapy is mainly used in adult patients with generalized vitiligo. Nowadays, this is considered to be a second-line therapy.
- A good dosimetry is required to avoid acute adverse effects like phototoxic reactions. The most important long term adverse event is cutaneous carcinogenesis.
- Topical photochemotherapy is recommended as a second line therapy in localized vitiligo in adults and children over 2 years.
- Topical PUVA produces less cumulative UVA doses but is less effective and promotes more phototoxic reactions.
- Patients should be motivated before initiating treatment since repigmentation can be delayed until 6 months and that 12-24 months of continuous therapy may be necessary to acquire maximal repigmentation.
- 8-MOP is the most used psoralen for oral PUVA but due to its lower phototoxicity TMP is preferred for treatment with sunlight as radiation source.
- Polypodium leucotomos (7.5mg/Kg) has shown efficacy in reducing the phototoxicity and hyperpigmentation associated to PUVA therapy
- In topical PUVA, psoralens should be formulated in creams or ointments at very low dosage since solutions produce hyperpigmented lines.
- Topical 3% khellin formulated in moisturizing cream combined with natural sunlight is safe and useful in good responding areas and especially in children.

References

1. Abdel-Naser MB, Hann SK, Bystryn JC (1997) Oral psoralen with UV-A therapy releases circulating growth factor(s) that stimulates cell proliferation. Arch Dermatol 133: 1530–1533
2. Abd el Fattah A, Aboul-Enein MN, Wassel GM (1982) An approach to treatment of vitiligo by khellin. Dermatologica 165:136–140
3. Alomar A (1992) Some new treatments of vitiligo vulgaris: phototherapy with topical khellin. dermatology: progress and perspectives. In: Burgdorf and Kats (eds.) Proceedings of the 18th World Congress of Dermatology. New York 517-19
4. Averbeck D (1989) Recent advances in psoralen phototoxicity mechanism. Photochem Photobiol 50:859–882
5. Bhatnagar A, Kanwar AJ, Parsad D et al (2007) Comparison of systemic PUVA and NB-UVB in the treatment of vitiligo: an open prospective study. J Eur Acad Dermatol Venereol 21:638–642
6. Norris G, Hawk JLM, Baker C (1994) British Photodermatology Group guidelines for PUVA. Br J Dermatol 130:246–255
7. Brücke J, Tanew A, Ortel B et al (1991) Relative efficacy of 335 and 365 nm radiation in photochemotherapy of psoriasis. Br J Dermatol 124:372–374
8. Carrascosa JM, Gardeazábal J, Pérez-Ferriols A et al (2005) Consensus document on phototherapy: PUVA therapy and narrow-band UVB therapy. Actas Dermosifiliogr 96: 635–658
9. Canton M, Caffieri S, Dall'Acqua F (2002) PUVA-induced apoptosis involves mitochondrial dysfunction caused by the opening of the permeability transition pore. FEBS Lett 522: 168–172
10. Chakrabarti SG, Grimes PE, Minus HR, Kenney JA Jr et al (1982) Determination of trimethylpsoralen in blood, ophthalmic fluids, and skin. J Invest Dermatol 79: 374–377
11. Collins P, Wainwright NJ, Amorim I et al (1996) 8-MOP PUVA for psoriasis: a comparison of a minimal phototoxic dose-based regimen with a skin-type approach. Br J Dermatol 135:248–254
12. Fischer T, Alsins J (1976) Treatment of psoriasis with trioxsalen baths and dysprosium lamps. Acta Derm Venereol 56:383–390
13. Freeman K, Watin AP (1984) Deterioration of liver function during PUVA therapy. Photodermatol Photoimmunol Photome 1:147–148
14. Grimes PE (1993) Vitiligo: an overview of therapeutic approaches. Dermatol Clin 11:325–338
15. Grimes PE, Minus HR, Chakrabarti SG (1982) Determination of optimal topical photochemotherapy for vitiligo. J Am Acad Dermatol 7:771–778
16. Anon (1994) Guidelines of care for phototherapy and photochemotherapy. American Academy of Dermatology Committee on Guidelines of Care. J Am Acad Dermatol 31: 643–648
17. Gupta AK, Anderson TF (1987) Psoralen photochemotherapy. J Am Acad Dermatol 17:703–734
18. Halpern SM, Anstey AV, Dawe RS (2000) Guidelines for topical PUVA: a report of a workshopof the British photodermatology group. Br J Dermatol 142:22–31
19. Henseler T, Wolff K, Hönigsmann H et al (1981) Oral 8-methoxypsoralen photochemotherapy of psoriasis. The European PUVA study: a cooperative study among 18 European centres. Lancet 1:853–857
20. Honig B, Morison WL, Karp D (1994) Photochemotherapy beyond psoriasis. J Am Acad Dermatol 31:775–790
21. Ibbotson SH, Dawe RS, Farr PM (2001) The effect of methoxsalen dose on ultraviolet-A-induced erythema. J Invest Dermatol 116:813–815
22. Ibbotson SH, Farr PM (1999) The time-course of psoralen ultraviolet A (PUVA) erythema. J Invest Dermatol 113: 346–350
23. Kenney JA Jr (1971) Vitiligo treated by psoralens: a long term follow-up study of permanency of repigmentation. Arch Dermatol 103:475–480
24. Kovacs SO (1998) Vitiligo. J Am Acad Dermatol 38: 647–666
25. McNeely W, Goa KL (1998) 5-Methoxypsoralen. A review of its effects in psoriasis and vitiligo. Drugs 56:667–690
26. Melski JW, Tanenbaum L, Parrish JA (1989) Oral methoxsalen photochemotherapy for the treatment of psoriasis: a cooperative clinical trial. 1977. J Invest Dermato l92:153S
27. Middelkamp-Hup MA, Pathak MA, Parrado C (2004) Orally administered *Polypodium leucotomus* extract decreases psoralen-UVA-induced phototoxicity, pigmentation, and damage of human skin. J Am Acad 50:41–49
28. Morison WL, Marwaha S, Beck L (1997) PUVA-induced phototoxicity: incidence and causes. J Am Acad Dermatol 36:183–185
29. Nijsten TE, Stern RS (2003) The increased risk of skin cancer is persistent after discontinuation of psoralen + ultraviolet A: a cohort study. J Invest Dermatol 121:252–258
30. Njoo MD, Westerhof W, Bos JD et al (1999) The development of guidelines for the treatment of vitiligo. Clinical Epidemiology Unit of the Istituto Dermopatico dell'Immacolata-Istituto di Recovero e Cura a Carattere Scientifico (IDI-IRCCS) and the Archives of Dermatology. Arch Dermatol 135(12): 1514–1521
31. Orecchia G, Perfetti L (1992) Photochemotherapy with topical khellin and sunlight in vitiligo. Dermatology 184:120–123
32. Ortel B, Tanew A, Höningsmann H (1986) Vitiligo treatment. Curr Probl Dermatol 15:256–271
33. Ortel B, Tanew A, Höningsmann H (1988) Treatment of vitiliogo with Khellin and ultraviolet A. J Am Acad Dermatol 18:693–701
34. Parrish JA, Fitzpatrick TB, Tanenbaum L et al (1974) Photochemotherapy of psoriasis with oral methoxsalen and longwave ultraviolet light. N Engl J Med 291:1207–1211
35. Parrish JA, Fitzpatrick TB, Shea C et al (1976) Photochemotherapy of vitligo. Arch Dermatol 112:1531–1534
36. Parsad D, Kanwar AJ, Kumar B (2006) Psoralen-ultraviolet A vs. narrow-band ultraviolet B phototherapy for the treatment of vitiligo. J Eur Acad Dermatol Venereol 20: 175–177
37. Pathak MA, Fitzpatrick TB (1992) The evolution of photochemotherapy with psoralens and UVA (PUVA): 2000 BC to 1992 AD. J Photochem Photobiol B 14:3–22
38. Pathak MA, Mosher DB, Fitzpatrick TB (1984) Safety and therapeutic effectiveness of 8-methoxypsoralen, 4,5′,8-trimethylpsoralen, and psoralen in vitiligo. Natl Cancer Inst Monogr 66:165–173
39. Schmutz JL, Jeanmougin M, Martin S et al (2000) [Recommendations of the French Society of Photodermatology for systemic PUVA therapy in psoriasis vulgaris. French Society of Photodermatology] Ann Dermatol Venereol 127:753–759

40. Shaffrali F, Gawkrodger D (2000) Management of vitiligo. Clin Exp Dermatol 25:575–579
41. Stern RS, Nichols KT, Vakeva LH (1997) Malignant melanoma in patients treated for psoriasis with methoxalen (psoralen) and ultraviolet A radiation (PUVA). N England J Med 336:1041–1045
42. Stolk L, Siddiqui AH, Cormane RH (1981) Serum levels of trimethylpsoralen after oral administration. Br J Dermatol 104:443–445
43. Taneja A (2002) Treatment of vitiligo. J Dermatolog Treat 13:19–25
44. Tanew A, Ortel B, Hönigsmann H (1999) Half-side comparison of erythemogenic versus suberythemogenic UVA doses in oral photochemotherapy of psoriasis. J Am Acad Dermatol 41:408–413
45. Westerhof W, Nieuweboer-Krobotova L (1997) Treatment of vitiligo with UV-B radiation vs topical psoralen plus UV-A. Arch Dermatol 133:1525–1528
46. Wu CS, Lan CC, Wang LF et al (2007) Effects of psoralen plus ultraviolet A irradiation on cultured epidermal cells in vitro and patients with vitiligo in vivo. Br J Dermatol 156:122–129
47. Yones SS, Palmer RA, Garibaldinos TM (2007) Randomized double-blind trial of treatment of vitiligo: efficacy of psoralen-UV-A therapy vs Narrowband-UV-B therapy. Arch Dermatol 143:578–584

UVB Total Body and Targeted Phototherapies

3.3.2

Giovanni Leone and Adrian Tanew

Contents

G. Leone (✉)
Phototherapy Unit, San Gallicano Dermatological Institute, IFO, Rome, Italy
e-mail: gleone@ifo.it

3.3.2.1 Narrowband-UVB (NB-UVB) Phototherapy

Introduction

The introduction of NB-UVB in the early eighties of the last century has evolved into one of the major accomplishments in the field of phototherapy. The concept of NB-UVB was derived from an action spectrum study in psoriasis published by Parrish et al. [68]. This study has indicated that wavelengths in the longer UVB range have superior antipsoriatic activity as compared to shorter UVB wavelengths. With the development of fluorescent tubes by the Philips Company emitting most of the light in a narrow range between 310 and 315 nm, with a peak at 311–313 nm (TL01 lamps), clinical researchers were able to put theory into practice. Subsequently, a wealth of studies indicated that NB-UVB has a better therapeutic index than conventional broadband UVB and appears to be nearly as effective as photochemotherapy (PUVA) in the treatment of psoriasis [21, 22, 47, 60]. Not surprisingly, NB-UVB was later on tried for a variety of other UV-responsive dermatoses including vitiligo.

Pioneer Studies on NB UVB for Vitiligo

The first trial on the use of NB-UVB in adult patients with vitiligo was published by in 1997 [86]. This study consisted of two parts. In the first part, the patients were treated for 4 months with topical PUVA using a 0.005% psoralen gel (n = 28) or NB UVB (n = 78). In the second part, patients were treated with NB-UVB for up to

M. Picardo and A. Taïeb (eds.), *Vitiligo*,
DOI 10.1007/978-3-540-69361-1_3.3.2,

12 months ($n = 51$). After 4 months of treatment, repigmentation occurred more frequently in the NB-UVB group (67%) than in the topical PUVA group (46%). After 1 year of twice weekly NB-UVB treatment, 32/51 (63%) of patients achieved a >75% repigmentation.

Following this pioneer study, the same group of authors conducted an open trial on NB-UVB for children with generalized vitiligo [59]. Fifty-one children with a mean age of 9.9.years were treated with NB-UVB twice weekly for up to 1 year. More than 75% repigmentation was seen in 27/51 (53%) of patients. Importantly, while 49/51 (96%) of the children had active disease at the onset of treatment, the disease had stabilized in 80% of the cases after therapy.

Confirmative Studies on NB UVB for Vitiligo

A retrospective analysis [78] of 7 adult patients treated with NB-UVB thrice weekly revealed > 75% of repigmentation in 5 patients after a mean number of 19 exposures. The mean disease duration among these 5 patients was 13 months. Two other patients had 50 and 40% repigmentation after 46 and 48 treatments, respectively. Their mean disease duration was 132 months, suggesting that patients with long-standing disease have a poorer prognosis. In a subsequent study [90], the same group of authors summarized the response rate of 71 vitiligo patients who had been treated with 15–123 sessions of NB-UVB. Significant (66–100%) repigmentation occurred in 39%, moderate (26–65%) or mild (10–25%) repigmentation in 22 and 21% of the patients, respectively, and less than 10% repigmentation in 10% of the patients. Patients with better improvement tended to have higher numbers of treatment indicating that the length of treatment is positively correlated with the degree of repigmentation. Again, patients with significant improvement had shorter disease duration than those with a less favorable response.

Another retrospective study from Thailand evaluated the efficacy of NB-UVB in 60 Asian patients with recalcitrant vitiligo (53 generalized, 7 segmental). After 36–175 treatments, 25/60 (42%) of cases had more than 50% and 20/60 (33%) more than 75% repigmentation on the face, trunk, arms, and legs. No patient had more than 25% repigmentation on the hands and feet. Nine patients with a response of >50% were followed up for an average period of 14.5 months. Four patients relapsed after 3–24 months [57].

Twenty-two patients with symmetric vitiligo were included in a randomized, right–left comparison trial involving NB-UVB exposure to half of the body (trunk and extremities) and no light exposure to the other half as a control. Treatment was given three times weekly for either 60 treatments or 6 months. The mean improvement due to NB-UVB was 42.9 versus 3.3% on the control side after 6 months of treatment. The lower extremities had the best response, followed by the trunk, arms, hands, and feet [34].

An open study reported the therapeutic outcome of generalized vitiligo to thrice weekly NB-UVB treatment for at least 1 year in 26 Indian children (aged 5–14 years). Fifteen children (75%) developed marked to complete (>75%) repigmentation, four a marked (50–75%) and one child a moderate (<25%) repigmentation. Fifty percent repigmentation was achieved by an average of 34 treatments. The mean duration of disease in patients with marked to complete repigmentation was significantly lower (12.5 months) than that of children with moderate or mild improvement (38.8 months). After 6 months of therapy, the disease had stabilized in all patients [41].

A retrospective study performed in Taiwan evaluated the response of 72 vitiligo patients (61 nonsegmental, 11 segmental) to 2–3 times weekly NB-UVB phototherapy given for a maximum period of 1 year. Excellent response (75–100% repigmentation) was obtained in 9 patients (12.5%), a good response (50–75% repigmentation) in 24 (33.3%), a moderate response (25–50% repigmentation) in 20 (27.8%), and a poor response in 19 (26.4%) of the patients [19].

The largest open trial conducted so far included 110 evaluable patients (97 nonsegmental, 13 segmental) who were treated with NB-UVB twice weekly for 2.5–19.5 months (mean 7.8 months). In the nonsegmental vitiligo group, 48% of patients showed a marked response (>75% repigmentation), 27% a moderate response (25–75% repigmentation), and 25% a mild response (<25% repigmentation). When analyzed by site, the best results were achieved in the face, followed by the trunk, and limbs. The poorest outcome was noted for acral lesions that at best showed a moderate response. As reported in earlier studies, the duration of the disease was inversely correlated with the degree of treatment-induced repigmentation. In the segmental vitiligo group, all but one patient showed no more than a mild response, whatever the site of the lesion was [5].

An open, uncontrolled study from Greece reported data on 70 patients with nonsegmental vitiligo who were treated over a maximum period of 1.5 years. Cosmetically acceptable repigmentation (>75%) was achieved in 34.4% of patients with lesions on the face after a mean treatment period of 6 months, but in only 7.4% of patients with lesions on the body after a mean treatment period of 9.2 months. Hand and feet vitiligo showed minimal or no pigmentation, and lesions on the elbows and knees responded less than lesions on the trunk, but better than acral vitiligo. Predictors of a good response were darker-skin types (III–V) and early initial repigmentation. Twenty-five patients were followed-up for up to 4 years. Seven patients (28%) remained stable over 1–4 years, whereas 18 patients (72%) relapsed after 1–3.5 years [58].

Variants influencing the clinical response of vitiligo were evaluated in an open, uncontrolled study from Italy on 60 patients treated for 6–24 months with NB-UVB. The best results were achieved in the face, followed by the neck, the trunk, and the limbs. The hands and feet were the sites with the poorest results. Lesions on the neck and the upper and lower extremities responded better in patients whose vitiligo had recent onset. Treatment outcome was also related to treatment duration in that better results were achieved with increasing numbers of NB-UVB exposure [13].

It is well-substantiated by all these studies that NB-UVB is a very useful treatment for patients with widespread or active vitiligo. Proper patient selection and information is crucial for the therapeutic outcome. It might require several months until appreciable repigmentation may become evident. Patients affected by large areas of vitiligo may need up to 2 years of continuous treatment to achieve a cosmetically satisfactory result. The following parameters were found in several, but not all studies to determine the clinical response: the localization of the affected skin, with the face and neck region almost invariably showing the greatest degree of repigmentation; the duration of vitiligo; the skin phototype; early repigmentation; and the duration of treatment. The extent of disease and the disease activity at the onset of treatment do not seem to have an effect on the likelihood of repigmentation.

One of the most essential questions pertains to the duration of the therapeutic effect. Relapse rates after discontinuation of treatment have as yet not been adequately delineated in large long-term follow-up studies. Observations from small samples of patients followed up for up to 3.5 years indicate that, depending on the length of the follow-up period, 20–70% of patients will again develop vitiligo in areas that had been repigmented by treatment [40, 41, 57, 58, 67, 89].

Efficacy of NB UVB vs. Other Phototherapeutic Modalities

After the establishment of NB-UVB for vitiligo, an obvious challenge was to assess its therapeutic efficacy relative to that of other phototherapeutic modalities, in particular, photochemotherapy. To this end, several trials have been performed so far.

In a bilateral comparison study, NB-UVB was compared with PUVA (oral 8-methoxypsoralen (8-MOP) and UVA) in 15 adult patients with symmetrical vitiligo. After 60 sessions, the clinical response to both treatments did not differ significantly [27].

A retrospective comparison of 38 patients treated with oral PUVA and 31 patients treated with NB-UVB showed a significantly better outcome for NB-UVB. In the PUVA group, marked to complete improvement was observed in 23.6%, moderate improvement in 36.8%, absent to mild repigmentation in 32.6%, and worsening in 7% of patients. The corresponding data for the NB-UVB group were 41.9, 32.2, 25.9, and 0%, respectively [67].

Another study compared randomly allocated treatment with thrice weekly NB-UVB and oral trimethylpsoralen (TMP) UVA in 50 consecutive nonsegmental vitiligo patients. The mean treatment duration was 6.3 months for NB-UVB and 5.6 months for TMP UVA. Both studies reported that NB UVB was superior to TMP UVA, in terms of stability achieved and efficacy in active and stable disease [10].

Recently, the first randomized, double-blind trial was published on the efficacy of NB UVB versus oral 8-MOP (or 5-MOP) UVA in 50 patients with nonsegmental vitiligo. Treatment was given twice weekly, and assessments were performed after every 16 sessions. At the end of the study, the PUVA group had received a mean number of 47 treatments as opposed to 97 treatments in the NB-UVB group. This difference was suspected to be due to differences in efficacy and adverse effects. Sixty-four percent of patients in the NB-UVB group had >50% improvement compared with 36% of patients in the PUVA group. Moreover, among the patients with more than 48 sessions of treatments, the reduction of the depigmented surface area was significantly greater for NB-UVB than for PUVA. The color match of repigmented skin was excellent in all patients treated with NB-UVB, but in only 44% of those treated with PUVA. The clear conclusion of this study was that NB-UVB is superior to oral PUVA in nonsegmental vitiligo [89].

A small, open, four-quarter comparative study evaluated NB-UVB versus BB-UVB in combination with topical calcipotriol versus placebo in nine patients with generalized symmetrical vitiligo. NB-UVB was delivered to the upper part of the body until the navel and BB-UVB to the lower part of the body. Irradiations were done thrice weekly during the first months and twice weekly thereafter. Additionally, calcipotriol was applied once in the evening on vitiligo lesions on the right side of the body and placebo ointment on lesions the left side. After six months of treatment, none of the patients showed repigmentation on the lower parts of the body, indicating that neither BB-UVB and calcipotriol separately nor their combination had been therapeutically effective. BB-UVB was then discontinued and NB-UVB applied to the whole body. At the end of the treatment (12 months), no difference in repigmentation was apparent between calcipotriol and placebo-treated sites, indicating that calcipotriol failed to enhance the response to NB-UVB [35].

In summary, the majority of comparative studies have shown that NB-UVB is more effective than other phototherapeutic modalities. Therefore, most treatment centers nowadays consider NB-UVB phototherapy as the first-line treatment for generalized vitiligo. Studies comparing NB-UVB with the excimer laser or the excimer light are discussed in the section on targeted phototherapy in vitiligo.

NB UVB in Combination with Topical Treatments

Corticosteroids, calcineurin inhibitors, Vitamin D analogs, and preparations containing pseudocatalase or a combination of catalase and superoxide dismutase have all been used as topical treatments for vitiligo with different and sometimes equivocal results. In addition to their use as a monotherapy, a number of studies have investigated the combination of these agents with NB-UVB with the aim to accelerate and increase the therapeutic response to phototherapy [53, 73, 75, 83].

A case report from India first described the use of thrice weekly NB-UVB in combination with calcipotriol cream on the right and placebo cream on the left lower limb. At the end of six months, repigmentation was almost complete over the right limb, whereas it was less than 50% on the placebo-treated side [26].

Subsequently, several other trials assessed the usefulness of a once or twice daily combination of NB-UVB with calcipotriol (calcipotriene) or tacalcitol in small patient cohorts. Most of these trials were open and uncontrolled and did not extend for more than six months. NB-UVB thrice weekly in combination with calcipotriol twice daily applied to all vitiligo lesions on the left side of the body gave better results in 6 out of 17 patients. The study period was not clearly specified, apparently up to 116 treatments (or more) of NB-UVB were received by some patients [45]. In another open trial on 24 patients, twice daily application of calcipotriol was found to potentiate the efficacy of NB-UVB. About two-third of the patients had an earlier onset of repigmentation with the combination. After 6 months of treatment the overall response rate was 51% for the combination and 39% for NB-UVB alone [30]. A greater extent of repigmentation and an increase in response rate was also reported for NB-UVB in combination with tacalcitol. In a randomized, investigator-blinded bilateral comparison study, 32 pairs of symmetrical vitiligo lesions were exposed twice weekly to NB-UVB. In addition, a standard dose of tacalcitol was applied once daily in the evening on one of the paired lesions. Throughout the whole observation period, the combination provided significantly higher repigmentation scores when compared with NB-UVB alone. Lesions treated with combination regimen repigmented both earlier and to a greater extent [49]. In contrast, no effect of twice daily application of calcipotriol cream in addition to NB-UVB twice or thrice weekly was observed in an investigator-blinded study on 20 patients after 6–12 months of treatment [2]. Two further studies also reported negative findings. Calcipotriol once daily did not enhance NB-UVB-induced repigmentation when given over a period of 1 year [35]. Another trial compared monotherapy with

NB-UVB (24 patients) with NB-UVB plus twice daily calcipotriol (13 patients). No significant difference in repigmentation was found between the two groups after 30 sessions of phototherapy [6].

A small, randomized, placebo-controlled double-blind trial compared NB-UVB plus tacrolimus versus NB-UVB, plus placebo in the treatment of generalized vitiligo. Paired vitiligo lesions in nine patients were treated thrice weekly with NB-UVB, plus twice daily with either 0.1% tacrolimus or petrolatum (filled in identical containers) over a total period of 12 weeks. Overall, both sides improved without a statistically significant difference between tacrolimus and placebo [53]. In an open, uncontrolled trial including 110 patients with 403 lesions, tacrolimus ointment once daily was combined with twice weekly NB-UVB for 16 weeks. More than 50% of repigmentation was observed in 42% of the lesions. Due to the uncontrolled nature of the study, it is not possible to assess the additional effect, if any, of tacrolimus to NB-UVB in this trial [29].

After an early report on the benefit of pseudocatalase in combination with short-term UVB exposure, no effect of the combination of pseudocatalase with NB-UVB was found in a later investigation. Both studies were small, open, and uncontrolled [73, 75]. A small, double-blind, intraindividual comparison study on NB-UVB in combination with either a gel containing catalase and superoxide dismutase (Vitix®) or placebo (the excipient only) suggested some effect of the verum preparation [43]. However, according to the last author of this communication, this new product has since then not proven to be effective in enhancing NB-UVB-induced repigmentation (Ortonne, personal communication). A recent, small study also described the use of this product in combination with NB-UVB. Due to the open, uncontrolled design of the study, the relevance of the data cannot be interpreted [44].

No studies have been performed so far on the combination of NB-UVB with topical corticosteroids.

In summary, there is some evidence that concurrent treatment with topical Vitamin D analogs might accelerate and increase the response to NB-UVB in a proportion of patients. However, controlled, larger scale trials with longer treatment periods are required to corroborate this contention. Available data on the combination of NB-UVB with other topical treatments are scarce and of low evidence level, thus precluding a well-founded assessment.

NB UVB in Combination with Vitamins and Antioxidants (See Part 3.4)

The addition of folic acid and vitamin B12 to NB-UVB was assessed. One cohort of 13 patients was exposed to NB-UVB twice weekly for 1 year, the other cohort of 14 patients additionally received twice daily treatment with 1000 μg vitamin B12 and 5 mg folic acid. After 12 months, maximum repigmentation rates did not differ significantly between the two groups [82].

The effect of supplementation with a balanced antioxidant pool (AP) containing α-lipoic acid, Vitamins C and E, and polyunsaturated fatty acids on the response to treatment with NB UVB was investigated in a double-blind placebo-controlled trial. AP or placebo, twice daily, was started eight weeks prior to initiation of phototherapy and was then continued together with twice weekly NB-UVB. After six months of combinatory treatment, 47% of patients in the AP group had >75% repigmentation and 23.5% had a 50–75% repigmentation. In the placebo group, 18% of the patients achieved >75 or 50–75% repigmentation. The average number of treatments required to induce 50% repigmentation was 18 in the AP group and 23 in the placebo group [23].

Polypodium leucotomos is an antioxidant and immunomodulatory plant extract. Given the pathogenic role of oxidative stress and autoimmunity in vitiligo, the therapeutic potential of *P. leucotomos* in combination with NB-UVB has been evaluated in a double-blind, placebo-controlled trial. Fifty patients were randomized to receive either 250 mg *P. leucotomos* capsules or placebo three times daily in conjunction with twice weekly NB-UVB. At Week 26, there was a body area-dependent trend toward more repigmentation in the *P. leucotomos* group. The mean cumulative NB-UVB dose was similar for both groups. Patients with Skin Type II and III appeared to benefit more from P. leucotomos than those with darker-skin types [56].

Practical Aspects of NB-UVB Treatment

NB-UVB is currently the first-line treatment for inducing repigmentation in generalized vitiligo affecting multiple or large areas of the body and can also be used to arrest disease progression in active vitiligo. NB-UVB phototherapy is basically easy to perform even if a proper dosimetry is mandatory to achieve optimum treatment results. In terms of photosensitivity, patients with vitiligo have traditionally been regarded as Skin Type I and consequently were treated with very low initial NB-UVB doses ranging from 150 to 250 mJ/cm^2 to avoid severe sunburn reactions. However, in a recent study, this approach has been challenged. It was shown that the erythemal sensitivity in vitiliginous skin depends on the skin type with darker-skin types tolerating higher UVB doses than subjects with a fair complexion. In addition, minimum erythema dose (MED) values in vitiligo skin were on average only 35% (95% CI = 31–39%) lower than in normal skin of the same individual [17]. Based on these results and empirical knowledge ,there is a trend nowadays to initiate vitiligo treatment with higher NB-UVB doses (50% of the MED of normal skin) (Part 3.10). Subsequently, the exposure dose should be adjusted regularly to the photoadaption of the irradiated skin by dose increments of 10–20% with the aim to induce and maintain a faint erythema reaction in lesional skin. Treatment is normally given twice or three times weekly and is continued as long as there is ongoing repigmentation. In case that NB-UVB phototherapy was initiated primarily to arrest disease progression, a 3-month course may often be sufficient. Discontinuation of treatment should be considered if no or minimal repigmentation is achieved within 4–6 months of treatment or if no further improvement occurs within 3 months of continuous NB-UVB exposure (Figs. 3.3.2.1–3.3.2.3).

Initiated by the Vitiligo European Task Force, there is currently a multicenter phototherapy study in progress that compares the conventional protocol of continuous treatment with repeated cycles of *on–off* treatment (2 months treatment, 1 month break). The protocol is based on theoretical considerations that intermittent irradiation might induce a more effective stimulation of melanocyte proliferation than a continuous treatment schedule.

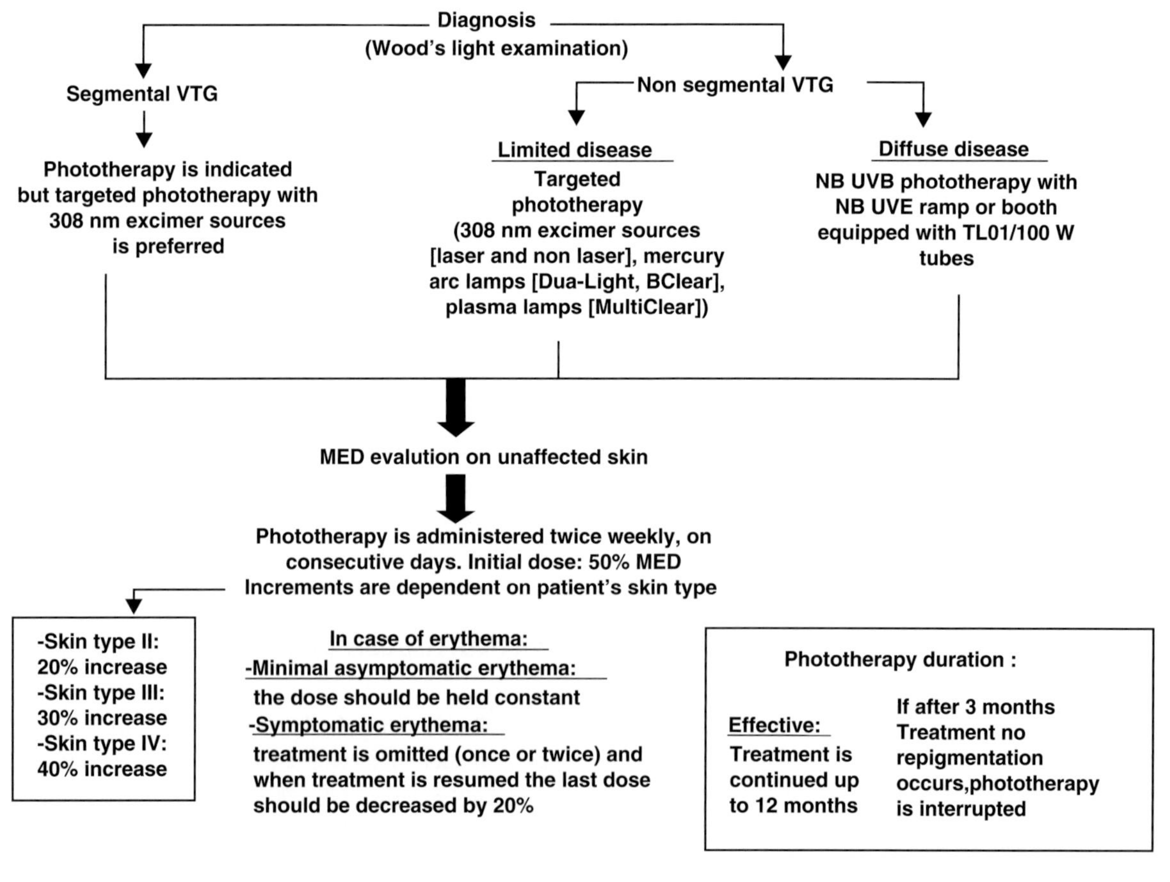

Fig. 3.3.2.1 Narrow band UVB protocol

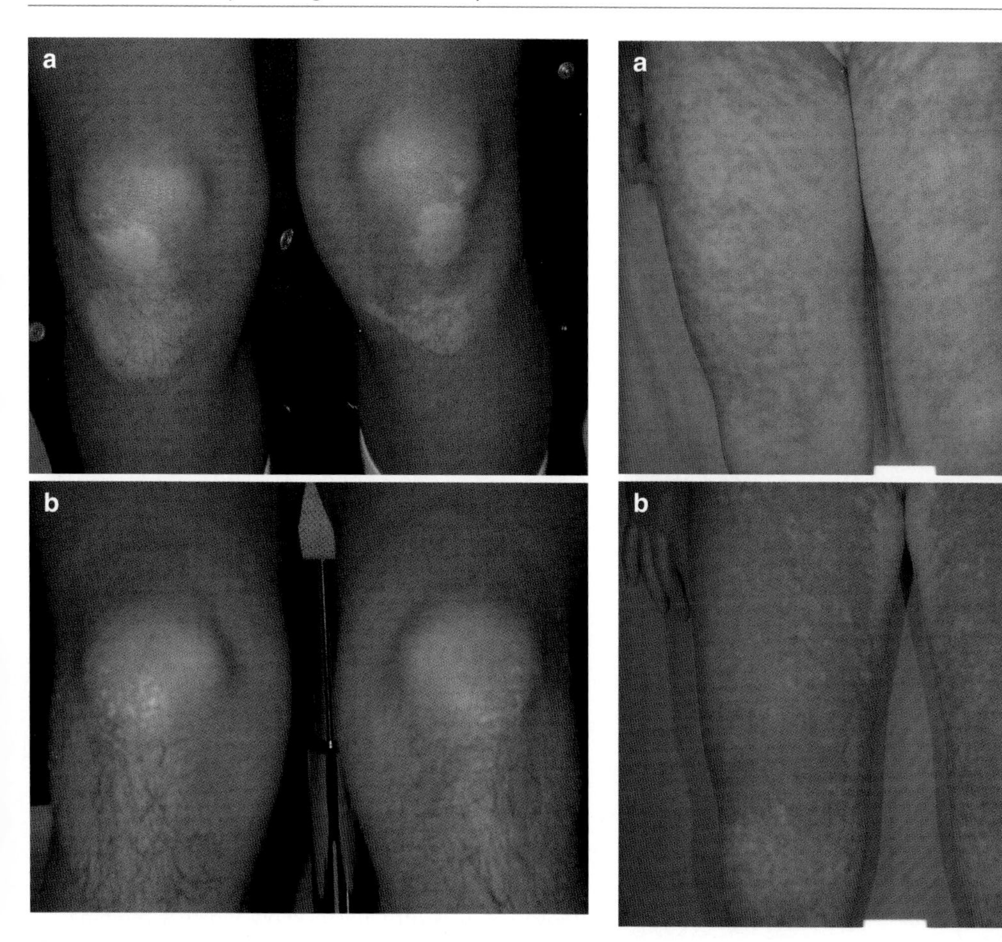

Fig. 3.3.2.2 Before (**a**) and after (**b**) 32 session treatments of NB UVB booth

Fig. 3.3.2.3 Before (**a**) and after (**b**) 32 session treatments of NB UVB booth

3.3.2.2 Targeted Phototherapy

Phototherapy, conventionally provided by psoralen plus UVA (PUVA), or, more recently, NB-UVB, is the mainstay for generalized vitiligo. However, these treatments are associated with burning and skin ageing when administered for a long-term. In conventional NB-UVB, as well as in any other type of phototherapy, ultraviolet radiation is delivered using a stand-up, whole-body unit: in such a manner that the normal, uninvolved skin is unavoidably exposed to UV radiation, resulting in several adverse effects.

Recently, efforts have been made to develop therapeutic devices that deliver light, both laser and incoherent one, selectively to the lesions. Using phototherapy units capable of emitting light in a more targeted manner with higher fluencies, the lesions can be selectively treated, while the normal skin is spared.

Lasers

Excimer Laser 308 nm

The excimer laser represents the latest advance in the concept of selective phototherapy (Fig. 3.3.2.4). It emits a wavelength of 308 nm and shares the physical properties of lasers: monochromatic and coherent beam of light, selective treatment of the target, and ability to deliver high fluences. The 308-nm excimer laser was first used in dermatology in 1997 for treating psoriasis [12]. Since then, many studies have evaluated this new device in a number of dermatologic disorders. Psoriasis and vitiligo have been further investigated, and the use of excimer lasers for both conditions is now approved by the US Food and Drug Administration [28, 32, 37, 48, 65, 84].

The excimer laser emits a wavelength of 308 nm, produced using xenon and chlorine gases. Transmission of

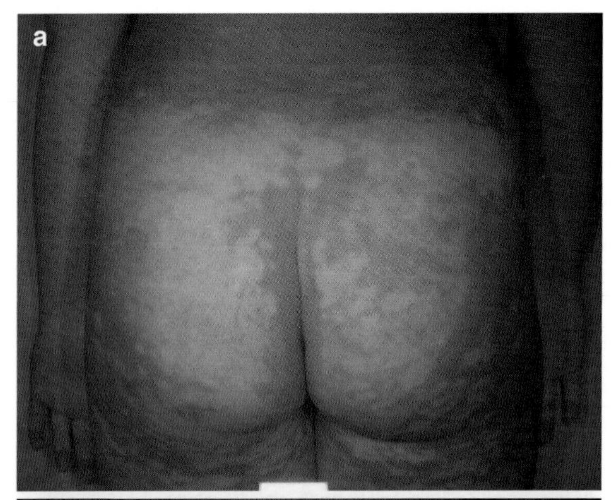

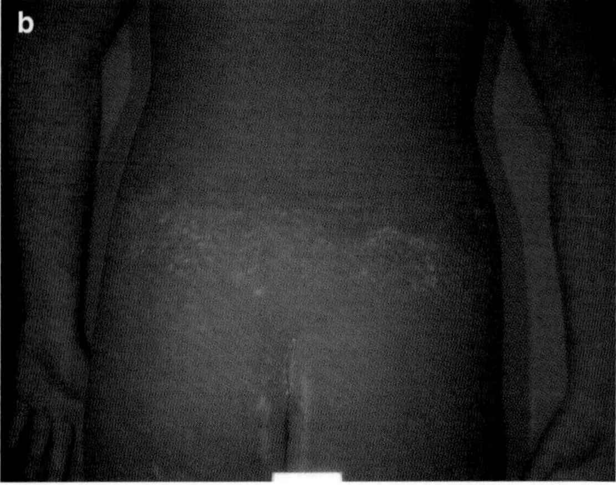

Fig. 3.3.2.4 Before (**a**) and after (**b**) 16 treatments with targeted phototherapy (monochromatic excimer light 308 nm). The repigmentation was almost complete

the beam of light is achieved by using a liquid light guide (LLG). Spot size is variable from 14 to 30 mm in diameter depending on the model used. These technical characteristics provide this laser with many advantages over conventional phototherapies. The high fluences emitted can be useful in thick plaques of psoriasis but not in vitiligo, where only low fluences are used. It is also possible to selectively turn off the beam of light and thus to treat the specific area involved, sparing healthy skin. In vitiligo, this selectivity limits the unsightly tanning of perilesional skin, which is commonly observed with other phototherapies. The LLG also makes it easier to reach areas that are usually difficult to treat, such as folds and mucosa. Disadvantages include the fact that the limited size of spots means that large surfaces (>20% of total surface body area) cannot be treated and that purchase and maintenance costs of these devices is rather expensive.

Due to its selectivity and propigmentary properties, the 308-nm excimer laser represents an interesting new approach for treating vitiligo. The use of the 308-nm excimer laser in treating vitiligo was first reported by Baltas et al. [9]. Subsequent studies have shown the efficacy of this laser for repigmenting vitiligo lesions. Spencer et al. [79] recently reported that the 308-nm excimer laser may represent a new therapeutic option for the management of vitiligo, resulting in repigmentation of vitiligo patches in less time than that required with other modalities. With this treatment, pigmentation can start after only five sessions and increase with continuation of treatment. Low fluences (from 50 to 200 mJ/cm^2) have been used in one to three sessions a week for 1–6 months, depending on the study.

Among factors that can influence the clinical response to treatment, localization of the lesions seems to play a crucial role. In their study, Taneja et al. [81] report repigmentation of at least 75% in all the lesions located on the face versus none on the hands and feet. The variability of some results reported certainly depends on the localization of target lesions.

Sessions can be performed once, twice, or three times a week. The repigmentation rate seems to be linked to the total number of sessions and not to their frequency. It is difficult to know whether repigmentation is stable because the follow-up of existing series is short or nonexistent. A recent study reports no depigmentation 1 year after the end of sessions. On the other hand, Passeron and Ortonne [71] in a recent review article report that in their series about 15% of new depigmentation is observed 1–3 years after the end of treatment.

Tolerance of treatment is usually very good, and immediate side effects are limited to erythema and in rare cases, blistering.

Topical tacrolimus has been shown to increase the efficacy of excimer laser phototherapy in vitiligo [42, 72]. Two pilot prospective studies have compared the efficacy of the excimer laser combined with 0.1% tacrolimus ointment with excimer laser monotherapy or laser associated with a placebo. In the first series, two versus three sessions per week were performed. In both cases, a total of 24 sessions were carried out and 0.1% tacrolimus ointment was applied twice a day. Results clearly showed a greater efficacy and shorter response to treatment with combined therapy as compared with excimer laser alone. Such a combination might not be used routinely since there is an ongoing debate whether tacrolimus might increase the risk for UV-induced cutaneous cancers. Another possible association could be with topical corticosteroids, but no study has still evaluated this combination.

Helium–Neon

Low-energy helium–neon (He–Ne) lasers (632.8 nm) have been employed in a variety of clinical treatments, including vitiligo management. Light-mediated reaction to low-energy laser irradiation is referred to have biostimulatory rather than thermal effect. A first report was published in 2003 [46]. This study investigated the effect of helium–neon laser both in vitro and in vivo. In vitro studies revealed a significant increase in basic fibroblast growth factor release from both keratinocytes and fibroblasts, as well as a significant increase in nerve growth factor release from keratinocytes. It has also been shown that melanocyte migration was enhanced either directly by helium–neon laser irradiation or indirectly by the medium derived from helium–neon laser treated keratinocytes. Furthermore, 30 patients with segmental-type vitiligo on the head and/or neck were enrolled in this study. Helium–Neon laser light was administered locally at 3 J/cm^2 with point stimulation once or twice weekly. The percentage of repigmented area was used for clinical evaluation of effectiveness. After an average of 16 treatment sessions, initial repigmentation was noticed. Marked repigmentation was observed in 60% of patients with successive treatments. These results suggest that helium–neon laser irradiation stimulates melanocyte migration and proliferation, release of mitogens for melanocytes, and may also rescue damaged melanocytes, therefore providing a microenvironment for inducing repigmentation in vitiligo. The same group demonstrated that Helium–Neon laser induced different physiological changes on melanoblasts at different maturation stages and recapitulated the early events during vitiligo repigmentation process brought upon by Helium–Neon laser in vitro [46].

Despite the interesting results obtained in cell cultures and the clinical improvement noted in patients with segmental vitiligo, further clinical studies are required on larger series, also including nonsegmental vitiligo, to confirm the indication for the use of Helium–Neon laser in vitiligo.

NonLaser Light Sources

Monochromatic Excimer Lamp or Light 308 nm (MEL 308 nm)

The 308 nm monochromatic radiation can be also delivered by excimer lamps. In this case, the effectiveness of a new 308 nm monochromatic excimer source, with emission close to that used in NB-UVB phototherapy, has been initially described in the treatment of recalcitrant palmoplantar psoriasis [11, 12, 15, 16].

A pilot study reports that 18 out of 37 vitiligo patients achieved 75% or more repigmentation after six months of treatment [62]. The source in this study was a 308 nm XeCl MEL device with a power density of 48 mW/cm^2 at a distance of 15 cm from the skin, and the irradiation field covered an area of 504 cm^2 with rectangular shape (36 × 14 cm). Interestingly, a satisfactory response on the hands was noted: two patients achieved Grade 3 repigmentation. This finding is not significant because of the limited number of patients that received treatment on the hands in this study ($n = 3$), but it may suggest the efficacy of higher NB-UVB fluences on these locations that usually respond poorly to conventional treatment [8, 11, 20]. The 308 nm MEL may present some advantages over the laser: lower power density and consequently reduced risk of accidents due to overexposure; larger irradiation field with the possibility to treat larger areas at a time, with shorter treatment duration. More recently, devices that deliver 308 nm MEL to the skin by means of an optic fiber or with a smaller hand piece have been introduced: this offers the possibility to treat both small and large lesions.

The cost of these devices is lower than that of lasers, and maintenance is less frequently required. Nevertheless, a comparative trial (308 nm excimer laser vs. 308 nm excimer lamp) in a larger population, in order to clarify whether the effectiveness of these two sources, is needed.

Excimer Laser/Lamp vs. NB-UVB

An important issue is to demonstrate the greater effectiveness of the high potency excimer sources as compared to conventional NB-UVB. Earlier studies have expounded on the safety and efficacy of treatment with the 308 nm excimer laser, possibly even more so than NB-UVB phototherapy [18, 22]. However, these reports involved comparisons of groups of patients, not direct comparisons within the same patient. A recent study demonstrated the greater efficacy of the 308 nm excimer laser treatment over NB-UVB phototherapy, because it produces a more rapid and profound repigmentation. The comparison has been made on 23 patients with symmetrical vitiligo patches treated with the 308 nm laser or NB-UVB on a 2-week sessions schedule for a maximum of 20 treatments [38]. A similar left–right study has been more recently performed comparing the therapeutic effectiveness of MEL 308 nm and conventional NB-UVB in 21 vitiligo patients with symmetrical vitiligo lesions. At the end of the study (6 months) 37.5% of lesions treated with 308 nm MEL and only 6% of lesions treated with NBUVB achieved an excellent repigmentation [38].

These two comparative studies suggest that the treatment with laser or MEL 308 nm may allow repigmentation within a shorter period of time, as compared to NB-UVB phototherapy, together with circuscribed exposure to the irradiation. Also, the rapid onset of repigmentation may play an important role in supporting patient motivation and compliance.

Mercury Arc Lamps

Different devices equipped with high-pressure mercury arc lamps are now available for targeted phototherapy. Usually the light is delivered to the skin by means of an optic fiber. On the basis of the emission spectrum of these lamps, this phototherapy is also referred as "targeted broad band UVB." Interesting results have been described in vitiligo, with a high-pressure mercury lamp capable of emitting either UVB or UVA (Dua-Light TheraLight Inc. Carlsbad, CA). The UVB spectral output of this light source includes peaks at 302 and 312 nm, with an average weighted erythemal wavelength of 304 nm. The high output of this device allows irradiation of 100 mJ/cm^2 of UVB to take place within approximately 0.7 s. Ultraviolet radiation is delivered through a square aperture sized 1.9 × 1.9 cm. Asawanonda et al. [7] reported their experience on 6 patients and 29 lesions were treated with targeted, broadband UVB phototherapy. Treatments were carried out twice weekly for 12 weeks. Some degree of repigmentation occurred in all subjects. Onset of repigmentation was as early as three weeks of treatment in some subjects. According to these results, broadband UVB could be an efficacious and safe modality for the treatment of localized vitiligo. A possible advantage is that these mercury arc lamps are less expensive and require minor maintenance as compared to the excimer devices.

Another device, similar to the aforementioned, has been used with interesting results in psoriasis, even if the current literature lacks a confirmation for the vitiligo. The BClear-Targeted PhotoClearing System (Lumenis Inc, Santa Clara, CA) uses a UVB lamp to deliver targeted broad-band UVB filtered incoherent pulsed or continuous UVB light at 290–320 nm. Peak irradiance occurs between 310 and 315 nm. The 16 × 16 mm spot size emits a pulse width of 0.5–2.0 s with a fluence range of 50–800 mJ/cm^2.

Plasma Lamps

A phototherapy device, the MultiClear® system (CureLight Ltd. USA), has been recently introduced. It is based on Selective Photo Clearing (SPC™), a proprietary technology generating, by means of high-power plasma light source, emission of different wavelengths: 296–315, 360–370, and 405–420 nm (blue light PDT). The system allows to select high intensity of UVB, UVA, and a blend of targeted UVB and UVA1 delivered by means of a flexible light guide, with a treatment spot of 23 × 23 mm. UVA1 and UVB could act synergistically to induce repigmentation in vitiligo according to the producer's specifications (unpublished data). The efficacy of this phototherapy device has to be demonstrated with clinical studies in vitiligo and in other indications.

Microphototherapy

Edited by Torello Lotti, Francesca Prignano, and Gionata Buggiani)

Microphototherapy is based on the photo-exposition limited to well-defined areas, avoiding the side effects associated to diffuse phototherapy (photoaging, erythema, and burns among the others) [31, 50, 51, 55, 59]. The NB-UVB lamp (Philips TL-10) with a wavelength peak at 311 nm, selectively delivers the light to the white patches. It allows one to obtain the drastic reduction of the total dose of radiation and thus the most common side effects related to the exposition to UV rays: excessive tanning of the nonaffected skin, photoaging, teleangectases, and the risk of neoplasms. Moreover, it reduces the chromatic contrast between normal and lesional skin. With this treatment, different doses of UVB radiation can be administered in different areas of the body, optimizing thus the treatment by tailoring it on each and every subject. The initial dose of radiation is 20% lower than the MED, which is evaluated through the exposition of affected and unaffected skin to increasing doses of UVB (80, 160, 240, 320, and 400 mW/cm2) at least three days before the beginning of the treatment. During the following sessions, every patch is uniformly irradiated. In "sensitive" areas (i.e., eyelids) 80% of MED doses are used. The radiation dose is increased by 20% at each session: when erythema occurs, the dose is lowered by 20% in the erythematous area only. Some skin areas can resist more to the photo-stimulation and are thus irradiated with higher regimens (up to twice the dose of the most sensitive areas). Irradiation sessions are repeated every 21–30 days, until repigmentation is reached, ranging from 2 months to 2 years [31, 50, 51]. A partial repigmentation is often seen after 3–6 sessions (63% of the cases), beginning just after 2 months of treatment as a pigment pitting around each follicular ostium (follicular repigmentation), usually accompanied by an evident interfollicular repigmentation. A photographic evaluation is useful to assess the clinical results, using Wood's light for lighter phototypes. In a recent study [55] 734 patients were irradiated using this protocol every 2 weeks for 12 consecutive months. At the end of the study period, 69.8% of patients ($n = 510$) achieved normal pigmentation on more than 75% of the treated areas, 21.12% ($n = 155$) achieved 50–75% repigmentation on the treated areas, and only 9.4% ($n = 69$) showed less than 50% repigmentation (in five subjects of this group vitiligo worsened), without statistical significance between segmental and nonsegmental vitiligo. The results of this study are similar to those obtained by total body UVB irradiation in international studies, enriched by the advantage of the lesser side effects.

Microphototherapy is particularly useful in patients affected by segmental vitiligo and bilateral symmetrical vitiligo, for which the total amount of body surface involved is less than 20%. The only side effect occasionally reported is transient erythema, rarely followed by desquamation [51].

Microphototherapy, as well as the other phototherapies, is not administered to subjects with actinic sensitivity (SLE, Xeroderma pigmentosum, porphyriasis, cutaneous viral infections) and in subjects treated with topical or systemic photosensitizing agents.

3.3.2.3 Mechanism of Action of Phototherapies

Narrow Band UVB

UVB induces several biological effects including erythema, hyperplasia, and nonmelanoma skin cancer. At the cellular level, UVB induces inflammation and DNA damage. Tissue/DNA repair mechanisms are employed during mild exposure to UVB. However, prolonged UVB exposure results in irreversible DNA damage leading to the programmed cell death. Thus, extensive UVB exposure leads to tissue damage through inflammation or apoptosis [3, 4, 14, 24, 33, 36, 66, 74, 85].

To date, little is known about the mechanism of action of NB-UVB in vitiligo. Melanocytes could be recruited from the outer root sheath of the hair follicle to repigment vitiliginous skin through the action of phototherapy. Recovery from vitiligo is initiated by the activation and proliferation of these melanocytes, followed by the upward migration to the nearby epidermis, accounting for the perifollicular pigmentation islands, and by subsequent downward migration to the hair matrices to produce melanin [39, 46, 69, 76, 77, 80, 87, 90]. PUVA acts by stimulating the outer root sheath melanocytes to migrate into the epidermis. The migration is thought to occur as a result of the release of cytokines and inflammatory mediators from the nearby keratinocytes [64, 67]. Abdel-Naser et al. [1] reported that PUVA treatment results in the release of growth factors into blood circulation that can stimulate the proliferation of melanocytes and of other cells. Similar to the phenomenon observed after PUVA therapy, vitiligo lesions treated with NB-UVB also repigmented in a perifollicular pattern and repigmentation is not seen in lesions with amelanotic hairs. NB-UVB also stimulates the melanocytes in the outer root sheath of the hair follicles. However, the exact mechanisms of NB–UVB in repigmentation are not still clarified.

There are evidences suggesting that focal adhesion kinase (FAK) plays a crucial role in transducing a variety of signals that modulate cell adhesion and cell migration. Increased expression of phosphorylated FAK ($p125^{FAK}$) may indirectly modulate cytoskeletal proteins necessary for cell migration. The matrix metalloproteinases (MPPs) are a family of enzymes involved in the degradation of extracellular matrix components such as collagen, gelatine and fibronectin. Expression of MMPs is markedly increased in situations involving active tissue remodeling and cell migration [66]. Both keratinocytes and fibroblasts, natural cellular neighbors of melanocytes, release putative melanocytes growth factors. Melanocytes proliferation is controlled by different classes of mitogens: leukotriene (LT) C4 and D4, endothelin-1 (ET-1) and tyrosine kinase growth factors-basic fibroblast growth factor (bFGF), stem cell factor, hepatocyte growth factor (HGF), melanotropin, epidermal growth factor, and platelet-derived growth factor.

Recently, Wu et al. [87] demonstrated that NB-UVB radiation stimulates the release of bFGF and ET-1 from keratinocytes, which induce melanocyte proliferation. In addition, NB-UVB irradiation stimulating the expression of $p125^{FAK}$ in melanocytes and inducing the expression of MMP2 in melanocyte supernatants may enhance melanocyte migration.

Targeted Phototherapy

Beside the mechanisms possibly responsible for the effectiveness of the conventional NB UVB, some additive ones can be considered in the case of the new potent monochromatic light sources, used for targeted phototherapy. Photobiologically, the wavelengths of the excimer laser (308 nm) and NB UVB (311 nm) are very close to one another, and the therapeutic effects may be similar. Recently, the mechanism of the excimer laser's high efficacy in psoriasis treatment has been investigated.

The mechanism of action in vitiligo has not been still fully understood. Stimulation of melanocyte migration and proliferation from progenitor niches located in hair follicles is certainly the major factor. This stimulation is not only due to the direct action of UV on melanocytes, but also due to the action of cytokines secreted by keratinocytes. Recent data on the autoimmune origins of vitiligo underline the probable implications of the immunosuppressive action of UV in treating vitiligo and thus the intense pro apoptotic effect of the 308 nm wavelength on T-cells could play a role also in vitiligo [62, 63]. The biological effects of coherent laser light may differ from those of incoherent light of the same wavelength. Conventional UVB sources emit polychromatic, continuous, incoherent light, whereas the excimer laser emits coherent, monochromatic, UVB light in short pulses. This permits the variation of some important phototherapeutic parameters, such as impulse frequency and intensity.

These optical properties of the excimer laser could make it more effective than NBU-VB in the treatment of vitiligo. Namely, the possibility to deliver high doses in a short interval of time may account for the differences in biological effects of conventional NB-UVB and those of the monochromatic excimer sources. Even if the Bunsen–Roscoe law (BRL) of reciprocity states that a certain biological effect is directly proportional to the total energy dose irrespective of the administered regimen, in some cases it has been shown not to hold and this could be also the case for the effects of the excimer laser.

3.3.2.4 Side Effects of Phototherapies

Narrowband UVB

Narrowband UVB, as well as targeted phototherapies is in general very well-tolerated. The most common acute adverse reaction is UV-induced erythema in vitiliginous skin, which is both skin type- (vitiligo patients with skin Type I or II have lower erythema threshold in lesional skin than darker-skin types) and UV dose-dependent. UVB erythema usually occurs 12–24 h after irradiation and subsides within another 24 h. Since patients are not treated on consecutive days, erythema induced by the last UVB exposure usually disappears before the next treatment session. Thus, it is essential to always ask the patients whether, and if so, to what extent, they had developed erythema in response to the previous irradiation. It should be kept in mind, however, that a slight erythema reaction in lesional skin is a good guideline for adequate dosimetry. Drug-induced photosensitivity is usually not an issue since most of the drugs have their action spectrum in the longer UV (UVA) wave range.

Other infrequent side effects include reactivation of herpes simplex infection, dryness of the skin, or elicitation of polymorphous light eruption in the initial phase of treatment. Considering that the doses used for the treatment of vitiligo are lower than those useful for the psoriasis management, low rate of acute, and possibly also long-term, adverse events occurs.

The main possible long-term hazards of NB-UVB treatment pertain to UV carcinogenicity. Sunlight and therapeutic ultraviolet radiation are known to potentially induce skin cancer, in particular, actinic keratoses, and squamous-cell carcinoma [54]. It is of special interest in this context that paradoxically skin cancer appears to occur rarely in vitiliginous skin despite the lack of protection by melanin from ultraviolet radiation [61,76]. This phenomenon could be related to epidermal upregulation of wild-type p53 in vitiligo [77].

Most part of the information about the carcinogenic risk of UVB irradiation comes from studies in psoriasis patients. Recent literature reviews suggest that the incidence of skin cancer due to UVB treatment is low; therefore UVB can be considered a relatively safe treatment with regard to potential long hazards [70]. A critical appraisal of the carcinogenic risk of NB-UVB compared to BB-UVB concluded that the cancer risk of comparable therapeutic doses of these two wavebands is presumably similar and much lower than that of PUVA. Two early follow-up studies on NB-UVB carcinogenicity have been published recently. The first one did not find any evidence for increased skin-cancer development, whereas the other indicates a slight increase in basal cell-cancer formation (but not squamous-cell carcinoma or malignant melanoma) [52]. However, it has to be pointed out that large multicenter studies with long follow-up periods will be required to accurately determine the carcinogenic potential associated with NB-UVB treatment [25].

Targeted Phototherapy

The major side effect of targeted phototherapies is erythema. Higher therapeutic doses are commonly applied when treating lesional skin only; therefore erythema reactions may occur more often and with greater intensity than with NB-UVB phototherapy. However, these reactions are confined to small areas of the treated skin and do not impair the general well-being of the patient.

Summary Messages

- Narrowband UVB (NB UVB) currently represents the phototherapy of choice for vitiligo. Side effects are less frequent than in PUVA therapy and efficacy is at least equivalent.
- Generalized active vitiligo is preferentially treated with total body narrowband UVB which can arrest disease progression and may induce significant repigmentation.
- Targeted phototherapy allows for selective treatment of lesional skin thereby avoiding unnecessary irradiation of healthy skin. Other known advantages, particularly when using high energetic monochromatic light sources, include rapid induction of repigmentation and the requirement of fewer treatments to achieve repigmentation as compared to traditional NB UVB.
- Although the carcinogenic risk associated with NB UVB phototherapy is still poorly defined, it can be assumed that the reduced cumulative UV dose needed to treat vitiligo with targeted phototherapy involves a lower potential for long-term hazards.
- The emission of some devices is filtered in order to eliminate erythemogenic radiation, but this process can considerably reduce the power density.

References

1. Abdel-Naser MB, Hann SK, Bystryn JC (1997) Oral psoralen plus UV A therapy releases circulation growth factors that stimulates cell proliferation. Arch Dermatol 133:1530–1533
2. Ada S, Sahin S, Boztepe G et al (2005) No additional effect of topical calcipotriol on narrow-band UVB phototherapy in patients with generalized vitiligo. Photodermatol Photoimmunol Photomed 21:79–83
3. Afaq F, Mukhtar H (2001) Effects of solar radiation on cutaneous detoxification pathways. J Photochem Photobiol B 63:61–69

4. Akasaka T, Leeuwen RL, Yoshinaga IG et al (1995) Focal adhesion kinase (p125FAK) expression correlates with motility of human melanoma cell lines. J Invest Dermatol 105:104–108
5. Anbar TS, Westerhof W, Abdel-Rahman AT et al (2006) Evaluation of the effects of NB-UVB in both segmental and non-segmental vitiligo affecting different body sites. Photodermatol Photoimmunol Photomed 22:157–163
6. Arca E, Tastan HB, Erbil AH et al (2006) Narrow-band ultraviolet B as monotherapy and in combination with topical calcipotriol in the treatment of vitiligo. J Dermatol 33:338–343
7. Asawanonda P, Charoenla M, Korkij W (2006) Treatment of localized vitiligo with targeted broadband UVB phototherapy: a pilot study Photodermatol Photoimmunol Photomed 22:133–136
8. Baltas E, Csoma Z, Ignacz F et al (2002) Treatment of vitiligo with the 308 nm xenon chloride excimer laser. Arch Dermatol 138:1619–1620
9. Baltas E, Nagy P, Bonis B et al (2001) Repigmentation of localized vitiligo with the xenon chloride laser. Br J Dermatol 144:1266–1267
10. Bhatnagar A, Kanwar AJ, Parsad D et al (2007) Psoralen and ultraviolet A and narrow-band ultraviolet B in inducing stability in vitiligo, assessed by vitiligo disease activity score: an open prospective comparative study. J Eur Acad Dermatol Venereol 21:1381–1385
11. Bianchi B, Campolmi P, Mavilla L et al (2003) Monochromatic excimer light (308 nm): an immunohistochemical study of cutaneous T cells and apoptosis related molecules in psoriasis. J Eur Acad Dermatol Venereol 17:408–413
12. Bonis B, Kemeny L, Dobozy A et al (1997) 308 nm UVB excimer laser for psoriasis. Lancet 350:1522
13. Brazzelli V, Antoninetti M, Palazzini S et al (2007) Critical evaluation of the variants influencing the clinical response of vitiligo: study of 60 cases treated with ultraviolet B narrow-band phototherapy. J Eur Acad Dermatol Venereol 21: 1369–1374
14. Cadet J, Douki T, Pouget JP et al (2001) Effects of UV and visible radiations on cellular DNA. Curr Probl Dermatol 29:62–73
15. Campolmi P, Mavilia L, Lotti TM et al (2002) 308 nm monochromatic excimer light for the treatment of palmoplantar psoriasis. Int J Immunopathol Pharmacol 13:11–13
16. Cappugi P, Mavilia L, Mavilia C et al (2002) 308 nm monochromatic excimer light in psoriasis: clinical evaluation and study of cytokine levels in the skin. Int J Immunopathol Pharmacol 13:14–19
17. Caron-Schreinemachers AL, Kingswijk MM, Bos JD et al (2005) UVB 311 nm tolerance of vitiligo skin increases with skin photo type. Acta Derm Venereol 85:24–26
18. Casacci M, Thomas P, Pacifico A et al (2007) Comparison between 308-nm monochromatic excimer light and narrow-band UVB phototherapy (311–313 nm) in the treatment of vitiligo – a multicentre controlled study. J Eur Acad Dermatol Venereol 21:956–963
19. Chen GY, Hsu MM, Tai HK et al (2005) Narrow-band UVB treatment of Vitiligo in Chinese. J Dermatol 32:793–800
20. Choi KH, Park JH, Ro YS (2004) Treatment of vitiligo with 308-nm xenon chloride excimer laser: therapeutic efficacy of different initial doses according to treatment areas. J Dermatol 31:284–292
21. Dawe RS, Cameron H, Yule S et al (2003) A randomized controlled trial of narrowband ultraviolet B vs. bath-psoralen plus ultraviolet A photochemotherapy for psoriasis. Br J Dermatol 148:1194–1204
22. Dawe RS (2003) A quantitative review of studies comparing the efficacy of narrow-band and broad-band ultraviolet B for psoriasis. Br J Dermatol 149:669–672
23. Dell'Anna ML, Mastrofrancesco A, Sala R et al (2007) Antioxidants and narrow band-UVB in the treatment of vitiligo: a double-blind placebo controlled study. Clin Exp Dermatol 32:631–636
24. De With A, Greulich KO (1995) Wavelength dependence of laser-induced DNA damage in lymphocytes observed by single-cell gel electrophoresis. J Photochem Photobiol B 30:71–76
25. Diffey BL, Farr PM (2007) The challenge of follow-up in narrowband ultraviolet B phototherapy. Br J Dermatol 157: 344–349
26. Dogra S, Parsad D (2003) Combination of narrowband UV-B and topical calcipotriene in vitiligo. Arch Dermatol 139:393
27. El Mofty M, Mostafa W, Esmat S et al (2006) Narrow band ultraviolet B 311 nm in the treatment of vitiligo: two right-left comparison studies. Photodermatol Photoimmunol Photomed 22:6–11
28. Esposito M, Soda R, Costanzo A et al (2004) Treatment of vitiligo with the 308 nm excimer laser. Clin Exp Dermatol 29:133–137
29. Fai D, Cassano N, Vena GA (2007) Narrow-band UVB phototherapy combined with tacrolimus ointment in vitiligo: a review of 110 patients. J Eur Acad Dermatol Venereol 21:916–920
30. Goktas EO, Aydin F, Senturk N et al (2006) Combination of narrow band UVB and topical calcipotriol for the treatment of vitiligo. J Eur Acad Dermatol Venereol 20:553–557
31. Grimes PE (2004) White patches and bruised souls: advances in the pathogenesis and treatment of vitiligo. J Am Acad Dermatol 51:s5–s7
32. Hadi SM, Spencer JM, Lebwhol M (2004) The use of the 308 nm excimer laser for the treatment of vitiligo. Dermatol Surg 30:983–986
33. Halaban R (2000) The regulation of melanocyte proliferation. Pigment Cell Res 13:4–14
34. Hamzavi I, Jaim H, McLean D et al (2004) Parametric modeling of narrowband UV-B phototherapy for vitiligo using a novel quantitative tool. Arch Dermatol 140:677–683
35. Hartmann A, Lurz C, Hamm H et al (2005) Narrow-band UVB311 nm vs. broad-band UVB therapy in combination with topical calcipotriol vs. placebo in vitiligo. Int J Dermatol 44:736–742
36. Hirobe T (1994) Keratinocytes are involved in regulating the developmental changes in the proliferative activity of mouse epidermal melanoblast in serum free culture. Dev Biol 161:59–69
37. Hofer A, Hassan AJ, Legat FJ et al (2005) Optimal weekly frequency of 308 nm excimer laser treatment in vitligo patients. Br J Dermatol 152:981–985
38. Hong SB, Park HH, Lee MH (2005) Short term effects of 308 nm xenon chloride excimer laser and narrow band ultraviolet B in the treatment of vitiligo: a comparative study. J Korean Med Sci 20:273–278
39. Jiang W, Ananthaswamy HN, Muller N et al (1999) P53 protects against skin cancer induction by UV B radiation. Oncogene 18:4247–4253

40. Kanwar AJ, Dogra S, Parsad D et al (2005) Narrow-band UVB for the treatment of vitiligo: an emerging effective and well-tolerated therapy. Int J Dermatol 44:57–69
41. Kanwar AJ, Dogra S (2005) Narrow-band UVB for the treatment of generalized vitiligo in children. Clin Exp Dermatol 30:332–336
42. Kawalek AZ, Spencer JM, Phelps RG (2004) Combined excimer laser and topical tacrolimus for the treatment of vitiligo: a pilot study. Dermatol Surg 30:130–135
43. Khemis A, Ortonne JP (2004) Study comparing a vegetal extract with superoxide dismutase and catalase activities (Vitix®) plus selective UVB phototherapy versus an excipient plus selective UVB phototherapy in the treatment of vitiligo vulgaris. Nouv Dermatol 23:7–11
44. Kostovic K, Pastar Z, Pasic A et al (2007) Treatment of vitiligo with narrow-band UVB and topical gel containing catalase and superoxide dismutase. Acta Dermatovenereol Croat 15:10–14
45. Kullavanijaya P, Lim HW (2004) Topical calcipotriene and narrowband ultraviolet B in the treatment of vitiligo. Photodermatol Photoimmunol Photomed 20:248–251
46. Lan CC, Wu CS, Chiou MH et al (2006) Low-energy helium-neon laser induces locomotion of the immature melanoblasts and promotes melanogenesis of the more differentiated melanoblasts: recapitulation of vitiligo repigmentation in vitro. J Invest Dermatol 126:2119–2126
47. Lapidoth M, Adatto M, David M (2007) Targeted UVB phototherapy for psoriasis: a preliminary study. Clin Exp Dermatol 32:642–645
48. Leone G, Iacovelli P, Paro Vidolin A et al (2003) Monochromatic excimer light 308 nm in the treatment of vitiligo: a pilot study. J Eur Acad Dermatol Venereol 17:531–537
49. Leone G, Pacifico A, Iacovelli P et al (2006) Tacalcitol and narrow-band phototherapy in patients with vitiligo. Clin Exp Dermatol 31:200–205
50. Lotti TM, Menchin G, Andreassi L (1999) UVB radiation microphototherapy. An elective treatment for segmental vitiligo. J Eur Acad Dermatol Venereol 13:102–108
51. Lotti TM (2002) Vitiligo: problems and solutions. Int J Immunopathol Pharmacol Dermatol 13(5):325–328
52. Man I, Crombie IK, Dawe RS et al (2005) The photocarcinogenic risk of narrowband UVB (TL-01) phototherapy: early follow-up data. Br J Dermatol 152:755–757
53. Mehrabi D, Pandya AG (2006) A randomized, placebo-controlled, double-blind trial comparing narrowband UV-B plus 0.1% tacrolimus ointment with narrowband UV-B plus placebo in the treatment of generalized vitiligo. Arch Dermatol 142:927–929
54. Melnikova V, Pacifico A, Peris K et al (2005) Fate of UVB-induced p53 mutations in SKH-hr1 mouse skin after discontinuation of irradiation: relationship to skin cancer development. Oncogene 24:7055–7063
55. Menchini G, Lotti T, Tsoureli-Nikita E, Hercogová J (2004) UV-B narrowband microphototherapy: a new treatment for vitiligo. In: Lotti T, Hercogová J (eds.) Vitiligo: problems and solutions. Marcel Dekker, New York
56. Middelkamp-Hup MA, Bos JD, Rius-Diaz F et al (2007) Treatment of vitiligo vulgaris with narrow-band UVB and oral *Polypodium leucotomos* extract: a randomized double-blind placebo-controlled study. J Eur Acad Dermatol Venereol 21:942–950
57. Natta R, Somsak T, Wisuttida T et al (2003) Narrowband ultraviolet B radiation therapy for recalcitrant vitiligo in Asians. J Am Acad Dermatol 49:473–476
58. Nicolaidou E, Antoniou C, Stratigos A et al (2007) Efficacy, predictors of response, and long-term follow-up in patients with vitiligo treated with narrowband UVB phototherapy. J Am Acad Dermatol 56:274–278
59. Njoo MD, Spuls PI, Bos JD et al (1998) Non surgical repigmentation therapies in vitiligo. Meta-analysis of the literature. Arch Dermatol 134:1532–1540
60. Njoo MD, Bos JD, Westerhof W (2000) Treatment of generalized vitiligo in children with narrow-band (TL-01) UVB radiation therapy. J Am Acad Dermatol 42:245–253
61. Nordlund J (2002) The paradox of hypopigmentation and decreased risk of skin cancer in vitiligo (abstract). Ann Dermatol Venereol 129:1S194
62. Novak Z, Bonis B, Baltas E et al (2002) Xenon chloride ultraviolet B laser is more effective in treating psoriasis and in inducing T cell apoptosis than narrow-band ultraviolet B. J Photochem Photobiol B 67:32–38
63. Ongenae K, Van Geel N, Naeyaert JM (2003) Evidence for an autoimmune pathogenesis of vitiligo. Pigment Cell Res 16:90–100
64. Ortonne JP, Schmitt D, Thivolet J (1980) PUVA induced repigmentation of vitiligo: scanning electron microscopy of hair follicles. J Invest Dermatol 74:40–42
65. Ostovari N, Passeron T, Zakaria W et al (2004) Treatment of vitiligo by 308-nm excimer laser: an evaluation of variables affecting treatment response. Lasers Surg Med 35: 152–156
66. Parks WC (1995) The production, role, and regulation of matrix metalloproteinases in the healing epidermis. Wounds 7:23A–73A
67. Parsad D, Kanwar AJ, Kumar B (2006) Psoralen-ultraviolet A vs. narrow-band ultraviolet B phototherapy for the treatment of vitiligo. J Eur Acad Dermatol Venereol 20:175–177
68. Parrish JA, Jaenicke KF (1981) Action spectrum for phototherapy of psoriasis. J Invest Dermatol 76:359–362
69. Parrish JA (1981) Phototherapy and photochemotherapy of skin diseases. J Invest Dermatol 77:167–171
70. Pasker-de Jong PCM, Wielink G, van der Valk PGM et al (1999) Treatment with UV-B for psoriasis and nonmelanoma skin cancer. A systematic review of the literature. Arch Dermatol 135:834–840
71. Passeron T, Ortonne JP (2006) Use of the 308-nm excimer laser for psoriasis and vitiligo Clin Dermatol 24:33–42
72. Passeron T, Ostovari N, Zakaria W et al (2004) Topical tacrolimus and the 308-nm excimer laser: a synergistic combination for the treatment of vitiligo. Arch Dermatol 140:1065–1069
73. Patel DC, Evans AV, Hawk JLM (2002) Topical pseudocatalse mousse and narrowband UVB phototherapy is not effective for vitiligo: an open, single-centre study. Clin Exp Dermatol 27:641–644
74. Petit-Frere C, Capulas F, Lyon DA et al (2000) Apoptosis and cytokine release induced by ionizing or ultraviolet B radiation in primary and immortalized human keratinocytes. Carcinogenesis 21:1087–1095
75. Schallreuter KU, Wood JM, Lemke KR et al (1995) Treatment of vitiligo with a topical application of pseudocatalase and calcium in combination with short-term UVB exposure: a case study on 33 patients. Dermatology 190: 223–229

76. Schallreuter KU, Tobin DJ, Panske A (2002) Decreased photodamage and low incidence of non-melanoma skin cancer in 136 sun-exposed Caucasian patients with vitiligo. Dermatology 204:194–201
77. Schallreuter KU, Behrens-Williams S, Khaliq TP et al (2003) Increased epidermal functioning wild-type p53 expression in vitiligo. Exp Dermatol 12:268–277
78. Scherschun L, Kim JJ, Lim HW (2001) Narrow-band ultraviolet B is a useful and well-tolerated treatment for vitiligo. J Am Acad Dermatol 44:999–1003
79. Spencer JM, Nossa R, Ajmeri J (2002) Treatment of vitiligo with the 308 nm excimer laser: a pilot study. J Am Acad Dermatol 46:727–731
80. Staricco RG (1962) Activation of amelanotic melanocytes in the outer root sheath of the hair follicle following ultraviolet exposure. J Invest Dermatol 39:163–164
81. Taneja A, Trehan M, Taylor CR (2003) 308 nm Excimer laser for the treatment of localized vitiligo. Int J Dermatol 42:658–662
82. Tijoe M, Gerritsen MJP, Juhlin L et al (2002) Treatment of vitiligo vulgaris with narrow band UVB (311 nm) for one year and the effect of addition of folic acid and vitamin B12. Acta Derm Venereol 82:369–372
83. Tran C, Lubbe J, Sorg O et al (2005) Topical calcineurin inhibitors decrease the production of UVB-induced thymine dimers from hairless mouse epidermis. Dermatology 211:341–347
84. Trehan M, Taylor CR (2002) High dose 308 nm excimer laser for the treatment of psoriasis. J Am Acad Dermatol 46:732–737
85. Weischer M, Blum A, Eberhard F et al (2004) No evidence for increased skin cancer risk in psoriasis patients treated with broadband or narrowband UVB phototherapy: a first retrospective study. Acta Derm Venereol 84:370–374
86. Westerhof W, Nieuweboer-Krobotova L (1997) Treatment of vitiligo with UV-B radiation vs topical psoralen plus UV-A. Arch Dermatol 133:1525–1528
87. Wu CS, Yu CL, Wu CS et al (2004) Narrow-band ultraviolet-B stimulates proliferation and migration of cultured melanocytes. Exp Dermatol 13:755–763
88. Yashar SS, Gielczyk R, Scherschun L et al (2003) Narrow-band ultraviolet B treatment for vitiligo, pruritus, and inflammatory dermatoses. Photodermatol Photoimmunol Photomed 19:164–168
89. Yones SS, Der D, Palmer RA et al (2007) Randomized double-blind trial of treatment of vitiligo. Efficacy of psoralen-UV-A therapy vs narrowband-UV-B therapy. Arch Dermatol 143:578–584
90. Yu HS, Wu CS, Yu CL et al (2003) Helium-neon laser irradiation stimulates migration and proliferation in melanocytes and induces repigmentation in segmental type vitiligo. J Invest Dermatol 120:56–64

Vitamins and Antioxidants: Topical and Systemic

3.4

Mauro Picardo and Maria Lucia Dell'Anna

Contents

3.4.1 Introduction

The therapeutic approach with antioxidants originated from the description of the occurrence of the oxidative stress and possible vitamin deficiency in vitiligo patients, involving melanocytes and other epidermal as well as nonepidermal cells [4, 14], and, besides possible nutritional aspects, of the mostly neurogenic view of vitiligo for vitamin B12 and folic acid (Chap 2.2.3). Increased production of H_2O_2, biopterins and catecholamines, defective expression and/or activity of the antioxidant enzymes catalase and glutathione peroxidise in addition to lipid peroxidation are the metabolic alterations reported by the literature [4, 14] (Chap. 2.2.6). In the last decade, several studies looking at the efficacy of vitamins and antioxidants in vitiligo have been performed. However, published studies are frequenly of poor quality, involving a small number of patients, or are lacking control groups, follow-up, or relevant clinical data.

3.4.2 Vitamin B 12 and folic acid, para-aminobenzoic acid

Vitamin B12 and folic acid have been found to be decreased in sera of vitiligo patients, even in the absence of specific clinical manifestations [7, 9], but the initial mechanism accounting for the reduction has been poorly evaluated. Folic acid and vitamin B12 derivatives interact in the one-carbon cycle, and the folate form N-N-methylene tetrahydrofolate gives the methyl group to homocysteine producing, in a vitamin B12-dependent manner, methionine. This enzymatic reaction determines the level of homocysteine, and the depigmentation may be due to the deficient methionine

M. Picardo (✉)
Istituto Dermatologico San Gallicano, via Elio Chianesi, 00144 Roma, Italy
e-mail: picardo@ifo.it

M. Picardo and A. Taïeb (eds.), *Vitiligo*,
DOI 10.1007/978-3-540-69361-1_3.4,

synthesis and homocysteine build-up. Taking into account that homocystinuria can cause the syndrome described as "pigmentary dilution," characterized by fair skin and hair, [14] and that the alteration of the homocysteine metabolism may be related to the catalase polymorphism, it is interesting to point out that in vitiligo patients, the serum level of homocysteine has been found increased and positively correlated with the activity of the disease [14]. The pteridine in folic acid might refund the pteridine deficiency, which inhibits the melanogenesis by lowering tyrosine; indeed, the pteridine from folic acid may stop the altered recycling of the reduced pterines (6BH4 and 7BH4) proved in vitiligo epidermis (Chap. 2.2.6). Finally, the mixed vitamins may support the UV-induced stimulation of melanocyte stem cells, through a mechanism independent on the serum level of folic acid [7]. An open study indicated that vitamin-induced repigmentation occurs mainly in patients with vitiligo more recent than 10 years, independently on the activity of the disease [7].

The combination of folic acid and vitamin B12 was assayed first with promising results in 1992 in the United States of America in an open study in association with vitamin C [9]. Another trial was conducted in Sweden, involving 100 patients treated with Folic acid (5 mg) and vitamin B12 (1 mg) for three months, and they were advised to expose their skin to the sun. Repigmentation was observed in 52 out of the treated patients [20].

It is known that *para*-aminobenzoic acid (PABA) induces hair and skin darkening, not specifically in vitiligo, but in patients treated for different diseases. The trial conducted in vitiligo patients indicated that 1000 mg/day, together with vitamin C and B12 and folic acid, enhanced the pigmentation in 12 out of 20 enrolled patients (Montes, personal communication). However, this study was limited by the low number of enrolled subjects as well as by the inconsistent outcome measures.

3.4.3 L-Phenylalanine

Four trials evaluated the effectiveness of L-phenylalanine (50–100 mg/kg up to 18 months) even as supporting therapy of UVA or UVB phototherapy, or of other approaches. All studies reported beneficial effects (30–90% of repigmentation in 25–60% of the patients) even if they show high patients dropout or inconsistent outcome measures [2, 13, 19].

3.4.4 Antioxidants

Background

The occurrence of a redox imbalance, both at epidermal and systemic levels, has been described in vitiligo patients (Chap 2.2.6). *In vitro* data suggest that the production of free radicals takes place within different cell types and cannot be due to a simple diffusion from the epidermis. Usually, the redox inbalance is thought to be caused by the hyperproduction of ROS not adequately associated with an increased activity of the corresponding detoxifying systems. Antioxidants form an integrated network in the cells, and the activity of each compound depends on the presence and function of the rest of the antioxidant molecules. The participating enzymes and non enzymatic molecules act as scavengers of the free radicals and are able to reduce the oxidised compounds within the network, allowing the restoration of a correct redox balance. The system glutathione/glutathione peroxidase, catalase, superoxide dismutases, a-lipoic acid, vitamin E, and vitamin C are the main components of this antioxidant network. a-lipoic acid is a lipophilic and hydrophilic compound acting as a fatty acids peroxyl and hydroxyl radical scavenger, lipoxygenase inhibitor and glutathione synthesis promoter. Moreover, a-lipoic acid is involved in recycling vitamins C and E. Vitamin C is a hydrophilic antioxidant; whereas vitamin E is a lipophilic free-radical scavenger and inhibits lipid peroxidation helping to maintain membrane integrity. According to physiological principles, a sound therapeutic approach should provide a balanced pool of antioxidant molecules, with the aim of restoring the correct intracellular network in the right place (Fig. 3.4.1).

Studies *(see also Chap 3.3.2)*

Starting from the occasional uncontrolled experience of an improvement of vitiligo lesions after topical application or systemic intake of antioxidant compounds, several trials have been conducted.

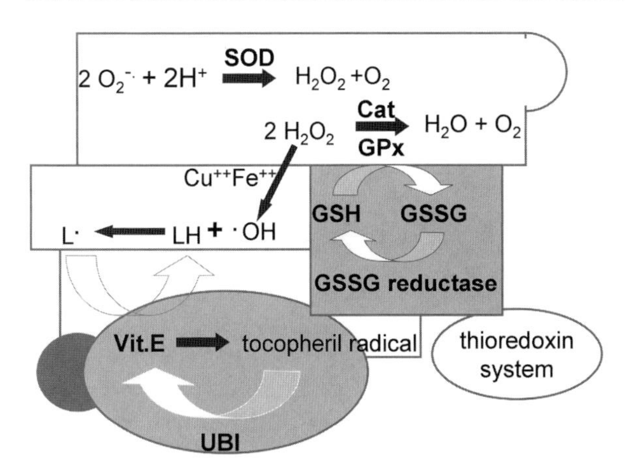

Fig. 3.4.1 A simplified schematic representation of the enzymatic and non-enzymatic antioxidant network.

Pseudocatalase has been developed by the group of Karin Schallreuter and Jim Wood based on *in vitro* and *in vitro* evidence of antioxidant status imbalance in vitiligo skin (Chap. 2.2.6). Antioxidant treatment was initially provided through the application, for 15 months, of a cream containing pseudocatalase and calcium chloride to 33 vitiligo patients during UVB phototherapy [18]. Pseudocatalase is a low molecular weight coordination complex [bis-manganese III-EDTA-$(HCO_3^-)_2$] more able than catalase itself to produce O_2 and H_2O from H_2O_2. The solar or UVB exposure is needed for its activation [18] but apparently not for its effects in therapy, since the amount of UV received was much lower than that used during phototherapy. 90% of the patients obtained excellent repigmentation, especially on the face and dorsum of hands. Pseudocatalase, with or without the Dead Sea bath that contains metals which can activate pseudoacatalase, has been reported to induce repigmentation better than UVB [17]. The combination pseudocatalase plus Dead Sea climatotherapy shortened the time required for the repigmentation onset (10 days versus some weeks) in nearly all the 39 patients. The clinical improvement was associated with a reduction of epidermal H_2O_2 level. The pseudocatalase-mediated H_2O_2 removal has also been associated with the recovery in the epidermis of catalase, tetrahydrobiopterin dehydratase, acetylcholinesterase, and dihydropteridine reductase activities. This last effect is strictly dependent on the elimination of H_2O_2, which causes the oxidation of methionine, cysteine, tryptophan of the enzymes allowing thus their inactivation, as suggested by computer simulation [5]. However, these studies have not confirmed by other authors, and the clinical efficacy of pseudocatalase is still controversial because large scale double-blinded versus placebo studies have not been published, and that another trial using a mousse based formulation was found ineffective [11]. The effectiveness of pseudocatalase seems to be dependent on the formulation of the cream, and the PC-KUS brand is reported to be the most successful. In all cases treated by Dr Schallreuter, the progression of the disease was stopped, without side effects (except a case of contact dermatitis due to the cream preservative) [16–18].

A clinical trial evaluated the effects of vitamin E on the recovery of skin lipid peroxidation induced by PUVA treatment [1]. The study reported 75% of repigmentation in 60% of patients treated with PUVA and vitamin E whereas the same percentage of repigmentation was obtained in 40% of the subjects exposed to PUVA only.

Taking into account the mechanisms underlying the correct synthesis and recycling of the antioxidant network, the administration of a balanced pool of antioxidants and molecules involved in their recycling may be useful. According to that, a pool of a-lipoic acid, vitamin E and C, has been evaluated in a double-blind placebo controlled trial in order to test its role in reducing the UV dosage and in improving repigmentation [3]. The oral intake of the balanced pool improved the effectiveness of NB-UVB phototherapy, increasing the extent of repigmentation and lowering the UV dosage. At the same time the activity of the catalase was partially restored and the intracellular level of ROS decreased (50% of the basal level). Excellent and good= repigmentation was reported in 40% and 20% of antioxidant-treated patients respectively (versus 22% and 10% respectively in patients exposed to NB-UV alone) (Fig 3.4.2). Clinical data have been also confirmed by digital photography performed at the beginning and at two different subsequent time points of the study.

A further antioxidant approach has been performed with *Polypodium leucotomos*. Polypodium leucotomos might represent a consistent support of other therapies on the basis of its antioxidant, photosensitizing and immunomodulatory activities [8, 12]. Two double-blinded placebo controlled trials have been performed to test its effectiveness. The association PUVA plus *Polypodium leucotomos* increases the percentage of patients with more than 50% of repigmentation [12]. A clear trend of repigmentation of neck and head in patients who where given *Polypodium leucotomos* during NB-UVB phototherapy (44% versus 27% out of the

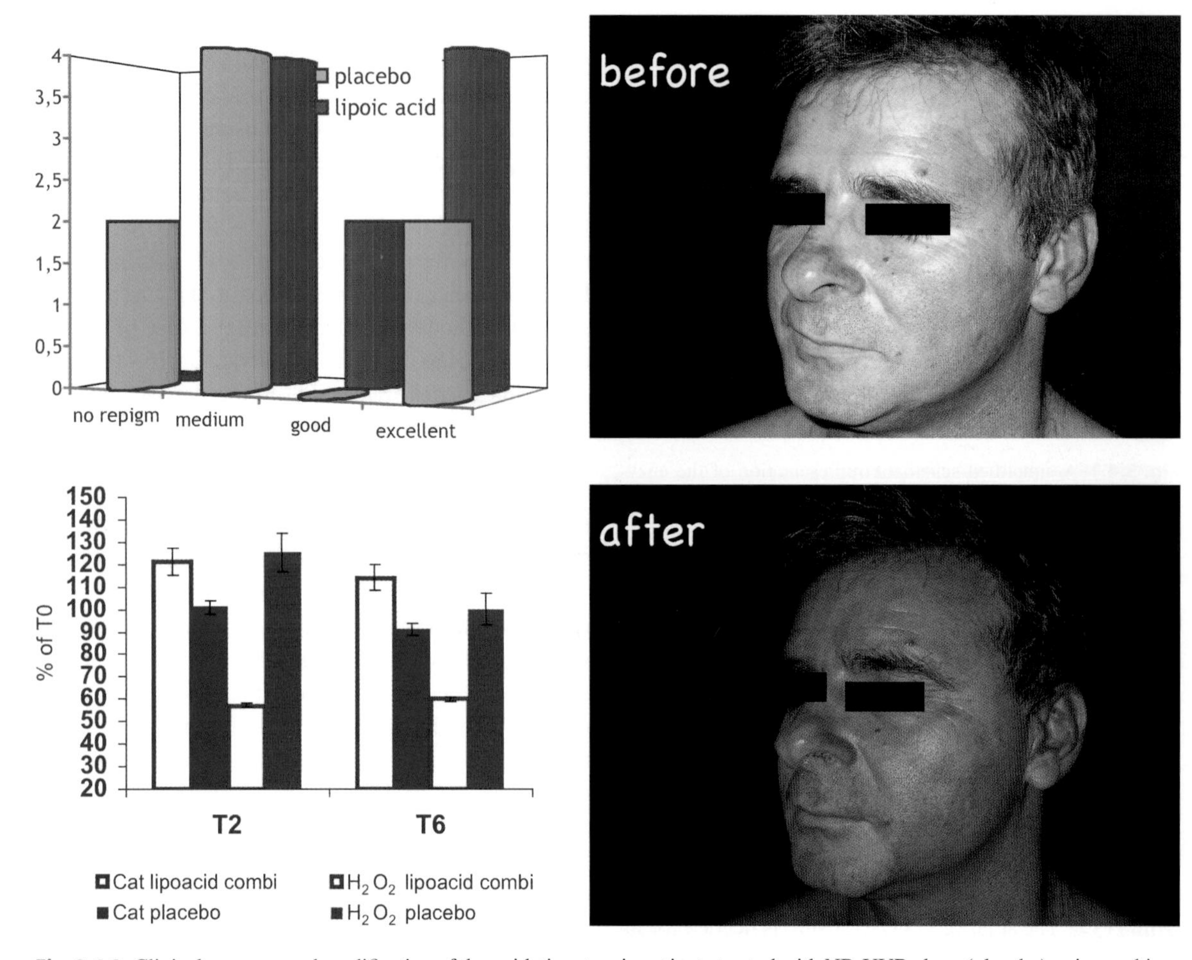

Fig. 3.4.2 Clinical response and modification of the oxidative stres in patients treated with NB-UVB alone (placebo) or in combination with lipoic acid

50 enrolled patients) has been described in the second study [8]. The trial has been also supported by digital photography performed at the beginning and at two different subsequent time points of the study. Finally, some studies have been conducted in order to test the possible therapeutical use of *Gingko biloba*; the extract of *Gingko biloba* contains polyphenol compounds, including terpenoids, flavonoids, and flavonol glycosides. Extracts have anti-inflammatory, immunomodulatory, and antioxidant activities. The terpenoids gingkolides B, C, J and M, as well as bilobalide appear to be the main contributors to the antioxidant and free radical-scavenger activities. Early *in vitro* data suggested the possible therapeutical role of *Gingko biloba* in vitiligo [10]. The trial was carried out for 12 months with promising results (90% patients stopped the progression and 70% of patients showed good or complete repigmentation). Further evidence in favor of the effectiveness of *G biloba* came from Chinese traditional medicines and from supporting results provided by an *in vitro* study. The main problems with the studies were the limited number of enrolled patients and the inconsistent outcome parameters. No side effects have been reported until now.

In summary, randomized controlled trials on systemic antioxidant supplements provide a moderate evidence of effectiveness, and therefore, no firm conclusion can be reached at this point (Table 3.4.1).

Safety concerns

The oral intake of antioxidants do not interfere with other drugs or other diseases (such as thyroiditis, diabetes possibly occuring in vitiligo patients). However, long-term follow-up studies have not been

Table 3.4.1 Studies with systemic antioxidants supplementation

Antioxidant Supplementation mg /day	Study	Enrolled patients	Duration months	effectiveness	Reference
Vitamin E	Controlled	30	6	Good improvement in 60% patients in the active group	[1]
Lipoic acid 50mg vitC 50 mg, cysteine monohydrate 50 mg, vitE 20 mg/PUFA12% plus NB-UVB vs NB-UVB Control: 18% excellent 18% good	RCT	35	8	Active 47% excellent 23% good	[3]
Polypodium Leucotomos 750 mg plus NB-UVB vs NB-UVB	RCT	50	4-5	Active 44% repigmentatioin head or neck lesions 30% extrmities Control: 27% haed and neck 26% extremities lesions	[8]
Polypodium Leucotomos 720 mg plus PUVA vs PUVA	RCT	19	3	Active 50% moderat 30% mild Control:40% mild	[12]
Ginko. Biloba 120 mg Vs placebo	RCT	52	6	Active: 20 stop progression 10 marked repigmentation Control: 8 stop progression 2 marked repigmentation	[10]

yet published. Even if the oral intake of antioxidants is usually not associated with the occurrence of side effects, recent data have raised some concerns for cancer prevention, and no definitive recommendations can be made at this point.

Summary Messages

- Oxidative stress takes part in vitiligo onset and progression.
- Decreased serum levels in vitamin B12 and folic acid may occur in vitiligo patients, which may interfere with melanogenesis
- Topical application of Pseudo catalase has been associated with the decrease of H2O2 level in the skin
- Based on the rationale that a balanced pool of antioxidants may help to recover a correct cellular redox status both at epidermal and systemic level, antioxidant supplementation combined or not with phototherapy has been advocated.
- However, most published studies show a moderate impact, with an adjuvant effect of phototherapies, and confirmative high quality controlled double-blinded trials are needed.
- The safety of prolonged administration of systemic antioxidants requires further evaluation.

References

1. Akyol M, Celik VK, Ozcelik S et al (2002) The effects of vitamin E on the skin lipid peroxidation and the clinical improvement in vitiligo patients treated with PUVA. Eur J Dermatol 12:24–26
2. Cormane RH et al (1985) Phenylalanine and UVA light for the treatment of vitiligo. Arch Dermatol Res 277:126–130
3. Dell'Anna ML, Mastrofrancesco A, Sala R et al (2007) Anrtioxidants and narrow band-UVB in the treatment of vitiligo: a double-blind placebo controlled trial. Clin Exp Dermatol 32:631–636
4. Dell'Anna ML, Picardo M (2006) A review and a new hypothesis for non-immunological pathogenetic mechanisms in vitiligo. Pigment Cell Res 19:406-411
5. Gibbons NCJ, Wood JM, Rokos H et al (2006) Computer simulation of native epidermal enzyme structure in the presence and absence of hydrogen peroxide (H_2O_2): potential and pitfalls. J Invest Dermatol 126:2576–2582
6. Hercberg S, Ezzedine K, Guinot C et al (2007) Antioxidant supplementation increases the risk of skin cancers in women but not in men. J Nutr 137:2098–2105
7. Juhlin L, Olsson MJ (1997) Improvement of vitiligo after oral treatment with vitamin B12 and folic acid and the importance of sun exposure. Acta Derm Venereol 77: 460–462
8. Middelkamp MA, Bos JD, Riuz-Diaz F et al (2007) Treatment of vitiligo vulgaris with narrow-band UVB and oral *Polypodium leucotomos* extract: a randomized double-blind placebo-controlled study. JEADV 21:942–950
9. Montes LF, Diaz ML, Lajous J et al (1992) Folic acid and vitamin B12 in vitiligo: a nutritional approach. Cutis 50:39–42
10. Parsad D, Pandhi R, Juneja A (2002) Effectiveness of oral Gingko biloba in treating limited, slowly spreading vitiligo. Clin Exp Dermatol 28:285–287
11. Patel DC, Evans AV, Hawk JL (2002) Topical pseudocatalase mousse and narrowband UVB phototherapy is not effective for vitiligo: an open single-center study. Clin Exp Dermatol 27:641–644

12. Reyes E, Jaen P, de las Heras E et al (2006) Systemic immunomodulatory effects of Polypodium leucotomos as an adjuvant to PUVA therapy in generalized vitiligo: a pilot study. J Dermatol Sci 41:213–216
13. Rojas-Urdaneta JE, Poleo-Romero AG (2007) Evaluation of an antioxidant and mitochondria-stimulating cream formula on the skin of patients with stable common vitiligo. Invest Clin 48:21–31
14. Shaker OG, El-Tahlawi SMR (2008) Is there a relationship between homocysteine and vitiligo? Br J Dermatol 159: 720–724
15. Schallreuter KU, Bahadoran P, Picardo M et al (2008) Vitiligo pathogenesis: autoimmune disease, genetic defect, excessive reactive oxygen species, calcuim imbalance, or what else. Exp Dermatol 17: 139-60
16. Schallreuter KU, Kruger C, Wurfel C et al (2008) From basic research to bedside: efficacy of topical treatment with pseudo catalase PC-KUS in 71 children with vitiligo. Int J Dermatol. 188:215-218
17. Schallreuter KU, Moore J, Behrens-Williams S, et al (2002). Rapid initiation of repigmentation in vitiligo with dead sea climatotherapy in combination with pseudocatalase (PC-KUS). Int J Dermatol 41: 482-487.
18. Schallreuter KU, Wood JM, Lemke KR et al (1995) Treatment of vitiligo with a topical application of pseudocatalase and calcium in combination with short-term UVB exposure: a case study on 33 patients. Dermatol 190: 223–229
19. Siddiqui AH, Stolk LML, Bhaggoe R et al (1994) L-phenylalanine and UVA irradiation in the treatment of vitiligo. Dermatol 188:215–218
20. Tjioe M et al (2002) Treatment of vitiligo vulgaris with narrow band UVB (311 nm) for one year and the effect of addition of folic acid and vitamin B12. Acta Derm Venerol 82: 369–372

3.5 Systemic Treatments

Corticosteroid Minipulses

3.5.1

Davinder Parsad and Dipankar De

Contents

D. Parsad (✉)
Department of Dermatology, Postgraduate Institute of Medical Education & Research, Chandigarh, India
e-mail: parsad@mac.com

3.5.1.1 Definition and Historical Background

Pulse therapy refers to the administration of large (supra-pharmacologic) doses of drugs in an intermittent manner to enhance the therapeutic effect and reduce the side effects of a particular drug [9]. The credit of the first use of corticosteroids in pulse form goes to Kountz and Cohn [4], who used them to prevent renal graft rejection. Subsequently, pulse corticosteroids have been used in various dermatological as well as nondermatological indications. Oral minipulse (OMP), that is, intermittent administration of betamethasone/ dexamethasone has been pioneered in India by Pasricha et al. [6] and was first published in 1989. They first used it in vitiligo. Subsequently, OMP has been used successfully in various other dermatoses such as extensive alopecia areata [8], cicatricial alopecia, extensive/bullous lichen planus [2], trachyonychia [5], infantile periocular haemangioma [11], etc. As the indications of use of OMP suggest, it is understandable that in dermatoses for which steroid therapy is effective, OMP may be used, for maintaining efficacy and cutting down on side effects.

Though controversy regarding the pathogenesis of vitiligo is far from being over, autoimmunity definitely plays a role, which has suggested to try various immunosuppressants, including topical or systemic corticosteroids. Though topical steroids are used extensively in the management of vitiligo, studies on systemic steroids have been sparingly reported in literature. The cause of this discrepancy is not known. Systemic steroids can arrest the activity of the disease, if they are used in sufficient doses for a sufficient period of time [1, 6]. They are in general not effective in repigmenting stable vitiligo. Moreover,

M. Picardo and A. Taïeb (eds.), *Vitiligo*,
DOI 10.1007/978-3-540-69361-1_3.5.1, © Springer-Verlag Berlin Heidelberg 2010

side effects associated with long-term use of daily systemic corticosteroids may act as deterrent against their common use.

3.5.1.2 Oral Corticosteroids Minipulses for Vitiligo

In the first reported study on OMP in vitiligo by Pasricha et al. [6], betamethasone/ dexamethasone was given as a single oral dose of 5 mg on 2 consecutive days per week. This dose of steroids did not have scientific reasoning and was decided upon arbitrarily. Progression of the disease was arrested in 91% of the patients. A degree of repigmentation was observed in a proportion of patients, and the side effects were either not significant or altogether absent.

In a subsequent trial of 40 patients, 36 with progressive and 4 with static disease, the same OMP regimen was used [7]. In children, the dose was proportionately reduced. In adults who did not respond to the standard dose of corticosteroids, the dose was increased to 7.5 mg/day and then reduced to 5 mg/day when disease progression was arrested. Within 1–3 months of starting treatment, 89% of the patients with progressive disease stabilized, while within 2–4 months, repigmentation was observed in 80% of total patient cohort. The area of repigmentation continued to progress as treatment continued, though none of the patients achieved complete repigmentation. Seventeen of 40 (42%) had at least one side effect, though they were not significant. The side effects profile included weight gain, dysgeusia, headache, transient mild weakness, acne, mild puffiness of the face alone, perioral dermatitis, herpes zoster, glaucoma, and amenorrhea. The authors presumed that the repigmentation observed was spontaneous and thus varied from patient to patient and lesion to lesion [7]. Subsequently Kanwar et al. [3] assessed the efficacy of OMP in vitiligo. They used dexamethasone in a dose of 5 mg/day on two consecutive days per week and the dose was halved in children 16 years of age or younger. Of thirty-seven patients included with actively spreading disease, 32 were evaluable at the end of the study. About 43.8% had mild to moderate repigmentation without appearance of new lesions. The pigmentation appeared in majority of patients within 15 weeks of starting treatment. No side effects were reported [3]. In the latest study by Radakovic-Fijan et al. [10], 29 patients, 25 with progressive disease and 4 with stable disease were included. The daily dose of dexamethasone was significantly increased to 10 mg/day and the treatment was continued for a maximum period of 24 weeks. In addition, plasma cortisol and corticotrophin levels were measured to detect hypothalamo–pituitary–adrenal axis suppression before and up to 6 days after the dexamethasone pulse in the first and fourth weeks of treatment in 14 patients. Disease activity was arrested in 88% of patients with progressive disease after an average treatment period of 18.2 weeks. Marked repigmentation was observed in 6.9% and moderate or slight repigmentation in 10.3%, while 72.4% had no response in repigmentation. A tendency toward better treatment results was observed with an increasing number of pulses. Side effects were observed in 69% patients, which included weight gain, insomnia, agitation, acne, menstrual disturbances, and hypertrichosis. Plasma cortisol and corticotrophin levels, though markedly decreased after one pulse, returned to normal before starting the next pulse. The authors observed that ethnic background may have an impact on therapeutic response [10].

3.5.1.3 Personal Remarks

The OMP with either betamethasone or dexamethasone can arrest progression of spreading disease. However, it is not usually suitable alone for repigmentation of vitiligo lesions. In patients with fast spreading vitiligo, disease progression is usually commenced after this intervention. There are no RCT confirming that either speed or magnitude of response to phototherapy and photochemotherapy in patients with generalized fast spreading vitiligo might be potentiated by concomitant administration of oral corticosteroid pulses. As noted earlier, the dosage of dexamethasone used in OMP has been arbitrarily chosen. In majority of the studies, a dose of 5 mg every day for two consecutive days per week has been used. For those who do not respond, 7.5 mg/day may be used, and then reduced to 5 mg/day if disease progression is arrested. In our institution, we use 2.5 mg in vitiligo patients (unpublished data) and overall this dose is sufficient in the majority of cases. The drug can be preferably given on weekends to increase the compliance as short-term side effects may be troublesome in some. If the 5 mg/day dose is used, the drug can be gradually tapered off over six months, after a desired response is achieved. However, when a lower dose is used (2.5 mg/day), the treatment can be stopped abruptly without any noticeable manifestation of hypothalamo–pituitary–adrenal axis suppression.

Summary Messages

- Oral minipulse (OMP), i.e. intermittent administration of betamethasone/dexamethasone, for the treatment of vitiligo, has been pioneered in India by Pasricha in 1989.
- In a first trial the progression of the disease was arrested in 91% of the patients without significant side effects.
- A tendency towards better treatment results was observed with an increasing number of pulses and of dexamethasone amount (from 5 to 10 mg/day).
- OMP is considered as not useful to repigment stable vitiligo
- Week end OMP starting with low doses (2.5 mg/ day) of dexamethasone is suggested based on the author's experience for fast spreading vitiligo, before starting photoherapy
- The benefit of adding OMP to phototherapy at onset of treatment in progressive vitiligo needs further assessment
- Optimal duration of OMP therapy to stop vitiligo progression is situated between 3 and 6 months

References

1. Farah FS, Kurban AK, Chaglassian HT (1967) The treatment of vitiligo with psoralens and triamcinolone by mouth. Br J Dermatol 79:89–91
2. Joshi A, Khaitan BK, Verma KK, Singh MK (1999) Generalized and bullous lichen planus treated successfully with oral mini-pulse therapy. Indian J Dermatol Venereol Leprol 65:303–304
3. Kanwar AJ, Dhar S, Dawn G (1995) Oral minipulse therapy in vitiligo. Dermatology 190:251–252
4. Kountz SL, Cohn R (1969) Initial treatment of renal allografts with large intrarenal doses of immunosuppressive drugs. Lancet 1:338–340
5. Mittal R, Khaitan BK, Sirka CS (2001) Trachyonychia treated with oral minipulse therapy. Indian J Dermatol Venereol Leprol 67:202–203
6. Pasricha JS, Seetharam KA, Dashore A (1989) Evaluation of five different regimes for the treatment of vitiligo. Indian J Dermatol Venereol Leprol 55:18–21
7. Pasricha JS, Khaitan BK (1993) Oral mini-pulse therapy with betamethasone in vitiligo patients having extensive or fast-spreading disease. Int J Dermatol 32:753–757
8. Pasricha JS, Kumrah L (1996) Alopecia totalis treated with oral mini-pulse (OMP) therapy with betamethasone. Indian J Dermatol Venereol Leprol 62:106–109
9. Pasricha JS (2003) Pulse therapy as a cure for autoimmune diseases. Indian J Dermatol Venereol Leprol 69:323–328
10. Radakovic-Fijan S, Firnsinn-Friedl AM, Honigsmann H et al (2001) Oral dexamethasone pulse treatment for vitiligo. J Am Acad Dermatol 44:814–817
11. Verma K, Verma KK (2001) Infantile periocular haemangioma treated with two days in a week betamethasone oral mini pulse therapy. Indian J Pediatr 68:355–356

Other Immunosuppressive Regimen 3.5.2

Markus Böhm

Contens

M. Böhm
Department of Dermatology, University of Münster, Münster, Germany
e-mail: bohmm@uni-muenster.de

3.5.2.1 Rationale for the Use of Systemic Immunomodulators

As described in Chap. 2.2.7, there is a large body of data indicating that vitiligo, at least distinct clinical subtypes of it, represents an immune-mediated inflammatory disease. This concept is further corroborated by those patients who experience beneficial effects from anti-inflammatory and/or immunomodulatory treatment, for example, after therapy with corticosteroids or topical calcineurin inhibitors. In this view, different systemic imunomodulators have been proposed for the treatment of the diffuse form of the disease. Recently, the widely used a new class of immunomodulators in dermatology defined as *biologics* have been tested in vitiligo in a limited number of patients to assess their therapeutic potential in this disease.

3.5.2.2 Traditional Systemic Immunosuppressants

Low dose of cyclophosphamide has been proposed in early 1980s. Cyclophosphamide inhibits the antibodies production by B lymphocytes [6]; the rational for its use in vitiligo derives from the experience in other autoimmune skin diseases, such as pemphigus, and it is supported by the identification of circulating melanocyte-specific antibodies in the serum of affected patients. Good response was seen at the dosage of 50 mg twice a day in about 27% of the cases, while no response was observed in 33% of the treated patients. Hematological toxicity, loss of hearing, and nausea have been reported as common side effects during the treatment, limiting its use [8, 9].

M. Picardo and A. Taïeb (eds.), *Vitiligo*,
DOI 10.1007/978-3-540-69361-1_3.5.2,

Experience with low-dose azathioprine (at the maximal dosage of 50 mg/day) in association with PUVA has been also proposed. Azathioprine is able to inhibit the cellular immune response, and it is currently used in the treatment of rheumatological and bowel diseases, and of autoimmune dermatoses, including pemphigus, at the dosage of 50–150 mg/day. A study performed on 60 patients randomized to receive either azathioprine (0.6–0.75 mg/kg/day) in association with PUVA or PUVA-therapy alone, has demonstrated a potential synergic effect of the two approaches, but true positive results have been demonstrated only in a minority of the subjects, with the limitation of lack of validated, standardized measures for vitiligo assessment [14]. Starting from the pathogenetic concept of a systemic over-activation of cellular immunity in vitiligo, systemic cyclosporine has been anecdotally used in patients with diffuse disease, at the dosage of 5 mg/kg/daily with no univocal response. Furthermore, existing ethical questions regarding the real safety of this therapy, and considering the kidney and liver toxicity of cyclosporine, further experience in this field was limited [11].

3.5.2.3 Anti-IFN-γ Strategy

The pioneering concepts and preclinical observations of Skurkovich et al. will be outlined [20, 21]. These scientists were among the first who proposed to remove not only certain types of IFNs, but also TNF-α to treat various autoimmune diseases.

Based on their longstanding research on the pathogenetic role of proinflammatory cytokines in immune-mediated inflammatory diseases, Skurkovich et al. initially proposed that IFN-γ should be removed in autoimmune diseases such as rheumatoid arthritis, multiple sclerosis, Type I diabetes, psoriasis vulgaris, alopecia areata, or vitiligo [23]. In this context, it is worth mentioning that IFN-γ mRNA levels are in fact increased in lesional and in adjacent uninvolved skin of patients with vitiligo [10]. It is also well-known that systemic administration of IFN-α can induce or aggravate vitiligo [3, 4, 18], although recently improvement of preexisting vitiligo under pegylated IFN-α-2A was reported in a patient with hepatitis C [26]. In a small preclinical case series, Skurkovich et al. injected polyclonal IFN-γ antibodies (both IgG and/or F(ab')2 antibody fragments) into patients with various Th-1-mediated autoimmune diseases [23]. The protein concentration of these antibodies was about 33 mg/mL with an IFN-γ neutralizing capacity of more 66 μg/mL [19]. Four patients with vitiligo (12–14 years old) received intradermal injections of F(ab')2 fragments (titer: 24 × 10^3 IU/mL) generated from goat antibodies to human IFN-γ [22]. Aliquots of 0.1 mL of the antibody were given perilesionally for 10 days. All treated patients experienced sustained erythema after three days of therapy followed by development of small, slightly infiltrated pinkish papules in the depigmented areas. On day 10, the authors observed a loss of the well-defined borders between the normal and depigmented skin in all treated patients [22]. In 2 of the intralesionally treated patients an additional course of anti-IFN-γ with intramuscular injections of the IFN-γ antibody was performed. A gradual diminishment of the border between the depigmented area and normal skin was seen according to the authors [19, 23]. In addition, the authors reported on the clearance of vitiligo in another patient with autoimmune polyglandular syndrome (APS)-4, that is, vitiligo and alopecia areata [23]. It is unclear for how long systemic anti-IFN-γ therapy was given in the latter individuals. Moreover, the aforementioned preclinical study lacks detailed description of patient characteristics, as well as any follow-up data. To the best of our knowledge, we are unaware of any further preclinical studies or clinical trials employing the anti-IFN-γ strategy in vitiligo. In the following section the so-far reported clinical experience of biologics approved in dermatology within

Table 3.5.1 Immunomodulating biologics approved for the treatment of skin diseases within the EU

Name	Target	Biochemical features	Indications
Etanercept (Enbrel)	TNF-α	TNF type II soluble receptor fusion protein	PSO, PA
Infliximab (Remicade)	TNF-α	Anti-human TNF-α chimeric (mouse-human) monoclonal antibody	PSO, PA
Adalimumab (Humira)	TNF-α	Humanized anti-human TNF-α monoclonal antibody	PSO, PA
Efalizumab (Raptiva)	LFA-1	Humanized LFA-1 monoclonal antibody	PSO

PSO psoriasis; *PA* psoriasis artritis

Europe for the treatment of Psoriasis (Table 3.5.1) will be reviewed. They include anti-TNF-a targeted therapies (etanercept, infi iximab, adalimumab) and efalizumab, the latter being an antibody that binds to the CD11 a sub-unit of LFA-1.

3.5.2.4 Targeting TNF-α by Antibodies

Since TNF-α protein expression and immunoreactivity is elevated in lesional skin of patients with vitiligo [5, 13], it is not surprising that biologics of the anti-TNF-α family just recently found their way into clinical pilot studies on patients with preexisting vitiligo. Notably, anti-TNF-α therapy, in accordance with its significant potential to trigger other autoimmune phenomena (such as alopecia and lupus erythematosus-like syndromes) can induce de novo vitiligo. Accordingly, a 61-year-old Caucasian suffering from rheumatoid arthritis was reported to develop vitiligo lesions on the dorsa of the hands 6 months after intravenous therapy with infliximab at 3 mg/kg [15]. Infliximab was not discontinued and the patient's vitiligo was subsequently treated with *Polypodium leucotomus* extracts and topical pseudocatalase to regain 50% of his pigmentation. In another recent case report, a 66-year-old white man receiving adalimumab for the treatment of his psoriasis is described [24]. Within 4 months of 40 mg adalimumab administered subcutaneously every other week, his psoriasis had cleared. However, depigmented skin was noticed in those areas previously affected by psoriasis, but not in unaffected areas. A skin biopsy specimen from the depigmented skin confirmed the presence of vitiligo. As discussed by the authors, concurrence of psoriasis and vitiligo appears to be a rare event (Chap. 1.3.8). In the present case, perhaps Koebnerization of vitiligo due to preexisting psoriasis rather than induction of vitiligo by the anti-TNF-α treatment had occurred since the vitiligo was not widespread [24].

Regarding the therapeutic potency of TNF-α antibodies in patients with preexisting vitiligo, the data are limited and controversial. Rigopoulos et al. [16] assessed the therapeutic potential of etanercept in a small open-label pilot study consisting of four male patients with vitiligo vulgaris (mean age: 29.3; mean duration of vitiligo: 7.5 months). All patients had progressive disease with development of new lesions within the previous three months. Most of the patients had an involvement of extremities including hands and feet. All the patients had not received any treatment for vitiligo for the last two months. The treatment protocol consisted of weekly injections of etanercept (50 mg subcutaneously) for 12 weeks followed by 25 mg of etanercept weekly for another 4 weeks. Treatment success was evaluated photographically and photometrically. Although the overall tolerability was good, none of the patients had any repigmentation. However, no aggravation of their vitiligo was noticed. The authors suggested that monotherapy with TNF-α antibodies should not be considered as a treatment option for vitiligo. In another case report, vitiligo improvement was observed in a patient with ankylosing spondylitis under infliximab treatment [17]. A 24-year-old male with ankylosing spondylitis since the age of 18 and generalized vitiligo for 11 years received 350 mg infliximab intravenously in weeks 0, 2 and 6, and then every other week for 10 months. Upon this treatment, the patient's disease activity score and functional indices as well as his vitiligo improved: six months after infliximab, spreading of two vitiligo spots at both axilla and pretibial areas was halted. In addition, several other vitiligo spots (located on the trunk, finger joints, face, and pretibial areas) revealed partial repigmentation or even disappeared. In accordance with the established role of TNF-α as an inhibitor of melanocyte proliferation [25] and melanogenesis [12], these data would suggest some potential in the treatment of vitiligo (Chap. 2.2.8).

3.5.2.5 Effects of Efalizumab

Two anecdotal reports describe a beneficial effect of efalizumab as a T-cell targeted recombinant antibody binding to the CD11a subunit of LFA-1 [7, 27]. As a consequence of such an approach, the influx of blood CD11a bearing leukocytes in limited, and the interaction between LFA-1 and intercellular adhesion molecule (ICAM)-1 expressed by numerous activated resident skin cells is blocked. Previously, expression of ICAM-1 was detected in perilesional melanocytes around active vitiligo patches [2]. Regarding expression of LFA-1 in vitiligo patients, it was shown that LFA-1 immunoreactivity in leukocytes is higher in minigrafts of nonresponders than in responders [1].

In the first report, a 52-year-old male Surinamese Hindustani man with plaque psoriasis and universal vitiligo received efalizumab [27]. He had been suffering from vitiligo since more than 10 years. His previous anti-psoriatic therapies had included ultraviolet

(UV) light and psoralen, methotrexate, cyclosporine and fumaric acid. Six weeks after starting subcutaneous efalizumab at 1 mg/kg once per week, the patient developed spotty repigmentation in vitiliginous areas of his face. Facial repigmentation continued in a perifollicular pattern under efalizumab treatment for further six weeks, when treatment had to be terminated due to deterioration of his psoriasis. Subsequently, his vitiligo on the face became worse. In the second casuistic report, Fernandez-Obregon [7] reported on a 43-year-old male Hispanic with a history for vitiligo vulgaris for more than 10 years and plaque psoriasis for 5 years. Before starting efalizumab he had received topical high-potency steroids, anthralin, and UVB phototherapy. However, UVB therapy was not tolerated and his vitiligo deteriorated. Upon systemic acitretin he developed pruritus and therapy had to be stopped. Efalizumab was started at 0.7 mg/kg and thereafter continued at 1 mg/kg/week. At the time when the patient's psoriasis improved (6–8 weeks after beginning with efalizumab) his vitiligo also showed some improvement, especially on his trunk and lower extremities. Later, the patient received systemic methyprednisolone followed by methotrexate upon which his vitiligo continued to improve.

These two reports indicate some therapeutic effect of efalizumab in even in long-standing vitiligo vulgaris. However, the overall number of treated patients is too low and clinical trials are needed to make definitive draw any clear-cut conclusion on the potential of efalizumab in vitiligo.

3.5.2.6 Concluding Remarks

Studies with classic immunosuppressants, such as methothrexate and azathioprine, are limited and it is probably still worth designing new trials in patients with proven immune/inflammatory vitiligo, based on more stringent inclusion criteria and using better assessment techniques.

Neutralization of IFN-γ, TNF-α and LFA-1 by antibodies (biologics) represents a novel systemic immunomodulatory approach in the treatment of vitiligo, even if limited data regarding its efficacy are available. Unfortunately, antibodies against IFN-γ are not yet available for the clinician in daily routine, whereas off-label use of TNF-α antibodies in vitiligo is possible. However, TNF-α antibodies should be considered with great caution. First, they may trigger de novo development of vitiligo. The data on the use of anti-TNF-α antibodies in preexisting vitiligo are controversial regarding the efficacy of these agents with no or only a limited repigmentary response at best. Our current knowledge on the use of efalizumab as a prototype of LFA-1 neutralization is likewise limited and cannot be generally advised.

Local administration of biologics could become an alternative option in selected vitiligo patients in future. Therapeutic future approaches in vitiligo should also include additional anti-cytokine therapies (e. g., targeting the inflammasomal pathway).

Summary Messages

- The imbalance between protective mediators/growth factors versus proinflammatory cytokines/pro-oxidant mediators is considered a rationale based on which systemic immunomodulators including biologics have been tested in a limited number of anecdotal reports and small pilot studies.
- The so-far tested systemic immunomodulators in vitiligo include antibodies against interferon (IFN)-γ and tumor necrosis factor-α (TNF-α) as well as efalizumab, a monoclonal antibody against the CD11a subunit of the lymphocyte function-associated antigen (LFA)-1.
- The anti IFN gamma strategy remains to be tested on a larger scale
- The development of antipsoriatic biologics (such as anti TNF agents and efalizumab) in the field of vitiligo is not warranted based on current knowledge of disease immunology, drug tolerance and limited off label use.
- There is room for more trials of classic immunosuppressants in inflammatory proven vitiligo

References

1. Abdallah M, Abdel-Naser MB, Moussa MH et al (2003) Sequential immunohistochemical study of depigmenting and repigmenting minigrafts in vitiligo. Eur J Dermatol 13:548–552

2. al Badri AM, Foulis AK, Todd PM et al (1993) Abnormal expression of MHC class II and ICAM-1 by melanocytes in vitiligo. J Pathol 169:203–206
3. Anbar TS, Abdel-Rahman AT, Ahmad HM (2007) Vitiligo occurring at site of interferon-alpha 2b injection in a patient with chronic viral hepatitis C: a case report. Clin Exp Dermatol 33:503
4. Bernstein D, Reddy KR, Jeffers L, Schiff E (1995) Canities and vitiligo complicating interferon therapy for hepatitis C. Am J Gastroenterol 90:1176–1177
5. Birol A, Kisa U, Kurtipek GS et al (2006) Increased tumor necrosis factor alpha (TNF-alpha) and interleukin 1 alpha (IL1-alpha) levels in the lesional skin of patients with non-segmental vitiligo. Int J Dermatol 45:992–993
6. Cara CJ, Pena AS, Sans M et al (2004) Reviewing the mechanism of action of thiopurine drugs: towards a new paradig in clinical practice. Med Sci Monit 10:247–254
7. Fernandez-Obregon AC (2008) Clinical management with efalizumab of a patient with psoriasis and comorbid vitiligo. J Drugs Dermatol 7:679–681
8. Gokhale BB (1979) Cyclophosphamide and vitiligo. Int J Dermatol 18:92
9. Gokhale BB, Parakh AP (1983) Cyclophosphamide in vitiligo. Indian J Dermatol 28:7–10
10. Grimes PE, Morris R, Avaniss-Aghajani E et al (2004) Topical tacrolimus therapy for vitiligo: therapeutic responses and skin messenger RNA expression of proinflammatory cytokines. J Am Acad Dermatol 51:52–61
11. Mahmoud BH, Hexsel CL, Hamzavi IH (2008) An update on new and emerging options for the treatment of vitiligo. Skin Therapy Lett 13:1–6
12. Martínez-Esparza M, Jiménez-Cervantes C, Solano F et al (1998) Mechanisms of melanogenesis inhibition by tumor necrosis factor-alpha in B16/F10 mouse melanoma cells. Eur J Biochem 255:139–146
13. Moretti S, Spallanzani A, Amato L et al (2002) New insights into the pathogenesis of vitiligo: imbalance of epidermal cytokines at sites of lesions. Pigment Cell Res 15: 87–92
14. Radmanesh M, Saedi K (2006) The efficacy of combined PUVA and low-dose azathioprine for early and enhanced repigmentation in vitiligo patients. J Dermatolog Treat 17: 151–153
15. Ramírez-Hernández M, Marras C, Martínez-Escribano JA (2005) Infliximab-induced vitiligo. Dermatology 210: 79–80
16. Rigopoulos D, Gregoriou S, Larios G et al (2007) Etanercept in the treatment of vitiligo. Dermatology 215:84–85
17. Simon J-A, Burgos-Vargas R (2008) Vitiligo improvement in a patient with ankylosing spondylitis treated with infliximas. Dermatol 216: 234–235
18. Simsek H, Savas C, Akkiz H, Telatar H (1996) Interferon-induced vitiligo in a patient with chronic viral hepatitis C infection. Dermatology 193:65–66
19. Skurkovich B, Skurkovich S (2003) Anti-interferon-gamma antibodies in the treatment of autoimmune diseases. Curr Opin Mol Ther 5:52–57
20. Skurkovich SV, Klinova EG, Eremkina EI, Levina NV (1974) Immunosuppressive effect of an anti-interferon serum. Nature 247:551–552
21. Skurkovich S, Skurkovich B, Bellanti JA (1987) A unifying model of the immunoregulatory role of the interferon system: can interferon produce disease in humans? Clin Immunol Immunopathol 43:362–373
22. Skurkovich S, Korotky NG, Shaova NM, Skurkovich B (2002) Successful anti-IFN therapy of alopecia, vitiligo, and psoriasis. Clin Immunol 103:S103
23. Skurkovich S, Skurkuvich B (2006) Inhibition of IFN-γ as a method of treatment of various autoimmune diseases, including skin disease. In: Numerof R, Dinarello CA, Asadullah K (eds) Cytokines as potential therapeutic targets for inflammatory skin diseases. Ernst Schering Research Foundation Workshop 58, Springer, Berlin, pp 1–27
24. Smith DI, Heffernan MP (2008) Vitiligo after the resolution of psoriatic plaques during treatment with adalimumab. J Am Acad Dermatol 58:S50–S51
25. Swope VB, Abdel-Malek Z, Kassem LM, Nordlund JJ (1991) Interleukins 1 alpha and 6 and tumor necrosis factor-alpha are paracrine inhibitors of human melanocyte proliferation and melanogenesis. J Invest Dermatol 96:180–185
26. Taffaro M, Pyrsopoulos N, Cedron H et al (2007) Vitiligo improvement in a hepatitis C patient after treatment with PEG-interferon alpha-2a and ribavirin: a case report. Dig Dis Sci 52:3435–3437
27. Wakkee M, Assen YJ, Thio HB, Neumann HA (2008) Repigmentation of vitiligo during efalizumab. J Am Acad Dermatol 59:S57–S58

Empirical, Traditional, and Alternative Treatments

3.6

Mauro Picardo and Alain Taïeb

Contents

M. Picardo (✉)
Istituto Dermatologico San Gallicano, via Elio Chianesi
00144 Roma, Italy
e-mail: picardo@ifo.it

3.6.1 Introduction

Trials have tested the use of alternative natural health products for the treatment of the vitiligo. Some prospective controlled, double-blind, and randomized studies have been performed. Traditional Chinese products and plant-derived photosensitizing agents have been proposed, and they are the most tested products.

3.6.2 Chinese Traditional Products

Traditional Chinese Medicine is currently attracting increasing interest in dermatological research looking at the possible development of new drugs. This approach is supported by the political strategies of the Western Countries toward China. The use of herb extracts as enhancers or modulators of the melanogenesis because of photosentizing substances or possessing antioxidant or anti-inflammatory properties is particular popoular. However the composition of the mixture used or the chemical characterization of the active compounds are not reported in all the studies.

An early randomized controlled study of Jin [6] performed on more than 200 subjects reported the effectiveness of an undefined mixture of Chinese herbs. The comparison was carried out versus oral corticosteroids alone (15 mg/day, eventually decreased to 5 mg, every 2–4 weeks), psoralen alone (topical, 30%), or corticosteroids combined with herbal products. After 2 months of treatment, the combination of corticosteroids plus herbal drugs was considered the best treatment since they produced a complete repigmentation in 31%, and more than 60% in 14% of treated patients. A more recent paper described the occurrence of complete or good (more than 50%)

M. Picardo and A. Taïeb (eds.), *Vitiligo*,
DOI 10.1007/978-3-540-69361-1_3.6,

repigmentation in 95% of the patients ($n = 41$) treated with Xiaobai mixture, versus 79% of control group ($n = 33$) treated with 10 mg 8-MOP definire. The mixture was an aqueous extract of walnut, red flower, black sesame, black beans, zhi bei fu ping, lu lu tong, and plums (1 mL contains 0.1 g of raw medication), which was administered (160 mL) every day for three months [9].

The main problem associated with these studies was the poor description and characterization of the compounds used, as well as the trials methods (group size, several arms, allocation of patients and outcome measure). Moreover, the rationale for a biochemical or molecular impact in vitiligo is clearly lacking within this empirical class of therapies. Overall, the evidence that traditional Chinese herbal medicines are effective in the treatment of vitiligo is inadequate, and none of the treatment published have been replicated and confirmed.

3.6.3 Plant-Derived Extracts

Some trials investigated the possible therapeutic activity of different plant extracts. Two trials utilized the photosensitizing effect of the derivative of the plant *Picorrhiza kurroa,* a khellin extract, and their effectiveness was assayed after oral intake in conjunction with UVA or UVB phototherapy. The treatments varied for 3–12 months and were characterized by a small number ($n = 32$) of patients [1]. A high rate of dropout was mentioned. Only 35% of patients obtained a repigmentation of more than 50% [2, 17]. One study investigated the effectiveness of the topical application of the extract of *Cucumis melo* [5], the orange melon, in combination with UVB, and no differences toward the placebo cream were detected. One study evaluated the effect of oral administration of *Ginko Biloba* itself and two *Polypodium leucotomos* in combination with photo or photochemiotherapy (Chap. 3.3.2 and Part 3.4) with the aim to increase the antioxidant activities of the skin (Chap. 2.2.6).

3.6.4 Melagenin

The extract of the human placenta, named "melagenin," has been reported to stimulate in vitr the proliferation of melanocytes and the synthesis of melanin. A pilot trial evaluated the effectiveness of the topical application of the melagenin in pediatric vitiligo patients. Sixty-two out of the 366 treated subjects, ranging from 4 to 15 years old were followed up for 1 year. Depigmentation involved more than 70% of the body. The authors reported that the treatment with melagenin proved to be effective in 83% of vitiligo patients [10]. Currently, the placenta extract is used in Cuba, even if the protocol lacks external validation. The possible mechanism of action of the extract has been reported to be associated with the presence of melanocyte growth factors in placenta, including endothelin and sphingolipids [9, 12].

3.6.5 Aspirn

The possible therapeutic use of acetylsalicylic acid (aspirin), based on its antioxidative and anti-lipoperoxidative effect associated with immune-modulator activities, has been advocated by Zailaie in Saudi Arabia [17-20]. A clinical study evaluated the ability of the daily intake of 300 mg of aspirin for 12 weeks to block the spreading of the lesions, as well as to induce repigmentation. The trial included 32 vitiligo patients: in all the enrolled subjects, the progression of the disease was arrested, and in 2 out of 32, a significant repigmentation was reported. The mechanism of action has been referred to the inhibition of leukotriene synthesis, and the induction of catalase/glutathione peroxidase activities [3]. Moreover, a reduction of the soluble IL-2 receptor (sIL2R) concentration, as well as of the serum IL-1β, IL-6, IL-8, and TNF-α levels and antimelanocytes antibodies has been reported to be associated with the treatment. In addition, low dose of aspirin appears to increase the proliferative capacity of the melanocytes and to counteract the lipoperoxidative process in vitro, whereas higher doses increased these phenomena. However, the main defect of these data results from the lack of external and further validation for both in vivo and in vitro results, and the limited number of patients evaluated [17–2].

3.6.6 Statins

Based on a "case report" and some in vitro studies,the possible therapeutic role of the statins has been suggested for the treatment of the vitiligo.

Statins are inhibitors of the hydroxy-3 methyl-3 glutaryl coenzyme A reductase (HMG-CoA reductase), the rate-limiting enzyme of the cholesterol synthesis pathway. The reported case describes a woman treated with statins for several months for her lipid abnormalities. The clinical effectiveness has been related to the immunomodulatory and anti-inflammatory activity of this family of compounds through the inhibition of TNF-a, IL-6, IL-8 release, and of the LFA1 and ICAM1 expression. Moreover, statins are also able to in vitro inhibit the expression of MHC class II molecules contributing thus to the immune suppression. A different possible mechanism of action has been related to the interference with the phosphatidylinositol-3-kinase/Akt signal transduction pathway [12, 13].

3.6.7 Dermabrasion combined with 5-Fluorouracil

Skin ablation with mechanical superficial dermabrasion combined with topical application of 5% 5-fluorouracil (5FU) was introduced in 1983. Only patients with stable disease and limited extension of the manifestations can be treated because of the risk of Koebner phenomenon or hyperpigmentation induced by the treatment. Overall, microdermabrasion studies have been overall conducted in small groups of patients. Protocols, units, and settings were not homogeneous. The mechanism of action is mainly based on the combination of epidermal removal and induction of irritation mediated by 5FU, which through the production of inflammatory cytokines and prostaglandins stimulate melanocyte migration and proliferation. A 5-fluorouracil dressing is applied for 7 days following dermabrasion and repigmention can be achieved in the following 6 months. Thirty patients were treated by Sethi et al [15]. After dermabrasion, a soframycin tulle dressing, a topical 5% 5-fluorouracil dressing, and a topical placentrex gel dressing were put on three different lesions for each patient. The authors described higher efficacy (73%) for the combinatory treatment dermabrasion plus 5-fluorouracil. Dermabrasion alone and dermabrasion combined with placentrex gel were reported to show similar efficacy [15]. The replacing of mechanical dermabrasion by erbium-YAG laser ablation has been proposed and better results have been reported at least in periungual lesions. Moreover the combination with short term NB-UVB phototherapy (4 months) evaluated in a left-right comparative study demonstrated that laser ablation and 5FU application improved the efficacy of phototherapy [1]. A trial, where 50 adult subjects with non segmental vitiligo were enrolled, has been carried out. All patients exhibited symmetrical lesions. One side was treated with ER:YAG laser ablation, followed by 5FU application before simultaneous NB-UVB therapy of both sides for a maximum period of 4 months. The overall response to therapy was better using the combination therapy. The authors reported for fifty patients (78%) a moderate-marked repigmentation response when the combinatory protocol was applied compared with 23% in the mono-therapy one. The response was reported to be significantly higher, except for foot lesions, which were better but not statistically significant. As regards to the possible side effects, tolerable pain during ablation or at sites of 5FU application was reported in all cases. Transient hyperpigmentation occurred in 30% of cases and 3% of lesions healed by a transient slate blue color. Half of the treated periungual lesions showed a temporary tiny brownish spot on nail plates and Köebnerization was not detected in any patient [1].

3.6.8 Others

Sporadic clinical and in vitro evaluations have been performed in order to select possible drugs which would be able to interfere with the physiological intracellular signal transduction pathways, affecting the melanocyte proliferation/migration/differentiation. Among these compounds, the secreted phospholipase A_2, component of the bee venom has been considered [7]. When tested in vitro, bee venom was able to promote melanocyte proliferation, dendriticity, and migration; the tyrosinase activity was also positively affected. The molecular mechanism has been supposed to be mediated by the activation of intracellular signal transduction mechanisms, and in particular, of protein kinase A (PKA), ERK, and PI3K/Akt. The water extract (0.1 mg/mL) of the *Piper nigrum*, and its alkaloid piperine, has been demonstrated to promote in vitro proliferative activity, probably mediated by activation of PKC, on melanocytes. Clinical observations confirmed the in vitro data [8]. The occasional report of PGE2-induced hyperpigmentation suggested its possible utilization for the topical treatment of the vitiligo lesions. In vitro PGE2 have been reported to induce the expression of mRNA for bFGF, and the oxidative stress-mediated GSH depletion may decrease PGE2 level in vitiligo epidermis. A recent trial indicates that the topical application of a gel containing PGE2 gives rise to repigmentation, mainly on the face, after six months of therapy. Segmental and focal vitiligo have been reported to be characterized by the highest percentage of repigmentation [4].

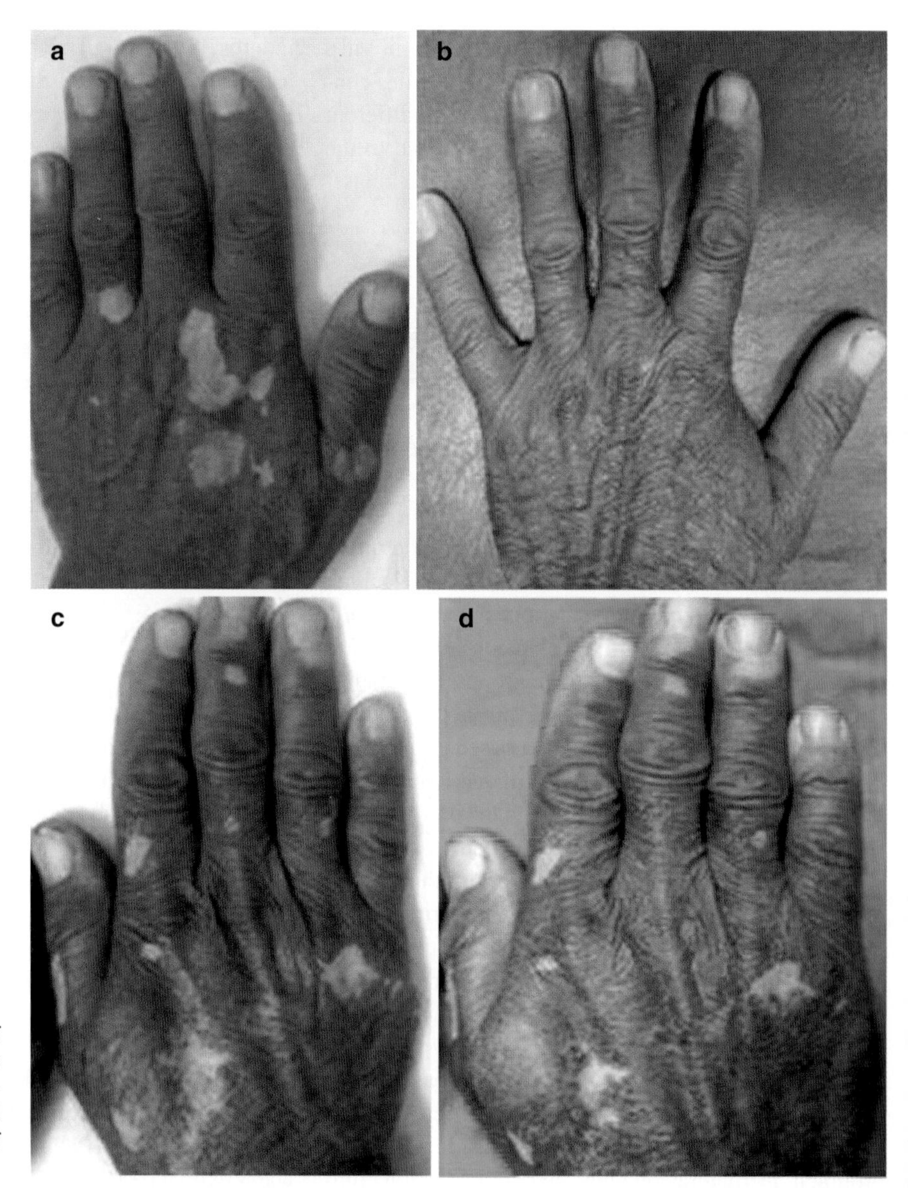

Fig 3.6.1 Vitiligo patches of the hand: (**a**) and (**b**) before the therapy, (**c**) after the treatment with UVB plus SFU, and (**d**) after UVB alone. Credit of figure: Dr Tag Anbar

Summary Messages

- Several alternative approaches for the treatment of vitiligo have been proposed originating from traditional Chinese medicine, alternative medicine and sporadic clinical observations.
- The in vitro properties of the possible active compounds include photosensitizing properties, induction of melanocyte proliferation, antioxidant activities, and immunoregulatory abilities.
- The current evidence of a therapeutic effect of traditional herbs derivatives as well as plant and animal extracts in vitiligo is at best limited.
- Based on other indications of alterative medicines, there is a potential for adverse effects which has not yet been reported in vitiligo.
- PGE2 based topical formulations may correspond to a significant therapeutic advance in the field, but safety concerns remain unanswered.
- The association laser dermabrasion-5 FU looks promising for recalcitrant locations (hands).

References

1. Anbar TS, Westerhof W, Abdel-Rahman AT, Ewiss AA, El-Khayyat MA (2008). Effect of one session of ERG:YAG laser ablation plus topical 5Fluorouracil on the outcome of short-term NB-UVB phototherapy in the treatment of non-segmental vitiligo: a left-right comparative study. Photodermatol Photoimmunol Photomed 24:322–329
2. Bedi KL, Zutshi U, Chopra CL et al (1989). Picrorhiza kurroa, an ayurvedic herb, may potentiate photochemotherapy in vitiligo. J Etnopharmacol 27: 347–352
3. Durak I, Karayvaz M, Cimen MY, et al (2001). Aspirin impairs antioxidant system and causes peroxidation in human erythrocytes and guinea pig myocardial tissue. Hum Exp Toxicol 20: 34–37
4. Kapoor R, Phiske MM, Jerajani HR (2009). Evaluation of safety and efficacy of topical prostaglandin E2 in treatment of vitiligo. Br J Dermatol. 2009;160:861–3
5. Khemis A, Ortonne JP (2004). Comparative study of vegetable extracts possessing active superoxide dismutase and catalase (Vitix) plus selective UVB phototherapy versus an excipient plus selective UVB phototherapy in the treatment of common vitiligo. Nouvelles Dermatologiques 23: 45–46
6. Jeon S, Kim NH, Koo BS, et al (2007). Bee venom stimulates human melanocyte proliferation, melanogenesis, dendriticity and migration. Exp Mol Med 39: 603–613
7. Jin QX, Wj M, Zs D et al (1983). Clinical efficacy observation of combined treatment with chinese traditional medicine and western medicine for 407 cases of vitiligo. 12: 9–11
8. Lin Z, Liao Y, Venkatasamy R, et al (2007). Amides from Piper nigrum L. with dissimilar effects on melanocytes proliferation in vitro. Pharm Pharmacol 59: 529–536
9. Liu ZJ, Xiang YP (2003). Clinical observation on treatment of vitiligo with xiaobai mixture. Chinese J Integr Trad Western Med 23: 596–598
10. Mal'tsev VI, Kaliuzhnaia LD, Gubko LM (1995). Experience in introducing the method of placental therapy in vitiligo in Ukraine. Lik Sprava 7-8: 123–125
11. Mallick S, Mandal SK, Bhadra R (2002). Human placental lipid induces mitogenesis and melanogenesis in B16F10 melanoma cells. J Biosci 27: 243–249
12. Namazi MR (2004). Statins: novel additions to the dermatologic arsenal? Exp Dermatol 13: 337–339
13. Noel M, Gagné C, Bergeron J, et al (2004). Positive pleiotropic effects of HMG-CoA reductase inhibitor on vitiligo. Lip Health Dis 3: 7–11
14. Pal P, Mallick S, Mandal SK, et al (2002). A human placental extract: in vivo and in vitro assessments of its melanocyte growth and pigment-inducing activities. Int J Dermatol 41: 760–767
15. Sethi S, Mahajan BB, Gupta RR, Ohri A (2007). Comparative evaluation of theraputic efficacy of dermoabrasion, dermoabrasion combined with topical 5% 5-fluorouracil cream, and dermoabrasion combined with topicalmplacentrex gel in localized stable vitiligo. Int J Dermatol 46:875–879
16. Szczurko O, Boon HS (2008). A systematic review of natural health product treatment for vitiligo. BMC Dermatol 8: 2–14
17. Zailaie MZ (2004a). Short- and long-term effects of acetylsalicylic acid treatment on the proliferation and lipid peroxidation of skin cultured melanocytes of active vitiligo. Saudi Med J 25: 1656–1663
18. Zailaie MZ (2004b). The effect of acetylsalicylic acid on the release rates of leukotrienes B4 and C4 from cultured melanocytes of active vitiligo. Saudi Med J 25: 1439–1444
19. Zailaie MZ (2005a). Aspirin reduces serum anti-melanocyte antibodies and soluble interleukin-2 receptors in vitiligo patients. Saudi Med J 26: 1085–1091
20. Zailaie MZ (2005b). Decreased proinflammatory cytokine production by peripheral blood mononuclear cells from vitiligo patients following aspirin treatment. Saudi Med J 26: 799–805

3.7 Surgical Therapies

Background and Techniques

3.7.1

Mats J. Olsson

Contents

M. J. Olsson
Department of Medical Sciences,
Dermatology and Venereology,
Uppsala University, Sweden
e-mail: mats.olsson@medsci.uu.se

3.7.1.1 Introduction

In many cases, medical therapies and ultraviolet light (UV) treatments may be of benefit for vitiligo patients. However, as in some other disorders characterized by a loss or lack of melanocytes in the epidermis as well as in the hair follicles (i.e., piebaldism, segmental vitiligo, and depigmentation after burn injury), these approaches may fail. Also, many cases of generalized vitiligo do not respond to medical or UV-treatment, especially those with lesions on hands, fingers, feet, and toes (i.e., areas with reduced numbers of hair follicles). To overcome this lack of response to noninvasive methods, a number of surgical techniques have been developed. Several of these techniques have been used in clinical practice for many years and are part of standard treatments used today.

The general aim in transplant treatment should always be to select a method that is likely to be the most effective for a particular patient. Limited technical/financial resources and lack of certain training are limiting factors.

While selecting a specific transplantation method, the following points should be taken into consideration:

- Location of the areas to be treated
- Extent
- Whether there are several small lesions or one large lesion
- Texture of the skin on the recipient area
- Presence of hairs in the recipient area
- Texture of the skin at possible harvest area
- Tangible limits in the surgery facility, ward, equipments, and laboratory
- Training and experience of the involved personnel

M. Picardo and A. Taïeb (eds.), *Vitiligo*,
DOI 10.1007/978-3-540-69361-1_3.7.1, © Springer-Verlag Berlin Heidelberg 2010

Once all these variables have been evaluated, the decision for the most suitable method can be made.

Treatment of leucoderma with transplantation of autologous melanocytes can be implemented in all types of vitiligo and piebaldism. Surgical methods are the treatment of choice in piebaldism, focal, and segmental vitiligo, and such methods can also be used to treat depigmentation after burns and chemical injuries. Selecting appropriate candidates among patients with generalized vitiligo (vitiligo vulgaris) is most difficult. Generalized vitiligo is progressive in most cases, and untreated lesions usually increase in size and number during the patient's lifetime. If not in stable condition, autoimmune components and/or other unknown factors decrease the chance for melanocyte be survival and attachment.

Patients should be counseled so they have realistic expectations about the outcome of the procedure. They should be informed that expecting a 100% cosmetically perfect result is unrealistic, even if the whole procedure is faultless. A small area may remain depigmented after the procedure, or a repigmented area may have a color slightly darker or lighter than the surrounding skin. Careful evaluation and experience will help predicting the outcome in individual patients.

Different techniques exist based on technical equipment, ways of collecting or processing the tissue/cells from the donor site, preparing the recipient site, and fixating the cells or tissue in place for a sufficient time. Some of the most used methods are reviewed on the basis of their advantages and limitations in different anatomical sites. A more in-depth discussion of all technical variants and applications of surgery in vitiligo can be found in a recent book [12].

3.7.1.2 Selection of Patients

In piebaldism and stable types of vitiligo (such as segmental unilateral vitiligo and focal vitiligo), the outcome of the transplantation treatments is usually excellent; in some cases, transplantation may indeed be the only effective treatment. But in the group of patients with generalized vitiligo, we know that an appropriate selection of patients is mandatory to achieve a good success rate. A careful history-taking is mandatory to make a decision. Long-term follow-ups have shown that patients with extensive generalized vitiligo and also those suffering from associated autoimmune hypothyroidism have a worse outcome compared to patients with more limited disease and those free from hypothyroidism [27]. In the group of patients with progressive vitiligo, it was more likely that some or all of the repigmentation obtained was lost during the follow-up period.

It is accordingly important that the condition is stable before surgery. In a "stable" condition, neither development of new lesions, nor enlargement of old lesions, nor Köbner phenomenon are identified for a certain period of time. There is no consensus definition of stability based on clinical grounds in nonsegmental vitiligo. I would recommend a minimum of one year total stability. Furthermore, no reliable test can be used to predict disease activity, future progress, or outcome of melanocyte transplantation treatment in patients with generalized vitiligo. Minigrafts have been used prior to more extensive treatments as an approach to predict stability. However, vitiligo can be active in one skin area, and inactive or even in regression in another one at the same time. The outcome of transplantation can be successful in one area and unsuccessful in another one in the same individual at the very same occasion. The disease can be stable for some time and then activated again in unpredicted cycles and length of the periods. Therefore, it is difficult to predict how long the disease will be stable, and similarly it is difficult to envisage when it will start to become unstable. Thus, it can be difficult to draw conclusions from the test spot.

Until we have a validated blood test to analyze vitiligo-specific markers/predictors of the activity of the disease, we are still dependent on a careful anamnesis. The final judgment will remain subjective, but based on long-term experience and logical applications of our clinical knowledge and progression pattern of the disease, some level of prediction is possible. But we have to keep in mind that many patients are willing to try almost anything, and might also believe that surgical treatment can stop the progression of the disease and correct the underlying causative factors. Sometimes, when patients know our selection criteria, their strong desire to undergo a transplantation treatment can lead them to adjust their history, which in turn, could bias our decision. We must therefore carefully explain how important it is to make the right treatment selection. We must also explain that we merely treat symptoms in the affected areas of the skin, and not the underlying disease.

Both written and oral information about the disease and planned procedure should be given to the patient in good time before the scheduled date of the surgery. Informed consent should also be used to ensure that the patient has taken part of the information and understood the content and possible outcomes of the treatment.

3.7.1.3 Surgical Methods

The surgical techniques can be grouped in techniques utilizing harvested tissue directly, without any processing, or techniques based on processing the donor tissue/cells before use. The most frequently used and scientifically proven methods are listed here, along with brief explanations and comments. Some of the more advanced techniques are explained later in greater detail.

Methods Based on Direct Transplantation of Unprocessed Tissue

- *Suction blisters.* Fluid-filled blisters are induced with the help of a suction machine connected to suction cups on the donor skin. The roof of the blister top is cut loose with the help of a small curved-eye scissors, slipped onto a saline moistened microscope slide to ensure that the thin tissue sheet will stay unfolded, and correctly oriented and then transferred to the prepared bed at the recipient site. The induction depth of the blisters is at the dermo–epidermal junction, and there is no risk of scaring at the donor site. This a relatively easy and inexpensive method [11, 13, 17, 18, 29, 31].

- *Split thickness grafts (Thiersch grafts.)* A method utilizing a special knife or dermatome to harvest a dermo–epidermal tissue biopsy to be directly transferred and fixed on the prepared wound bed at the recipient site. Too thick biopsy may results in a stuck on appearance and persistent hyperpigmentation. The risk of scaring at the donor site is surgeon-dependent based on the depth of the shave. This is a relatively easy and inexpensive method [1, 2].

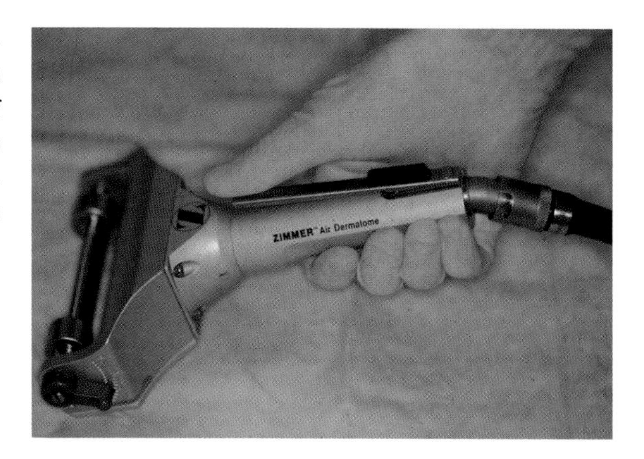

Fig. 3.7.1.1 The Zimmer air dermatome with the high-pressure hose. Make sure you use medical grade air from a tube or hospital outlet

- *Minigrafting.* Small punch biopsies (minigrafts) are harvested with a punch biopsy needle of Ø 2 mm or smaller. The grafts are transferred to the recipient site and planted into perforations done with the same biopsy needle and secured in place with an adhesive tape and pressure bandage. Smaller minigrafts give less risk of cobblestone side effect at the recipient site or scarring at the donor site. Easy and inexpensive method [3, 7].

- *Ultra-thin grafting.* An air-driven high speed dermatome is used to harvest ultra-thin shaves. The machine used is very precise and renders possible a minimally thin harvest of a donor tissue. The result is in the hand of the surgeon, and if the equipment is used properly, scarring at the donor site is very rare. Medium advanced method [16, 25] Fig 3.7.1.1.

Methods Based on Transplantation of Processed Cells or Tissue

- *Cultured epithelial sheets carrying melanocytes.* This is a technique similar to that used at advanced burn clinics to cover large areas of burn-injured skin [4, 8, 10, 19]. A shave-biopsy is taken from a normally pigmented location. The epidermis is removed from the dermis with the help of the enzymes dispase or trypsin. The basal layer cells are detached and put in culture. Enough melanocytes should survive in the keratinocyte culture to be able to repigment the recipient site. Preseeding of the culture flasks with feeder-layers of 3T3-mouse fibroblasts or precoating of the culture flasks with collagen or fibronectin enhance cell attachment and growth potential. For a more defined and safe system, culture systems free from serum and feeder-layers [19] or defined systems with a coating of plasma-polymers to enhance the growth of both keratinocytes and melanocytes [5, 6] are recommended. These advanced methods require cell-culture facilities and trained personnel.

- *Cultured melanocyte suspension.* Technique utilizing cell-culture expansion of melanocytes from a shave-biopsy [22–24]. The method gives a high yield of melanocytes from a relatively small donor site. The exchange rate between the donor and recipient site is high, and large areas of skin can be treated in one session; alternatively, the treatment can be divided into several smaller sessions and in that case the melanocytes need to be cryo-preserved [28]. The possibility to cryo-store the cells makes it possible to plan and schedule the treatment, when it is most suitable for all parties, and to ship cells on dry-ice to be used at another hospital. It is possible to establish 100% pure melanocyte cultures, but that is not the main purpose for the use in transplantation treatments. Some accompanying keratinocytes and fibroblasts do not disturb the outcome of the treatment. This method allows to control the exact number of melanocytes applied per square millimeter. These advanced methods require cell-culture facilities and trained personnel.
- *Basal cell-layer suspension.* Technique utilizing melanocytes and keratinocytes from a shave-biopsy, which has several steps in common with the methods described earlier (Cultured epithelial sheets carrying melanocytes and Cultured melanocyte suspension), but instead of expanding basal cells in culture, the cell suspension is used immediately for transplantation [9, 26]. It is important to wash the cells and concentrate them in a centrifugation step to achieve the right (small) volume and viscosity [26]. Without centrifugation steps, there is a risk of too high seeding volume, allowing transplant cells to leak out of the recipient area. Methods trying to bind a larger volume of excess liquid with the help of hyaluronic acid, serum, collagen, etc. carry the risk for limiting cell attachment to the recipient surface (i.e., the cells may remain in suspension in the viscous solution/matrix). These advanced method require minimal cell-culture facilities and trained personnel.
- *Melanocytes and keratinocytes cultured and delivered on defined membranes* Attempt to make one of the advanced methods fully defined, reproducible, more user-friendly and efficient in the hands of the clinicians [5, 6]. A specialized laboratory performs all the technical steps until the delivery to the bedside. To achieve optimal growth and transfer conditions, cells have been seeded on chemically defined substrates (produced by plasma polymerization of carboxylic acid, acrylic acid, allylamine, or a mixture of these monomers) either as mono- or co-cultures and using media combinations [5]. The ability of keratinocytes and melanocytes to be transferred from cell carriers under different media conditions to an in vitro human wound bed model has been tested. The number of melanocytes transferred, their location within the neoepidermis, and their ability to pigment were evaluated as preclinical end points. A highly efficient and reproducible transfer of physiologically relevant numbers of melanocytes capable of pigmentation from the coculture of melanocytes and keratinocytes was obtained using M2 medium, and a silicone carrier pretreated with 20% carboxylic acid deposited by plasma polymerization [6]. Very advanced method requiring cell-culture facilities and special trained personnel, which needs a further industrial step to reach the clinic.

3.7.1.4 Preparation of the Recipient Area

There are a variety of methods for preparing a wound bed at the recipient site:

- Suction blisters
- Friction blisters
- Heat separation
- Chemically induced blisters, e.g., NaOH
- PUVA induced blisters [18]
- Liquid nitrogen induced blisters [7, 9, 14, 31]
- Ultrasonic abrasion [32]
- Dermatome shaves [2]
- Flip-top [20]
- Laser ablation [10, 13, 16, 17, 29, 33]
- Dermabrasion [1, 4, 11, 23, 25]

Some of these methods utilize the creation of a blister are not always practical to use. It usually takes a considerable time to create a blister and most often, due to the delay of outcome, it is difficult to know beforehand if a sufficient blister will be induced or not. Further vitiligo lesions can have any kind of shape and location and it is not easy to follow the borders with a blister formation.

In practical terms, dermabrasion and laser ablation are today the most effective, quickest, and most controllable means of removing the epidermis. The drawback with dermabrasion and laser ablation is that it creates an aerosol (airborne small fragments of epidermal cells and blood particles or smoke). A face visor, face mask class FFP2 is needed for the operator. For laser surgery, a suction fan with a particle filter is mandatory (Fig. 3.7.1.2).

In future, we can expect the development of chemical removal of epidermis (i.e., quick and effective peelings) as well as faster and more easy to maneuver devices for ultrasonic and heat de-epidermization.

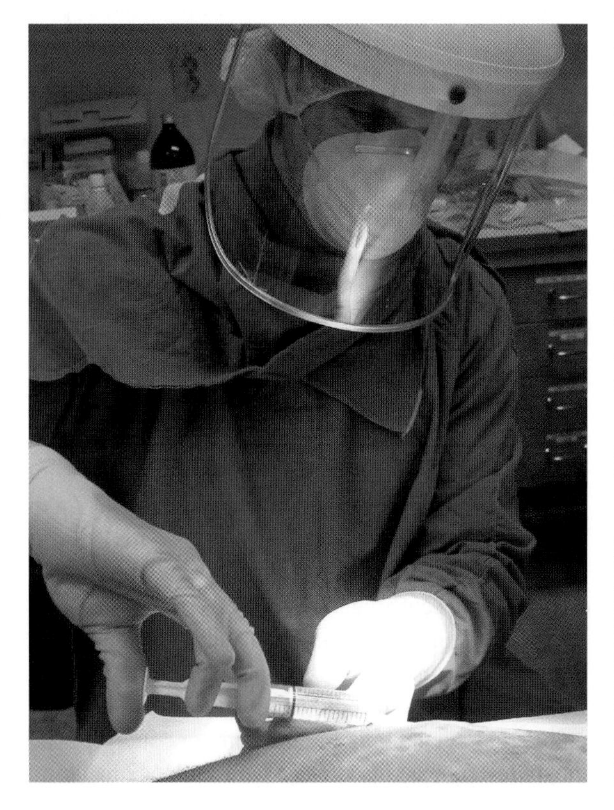

Fig. 3.7.1.2 Usage of gears protecting eyes and airways such as face mask class FFP2 and face visor is important when a deepithelialisation method that spreads airborne particles is used

3.7.1.5 A Guide to Advanced Methods for Vitiligo Surgery

The three methods described hereafter are the most efficient in terms of the capacity to cover large areas in one session and to produce an even repigmentation.

Basal Cell-Layer Suspension and Cultured Melanocyte Suspension

The donor cells are derived from a superficial shave biopsy, taken from a hidden area such as the buttocks.

Donor Tissue

A normally pigmented area of about $2 \times 4\,\mathrm{cm}^2$ (or somewhat larger if the cells are going to be used directly and not expanded in culture) is marked, surgically cleansed and anaesthetized with a solution containing equal amounts of 10 mg/mL lidocaine and Tribonat® (bicarbonate solution from Fresenius Kabi, Uppsala, Sweden). The bicarbonate solution neutralizes the pH, and makes the injection less painful. A long thin needle is inserted from outside the marked donor area to avoid bleeding at the site from needle stick, and then parallel to the skin surface thrust into the marked area. Avoid the use of adrenaline due to the risk of a buckled area with embossed lesions. Press the area with a sterile gauze-pad to facilitate the diffusion. An even surface makes it easier to harvest a very thin and coherent sheet.

A shallow shave biopsy is taken with a Goulian skin graft knife (Edward Weck & Company, Inc, Research Triangle Park, NC) (Fig. 3.7.1.3). The Goulian knife

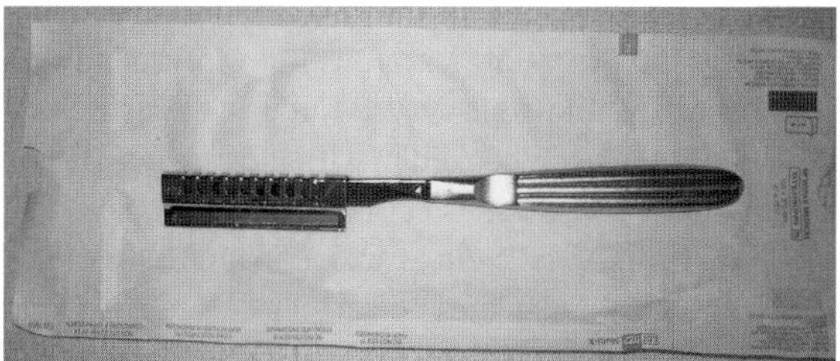

Fig. 3.7.1.3 A dermatome such as the Goulian Weck knife, equipped with a 006 blade shield, can be used to harvest shallow shave biopsies

should be equipped with a 006 shield, to ensure very shallow biopsies. The specimen is put in a 15 mL test tube containing, Joklik's modified minimal essential medium (s-MEM, GIBCO BRL, Life Technology, Gaitersburg, MD) completed with antibiotics (e.g., 50 U/mL penicillin and 0.05 mg/mL streptomycin) and transferred to the laboratory for preparation. After surgery, the donor area is covered for eight days with semipermeable Tegaderm™ (3M, St Paul, MN), plus a layer of the air and water vapor permeable stretch fabric tape Fixomull®, extending a few cm beyond the margins of the Tegaderm. Both the Tegaderm and the Fixomull are glued in place with Mastisol® (Ferndale Laboratories, Inc, Ferndale, Michigan). The glue ensures that the dressing will remain secured for the whole bandage period.

If the lag period between harvesting and the preparation of the cells is more than 3 h, the biopsy should be put in complete M2 melanocyte medium (PromoCell, Heidelberg, cat. No. C-24300) and kept at 8°C to ensure the highest possible survival rate of the melanocytes.

Release and Preparation of Free Cells

The donor sample is kept in a Ø 6 cm or Ø 10 cm Petri dish inside a laminar flow-hood and washed once with 4 or 8 mL 0.25% (w/v) trypsin and 0.18% (w/v) EDTA in PBS (SVA, Uppsala, Sweden) and refurnished with 5 or 10 mL (depending on the size of the Petri dish) of the trypsin/EDTA solution. The sample is turned back and forth with the help of jeweller's forceps to ensure that it comes in complete contact with the solution (Fig. 3.7.1.3), and finally, with the epidermis side facing upward, torn or cut into pieces of about 4 cm^2. The air-bubbles under the thin fragments should be removed by gently pressing and scraping the surface with a curved forceps. The Petri dish is incubated at 37°C for about 55 min (Fig. 3.7.1.4). At about half the incubation time the pieces are carefully moved around and lightly pressed on with a curved forceps, to ensure that the whole tissue gets soaked in with the trypsin/EDTA-solution.

It is difficult to give the exact incubation time, since it depends on the thickness of the sheet, but by gentle pressing with a curved forceps on the epidermis, the impression pattern of slightly lighter streaks (indicating "ready to be processed") will make the operator able to detect if the incubation time is sufficient or not.

Fig. 3.7.1.4 The thin shave biopsy is with the help of fine forceps stretched out and worked up in Trypsin-EDTA solution. All air bubbles are removed by scraping the biopsy with a curved jeweller's forceps

After incubation, the trypsin/EDTA solution is removed with the help of a pipette and 3 mL (15°C) of 0.5 mg/mL trypsin inhibitor (Sigma, St Louis, MO) in PBS is added to the Petri dish to terminate the trypsin digestion. The epidermis is removed from the dermis with the help of forceps and the dermis is transferred to a 15 mL test tube containing 5 mL of serum-free, melanocyte medium M2 (PromoCell, cat. no. C-24300) and vortex-mixed for 5 s. The dermal pieces are then removed with the tip of a Pasteur pipette or with the help of a hooked forceps and then discarded. The epidermal pieces are scraped with a curved jeweler's forceps to free basal cells, minced to smaller fragments, and then transferred, together with the trypsin inhibitor, to a test tube. The tube is vortex-mixed for 30 s. The Petri dish is rinsed twice with a small volume of medium, which is also transferred to the test tube, run up end down a few times in a Pasteur pipette to ensure a homogenous solution of individual cells, filled up to 14 mL with medium, and then centrifuged for 7 min at 180 × g.

After the centrifugation the supernatant and the floating stratum corneum-granulosum fragments are removed and the pellet is resuspended in 7 mL M2 melanocyte medium and transferred to a 75 or 150 cm^2 culture flask containing M2 medium for culturing. The empty test tube is rinsed twice with 1 mL M2 medium, which is also transferred to the culture flask, to ensure that no cells get wasted. The culture flask should be kept lying flat when transferring the cell-suspension, so that the cells do not stick to the sides or ceiling of the flask. This is to ensure a maximum exchange rate. A 75 cm^2 flask should have a total of about 15 mL medium and a 150 cm^2 flask should contain about 30 mL of culture medium.

Cells to be used for direct transplantation without preceding culture expansion should be washed and centrifuged a second time in 8 mL medium. The cell-pellet is then dissolved in a very small volume (a total of 200–600 μL), depending on the size of the area to be treated, and then ready to be seeded on the recipient area.

Culturing of Melanocytes

The cells should be microscopically controlled every day to ensure that the cells are morphologically looking healthy, and that there are no signs of infection (Fig. 3.7.1.5). The medium in the culture-flasks should be changed every third day. When the culture becomes confluent, the cells are detached with trypsin and subcultured. For therapeutic purposes, there is no need to eliminate a possible incorporation of some fibroblasts and keratinocytes, as they may support the melanocytes and, theoretically also may enhance the healing process in the treated area. When a 100% pure melanocyte culture is needed for a research purpose, the level of Ca^{2+} may be elevated to 1.6 mM for 2 days to differentiate and eliminate keratinocytes and a 3 days

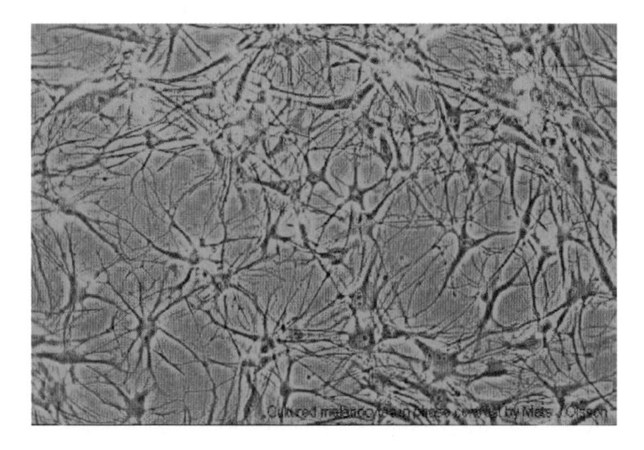

Fig. 3.7.1.5 Human adult melanocytes cultured in serum free and phorbolester free conditions seen in phase contrast microscopy. The morphology of the cells varies with the culturing conditions, but is also to some extent unique for the individual patient. The cells are looked on every day to ensure that they are free from fungus and bacteria

incubation starting from the tenth day in culture with 45 μg/mL geniticin (G418) is needed to selectively get rid of the fibroblasts.

After 2 weeks, the number of cells cultured from one biopsy varies from 10×10^6 to 50×10^6 depending on the size of the biopsy and the age of the donor. At this stage the cells are ready for transplantation.

When the cells in the flasks are confluent, they are harvested immediately before the transplantation. The culture medium is removed from the flask and about 5 mL of 37°C trypsin/EDTA solution is added to each 150 cm^2 flask. The flask is tilted back and forth a few times to ensure that the solution comes in contact with all cells and then incubated in 37°C for about 2 min. After the incubation the flask is tapped on the side with the palm of the hand, while holding the flask with the other hand. This gives the flask a jerking acceleration sideways, releasing the cells from the plastic surface. Transfer the free cells quickly to a 15 mL test tube containing 4 mL 15°C trypsin inhibitor with the help of a pipette. The remaining cells in the flask are washed with additional 5 mL 15°C trypsin inhibitor and transferred to the same test tube. A gentle spin of 170 × g for 6 min will settle the cells down into a pellet. The supernatant is discarded, and the cells are resuspended in 6 mL room-tempered s-MEM medium without any additives. Resuspension/wash is done twice if cells are to be used for transplantation and only once if the cells are to be seeded into new flasks for subculture. At resuspension the cells are only centrifuged for 5 min at 170 × g, the supernatant is discarded, and the cells to be used for transplantation are resuspended in a very small volume of s-MEM (about 0.2–0.4 mL), to be used for immediate transplantation. The small volume of s-MEM resuspended cell-pellet is kept in the test tube for direct application onto the skin with the help of a pipette.

Premedication

The patient should remain calm during the transplantation procedure. Therefore, it is recommended to give 5–10 mg of diazepam and/or 10 mg of ketobemidone orally 1 h before the transplantation, plus 500–1,000 mg of paracetamol 40–50 min before the surgery. Antibiotics such as penicillin M, are given for eight days, starting on the day of surgery. Ask beforehand for a history of known allergic reaction against penicillin and derivatives and avoid medicines containing salicylates (e.g., Aspirin) for 10 days before surgery and 1 week after the surgery.

Anesthesia of the Recipient Site

Large recipient areas are anesthetized with EMLA® cream (Astrazeneca, Södertälje, Sweden) applied under plastic foil occlusion (e.g., saran wrap) for 1–2 h [24, 15] and then also locally anaesthetized with a mixture of equal parts of 1% lidocaine and Tribonate®-buffer (Fresenius Kabi) immediately prior to the surgery. A thin and long needle is used, which is inserted outside the depigmented lesion and then thrust parallel to the skin surface into the lesion. This is to avoid bleeding from needle stick in the area to be treated. In a large area, it may not be possible to reach the center of the lesions from the borders, but a combination of "ring-block" effect of peripheral injections and EMLA will provide satisfactory anesthesia also in the center of the lesion. In small areas, local anesthetic can be infiltrated to the whole lesion, and therefore EMLA is not needed.

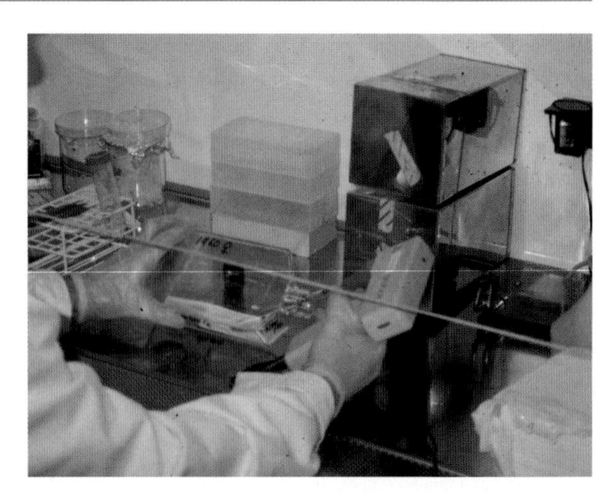

Fig. 3.7.1.6 Dermabrasion. The handle should be kept parallel to the surface and with a light downward pressure moved back and forth in the direction of the extension of the handle in a filling movement at the same time as the handle is moved steady and slowly in a 90° course. It is important that the lesion is worked over in at least two different directions. Hold the handle steady with one of the hands and use a finger from the other hand to support the handle at the same time as you keep a couple of the other fingers on the patient. This is to minimize the risk of uncontrolled jerking movements in the direction of the rotation

Transplantation and Aftercare

The recipient area is cleaned with alcohol, outlined with a sterile surgical marker, and the epidermis is removed down to the dermal–epidermal junction, using a high-speed dermabrader, fitted with a diamond fraise (Fig. 3.7.1.6). A wheel, pear, or cone suitable for the size and location of the area to be treated is used. Normally, you can dermabrade most of the lesions with a 6 mm wide regular fraise wheel but on rough skin, such as that on the knees, you may first need a coarse wheel and on delicate areas such as around the nostrils, corners of the mouth and on the eyelids, a small pear-shaped fraise may be needed. For eyelids, a specially made hand tool fitted with a regular cone ensures prevention of damage to the eyelid or the eye during the procedure.

The dermabrasion should be performed in at least two different directions until a uniform punctate capillary bleeding from the dermal papilla can be seen (Fig. 3.7.1.7). Light freezing with fluoro-ethyl spray can be used to reveal if there are any islands of epidermis left. The denuded area is washed with saline solution and kept under moistened gauze for few minutes to ensure that bleeding has stopped. If some puncture bleeding points persist, electrocoagulation is performed with a fine-pointed electrocate-needle.

The cell-suspension is applied (with an exchange rate of about 10 times donor:recipient area if direct transplantation of noncultured basal cell-layer suspension is used and a seeding density of 700–1,000 cells/mm² if cultured melanocytes are used) to the denuded areas and spread with the tip of the pipette. The cells are secured with a silicone netting (Mepitel®, Mölnlycke AB, Mölnlycke, Sweden) extending about 1 cm onto the dry non-dermabraded surrounding skin, which locks it in place so the netting will not slide over the slippery wound-bed. Then two layers of saline-moistened woven gauze compresses, and a semipermeable Tegaderm™-film are applied. The later is glued in place onto the surrounding skin with Mastisol® (Ferndale Laboratories, Inc). The glue ensures that it is secure in place for the period needed.

In all cases, the patients should rest a minimum of 4 h, preferably longer, in a hospital bed after the procedure has been completed. The patient is informed that the first 48 h are most critical for the outcome, and that it is important to restrict the physical activities and avoid tight clothing for the first 2 weeks. The dressing is removed after 8–9 days.

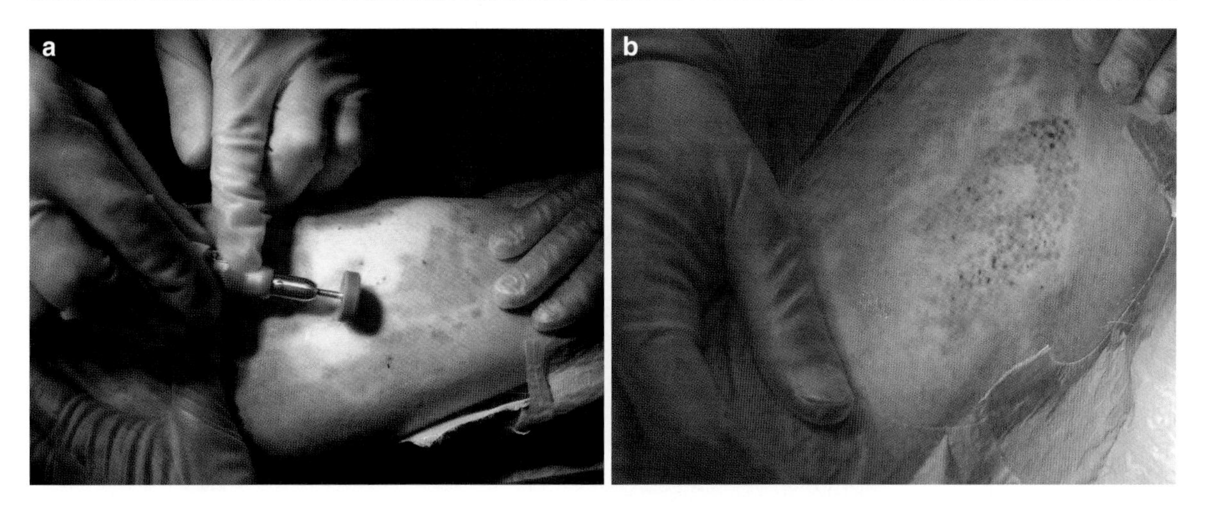

Fig. 3.7.1.7 Light and uniform capillary bleeding should be seen directly after the dermabrasion. The abraded surface is then kept a few minutes under saline moistened compresses before the cells are seeded. This is to ensure that the bleeding has stopped

The Tegaderm covering the donor site is also removed on the same day as the removal of the dressing. Usually the skin heals nicely at both the recipient and donor sites and a careful flush with sterile saline solution and light padding with dry gauze compresses is usually sufficient. This is followed by an application of a layer of pure vaseline with a spatula. On skin exposed to friction or trauma, such as elbows, hands, and feet, the area is after the application of vaseline covered with a gauze compress, which is secured in place with a Micropore™ tape to be worn for another two days. The patient is advised to apply a thin layer of vaseline once a day for seven days after the removal of the dressing. This is to minimize the frictional trauma and desiccation to the still fragile surface. A layer of vaseline is also applied to the donor area.

The patients are advised to expose themselves to midday sunlight for a few minutes 2 times a week for about 2 month, commencing at 1 week after the removal of the bandage.

Follow-Up Evaluation

About 5–8 months after transplantation, an overall follow-up evaluation is advised.

During the first few weeks, the transplanted area is erythematous, but pigmentation can be seen as early as 2–3 weeks posttransplantation using Wood's light or diascopy.

During the first year it is not uncommon to see a slight hyper- or hypopigmentation. But the color gradually matches with that of the surrounding skin and after a year it most often blends well with the surrounding skin.

Sometimes, a 1–2 mm depigmented halo persists around the transplanted area. This is seen more often in patients with generalized vitiligo than in those with piebaldism or segmental vitiligo. Generally, this halo is repigmented with repeated sun-exposures.

The outcome can be predicted early, but the final result will usually first be seen about 1.5 years after the surgery (Fig. 3.7.1.8).

Ultra-Thin Grafting

Treatment of leucoderma with ultra-thin epidermal sheets can be implemented in all types of vitiligo and piebaldism. Ultra-thin epidermal sheet transplantation is quick, effective and appropriate in flat, large, coherent areas that are free from stretching or can be immobilized for some time. Areas such as those over the joints, eyelids, and the corners of one's mouth are more difficult to treat with epidermal sheets than with cells free in a suspension. Areas in the face should be avoided due to the increased risk of hyperpigmentation [27]. Lesions in hairy locations should also be avoided due to the simple reason that the growing hair will elevate the sheet and reduce the chance of a graft uptake.

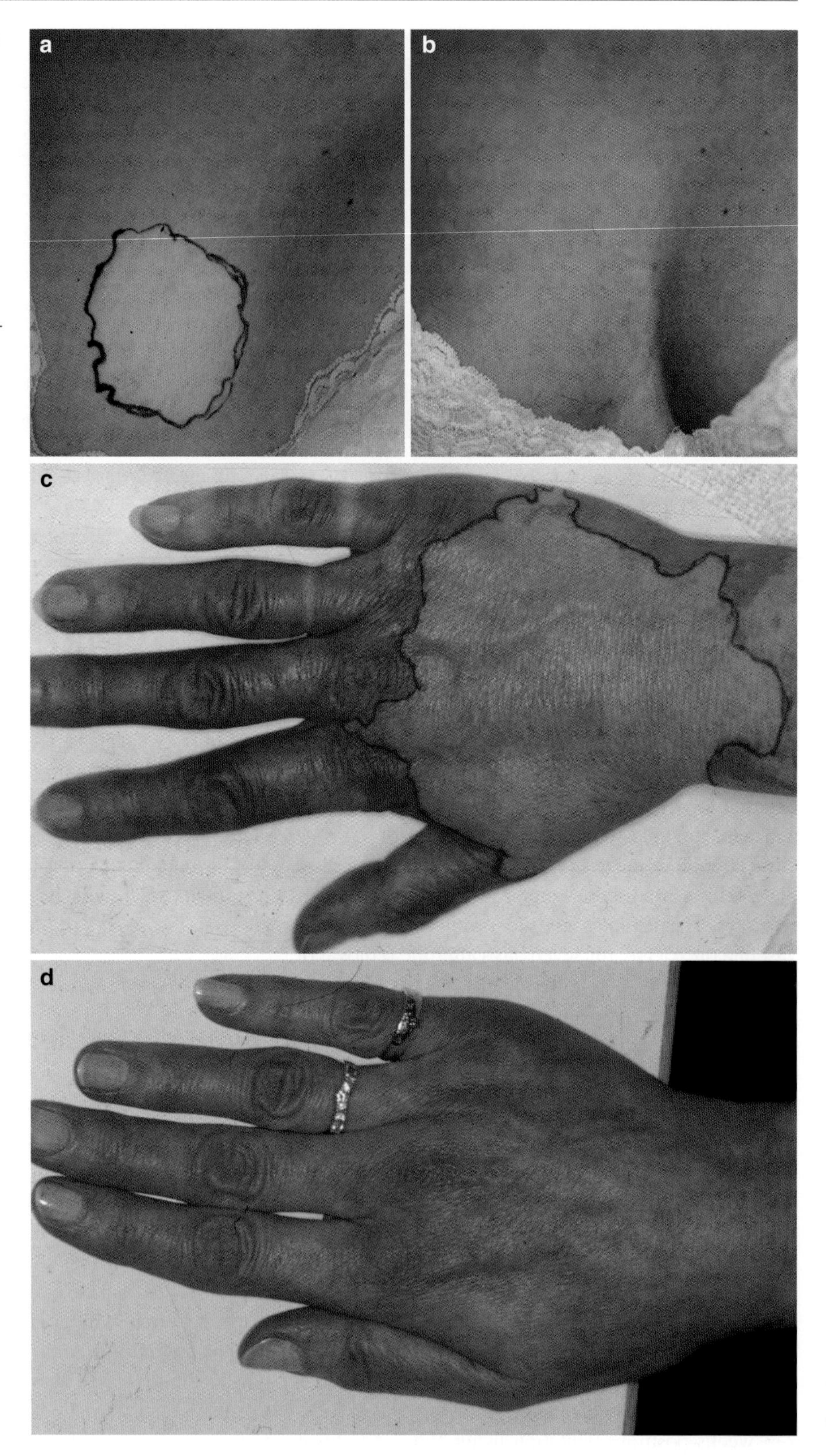

Fig. 3.7.1.8 A vitiligo lesion on the chest on a 24-year-old woman with generalized vitiligo before the treatment (**a**). The same woman 1 year after transplantation with autologous basal cell-layer suspension (**b**). A vitiligo lesion on the hand on a 47-year-old woman with generalized vitiligo before the treatment (**c**). The same woman 1 year after transplantation with cultured autologous melanocytes (**d**)

Calculation of Areas

The borders of the chosen recipient areas are marked with a pen (use Wood's-light on fair skin). A transparent film (overhead-projector plastic sheets) is put over the marked areas and the markings are transferred to the film. The film is copied in a copy machine (Xeroxed) and the outlined areas on the paper are cut out and weighed on an electronic balance. The total area is calculated. This is an important maneuver that will ensure that excessive donor tissue is not harvested or excessive recipient area is not denuded.

Premedication

A still and calm patient is important for success of the procedure. Therefore, we usually recommend 5–10 mg of diazepam and/or 10 mg of ketobemidone, orally, 60–80 min before transplantation, plus 500–1,000 mg of paracetamol 40–50 min before the surgery. Heracillin (Flukoloxacillin), or other preferable antibiotic, is given for eight days, starting on the day of surgery. Always ask for a history of known allergic reaction against penicillin and ask to avoid any medicine containing salicylates (e.g., Aspirin) for 10 days before surgery and 1 week after the surgery.

Anesthesia of Recipient Site

The recipient area is anesthetized as mentioned earlier. Local anesthesia with a freezing spray such as ethyl chloride or fluor-ethyl can also be used immediately prior to the dermabrasion. The sprays provides some anesthesia but not sufficient to give total pain-relief. However, when skin is chilled it also becomes more firm and easy to dermabrade. It is also much easier for an inexperienced eye to detect remaining epidermal remnants on a chilled denuded surface.

Preparation of the Recipient Site (as Above)

The recipient area is cleaned with alcohol, outlined with a sterile surgical marker pen, and the epidermis is removed down to the dermal–epidermal junction, using a high-speed dermabrader (20,000 rpm), fitted with a diamond fraise.

Donor Area

Normally, pigmented gluteal or thigh skin is shaved and sterilized with alcohol. An area of the size of the recipient patch(es) is marked with a sterile surgical marking pen. The marked area is anesthetized as mentioned earlier. The skin is stretched hard by an assistant and a very shallow sheet of skin is taken with a high speed Zimmer®-air driven ultra-dermatome (Zimmer Inc. Warsaw, IN). Depending on the anatomical area to be harvested, the nature of the skin and the total graft size needed, the machine is fitted with a special shield plate, controlling the width of the sheet. The machine is delivered in a special autoclavable metal box, including a set of four different bottom plates with width gaps from 1 to 4 in. The level control is set at a shallow marking. In principle, it is possible to harvest thinnest possible grafts, but in practice, a coherent melanocyte containing sheet, strong enough to be transferred to and fastened onto the recipient area, is needed. This is craftsmanship, and machines and operators have their own small individual peculiarities and nature, and therefore, the user has to set right his/her own individual settings, depending on personal angle approach (about 25°), forward speed, and downward pressure. The harvested sheet is immediately moistened with s-MEM, that is, Joklik's modified MEM (GIBCO BRL, Life Technology), so as not to dry out before it is grafted to the recipient site.

The donor area is covered for 8–9 days as mentioned earlier.

Transplantation Phase

The thin epidermal sheets are applied onto the dermabraded recipient areas. For large areas the sheets are cut into somewhat smaller pieces to allow the drainage of the exudate. The sheet can be moistened and stretched out on the machine itself or in a large Petridish there, and can easily be cut in suitable pieces with an ordinary scalpel. The epidermal sheet is transferred to the recipient area with the help of two forceps holding the corners or for smaller pieces on the blade of a scalpel or sterilized microscope slide. To ensure that no bubbles or fluid are left within between the wound-bed and the epidermal sheet, a metal spatula or curved iris-scissors is used to scrape the surface of the sheet. The grafts are secured with a silicone netting as described earlier.

In all cases, the patient should stay at least for 4 h in a hospital bed after the procedure has been completed, if possible longer. Immobilization improves the chances for a good outcome. Hands should be splinted, and if legs or feet are treated, the patient should be transported in a wheelchair to the car for home transportation. The bandages are removed after 8–9 days.

Follow-Up Inspection

An overall follow-up evaluation at 5–8 months after transplantation is recommended.

During the first two weeks, the transplanted areas are slightly erythematous. Some crust formations of overlapping skin or dry oozed serum in the corners of the areas and small punctual bleedings under the sheets can be seen during this period. Often the pigmentation can be seen immediately, but in fair-skinned individuals it may take some time before the outcome can be judged. Woods-light or diascopy pressure can help when evaluating at an early stage.

During the first year, a slight hyper- or hypo-pigmentation in some of the treated areas is not uncommon, but the color gradually matches with the complexion of the surrounding skin and most often blends in well in about 1–2 years. Long-lasting hyperpigmentations can occur.

A white, 1–2 mm halo between the transplant and normally pigmented skin can sometimes be seen. This is seen less often in patients with piebaldism or segmental vitiligo, compared with generalized vitiligo. Most often this achromic border is filled after some time of repeated sun-exposure.

Evaluation and Documentation

The final result can be seen about one year after the surgery. The transparent plastic film, used prior to the surgery to calculate the recipient areas is stored in the patient's file and used for evaluation at follow-up inspection. The film is put over the lesions to evaluate the outcome/progress. It is important to mark the angle of an elbow, knee, or neck on the plastic sheets since a change in position (degrees) can alter the size.

Total body chart drawings and photos of the treated lesions shall be taken before the treatment and at follow-ups.

An estimation of the total extension of the white areas (treated and not treated) is also important to document. This is important for planning of future sessions in patients with extensive areas and for scientific evaluation of outcome, when comparing the results in extensive versus nonextensive generalized vitiligo [27]. Files with anamneses and treatment charts are kept in the hospital's central records.

Summary Messages

- To overcome the lack of response in many patients to pharmacological and UV-treatments a number of surgical techniques have been developed.
- In segmental vitiligo, transplantation methods are often the only effective means to restore the pigmentation in the affected areas.
- The condition should be stable before surgery is performed.
- Surgical transplantation is the fastest way of restoring a loss of pigment cells.
- The surgical techniques can roughly be grouped in techniques utilising harvested tissue direct, without any processing, or techniques based on processing the donor tissue/cells before the use.
- In generalised vitiligo it is important to let the patient understand that the treatment is not a cure for the underlying disease, merely a treatment of the symptoms.
- Always be realistic to yourself and the patient when you predict the outcome.
- If you or the patient is not sure, in this specific case, that surgical intervention is the treatment of choice, avoid treating.
- Avoid treating young children. The expressed desire of undergoing a treatment should come from the patient and not from the parents.
- It is important to protect both the patient and yourself by working in sterile conditions such as lamina flow hood and using protecting gears such as face visor and FFP2 class face mask.
- A successful treatment can change the life for the individual and his/her family.

References

1. Agrawal K, Agrawal L (1995) Vitiligo: repigmentation with dermabrasion and thin split-thickness skin graft. Dermatol Surg 21:295–300
2. Behl PN, Bhatia RK (1973) Treatment of vitiligo with autologous thin Thiersch's grafts. Int J Dermatol 12:329–331
3. Boersma BR, Westerhof W, Bos JD (1995) Repigmentation in vitiligo vulgaris by autologous minigrafting: results in nineteen patients. J Am Acad Dermatol 33:990–995
4. Brysk MM, Newton RC, Rajaraman S et al (1989) Repigmentation of vitiliginous skin by cultured cells. Pigment Cell Res 2:202–207
5. Eves PC, Beck AJ, Shard AJ et al (2005) A chemically defined surface for the co-culture of melanocytes and keratinocytes. Biomaterials 26:7068–7081
6. Eves PC, Bullett NA, Haddow D et al (2008) Simplifying the delivery of melanocytes and keratinocytes for the treatment of vitiligo using a chemically defined carrier dressing. J Invest Dermatol 128:1554–1564
7. Falabella R (1983) Repigmentation of segmental vitiligo by autologous minigrafting. J Am Acad Dermatol 9:514–521
8. Falabella R, Escobar C, Borrero I (1989) Transplantation of in vitro-cultured epidermis bearing melanocytes for repigmenting vitiligo. J Am Acad Dermatol 21:257–264
9. Gauthier Y, Surleve-Bazeille JE (1992) Autologous grafting with noncultured melanocytes: a simplified method for treatment of depigmented lesions. J Am Acad Dermatol 26: 191–194
10. Guerra L, Primavera G, Raskovic D et al (2003) Erbium:YAG laser and cultured epidermis in the surgical therapy of stable vitiligo. Arch Dermatol 139:1303–1310
11. Gupta S, Goel A, Kanwar AJ, Kumar B (2006) Autologous melanocyte transfer via epidermal grafts for lip vitiligo. Int J Dermatol 45:747–750
12. Gupta S, Olsson MJ, Kanwar AJ, Ortonne JP (2007) Surgical management of vitiligo. Blackwell, Oxford
13. Hasegawa T, Suga Y, Ikejima A et al (2007) Suction blister grafting with CO(2) laser resurfacing of the graft recipient site for vitiligo. J Dermatol 34:490–492
14. Issa CM, Rheder J, Taube MB (2003) Melanocyte transplantation for the treatment of vitiligo: effects of different surgical techniques. Eur J Dermatol 13:34–39
15. Juhlin L, Olsson MJ (1995) Optimal application times of a eutectic mixture of local anaesthetics (EMLA) cream before dermabrasion of vitiliginous skin. Eur J Dermatol 5: 368–369
16. Kahn AM, Cohen MJ (1998) Repigmentation in vitiligo patients. Melanocyte transfer via ultra-thin grafts. Dermatol Surg 24:365–367
17. Kim HU, Yun SK (2000) Suction device for epidermal grafting in vitiligo: employing a syringe and a manometer to provide an adequate negative pressure. Dermatol Surg 26: 702–704
18. Koga M (1988) Epidermal grafting using the tops of suction blisters in the treatment of vitiligo. Arch Dermatol 124: 1656–1658
19. Lontz W, Olsson MJ, Moellmann G et al (1994) Pigment cell transplantation for treatment of vitiligo: a progress report. J Am Acad Dermatol 30:591–597
20. McGovern TW, Bolognia J, Leffell DJ (1999) Flip-top pigment transplantation: a novel transplantation procedure for the treatment of depigmentation. Arch Dermatol 135: 1305–1307
21. Olsson MJ (2004) What are the needs for transplantation treatment in vitiligo, and how good is it? Arch Dermatol 140:1273–1274
22. Olsson MJ, Juhlin L (1992) Melanocyte transplantation in vitiligo. Lancet 340:981
23. Olsson MJ, Juhlin L (1993) Repigmentation of vitiligo by transplantation of cultured autologous melanocytes. Acta Derm Venereol 73:49–51
24. Olsson MJ, Juhlin L (1995) Transplantation of melanocytes in vitiligo. Br J Dermatol 132:587–591
25. Olsson MJ, Juhlin L (1997) Epidermal sheet grafts for repigmentation of vitiligo and piebaldism, with a review of surgical techniques. Acta Derm Venereol 77:463–466
26. Olsson MJ, Juhlin L (1998) Leucoderma treated by transplantation of a basal cell layer enriched suspension. Br J Dermatol 138:644–648
27. Olsson MJ, Juhlin L (2002) Long-term follow-up of leucoderma patients treated with transplants of autologous cultured melanocytes, ultrathin epidermal sheets and basal cell layer suspension. Br J Dermatol 147:893–904
28. Olsson MJ, Moellmann G, Lerner AB, Juhlin L (1994) Vitiligo: repigmentation with cultured melanocytes after cryostorage. Acta Derm Venereol 74:226–228
29. Sachdev M, Krupashank DS (2000) Suction blister grafting for stable vitiligo using pulsed erbium:YAG laser ablation for recipient site. Int J Dermatol 39:471–473
30. Stromberg S, Bjorklund MG, Asplund A et al (2008) Transcriptional profiling of melanocytes from patients with vitiligo vulgaris. Pigment Cell Res 21:162–171
31. Suvanprakoren P, Dee-Ananlap S, Pongsomboon C, Klaus SN (1985) Melanocyte autologous grafting for treatment of leukoderma. J Am Acad Dermatol 13:968–974
32. Tsukamoto K, Osada A, Kitamura R et al (2002) Approaches to repigmentation of vitiligo skin: new treatment with ultrasonic abrasion, seed-grafting and psoralen plus ultraviolet A therapy. Pigment Cell Res 15:331–334
33. van Geel N, Ongenae K, De Mil M et al (2004) Double-blind placebo-controlled study of autologous transplanted epidermal cell suspensions for repigmenting vitiligo. Arch Dermatol 140:1203–1208

3.7.2 The Outcomes: Lessons About Surgical Therapy for Vitiligo in the Past Two Decades

Rafael Falabella

Contents

R. Falabella
Universidad del Valle and Centro Dermatologico de Cali, Cali, Colombia
e-mail: rfalabella@uniweb.net.co

3.7.2.1 Introduction

Melanocyte transplantation has become a standard procedure for depigmented skin in vitiligo. For this purpose, different surgical approaches have been described, most of them providing remarkable restoration of affected skin with acceptable or minimal side effects. However, these therapies need to be performed under rigorous conditions to avoid repigmentation failure, side effects, and undesirable pigmentation abnormalities following surgery, although minor pigmentary changes are normally expected.

During the last 20 years, much has been learned in regard to surgical treatments with diverse methods allowing many patients to recover from depigmentation, with good and excellent cosmetic results, minimizing unwanted complications [8, 14, 16].

This chapter summarizes the most important aspects learned during the last two decades in regard to melanocyte transplantation.

3.7.2.2 Melanocytes Can Be Effectively and Safely Transplanted for Repigmentation of Depigmented Skin

From early trials with melanocyte transplantation [2, 6], it was learned that pigment cells could be "moved" from donor sites at their original location and placed on a new recipient depigmented bed, where they continued the normal pigmentation process. No reports of specific complications have been reported thereafter, when techniques are performed within established parameters as evidenced in numerous publications

M. Picardo and A. Taïeb (eds.), *Vitiligo*,
DOI 10.1007/978-3-540-69361-1_3.7.2, © Springer-Verlag Berlin Heidelberg 2010

done so far. A slight degree of hyperpigmentation may be expected in a small percentage of patients, but improvement far exceeds this minor side effect [18].

3.7.2.3 Melanocyte Transplantation Is Mainly Successful in Stable Vitiligo

When selecting patients with depigmentation for melanocyte transplantation, one of the important issues is to define the stability of the depigmented area to be treated. There are two clinical situations in which melanocyte transplantation is successful: (a) segmental (unilateral) vitiligo [7] and (b) vitiligo vulgaris (bilateral) of long duration, where stability is achieved after a prolonged course of the disease. But in such a case, vitiligo stability should be carefully evaluated [12, 17]. A minigrafting test with 4–5 minigrafts of 1.0–1.2 mm implanted within the area to be treated may yield important information about vitiligo stability (Fig. 3.7.2.1).

3.7.2.4 With Less Invasive Methods, Better Results Are Achieved

From different techniques, those methods in which the epidermis is only manipulated without harming the dermis with invasive procedures, as in epidermal grafting [15, 24, 22], epidermal suspensions [19, 30], in vitro cultured epidermis with melanocytes [29, 37], or melanocyte suspensions [5], excellent cosmetic results without scarring are achieved, since the surgical trauma is minimal and the healing process occurs faster.

In addition, when dermo–epidermal grafts performed with very thin epidermal sheets [25] and minigrafts are harvested and implanted with a very small punch (1.0–1.2 mm), dermal disturbance is also minimal, and repigmentation outcomes are very acceptable [14].

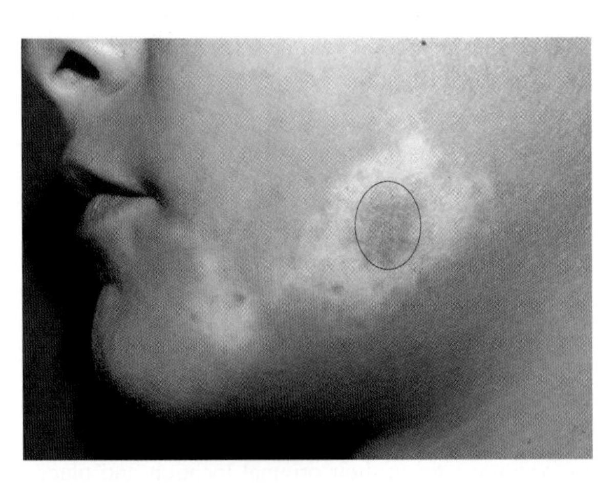

Fig. 3.7.2.1 Minigrafting test. This 12-year-old boy has had segmental vitiligo on his cheek for 2 years. Five minigrafts implanted within the depigmented lesion repigmented an area of about 1.3 cm in diameter (area within *circle*) indicating high possibilities for repigmentation of the remaining affected skin with additional treatment

3.7.2.5 PUVA, NB-UVB, or Sunlight Exposure Enhance Repigmentation Rates

After healing, the grafted area slowly recovers, and repigmentation occurs several weeks after surgery, although color matching is a more slow process, which gradually improves during several months. Several publications indicate that with either PUVA, or NB-UVB, or plain sunlight exposure starting 2–3 weeks after healing, repigmentation will be faster, with homogeneous spreading, and pigmentation will be similar to normal surrounding skin [34, 26]. Depigmented spaces between grafts repigment more easily when light therapy is used.

3.7.2.6 The Best Repigmentation Results Are Achieved in Segmental Vitiligo

With melanocyte transplantation procedures in segmental (unilateral) vitiligo, high repigmentation rates, above 80% of treated patients [12, 29] are achieved (Fig. 3.7.2.2); probably, the pathogenic mechanisms in this clinical form come to a halt and repigmentation occurs in absence of a depigmenting and active noxa (Fig. 3.7.2.5). On the other hand, in vitiligo vulgaris, excellent repigmentation rates around 50% in stable disease [17, 30] are observed.

The crucial point to get a favorable response after melanocyte transplantation is vitiligo activity and hence the most accurate evaluation in this regard is recommended [31].

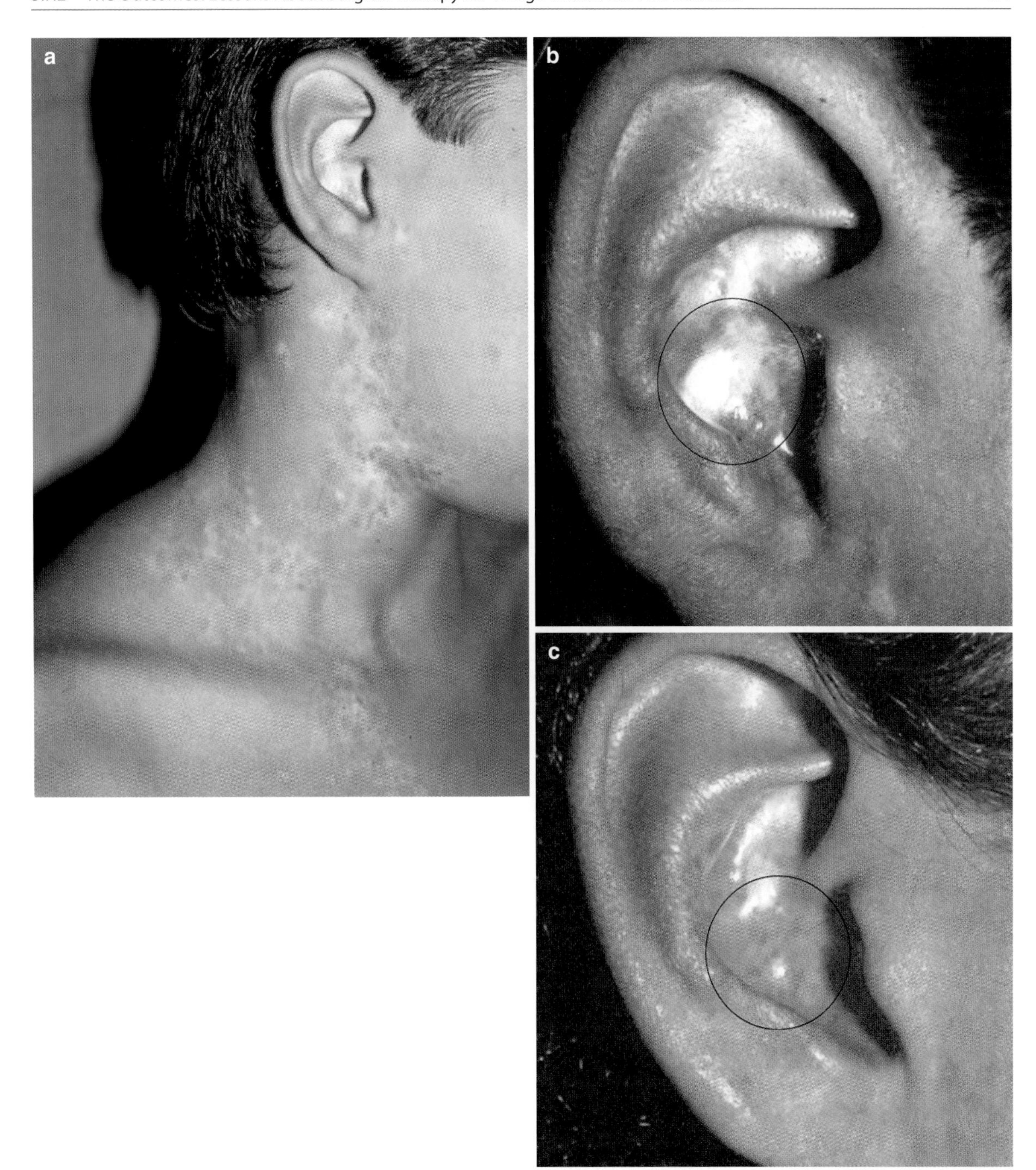

Fig. 3.7.2.2 (**a**) Segmental vitiligo. Repigmentation by medical therapy. Same patient of (**b**, **c**) disclosing the affected neck area which became irregularly repigmented after treatment with topical corticosteroid therapy during 2 years. (**b**) Segmental vitiligo before surgical treatment. This 15-year-old boy has had segmental vitiligo affecting a small area of the right ear, next to opening of the external auditory canal in a difficult location for grafting, where a depigmented spot of about 1 cm can be observed (lesion within *circle*); an extensive area of the right side of his neck was also involved (see (**a**)). (**c**) Segmental vitiligo after surgical treatment The lesion was completely repigmented after treatment with 5 minigrafts of 1.0 mm and complete repigmentation was achieved as observed 12 months later (repigmented area within *circle*)

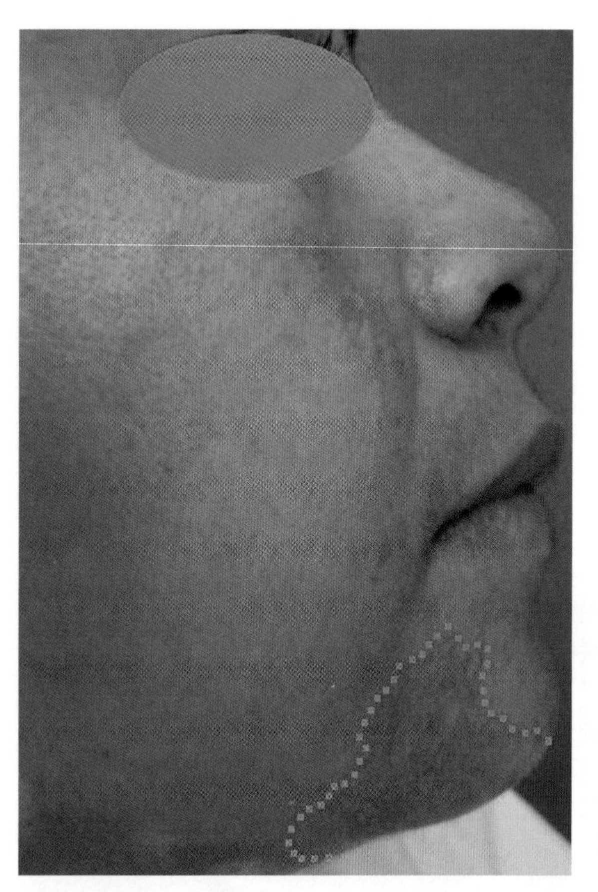

Fig. 3.7.2.5 Segmental vitiligo and long-term results. This 38-year-old woman had segmental vitiligo on her chin at age 18, which was completely repigmented with mingrafting at that time (area inside of *dotted line*). Twenty years have elapsed since original treatment and repigmentation remains unchanged indicating definitive permanency of melanocytes and pigmentation

3.7.2.7 Lasers Are Useful to Denude Recipient Sites

One of the most important steps during melanocyte transplantation is preparing the recipient site. Since melanocytes are located in the basal-cell layer, removal of epidermis alone is needed, and every effort should be done to avoid invasive procedures that may induce scarring.

Removing the epidermis and the most superficial layers of the papillary dermis with modern lasers such as pulsed erbium:YAG [20] and short-pulsed CO_2 lasers [4, 35], excellent repigmentation without scarring has been achieved. When feasible, the use of lasers for this purpose is highly encouraged.

3.7.2.8 Frequent Failures with Surgery in Acral Vitiligo Are Seen

As it occurs with medical treatment of vitiligo, acral areas are refractory and highly resistant to all types of melanocyte transplantation procedures [18, 21, 34]. Inappropriate immobilization of grafts on recipient sites may be important for this outcome in addition to other unknown factors.

However, in very stable vitiligo, that is segmental (unilateral) or long-standing quiescent vitiligo vulgaris (bilateral), repigmentation is possible, but several procedures may be needed, possibly because of difficulties with graft survival [11].

3.7.2.9 Surgical Repigmentation Is Similar to Medical Treatment

There are frequent concerns in regard to side effects and unsightly appearance after surgical procedures, but this is only true with very invasive procedures. Most of reported complications and unacceptable side effects have been published with punches larger than 1.2 mm and after dermo–epidermal grafts are performed without the thin-graft technique. If properly done, surgical procedures initiate similar results as medical therapy with insignificant side effects, if any [18, 30]; sometimes, repigmentation achieved by medical treatment is irregular and less homogeneous than that obtained with surgical methods (Fig. 3.7.2.4).

3.7.2.10 Surgical Repigmentation Is Permanent

This is a statement that depends on vitiligo stability, an appropriate technique and a demonstration of stability such as the minigrafting test described earlier. If stable, vitiligo repigmentation is usually permanent, but if unstable, vitiligo depigmentation occurs shortly after surgery [15]. We have observed 3 patients after 7 years with permanent repigmentation, and 5 others over 15 years without depigmentation (Fig. 3.7.2.3).

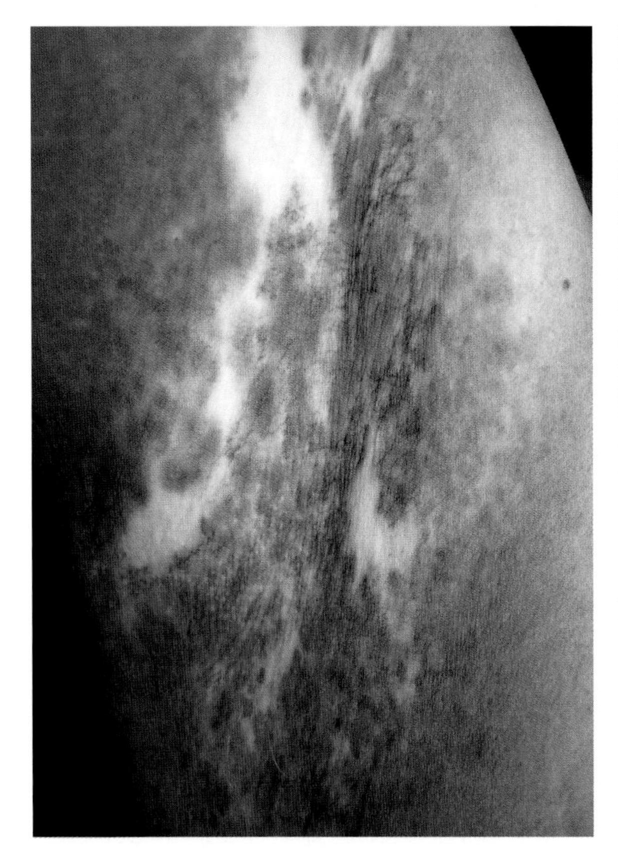

Fig. 3.7.2.4 Axilla after medical treatment. This 56-year-old woman has had vitiligo for over 30 years. Both axillary regions were affected but after 2 years repigmentation occurred in most of involved areas with topical corticosteroids. Notice the irregularity of repigmentation, lack of complete pigment coalescence and diverse skin color tonalities compared with the homogeneous repigmentation achieved by surgical methods as depicted in other illustrations of this article

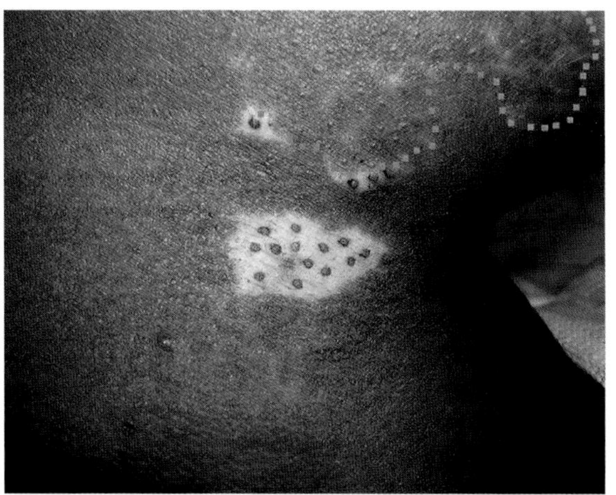

Fig. 3.7.2.3 Segmental vitligo. This 13-year-old boy had segmental vitiligo on his right cheek and neck for the past 4 years. Previous surgical treatment with 1 mm minigrafts successfully repigmented the upper cheek (within *dotted line*). Additional minigrafts to complete repigmentation have been implanted next and below the already treated area

3.7.2.11 Combination Therapy Is Useful to Enhance Repigmentation Rates

Different repigmentation procedures in patients with vitiligo are followed by adequate repopulation of melanocytes and pigmentation recovery is usually achieved. However, in multiple occasions, small areas, randomly distributed, measuring a few cm^2 or mm^2 may still show persistent depigmentation which remains indefinitely. For completing repigmentation, depigmented spots may be treated with a few minigrafts (1 or 1.2 mm) instead of performing the most time-consuming methods originally used [13], such as epidermal grafting, epidermal suspensions, and in vitro cultured grafts.

3.7.2.12 In Vitro Cultured Epidermal Sheets and Melanocyte Suspensions Are Important Options for Repigmenting Stable Vitiligo

Since first reported in 1989 [3, 9], cultured melanocytes in vitro have been shown to be useful in the management of refractory and stable vitiligo. Several methods with epidermal sheets bearing melanocytes [10, 11, 33, 35] and melanocyte cell suspensions [27, 28, 32] have been described with high rates of repigmentation.

The most important indication for these methods is the extensive form of vitiligo, where other methods have a limited success, since from small donor site samples large amounts of cells can be harvested and extensive areas may be treated.

In addition, 19 years have elapsed since the first transplantation trials of in vitro cultures [3,9], and so far no important complications have been reported in spite of numerous publications with diverse methods.

These techniques may be the therapy of the future when simplified and become cost effective.

3.7.2.13 Repigmentation of Leukotrichia May Be Achieved by Melanocyte Transplantation

During surgical repigmentation with in vitro cultured grafts [11], epidermal grafts [23], and dermoepidermal grafts [1], several authors noticed that hairs within vitiligo lesions became repigmented. A reversal mechanism of repigmentation in which the transplanted melanocytes migrate from the epidermis toward the outer root sheath and thereafter to the hair bulb, as opposed to the normal pathway of perifollicular melanocytes migrating toward the epidermis during vitiligo repigmentation, has been proposed as the mechanism of hair-shaft repigmentation [11] (see Chap. 1.3.7, hair melanocytes).

3.7.2.14 What to Avoid when Performing Melanocyte Transplantation

There are certain important principles that should be followed in order to achieve good and excellent results and avoid complications. For this purpose it is recommended not to:

1. Perform procedures in patients with a hyperpigmentation diathesis, which may be observed in areas of recent superficial skin trauma.
2. Harvest thick dermo–epidermal grafts with knife or dermatome.
3. Use punch grafts larger than 1.2 mm (1.0 mm ideal for facial areas).
4. Treat patients with unstable, active vitiligo.
5. Attempt to repigment extensive areas without planning full treatment.
6. Offer therapeutic success without previous repigmentation testing or at least warning about expected results.
7. Manipulate recipient sites with excessively invasive methods.

Summary Messages

- Surgical therapy is not a first option in vitiligo; medical treatment should always be first
- One of the most important recommendations when surgical therapy is an option, is to define stable disease as close as possible
- Preparation of recipient sites by removal of epidermis with lasers is a major improvement of techniques for avoiding excessive dermal manipulation during surgery
- Stable segmental vitiligo treated with minimally invasive methods offers the best repigmentation outcome which is usually permanent
- A technique well performed, regardless the selected method may offer appropriate repigmentation, but knowledge of the multiple factors involved in these procedures is essential for achieving acceptable results
- A simple method is minigrafting, but small minigrafts (1.0 mm for facial areas and 1.2 mm for remaining areas) are required
- For better and less expensive results, epidermal suspensions and epidermal grafting yield optimal results
- Completely depigmented lesions yield better results than partially hypopigmented defects after melanocyte transplantation
- When performing surgical therapy, a variable proportion of grafts may not survive and additional grafting or "touch ups" may be anticipated
- UV light therapy and combination procedures may enhance repigmentation in treated patients

References

1. Agrawal K, Agrawal A (1995) Vitiligo: surgical repigmentation of leukotrichia. Dermatol Surg 21:711–715
2. Behl PN, Bhatia RK (1973) Treatment of vitiligo with autologous thin Thiersch's grafts. Int J Dermatol 12:329–331
3. Brysk MM, Newton RC, Rajaraman S et al (1989) Repigmentation of vitiliginous skin by cultured cells. Pigment Cell Res 2:202–207
4. Chen YF, Chang JS, Yang PY et al (2000) Transplant of cultured autologous pure melanocytes after laser-abrasion for the treatment of segmental vitiligo. J Dermatol 27:434–439

5. Chen YF, Yang PY, Hu DN et al (2004) Treatment of vitiligo by transplantation of cultured pure melanocyte suspension: analysis of 120 cases. J Am Acad Dermatol 51:68–74
6. Falabella R (1971) Epidermal grafting. An original technique and its application in achromic and granulating areas. Arch Dermatol 104:592–600
7. Falabella R (1988) Treatment of localized vitiligo by autologous minigrafting. Arch Dermatol 124:1649–1655
8. Falabella R (1989) Grafting and transplantation of melanocytes for repigmenting vitiligo and other types of leukoderma. Int J Dermatol 28:363–369
9. Falabella R, Escobar C, Borrero I (1989) Transplantation of in vitro-cultured epidermis bearing melanocytes for repigmenting vitiligo. J Am Acad Dermatol 21:257–264
10. Falabella R, Borrero I, Escobar C (1989) In vitro culture of melanocyte-bearing epidermis and its application in the treatment of vitiligo and the stable leukodermas. Med Cutan Ibero Lat Am 17:193–198
11. Falabella R, Escobar C, Borrero I (1992) Treatment of refractory and stable vitiligo by transplantation of in vitro cultured epidermal autografts bearing melanocytes. J Am Acad Dermatol 26:230–236
12. Falabella R, Arrunategui A, Barona MI et al (1995) The minigrafting test for vitiligo: detection of stable lesions for melanocyte transplantation. J Am Acad Dermatol 32:228–232
13. Falabella R, Barona M, Escobar C et al (1995) Surgical combination therapy for vitiligo and piebaldism. Dermatol Surg 21:852–857
14. Falabella R (1997) Surgical therapies for vitiligo Clin Dermatol 15:927–939
15. Falabella R (2001) Surgical therapies for vitiligo and other leukodermas, part 1: minigrafting and suction epidermal grafting. Dermatol Ther 14:7–14
16. Falabella R (2003) Surgical treatment of vitiligo: why, when and how. J Eur Acad Dermatol Venereol 17:518–520
17. Falabella R (2004) The minigrafting test for vitiligo: Validation of a predicting tool. J Am Acad Dermatol 51: 672–673
18. Falabella R (2005) Surgical approaches for stable vitiligo. Dermatol Surg 31:1277–1284
19. Gauthier Y, Surleve-Bazeille JE (1992) Autologous grafting with noncultured melanocytes: a simplified method for treatment of depigmented lesions. J Am Acad Dermatol 26: 191–194
20. Guerra L, Primavera G, Raskovic D et al (2003) Erbium:YAG laser and cultured epidermis in the surgical therapy of stable vitiligo. Arch Dermatol 139:1303–1310
21. Guerra L, Capurro S, Melchi F et al (2000) Treatment of "stable" vitiligo by Timedsurgery and transplantation of cultured epidermal autografts. Arch Dermatol 136:1380–1389
22. Gupta S, Kumar B (1992) Epidermal grafting in vitiligo: influence of age, site of lesion, and type of disease on outcome. J Am Acad Dermatol 49:99–104
23. Hann SK, Im S, Park YK (1992) Repigmentation of leukotrichia by epidermal grafting and systemic psoralen plus UV-A. Arch Dermatol 128:998–999
24. Hann SK, Im S, Bong HW (1995) Treatment of stable vitiligo with autologous epidermal grafting and PUVA. J Am Acad Dermatol 32:943–948
25. Kahn AM, Cohen MJ (1998) Repigmentation in vitiligo patients. Melanocyte transfer via ultra-thin grafts. Dermatol Surg 24:365–367
26. Lahiri K, Malakar S, Sarma N et al (2006) Repigmentation of vitiligo with punch grafting and narrow-band UV-B (311 nm) – a prospective study. Int J Dermatol 45:649–655
27. Liu JY, Hafner J, Dragieva G et al (2004) Bioreactor microcarrier cell culture system (Bio-MCCS) for large-scale production of autologous melanocytes. Cell Transplant 13: 809–816
28. Löntz W, Olsson MJ, Moellmann G (1994) Pigment cell transplantation for treatment of vitiligo: a progress report. J Am Acad Dermatol 30:591–597
29. Mulekar SV (2005) Long-term follow-up study of 142 patients with vitiligo vulgaris treated by autologous, non-cultured melanocyte-keratinocyte cell transplantation. Int J Dermatol 44:841–845
30. Olsson MJ, Juhlin L (2002) Long-term follow-up of leucoderma patients treated with transplants of autologous cultured melanocytes, ultrathin epidermal sheets and basal cell layer suspension. Br J Dermatol 147:893–904
31. Olsson MJ, Juhlin L (1995) Transplantation of melanocytes in vitiligo. Br J Dermatol 132:587–591
32. Pianigiani E, Andreassi A, Andreassi L (2005) Autografts and cultured epidermis in the treatment of vitiligo. Clin Dermatol 23:424–429
33. Skouge J, Morison WL (1995) Vitiligo treatment with a combination of PUVA therapy and epidermal autografts. Arch Dermatol 131:1257–1258
34. Toriyama K, Kamei Y, Kazeto T et al (2004) Combination of short-pulsed CO_2 laser resurfacing and cultured epidermal sheet autografting in the treatment of vitiligo: a preliminary report. Ann Plast Surg 53:178–180

Combined Therapies for Vitiligo

3.8

Thierry Passeron and Jean-Paul Ortonne

Contents

T. Passeron (✉)
Department of Dermatology, University Hospital of Nice, Nice, France
e-mail: passeron@unice.fr

3.8.1 Introduction

Vitiligo is a challenging disorder to treat. To date, no treatment gives truly satisfactory results. Some areas, such as bone prominences and extremities are almost impossible to fully repigment. However, from molecules that target the immune system, to therapies that stimulate the proliferation and differentiation of melanocytes and/or melanocyte progenitors, and others that fight against radical species or bring new melanocytes to the affected areas, we now have many strategies to repigment vitiligo patches. Literature is surprisingly very poor concerning the evaluation of the combination of those repigmenting strategies. However, in the light of existing results, combination therapies appear to provide higher rates of repigmentation as compared with traditional monotherapies.

3.8.2 Combination of Surgical Therapies and Phototherapy

Surgical therapies of vitiligo such as punch, blister, split thickness grafting, or more recently transplanted autologous melanocyte or epidermal-cell suspensions have been the first therapeutic options to be associated with phototherapy. In 1983, Bonafé et al. [6] first reported the combined use of skin grafts followed with psoralen plus ultraviolet A (PUVA) therapy. The combination of several surgical therapies of vitiligo with PUVA has been then further studied and remains today the best tested combination in vitiligo treatment. Indeed, several studies have shown that adjunction of PUVA following surgical treatment enhances the repigmentation rate of vitiligo patches [3, 5, 15, 16,

M. Picardo and A. Taïeb (eds.), *Vitiligo*,
DOI 10.1007/978-3-540-69361-1_3.8,

22, 37, 39, 46]. Most of the authors add phototherapy several weeks after the surgical procedures (usually 3 or 4 weeks after) and use standard phototherapy protocols for vitiligo treatment. The interest of combining phototherapy and surgical procedures has been proven in a prospective, randomized, double-blind study that clearly showed that autologous transplanted epidermal-cell suspensions followed by narrow-band UVB (NB-UVB) or PUVA was clearly superior to phototherapy alone for repigmenting vitiligo [46].

Combination of NB-UVB with surgical therapies has been less studied, while it is a fact that this enhances repigmentation [21, 33, 44]. No direct comparison between NB-UVB and PUVA, as a synergic agent for surgical therapies, is available yet. A series of five patients who failed to repigment after the combination of grafting plus PUVA has been reported to finally repigment when the surgical procedure was combined with NB-UVB [20]. Although interesting, this report is not sufficient to prove the superiority of NB-UVB in this indication, and further studies are required.

3.8.3 Combination of Surgical Therapies and Corticosteroids

Although most combined approaches using surgical therapies have been performed with phototherapy, topical and systemic steroids have also been evaluated. In a prospective study performed on 50 vitiligo patients, the combination of punch grafting with topical steroids (fluocinolone acetonide 0.1%) has been shown to be as effective as punch grafting followed by PUVA [4].

More recently, an open series and a case report have suggested that low doses of oral steroids might be beneficial in addition to surgical procedures [24, 29]. This needs to be confirmed in a controlled trial.

3.8.4 Combination of Phototherapy and Topical Steroids

The combined use of UVA with topical steroids has been studied by the Westerhof group [45]. In a prospective, randomized, controlled, left–right comparison study, it was shown that the combination of UVA and fluticasone propionate was much more effective than UVA or topical steroid alone.

To the best of our knowledge, the combination of UVB therapy with topical steroids has not been still evaluated, although some series have studied the association of UVB and other synergistic drugs. In our experience, adjunction of Class 3 topical steroids with 308 nm excimer laser increases its efficiency for treating vitiligo (Fig. 3.8.1). Association of daily application of topical steroid increases the speed and the rate of repigmentation [14]. This is particularly helpful in difficult to treat areas such as bone prominences. Indeed, prospective studies are certainly lacking, but the combination of topical steroids and UVB sources (NB-UVB and 308 nm

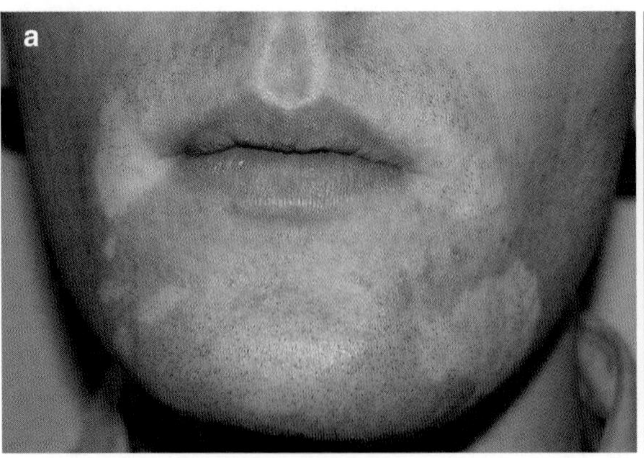

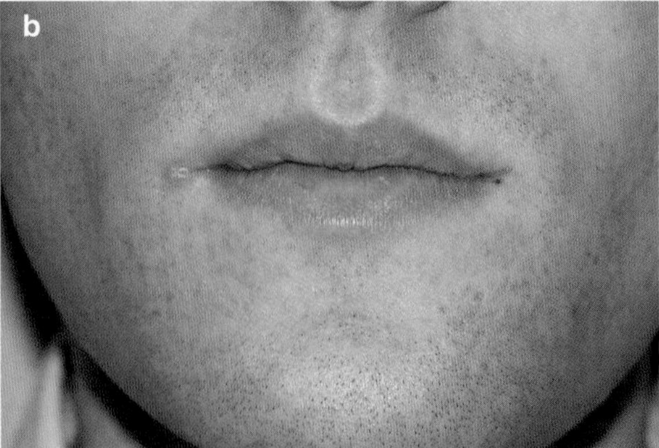

Fig. 3.8.1 (**a**) Vitiligo patches of the face. (**b**) Clinical aspect after 12 weeks of twice weekly 308 nm excimer laser combined with daily application desonide 0.05% cream.

excimer lasers or lamps) should provide a useful therapeutic approach for difficult to treat areas.

3.8.5 Combination of Phototherapy and Topical Calcineurin Inhibitors (TCI)

TCI, such as tacrolimus and pimecrolimus have shown some interesting results in the treatment of vitiligo, but the best results were achieved in sun-exposed area, especially on the face [13, 25, 38, 40, 42]. Treatment with such drugs in monotherapy for other areas provided disappointing results [29]. Two studies evaluated if the combination of 308 nm excimer laser and topical tacrolimus could be synergistic. They have compared the efficiency of 308 nm excimer combined with tacrolimus ointment to 308 nm excimer laser monotherapy [31] or associated with placebo ointment [18]. In both cases, a total of 24 sessions were evaluated and tacrolimus ointment was applied twice a day. The results were similar and showed a greater efficiency with the combined treatment as compared to laser alone. Tolerance was good and side effects were limited to constant erythema, sticking due to ointment, and rare bullous lesions. These encouraging results are corroborated by two other reports associating UVB light and topical tacrolimus [6, 41]. More recently, a large retrospective study reported the synergistic effect of the combination of topical tacrolimus with NB-UVB [10]. However, the increased risk of skin cancers promoted by the association of two immunosuppressive treatments remains to be taken into consideration. Caution is thus necessary before embarking on long-term protocols, and registries of treated patients should be implemented.

3.8.6 Combination of Phototherapy and Topical Vitamin D

The occurrence of repigmentation of vitiligo in patients treated with calcipotriol (a Vitamin D3 analog) for psoriasis has suggested that it might be efficacious in treating vitiligo. It appears now quite clear that calcipotriol in monotherapy is useless for treating vitiligo [8]. The use of calcipotriol with sun or PUVA therapy has provided some interesting rates of repigmentation. However, the efficacy of this combination is not clearly established [4, 29, 30]. Combination of calcipotriol and UVB also provides controversial data. About half the studies show more potency of UVB when combined with calcipotriol or other topical Vitamin D analogs, whereas the other half reports no enhancing effect on repigmentation [1, 2, 11, 17, 19, 24, 26].

Calcipotriol treatment associated with 308 nm excimer laser was also recently evaluated in a short prospective study [12]. Results show that the adjunction of topical calcipotriol does not increase the efficacy of the 308 nm excimer laser. Thus, if vitamin D analogs increase the pigmentation induced by UVB, their effects appear to be at best very limited.

3.8.7 Combination of Phototherapy and Antioxidants

Oxidative stress has been shown to be involved in the pathogenesis of vitiligo. Pseudocatalase has the ability to remove hydrogen peroxide and so could be interesting in the treatment of vitiligo. The combination of topical pseudocatalase with UVB has shown very promising results in a pilot non-RCT study (complete repigmentation on the face and the dorsum of the hands in 90% of patients) [35]. Unfortunately, these results were not confirmed in another study [33].

More recently, oral antioxidant supplementation was reported to increase the effectiveness of UVB phototherapy in a prospective double-blind placebo-controlled study [8]. Although this study concerned a relatively small number of patients with only 80% having completed the entire treatment, the results are certainly encouraging. This potential interest of oral antioxidants was also supported by a recent prospective double-blind series that study the combination of *Polypodium leucotomos* extract with NB-UVB [27]. Better repigmentation was observed with combination of *Polypodium leucotomos* extract and UV, but the difference was slight and only observed in a subgroup of patients. Thus, larger studies are surely required to confirm the interest of oral antioxidants combined with phototherapy.

Summary Messages

- Most of the studies on new combinatory appraches confirm the synergistic effect of targeting different ways for promoting repigmentation.
- Up to now, only the combination of surgical procedures with phototherapy has been clearly shown to increase repigmentation rates.
- Some prospective randomized studies support the association of topical calcineurin inhibitors with phototherapy.
- Such combinations will probably have limited effects for vitiligo lesions relatively easy to repigment such as those of the face, but should provide better therapeutic approaches on difficult to treat areas.
- Optimal combination of those therapies has to be further studied to maximize repigmentation.

References

1. Ada S, Sahin S, Boztepe G et al (2005) No additional effect of topical calcipotriol on narrow-band UVB phototherapy in patients with generalized vitiligo. Photodermatol Photoimmunol Photomed 21:79–83
2. Arca E, Tastan HB, Erbil AH et al (2006) Narrow-band ultraviolet B as monotherapy and in combination with topical calcipotriol in the treatment of vitiligo. J Dermatol 33:338–343
3. Barman KD, Khaitan BK, Verma KK (2004) A comparative study of punch grafting followed by topical corticosteroid versus punch grafting followed by PUVA therapy in stable vitiligo. Dermatol Surg 30:49–53
4. Baysal V, Yildirim M, Erel A et al (2003) Is the combination of calcipotriol and PUVA effective in vitiligo? J Eur Acad Dermatol Venereol 17:299–302
5. Bonafé JL, Lassere J, Chavoin JP et al (1983) Pigmentation induced in vitiligo by normal skin grafts and PUVA stimulation: a preliminary study. Dermatologica 166:113–116
6. Castanedo-Cazares JP, Lepe V, Moncada B (2003) Repigmentation of chronic vitiligo lesions by following tacrolimus plus ultraviolet-B-narrow-band. Photodermatol Photoimmunol Photomed 19:35–36
7. Chiaverini C, Passeron T, Ortonne JP (2002) Treatment of vitiligo by topical calcipotriol. J Eur Acad Dermatol Venereol 16:137–138
8. Dell'Anna ML, Mastrofrancesco A, Sala R et al (2007) Antioxidants and narrow band-UVB in the treatment of vitiligo: a double-blind placebo controlled trial. Clin Exp Dermatol 32:631–636
9. Ermis O, Alpsoy E, Cetin L et al (2001) Is the efficacy of psoralen plus ultraviolet A therapy for vitiligo enhanced by concurrent topical calcipotriol? A placebo-controlled double-blind study. Br J Dermatol 145:472–475
10. Fai D, Cassano N, Vena GA (2007) Narrow-band UVB phototherapy combined with tacrolimus ointment in vitiligo: a review of 110 patients. J Eur Acad Dermatol Venereol 21:916–920
11. Goktas EO, Aydin F, Senturk N et al (2006) Combination of narrow band UVB and topical calcipotriol for the treatment of vitiligo. J Eur Acad Dermatol Venereol 20:553–557
12. Goldinger SM, Dummer R, Schmid P et al (2007) Combination of 308-nm xenon chloride excimer laser and topical calcipotriol in vitiligo. J Eur Acad Dermatol Venereol 21:504–508
13. Grimes PE, Soriano T, Dytoc MT (2002) Topical tacrolimus for repigmentation of vitiligo. J Am Acad Dermatol 47:789–791
14. Gupta S, Olson M, Kanwar AJ, Ortonne JP (eds) (2007) Medical treatment of vitiligo. Surgical management of vitiligo. Blackwell, Oxford
15. Hann SK, Im S, Park YK et al (1992) Repigmentation of leukotrichia by epidermal grafting and systemic psoralen plus UV-A. Arch Dermatol 128:998–999
16. Hann SK, Im S, Bong HW et al (1995) Treatment of stable vitiligo with autologous epidermal grafting and PUVA. J Am Acad Dermatol 32:943–948
17. Hartmann A, Lurz C, Hann H et al (2005) Narrow-band UVB311 nm vs. broad-band UVB therapy in combination with topical calcipotriol vs. placebo in vitiligo. Int J Dermatol 44:736–742
18. Kawalek AZ, Spencer JM, Phelps RG (2004) Combined excimer laser and topical tacrolimus for the treatment of vitiligo: a pilot study. Dermatol Surg 30:130–135
19. Kullavanijaya P, Lim HW (2004) Topical calcipotriene and narrowband ultraviolet B in the treatment of vitiligo. Photodermatol Photoimmunol Photomed 20:248–251
20. Lahiri K, Malakar S, Sarma N et al (2004) Inducing repigmentation by regrafting and phototherapy (311 nm) in punch grafting failure cases of lip vitiligo: a pilot study. Indian J Dermatol Venereol Leprol 70:156–158
21. Lahiri K, Malakar S, Sarma N et al (2006) Repigmentation of vitiligo with punch grafting and narrow-band UV-B (311 nm) – a prospective study. Int J Dermatol 45: 649–655
22. Lee AY, Jang JH (1998) Autologous epidermal grafting with PUVA-irradiated donor skin for the treatment of vitiligo. Int J Dermatol 37:551–554
23. Lee KJ, Choi YL, Kim JA et al (2007) Combination therapy of epidermal graft and systemic corticosteroid for vitiligo. Dermatol Surg 33:1002–1003
24. Leone G, Pacifico A, Iacovelli P et al (2006) Tacalcitol and narrow-band phototherapy in patients with vitiligo. Clin Exp Dermatol 31:200–205
25. Lepe V, Moncada B, Castanedo-Cazares JP et al (2003) A double-blind randomized trial of 0.1% tacrolimus vs 0.05% clobetasol for the treatment of childhood vitiligo. Arch Dermatol 139:581–585

26. Lu-yan T, Wen-wen F, Lei-hong X et al (2006) Topical tacalcitol and 308-nm monochromatic excimer light: a synergistic combination for the treatment of vitiligo. Photodermatol Photoimmunol Photomed 22:310–314
27. Middelkamp-Hup MA, Bos JD, Rius-Diaz F et al (2007) Treatment of vitiligo vulgaris with narrow-band UVB and oral *Polypodium leucotomos* extract: a randomized double-blind placebo-controlled study. J Eur Acad Dermatol Venereol 21:942–950
28. Mulekar SV (2006) Stable vitiligo treated by a combination of low-dose oral pulse betamethasone and autologous, non-cultured melanocyte-keratinocyte cell transplantation. Dermatol Surg 32:536–541
29. Ostovari N, Passeron T, Lacour JP et al (2006) Lack of efficacy of tacrolimus in the treatment of vitiligo in the absence of UV-B exposure. Arch Dermatol 142:252–253
30. Parsad D, Saini R, Verna R (1998) Combination of PUVAsol and topical calcipotriol in vitiligo. Dermatology 197: 167–170
31. Passeron T, Ostovari N, Zakaria W et al (2004) Topical tacrolimus and the 308-nm excimer laser: a synergistic combination for the treatment of vitiligo. Arch Dermatol 140: 1065–1069
32. Patel DC, Evans AV, Hawk JL (2002) Topical pseudocatalase mousse and narrowband UVB phototherapy is not effective for vitiligo: an open, single-centre study. Clin Exp Dermatol 27:641–644
33. Pianigiani E, Risulo M, Andreassi A et al (2005) Autologous epidermal cultures and narrow-band ultraviolet B in the surgical treatment of vitiligo. Dermatol Surg 31:155–159
34. Schallreuter KU, Wood JM, Lemke KR et al (1995) Treatment of vitiligo with a topical application of pseudocatalase and calcium in combination with short-term UVB exposure: a case study on 33 patients. Dermatology 190: 223–229
35. Shenoi SD, Srinivas CR, Pai S (1997) Treatment of stable vitiligo with autologous epidermal grafting and PUVA. J Am Acad Dermatol 36:802–803
36. Skouge JW, Morison WL, Diwan RV et al (1992) Autografting and PUVA. A combination therapy for vitiligo. J Dermatol Surg Oncol 18:357–360
37. Skouge J, Morison WL (1995) Vitiligo treatment with a combination of PUVA therapy and epidermal autografts. Arch Dermatol 131:1257–1258
38. Smith DA, Tofte SJ, Hanifin JM (2002) Repigmentation of vitiligo with topical tacrolimus. Dermatology 205:301–303
39. Suga Y, Butt KI, Takimoto R et al (1996) Successful treatment of vitiligo with PUVA-pigmented autologous epidermal grafting. Int J Dermatol 35:518–522
40. Tanghetti EA (2003) Tacrolimus ointment 0.1% produces repigmentation in patients with vitiligo: results of a prospective patient series. Cutis 71:158–162
41. Tanghetti EA, Gillis PR (2003) Clinical evaluation of B Clear and Protopic treatment for vitiligo. Lasers Surg Med 32(S15):37
42. Travis LB, Weinberg JM, Silverberg NB (2003) Successful treatment of vitiligo with 0.1% tacrolimus ointment. Arch Dermatol 139:571–574
43. Tsukamoto K, Osada A, Kitamura R et al (2002) Approaches to repigmentation of vitiligo skin: new treatment with ultrasonic abrasion, seed-grafting and psoralen plus ultraviolet A therapy. Pigment Cell Res 15:331–334
44. Van Geel N, Ongenae K, De Mil M et al (2004) Double-blind placebo-controlled study of autologous transplanted epidermal cell suspensions for repigmenting vitiligo. Arch Dermatol 140:1203–1208
45. Westerhof W, Nieuweboer-Krobotova L, Mulder PG, et al (1999) Left-right comparison study of the combination of fluticasone propionate and UV-A vs. either fluticasone propionate or UV-A alone for the long-term treatment of vitiligo. Arch Dermatol 135:1061–1066

Camouflage

3.9

Thomas Jouary and Alida DePase

Contents

T. Jouary (✉)
Service de Dermatologie, Hôpital Saint André,
CHU de Bordeaux, France
e-mail: thomas.jouary@chu-bordeaux.fr

3.9.1 Introduction

The World Health Organisation defines 'health' as physical and psychological well-being. Considering all physical and psychological aspects of vitiligo, dermatologists should suggest a therapy when feasible, and should also recommend psychological approaches and the use of camouflage [1–7]. Benefits can be obtained by the skilled use of corrective cosmetics, which are presented in the chaps 3.9.6–3.9.7.

3.9.2 Why Camouflage and Cosmetic Rehabilitation Are Needed

Camouflage refers to a range of special products, specially developed to disguise aesthetic skin disfigurement of any kind, requiring special application techniques. Cosmetic rehabilitation helps patients to achieve a positive image of the self.

Vitiligo is a disfiguring disease, pervasive to the patient. In an attempt to cope, patients may drastically change their way of life. They choose clothes with the sole aim of covering the patches. They feel obliged to wear long sleeves and long trousers even in the hottest summers. Patients presenting face and hands lesions demonstrate usually a higher impairment of their quality of life and self-body image than others (Chap 3.13).

The expressed fear of many vitiligo patients is that the disease may spread to visible areas. Indeed, patients with visible vitiligo are usually seeking for a treatment more actively than others. They are logically searching help first for patches in the visible areas, and only secondly for areas of lesser aesthetic importance.

M. Picardo and A. Taïeb (eds.), *Vitiligo*,
DOI 10.1007/978-3-540-69361-1_3.9, © Springer-Verlag Berlin Heidelberg 2010

Standard and experimental vitiligo treatment options are demanding from the psychological point of view, and may be disappointing for both patient and dermatologist. Moreover, repigmentation is often partial, and relapses may occur after stopping treatment. The treatment of hand and face lesions is particularly difficult, because of Koebnerisation (Sect. 2.2.2.1). Segmental vitiligo of the face starts frequently in childhood. At this age, melanocyte cell grafts are rarely considered for ethical reasons and because some repigmentation may appear spontaneously in the future. In all these situations, when classical medical treatments fail or cannot be considered, efficient camouflage techniques should be proposed to the patient (Fig. 3.9.1).

3.9.3 Camouflage as a Medical Intervention?

Dermatologists as a rule do not know much about camouflage; in the past, they have been dismissive of what they considered merely make-up. They are uninformed about the wide range of available camouflage products and the different techniques of application. They are equally unaware of the benefits that can be obtained by the skilled use of corrective cosmetics, such as self-tanners, stains, dyes, whitening lotions, tinted cover creams, compact, liquid and stick foundations, fixing powders, fixing sprays, cleansers, semipermanent and permanent tattoos, and dyes for facial and head-white hair.

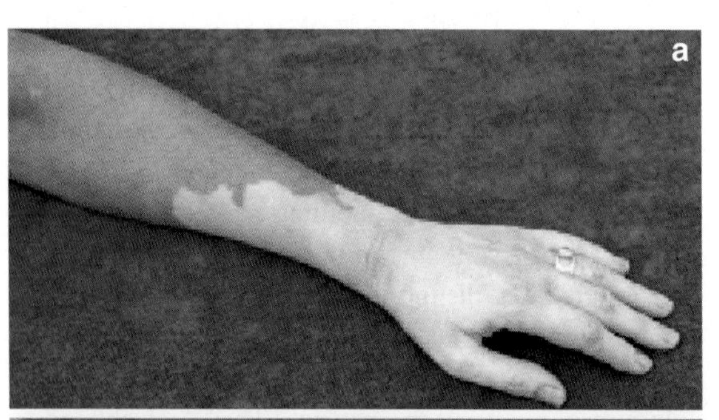

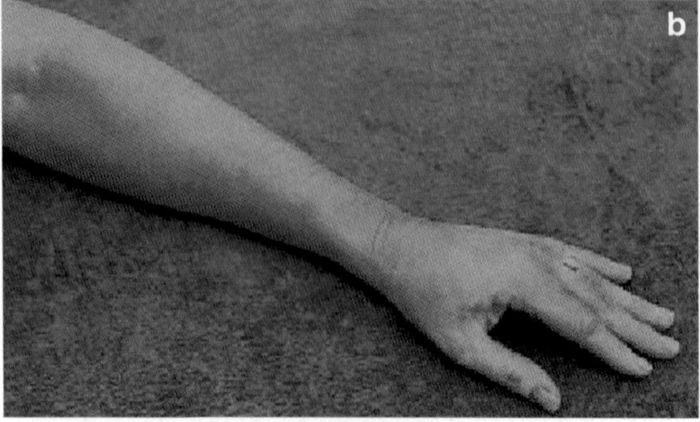

Fig. 3.9.1 (**a**, **b**) Camouflaging of hands and forehands

It is only recently that camouflage has been recognised as being equally worthy of consideration as a medical intervention, when there are no other satisfactory options to really help the patient and when the disease is recalcitrant to all standard and alternative therapies (Fig. 3.9.2). Camouflage consultations have been developed in some dermatology departments. The aim of these consultations is to educate patients about camouflage (Table 3.9.1 details the application method of commonly used products). These multidisciplinary consultations, composed of camouflage trained nurses, dermatologists, and camouflage specialised persons, have had a favourable impact on patients. Camouflage therapy or education means also to enable patients to apply the cosmetics themselves. It is worthwhile when the patient can master the correct procedure with the products chosen and the various techniques of application. Patients should also be informed on where and how to obtain the camouflage products that suit their individual needs.

Candidates for vitiligo camouflage are patients of both genders, and of all ethnic origins and ages. As with all patients, it is important for the camouflage practitioner to learn about their prior history, current medical situation, emotional state, and attitude towards camouflage. The patient's ability and desire to perform the various camouflage techniques should be discussed and the patient should be asked whether he/she has experienced allergic reactions to cosmetics in general in the past. It is important to discuss with the patient his or her lifestyle to choose products (self-tanners, cover creams, etc.) suitable for his/her case.

3.9.4 A Brief History of Camouflage

Since ancient times, an unblemished face has universally been considered a symbol of beauty and therefore sought after at all costs. In ancient Roman times, camouflage was used by the slaves who had gained their freedom, become rich, and determined to leave their

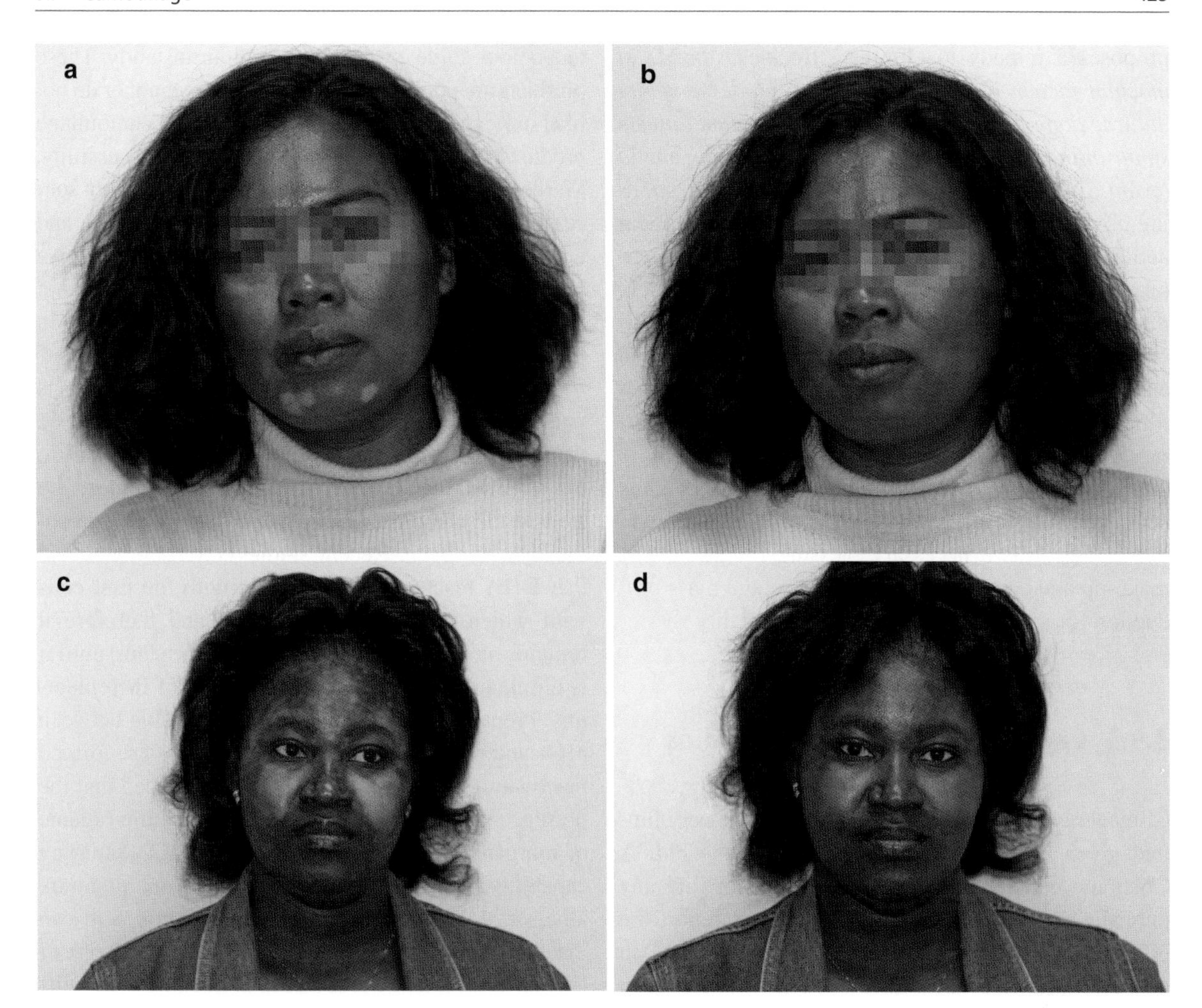

Fig. 3.9.2 (**a–d**) Camouflaging of facial vitiligo in two dark skinned patients

Table 3.9.1 Self-tanner and cover creams: application method

- A cotton washcloth to exfoliate the skin, or a commercial defoliator
- Soap, skin lotion (an ordinary skin lotion should be applied to feet and ankles, knees, elbows, hands, wrists, and any area with fine lines, such as neck and face). The chosen self tanner (cream, gel, lotion or spray only for large areas)
- Gloves, the inexpensive medical supply kind which come in a box
- Facilities to wash hands frequently, a band or tie to keep hair off of the face, and a fingernail scrub brush
- Baby Wipes, Quickies, or similar products to clean the palms of hands after products application
- The room or the bathroom should not be steamy or warm, as sweating is not good
- If self tanner is applied on the back, a sponge paintbrush is needed ,with a handle of three inches or more. If applied on large areas or the whole body, it's good to wear a loose outfit, while the product is drying, which takes from one to two hours
- Dark-coloured bathing suits are the best choice if the product is applied on the whole body . If the sunless tanner needs to be applied to patches on face, arms, and legs, the best choice are a dark T-shirt and shorts
- Light to medium-coloured fabrics which contain nylon are not indicated, because sunless tanners stain them permanently
- A moisturizer, to be applied every day on the fake tan

past behind. This consisted of an ochre-coloured paste made of clay and helped to cover the mark made on their forehead with a hot iron, the ignominious stigma of slavery. In the chapter Cutis et faciei vitiis propellendis of his Liber Medicinalis, Quinto Sereno Sammonico, a doctor in the Second Century AC, proposes a remedy to eliminate freckles: '*Invida si maculat faciem lentigo decoram nec prodesse valent naturae dona benignae, erucam atque acidum laticem simul inline mali Saepiolae cineres ex ossibus omnia levant…*' (If the horrible ephelids spoil the skin, spoiling its natural beauty, spread a lotion made of vinegar and rocket on the skin). Until the early twentieth century, make-up in general was used only by the rich, theatre actors, or prostitutes. It was only in the twentieth century, with the birth of the film industry, that camouflage and make-up in general became a must for film stars. Products resistant to the effect of stage lighting were requested for actors with imperfections to hide. After the First World War, many soldiers came back from the front severely burnt or disfigured, and camouflage was a necessary blessing for them. The make-up and camouflage era was born, and mass production for the general public became a reality.

3.9.5 Camouflage Controlled Studies

Clinical research on camouflage in vitiligo is very limited given the practical importance of this field. A Cochrane database review in 2006 did not find any published trial in this field. Some authors report the efficacy and safety of dihydroxyacetone (DHA) in healthy volunteers and vitiligo patients. These open and/or retrospective studies have compared different DHA concentrations in patients of various phototypes. The higher the concentration, the better the response observed particularly in darker phototypes [9]. Only one study demonstrated the positive impact of self-tanning interventions on the quality of life in a cohort of vitiligo patients [8]. These studies were principally conducted with DHA-based products, which are described in detail hereafter.

3.9.6 Self-Tanning Creams, Lotions, and Sprays

Self-tanners in gel, cream, lotion, or spray give the skin a brown colour that resembles a natural tan, and normally lasts 3–5 days. The tanning of the skin develops in about 3–24 h after the application. Instant colour self-tanners, available by some manufacturers, thanks to a colour guide, give a tanned colour instantly. These products are popular among those who cannot or do not like sun exposure, and can be considered camouflage products as they disguise depigmentation successfully. Marketed for about 40 years, self-tanners were not successful at first because they provided a yellowish and uneven colour, while today the latest formulas give excellent aesthetic results. Unfortunately, they are not suitable for coloured people, and the best results are in Caucasians of Phototypes I–III.

The active ingredient is DHA, a sugar that reacts with the proteins of the stratum corneum, and gives a tan resembling the solar UV-induced tan. This is due to the so-called Maillard's reaction, after the author who studied the chemical reaction that golden the crust of bread in the oven. Recent studies have demonstrated that DHA reacts totally and only with the first cells with which it comes into contact, and therefore it remains on the surface of the stratum corneum, until it is eliminated with the normal turnover of the epidermis. Preparations containing DHA are stable between pH 4 and 6. At neutral pH, brown grumes are formed inactivating pigmentation. It has been noticed that the presence on the skin or in the product of little organic or inorganic molecules may alter the DHA colouring capability. Traces of metals such as iron, titanium, zinc, or alpha-hydroxy-acids such as lactic acid can inactivate the product. So to avoid the frequent risk of lack of uniformity, a good rule is not to utilise self-tanners after using creams with zinc or titanium or after washing the skin with alkaline or lactic acid-based soaps largely present in products being defined by manufacturers as 'at physiologic pH' [1]. They can be used throughout the year, they are waterproof, and the fake tan developed does not stain clothes or sheets. However, sea water makes them fade away quickly, whilst swimming pool water does not.

No sunless tanner currently available contains adequate sunscreen, so sun-shielding products as well as moisturisers may be applied during the day, but only after the desired colour intensity has been obtained. Before applying self-tanners, it is advisable to gently rub the skin with a very soft brush to eliminate dead cells, especially on elbows, knees, and knuckles, in order to obtain an even skin colouring. It is advisable not to apply these products during the hot hours of the day in summer, because excessive sweat can result in uneven application and may prevent the active substance to develop colour properly. Only a small

Table 3.9.2 DHA tanning camouflage, practical points

-the products are to be used on perfectly dry skin -on large areas, they should be developed with circular movements to obtain a homogeneous result, - use a very spare quantity on hands, elbows, knuckles, and knees - application on eyebrows or near the forehead hairline should be avoided - apply sparingly to the face and neck area, because this part of the skin takes to self tanners quite well, and especially sparingly to the forehead hairline area. If the hair is short, the product should be applied behind ears. - Application on mucosa or near the eyes should be avoided. - After applying the product, avoid washing for 3 hours. No swim, bath, or anything that will make sweat for one hour; no tight jeans, belts, shoes, and bras for one hour, if the products have been used on the body. - wash hands thoroughly, use a nail scrub brush to clean the nails, but be prudent because of Koebner's phenomenon if this area is not already depigmented. The palms of the hands, between the fingers, the knuckles have to be cleaned well.

quantity of the product should be applied first, and if the desired colouring is not achieved, it is possible to intensify with additional daily applications. If too much of the product is used, the result will be unnatural.

For small-sized vitiligo lesions, it is advisable to use a self-tanner in lotion (not in cream) with a q-tip. The lotion is spread from the centre of the lesion towards the outside, up to 1–2 mm from the lesion edge. A different technique is to spread the product over the entire area, for instance hands and arms, including the normal pigmented areas and subsequently repeat over only in the white areas using a q-tip dipped in the product.

Most sunless tanners' instructions for using the product advise to rub it well until it is absorbed, but this takes too much time for hands that are going to be orange. Instead, the product should be applied quickly, but thoroughly, spreading it in a circular motion to avoid streaking. Other important points are outlined in Table 3.9.2.

3.9.7 Cover Creams, Foundations, Sticks

Highly pigmented creams come in compact, liquid, or stick formulations and are available from several manufacturers in a wide range of natural skin colours.

Table 3.9.3 Tips on cover creams application technique

-The cover-cream mixture is applied with the ball of the middle finger, and smoothed over the discoloured area with a light-pressing motion. A flat brush can be used to feather out the outer edges of the thick, opaque cover-cream solution until it is so well-blended to become undetectable. -Fixing powder: with a small cotton pad apply the translucent, colorless powder to stabilize the foundation. This procedure makes the application waterproof and resistant to smudges and friction. A few minutes for the talc to be absorbed, then the excess is dusted off. -Fixing spray: maintains the corrective make-up for a whole day. Vaporized at a distance of 40 cm, it creates, thanks to its silicon and polymers formula, an elastic film that guarantees coverage.

They are lightweight and easy to apply, usually free of contact allergens, but dermatitis and allergic reactions may occur, due to fragrances and preservatives present in some brands. The texture is denser than traditional foundation creams used by the general public, as their aim is to provide effective cover. They may contain up to 50% mineral oils and wax. The texture, different from normal foundation products, is also due to titanium dioxide, used as a thickening and shielding agent, offering sun protection, while the colours are provided by iron oxides. Blended to complement the individual's particular skin colour, they can conceal disfiguring vitiligo patches of exposed areas. Suitable for men, women, and children, their waterproof characteristics allow shower and swimming. They should be applied on the face and removed every day. It is crucial to choose one's basic colour, and then the various shades that can vary indefinitely, according to the different parts of the body, to the seasons or simply from one day to another. Sometimes, it is necessary to mix more colours of different brands.

For a correct application, the skin should be cleaned. Any previous makeup has to be gently removed. The patient has to learn proper makeup-removal techniques, that is, gentle movements to avoid the Koebner phenomenon. Using a sample palette of cover creams, two or three shades can be mixed to achieve the desired colour match. More than three colours will make the procedure too complicated and expensive. An optimal blend of colours should be created, and when applied, this should match the colour of the surrounding skin as closely as possible (Table 3.9.3).

3.9.8 Vitiligo of the Lips

Transfer-resistant lip colour lasting up to 12h of wear, in matte formulas, is available in many shades that duplicate the natural colour of the lips, suitable also for male patients. They are an alternative to lip tattoos. These very popular cosmetics are easily available in almost all departmental stores and pharmacies.

3.9.9 Leukotrichia

Leukotrichia can affect visible areas on the beard, moustaches, eyebrows, and eyelashes. In these cases, it is advisable to dye the white hair in a hairdressing salon the first time, and subsequently the dyeing can be practiced at home with products formulated for these delicate areas.

3.9.10 Permanent and Semi-Permanent Camouflage

Also referred to as micro-pigmentation, cosmetic tattooing or dermal pigmentation, are techniques that require an experienced technician, qualified to offer these services. Results vary according to the skill of the practitioner and his experience with colours. Cosmetic tattoo may be suitable for depigmented lips, especially in black people, and for depigmented nipples. As regards other vitiligo areas, results may be very disappointing. Pigments specially formulated for cosmetic application are implanted beneath the epidermis into the dermal layer, by microinsertion. Topical anaesthetics are used to minimise discomfort; allergic reactions are rare; sterile needles and surgical gloves are used for each procedure. Semi-permanent camouflage or cosmetic tattooing differs from traditional tattooing, where permanent skin inks and dyes are placed into the skin. It can be considered 'permanent' if compared to normal make-up, but as it will fade with time – usually it lasts from 2 to 5 years – it is actually "semi-permanent.'

3.9.11 Precautions of Use

Some precautions should be considered when a camouflage technique is used or proposed to the patients. Firstly, the patient should be taught that camouflage has to be removed smoothly avoiding intensive friction or use of washcloths. This is of importance as the Koebner's phenomenon is observed in areas where an adherent camouflage needs an intense frictional washing to pull it off (Sect. 2.2.2.1). Similarly, fixing powder should be used cautiously and avoided whenever possible. Smooth, liquid, and light instant colour self-tanner and stains should be preferred for these reasons.

Another point of importance is the risk with permanent camouflage. Indeed, vitiligo course is highly unpredictable. Even after many years, stable large patches can resolve spontaneously. Given these considerations, permanent camouflage and tattoos should be considered with particular caution. The development of a depigmented patch around an area previously treated with these techniques can lead to inaesthetic results. Thus, if performed in vitiligo, the colour of the tattoo has to be very close to the natural pigmentation of the patient, which is very difficult. Since the colour of the tattoo is definitive and does not follow the UV-induced tanning changes, the contrast between tattooed skin and natural pigmented areas may cause problems.

3.9.12 Conclusion: Camouflage as a Balm for 'Bruised' Souls

Camouflage increases the patient's confidence and improves his quality of life. It is readily accepted by women, who, unlike children and men, may already be accustomed to use make-up products. But both men and adolescents can easily learn how to apply the products.

There are many ways to conceal small or large areas of vitiligo, and the natural result increases the patient's confidence. No more embarrassing questions or intrusive staring, wearing a short-sleeved shirt or shorts in the summer, no hands hidden in pockets, greeting with a handshake without fear; this is what 'camouflage therapy' can do to improve the quality of life of vitiligo patients.

Summary Messages

> - It is only recently that camouflage has been recognised as being a medical intervention, when there are no other satisfactory options to really help the patient.
> - It is important to discuss with the patient his or her lifestyle to choose products (self-tanners, cover creams, etc.) suitable for his/her case.
> - Non permanent techniques such as self tanners should be usually preferred
> - The Koebner's phenomenon can be observed in areas where an adherent camouflage needs an intense frictional washing to pull it off
> - Cosmetic tattoo may be suitable for depigmented lips, especially in black people, and for depigmented nipples
> - Camouflage can improve the quality of life of vitiligo patients

References

1. DePase A (2000) Importanza del Camouflage in Pazienti con Vitiligine. In: Lotti T (ed) La Vitiligine, nuovi Concetti e Nuove Terapie, Ed Utet
2. DePase A (2003) La Voce dei Pazienti. EBD evidence based dermatology. Masson, Milano
3. DePase A, Naldi L (2004) Vitiligo, some reflections on clinical research in a rather elusive disorder Centro Studi Gised. Ospedali Riuniti, Bergamo
4. DePase A (2004) Vitiligo, au-delà de la maladie I" In: Ortonne JP (ed) Cutis and Psyche. France
5. DePase A (2004) The view of the cosmetologist. In: Lotti T, Hercegova L (eds) Vitiligo: problems and solutions. Marcel Dekker, New York
6. DePase A (2006) Il Camouflage. In: Naldi L, Rebora A (eds) Dermatologia Basata sulle prove di Efficacia, chapter 69. Masson, Milano
7. Ongenae K, Dierckxsens L, Brochez L et al (2005) Quality of life and stigmatization profile in a cohort of vitiligo patients and effect of the use of camouflage. Dermatology 210:279–285
8. Rajatanavin N, Suwanachote S, Kulkollakarn S (2008) Dihydroxyacetone: a safe camouflaging option in vitiligo. Int J Dermatol 47:402–406
9. Whitton ME, Ashcroft DM, Barrett CW, González U (2006) Interventions for vitiligo. Cochrane Database Syst Rev (1):CD003263

Further Reading

www.skin-camouflage.net < http://www.skin-camouflage.net >
http://www.redcross.org.uk/standard.asp?id = 49354

Photoprotection Issues

3.10

Alessia Pacifico, Giovanni Leone, and Mauro Picardo

Contents

A. Pacifico (✉)
Phototherapy Unit, San Gallicano Dematological Institute, IFO, Rome, Italy
e-mail: alessia.pacifico@tiscali.it

3.10.1 Normal and Vitiligo Skin UV Sensitivity

A number of mechanisms have been developed during evolution to protect human skin from excessive UV radiation. To classify the susceptibility of human subjects to develop erythema following UV exposure, Fitzpatrick's system differentiates six sun reactive skin phototypes (SPT) [9] (Table 3.10.1). In this system, the capacity to tan is equally important to help to categorise individuals of any colour or ethnic background. Using this method, the physician can estimate the relative risk of developing acute and chronic changes related to UV exposure. A study by Carrettero-Mangolis and Lim confirms that there is a correlation between minimal erythema dose (MED) and SPT, that is, the higher the SPT, the higher the MED [6].

There is evidence that the stratum corneum (SC) protects against UV mainly because of reflection of radiation, absorption of the radiant energy, and internal scattering. In the palms where the SC is very thick, the minimal erythema dose is 16 times higher than that on the back. It was observed that in areas stripped from SC, sunburn develops more easily than in the unstripped areas of the skin [23].

Melanin is a potent UV absorber, and thus skin pigmentation is another natural protective factor against UV radiation. It is already well-known that pigmented skin is more resistant to sunburn than poorly pigmented skin [14, 18]. Delayed effects of chronic UV exposure, such as development of skin cancer and photoageing are also more pronounced in individuals with fair-skin colour [28]. Skin responds to repeated exposure by tanning, especially in Skin Types III and IV who tan easily and burn rarely, which is the result of melanogenesis and associated SC thickening. A study performed by Cario-André et al. assessed the photoprotective role of

M. Picardo and A. Taïeb (eds.), *Vitiligo*,
DOI 10.1007/978-3-540-69361-1_3.10, © Springer-Verlag Berlin Heidelberg 2010

Table 3.10.1 Fizpatrick SPT classification

SPT I	Always burns, never tans
SPT II	Burns easily, tans minimally
SPT III	Burns moderate, tans gradually to light brown
SPT IV	Burns minimally, alias tans well to moderate brown
SPT V	Rarely burns, tans profuse to dark
SPF VI	Never burns, deeply pigmented

melanocytes in the epidermis and studied the effects of UVB on epidermis reconstructed with and without melanocytes. To address more specifically the role of melanin in fair-skinned individuals, experiments were done with cells obtained from human skin of low-skin types (II–III). In order to study the effect of constitutive melanin and possibly that of newly synthesised melanin precursors, a single dose of ultraviolet B (UVB) (4–5 MED) was administered to reconstructs and the effects were then monitored over the first 24h. The results clearly showed that low phototype melanocytes protect epidermal basal cells against UVB-induced apoptosis as well as necrosis, and may thus preserve epidermis integrity after UVB irradiation. On the other hand, such melanocytes do not seem to have a protective role against DNA damage and consequently may not prevent skin cancer [4].

In a recent study by Yamaguchi et al., DNA damage and apoptosis in different skin types before and after UV exposure have been reviewed. Their results, along with other published reports, indicate that UV-induced DNA damage is more effectively prevented in darker-skin types. This study also demonstrated that rates of repair of DNA damage may differ greatly among different individuals and that UV-induced apoptosis is significantly greater in darker skin. These results suggest that UV-damaged cells are more efficiently removed in darker skin [29].

In patients with vitiligo, it is generally assumed that lesional skin of SPT I–VI would be very sensitive to UV radiation owing to the fact that pigment is absent in such lesions, but it has been recently established that there are several mechanisms that contribute to modulate UV sensitivity in vitiliginous skin. Depigmented area of vitiligo patients does not contain pigment, and thus protection against UV is afforded by the SC and the rest of epidermis. It might thus be suspected that hyperkeratosis and epidermal hyperplasia would develop in vitiligo exposed to UV to compensate for the lack of pigment. The importance of hyperplasia in photoprotection was validated in studies looking at vitiligo skin which is characterised by a lack of melanocytes. Gniadecka et al. assessed solar simulated radiation (SSR) MED on amelanotic and adjacent normally pigmented skin of 14 vitiligo patients. Erythema and melanin content were quantified using a reflectance device, and SC thickness was determined from frozen skin samples. These authors reported that the predominant protective factor in both vitiligo and pigmented skin is the SC where it accounted for 57% of the total photoprotection in pigmented skin. Repeated exposures to this skin were shown to elicit a protection factor of 15 in the absence of melanocytes. In contrast, selective activation of melanocytes by ultraviolet A (UVA) that does not induce SC thickening only gave a protection factor of 2–3. Gniadecka et al. suggested that SC accounted for over two-thirds of photoprotection observed in normal skin and hence was far more significant in this role than induced tanning [10]. The significance of SC in photoprotection may be greater in fair-skinned individuals than in pigmented individuals, especially in vitiligo skin, where the SC may represent the only source of protection. However, studies performed by Sheehan et al. did not support a significant photoprotective role for SC thickening [27]. The different results obtained by Gniadecka and Sheehan may be due to several variables. It is known in fact that UVR sensitivity vary between body sites on the same person.

Kaidbey et al. investigating epidermal UVR transmission in black skin (Skin Type VI) as well as Caucasian skin (Skin Types I–III), found that in Skin Types I, II, and III the SC is the main site of UVR screening, absorbing over 50% of the incident radiation. However, samples were taken from previously sun-exposed sites (abdominal skin) [14]. Kaidbey and Klingman also reported in an earlier study that melanogenesis (without appreciable thickening of the SC) induced by repeated UVA exposure afforded a protection factor of 2–3 [15].

3.10.2 Photoadaptation of Vitiliginous Skin to UV Irradiation

A recent retrospective study showed that most patients with vitiligo treated with narrow band UVB (NB-UVB) phototherapy did not develop phototoxicity in their skin despite increasing doses of UV radiation [13].

Thus, these patients developed photoadaptation, a frequently observed phenomenon. Photoadaptation has been described in different SPT, and it is probably due to both pigmentary and non-pigmentary influences. Several studies have indicated that factors involved in photoadaptation include hyperkeratosis, acanthosis, melanogenesis as well as an unknown factor that may be due to DNA repair or an immune-related process. Photoadaptation was described by Oh et al. as a 'number of changes occurred that are adaptive, in the sense that they result in a diminished future response to equivalent doses of radiation' [21]. One of the methods to measure photoadaptation is MED phototesting.

In phototherapy protocols, vitiliginous skin has been always classified as Fitzpatrick's SPT I or II because of the lack of pigment. Caron–Schreinmachers et al. have recently showed that by elicitation of the MED on areas of vitiliginous skin, UVR sensitivity varied with total body skin type even in skin without pigment. Unlike earlier studies in which we only had information about the relation between MED and SPT from normal skin of different SPT, in this study, it has been proved that there is a linear relationship between the SPT of non-affected skin of vitiligo patient and the sensitivity to NB-UVB irradiation of lesional skin (Fig. 3.10.1) [5]. Since melanocytes are absent or scarcely present in lesional vitiligo skin, this difference in photosensitivity cannot be due to melanin, but this protection must be based on other mechanisms. We know that epidermis thickness increases after UV irradiation. In the study performed by Caron–Schreinmachers et al. however, patients' skin had not been exposed to UV for at least three months, and tests were carried out in regions that had not been exposed to sun so skin thickening could not be responsible for the observed differences. One mechanism that possibly accounts for the SPT-related differences is the antioxidant status of the skin. It has been already well-established that antioxidants provide photoprotection. Bessou-Touya et al. have shown that melanocytes of Caucasian subjects which have a higher content of unsaturated fatty acids in their cell membrane, are more prone to the peroxidative effects of UV light and that keratinocytes participate in photoprotection via phototype-dependent antioxidant enzyme activities [1]. Picardo et al. demonstrated that there are significant differences in antioxidant status in normal skin between people with high (III–V) and low (I–II) SPT [24]. Possibly, the same differences in antioxidant status between the different SPT exist in vitiligo skin as well.

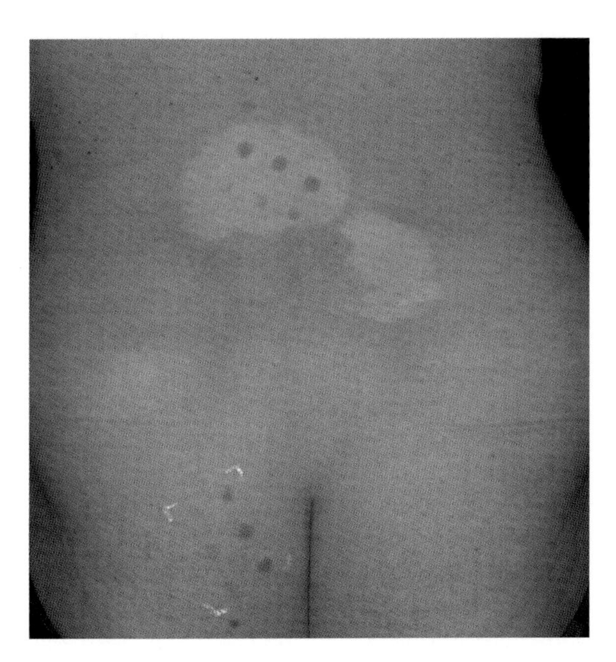

Fig. 3.10.1 Unmodified MED on lesional and non lesional skin in NS vitiligo

Overall, the studies performed indicate that photoadaptation occurs following UVR exposure also in depigmented areas of vitiligo subjects and that the normally pigmented skin phototype should be taken in account when a photoprotection strategy (or phototherapy) is recommended.

3.10.3 Vitiligo and Skin Cancer

Sun exposure is the main cause of photocarcinogenesis, photoageing, and photosensitivity. Unprotected exposure to ultraviolet radiation is a major causal factor in the development of skin cancer. Non-melanoma skin cancers (NMSC) are initiated for the most part by chronic sunlight exposure and can readily be produced by experimental exposure to ultraviolet radiation in animal models [8, 17]. UVB is the major active waveband region that causes direct photochemical damage to DNA, from which gene mutations arise. Unlike UVB, UVA could have more indirect effects on DNA via the generation of reactive oxygen species. In contrast with NMSC, cutaneous melanoma is more commonly associated with sporadic burning exposure to sunlight, especially early in life, but the wavelengths responsible have not been clearly identified. There are several indications that UVA might have an important role in the pathogenesis of melanoma [30]. However, this involvement has recently been questioned, since only UVB could induce melanoma in a transgenic mouse model [7].

The action spectrum for UV-induced tanning and erythema are almost identical. Indirect evidence suggests that UVA has a greater role in long-term sun damage than it does in acute effects such as sunburn or Vitamin D synthesis, which are overwhelmingly attributable to UVB [16, 20].

Considering that patients with vitiligo have patches lacking pigment and that a group of patients develop the disease during childhood, it would be expected that these subjects with the involvement of the head, neck, and hands, which are sites at constant risk of sun damage, should develop an increased risk for NMSC [3]. A review of the literature indicates even a lower risk of NMSC. The rarity of NMSC in vitligo was noted earlier in a study by Calanchini-Postizzi and Frank on 23 patients who had a mean duration of vitiligo of 15.1 years. These investigators found only 3 actinic keratosis in light-exposed vitiligo patches in those 23 patients. In addition, the number of 'sunburn cells' commonly considered as the morphological sign of UV damage to keratinocytes, was found significantly lower compared to healthy controls [21]. Schallreuter et al. more recently hypothesised that the low incidence for actinic damage, basal and squamous cell carcinomas as documented in vitiligo could be due to a protective function of up-regulated wild type p53 induced by the constant H_2O_2 stress existing in all the epidermis of vitiligo patients [25, 26].

The association between vitligo and melanoma is interesting because both diseases affect melanocytes and that immunological mechanisms play a part in both conditions. Different clinical studies report the connection between malignant melanoma and vitiligo, and also several authors suggest that the appearance of depigmentation during the course of malignant melanoma or its treatment with interferon can be considered a good prognostic sign [2, 11, 19].

Thus a series of arguments indicate that either the chronic adaptation to UV stress via increased antioxidant responses or reinforced disease-driven immunosurveillance against melanocyte antigens may naturally protect vitiligo patients. However, large-scale epidemiological studies are needed to clarify this important issue.

3.10.4 Practical Photoprotection

All patients with vitiligo, but particularly those with fair skin, should have a photoprotective counselling adapted to their phototype, environmental exposure risk, and therapy plan. Indeed, repigmentation needs UV stimulation and an exceedingly high photoprotection could be counterproductive. However, for photoprotection in sunny and more risky areas such as tropical climates, avoidance of peak sunny hours and clothing are the major steps. It is thus most important to explain to the patient the importance of wearing sun-protective clothing (such as wide brimmed hats) to help prevent tanning and sunburns. Sunblockers come as second in line. Sunscreen that provides protection from both UVA and UVB should be used. Besides helping to protect the skin from sunburn and resulting koebnerisation, long-term damage of susceptible depigmented areas, sunscreens also minimises tanning, which makes the contrast between normal and depigmented skin less noticeable [12]. A waterproof, broad-spectrum sunscreen with a sun protection factor of at least 15, should be used on all exposed skin, pigmented and depigmented. Particular attention should be given to the formulation of the sunscreen to allow easy spreading (and removal if needed) without risk of skin trauma.

Sunscreen act by one of the two mechanisms, either by absorbing UV rays or by blocking or/and scattering these rays. Chemical sunscreens function by absorbing UVB and/or UVA. Nowadays, protection against both UVA and UVB is very common and as a result, sunscreens often contain a mixture of light absorbing chemicals. Sunscreen's efficacy in absorbing UVB is measured by the sun-protective factor (SPF). The UVB absorbers have been commonly used worldwide for decades, whereas most UVA and broadband absorbers have been developed in recent years. Since a sunscreen has to protect against the entire UV spectrum, different filters have to be combined in the same product (Table 3.10.2). The cinnamates (2-ethyl *p*-methoxycinnamate) are by far the most popular UVB absorbers in both United States ofAmerica and Europe and they are used in combination with other UVB absorbers to achieve a high SPF. The second most popular filters during the recent past are camphor derivatives. Salycilates and *para*-aminobenzoic acid (PABA) and its derivatives are among the oldest commercially available UVB filters and they are still used worldwide. The increasing need

Table 3.10.2 Sunscreen agents permitted as active ingredients in the European Committee (EU), USA and Australia (AUS)

UVB FILTERS	Synonims, abbreviations, trade names	PERMITTED IN
PABA derivatives		
4-Aminobenzoic acid	PABA	EC, USA, AUS
Cinnamates		
Ethylhexyl methoxycinnamate	Octyl methoxycinnamate, Eusolex 2292, Parsol MCX	EC, USA, AUS
Salycilates		
2-Ethylhexyl salicylate	Octyl salycilate, octisalate, Escalol 587	EC, USA, AUS
Camphors		
Benzylydene camphor sulfonic acid	Mexoryl SD-20, Unisol S22	EC, USA, AUS
UVA FILTERS	**Synonims, abbreviations, trade names**	**PERMITTED IN**
2,2′-Methylene-bis-6-(2H-benzotriazol-2yl)-4-(tetramethyl-butyl)-1,1,3,3-phenol	Tinosorb M	EC, AUS
Phenol,2(2H-benzotriazol-2yl)-4-methyl-6[2-methyl-3-[1,3,3,3-tetramethyl-1-[(trimethylsilyl)oxy]disiloxanyl]propyl	Mexoryl XL	EC, AUS
Terephthalylidene dicamphor sulfonic acid	Mexoryl SX	EC, AUS
(1,3,5)-Triazine-2,4-bis((4-(2-ethyl-hexyloxy)-2-hydroxy)-phenyl)-6-(4-methoxyphenyl)	Tinosorb S	EC, AUS

for broadband agents and improved photostability has led to the introduction of a new generation of filters, including methylene bis-benzotriazolyl tetramethylbutylphenol (Tinosorb M) and bis-ethylhexyloxyphenol methoxyphenol triazine (Tinosorb S), both manufactured by CIBA Specialty (Basel, Switzerland), as well as terephthalylidene dicamphor sulfonic acid (Mexoryl SX) and drometrizol trisiloxane (Mexoryl XL), produced by L'Oreal (Clichy, France). The mexoryls and tinosorbs are not licensed in the United States of America and Japan [22].

Physical sunscreens act by blocking or scattering UV rays. Micronised formulations of zinc oxide and titanium dioxide are gaining in popularity. Mixtures of these minerals, along with chemical sunscreens, have led to a marked reduction in the transmission of both UVB and UVA. In particular, the use of broad-spectrum sunscreens alone limiting tan contrast may be an effective therapy in vitiligo patients with Skin Type I or II. Many of these agents are also used in cosmetic products such as eye shadow, foundation, and powders. Zinc oxide, titanium dioxide, talc, kaolin, and calamine are examples of physical agents. In the past, formulations containing these agents were often opaque, and as a result, patients found them cosmetically unacceptable. More recently, brown colours, such as iron oxide, were added as ingredients; not only does this make the physical sunscreen more appealing cosmetically, it also helps to scatter the UV rays making the formulation more effective.

It is recommended that the sunscreen be reapplied approximately every 90 min. In patients with darker-skin types (>III), we usually recommend the use of two different sunscreens: the one on the vitiliginous lesions with SPF around 15 and another one with SPF 50+ on surrounding healthy skin, if possible with a high UVA protection factor. We find this strategy particularly useful in minimising the contrast between healthy and vitiliginous skin and nevertheless allowing part of the UV spectrum to stimulate pigmentation on affected skin. New long lasting sun blockers (Daylong actinica, Spirig) may limit difficult to follow reapplication schemes.

Summary Messages

- Using Fitzpatrick's grading system for phototype, the physician can estimate the relative risk of developing acute and chronic changes related to UV exposure.
- Stratum corneum protects against UV mainly because of reflection of radiation, absorption of the radiant energy and internal scattering.
- Melanin is a potent UV absorber and a protective factor against UV radiation.

> Hyperkeratosis and epidermal hyperplasia develop in vitiligo exposed to UV to compensate for the lack of pigment.
> Epidermal antioxidant status is also important in photoprotection.
> Low incidence of actinic damage, basal and squamous cell carcinomas in vitiligo could be due to a protective function of up-regulated wild type p53.
> For photoprotection in sunny and tropical climates, avoidance of peak sunny hours and clothing is the major step.
> In patients with skin types >III the use of two different sunscreens (the first one on the vitiliginous lesions with SPF around 15 and another one with SPF 50+ on surrounding healthy skin) is recommended.
> New long lasting sun blockers may limit difficult to follow reapplication schemes.

References

1. Bessou-Touya S, Picardo M, Maresca V et al (1998) Chimeric human epidermal reconstructs to study the role of melanocytes and keratinocytes in pigmentation and photoprotection. J Invest Dermatol 111:1103–1108
2. Buljan M, Situm M, Lugovic L, Vucic M (2006) Metastatic melanoma and vitiligo: a case report. Acta Dermatovenereol Croat 14:100–103
3. Calanchini-Postizzi E, Frank E (1987) Long term actinic damage in sun exposed vitiligo and normally pigmented skin. Dermatologica 174:266–271
4. Cario-André M, Pain C, Gall Y et al (2000) Studies on epidermis reconstructed Wight and without melanocytes: melanocytes prevent sunburn cell formation but not appearance of DNA damaged cells in fair skinned Caucasians. J Invest Dermatol 115:193–199
5. Caron-Schreinemachers ALDB, Kingswijk MM, Bos JD, Westerhof W (2005) UVB 311 nm tolerance of vitiligo skin increases with skin photo type. Acta Derm Venereol 85:24–26
6. Carretero-Magolis C, Lim HW (2001) Correlation between skin types and minimal erythema dose in narrow band UVB (TL-01) phototherapy. Photodermatol Photoimmunol Photomed 17:244–246
7. De Fabo EC, Noonan FP, Fears T, Merlino G (2004) Ultraviolet B but not ultraviolet A radiation initiates melanoma. Cancer Res 64:6372–6376
8. Dumaz N, van Kranen HJ, de Vries A et al (1997) The role of UVB light in skin carcinogenesis through the analysis of p53 mutations in squamous cell carcinomas of hairless mice. Carcinogenesis 18:897–904
9. Fitzpatrick TB (1988) The validity and practicality of sunreactive skin types I through VI. Arch Dermatol 124:869–871
10. Gniadecka M, Wulf HC, Mortensen NN, Poulsen T (1996) Photoprotection in vitiligo and normal skin. A quantitative assessment of the role of stratum corneum, viable epidermis and pigmentation. Acta Dermatol Venereol 76: 429–432
11. Gogas H, Ioannovich J, Dafni U et al (2006) Prognostic significance of autoimmunity during treatment of melanoma with interferon. N Engl J Med 354:709–718
12. Halder RM, Brooks HL (2001) Medical therapies for vitiligo. Dermatol Ther 14:1–6
13. Hamzavi I, Deleon S, Yue K, Murakawa G (2004) Repigmentation does not affect tolerance to NB UVB light in patients with vitiligo. Photodermatol Photoimmunol Photomed 20:117
14. Kaidbay KH, Agin PP, Sayre RM, Kligman AM (1979) Photoprotection by melanin-a comparison of black and Caucasian skin. J Am Acad Dermatol 1:249–260
15. Kaidbay KH, Kligman AM (1978) Sunburn protection by longwave ultraviolet radiation induced pigmentation. Arch Dermatol 114:46–48
16. Lim HW, Naylor M, Honigsman H et al (2001) American Academy of Dermatology consensus conference on UVA protection of sunscreens: summary and recommendations. J Am Acad Dermatol 44:505–508
17. Madan V, Hoban P, Strange RC et al (2006) Genetics and risk factors for basal cell carcinoma. Br J Dermatol 154 (Suppl 1):5–7
18. McFadden AW (1961) Skin disease in the Cuna Indians. Arch Dermatol 84:1013–1023
19. Michail M, Wolchock J, Goldberg SM et al (2008) Rapid enlargement of a malignant melanoma in a child with vitiligo vulgaris after application of topical tacrolimus. Arch Dermatol 144:560–561
20. Morison WL (2004) Clinical practice. Photosensitivity. N Engl J Med 350:1111–1117
21. Oh C, Hennessy A, Ha T et al (2004) The time course of photoadaptation and pigmentation studies using a novel method to distinguish pigmentation from erythema. J Invest Dermatol 123:965–972
22. Palm MD, O'Donoghue MN (2007) Update on photoprotection. Dermatol Ther 20:360–376
23. Pathak MA, Fitzpatrick TB (1974) The role of natural photoprotective agents in human skin. In: Pathak MA, Harber LC, Seljl M, Kukita A (eds) Sunlight and man, normal and abnormal photobiologic responses. University of Tokyo Press, Tokyo, pp 725–750
24. Picardo M, Maresca V, Eibenschutz L et al (1999) Correlation between antioxidants and photypes in melanocytes cultures. A possible link of physiologic and pathologic relevance. J Invest Dermatol 113:424–425
25. Schallreuter KU, Tobin DJ, Panske A (2002) Decreased photodamage and low incidence of non melanoma skin cancer in 136 sun exposed Caucasian patients with vitiligo. Dermatology 204:194–201
26. Schallreuter KU, Behrens-Williams S, Khaliq TP et al (2003) Increased epidermal functioning wild-type p53 expression in vitligo. Exp Dermatol 12:268–277

27. Sheehan JM, Potten CS, Young AR (1998) Tanning in human skin types II and III offers modest photoprotection against erythema. Photochem Photobiol 68:588–592
28. Taylor ChR, Stern RS, Leyden JJ, Gilchrest BA (1990) Photoaging/photodamage and photoprotection. J Am Acad Dermatol 22:1–15
29. Yamaguchi Y, Beer JZ, Hearing VJ (2008) Melanin mediated apoptosis of epidermal cells damaged by ultraviolet radiation: factors influencing the incidence of skin cancer. Arch Dermatol Res 300:s43–s50
30. Wang SQ, Setlow R, Berwick M et al (2001) Ultraviolet A and melanoma: a review. J Am Acad Dermatol 44:837–846

Depigmenting Agents

3.11

Mauro Picardo and Maria Lucia Dell'Anna

Contents

M. Picardo (✉)
Istituto Dermatologico San Gallicano,
via Elio Chianesi, 00144 Roma,
Italy
e-mail: picardo@ifo.it

3.11.1 Introduction

The depigmenting approach is quite recent, deriving from the observation of unwanted depigmenting action of the phenol derivatives [6] (Sect. 2.2.2.2). On the basis of this clinical observation, the researchers aimed to define the possible mechanisms of action of this class of compounds. The first suggested target was the enzyme tyrosinase, and the capability of different phenol derivatives to act as alternative substrate of the enzyme or as competitive inhibitor was evaluated. Consequently, it was hypothesized that this class of substances, or some of them, may be used for the treatment of the skin disorders due to hyperpigmentation or melanocyte hyperproliferation.

Structural studies have indicated the role of the position and of the type of substitutes in the phenolic ring to allow the compound to be hydroxylated or oxidated by tyrosinase [8]. Hydroquinone (HQ) belongs to the phenol/catechol class of chemical agents (Sect. 2.2.2.2). HQ inhibits tyrosinase through the interaction with the copper at the active site, as well as decreases the amount of intracellular glutathione and induces the production of oxygen-reactive species. HQ acts as alternative substrate, according to most part of phenol/catechol compounds, because it is similar to tyrosine. HQ can be thus oxidized by the enzyme without generating the pigment. In addition, the produced quinones are able to react with the sulphydryl residues of the proteins generating oxidative damage and affecting the cell growth. The oxidative damage, involving both lipids and proteins of the cellular membranes, may thus account for the depigmenting action. Functional studies (see Sect. 2.2.2.2) have demonstrated that HQ and other phenolic compounds, such as *tert*-buthyl-phenol, may act even through different mechanisms, including the oxidation of TRP1, and by

M. Picardo and A. Taïeb (eds.), *Vitiligo*,
DOI 10.1007/978-3-540-69361-1_3.11, © Springer-Verlag Berlin Heidelberg 2010

interfering with RNA and DNA synthesis. HQ has been identified as the main depigmenting agent, whereas among the several phenolic derivatives, the monobenzyl ether of hydroquinone (MBEH) appeared as the more handful one (Fig. 3.11.1).

3.11.2 Chemical Agents

In patients with extensive and refractory vitiligo, depigmenting the remaining islands through chemical or physical methods of normal skin may be more cosmetically acceptable. The MBEH, or monobenzone, is a derivative of HQ. It is metabolized by the cells, producing the corresponding quinone, responsible for the depigmenting effect. Unlike HQ, MBEH almost always causes nearly irreversible depigmentation of the skin (Fig. 3.11.2). The suggested mechanisms of depigmentation by MBEH are the selective melanocyte destruction through free-radical formation, the competitive inhibition of the tyrosinase enzyme system, and the increased release of melanin [2, 3, 8]. According to the most part of diphenol compounds, it acts also as cytotoxic agent beside the inhibitor of tyrosinase. The in vitro data underlined (Sect. 2.2.2.2) that the toxic effect of the *tert*-butyl-phenol may be mediated by the TRP1-dependent production of quinones inside the cells.

A clinical observation suggested the potential depigmenting action of the imatinib mesilate, a selective tyrosine kinase inhibitor [3]. It was used at 300 mg/ day for the treatment of the chronic myeloid leukemia in a woman with vitiligo, giving rise to a speedy progressive depigmentation. The inhibition of c-Kit/SCF has been suggested to be responsible for the depigmenting effect (Chap. 1.2.1).

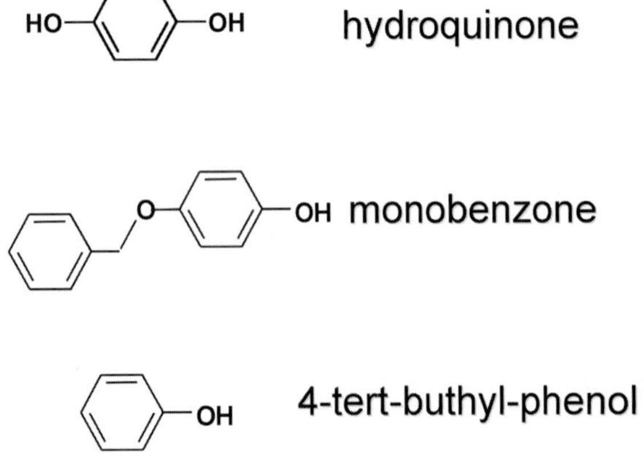

HO—⬡—OH hydroquinone

⬡—O—⬡—OH monobenzone

⬡—OH 4-tert-buthyl-phenol

Fig. 3.11.1 Chemical structure of some hydroquinone-related compounds with depigmenting activity

3.11.3 Patient's Selection

Only patients with extensive disfiguring vitiligo should be treated, and this must be done only after exploring other possible therapies. The patient should be advised that monobenzone is a potent depigmenting agent and not a cosmetic skin bleach (Fig 3.11.2) [4, 6, 8].

The subjects with highest phototype (V and VI), for which the contrast between dark-pigmented skin and white vitiligo areas is actually disfiguring, mainly when involving exposed areas, may be candidate to this approach. For the opposite reason, the lowest phototypes (I and II) may be cosmetically improved by the depigmenting agents because even the possible spontaneous repigmentation of the vitiligo lesions did not allow any difference with respect to the drug-depigmented skin. Moreover, incomplete or trichrome repigmentation (e.g., when using UV light) may cause more disfigurement. The patients should be indeed advised that the repigmentation may occur in vitiligo lesions, causing thus further depigmenting cycles. Alternatively, the depigmenting approach may be useful for high phototype vitiligo patients with limited, but highly disfiguring lesions involving the face or the hands. Considering that most of the used approaches lead to a definitive irreversible depigmentation, the patients must be extensively informed.

3.11.4 Protocol

Monobenzone is applied topically in the form of a 20% cream. A thin layer of monobenzone cream should be applied uniformly and rubbed into the pigmented area 2–3 times daily. Prolonged exposure to sunlight should be avoided during treatment with the drug or a sunscreen should be used, as exposure to sunlight reduces the depigmenting effect of the drug. Depigmentation is usually accomplished after 1–4 months of treatment; if satisfactory results are not obtained after 4 months of treatment, the drug should

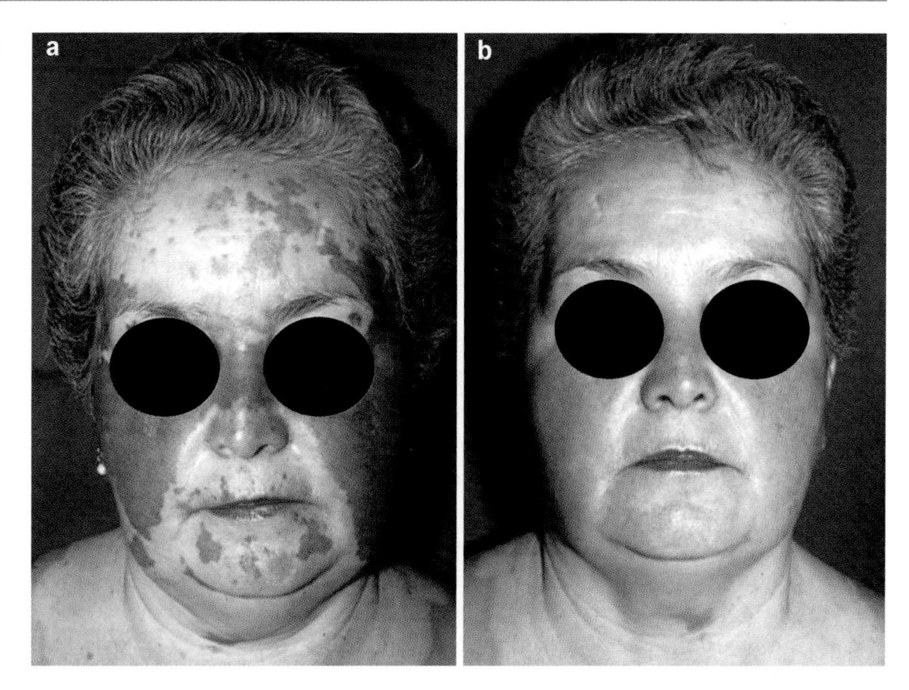

Fig. 3.11.2 Residual patches of pigmentation (**a**) and (**b**) final results after monobenzone (courtesy Dr Rafael Falabella)

be discontinued. When the desired degree of depigmentation is obtained, monobenzone should be applied as often as needed to maintain depigmentation (usually only two times weekly) [4].

3.11.5 Side Effects

Among the possible adverse effects, mild, transient skin irritation and sensitization, including erythematous and eczematous reactions, have occurred following topical application of monobenzone. Although these reactions are usually transient, the treatment should be discontinued if irritation, burning sensation, or dermatitis occurs. Sometimes ocular side effects have been reported. Areas of normal skin distant to the site of monobenzone application frequently have become depigmented.

3.11.6 Combinatory Chemical Approaches

Monobenzone has been proposed in association with retinoic acid in order to overcome the resistance to the treatment of a considerable number of vitiligo patients. Retinoic acid is a derivative of Vitamin A, able to in vitro promote or inhibit melanogenesis on the basis of the experimental condition and cell type. In guinea pig model, it has been demonstrated to potentiate the depigmenting effect of the monobenzone. The depigmentation obtained by 0.025% retinoic acid plus 10% monobenzone was omogeneous, corresponding to the complete disappearance of epidermal melanocytes. Initially, the activity of retinoic acid was attributed to the promotion of keratinocye turnover, leading to the accelerated melanin loss. As alternative mechanism of action for retinoic acid, the impairment of the glutathione-dependent defense has been suggested. However, considering the skin irritation and exfoliation induced by retinoic acid, other retinoic derivative may be used with more successful results [2].

3.11.7 Physical Approaches

The Q-switched 755-nm laser has been applied in order to induce persistent depigmentation [1]. The Q-switched ruby laser (QSR) alone or in combination with methoxyphenol has been also used [5, 7]. The QSR has been reported to be able to destroy melanin and melanin-bearing cells.

However, up to now only few reports have been provided. Consequently, the effective therapeutic role and the stability of the result need further evaluation. QSR was reported as useful treatment for MBEH-unresponsive patient. Ten treatments were performed in order to obtain the depigmentation, and after 12 months, a few non-relevant repigmentation occurs. Further QSR treatments have been carried out on the minimal residual pigmentation without side effects.

When the combinatory therapy has been chosen, a cream containing 4-methoxyphenol was applied. The trial included 16 patients with vitiligo universalis, for which QSR or cream, never simultaneously, was used. Total depigmentation has been described in 70% of the patients after the application of the cream. One or two weeks were the time requested to start the depigmentation, whereas a time of 4–12 months has been reported as useful in order to reach complete depigmentation. The utilization of the QSR for the patients not responding to 4-methoxyphenol produced the depigmenting effect. Four out of the cream-treated patients reported itching. A relapse rate of repigmentation of 36% (4 out of 11 responder subjects) has been described in the studied population during the next 36 months of treatment-free period. No side effects have been described. However, the authors underlined the possible repigmentation after interruption of the treatment [1, 5, 7].

Summary Messages

- In patients with extensive and refractory vitiligo, depigmenting the remaining islands may be more cosmetically acceptable.
- The suggested mechanisms of depigmentation of monobenzone are the selective melanocytic destruction through free-radical formation, the competitive inhibition of the tyrosinase enzyme system, and the increased release of melanin.
- Only patients with extensive/disfiguring vitiligo should be treated and only after exploring other possible therapies.
- The patient should be advised that monobenzone is a potent depigmenting agent and not a cosmetic skin bleach, and that results are usually irreversible
- Unsighty repigmentation can be a major problem with this technique
- Depigmentation has be also obtained by the Q-switched ruby laser, alone or in combination with methoxyphenol.

References

1. Kim YJ, Chung BS, Choi KC (2001) Depigmtation therapy with Q-switched ruby laser after tanning in vitiligo universalis. Dermatol Surg 27:969–970
2. Kasraee B, Fallahi MR, Ardekani GS et al (2006) Retinoic acid synergisitcally enhances the melanocytotoxic and depigmenting effects of monobenzylether of hydroquinone in black guinea pig skin. Exp dermatol 15:509–514
3. Legros L, Cassuto JP, Ortonne JP (2005) Imatinib mesilate (Glivec): a systemic depigmenting agent for extensive vitiligo? Br J Dermatol 153:691–692
4. Mosher DB, Parrish JA, Fitzpatrick TB (1977) Monobenzylether of hydroquinone. A retrospective study of treatment of 18 vitiligo patients and a review of the literature. Br J Dermatol 97:669–679
5. Njoo MD, Vodegel RM, Westerhof W (2000) Depigmentation therapy in vitiligo universalis with topical 4-methoxyphenol and the Q-switched ruby laser. J Am Acad Dermatol 42: 760–769
6. Nordlund JJ (2000) Depigmentation for the treatment of extensive vitiligo. In: Hann SK, Nordlund JJ (ed) Vitiligo. Blackwell Science, Lucon, pp 207–213
7. Rao J, Fitzpatrick RE (2004) Use of the Q-switched 755-nm alexandrite laser to treat recalcitrant pigment after depigmentation therapy for vitiligo. Dermatol Surg 30: 1043–1045
8. Solano F, Briganti S, Picardo M et al (2006) Hypopigmenting agents: an updated review on biological, chemical and clinical aspects. Pigment cell Res 19:550–571

Therapy Adapted for Age, Gender, and Specific Locations

3.12

Alain Taïeb and Ludmila Nieuweboer-Krobotova

Contents

A. Taïeb (✉)
Service de Dermatologie, Hôpital St André,
CHU de Bordeaux, France
e-mail: alain.taieb@chu-bordeaux.fr

In the earlier sections, therapeutic issues have been discussed for mainstream adult patients. In clinical practice the extreme ages, some specific gender-related problems, and special locations need a more detailed coverage.

3.12.1 Age and Gender Issues

Children

It is crucial to avoid under-treatment of common conditions in childhood, especially skin disorders such as atopic dermatitis and psoriasis, and also in vitiligo, where early intervention is helpful before a definitive loss of skin melanocytes. The reasons for under-treatment of skin disorders in children are not always clear, but the reluctance to treat because of age first, assuming that children are less hardy than adults, or because of interference with growth and development, is usually not scientifically founded. Parents may also not adhere to the management plans due to personal beliefs or neglect. In contrast, aggressive parents seeking to have a 'perfect child,' may push the physician to intervene because of a minor cosmetic disfigurement (e.g. focal vitiligo). The dictum: '*primum non nocere*' should always be kept in mind. Overtreatment of benign cutaneous conditions is a difficult problem, but physicians must be persuasive, and parent management through good quality information is a key issue.

Treatment request by the child is usually not common until the age of 6, when entering primary school.

M. Picardo and A. Taïeb (eds.), *Vitiligo*,
DOI 10.1007/978-3-540-69361-1_3.12, © Springer-Verlag Berlin Heidelberg 2010

However, earlier unformulated harm to the building of self-image may remain undetected. As indicated before, early intervention, whatever the type of vitiligo in the child, is preferable to limit disease extension. However, the benefits/risks of the treatment should be weighted cautiously, in terms of time needed to apply topicals or more importantly, in case of deciding for phototherapy. The chronic use of calcineurin inhibitors is debated in children, but there is little evidence of induction of local/systemic immunosuppression, especially in a disorder such as vitiligo which is not associated with increased skin permeability.

The decision to start a phototherapy is based on medical grounds, when other especially topical options are unrealistic, and when the child is able to stand alone in the cabin. The duration of the sessions is usually not a problem, but the distance and time to treatment station can be very limiting. For counselling issues, the importance of the Koebner's phenomenon should be clinically evaluated with respect to the activities of the child. In case of detection of a relevant trauma-prone activity, the decision for cessation should be taken in accordance with the child if age permits, at least for a few months to assess repigmentation. Particular attention should be given to uncontrolled, repeated movements and tics which can be avoided. Guidance through educative slides is helpful. For segmental vitiligo, there is some clinical evidence that neurogenic influences may play an aggravating role. A clinical and X-ray search for a spine trauma or deformation according to segment location can be indicated, and may require physical therapy or orthopaedic measures (Y Gauthier, personal communication). Surgical procedures are rarely attempted in children. Most would need repeated general anaesthesia, which is questionable if the disfigurement is limited. Generally, surgery is started at adolescence, when appropriate. Camouflage should be discussed in children handicapped by disfiguring lesions, especially when bullying at school is noticed, and when surgery is not yet possible. Camouflage workshops are organised at some centres and nurses can provide information following the visit.

Elderly Patients

Vitiligo in light-skinned elderly patients gets frequently unnoticed. Those patients usually have no or minimal aesthetic complaints. In case of darker complexion, ambulatory elderly patients seek advice as younger adults, and should be treated accordingly (Chap.1.3.13). Phototherapies should be limited in case of chronic actinic damage or prior exposure to immunosuppressive drugs, which may be the case if vitiligo accompanies a multiple autoimmune syndrome.

Gender Issues

Women in vitiligo clinics outnumber men. The question of camouflage is usually more central in the counselling. An expert view on the subject is needed because the best solutions need an accurate assessment of the individual situation of the patient (Part 3.9). Camouflage workshops are organised at specialised institutions.

3.12.2 Particular Locations

Mucosae

Vitiligo lesions involving the lips, oral, and genital mucosa are more resistant to medical therapies, since no hair follicle pigment cells can be mobilised. Topical calcineurin inhibitors can be effective in lip and penile vitiligo [8]. Localised-UVB devices are a good indication in a separate or combined approach. For lips, tattooing is a possible option in dark-skinned individuals, but the results may deteriorate with time [2, 3]. The surgical approach [5, 6] has been reviewed in Part 1.3.4. Autologous minigrafting (punch grafting) is one of the treatments for stable vitiligo spots that are recalcitrant to medical local therapy, which can be adapted to lip vitiligo. Punch grafts of 3 mm full thickness from a normally pigmented donor site are placed in the acceptor site at 5–10 mm distance from each other. They are secured with absorbable sutures. The grafting is followed by narrow-band UVB irradiation twice a week. Post-treatment observation shows repigmentation around the grafts without side effects including cobblestone-like texture or scarring at the acceptor site. This technique is an interesting treatment option for the difficult areas of

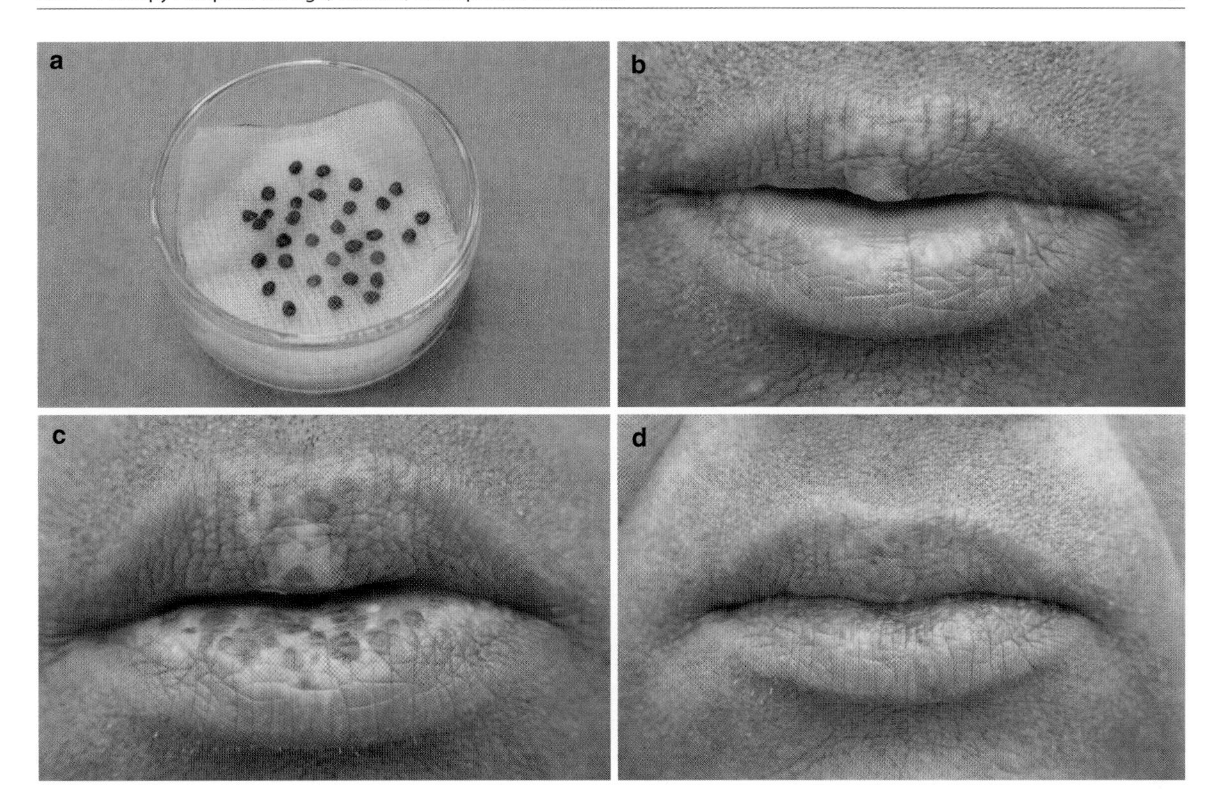

Fig. 3.12.1 Mucosal minigrafting technique (Pigment Cell Melanoma Res. 2008;21:277) Donor site location = hip; 7 sessions of minigrafting plus UVB 311nm during one year. The biopsies are done with a 3mm Stiefel punch and sutures with Monocryl 6-0 (Ethicon, Johnson & Johnson). (**a**) Petri dish with 3 mm biopsies; (**b**) acceptor site before transplantation; (**c**) acceptor site during treatment period; (**d**) acceptor site after treatment one year later.

stable vitiligo spots of not only the lips, but also for fingers, arm-pit, and scalp (Fig. 3.12.1).

Hairs

The therapeutic approach has been discussed in Part 1.3.6

Hands

The repigmentation of the dorsum of hands and digits is particularly challenging, mostly because of the prevalence of a difficult-to-prevent Koebner's phenomenon. Surgery can be proposed in stable cases [7] and camouflage is useful [9]. Occlusive tacrolimus dressings [4] and the combination of laser and 5-Fluorouracil [1] are new treatment options, which deserve confirmation studies. The sutured minigrafts technique can be used (see above).

Summary Messages

- Objective and perceived severity of disease is more important than age to make a difficult therapeutic decision
- In children, the primary aim is to treat the child, not the parents
- Children: always consider lack of observance from the parents in case of failure of treatment
- Children : always consider the opinion of the child when in age of giving one (usually from 6–7)
- Elderly patients: phototherapies should be limited in case of chronic actinic damage or prior exposure to immunosuppressive drugs
- Females but also males require professional camouflage counselling
- Particular locations: mucosal, hair and hand involvement pose specific and particular challenging problems and possible solutions consist in combining medical and surgical approaches, such as the sutured minigraft technique plus targeted phototherapy.

References

1. Anbar T, Westerhof W, Abdel-Rahman A et al (2007) Treatment of periungual vitiligo with erbium-YAG-laser plus 5-flurouracil: a left to right comparative study. J Cosmet Dermatol 5:135–139
2. Centre JM, Mancini S, Baker GI et al (1998) Management of gingival vitiligo with use of a tattoo technique. Br J Dermatol 138:359–360
3. Hann SK, Nordlund JJ (2000) Clinical features of Generalized Vitiligo. In: Hann Sk, Nordlund JJ (eds) Vitiligo. Blackwell Science, Oxford, pp 35–48
4. Hartmann A, Bröcker EB, Hamm H (2008) Occlusive treatment enhances efficacy of tacrolimus 0.1% ointment in adult patients with vitiligo: results of a placebo-controlled 12-month prospective study. Acta Derm Venereol 88:474–479
5. Gupta S, Sandh K, Kanwar A et al (2004) Melanocyte transfer via epidermal grafts for vitiligo of labial mucosa. Dermatol Surg 30:45–48
6. Malakar S, Lahiri K (2004) Punch grafting in lip leucoderma. Dermatology 208:125–128
7. Parsad D, Gupta S, IADVL Dermatosurgery Task Force (2008) Standard guidelines of care for vitiligo surgery. Indian J Dermatol Venereol Leprol 74(Suppl):S37–S45
8. Souza Leite RM, Craveiro Leite AA (2007) Two therapeutic challenges: periocular and genital vitiligo in children successfully treated with pimecrolimus cream. Int J Dermatol 46:986–989
9. Tanioka M, Miyachi Y (2008) Camouflaging vitiligo of the fingers. Arch Dermatol 144:809–810

Psychological Interventions

3.13

Panagiota Kostopoulou and Alain Taïeb

Contents

P. Kostopoulou (✉)
Service de Dermatologie, Hôpital Saint André, CHU de Bordeaux, France
e-mail: p_kostop@yahoo.

3.13.1 Why Psychological Support Is Important

Although vitiligo does not lead to severe physical illness, patients experience a variable degree of psychosocial impairment. The psychological impact of vitiligo has been shown in different studies all over the world. Patients with vitiligo suffer from poor body image, low self-esteem, and social isolation, caused by feelings of embarrassment, and they experience a considerable level of disability [1–7]. The prevalence of psychiatric morbidity associated with vitiligo ranges from 25 to 30% in western Europe [2, 8] and from 56 to 75% in India [3, 9, 10]. A low self-esteem and high levels of perceived stigma seem to be important factors for quality of life impairment in vitiligo patients [9, 11–16] (Part 1.5). The majority of patients with vitiligo found their disfigurement moderately or severely intolerable [14], and most of them said that vitiligo had affected their lives recently [9]. Psychological interviews with patients at our department confirm that vitiligo has an important impact on their daily lives. Visible or not directly visible lesions influence not only their personal relationships, their professional career, but also their social life. Patients said they avoided daily activities and different social events in order to protect themselves from embarrassing comments [17]. In the same study, we found that perceived severity of the disease and patient's personality are important factors to consider when assessing the psychological impact of vitiligo. If self-body image is more influenced by gender, perceived severity is more influenced by patient's personality than by objective criteria of the disease (Fig. 3.13.1) [17].

M. Picardo and A. Taïeb (eds.), *Vitiligo*,
DOI 10.1007/978-3-540-69361-1_3.13, © Springer-Verlag Berlin Heidelberg 2010

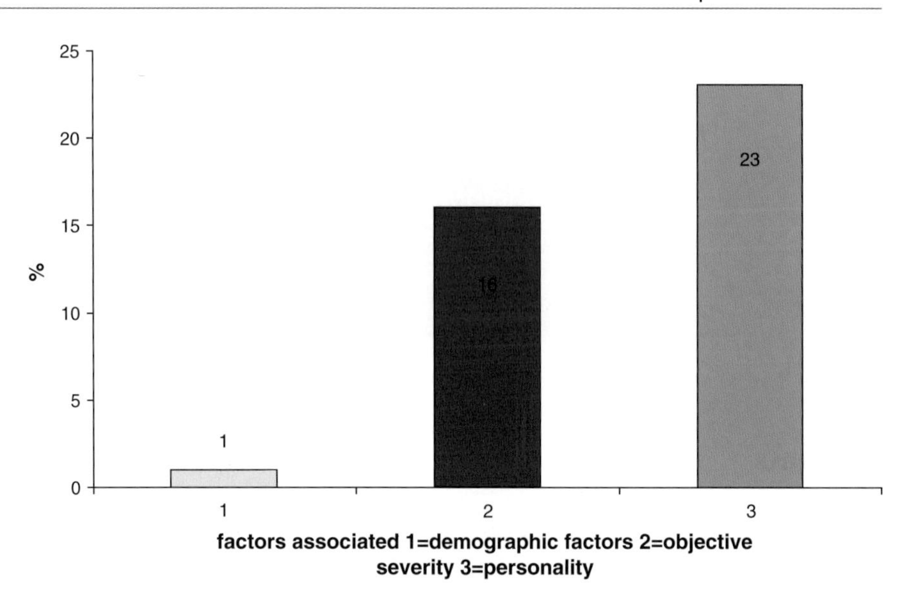

Fig. 3.13.1 Relative importance of factors affecting the variance of perceived severity.

3.13.2 Screening Patients in Need of Psychological Support

Objective criteria like percentage of the body area affected and staging score [15] are important, but not enough for patient's assessment and follow-up. A psychologically oriented interview is essential in order to evaluate the perceived impact of the skin disorder. This perceived impact is important to tailor the management to the patients' needs and difficulties. A simple visual analogue scale from 0 to 10 with the notice 'How much does your skin disease bother you currently'? helps to measure coarsely, the way patients feel about their disease, independent of medical findings. A simple questionnaire of quality of life such as the DLQI [18, 19, 20] which is very easy and rapid to use can also be used to measure patients' daily difficulties associated with their skin disorder.

3.13.3 Psychological Interventions

Different types of psychological support and help can be proposed. Most published articles concentrate on the type of psychological difficulties and the factors that influence them, but do not mention specific interventions. The only psychological intervention published on vitiligo is that of Papadopoulos et al. [12], who have used a cognitive–behaviour therapy (CBT). They indicated that it can help improving the quality of life, self-esteem, and perceived body-image. They also suggested that CBT may influence the progression of the condition itself. Even if their findings are based on a very small sample of patients and if their conclusions are difficult to acknowledge without reservation, this work offers new perspectives. Dermatologists might consider adding psychosocial interventions to standard medical treatment [13].

In addition, based on our experience, a simple psychological interview and a supportive attitude can also be beneficial. Patients are happy to have the opportunity to express difficulties resulting from their disease. Being understood and listened to is a part of the global management of vitiligo. Sometimes, a simple discussion with a professional can help the patient become free from the complexes caused by the disease and embarrassing comments he has to endure. Throughout this process, patients may evolve from the status of patients to that of near normal individuals.

During the interview, we can also evaluate the patient's socio–professional context and the stigmatisation caused by the disorder either at work or during daily activities. Different strategies to be adopted to enjoy better social life can be proposed. Patients who can benefit from a psychological follow-up can receive more attention and may just feel better because of that. The objective is that patients' negative thoughts related to the disease should be gradually replaced by other more positive thoughts based on their personality and qualities. By doing so, patients should be able to improve their self-esteem and socialisation.

3.13.4 Concluding Remarks

Psychological support primarily has the intention to help the patients express their psychological suffering and the negative feelings associated with the disease. It is also aimed at helping the patients accept themselves as they are, with or without the disease. In case of beneficial impact of the intervention, patients may be able, not only to resume their activities and to participate in their social life without feeling handicapped by their skin disorder, but also to develop their own personality without being restricted by disease considerations.

Summary Messages

- A low self-esteem and high levels of perceived stigma seem to be important factors for quality of life impairment in vitiligo patients
- A simple visual analogue scale from 0 to 10 with the notice "How much do your skin disease bother you?" help to measure coarsely how patients feel about their disease independently of medical findings
- Management should be tailored to self perceived disease severity
- Cognitive-behavioural therapy might help improving the quality of life, self-esteem and perceived body-image

References

1. Firooz A, Bouzari N, Fallah N et al (2004) What patients with vitiligo believed about their condition. Int J Dermatol 43:811–814
2. Kent G, Al Abadie M (1996) Factors affecting responses on dermatology life quality index items among vitiligo sufferers. Clin Exp Dermatol 21:330–333
3. Matoo SK, Handa S, Kaur I et al (2002) Psychiatric morbidity in vitiligo: prevalence and correlates in India. J Eur Acad Dermatol Venereol 16:573–578
4. Porter JR, Beuf AH, Lerner A, Nordlund J (1986) Psychosocial effect of vitiligo: a comparison of vitiligo patients with 'normal' control subjects, with psoriasis patients and patients with other pigmentary disorders. J Am Acad Dermatol 15:220–224
5. Porter JR, Beuf AH, Lerner A, Nordlund J (1979) Psychological reaction to chronic skin disorders: a study of patients with vitiligo. Gen Hosp Psychiatry 1:73–77
6. Porter JR, Beuf AH, Lerner A, Nordlund J (1990) The effect of vitiligo in sexual relationships. J Am Acad Dermatol 22:221–222
7. Sampogna F, Raskovic D, Guerra L et al (2008) Identification of categories at risk for high quality of life impairment in patients with vitiligo. Br J Dermatol 159:351–359
8. Picardi A, Abeni D, Melchi CF et al (2000) Psychiatric morbidity in dermatological outpatients:an issue to be recognised. Br J Dermatol 143:983–991
9. Kent G, Al'Abadie M (1996) Psychologic effects of vitiligo: a critical incident analysis. J Am Acad Dermatol 35:895–898
10. Matoo SK, Handa S, Kaur I et al (2002) Psychiatric morbidity in vitiligo and psoriasis: a comparative study from India. J Dermatol 28:424–432
11. Harlow B, Poyner T, Finlay AY, Dykes PJ (2000) Impaired quality of life of adults with skin disease in primary care. Br J Dermatol 143:979–982
12. Papadopoulos L, Bor R, Legg C (1999) Coping with the disfiguring effects of vitiligo: a preliminary investigation into the effects of Cognitive-Behavioural Therapy. Br J Med Psychol 72:385–396
13. Picardi A, Abeni D (2001) Can Cognitive-Behavioural Therapy help patients with vitiligo? Arch Dermatol 137: 786–788
14. Salzer B, Schallreuter KU (1995) Investigation of the personality structure in patients with vitiligo and a possible association with catecholamine metabolism. Dermatology 190:109–115
15. Taieb A, Picardo M (2007) The definition and assessment of vitiligo: a consensus report of the Vitiligo European Task Force. Pigment Cell Res 20:27–35
16. Whitton ME, Ashcroft DM, Barrett CW, González U (2006) Interventions for vitiligo. Cochrane Database Syst Rev (1):CD003263
17. Kostopoulou P, Jouary T, Quintard B et al (2009) Objective vs subjective factors in the psychological impact of vitiligo: the experience from a French referral centre. Br J Dermatol 161:128–133
18. Finlay AY, Khan GK (1994) Dermatology Life Quality Index (DLQI). A simple practical measure for routine clinical use. Clin Exp Dermatol 19:210–216
19. Finlay AY(1997) Quality of life measurement in dermatology: a practical guide. Br J Dermatol 136:305–314
20. Ongenae K, Van Geel N, De Schepper, Naeyaert JM (2005) Effect of vitiligo on self-reported health-related quality of life. Br J Dermatol 152:1165–1172

The Patient Perspective

3.14

Maxine E. Whitton and Alida DePase

Contents

M. E. Whitton (✉)
Wanstead, UK
e-mail: mewhitton@googlemail.com

3.14.1 Introduction

Variations in skin colour, cultural background, access to treatment, and severity of the disease contribute to the degree to which vitiligo impacts on people's lives. While some of those with the disease manage to develop strategies to cope, others succumb to depression and despair. Far from being a mere cosmetic problem, vitiligo can sometimes ruin people's lives.

While we cannot speak for all patients, there are certain common concerns experienced by all those who have the disease. We sincerely hope that researchers, clinicians and health professionals will have a clearer understanding of the patient perspective after reading this chapter.

3.14.2 Impact of the Disease on Patients and Families

Adults

For adults, especially when exposed areas of the body or genitals are affected, vitiligo can be devastating and can severely affect the quality of life. The unpredictable course of the disease as well as its tendency to spread only adds to the anxiety and stress as the patient struggles to adjust to impaired appearance and increasing disfigurement and, in some cases, the possible loss of identity [14]. Vitiligo can undermine self-confidence and self-esteem, leading in some cases to severe depression, and even to suicidal thoughts. It can cause a feeling of isolation and can influence many aspects of everyday life and social interaction (Table 3.14.1).

M. Picardo and A. Taïeb (eds.), *Vitiligo*,
DOI 10.1007/978-3-540-69361-1_3.14,

Table 3.14.1 Impact of Vitiligo

Choice of occupation	It may deter people from entering certain professions or jobs if their vitiligo is visible (e.g. catering, musicians, nursing)
Isolation	People can be ostracised or feel afraid to leave the house for fear of the reaction to their disease
Style of dress	Many people whose vitiligo is on exposed areas wear long sleeves throughout the year
Leisure pursuits and holidays	Avoiding the beach, not joining in sports such as swimming
Relationships	Establishing friendships and finding a partner can be difficult
Marriage prospects	Particularly in some cultures the lives of young people can be blighted if vitiligo is in the family
The decision to have children	Many fear passing on the disease

Children

Around half of the people with vitiligo develop the condition before the age of 20. This means that a large proportion of patients are children with the prospect of living with a chronic, incurable, lifelong disease. Although physical symptoms are not severe, restricted to susceptibility to sunburn on affected areas or itching in some sufferers, the disease can have a devastating effect.

Children can be the butt of cruel teasing and bullying at school, leading to withdrawal, loneliness, poor self-esteem, lack of confidence and underachieving in school subjects. Transition from primary to secondary school can be a particularly difficult time for children when they have to adjust to a change and make new friends. The onset of vitiligo during adolescence can be devastating, as this is a crucial stage in the child's development in the journey towards adulthood. At this stage, young people want to be part of the crowd and belong to a group, but vitiligo makes them stand out. There are anecdotal reports of some adolescents who avoid forming relationships or having sexual intercourse for fear of having to reveal affected genitals.

Families

Parents sometimes go to great lengths and pay a lot of money to try and cure their child. Some parents can unknowingly transfer their anxieties to their child by relentlessly seeking a cure instead of offering support, understanding, and the security of love in spite of the way the child looks. Having a child with vitiligo can put a strain on the parents, on marriages, and also the family unit as a whole. In a study by Basra and Finlay [1], the majority of relatives in their sample reported that having a family member with a skin condition had a detrimental effect on their lives. The sample included people with different skin diseases and did not just target parents, but other family members. However, the majority of cases were eczema sufferers with only one patient with vitiligo included in the study. Studies to ascertain how vitiligo affects families could well prove to be of value if the results lead to a method of helping families to cope. As vitiligo is such a difficult disease to treat, this could be an important element in the care and support of patients and their families. Family impact should be taken into consideration when assessing and treating patients and also in study design.

3.14.3 Role of Psychotherapy

Acceptance of the disease can vastly improve the patient's quality of life. There are clearly some patients who could benefit from psychological interventions, but few psychotherapists are trained or experienced to work with patients with chronic skin disease. Despite the fact that many clinicians acknowledge the need to address the psychosocial impact of skin disease, there is no provision for it. Quite apart from the cost to the patient, unless it is accepted as a useful adjunct to conventional treatment, it will continue to be talked about, but not implemented. The lack of evidence for this type of intervention is also a concern and may also make dermatologists reluctant to suggest it. When patients request it, not many dermatologists can recommend suitably qualified practitioners.

It is an indisputable fact that emotional and psychological distress are the main symptoms of vitiligo which is probably why it has always been considered merely a cosmetic problem and therefore easily dismissed as trivial [3]. The failure of medical interventions to provide a lasting improvement makes it imperative that this aspect of the disease is given more priority until a cure or more effective treatments can be found.

Stigma and Vitiligo

Although vitiligo is not life-threatening, it is certainly life-altering as people with the disease have to cope with a visible disfigurement, which can get worse over time and which may last a lifetime. People with vitiligo often have to face stares, adverse comments, and recoil from touch on a regular basis. The stigma of vitiligo is felt by everyone with the disease, but women in particular bear the greater burden of stigmatisation because of societal perceptions of beauty. Sometimes, stigma is not actual, but perceived. It is nonetheless very real to the patient who can misinterpret the reasons for the stares, which may have nothing to do with the presence of vitiligo.

People whose face and hands are affected by vitiligo are likely to experience more adverse reactions to their appearance than those whose vitiligo is not visible. Assessment of the severity of the disease should take this into consideration as well as the degree of psychological distress, not just the number of lesions and extent of affected areas [11].

Stress and Vitiligo

Vitiligo, like alopecia areata and psoriasis, can be triggered and aggravated by stress. Many patients link the onset of their disease to bereavement or other distressing life event. In a study in which vitiligo patients and matched controls completed the 12-month version of the Schedule of Recent Experience – a questionnaire measuring the number and frequency of stressful life events – Papadopoulos concluded that vitiligo patients experienced more stressful events than controls [7]. In another more recent study of 45 alopecia patients and 32 vitiligo patients, Manolache [6] found that in the vitiligo group more patients than controls reported stressful events (65.62 compared to 21.87%). In both the alopecia and vitiligo groups, one specific event occurring just before onset of the disease seems to have greater impact than several stressful situations. In both groups, personal problems, including exams and job/financial problems were significant.

Cognitive–Behavioural Therapy (CBT)

A recent review of psychosocial interventions for people with visible difference, including vitiligo, did not find convincing evidence for clinical effectiveness in the twelve studies evaluated [2]. There are two studies which explore the possible value of psychotherapy for vitiligo patients [8, 9]. The first study published in 1999 was small with only 16 participants randomised into two groups of eight. One group received individual sessions of CBT and the other group did not receive any counselling, but in both treatment and controlled arms two patients continued to receive PUVA treatment. Groups were well-matched at baseline and the study lasted for eight weeks with a follow-up assessment after five months. The results after treatment showed an improvement in quality of life, self-esteem, and body image in the treatment group compared to no change in the low scores recorded before and after the duration of the study in the control group. Results of the second study comparing CBT with person-centred therapy did not support the findings of the first study. This may be because it was designed to use a group-therapy approach, which did not work as well as individual sessions described in the first study. More and better-powered studies are needed to establish whether this type of therapy is a useful adjunct to conventional treatment.

3.14.4 Patients' Stories

Patients' stories give an insight into what it is like to live with vitiligo. In this section we quote from the patient's own words. Lee's story, which is given in more detail, and Gurdeep's narration show how people with vitiligo gain strength from speaking out, showing their disease instead of hiding it and ultimately helping others with it to lead fulfilling lives.

Maria, 55 years old

> 'I have vitiligo, and my son has inherited the same disease. I know he is negative about sexual relations, being ashamed to show his body as his most intimate parts are affected. He tries to avoid making love with his girlfriend for this very reason, and does not dare to turn out the light during love-making since he believes this gesture could give rise to who knows what suspicions. He ends up avoiding any sort of relationship.
>
> As regards me, I hide with clothing, as far as I can, the parts of my body affected, and am bothered by the condition only at the swimming pool or on the beach. I have to confess that I am overdressed even in the hottest summers. I have given up going to the gym, as the fear of being seen with my skin uncovered keeps me in a

continuous state of feeling guilty, threatened and vulnerable I know that somebody with a visible skin disease was refused admission to the gym where I used to go.

After joining other patients of the Vitiligo Group, I have come to the conclusion that I live best and with least distress if I openly talk about my condition, explaining clearly what it is about. In the long run, hiding myself and hiding it leads to isolation and anxiety.........'

Alfredo, 60 years old

'I am a middle-aged estate agent living in a sea-side town in Southern Italy, and each day I have appointments with new clients I take to view the flats I sell. Immediately I have to hold out my hand, patchy like my arms, to a stranger, I am filled with stress and anxiety. I also experience a sense of panic during meetings with my bosses and colleagues around a crowded table when it seems to me that all eyes are on my very obvious patches. This feeling of isolation is even stronger when I need, in a new situation, to feel myself part of a group, accepted and protected for being equal, and not rejected for being different.........'

Robert, 11 years old

'When the creams had no dramatic effect on my vitiligo this was another blow to me, I felt a sense of despair as if there was no chance for me to ever rid myself of vitiligo and look normal. I was still very depressed and was still crying myself to sleep......I did not believe that anyone understood how I felt and if I tried to explain then people would always start by saying. well, it could be worse you know or what about people living with cancer....I was tired of explaining what I had to strangers and friends, tired of trying to cover up and tired of trying to find ways of reducing the patches.'

Henrietta, 40 years old

'I had heard of the condition but knew very little about it until two and a half years ago when, during drug induced menopause and HRT (hormone replacement therapy) I started developing white spots on the insides of my wrists. After a few visits to medical professionals I was given a diagnosis and told to accept it. Easy to say when it's not happening to you but post-hysterectomy the spread of my patches, which increased on a weekly basis, had such a profound effect on me mentally that I did not see how I would every accept it.'

Gurdeep

'I have had vitiligo since I was 10 years old and see my condition as having taken me through a journey. I used to hide my white patches by wearing long sleeved clothes and sometimes sitting on my hands! I then realised that this was something I needed to accept before anyone else would... I have built a lot of strength of self confidence which has helped me to deal with having the condition. My friends and family have played a huge role in helping me through this journey. I now help others to deal with the condition and try to encourage them not to hide it. We have to educate others and raise awareness of vitiligo so we can eliminate the ignorance which people have. Having vitiligo is part of my life now and many people don't even see it on me as they see someone with confidence. This is how I want others with vitiligo to feel. You can lead a happy life with vitiligo and it should not stop us from doing anything.'

Lee

Lee Thomas is a feature/entertainment reporter who works for an American TV company. He developed vitiligo in 1997, when he was twenty-five, and it has since spread relentlessly to cover half his face and a lot of his body. He spoke openly about his condition on TV and wrote a book 'Turning White: a memoir of change' [12] some of the proceeds of which go towards funding a vitiligo support group, The Turning White Foundation. The following are excerpts from the book, which tells of his journey with vitiligo.

'Even people who have known me for years avoid eye contact when they see my face without makeup for the first time.'

'So many times, I thought I should just quit and go hide somewhere. Other times, I just hid out. There are times of weakness. But I have vowed to stay engaged in life no matter what. I had no idea "what" would mean watching myself turn white. I never envisioned this. I call it the phobia of me. I look like a monster.'

'...sometimes I want to scream when I see myself in the mirror. I want to cry until it's all better. Other times, I don't even notice it because my mind is somewhere else.'

In the book Lee describes two encounters with two little girls which had a profound effect on him. The first was at a playscape launch when a little girl ran screaming from him, terrified at the sight of his hands. He tried to imagine how much worse it could have been if she had seen his face and not just his hands. This experience stopped him from leaving his house for several weeks.

The other story is about a little girl he met is a store while reaching for a product. This time the little girl did not run away but looked concerned, touched his face and asked him if it hurt. He was so moved by her concern that he was able to go out again in public without make-up.

"Both little girls helped me to understand myself. They helped me to clarify what I call the duality of me. They exhibited the two extremes of reactions that I get every day – plus everything in between – in relation to how people feel about seeing me with this shocking, visual disease.

One reaction was based on ignorance. And that's not a negative thing. The little girl had never seen anything like me before; sometimes ignorance can cause fear. But the other little girl in the grocery store thought she knew what was wrong. Because of that, she had no fear. She had compassion and concern and displayed a kind heart".

3.14.5 Patient Support Organisations

If proof were needed of the importance of emotional support, this can be amply supplied by the existence of Patient Support Organisations (PSOs). Most were set up by people with vitiligo (many of whom are volunteers) who want to share experiences, support each other, and gather information about the disease. Organisations vary in size and resources as well as in membership, with many in Europe struggling to survive. Some, unfortunately, have lost the battle. The list of PSOs in Table 3.14.2 is not comprehensive but gives some idea of the variety and achievements of patient support organisations which can be a lifeline for people with vitiligo.

3.14.6 Inter-Organisational Cooperation

In order to get more recognition for the disease and have more influence on health policy, some organisations have formed alliances with other skin groups. Individual patient organisations may find it difficult to make a significant contribution to research, whereas an alliance is likely to have a bigger voice and a stronger influence and raise more money for research.

In the United Kingdom, the Vitiligo Society is a member of the All Party Parliamentary Group on Skin (APPGS) whose aims include raising awareness of skin issues in Parliament and making recommendations to the government to improve dermatology services. Over the years, they have produced reports on issues such as training for health professionals and the influence of skin disease on patient's lives. The Society provided written evidence for these reports and representatives from the Society gave evidence in the House of Commons and sat on some of the Expert Committees. The Vitiligo Society is also a founder member of the Skin Care Campaign (SCC), an umbrella organisation that represents the interests of all people with skin diseases in the United Kingdom. The SCC also gets advice and support from dermatologists, pharmacists, and other health professionals as well as sponsorship from a number of pharmaceutical companies.

Both, the National Vitiligo Foundation (NVFI) and the Vitiligo Support International (VSI) are members of the Coalition of Skin Diseases (CSD), a voluntary coalition of patient groups based in the United States of America. They also work closely with the National Institute of Arthritis and Musculoskeletal and Skin Diseases (NIAMS) to coordinate research in skin diseases. The VSI is also involved with The National Coalition of Autoimmune Patient Groups (NCAPG). The NCAPG works to consolidate the voice of autoimmune disease patients and to promote increased education, awareness, and research into all aspects of autoimmune diseases through a collaborative approach.

3.14.7 Other Issues

Treatments

Current treatments are not satisfactory and often require great commitment on the part of the patient and may result in only partial repigmentation. Even when normal colour is restored, it can often be lost again in the same treated areas, and the disease can spread more aggressively than before. There is no strategy for maintenance of improvement once treatment has stopped and for all these reasons some patients are unwilling to undergo therapy. Many practitioners recommend cosmetic camouflage as a useful way of disguising the white patches on the face and other exposed areas, thus improving confidence.

Family doctors, Misdiagnosis, Early Diagnosis, and Treatment

As with all chronic illnesses where the cause is unknown, the fruitless pilgrimage of the patient from one specialist to another can be as discouraging as the treatments prescribed. It is not unusual for patients at their first visit to their family doctor to be given no information and offered no treatment or advice, just

Table 3.14.2 Patient support organisations

Organisation details	Aims and mission	Activities	Publications	Status & professional advice
ARIVONLUS (Associazione Italiana Ricerca e Informazione per la Vitiligine) 2006 (Italy) www.arivonlus.it	To improve the quality of life of patients and families, helping them to cope with the disease and its disfiguring skin changes. To increase the government's investment in research, educate decision-makers about vitiligo, improve access to effective treatments, help researchers make real progress	The ARIV organises successful patients/ researchers/ clinicians meetings with a high degree of participation. Social and educational events such as walks, forums and fundraising initiatives such as the innovative annual ARIV Golf Trophy, a series of golf tournaments in different Italian regions	A magazine, printed in 12,000 copies, is distributed to patients, dermatologists, pharmacies and the charity sector. A monthly electronic newsletter is sent to all on-line contacts	Not-for profit organization Scientific Committee
AVRF (American Vitiligo Research Foundation) www.avrf.org	Provides public awareness about vitiligo through dedicated work, education and counselling. Seeks to make a difference worldwide to those afflicted by the disease, focusing on children and their families. The AVRF embraces diversity and encourages acceptance. It also encourages higher ethical standards in research, and therefore supports finding a cure through alternatives to animal testing	Children's Dream Program which arranges for children to meet their favourite personality. Vitiligo Walk-a-Thons, Skate-a-Thons, Swim-a-Thons, Skip-a-Thons and BBQs. Yearly Medical/Research/ Patient Seminars	annual children's calendar Medical Newsletters	Non-profit, tax-exempt charity Medical Advisory Board
NVFI (National vitiligo Foundation Inc.) 1985 (USA) www.nvfi.org	To educate and help the world to understand and accept people with Vitiligo with unquestionable love and respect; while also helping the find a cure	Annual national convention, regional vitiligo conferences (Meet and Greet) where NVFI members get the opportunity to meet other patients, dermatologists, researchers and company representatives and to discuss vitiligo issues	The NVFI website also provides a vetted list of dermatologists and current research information, mainly projects carried out in the U.S.	Non-profit tax exempt charity Medical Advisory Board

Table 3.14.2 (continued)

Organisation details	Aims and mission	Activities	Publications	Status & professional advice
Vitiligo Society 1985 (UK) www.vitiligosociety.org.uk	To beat vitiligo by eradicating the psychological, social and physical effects it has on people's lives, and by finding effective treatments and a cure	Living with "Vitiligo" workshops for adults. Workshops for parents of children with vitiligo	Magazine, "Dispatches" published regularly, information leaflet for health professionals, children's DVD	Registered charity Medical and Scientific Advisory Panel
VSI Vitiligo Support International 2002 (USA) www.vitiligosupport.org	A global community organization, committed to supporting those affected by vitiligo, raising awareness, promoting research, and discussing effective treatments until a cure is found	Annual conference	Online magazine "Spotlight"	Non-profit tax- exempt charity Medical and Scientific Advisory Committee

told the oft repeated mantra 'think yourself lucky that you don't have cancer, there's nothing to be done, learn to live with it.' If and when the hoped for referral comes to a specialist, more disappointments often follow.

Very often, in fair-skinned individuals, the best treatment for vitiligo is no treatment at all and aggressive treatment is generally not used in children. For the majority of patients, early diagnosis is the key to halting the progress of the disease, and there are still some family doctors who misdiagnose vitiligo. There are many anecdotal reports of patients who have been prescribed Selsun shampoo, for example, in the mistaken belief that they have pityriasis versicolor. Because topical creams, including potent corticosteroids and the newer calcineurin inhibitors are not licensed for vitiligo, general practitioners are, understandably, reluctant to prescribe them. The result is that the patient's vitiligo is often very advanced by the time they get to see a dermatologist, which could make it much harder to treat the disease. In the case of children it is even more important to start treatment at early onset as in general they appear to respond better to treatment than adults.

Internet and Alternative Treatments

If we search for vitiligo on the Internet, we are inundated with thousands of scientific publications and hundreds of thousands of pages on the disease, treatments, and cures of every description including miracle cures from all over the world [4]. It is difficult to know which of these cures, if any, have any scientific basis for their claims. In some countries without national health systems, treatment can be costly and vulnerable patients, desperate to find a cure, can be exploited. Many individuals with vitiligo decide to try alternative and unproven therapies but what is the real cost to them, not just in terms of money but of the effect on their lives? Clinicians can be offhand, belittling the impact of vitiligo, trivialising the condition, and sending patients away with no hope. It is not at all surprising that in these circumstances people will go to any lengths to find help and clutch at any straws that offer them hope of a cure.

What patients really want is an effective treatment that is easy to use, and that which will not require many time-consuming weekly trips to hospital.

Research

Although vitiligo is not a rare disease, investment in vitiligo research is poor, partly because it is not considered a proper disease in spite of being classified as such by the WHO (International Classification of Diseases, ICD, L80). In addition, there is no apparent strategy to develop clinical research in this area.

The last 10 years have seen an increase in vitiligo research and a better understanding of the underlying mechanisms of the disease, for example, genetic susceptibility and the link to autoimmunity. It is also encouraging to note that researchers around the world are beginning to share their knowledge and expertise to improve their understanding of the pathogenesis of vitiligo [10].

Involvement of Patients and Patient Support Organisations in Research

Recognising the paucity of good research and the lack of effective, evidence-based treatments, all Patient Support Groups strive to support, fund, and encourage research (Fig. 3.14.1). The degree to which they can do this is largely dependent on their resources. The Vitiligo Society part funded the study by Gottumukkala et al., investigating auto-antibodies in vitiligo [5], and the pilot study by Papadopoulos et al. [8]. Patient Groups around the world – including the Associazione Italiana Ricerca e Informazione per la Vitiligine (ARIV), VSI, the NVFI in the United States of America, and the UK Vitiligo Society, have participated in the VitGene Consortium which includes 23 physicians and researchers from the United States of America, Columbia, Japan, Korea, Pakistan, England, France, Holland, Belgium, and Italy, all pooling their resources to discover the genes that are involved in vitiligo. The ARIV also takes an active part in the European Society for Pigment Cell Research, an established association of scientists and researchers that has lately focused on vitiligo. The ARIV and the Vitiligo Society support the work of the Vitiligo European Task Force (VETF), the largest European group to work in partnership with patients and scientists on vitiligo. The task force acknowledges the importance of patient input in devising the patient assessment form.

Many patient support organisations fund specific research projects, reporting on their progress to members and the general public via newsletters and their websites. The growing movement of patient involvement in health care and research means that patients are no longer confined to being only participants in studies, but now have the opportunity to work in partnership with investigators and healthcare professionals to drive the agenda for research that is more relevant to patients' needs. The following are some examples of this.

INVOLVE is a national advisory group funded by the Department of Health in the United Kingdom which aims to promote and support active public involvement in public and social-health research. The Cochrane Skin Group, part of the international Cochrane Collaboration, encourages patient involvement at all levels of developing a systematic review, including commenting on protocols and reviews. The Cochrane Systematic Review of interventions for Vitiligo is a patient-led review, which highlights the gaps in research and the lack of robust evidence for current treatments [13]. The British Association of Dermatologists had patient representation on the vitiligo guideline committee, which has

Fig. 3.14.1 This illustration was used to communicate by a patient's support group

developed an evidence-based guideline for the disease. The UK Dermatology Clinical Trials Network (UKDCTN) actively seeks the involvement of consumers and patient representatives at all stages in the process of trial development.

Useful Websites

WHO ICD classification for vitiligo: http://www.who.int/classifications/apps/icd/icd10online/
INVOLVE www.invo.org.uk
Cochrane Skin Group: http://www.nottingham.ac.uk/~muzd
UKDCTN: http://www.ukdctn.org

Summary Messages

- Patients with vitiligo should be treated with respect as people suffering from a genuine disease and not just a cosmetic problem.
- Vitiligo should be a compulsory element in training for health professionals, including specialist dermatology nurses and general practitioners.
- The psychosocial nature of the disease needs to be acknowledged and addressed. Supportive psychotherapy should be more widely available.
- Patients need treatments to be easy to use and effective. Thought should be given to ways of maintaining improvement once achieved.
- Patients should be offered cosmetic camouflage, routinely, as part of the management of the disease.
- Patient Organisations are a valuable resource for patients and health professionals and an excellent source of support.
- Patients and patient organisations can play a useful role in the research process.
- There should be a holistic, multi-disciplinary approach to the management of vitiligo which includes information about the disease, as well as patient support groups, cosmetic camouflage and referral to psychotherapy where appropriate.
- Funding bodies and pharmaceutical companies should invest more in the work of patient support groups and vitiligo research.

References

1. Basra MKA, Finlay AY (2007) The family impact of skin diseases: the Greater Patient concept. Br J Dermatol 156:929–937
2. Bessell A, Moss TP (2007) Evaluating the effectiveness of psychosocial interventions for individuals with visible differences: a systematic review of the empirical literature. Body Image 4:227–238
3. DePase A (2000) Importanza del Camouflage in Pazienti con Vitiligine. In:Lotti T (eds) La Vitiligine: Nuovi Concetti e Nuove Terapie. Utet, Milan
4. DePase A (2003) La Vitiligine: Il Patiente Oltre La Malattia. EBD Evid Based Dermatol (Quarterly Rev) 3:47–48
5. Gottumukkala RV, Waterman EA, Herd LM et al (2003) Autoantibodies in vitiligo patients recognize multiple domains of the melanin-concentrating hormone receptor. J Invest Dermatol 121(4):765–770
6. Manolache L, Benea V (2007) Stress in patients with alopecia areata and vitiligo. J Eur Acad Dermatol Venereol 21:921–928
7. Papadopoulos L, Bor R, Legg C et al (1998) Impact of life events on the onset of vitiligo in adults: preliminary evidence for a psychological dimension in aetiology. Clin Exp Dermatol 23:243–248
8. Papadopoulos L, Bor R, Legg C (1999) Coping with the disfiguring effects of vitiligo: a preliminary investigation into the effects of cognitive-behavioural therapy. Br J Med Psychol 72:385–396
9. Papadopoulos L, Walker c, Anthis L (2004) Living with vitiligo: a controlled investigation into the effects of group cognitive-behavioural and person-centred therapies. Dermatology 5:172–177
10. Schallreuter KU, Bahadoran P, Picardo M et al (2008) Vitiligo pathogenesis: autoimmune disease, genetic defect, excessive reactive oxygen species, calcium imbalance, or what else? Exp Dermatol 17:139–160
11. Schmid-Ott G, Künsebeck H-W, Jecht E et al (2007) Stigmatization experience, coping and sense of coherence in vitiligo patients. J Eur Acad Dermatol Venereol 21: 456–461
12. Thomas L (2007) Turning White: a memoir of change. Momentum Books LLC, Canada
13. Whitton ME, Ashcroft DA, Barrett CM et al (2006) Interventions for vitiligo. Cochrane Database Syst Rev (1):CD003263
14. Whitton M (2007) A patient's experience of vitiligo and its treatment. Dermatological Nursing 6(1):43

Evidence-Based Medicine Perspective 3.15

Jean-Paul Ortonne, Thierry Passeron, DJ Gawkrodger, Davinder Prasad, and Somesh Gupta

Contents

3.15.1 History of EBM Approaches in Vitiligo

The first meta-analysis of nonsurgical repigmentation therapies and/or systemic review of autologous transplantation methods in vitiligo were published in 1998 [10, 11]. In 2006, the Cochrane Collaboration published a review of interventions for vitiligo [17]. This document was based upon electronic databases (Cochrane Skin Group Register, the Cochrane Central Register of Controlled Trials, Medline, EMBASE, AMED) and other databases (last search on September 2004). The last evidence-based analysis of the literature on vitiligo therapy was published in 2007 by a group of German dermatologists [3]. In 2008, guidelines for the diagnosis and management of vitiligo were produced by the British Association of Dermatologists, including the Vitiligo Society, the Cochrane Skin Group and the Royal College of Physicians [4], and guidelines for the surgical management of vitiligo were produced by an ad hoc task force [13]. Tables 3.15.1 and 3.15.2 summarize the grading of evidence and quality used in this chapter.

3.15.2 Meta-Analyses

- The 1998 meta-analysis

The first meta-analysis included 63 studies on therapies for localized vitiligo and 117 studies for generalized vitiligo [10]. A total of 33 randomized controlled trials (RCTs) (11 for localized vitiligo and 22 for generalized vitiligo) were found. Among RCTs on localized vitiligo, the pooled adds ratio (OR) vs. placebo was significant for topical class 3 corticosteroids. Side effects were reported mostly for

J.-P. Ortonne (✉)
Department of Dermatology, Archet-2 hospital, Nice, France
e-mail: Jean-Paul.ORTONNE@unice.fr

M. Picardo and A. Taïeb (eds.), *Vitiligo*,
DOI 10.1007/978-3-540-69361-1_3.15, © Springer-Verlag Berlin Heidelberg 2010

Table 3.15.1 Grades of recommendations as used in the UK Guidelines

Grades of recommendation	Levels of evidence
A: At least one meta-analysis, systematic review, or RCT rated as 1++, and directly applicable to the target population; or a systematic review of RCTs or a body of evidence consisting principally of studies rated as 1+, directly applicable to the target population, and demonstrating overall consistency of results	1++: High-quality meta-analyses, systematic reviews of RCTs, or RCTs with a very low risk of bias 1+: Well-conducted meta-analyses, systematic reviews of RCTs, or RCTs with a low risk of bias
B: A body of evidence including studies rated as 2++, directly applicable to the target population, and demonstrating overall consistency of results; or extrapolated evidence from studies rated as 1++ or 1+	1−: Meta-analyses, systematic reviews of RCTs, or RCTs with a high risk of bias 2++: High-quality systematic reviews of case–control or cohort studies; High-quality case–control or cohort studies with a very low risk of confounding, bias, or chance and a high probability that the relationship is causal
C: A body of evidence including studies rated as 2+, directly applicable to the target population and demonstrating overall consistency of results; or extrapolated evidence from studies rated as 2++	2+: Well-conducted case–control or cohort studies with a low risk of confounding, bias, or chance and a moderate probability that the relationship is causal
D: Evidence level 3 or 4; or extrapolated evidence from studies rated as 2+	2−: Case–control or cohort studies with a high risk of confounding, bias, or chance and a significant risk that the relationship is not causal 3: Nonanalytical studies, e.g., case reports, case series 4: Expert opinion

RCT randomized controlled trial

Table 3.15.2 Levels of evidence for surgical approaches (as used by the Indian Surgery Task Force)

Level	Research-based evidence	Studies and results
A	Strong	High quality, homogeneous
B	Moderate	Multiple adequate studies
C	Limited	At least one adequate study
D	Absent	Expert panel evaluation or other information

class 4 corticosteroids. For generalized vitiligo, oral methoxsalen plus sunlight and oral trioxsalen plus sunlight RCT were reviewed but no RCT was found on narrow-band and broad-band UVB therapy. Oral methoxsalen plus UVA was associated with the highest rates of side effects. On the basis of this review, several recommendations were made concerning the choice of the most effective and safest therapy. For localized vitiligo, class 3 corticosteroid was advised as the first-choice treatment. Although, there were at that time still insufficient data to provide strong evidence-based guidelines concerning narrow-band UVB therapy, the authors concluded that UVB (narrow-band or broad-band) therapy or oral methoxsalen plus UVA was recommended for the treatment of generalized vitiligo. However, the authors stated that oral methoxsalen plus UVA was associated with the highest rates of side effects and recommended that guidelines for maximum cumulative PUVA doses in vitiligo should follow those recommended for psoriasis. This meta-analysis included a systematic review of the effectiveness, safety, and applicability of autologous transplantation methods in vitiligo. Sixty-three studies were obtained, of which 16 reported on minigrafting, 13 on split-thickness, 15 on grafting epidermal blisters, 17 on grafting of cultured melanocytes, and 2 on grafting of noncultured epidermal suspension. However, noncontrolled trials were included. Split thickness and epidermal blister grafting were recommended as the most effective and safest techniques. No definite conclusions could be drawn about the effectiveness of culturing techniques because only a small number of patients had been studied.

- The 2006 Cochrane Skin Group meta-analysis

The Cochrane group assessed interventions used to manage vitiligo [17]. The material for this review was obtained by a search of several databases (the Cochrane Skin Group, the Cochrane Central Register of Controlled Trials, Medline, EMBASE, AMED, and other databases). The search ended on September 2004. Only RCTs were considered. Initially, 22 studies were identified, of which 19 with a total of 1,350 participants were included. All studies assessed outcomes as the presence of repigmentation. None of the trials had similar enough interventions to allow data pooling. The results are summarized as follows:

The use of a combination of a light source and a photoactive chemical (i.e., psoralens or khellin) administered either orally or topically was the most common intervention among the included trials (12 RCTs). The Cochrane group search uncovered limited to moderate evidence to various types and regimens of phototherapy (UVA and UVB) used alone or in combination with psoralens, calcipotriol, folic acid and vitamin B12, oral L-phenylalanine, and topical pseudocatalase. Topical khellin combined with UVA is commonly used, but there was a lack of available evidence of benefit. There was also limited evidence of the benefit of topical immunosuppressants and topical calcipotriol, used in conjunction with ultraviolet. Several studies with topical steroids showed a good evidence of efficacy. Results of individual studies showed that clobetasol propionate was better than PUVAsol, bethametasone valerate was better than placebo; there was no difference between tacrolimus and clobetasol propionate, and flucasone propionate was better when used in combination with UVA than either UVA or fluticasone alone. Intralesional or oral corticosteroids had been evaluated in trials of limited quality. Topical tacrolimus appeared to have similar effects to topical steroids (but a better safety profile), but required further evaluation (Chap. 3.2.2).

Sea climatotherapy and *gingko biloba* extract were found to give some evidence of repigmentation. Another unconventional intervention, melagenina, showed no statistical difference with placebo. Grafting appeared to work best on stable nonsegmental vitiligo and on segmental forms using split-thickness/suction grafts. The authors concluded that some evidence support existing therapies for vitiligo. However, they found that the different design and outcome measurements, lack of quality of life measures, and adverse effect reporting in the reported studies limited the usefulness of their findings. Another main conclusion was that high-quality randomized trials using standardized measures of repigmentation and addressing relevant clinical outcomes, including quality of life, were needed.

No RCT studies for depigmentation were available. Only an open retrospective study of 18 patients treated with monobenzyl ether of hydroquinone was published. Eight of these patients achieved 100% depigmentation. Guidelines for the use of MBEH were suggested: (a) desire of permanent depigmentation in patient with vitiligo; (b) age over 40; (c) more than 50% of the skin depigmented; and (d) willingness to accept the fact that the repigmentation will no longer be possible (this statement is not fully true since patients depigmented by MBEH can repigment).

One study reported that the combination treatment topical 4-methoxyphenol + Q-switched ruby laser can be useful for depigmentation. Cryotherapy has also been demonstrated to achieve complete and permanent depigmentation.

3.15.3 Lim and Hexsel's Algorithm

An editorial published in 2007 summarized the present status of vitiligo treatments and proposed a treatment algorithm (Fig. 3.15.1). This review focused primarily on the more recent studies (n = 15) with a high level of evidence. The main statement in this paper is that the question when managing patients with vitiligo, especially that involving exposed sites, is no longer whether to treat or not to treat, but to decide which treatment method is most appropriate for the individual patients [9].

Based on this algorithm, on exposed areas, if the body surface area (BSA) involved is superior to 11%, the first-choice treatment is narrow-band UVB with or without calcipotriene or oral PUVA with or without calcipotriol. Oral PUVA should be second line as evidence exists that UVB irradiation is not associated with an increased risk of skin cancer while it is definitely increased in PUVA therapy. On exposed areas, if

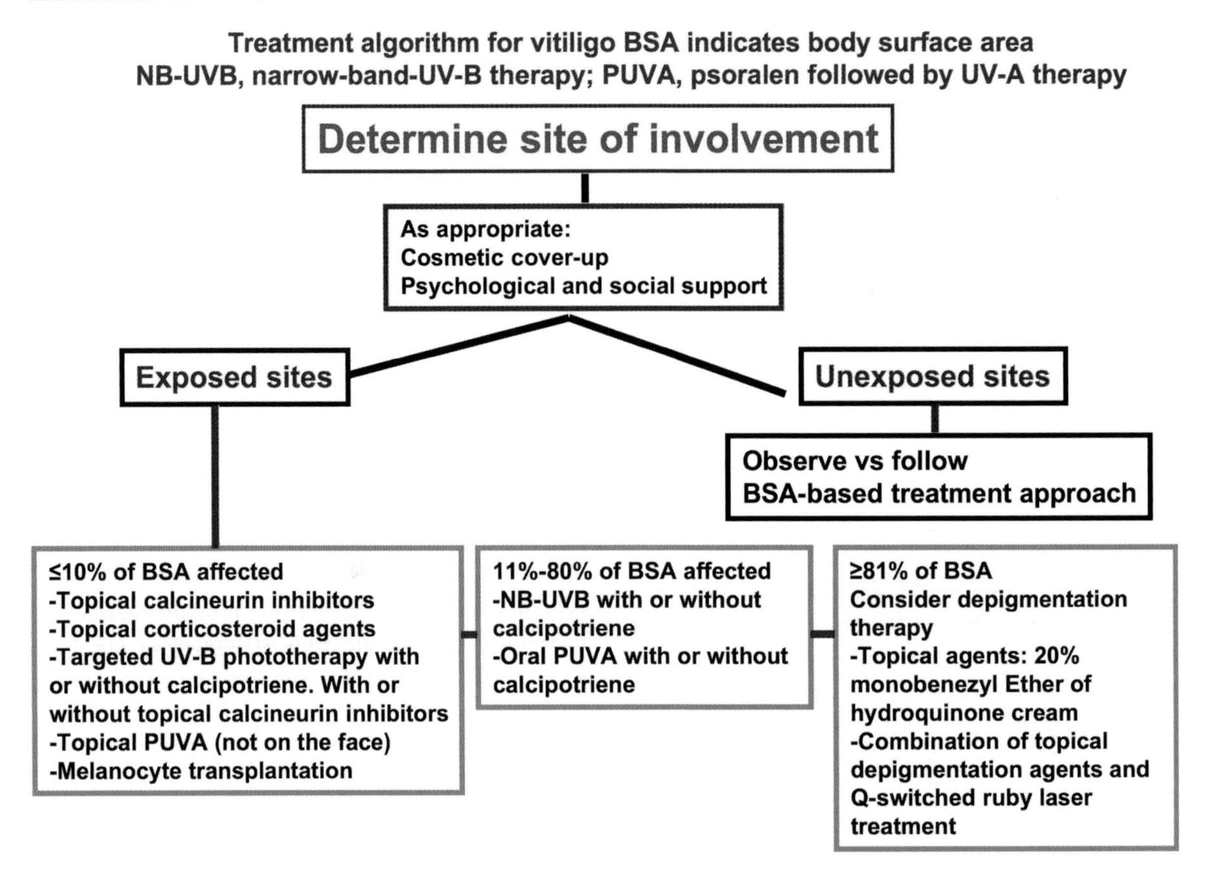

Fig. 3.15.1 The vitiligo treatment algorithm

the BSA is equal or inferior to 10%, several treatment options should be considered: topical calcineurin inhibitors with or without natural sun-exposure; topical corticosteroids with or without normal natural sun-exposure; targeted UVB exposure (311 nm) phototherapy (308 nm excimer lamp or laser) with or without topical calcipotriol or with or without inhibitors. PUVA therapy can be considered, but not first line and not on the face.

For surgical repigmentation, the authors concluded that maximal repigmentation occurred with split-thickness grafting when compared with minigrafting. Grafting of cultured autologous melanocytes has the unique advantage of allowing treatment of large depigmented areas. However, the major disadvantages are the cost and the required infrastructure.

When more than 81% of BSA is involved, depigmentation therapy should be considered either using depigmenting agents (20% monobenzyl ether of hydroquinone) or a combination of topical depigmenting agents and Q-switched Ruby laser treatment.

3.15.4 The 2007 German Evidence-Based Analysis

This analysis [3] includes clinical trials published from 1997 to 2006. It includes modern therapeutic options such as topical immunomodulators (tacrolimus, pimecrolimus), excimer laser, and surgery/transplantation.

The results can be summarized as follows: phototherapy achieves good results, especially in generalized vitiligo with narrow-band UVB appearing superior to broad-band UVB. Vitamin D3 analogues, such as calcipotriol, do not show convincing either in monotherapy or in combination with UVB. In comparison to conventional narrow-band UVB, excimer laser treatment achieves significantly better results. Combination of tacrolimus and excimer laser increases efficacy of times and facilities, thus limiting its widespread use. Topical immunomodulators are considered as better suitable for use on the face.

3.15.5 The 2008 Systematic Review of Natural Health Product Treatments

A systematic review of natural health product treatment of vitiligo has been published in May 2008 [16]. Fifteen clinical trials were identified and organized into four categories based on the natural health product, L. phenylalanine, traditional Chinese medicine, plants (photosensitizers, *ginkgo biloba*), and vitamins (cobalamin, folic acid, vitamin E). Most of the clinical trials identified were uncontrolled. Of the few controlled clinical trials reviewed, most were of poor quality and contained significant reporting flaws. The authors concluded that phenylalanine used with phototherapy produces beneficial effects. Among the plants, the studies provided weak evidence that photosensitizing plants can be effective in conjunction with phototherapy, and moderate evidence that *gingko biloba* monotherapy can be useful for vitiligo (Chap. 3.6).

3.15.6 The British Guidelines for the Treatment of Vitiligo

Communicated by D. J. Gawkrodger on behalf of the UK guidelines group.

- Therapeutic algorithm for children (summarized in Table 3.15.3)
- Therapeutic algorithm in adults

No Treatment Option

In adults with skin types I and II, in the consultation, it is appropriate to consider, after discussion, whether the initial approach may be to use no active treatment other than use of camouflage cosmetics and sunscreens (D/4).

Topical Treatment

- In adults with recent onset of vitiligo, treatment with a potent or very potent topical steroid should be considered for a trial period of no more than 2 months. Skin atrophy has been a common side effect (B/1+).
- Topical pimecrolimus should be considered as an alternative to a topical steroid, based on one study. The side-effect profile of topical pimecrolimus is better than that of a highly potent topical steroid (C/2+).
- Depigmentation with *p*-(benzyloxy)phenol (monobenzyl ether of hydroquinone) should be reserved for adults severely affected by vitiligo (e.g., more than 50% depigmentation or extensive depigmentation on the face or hands) who cannot or choose not to seek repigmention and who can accept permanently not tanning (D/4).

Phototherapy

Narrow-band UVB phototherapy (or PUVA) should only be considered for treatment of vitiligo in adults

Table 3.15.3 Algorithm for children management (UK Guidelines)

Option	Who or what	Alternative suggestion
No treatment (D/4)	Skin types I and II	Camouflage, sunscreens
Topical treatment (B/1+)	Potent and very potent steroids for 2 months; skin atrophy as side effect	Pimecrolimus or tacrolimus; better safety profile
Phototherapy (D/4)	Widespread lesions without any alternative approach and with high impact on quality of life	PUVA (A/1+)
Systemic treatment	High unacceptable risk of side effects (B/2++)	–
Surgical treatment	Studies are not performed until now	–
Psychological treatment (C/2++)	Children with coping defects	Parents of children should be also supported

who cannot be adequately managed with more conservative treatments (D/4), who have widespread vitiligo, or localized vitiligo with a significant impact on QoL. Ideally, this treatment should be reserved for patients with darker skin-types and monitored with serial photographs every 2–3 months (D/3). Narrowband UVB should be used in preference to oral PUVA in view of evidence of greater efficacy (A/1+).

Systemic Therapy

The use of oral dexamethasone to arrest progression of vitiligo cannot be recommended because of an unacceptable risk of side effects (B/2++).

Surgical Treatments

- Surgical treatments are reserved for cosmetically sensitive sites where there have been no new lesions, no Koebner phenomenon, and no extension of the lesion in the previous 12 months (A/1++).
- Split skin grafting gives better cosmetic and repigmentation results than minigraft procedures and utilizes surgical facilities that are relatively freely available (A/1+). Minigraft is not recommended, due to a high incidence of side effects and poor cosmetic results (A/1+). Other surgical treatments are generally not available.

Psychological Treatments

Clinicians should make an assessment of the psychological and QoL effects of vitiligo on patients (C/2++). Psychological interventions should be offered as a way of improving coping mechanisms in adults with vitiligo (D/4).

3.15.7 The Indian Evidence-Based Practice Guidelines for Surgical Management of Vitiligo

Communicated by D. Parsad and S. Gupta on behalf of the ad hoc Indian Task Force

These guidelines provide minimal standards of care for various surgical methods of treatment of vitiligo, with a brief description of the procedures as well as their advantages and disadvantages.

- Facility

Vitiligo surgery can be performed safely in an outpatient day care dermatosurgical facility under local anesthesia. The day care operating theater should be equipped with facilities for monitoring and handling emergencies. Transplantation for extensive areas of vitiligo may need general anesthesia, and in such cases, an operation theater facility in a hospital setting and the presence of an anesthetist are recommended.

- Indications for surgery and patient selection

Surgery is indicated for all types of stable vitiligo, including segmental, generalized, and acrofacial types, that do not respond to medical treatment. Test grafting may be performed in doubtful cases to detect stability. The choice of surgical intervention should be individualized according to the type of vitiligo, stability, localization of lesions, and cost-effectiveness of the procedure.

- Consensus recommendation of the task force on stability

The stability status of vitiligo is the single, most-important prerequisite in case selection. However, there is no consensus regarding the minimum required period of stability. The recommended period of stability in different studies has varied from 4 months to 3 years. Although there is no consensus on definitive parameters for stability, various recommendations suggest a period of disease inactivity ranging from 6 months to 2 years. The task force agrees on a year of disease inactivity as the cut-off period for defining stability (Level D). The parameters include absence of new lesions, no extension of old lesions, and absence of Koebner phenomenon during the past 1 year. If there is a doubt in the patient's history, test minigrafting may be done as proposed by Falabella et al. [1]. The test was considered positive if unequivocal repigmentation took place beyond 1 mm from the border of the implanted graft over a period of 3 months. Although this test has been considered as the gold standard for establishing the stability and success of repigmentation, doubts have been expressed over its utility. Spontaneous repigmentation should be considered as a favorable sign for the transplantation procedure. A test graft may be considered whenever there is a doubt about the stability, or the patient is unable to give a clear history on the stability. It needs to be stressed here that the treating physician should always consider each patient individually and exercise his/her judgment (Level D).

- The age of the patient for surgery

Vitiligo surgery is generally performed under local anesthesia, which would be difficult in children. General anesthesia for vitiligo surgery in a young child poses unacceptable risks, and the progress of the disease is difficult to predict in children. Hence, many dermatologists feel that surgical procedures should not be performed in children. However, studies have suggested that results of transplantation procedures were better in younger individuals than in older ones [5, 7]. Thus, no consensus exists in this aspect, and physicians should exercise their judgment after taking all aspects of the individual patient into consideration (Level C).

- Preoperative counseling and informed consent

Proper counseling is essential; the nature of the disease, procedure, expected outcome, and possible complications should be clearly explained to the patient. The need for concomitant medical therapy should be emphasized. Patients should understand that proper results may take time to appear (few months to 1 year). A detailed consent form, describing the procedure and possible complications, should be signed by the patient. The consent form should specifically state the limitations of the procedure, possible future disease progression, and whether more procedures will be needed for optimal outcome.

- Anesthesia

The recipient and donor sites are locally anesthetized by infiltration of 2% xylocaine, the pain of which can be reduced by prior application of EMLA® cream applied under occlusion for 1–2 h. Adrenaline should not be used on the recipient site as it makes the judgment of adequacy of the denudation to the required depth difficult. Tumescent anesthesia and nerve blocks may be used in larger areas. If grafting is planned for extensive areas, general anesthesia may be needed in a hospital setting (Level D).

- Methods of surgical modalities

Tissue grafts include punch grafts, suction blister epidermal grafts (SBEGs), split thickness skin grafts, hair follicle grafts, and flip-top grafts. Cellular grafts include autologous noncultured epidermal cell suspension (NCES), Autologous cultured melanocytes, and autologous cultured epithelial grafts.

- Evidence-based practice guidelines for the choice of method

As already highlighted, there is no uniformly acceptable objective measurement tool to compare the surgical outcome of a given modality (see Chaps. 1.4). A scoring system has been suggested to evaluate the outcome of transplantation procedures in vitiligo, although it is mostly subjective and has not been validated for interobserver bias in a large sample size [6] (Level C).

Various meta-analyses of published studies have shown that tissue grafting methods split-thickness grafting, suction blister epidermal grafting, and punch grafting have comparable success rates of repigmentation [2, 11, 14, 15] (Level A) (Fig. 3.15.2).

Similarly, all three cellular grafting techniques (NCES, cultured melanocytes, and cultured epidermis) were found to be equally effective [8, 12]. However, in comparison to tissue grafts, cellular grafts showed slightly lower success rates (Level A). One explanation may be that cellular grafts are generally used to treat larger areas in comparison to tissue grafts, and the success rates are lower when transplantation is done in patients with large surface area of involvement than in patients with smaller surface area of involvement regardless of stability.

When compared on the basis of adverse events, the highest incidence of adverse events was reported with punch grafting followed by split-thickness grafting and suction blister epidermal grafting. Cellular grafting methods were associated with the lowest incidence of adverse events (Level B) [11, 15].

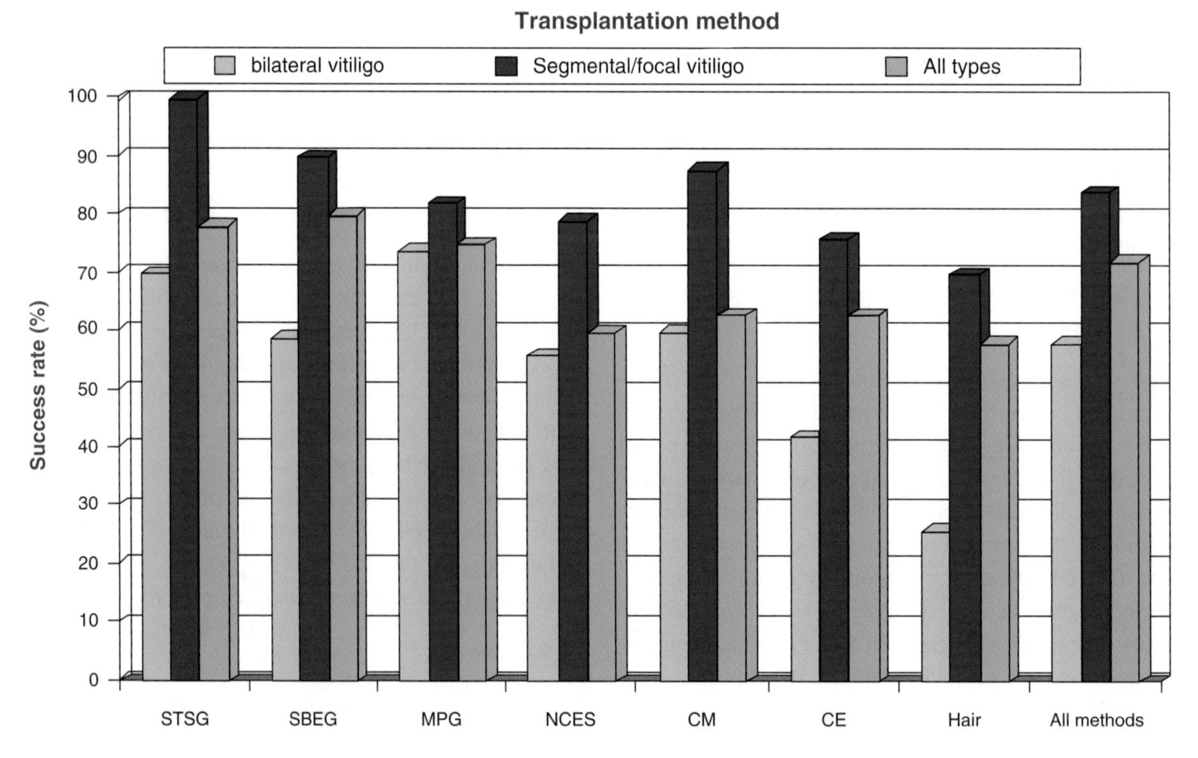

Fig. 3.15.2 Success rates of various transplantation methods in vitiligo, including its subtypes. The values given on each bar represent numbers of successful patients/total number of patients (From [8])

Table 3.15.4 Guidelines for surgical management of vitiligo

Vitiligo site	First choice	Alternative
Acral (fingers and toes)	CM, SBEG, STSG	MPG, NCES, CE
Palms	MPG	–
Lips	SBEG, STSG	MPG, NCES
Eyelids	SBEG, NCES, STSG	MPG
Nipple and areola	SBEG, NCES	
Genitals	NCES, SBEG, CM	–
Vitiligo area		
Small	SBEG, STSG	MPG, hair
Moderate	NCES, STSG	MPG
Extensive	CM, CE, NCES	STSG

STSG thin and ultra-thin split-thickness skin grafts; *SBEG* suction blister epidermal grafts; *MPG* mini-punch grafts; *NCES* noncultured epidermal cell suspension; *CM* cultured "pure" melanocytes; *CE* cultured epithelial grafts (From [8])

The choice of the procedure is dependent upon the site, area, availability of infrastructure (for cellular grafts), expertise of the dermatological surgeon, cost of the procedure, and the patient's preference (Table 3.15.4).

Summary Messages

- To treat or not to treat is a possible question in patients with a fair skin complexion
- Some medical or surgical possibilities have shown efficacy and can be advocated based on EBM principles
- However, the level of evidence and quality for studies is generally low, and EBM recommendations reach overall a low grade for vitiligo interventions.

References

1. Falabella R, Arrunategui A, Barona MI, Alzate A (1995) The minigrafting test for vitiligo: detection of stable lesions for melanocyte transplantation. J Am Acad Dermatol 32: 228–232
2. Falabella R (2003) Surgical treatment of vitiligo: why, when and how. J Eur Acad Dermatol Venereol 17:518–520
3. Forschner T, Bucholtz S, Stockfleth E (2007) Current state of vitiligo therapy-evidence based analysis of the literature. J Dtsch Dermatol Ges 5:467–475
4. Gawkrodger DJ, Ormerod AD, Shaw L et al (2008) Guideline for the diagnosis and management of vitiligo. Br J Dermatol 159:1051–1076
5. Gupta S, Kumar B (2002) Epidermal grafting for vitiligo in adolescents. Pediatr Dermatol 19:159–162
6. Gupta S, Handa S, Kumar B (2002) A novel scoring system for evaluation of results of autologous transplantation methods in vitiligo. Indian J Dermatol Venereol Leprol 68: 33–37
7. Gupta S, Kumar B (2003) Epidermal grafting in vitiligo: influence of age, site of lesion, and type of disease on outcome. J Am Acad Dermatol 49:99–104
8. Gupta S, Narang T, Olsson MJ, Ortonne JP (2007) Surgical management of vitiligo and other leukodermas: evidence-based practice guidelines. In: Gupta S, Olsson MJ, Kanwar AJ, Ortonne JP (eds) Surgical management of vitiligo. Blackwell, Oxford, pp 69–79
9. Lim HW, Hexsel CL (2007) Vitiligo: to treat or not to treat. Arch Dermatol 143:643–646
10. Njoo MD, Spuls PI, Bos JD et al (1998) Nonsurgical repigmentation therapies in vitiligo. Meta-analysis of the literature. Arch Dermatol 134:1532–1540
11. Njoo MD, Westerhof W, Bos JD et al (1998) A systematic review of autologous transplantation methods in vitiligo. Arch dermatol 134:1543–1549
12. Olsson MJ, Juhlin L (2002) Long-term follow-up of leucoderma patients treated with transplants of autologous cultured melanocytes, ultrathin epidermal sheets and basal cell layer suspension. Br J Dermatol 147:893–904
13. Parsad D, Gupta S; IADVL Dermatosurgery Task Force (2008) Standard guidelines of care for vitiligo surgery. Indian J Dermatol Venereol Leprol 74(Suppl):S37–S45
14. Rusfianti M, Wirohadidjodjo YW (2006) Dermatosurgical techniques for repigmentation of vitiligo. Int J Dermatol 45:411–417
15. Savant SS (1992) Autologous miniature punch grafting in vitiligo. Indian J Dermatol Venereol Leprol 58:310–314
16. Szczurko O, Boon H (2008) A systematic review of natural health product treatment for vitiligo. BMC Dermatol 8:2
17. Whitton ME, Aschcroft DM, Barrett CW et al (2006) Interventions for vitiligo. Cochrane Database Syst Rev (1):CD003263

Editor's Synthesis and Perspectives

3.16

Alain Taïeb and Mauro Picardo

Contents

A. Taieb(✉)
Service de Dermatologie, Hôpital St André,
CHU de Bordeaux, France
e-mail: alain.taieb@chu-bordeaux.fr

3.16.1 From EBM Guidelines to Clinical Practice

The major interest of EBM reviews and meta-analyses is to point to potentially harmful interventions or those with limited interest. Given the importance of charlatanism in the vitiligo field, counselling patients to avoid some therapies of dubious efficacy is indeed a major step.

As partly stated in the Cochrane review [12], there are many limitations to derive a valuable algorithm of treatment for all vitiligo patients based on RCTs. First, RCTs are rare and often lacking important methodological steps or details. Second, studies have often been conducted in heterogeneous groups in terms of vitiligo duration or progression, if not mixing localized, segmental and non-segmental forms. Third, confounding factors are many, e.g. light exposure in long term interventions, for which light sources may influence outcome, nutritional intake if antioxidant status is considered, or awareness of the limitation of the Koebner's phenomenon, which is rarely taken into account.

Nevertheless, a stepwise treatment approach divided by the type of vitiligo and the extent, which needs modulation by visibility, age and coping, is outlined in Table 3.16.1 [11]. A zero line is always possible, meaning no treatment if the disease is not bothering the patient. The environmental factors (occupation, Koebner's phenomenon, sustained stress or anxiety) should be always discussed in the management plan. For segmental vitiligo (SV), triggering neurogenic factors are usually envisaged but good studies to prove this point are lacking. This stepwise approach should be considered as a proposal, based mostly on EBM data. However, there is much room for modulation and innovation based on this scheme.

M. Picardo and A. Taïeb (eds.), *Vitiligo*,
DOI 10.1007/978-3-540-69361-1_3.16, © Springer-Verlag Berlin Heidelberg 2010

Table 3.16.1 General outline of management for vitiligo (adapted from Taieb and Picardo [11]).

Type of vitiligo	Usual management
Segmental and limited non-segmental <2–3% body surface involvement	First line: Avoidance of triggering factors, local therapies (corticosteroids, calcineurin inhibitors). Second line: Localized NB-UVB therapy, especially Excimer monochromatic lamp or laser. Third line: Consider surgical techniques if repigmentation is cosmetically unsatisfactory on visible areas
NSV	First line: Avoidance of triggering/aggravating factors. Stabilization with NB-UVB therapy, at least 3 months. Optimal duration at least 9 months after response. Combination with systemic/topical therapies, including reinforcement with localized UVB therapy, if possible Second line: Consider systemic steroids (e.g. 3–4 month minipulse therapy) or immunosuppressants in case of rapidly progressing disease or absence of stabilization under NB-UVB. Third line: Consider surgical techniques in non responding areas especially with high cosmetic impact. However, Koebner phenomenon limits the persistence of grafts. Relative contraindication in areas such as dorsum of hands Fourth line: Consider depigmentation techniques (hydroquinone monobenzyl ether or 4-methoxyphenol alone or associated with Q switch ruby laser) in non responding widespread (>50%) or highly visible recalcitrant facial/hands vitiligo

A no treatment option (zero line) can be considered in patients with a fair complexion after discussion. For children, phototherapy is limited by feasibility in the younger age group and surgical techniques rarely proposed before prepubertal age. There is no current recommendation applicable to the case of rapidly progressive vitiligo, not stabilized by UV therapy. For all subtypes of the disease or the lines of treatment, psychological support and counselling, including access to camouflage instructors is needed

3.16.2 Perspectives

Part 2 of this book has put the emphasis on numerous potential therapeutic resources, based on a better understanding of the crucial steps of melanocyte mechanisms of disappearance and repigmentation schemes. For a common disorder like vitiligo, there are probably subtypes in terms of mechanisms of melanocyte loss. We still do not know if cutaneous inflammation, which seems more common than previously envisaged in progressive disease, is a shared feature in all cases. If it is the case, a more aggressive anti-inflammatory therapy would probably be helpful, and there is clearly a need to assess more precisely the use of systemic immunosuppressants in this group. It is surprising that a drug widely used in other chronic inflammatory skin disorders such as methotrexate has not yet been tested. If the initial step preceding inflammation comes from a local predisposition of melanocytes to attach poorly to the basement membrane, there are possible targets to improve calcium-dependent adhesion mechanisms, but a more precise description of the basic impairment is obviously needed. If some vitiligo cases are due to the impairment of melanocyte survival mechanisms, growth factor supplementation, which is possible (MSH analogs, [3]), could be tested. The issue of self-renewal (stemness) aptitude of melanocytes has been raised especially for SV, which clearly benefits from autologous grafting [10]. It is yet unclear if this is a real issue in common NSV, and if this is related to the cellular environment rather than directly to melanocyte themselves, as suggested by Guerra's group [1]. However, improving in a symptomatic approach the antioxidant status of the epidermis has been attempted [2, 9], but more powerful tools using gene transfer might be used in the future [7, 8].

When melanocyte loss has been stopped, therapy tries to address repigmentation. New repigmenting therapies such as He–Ne lasers [5] and prostaglandin E2 [4] are emerging. Recent development in the field of melanocyte precursors in the hair follicle are promising (Chap. 2.3.4). If we can better stimulate the emigration of those stem cells towards the epidermis and understand why they usually stop migrating when becoming pigmented, a major step would be achieved. There is also the issue of dormant residual (de-differentiated) melanocytes which could be resuscitated in interfollicular zones and opening some new therapeutic avenues. A better appreciation of this issue is clearly needed to consider newer possibilities for medical treatments. Surgical treatments for limited and disfiguring diseases have dramatically improved in the last 20 years. Newer technologies derived from progenitors or reprogrammed skin cells [6] will probably further increase our surgical possibilities of intervention.

Summary Messages

- The major interest of EBM reviews and meta-analyses is, besides grading the efficacy and tolerance of interventions, to point to potentially harmful interventions or to those of limited interest
- We recommend a stepwise treatment approach by type and extent of vitiligo, which needs to be modulated by visibility, age and coping
- Subtyping the mechanisms of melanocyte loss may lead to improved pathophysiologic schemes of treatment.
- If cutaneous inflammation is a common feature in vitiligo, staging inflammation needs more attention to target appropriately anti-inflammatory therapy in future trials
- Emerging repigmenting therapies such as He-Ne lasers and prostaglandin E2 and newer technologies derived from progenitors or reprogrammed skin cells will increase our possibilities of intervention.

References

1. Bondanza S, Maurelli R, Paterna P et al (2007) Keratinocyte cultures from involved skin in vitiligo patients show an impaired in vitro behaviour. Pigment Cell Res 20:288–300
2. Dell'Anna ML, Mastrofrancesco A, Sala R et al (2007) Antioxidants and narrow band-UVB in the treatment of vitiligo: a double-blind placebo controlled trial. Clin Exp Dermatol 32:631–636
3. Harms J, Lautenschlager S, Minder CE, Minder EI (2009) An alpha-melanocyte-stimulating hormone analogue in erythropoietic protoporphyria. N Engl J Med 360: 306–307
4. Kapoor R, Phiske MM, Jerajani HR (2009) Evaluation of safety and efficacy of topical prostaglandin E2 in treatment of vitiligo. Br J Dermatol 160:861–863
5. Lan CC, Wu CS, Chiou MH et al (2006) Low-energy helium-neon laser induces locomotion of the immature melanoblasts and promotes melanogenesis of the more differentiated melanoblasts: recapitulation of vitiligo repigmentation in vitro. J Invest Dermatol 126:2119–2126
6. Takahashi K, Tanabe K, Ohnuki M (2007) Induction of pluripotent stem cells from adult human fibroblasts by defined factors. Cell 131:861–872
7. Rezvani HR, Cario-André M, Pain C et al (2006) Protection of normal human reconstructed epidermis from UV by catalase overexpression. Cancer Gene Ther 14:174–186
8. Rezvani HR, Ged C, Bouadjar B et al (2008) Catalase overexpression reduces UVB-induced apoptosis in a human xeroderma pigmentosum reconstructed epidermis. Cancer Gene Ther 15:241–251
9. Schallreuter KU, Kruger C, Wurfel C et al (2008) From basic research to the bedside: efficacy of topical treatment with psudocatalase PC-KUS in 71 children with vitiligo. Int J Dermatol 47:743–753
10. Taieb A (2000) Intrinsic and extrinsic pathomechanisms in vitiligo. Pigment Cell Res 13:41–47
11. Taieb A, Picardo M (2008) Clinical practice. Vitiligo. N Engl J Med 360:160–169
12. Whitton ME, Ashcroft DM, Barrett CW, Gonzalez U (2006) Interventions for vitiligo. Cochrane Database Syst Rev (1):CD003263

Index

J

K

L

M

Printing and Binding: Stürtz GmbH, Würzburg